The Biology of Dendritic Cells and HIV Infection

The Biology of Dendritic Cells and HIV Infection

Edited by

Sandra Gessani
Istituto Superiore di Sanità
Rome, Italy

Filippo Belardelli
Istituto Superiore di Sanità
Rome, Italy

 Springer

Sandra Gessani
Istituto Superiore di Sanità
Department of Cell Biology
   and Neurosciences
Viale Regina Elena, 299
00161 Rome
Italy
gessani@iss.it

Filippo Belardelli
Istituto Superiore di Sanità
Department of Cell Biology
   and Neurosciences
Viale Regina Elena, 299
00161 Rome
Italy
belard@iss.it

Cover illustration: Illustration of dendritic cells and T cells. The education of killer T cells by dendritic (Langerhans) cells occurs in a lymph node. These dendrites have processed antigenic information obtained from an autologous vaccine made with proteins purified from the patient's own cancer cells. Killer T cells then target all cells tagged with the antigen.

Library of Congress Control Number: 2006925170

ISBN-10: 0-387-33784-9   e-ISBN-10: 0-387-33785-7
ISBN-13: 978-0-387-33784-5   e-ISBN-13: 978-0-387-33785-2

Printed on acid-free paper.

9 8 7 6 5 4 3 2 1

springer.com

We dedicate this book in memory
of the late Prof. Giovanni Battista Rossi,

who was for many years chairman of the Department of Virology of the Italian National Institute of Health until he passed away (February 20, 1994). Prof. Rossi, Giovanni to his friends, was the promoter and scientific coordinator of the Italian National Project on AIDS (1987–1993) and the organizer of the VII International Conference on AIDS (Florence, June 1991). All these activities had earned him the admiration of the international scientific community. Giovanni was both a mentor and a friend. We commenced our scientific careers in his laboratory working on the differentiation of Friend virus–transformed mouse leukemia cells and on the antiviral and antitumoral effects of interferons. He inspired us to start research projects on AIDS, with the vision that the knowledge of cytokines could be instrumental in the understanding of complex aspects of the pathogenesis and immunology of HIV infection. Giovanni at the time knew little about dendritic cells, because it took several years to understand the crucial importance of these cells in the regulation of the immune response after the pioneering work of Ralph Steinman and Zanvil Cohn in 1973. However, having personally known his great curiosity and interest in cell differentiation, we are certain that he would have been captivated by the recent knowledge of the biology of dendritic cells and its potential importance in the development of novel strategies of immunotherapy of cancer and HIV infection. Giovanni's major interest was in the research of cell differentiation, HIV, and cancer. It is now becoming evident that differentiation of dendritic cells can play an important role in pathogenesis and control of both cancer and HIV infection. It is a pity that Giovanni had not had the chance to witness the recent exciting progress in dendritic cell biology! Dedicating this book to him is in strong recognition not only of his commitment to AIDS research but also to his interdisciplinary vision of biomedical research.

Preface

This book represents a special effort to bring together the most recent information on the biology of dendritic cells (DCs) focusing on the role of these cells in the pathogenesis and immunity of HIV-1 infection. The recent progress in immunology has revealed the key importance of DCs in the regulation of the immune response to infections as well as to certain malignant, allergic, and autoimmune diseases. In particular, research on DCs has recently emerged as a fundamental aspect for the comprehension of the mechanisms underlying the pathogenesis of viral diseases, as well as for the progress on the development of prophylactic and therapeutic vaccines. Recent findings have led to substantial conceptual advances in our understanding of the role of DCs in HIV infection and clarified that DCs are important targets and reservoirs of HIV, playing a crucial role in several aspects of viral pathogenesis. Interestingly, HIV can exploit many of the cellular processes responsible for the generation and regulation of the adaptive immune responses to gain access to its main target cells, namely the $CD4^+$ T lymphocytes. Hence, the central role of DCs in stimulating T-cell activation not only provides a route for viral transmission but also represents a vulnerable point where HIV-1 can interfere with the initiation of T cell–mediated immunity. Recent studies have revealed that several HIV proteins can profoundly influence the phenotype and functions of DCs even in the absence of a productive infection, often resulting in an abnormal immune response. This knowledge has resulted in the identification of some major mechanisms involved in the pathogenesis of HIV-1 infection. In addition, the recent advances in DC biology have opened perspectives in the research on new adjuvants and novel strategies for the *in vivo* targeting of antigens to DCs, which are instrumental in the development of HIV vaccines.

This book is of special interest to researchers in the fields of basic and applied immunology, virology, cell biology, and vaccine development and to practitioners and students. It offers the reader 14 chapters written by international experts on DCs and the pathogenesis and immunology of HIV infection. The text is divided into three parts. Part I includes a detailed overview of the biology, ontogeny, function, and therapeutic potential of human DC subsets, which is intended to give readers an introduction to the field of DCs and to their possible clinical exploitation, as well as provide a framework for a fuller understanding of the content of subsequent chapters. Part II focuses on some general aspects of the immunopathogenesis of HIV infection and provides a comprehensive overview of cells and factors involved in the innate and adaptive immune response to HIV-1. These general concepts are of relevance for a deeper understanding of the functional consequences of HIV-1 interaction with DCs. This then becomes the specific topic of Part III, which also focuses on the mechanisms leading to immune evasion and dysfunction and on the importance of DCs as targets and tools for the immunotherapy of HIV infection. We have now begun to understand the complex interactions between HIV and DCs in the pathogenesis of AIDS and have learned lessons about how to prepare potentially effective DCs for cancer vaccines. We assume that in the future, DC-based therapeutic vaccines can represent a topic of increasing interest for the immunotherapy of HIV-1–infected patients. Although DC-based vaccines will certainly not solve the drastic needs of HIV-infected individuals in the developing countries, the progress of the research in this field will help us to identify novel and practical strategies for the *in vivo* targeting of HIV antigens to DCs. We thank all the authors for their generous participation in this initiative, which permitted us to provide the reader with a comprehensive overview on the biology of DCs and their role in the pathogenesis and immunity of HIV-1 infection.

Rome, Italy

Sandra Gessani
Filippo Belardelli

Acknowledgments

Although many individuals have offered help or suggestions in the preparation of this book, several deserve special mention. We are grateful to Cinzia Gasparrini, Alessandro Spurio, Barbara Varano, and Paola De Castro for their invaluable editorial assistance. We would also like to give great thanks to all the people who read parts or all of the chapters and advised us on how to make them better.

Rome, Italy   Sandra Gessani
                                                        Filippo Belardelli

Contents

Preface..vii

Acknowledgments...ix

Contributors...xxi

Part I   **General Aspects of Dendritic Cell Biology,
Functions, and Clinical Application............................1**

Chapter 1   **Dendritic Cell Biology: Subset Heterogeneity
and Functional Plasticity..3**
Vassili Soumelis, Yong-Jun Liu, and Michel Gilliet

1.1 Introduction 3
1.2 Origin and Development of Human DC Subsets 4
1.3 Anatomic, Phenotypic, and Functional Features of DC Subsets 7
    1.3.1 Thymic DCs...7
    1.3.2 Blood DCs...8
    1.3.3 Skin and Mucosal DCs..9
    1.3.4 DCs in Solid Organs 9
    1.3.5 DCs in Secondary Lymphoid Organs 10
1.4 Innate Functions of DC Subsets...11
    1.4.1 Invariant Receptors on DCs 11
    1.4.2 Antigen Uptake and Processing by DC Subsets 14
    1.4.3 Cytokine and Chemokine Production by DCs 15
    1.4.4 Cross-talk Between DCs and Other Innate Immunity Cell Types 16
    1.4.5 DC Maturation and Migration to the Secondary Lymphoid Organs............17
1.5 Role of DC Subsets in Adaptive Immunity..18
    1.5.1 MDCs in Adaptive Immune Responses......................................18
    1.5.2 pDCs in Adaptive Immune Responses 20
    1.5.3 Molecular Basis for DC-mediated Th1/Th2 Polarization...........................22
1.6 DCs and Tolerance 23

1.7 Concluding Remarks...25
References..25

**Chapter 2   Dendritic Cells and Their Role in Linking Innate
                and Adaptive Immune Responses.....................................45**
                Mary F. Lipscomb, Julie A. Wilder, and Barbara J. Masten

2.1 Introduction...45
2.2 DC Origins, Subsets, and Differentiation 46
2.3 Plasticity and Trafficking Pathways...51
2.4 DC–T Cell Interactions 55
    2.4.1 Antigen Uptake 56
    2.4.2 Antigen Processing..56
        2.4.2.1 MHC Class II Presentation 56
        2.4.2.2 MHC Class I Presentation..58
    2.4.3 Costimulation and Suppression Molecules.....................................59
    2.4.4 Events at the DC–T Cell Interface 60
2.5 DCs Link Innate and Acquired Immunity...61
    2.5.1 Influence of the Microenvironment on DC Phenotype and Function..........61
        2.5.1.1 Exogenous Signals..64
        2.5.1.2 Endogenous Signals 64
    2.5.2 DCs in T-Cell Activation and Differentiation..................................66
    2.5.3 Regulation of Pulmonary Immunity by Lung DCs...........................68
    2.5.4 DCs and B-Cell Function 69
2.6 Concluding Remarks...69
References..70

**Chapter 3   Dendritic Cell and Pathogen Interactions
                in the Subversion of Protective Immunity. 85**
                John E. Connolly, Damien Chaussabel, and Jacques Banchereau

3.1 Introduction...85
3.2 Dendritic Cell Biology 86
    3.2.1 Dendritic Cells Are Composed of Subsets....................................86
        3.2.1.1 Distinct DC Subsets Are Endowed with Distinct
                 Functional Properties 86
    3.2.2 Dendritic Cell Ontogeny 88
        3.2.2.1 DC Progenitors and Precursors...88
    3.2.3 Antigen Capture, Processing, and Presentation 88
        3.2.3.1 Presentation via MHC Class I 90
        3.2.3.2 Presentation via MHC Class II...91
        3.2.3.3 Presentation via CD1 Family of MHC Molecules 91
    3.2.4 Dendritic Cell Maturation...92

3.2.4.1 Maturation Signals 92
3.2.5 DC Migration 94
3.2.6 Dendritic Cells and Tolerance...96
    3.2.6.1 Central Tolerance 96
    3.2.6.2 Peripheral Tolerance Through DCs 96
    3.2.6.3 Inhibitory Receptors 97
3.2.7 Dendritic Cell Death 98
3.3 Subversion of Dendritic Cell Function by Pathogens...........................99
    3.3.1 Pathogens Evade Dendritic Cell Antigen Processing 99
        3.3.1.1 Evasion of DC Uptake 99
        3.3.1.2 Inhibition of Classical Class I Processing and Presentation..........100
        3.3.1.3 Upregulation of Inhibitory Nonclassical Class I Molecules..........102
        3.3.1.4 Inhibition of Class II Processing and Presentation 102
        3.3.1.5 Inhibition of CD1 Processing and Presentation 104
    3.3.2 Selective Targeting of DC Subsets by Pathogens.......................104
    3.3.3 Pathogens Target DC Precursors 105
    3.3.4 Inhibition of DC Maturation by Pathogens.............................106
        3.3.4.1 Inhibition of DC Activation 106
        3.3.4.2 Pathogen Modulation of DC Maturation 107
    3.3.5 Pathogen-mediated Inhibition of DC Migration 108
    3.3.6 Viral Modulation of DC Tolerance...109
    3.3.7 Pathogen-mediated Dendritic Cell Death 109
3.4 Concluding Remarks...110
References 110

Chapter 4   Dendritic Cells as Keepers of Peripheral Tolerance 129
Sabine Ring, Alexander H. Enk, and Karsten Mahnke

4.1 Introduction 129
4.2 The Concept of "Steady State" versus "Activated" DC.......................130
4.3 Factors That Affect Tolerogenic DCs 136
    4.3.1 TNF-α and Semimature DCs 136
    4.3.2 Vitamin D_3 Affects DC Maturation 139
    4.3.3 IL-10 Modulates DCs for Tolerance Induction.........................143
4.4 Molecular Mechanisms Involved in Tolerance Induction 144
    4.4.1 Costimulatory Molecules..144
    4.4.2 RelB Translocation Is Crucial for DC Maturation.....................146
    4.4.3 The Role of Indoleamine 2,3-dioxygenase (IDO)
        in Tolerance Induction 148
4.5 Are There Specialized Subsets of Tolerogenic DCs? 151
    4.5.1 Surface Marker Expression 151
    4.5.2 Cytokine Expression 153
4.6 "Designer DC": Tailored for Tolerance Induction................................154

4.7 Concluding Remarks..158
References...159

**Chapter 5   Adjuvants, Dendritic Cells, and Cytokines:
                   Strategies for Enhancing Vaccine Efficacy..................171**
Paola Rizza, Imerio Capone, and Filippo Belardelli

5.1 Introduction 171
5.2 Brief Historical Background on Adjuvants...172
5.3 The Need of New Adjuvants for Vaccine Development and crucial
     importance of Adjuvant–Dendritic Cell Interactions for the Polarization
     of Immune Response...175
5.4 New Classification of Vaccine Adjuvants 179
5.5 Ligands for TLRs and Their Potential Role as Adjuvants.......................183
     5.5.1 PAMP-dependent Targeting of DCs...183
     5.5.2 CpG Oligonucleotides 184
     5.5.3 PAMP-independent Activation of DCs 186
     5.5.4 Small-Molecule Immune Potentiators: Imidazoquinolines 187
5.6 Interaction of Cytokines with DCs: Importance for Vaccine Development........187
5.7 Concluding Remarks...194
References...195

**Chapter 6   *Ex Vivo*–Generated Dendritic Cells for Clinical
                   Trials versus *In Vivo* Targeting to Dendritic Cells:
                   Critical Issues 203**
*Joannes F.M. Jacobs, Cândida F. Pereira, Paul J. Tacken,
I. Jolanda M. de Vries, Cornelus J.A. Punt, Gosse J. Adema,
and Carl G. Figdor*

6.1 Introduction 203
6.2 DC Culture, from Bench to Clinical-Grade Product 205
     6.2.1 Introduction 205
     6.2.2 Precursor Isolation..205
     6.2.3 Differentiation into DC Phenotype..206
     6.2.4 Maturation 207
          6.2.4.1 Maturation Signals...207
          6.2.4.2 Effect of Maturation on the Function of DCs 208
     6.2.5 The Immune Target 209
          6.2.5.1 MHC Class I and Class II Loading 209
          6.2.5.2 Antigen Source...210
               6.2.5.2.1 Loading DCs with Peptides...210
               6.2.5.2.2 Loading DCs with Proteins 211

6.2.5.2.3 Loading DCs with Whole Target 212
6.2.5.2.4 Loading DCs with RNA or DNA 212
6.2.5.3 Methods for Loading DCs with Antigen 213
6.2.5.4 Loading of DCs with Antigens: Summary 214
6.2.6 Storage 214
6.3 DC Quality Check 214
6.3.1 Introduction 214
6.3.2 DC Phenotype and Purity 215
6.3.3 Function 215
6.4 Vaccine Administration 216
6.5 Immunomonitoring 216
6.6 Targeting DCs *In Vivo* 218
6.6.1 Introduction 218
6.6.2 Maturing DCs *In Vivo* 219
6.6.3 Targeting Antigens to the MHC Class I and II Pathway 219
6.6.4 Drug Delivery Systems 220
6.6.4.1 Live Vectors to Deliver DNA to DCs 221
6.6.4.2 Microparticles to Deliver DNA, Protein, or Peptides
        to DCs 221
6.6.4.3 Receptor Ligands to Deliver Protein or Peptides to DCs 223
6.6.5 Targeting DC Surface Receptors 224
6.6.5.1 Mac-1 225
6.6.5.2 Gb3 225
6.6.5.3 CD40 225
6.6.5.4 Fc Receptors 226
6.6.5.5 C-type Lectin Receptors 227
6.7 Concluding Remarks 229
References 229

Part II   General Aspects of the Pathogenesis and Immune Response to HIV-1 Infection 243

Chapter 7   Immunopathogenesis of HIV Infection 245
Elisa Vicenzi, Massimo Alfano, Silvia Ghezzi, and Guido Poli

7.1 Introduction 245
7.2 The Life Cycle of HIV-1 in CD4[+] T Cells and Mononuclear Phagocytes 246
7.3 CCR5- versus CXCR4-dependent HIV-1 Infections 252
7.4 Intrinsic Resistance to HIV Infection 259
7.5 HIV Cytopathicity 261
7.6 HIV Replication in Lymphoid Organs and Central Nervous System 265
7.7 Host Determinants of HIV Propagation: Cytokines and Chemokines 268

7.8 Concluding Remarks..273
References..274

**Chapter 8   Innate Cellular Immune Responses in HIV
             Infection 297**
             *Barbara Schmidt, Nicolai A. Kittan, Sabrina Haupt,
             and Jay A. Levy*

8.1 Characteristics of Innate Immunity..297
8.2 Cells and Soluble Factors Involved in Innate Immunity Against HIV...............299
8.3 Plasmacytoid and Myeloid Dendritic Cells and Their Association
    with HIV Infection 301
    8.3.1 Overview..301
    8.3.2 PDC Characteristics..302
    8.3.3 Relationship of Dendritic Cells to the HIV Clinical State..........................303
    8.3.4 HIV Interaction with Dendritic Cells and IFN Production......................306
8.4. The Role of Noncytotoxic CD8$^+$ T Cells in Anti-HIV Responses 307
    8.4.1 Overview 307
    8.4.2 The CD8$^+$ T-Cell Antiviral Factor (CAF) 309
    8.4.3 Mechanism of Action 310
    8.4.4 Characteristics of CAF 310
    8.4.5 Relationship of CNAR to Clinical State.......................................311
8.5 Other Innate Immune Cells 312
    8.5.1 Natural Killer Cells...312
    8.5.2 NK-T Cells 313
    8.5.3 γδ T Cells 315
8.6 Concluding Remarks 316
References..317

Chapter 9   Adaptative Immune Responses in HIV-1 Infection 333
             Mara Biasin and Mario Clerici

9.1 Introduction to the Characteristics of Specific Immunity.....................................333
9.2 Functional Immune Disregulation in the Different Phases
    of HIV-1 Infection..334
9.3 Cellular Immune Response to HIV-1..336
    9.3.1 Differentiation and Functions of CTLs 337
    9.3.2 CTLs in HIV-1 Infection 339
    9.3.3 Viral Escape from CTLs...341
    9.3.4 Antiviral Soluble Factors...344
    9.3.5 Alteration of T-Helper Functions in HIV-1 Infection 346
    9.3.6 Immunologic Profile of People Living in Different
          Areas of the World..350

9.4 Humoral Immune Responses in HIV-1 Infection 351

    9.4.1 Differentiation and Functions of B Cells 352

    9.4.2 Antibody Responses in HIV-1 Infection..354

    9.4.3 Viral Escape from Antibodies..356

9.5 Immune Responses in HIV-Exposed Seronegative Individuals.........................357

9.6 Immune Responses in Long-Term Nonprogressors...359

9.7 Concluding Remarks...361

References 362

Part III   Dendritic Cells and HIV Interactions and Their Role in Pathogenesis and Immunity 379

Chapter 10 Binding and Uptake of HIV by Dendritic Cells and Transfer to T Lymphocytes: Implications for Pathogenesis..381

Anthony L. Cunningham, John Wilkinson, Stuart Turville, and Melissa Pope

10.1 Introduction 381

10.2 Transmission of HIV 381

10.3 HIV Infection of Female Genital Tract...382

10.4 The Role of Dendritic Cells in HIV Infection...384

    10.4.1 Immature versus Mature DCs..385

    10.4.2 HIV Receptors on DCs...386

        10.4.2.1 C-type Lectin Receptors...386

        10.4.2.2 Diversity of CLR Expression on DCs 388

10.5 Transmission of HIV to T Cells by DCs...390

    10.5.1 Effect of Maturation 393

    10.5.2 HIV, CLRs, and Nonepithelial DCs...394

    10.5.3 Control of HIV Spread and Vaginal Microbicides................................394

    10.5.4 Targeting the Virus..394

    10.5.5 Targeting the Cell...395

10.6 Models for Examining HIV Infection and Testing of Vaginal Microbicides in the Genital Mucosa 396

    10.6.1 Monocyte-derived Dendritic Cells and Monocyte-derived Langerhans Cells..396

    10.6.2 *Ex Vivo* Cervical Explants..397

    10.6.3 Macaque Models 397

10.7 Concluding Remarks..398

References 399

**Chapter 11 Loss, Infection, and Dysfunction of Dendritic
                 Cells in HIV Infection 405**
Steven Patterson, Heather Donaghy, and Peter Kelleher

11.1 Overview 405
11.2 Introduction to Myeloid and Plasmacytoid DCs 406
    11.2.1 Myeloid DCs 406
        11.2.1.1 Phenotype and Location 406
        11.2.1.2 Maturation 407
        11.2.1.3 Development 408
        11.2.1.4 Function 408
    11.2.2 Plasmacytoid DCs 409
        11.2.2.1 Location and Development 409
        11.2.2.2 Maturation and Function 410
11.3 Practical Issues in DC Research 411
11.4 Differentiation of Blood Myeloid DCs 412
11.5 DCs and Transmission of HIV 415
11.6 Loss of DCs in HIV Infection 416
    11.6.1 Tissue Langerhans Cells 416
    11.6.2 Blood DCs 417
    11.6.3 Anti-retroviral Drugs and DC Numbers 419
11.7 Infection of DCs by HIV 420
    11.7.1 *In Vitro* Studies 420
        11.7.1.1 HIV Receptor Expression on DCs 420
    11.7.2 *In Vitro* Infection Studies 422
        11.7.2.1 Blood Plasmacytoid DCs 422
        11.7.2.2 Blood Myeloid DCs 424
        11.7.2.3 Monocyte-derived DCs 424
        11.7.2.4 Langerhans Cells 426
    11.7.3 DC Infection *In Vivo* 426
        11.7.3.1 Blood DCs 426
        11.7.3.2 Langerhans Cells 427
11.8 Dysfunction of DCs in HIV Infection 427
    11.8.1 Blood Myeloid DCs 428
    11.8.2 Monocyte-derived DCs 430
    11.8.3 Langerhans Cells 431
    11.8.4 Plasmacytoid DCs 432
11.9 Concluding Remarks 432
References 433

**Chapter 12   HIV Exploitation of DC Biology to Subvert
the Host Immune Response**...**447**
*Manuela Del Cornò, Lucia Conti, Maria Cristina Gauzzi,
Laura Fantuzzi, and Sandra Gessani*

12.1 Introduction...447
12.2 The Host Response to Viral Infection 448
    12.2.1 Innate Immune Response 448
    12.2.2 Adaptive Immune Response 451
12.3 Virus-induced Phenotypic and Functional Alterations of Human DCs 454
    12.3.1 Interference with DC Generation and Survival 456
    12.3.2 Loss of DC Morphology 456
    12.3.3 Interference with DC Maturation 457
    12.3.4 Modulation of DC Migration 458
12.4 DCs in HIV Pathogenesis: Protective or Defective? 459
    12.4.1 HIV-1 Effects on DC Differentiation/Maturation 464
    12.4.2 Modulation of Cytokine/Chemokine Secretion 469
    12.4.3 Regulation of DC Chemotactic Functions........................472
    12.4.4 Effect on DC Survival...473
    12.4.5 Effect on DC Cytotoxic Activity.....................................473
12.5 Concluding Remarks 474
References...474

**Chapter 13   Cross-Presentation by Dendritic Cells: Role
in HIV Immunity and Pathogenesis** **485**
*Concepción Marañón, Guillaume Hoeffel, Anne-Claire
Ripoche, and Anne Hosmalin*

13.1 Introduction...485
13.2 Antigen Presentation Pathways 485
    13.2.1 MHC Class II–restricted Antigen Presentation 486
    13.2.2 MHC Class I–restricted Antigen Presentation 486
13.3 Direct Presentation of HIV from Infected DCs 489
13.4 Cross-Presentation of HIV 490
    13.4.1 Cross-Presentation of Defective HIV 490
    13.4.2 Cross-Presentation from Apoptotic, HIV-infected Cells.......491
    13.4.3 Cross-Presentation from Live, HIV-infected Cells 492
        13.4.3.1 Evidence of Cross-Presentation from Live,
HIV-infected Cells...492
        13.4.3.2 Mechanisms of Antigen Internalization from Live
Cells into DCs 494
13.5 Comparison of Cross-Presentation from Live or from Dead Cells.................496
    13.5.1 Uptake and Antigen-processing Pathways 496
    13.5.2 Role of DC Maturation 497

13.5.3 Is Death Necessary to Cross-Presentation?... 498
13.6 Role of Cross-Presentation from Infected Cells in HIV Immunity
    and Pathogenesis 499
13.7 Concluding Remarks.. 501
References.. 502

Chapter 14 Immunotherapy of HIV Infection: Dendritic Cells as Targets and Tools... 515

Imerio Capone, Giuseppe Tambussi, Paola Rizza, and Adriano Lazzarin

14.1 Introduction.. 515
14.2 The Rationale for Combining HAART with Immunotherapy
    in the Treatment of HIV-1 Infection 516
    14.2.1 Drug Therapy Limitations 516
    14.2.2 Immune Control of HIV-1 517
14.3 Overview of the Clinical Trials of Immunotherapy for the Treatment
    of HIV-1 Infection 519
    14.3.1 Introduction.. 519
    14.3.2 IL-2 Treatment in HIV-infected Patients................................. 520
    14.3.3 The Use of Immunomodulatory Drugs in Primary HIV Infection 521
        14.3.3.1 Use of Cyclosporin A Alongside HAART............................... 522
        14.3.3.2 Mycophenolic Acid in Patients Undergoing Supervised
            Interruption of Therapy 523
14.4 The Dendritic Cell as Target for the Development
    of HIV-1 Vaccines 524
14.5 Dendritic Cells as Tools for the Development of Therapeutic
    Vaccines Against HIV 526
    14.5.1 Introduction.. 526
    14.5.2 Studies in Hu-PBL-SCID Mouse Model.................................. 529
    14.5.3 Studies in Non-Human Primate Model 530
    14.5.4 Clinical Studies with DC-based Vaccines
        in HIV-infected Individuals.. 531
14.6 Concluding Remarks.. 533
References.. 534

Index...541

Contributors

Gosse J. Adema
Tumor Immunology, Medical Centre Nijmegen,
Nijmegen, The Netherlands

Massimo Alfano
AIDS Immunopathogenesis Unit, San Raffaele Scientific Institute,
Center of Excellence on Physiopathology of Cell Differentiation,
Milan, Italy

Jacques Banchereau
Baylor Institute for Immunology Research,
Dallas, Texas, USA

Filippo Belardelli
Department of Cell Biology and Neurosciences, Istituto Superiore di Sanità,
Rome, Italy

Mara Biasin
Chair of Immunology, DISP LITA Vialba, Milan University Medical School,
Milan, Italy

Imerio Capone
Department of Cell Biology and Neurosciences, Istituto Superiore di Sanità,
Rome, Italy

Damien Chaussabel
Baylor Institute for Immunology Research,
Dallas, Texas, USA

Mario Clerici
Chair of Immunology, DISP LITA Vialba, Milan University Medical School,
Milan, Italy

John E. Connolly
Baylor Institute for Immunology Research,
Dallas, Texas, USA

Lucia Conti
Department of Cell Biology and Neurosciences, Istituto Superiore di Sanità,
Rome, Italy

Manuela Del Cornò
Department of Cell Biology and Neurosciences, Istituto Superiore di Sanità,
Rome, Italy

Anthony L. Cunningham
Centre for Virus Research, Westmead Millennium Institute,
Westmead; and University of Sydney,
Sydney, Australia

I. Jolanda M. de Vries
Departments of Pediatric Hemato-Oncology, Medical Centre Nijmegen;
and Tumor Immunology, Medical Centre Nijmegen,
Nijmegen, The Netherlands

Heather Donaghy
Centre for Virus Research, Westmead Millennium Institute,
Westmead, New South Wales, Australia

Alexander H. Enk
Department of Dermatology, University of Heidelberg,
Heidelberg, Germany

Laura Fantuzzi
Department of Cell Biology and Neurosciences, Istituto Superiore di Sanità,
Rome, Italy

Carl G. Figdor
Tumor Immunology, Medical Centre Nijmegen,
Nijmegen, The Netherlands

Maria Cristina Gauzzi
Department of Cell Biology and Neurosciences, Istituto Superiore di Sanità,
Rome, Italy

Sandra Gessani
Department of Cell Biology and Neurosciences, Istituto Superiore di Sanità,
Rome, Italy

Silvia Ghezzi
AIDS Immunopathogenesis Unit, San Raffaele Scientific Institute,
Center of Excellence on Physiopathology of Cell Differentiation,
Milan, Italy

Michel Gilliet
Department of Immunology, MD Anderson Cancer Center,
Houston, Texas, USA

Sabrina Haupt
Institute of Clinical and Molecular Virology,
German National Reference Centre for Retroviruses,
Erlangen, Germany

Guillaume Hoeffel
Institut Cochin, Département d'Immunologie, INSERM U567, CNRS,
UMR 8104, IFR 116, Université Paris V René Descartes,
Paris, France

Anne Hosmalin
Institut Cochin, Département d'Immunologie, INSERM U567, CNRS,
UMR 8104, IFR 116, Université Paris V René Descartes,
Paris, France

Joannes F.M. Jacobs
Departments of Pediatric Hemato-Oncology, Medical Centre Nijmegen,
Nijmegen, The Netherlands

Peter Kelleher
Department of Immunology, Imperial College School of Medicine,
Chelsea and Westminster Hospital,
London, United Kingdom

Nicolai A. Kittan
Institute of Clinical and Molecular Virology,
German National Reference Centre for Retroviruses,
Erlangen, Germany

Adriano Lazzarin
Infectious Disease Clinic, San Raffaele Scientific Institute,
Milan, Italy

Jay A. Levy
Department of Medicine, Division Hematology/Oncology,
University of California,
San Francisco, California, USA

Mary F. Lipscomb
Department of Pathology, University of New Mexico School of Medicine,
Albuquerque, New Mexico, USA

Yong-Jun Liu
Department of Immunology, MD Anderson Cancer Center,
Houston, Texas, USA

Karsten Mahnke
Department of Dermatology, University of Heidelberg,
Heidelberg, Germany

Concepción Marañón
Institut Cochin, Département d'Immunologie, INSERM U567, CNRS,
UMR 8104, IFR 116, Université Paris V René Descartes,
Paris, France

Barbara J. Masten
Department of Pathology, University of New Mexico School of Medicine,
Albuquerque, New Mexico, USA

Steven Patterson
Department of Immunology, Imperial College School of Medicine,
Chelsea and Westminster Hospital,
London, United Kingdom

Cândida F. Pereira
Tumor Immunology, Medical Centre Nijmegen,
Nijmegen, The Netherlands

Guido Poli
Vita-Salute San Raffaele,
Milan, Italy

Melissa Pope
Center for Biomedical Research, Population Council, New York,
New York, USA

Cornelus J.A. Punt
Medical Oncology, Radboud University, Medical Centre Nijmegen,
Nijmegen, The Netherlands

S. Ring
Department of Dermatology, University of Heidelberg,
Heidelberg, Germany

Anne-Claire Ripoche
Institut Cochin, Département d'Immunologie, INSERM U567, CNRS,
UMR 8104, IFR 116, Université Paris V René Descartes,
Paris, France

Paola Rizza
Department of Cell Biology and Neurosciences, Istituto Superiore di Sanità,
Rome, Italy

Barbara Schmidt
Institute of Clinical and Molecular Virology, German National Reference
Centre for Retroviruses, Erlangen, Germany; and Department of Medicine,
Division Hematology/Oncology, University of California,
San Francisco, California, USA

Vassili Soumelis
Inserm U520, Institut Curie,
Paris

Paul J. Tacken
Tumor Immunology, Medical Centre Nijmegen,
Nijmegen, The Netherlands

Giuseppe Tambussi
Infectious Disease Clinic, San Raffaele Scientific Institute,
Milan, Italy

Stuart Turville
Center for Biomedical Research, Population Council, New York,
New York, USA

Elisa Vicenzi
AIDS Immunopathogenesis Unit, San Raffaele Scientific Institute,
Center of Excellence on Physiopathology of Cell Differentiation,
Milan, Italy

Julie A. Wilder
The Lovelace Respiratory Research Institute,
Albuquerque, New Mexico, USA

John Wilkinson
Centre for Virus Research, Westmead Millennium Institute,
Westmead; and University of Sydney,
Sydney, Australia

Part I

General Aspects of Dendritic Cell Biology, Functions, and Clinical Application

Chapter 1

Dendritic Cell Biology:
Subset Heterogeneity and Functional Plasticity

Vassili Soumelis, Yong-Jun Liu, and Michel Gilliet

1.1 Introduction

The immune system is central to the homeostasis of living organisms and has evolved to protect us against a broad variety of microbial pathogens. Cells of the innate immune system represent the frontline of the defense against invading pathogens and play a key role in controlling infections during the time needed for the induction of an adaptive immune response. In addition, innate immune responses are essential to the efficient priming of naïve T and B cells and the generation of an appropriate adaptive immune response. In this respect, dendritic cells (DCs) have the unique capacity to link innate and adaptive immunity, since they sense invading pathogens at the site of infection, capture and process foreign antigens (Ag), and then migrate to the secondary lymphoid organs where they activate naïve T cells in an Ag-specific manner. The three major defining criteria of mature DCs were proposed by Steinman in 1991 (Steinman, 1991): (1) dendritic morphology; (2) constitutive expression of high levels of major histocompatibility complex (MHC) class II molecules; and (3) capacity to induce strong proliferation of naïve $CD4^+$ T cells in a mixed lymphocyte reaction (MLR). Other important features of DCs include the lack of lineage-specific markers, the expression of a variety of costimulatory molecules that will provide signals for the activation and the polarization of T cells (Lanzavecchia and Sallusto, 2001), and the ability to interact, either directly or indirectly, with various immune cell types of the innate or adaptive immune systems (Banchereau and Steinman, 1998; Banchereau et al., 2000; Pulendran et al., 2001a).

DCs form a dynamic "system" with several levels of complexity and diversity according to their lineage origin, stage of differentiation, as well as anatomic location. In each of these different states, DCs exhibit great plasticity, which allows them to adjust their functional properties depending on the nature of the pathogens (Liu *et al.*, 2001).

In this chapter, we summarize current knowledge of the basic developmental, phenotypic, and functional characteristics of DC subsets, trying to underline their differences and complementarities. We chose to focus on human DCs, given the scope of this book. Studies on mouse DCs are only described briefly when bringing important knowledge on the general physiology of DCs. DCs from non-human primates are not specifically discussed but share most major characteristics of human DCs, including subset diversity, phenotype, and function. This general overview should serve as a basis to understand more specific chapters on the role of DCs in HIV infection.

1.2 Origin and Development of Human DC Subsets

DCs are continuously produced from hematopoietic stem cells within the bone marrow, and FLT-3 ligand represents the key DC growth and differentiation factor *in vivo* (Pulendran *et al.*, 2001a). The principal developmental pathways of human DCs from hematopoietic stem cells of the bone marrow are illustrated in Figure 1.1. $CD34^+$ stem cells differentiate into common lymphoid progenitors (CLPs) and common myeloid progenitors (CMPs). CMPs appear to differentiate into CLA^+ and CLA^- populations, which subsequently differentiate into $CD11c^+CD1a^+$ and $CD11c^+CD1a^-$ DC, respectively (Strunk *et al.*, 1997). Whereas $CD11c^+CD1a^+$ DCs migrate into the skin epidermis and become Langerhans cells, $CD11c^+CD1a^-$ DCs migrate into the skin dermis and other tissues and become interstitial DCs (Ito *et al.*, 1999). TGF-β plays a critical role in Langerhans cell (LC) development as demonstrated by the fact that the *in vitro* generation of Langerhans cells from $CD34^+$ progenitors can be greatly enhanced by TGF-β (Caux *et al.*, 1999) and Transforming growth factor (TGF)-β knockout mice lack Langerhans cells (Borkowski *et al.*, 1996). In their peripheral locations, both Langerhans cells and interstitial DCs are immature DCs that are readily activated by products of microbial invasion. Langerhans cells and interstitial DCs display different phenotypes and functions (Caux *et al.*, 1997). While Langerhans cells express CD1a, Lag-antigen, E-cadherin, and Birbeck granule-associated Ag langerin (Romani *et al.*, 2003), interstitial DCs express CD2, CD9, CD68, and factor XIIIa. Interstitial DCs, but not Langerhans cells, have the ability to take up large amounts of antigens by the mannose receptors and to produce IL-10, which may contribute to naïve B-cell activation and IgM production in the presence of CD40-ligand (CD40L) and IL-2.

In addition to these two subsets of immature myeloid DCs (MDCs), stem cells also give rise to two types of DC precursors: monocytes and plasmacytoid DC precursors (pDCs) (Liu *et al.*, 2001). DC precursors are defined by their low expression levels of costimulatory molecules, their failure to induce significant naïve T-cell activation, the lack of DC morphology and mobility in culture, and the ability to colonize non lymphoid tissues in the absence of stimulation. Monocytes are of myeloid origin, and express the myeloid antigens CD11b, CD11c, CD13, CD14, and CD33 (Liu *et al.*, 2001), mannose receptors, and CD1a, b, c, and d.

Stem cells differentiate into CLPs and CMPs. CMPs appear to differentiate into CLA$^+$ and CLA$^-$ populations, which subsequently differentiate and migrate into the skin epidermis to become Langerhans cells and the skin dermis to become interstitial DCs, respectively. In their peripheral locations, both Langerhans cells and the interstitial DCs are immature DCs that can be readily activated by products of microbial invasion. Stem cells also give rise to two types of DC precursors: monocytes and pDCs. Monocytes differentiate into immature DCs in culture with granulocyte-macrophage colony-stimulating factor (GM-CSF) and Interleukin (IL)-4 and can be further activated into mature DCs by stimulating with proinflammatory cytokines such as Tumor Necrosis Factor (TNF)-α, microbial products such as Lipopolysaccharide (LPS) or T cell–derived CD40L. Monocytes may also differentiate into Langerhans cells *in vivo* (Fig. 1.1, dashed line). pDCs represent the natural type 1 interferor (IFN)-producing cell (IPC) and have the ability to differentiate into mature DCs in culture with IL-3 or upon viral activation through Toll-like receptor (TLR)7 or 9.

Whereas the presence of macrophage colony-stimulating factor (M-CSF) leads to the differentiation of monocytes into macrophages, culture with GM-CSF and IL-4 leads to their differentiation into immature DCs with the ability to produce large amounts of IL-12 upon subsequent trigger with CD40L. These immature monocyte-derived DCs (MoDCs) resemble interstitial DCs and can be further activated into mature DCs by stimulating with proinflammatory cytokines such as TNF-α, microbial products such as LPS or T cell–derived CD40L. Addition of TGF-β to GM-CSF and IL-4 can drive the *in vitro* differentiation of monocytes into langerin$^+$, Birbeck granules$^+$ Langerhans cells (Geissmann *et al.*, 1998). This suggests that monocytes form a circulating pool of precursors capable of differentiating into various phagocytic cells in the tissue depending on the local factors they will encounter. *In vivo* monocyte-to-DC transformation may occur after transmigration from the blood to tissues across the endothelial barrier (Randolph *et al.*, 1998; Randolph *et al.*, 1999).

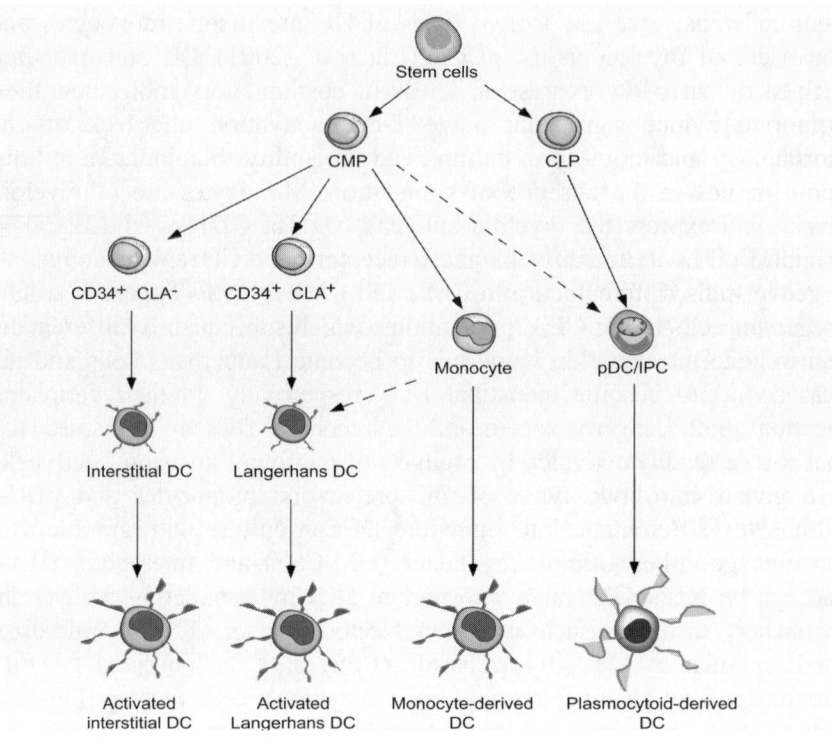

FIGURE 1.1. Pathways of human DC development.

The second DC-precursor subset generated from stem cells is the pDC subset. pDCs represent the natural type 1 IFN-producing cell (IPC) (see Section 1.4) and have the ability to differentiate into mature DCs in culture with IL-3 or upon viral activation through TLR7 or 9 (see Section 1.5). pDCs have plasmacytoid morphology and are lineage-CD4$^+$CD11c$^-$ cells that express BDCA2 (a novel C-type lectin receptor) and BDCA4 (which is identical to neuropilin-1), two specific markers of blood pDCs (Dzionek *et al.*, 2000; Dzionek *et al.*, 2001). pDC also express CD45RA, lack myeloid markers, and express high levels of the IL-3 receptor (CD123). The developmental path and molecular regulation of pDCs are not fully understood. To date, FLT3L ligand is the only known cytokine that is critical for pDC development from hematopoietic stem cells (HSCs) in humans and mice (Blom *et al.*, 2000; Chen *et al.*, 2004; Gilliet *et al.*, 2002). The ability of FLT3L to promote pDC development *in vivo* was confirmed by experiments showing that administration of FLT3L into human volunteers led to an increase in the number of peripheral blood pDCs (Pulendran *et al.*, 2000) and that FLT3L-transgenic mice have increased numbers of pDC, whereas FLT3L-deficient mice have less pDCs (Manfra *et al.*, 2003). Whereas the other human DC subsets described above are of myeloid origin (called myeloid DCs; MDCs), it has been suggested that pDC/IPC are of lymphoid origin. This notion has been supported by findings such as their expression of many lymphoid markers, the lack of surface myeloid markers, and the production of mRNA for germ-line IgK and for pre–T cell receptor (Grouard *et al.*, 1997; Rissoan *et al.*, 1999c). Moreover, two separate studies further support the lymphoid origin of pDCs in humans and mice: (a) overexpression of the dominant-negative transcription factors Id2 or Id3 in human $CD34^+$ hematopoietic progenitor cells blocks development of pDCs, T cells, and B cells, but not of myeloid DCs (Spits *et al.*, 2000); (b) knock down of Spi-B mRNA in human $CD34^+$ hematopoietic progenitor cells strongly inhibits their potential to differentiate into pDCs (Schotte *et al.*, 2004). However, more recent studies revealed that $FLT3^+$ cells within either CLPs or CMPs could differentiate into both MDCs and pDCs in cultures and *in vivo* (Chicha *et al.*, 2004). As a result of these seemingly divergent findings, several different hypotheses have been proposed regarding the developmental origin of pDCs, including the existence of a common DC precursor in blood that can give rise to all DC subsets, pDCs arising as a branch of the committed lymphoid lineage (Corcoran *et al.*, 2003), and lineage conversion (Zuniga *et al.*, 2004).

1.3 Anatomic, Phenotypic, and Functional Features of DC Subsets

Although the *in vitro* differentiation pathways of human DC subsets from different precursors are being well characterized, the correspondence with DCs observed *in vivo* or *in situ* in the tissue (resident or "naturally occurring" DCs) is not always clear (Shortman and Liu, 2002). However, both human and mouse studies have greatly advanced our understanding of the heterogeneity of DCs in peripheral tissue, which is based on the subset, activation state, and distinct microenvironments depending on the anatomic localization.

1.3.1 Thymic DCs

The thymus is a primary lymphoid tissue where T-cell differentiation and selection occurs and leads to the generation of naïve $CD4^+$ and $CD8^+$

T cells with a diverse TCR repertoire, and also naturally occurring CD4[+]CD25[+] regulatory T cells (Apostolou *et al.*, 2002; Jordan *et al.*, 2001; Watanabe *et al.*, 2005) as well as some of the double-negative invariant T-cell subsets, such as NKT cells (Benlagha *et al.*, 2005; Tilloy *et al.*, 1999) or mucosa-associated invariant T (MAIT) cells (Treiner *et al.*, 2003). Human thymus contains pDCs and two subsets of mature CD11c[+] MDCs: CD11b[-]CD45RO[low] DC that lack myeloid markers, and a minority of CD11b[+]CD45RO[high] DCs expressing many myeloid markers (Bendriss-Vermare *et al.*, 2001; Vandenabeele *et al.*, 2001). Thymic pDCs were shown to produce type I IFN in HIV-1–infected thymus, which exerts antiviral effects (Gurney *et al.*, 2004) and upregulates MHC class I expression on thymocytes (Keir *et al.*, 2002). Whether thymic pDCs play a role in the differentiation and/or selection of T-cell subsets is not known.

Thymic MDCs differ from other peripheral MDC subsets in two major ways: (1) they derive from an intrathymic precursor and die within the thymus, suggesting a nonmigratory behaviour (Ardavin *et al.*, 1993); (2) they mostly present self-Ag rather than foreign Ag (Steinman *et al.*, 2003). Thymic MDCs may be involved in the induction of central tolerance through the process of negative selection as well as the generation of the naturally occurring CD4[+]CD25[+] regulatory cells (Watanabe *et al.*, 2005) (see Section 1.6).

1.3.2 Blood DCs

Human DC subsets in the blood are well characterized because of tissue accessibility. Human blood contains two types of DC precursors, monocytes and pDCs, which can be induced to differentiate into DCs after *ex vivo* culture (Grouard *et al.*, 1997; Rissoan *et al.*, 1999a; Sallusto and Lanzavecchia, 1994) and have been described in Section 1.2. In addition, human blood contains a subset of immature CD11c[+] MDCs (O'Doherty *et al.*, 1994). Blood CD11c[+] MDC subsets are considered naïve cells that are migrating from the bone marrow to the peripheral tissue. This assumption is based on their immature phenotype and on the fact that DCs do not recirculate from peripheral tissue to blood, as suggested by mouse studies (Austyn *et al.*, 1988; Kupiec-Weglinski *et al.*, 1988). It is currently believed that blood CD11c[+] MDCs locate to the secondary lymphoid organs and peripheral tissues as resting interstitial DCs and that they are related to the *in vitro* generated monocyte-derived or CD34-derived interstitial DCs (Shortman and Liu, 2002). MDCs are lineage-CD4[+/-] CD11c[+] cells expressing CD45RO, myeloid markers, such as CD13 and CD33, and the MHC-like molecule CD1c. These phenotypic markers allow clear distinction of blood MDCs from blood pDCs (which lack CD11c but express CD123 and BDCA2) and can be used to purify the two subsets for *in vitro* studies (Duramad *et al.*, 2003; Rissoan *et al.*, 1999a; Soumelis *et al.*, 2002).

Blood pDCs express L-selectin and migrate to the secondary lymphoid organs through the high endothelial venules (Yoneyama *et al.*, 2004).

1.3.3 Skin and Mucosal DCs

In the skin and mucosa, DCs form a dense network of resident cells, both within pluri-stratified epithelia (epidermis, anogenital and oropharyngeal epithelium) as well as in subepithelial areas. The best-studied epithelial DC is the Langerhans cell (LC) of the epidermis. As described in Section 1.2, LCs are immature MDCs expressing CD1a, Lag-antigen, E-cadherin, and Birbeck granule-associated Ag langerin (Romani *et al.*, 2003). LC activation induces their migration out of the epidermis, as it is observed *in vivo* in mouse studies (Baldwin *et al.*, 2004; Ratzinger *et al.*, 2002; Romani *et al.*, 2003), *ex vivo* in cultures of human skin explants (Larsen *et al.*, 1990; Schuler *et al.*, 1993), and *in situ* in skin inflammatory diseases, where LC density is markedly decreased in areas of inflammation (Soumelis *et al.*, 2002).

DCs in subepithelial areas, such as the dermis, have a phenotype of interstitial DCs (Shortman and Liu, 2002). In the mucosa of the intestinal tract, DCs are absent in the epithelium but are abundant in the subepithelial region. Here DCs are lined-up under the basal membrane and protrude thin dendrites through the epithelium into the intestinal lumen to sample for foreign Ag (Kelsall and Rescigno, 2004; Rescigno *et al.*, 2001).

In contrast with these MDC subsets, pDCs are not resident cells of normal skin and mucosa (Gilliet *et al.*, 2004; Wollenberg *et al.*, 2002) but are present in HPV-related cervical cancer (Bontkes *et al.*, 2005), skin melanoma lesions (Salio *et al.*, 2003), lupus erythematosus (Farkas *et al.*, 2001), psoriasis (Nestle *et al.*, 2005), allergic contact dermatitis (Bangert *et al.*, 2003) and in the nasal mucosa as early as 6 h after allergen challenge (Jahnsen *et al.*, 2000), suggesting an active recruitment of blood pDCs to the site of peripheral inflammation. Furthermore, pDC recruitment to the skin has been observed in a therapeutic setting in which skin tumors were treated topically with TLR7 agonist imiquimod (Urosevic *et al.*, 2005). As will be discussed further, pDC trafficking has many similarities with T cells, both being attracted to the site of inflammation by chemokines (SDF-1/CXCR3-ligands).

1.3.4 DCs in Solid Organs

DCs are found in small numbers in all organs except brain (Hart and Fabre, 1981). They have characteristics of interstitial DCs. In each organ, the properties of resident DCs are greatly influenced by the local microenvironment. For example, lung DCs are biased for priming T helper 2 (Th2)-type responses (Holt, 2001; Lambrecht, 2001). Recent studies suggest that epithelial cell–derived thymic stromal lymphopoietin (TSLP) could participate in this process by endowing the DCs with a Th2 priming potential (Al-Shami *et al.*, 2005; Soumelis *et al.*, 2002a). In the liver, pDCs and MDCs were both observed in the mouse (Jomantaite *et al.*, 2004). Liver DCs were described to have a tolerogenic potential and can be used to induce tolerance to allografts (Thomson and Lu, 1999). Whether these functional differences are only due to environmental factors, to DC subset diversity, or both is not clearly established.

1.3.5 DCs in Secondary Lymphoid Organs

Secondary lymphoid organs (SLOs) are the anatomic site of naïve T- and B-cell activation and selection of Ag-specific lymphocyte populations. Three subsets have been described in human tonsils (Liu *et al.*, 2001): interstitial DCs located in the T-cell areas and having an immature DC phenotype similar to dermal DCs; pDCs, also located in the T-cell area, and comparable with blood pDCs; germinal center DCs (GCDCs) located in the germinal centers and similar to blood CD11c[+] MDCs (Grouard *et al.*, 1996). An additional subset was described in the marginal zone of mouse spleen that expresses high levels of the endocytic receptor DEC205 (De Smedt *et al.*, 1996). In inflammatory conditions, most of these DC subsets increase in numbers after recruitment from blood and peripheral tissue. Marginal zone DCs rapidly migrate into the T-cell zone after LPS stimulation (De Smedt *et al.*, 1996). Interstitial DCs from the tonsil and lymph nodes correspond with DCs that have captured Ag in peripheral tissues and have migrated to secondary lymphoid organs via the afferent lymph. pDC numbers are also dramatically increased in inflamed lymph nodes (Yoneyama *et al.*, 2004), although the exact migration pattern of pDCs is not clearly established, especially to what extent pDCs from the blood or from inflammatory sites account for this increase.

Migration of DCs from peripheral sites of inflammation to the SLO is an essential step to bring the APC loaded with foreign Ag in close contact with naïve CD4[+] T cells and initiate primary immune responses (Randolph *et al.*, 2005). However, the view that the same DC captures Ag in the periphery and presents this Ag to T cells within SLO was challenged by a recent study suggesting that T cells could be primed by resident SLO DCs that were different from the Ag-loaded DCs (Smith and Fazekas de St Groth, 1999). It was also suggested that free soluble Ag could migrate to the SLO via the afferent lymph and that uptake, processing, and presentation would be performed by resident SLO DCs (Sixt *et al.*, 2005). All three scenarios could be involved *in vivo* in this complex process, depending on the type of inflammation and the type of Ag. The result in all cases is Ag presentation to naïve T cells by a mature DCs that can provide the adequate costimulatory signals to induce the generation of effector T cells.

1.4 Innate Functions of DC Subsets

Although the major defining function of DCs is Ag presentation to T cells, they also perform a number of innate effector functions, in particular at the precursor or immature stage. DCs are particularly abundant at the sites of pathogen entry, such as the skin, the respiratory tract, lungs, and gut (Foti *et al.*, 2004). Before accomplishing their function in T-cell stimulation, they play a key role in orchestrating the recruitment and activation of various innate cell types, such as granulocytes, NK cells, NKT cells, and other DCs. They also directly participate in the control of infectious pathogens, mostly by producing antiviral cytokines such as type I IFNs. Many of these innate DC functions are critical to control infection and can greatly influence the subsequent adaptive immune response (Medzhitov and Janeway, 1997).

1.4.1 Invariant Receptors on DCs

DCs express a broad set of receptors that they use to sense the environment for danger signals, most of which are associated with infection or tissue damage. These innate receptors are germ-line encoded and can be categorized based on the type of structures that they recognize (Janeway and Medzhitov, 2002). Pattern-recognition receptors (PRRs) recognize conserved components that are specifically associated with microbial pathogens, also termed pathogen-associated molecular patterns (PAMPs). PPRs include Toll-like receptors (TLRs) 1-11 (Akira, 2003; Beutler, 2002; Medzhitov and Janeway, 2000) and a variety of C-type lectins (Gordon, 2002) (Table 1.1). Their ligands range from glycoproteins and polysaccharides to DNA oligonucleotides and single- or double-stranded RNA. Other invariant receptors include cytokine and chemokine receptors that deliver activating or inhibitory signals to the DCs depending on the inflammatory environment. Fc receptors and complement receptors mediate the opsonization of microbes or apoptotic cells (Verbovetski

TABLE 1.1. Phenotypic and functional properties of human pDC precursors and DC subsets.

Subsets of human DC-precursors

Phenotype	Monocytes	pDC/IPC
Myeloid marker		
CD11b	+	−
CD11c	+	−
CD13	+	−
CD14	+	−
CD33	+	−
Lymphoid marker		
Pre-T&	−	+
lg λ-like 14.1	−	+
Spl-B	−	+
Pattern-recognition receptors		
Mannose R	+/−	−
BDCA2	−	+
TLR2	++	−
TLR4	++	−
TLR7	−	++
TLR8	−	++
CD1a, b, c, d	+/−	−
DC-sign	−	−
Other differentially expressed antigens		
CD4	+	++
CD16RA	−	+
CD45RO	+	−
IL3R	+	+++
GM-CSFR	++	+
Function		
Phagocytosis & kill bacteria	++	−
IFN-α/β production	+	++++

DC Differentiation

	Monocytes	pDC/IPC	
In vitro	OM/IL4 CD40L	IL3 CD40L	Virus
In vivo	endothelial transmigration	?	?
DC function			
IL-12	+++	−	−
IFN-α/β	−	−	+++
TH1	++	+/−	++
TH2	−	++	−
CTL	++	−	?
IL-10 T suppressors	−	++	+

Subsets of human SC-derived immature DC

	Langerhans cells	Interstitial DC
Phenotype		
CD1a	+	−
CD2	−	+
CD9	−	+
CD68	−	+
Factor XIIa	−	+
E-cadherin	+	−
Birback granular		
Lag-antigen	+	−
DG sign	−	++
Function		
Macropinocytosis	−	+
IL-10 production	−	+
B-cell activation	+/−	+
COB T-cell priming	−	+

et al., 2002) and have key roles in the phagocytosis process. They also deliver specific activating or inhibitory signals and condition the processing and presentation pathways for a given Ag (Amigorena, 2002).

Expression of invariant receptors on DC subsets reveals a remarkable complementarity between MDCs and pDCs, suggesting that these two subsets evolved to perform distinct functions (Liu *et al.*, 2001). In the human, MDCs express TLR2 and TLR4 (Hornung *et al.*, 2002; Jarrossay *et al.*, 2001; Kadowaki *et al.*, 2001b), which mediate the response to ligands such as glycoproteins from mycobacteria and Gram-positive bacteria (Aliprantis *et al.*, 1999; Hirschfeld *et al.*, 1999; Poltorak *et al.*, 1998; Yoshimura *et al.*, 1999). TLR3 is found in the intracellular compartments of both monocyte-derived (MoDCs) and MDCs (Matsumoto *et al.*, 2003) and binds to double-stranded RNA, a bystander product of viral replication. MDCs also express high amounts of C-type lectin receptors that bind specific carbohydrate structures from pathogens and from self glycoproteins (Figdor *et al.*, 2002; Gordon, 2002). The mannose receptor (CD206) mediates the rapid uptake of large amounts of polysaccharide antigens such as fluorescein isothiocyanate-dextran (Grouard *et al.*, 1997). Although mostly considered as an Ag uptake receptor, signaling through the mannose receptor could inhibit LPS-induced IL-12 production by MDCs (Nigou *et al.*, 2001). DC-SIGN, another C-type lectin, is expressed by immature dermal and interstitial DCs and facilitates the endocytosis of several microbial pathogens, including HIV (van Kooyk and Geijtenbeek, 2003), *Schistosoma mansoni* (van Kooyk and Geijtenbeek, 2003), myco-bacteria (Geijtenbeek *et al.*, 2003; Tailleux *et al.*, 2003), and *Helicobacter pylori* (Bergman *et al.*, 2004). Binding of these ligands to DC-SIGN can induce the production of IL-10 by DCs and could favor the establishment of persistent infection (van Kooyk and Geijtenbeek, 2003).

Contrary to MDCs, pDCs and plasmacytoid-derived DCs express TLR7 and TLR9 (Jarrossay *et al.*, 2001; Kadowaki *et al.*, 2001b) and respond to their respective ligands, imidazoquinolines (Gibson *et al.*, 2002; Hemmi *et al.*, 2002) and single-stranded RNA (Diebold *et al.*, 2004; Heil *et al.*, 2004; Lund *et al.*, 2004) for TLR7, CpG-containing oligonucleotides (ODN) (Hemmi *et al.*, 2000; Krug *et al.*, 2004) for TLR9. They do not express TLR2, TLR3, and TLR4 and do not respond to such ligands as peptidoglycan, LPS, or double-stranded RNA (Jarrossay *et al.*, 2001; Kadowaki *et al.*, 2001b). Also contrary to MDCs, pDCs do not express mannose receptor and DC-SIGN but express the lectin BDCA2 (Dzionek *et al.*, 2000; Dzionek *et al.*, 2001). BDCA2 mediates Ag uptake and inhibits pDC production of type 1 IFN induced by influenza virus (Dzionek *et al.*, 2001).

Another PRR expressed by MDCs is NOD2, a member of the NOD-leucine-rich repeat protein family, which recognizes the muramyl dipeptide of peptidoglycan and was shown to be a negative regulator of TLR2-mediated T helper 1 (Th1) responses (Watanabe *et al.*, 2004). NOD2 is mutated in patients with inflammatory bowel disease, potentially inducing an over-reaction to endogenous commensal bacteria (Hugot *et al.*, 2001). Whether pDCs express NOD2 is currently unknown.

Finally, MDCs and pDCs also differ in the expression of Fc receptors. MDCs express the Fc-γ receptors I (CD64) and II (CD32), which mediate phagocytosis of immune complexes (Fanger *et al.*, 1996), but also DC maturation and MHC class I Ag presentation (Banki *et al.*, 2003; Regnault *et al.*, 1999), as well as tumor immunity in a mouse melanoma model (Kalergis and Ravetch, 2002). On the contrary, pDCs only express low levels of FcγRIIa (CD32), which was shown to mediate the IFN-α production of pDCs in response to apoptotic cells combined with lupus IgG or DNA-containing immune complexes (Bave *et al.*, 2003; Means *et al.*, 2005).

The broad variety of innate receptors expressed by DCs enables the DCs to receive a large number of signals from a complex inflammatory milieu. These signals will induce and/or modulate key innate functions of DCs, such as Ag uptake and processing, cytokine and chemokine production, and ultimately maturation and migration of the DC to the secondary lymphoid organs. During this process, the DCs will integrate all the information received at the site of Ag challenge and translate it into the appropriate adaptive immune response.

1.4.2 Antigen Uptake and Processing by DC Subsets

DCs continuously sample their environment in peripheral tissue and use a variety of specific or nonspecific mechanisms for Ag uptake. MDCs constitutively use macropinocytosis, which is facilitated by the expression of certain aquaporins (de Baey and Lanzavecchia, 2000). They also use receptor-mediated endocytosis. This process is linked to the broad variety of innate receptors expressed by DCs, as detailed in the above section. Among all these receptors, C-type lectins are mostly involved in endocytosis of their respective glycosylated Ag ligand, especially DEC205, mannose receptor, and DC-SIGN (Thery and Amigorena, 2001). DCs also uptake Ag through Fc receptors that mediate internalization of immune complexes (Maurer *et al.*, 1998; Regnault *et al.*, 1999), while receptors for the heat-shock proteins (hsp) gp96 and hsp70 mediate the internalization of hsp-peptide complexes (Arnold-Schild *et al.*, 1999; Binder *et al.*, 2000b; Binder *et al.*, 2000a; Castellino *et al.*, 2000; Singh-Jasuja *et al.*, 2000). Another Ag uptake pathway is through phagocytosis of apoptotic bodies, which is mediated by surface CD36 and the alphavbeta5 integrin (Albert *et al.*, 1998, 2000). This pathway was shown to be important for the maintenance of peripheral tolerance (Steinman *et al.*, 2003), whereas phagocytosis of necrotic cells induced maturation of MDCs and could favor immunity (Sauter *et al.*, 2000).

As opposed to MDCs, pDCs and pDC-derived DCs lack expression of most C-type lectins and Fc receptors that are known to mediate endocytosis of specific ligands (Liu, 2005). They also have very weak macropinocytosis (Grouard *et al.*, 1997), suggesting that their role in sampling the environment is different from MDCs. It was suggested that pDCs could present endogenous self-Ag that do not require uptake or that their function in Ag presentation could be restricted to viral Ag (Liu, 2005). Indeed, pDCs were shown to be able to process and present viral Ag *in vitro* (Fonteneau *et al.*, 2003) and *in vivo* (Salio *et al.*, 2004) when the mechanisms mediating viral entry into the cell were intact. Interestingly, stimulation of virus-specific CD8$^+$ T cells was abolished when pDCs were stimulated with boiled influenza virus that had lost its fusogenic properties and could not actively enter the cell (Fonteneau *et al.*, 2003).

In MDCs, Ag uptake capacities are down-modulated upon maturation (Thery and Amigorena, 2001). This is partly due to a decreased level of expression of endocytic receptors but not to an intrinsic defect of mature DCs in endocytosis (Garrett *et al.*, 2000; West *et al.*, 2000). However, the overall levels of phagocytosis and macropinocytosis were shown to be down-modulated in mature DCs (Garrett *et al.*, 2000; West *et al.*, 2000). This down-modulation is believed to restrict T-cell stimulation to those Ag that the DC encounters in the periphery at the site of antigenic challenge (Guermonprez *et al.*, 2002).

1.4.3 Cytokine and Chemokine Production by DCs

Both TLR-dependent and -independent DC activation results in the production of cytokines and chemokines, although some cytokines, such as pDC-derived type I IFN, show a strict TLR dependency (Asselin-Paturel and Trinchieri, 2005; Colonna *et al.*, 2002; Liu, 2005; Reis e Sousa, 2004). MDCs have the ability to produce a variety of proinflammatory (TNF-α, IL-1, IL-6) and anti-inflammatory (IL-10) cytokines. They are the main source of the Th1-polarizing cytokine IL-12, together with macrophages that produce much lower levels. Although MDCs can produce IL-12 *in vitro* after CD40-ligand stimulation (Cella *et al.*, 1996), TLR triggering was shown to be required for IL-12 production *in vivo* (Sporri and Reis e Sousa, 2005). Autocrine type-I IFN could play a key role in this process as it was shown to be required *in vitro* for the production of IL-12 by DCs (Gautier *et al.*, 2005). Another important TLR-induced cytokine is IL-6, which inhibits regulatory T cell–mediated immune suppression (Pasare and Medzhitov, 2003). Contrary to MDCs, pDCs produce no or very low levels of IL-12 but are specialized in the production of type I IFN (Asselin-Paturel and Trinchieri, 2005; Colonna *et al.*, 2002; Liu, 2005). They are the principal source of type I IFN in human blood and very rapidly produce all type I IFN isoforms in response to microbial stimuli, such as virus (Cella *et al.*, 1999a; Diebold *et al.*, 2004; Siegal *et al.*, 1999), CpG-containing oligonucleotides (Hartmann *et al.*, 1999; Kadowaki *et al.*, 2001b), or the synthetic molecules imidazoquinolines (Gibson *et al.*, 2002). pDC-derived type I IFN has direct antiviral activity against a variety of virus, including HIV, and has important adjuvant functions on other immune cell-types, such as NK cells, T cells, macrophages, and DCs (Bogdan, 2000).

Chemokines are also differentially produced by MDCs and pDCs during an immune response. MDCs produce high levels of the homeostatic chemokines CCL17 and CCL22, while pDCs mostly produce the inflammatory chemokine CCL3 (Bendriss-Vermare *et al.*, 2005; Penna *et al.*, 2002). CCL4 and CXCL8 are produced by both subsets. This suggests the recruitment of different T-cell subsets at the site of inflammation.

1.4.4 Cross-talk Between DCs and Other Innate Immunity Cell Types

In addition to T cells and B cells, DCs were shown to interact with cells of the innate immune system. First evidence of this cross-talk relevant to innate immunity came from *in vivo* mouse studies showing that DCs are critical for the function of NK cells (Fernandez *et al.*, 1999). MDCs can establish cellular interactions with NK cells with characteristics of an immune synapse (Borg *et al.*, 2004). The molecular basis of this interaction is being actively studied. DC-derived IL-2 was shown to play a crucial role in NK cell activation (Granucci *et al.*, 2004). IL-2–activated NK cells in turn could promote the maturation of MDCs via IFN-γ production (Gerosa *et al.*, 2002; Martin-Fontecha *et al.*, 2004), suggesting a bidirectional cross-talk between these two cell-types. Other DC-derived cytokines important in this cross-talk include IL-12 for IFN-γ priming and IL-15 for NK cell survival and proliferation (Ferlazzo *et al.*, 2004). PDCs induce NK cell activation by producing type I IFN, but other mechanisms could also be involved (Gerosa *et al.*, 2005).

MDCs express the MHC class I-like molecule CD1d and are able to activate NKT cells via presentation of the glycolipid Ag α-galactosylceramide (Ikarashi *et al.*, 2001; Kadowaki *et al.*, 2001a), the endogenous lysosomal glycosphingolipid isoglobotrihexosylceramide (iGb3) (Zhou *et al.*, 2004), or glycosylceramides from the cell wall of LPS-negative bacteria (Mattner *et al.*, 2005). Contrary to MDCs, pDCs do not express CD1d and do not directly interact with NKT cells (Kadowaki *et al.*, 2001a). Recently, DCs were shown to interact with neutrophils via the lectin receptor DC-SIGN and the neutrophil β(2)-integrin Mac-1. This interaction together with neutrophil-derived TNF-α induces DC maturation and favours a Th1 polarization of naïve CD4[+] T cells (van Gisbergen *et al.*, 2005). During the effector phase of allergy, mast cells could influence pDC function through histamine-mediated down-regulation of type I IFN production (Mazzoni *et al.*, 2003). Other interactions between cells of the innate immunity could take place *in vivo* given the complexity of the cellular infiltrate.

1.4.5 DC Maturation and Migration to the Secondary Lymphoid Organs

Maturation and migration of DC subsets occur simultaneously in response to microbial and inflammatory stimuli. TLR ligands induce a strong DC maturation, both by direct effects mediated by NFkB activation and via the induction of cytokines, such as TNF-α, which act in an autocrine or paracrine fashion (Kaisho and Akira, 2001; Medzhitov and Janeway, 2000). Inflammatory cytokines, such as TNF-α, GM-CSF, IL-1, and IL-6, could participate in this maturation process, in collaboration with other signals, such as those mediated by Fc receptors (Amigorena, 2002). IL-3 specifically and strongly activates pDC differentiation into DCs (Grouard et al., 1997; Rissoan et al., 1999b) and potentiates TLR-induced pDC differentiation and type I IFN production (Gary-Gouy et al., 2002). Inflammatory cytokines, such as TNF-α and IL-1β (autocrine or paracrine), correlated with the expression of the CCR7-ligand SLC by the lymphatic endothelium (Martin-Fontecha et al., 2003) and play a role in DC migration to the lymph node (Austyn et al., 1988; Martin-Fontecha et al., 2003; Stoitzner et al., 1999). Other factors were critical for the induction of DC migration to the SLO, including IL-16 (Stoitzner et al., 2001), 6 integrins (Price et al., 1997), and the metalloproteinases MMP9 and MMP2 (Ratzinger et al., 2002). In terms of chemokine receptors, blood MDCs and pDCs express a similar pattern, including CXCR4, CCR5, CCR7, and CXCR3. CCR7 on MDCs is upregulated upon maturation and is essential for the migration of MDCs to the lymph nodes (Dieu et al., 1998; Ngo et al., 1998; Sozzani et al., 1998). Although blood pDCs express higher levels of CCR7 than MDCs, they become responsive to CCR7 ligands only after activation through CD40 ligation (Penna et al., 2001). Blood pDCs also show an impaired capacity to migrate in response to inflammatory chemokines suggesting a propensity to migrate to secondary lymphoid organs (Penna et al., 2001). Although the expression of the skin homing Ag CLA indicates a capacity to migrate to the skin, the precise mechanisms orchestrating pDC migration to the inflamed tissues remain unclear. A recent study indicates that the synergistic action of the constitutive CXCR4 ligand CXCL12 and the inducible CXCR3 ligands CXCL9 and CXCL1 greatly enhances pDC migration to CXCL12 and could regulate pDC recruitment either in lymph nodes or at sites of inflammation (Vanbervliet et al., 2003).

1.5 Role of DC Subsets in Adaptive Immunity

The human immune system has evolved to have two distinct mechanisms for protection against microbes. In response to intracellular microbes (bacteria, viruses, fungi, and intracellular parasites), DC subsets produce IL-12 or type I IFN (MDCs and pDCs, respectively), which stimulate CD4$^+$ T helper cells to differentiate into IFN-γ–producing T helper (Th) 1 cells (Lanzavecchia and Sallusto, 2001). Th1 cells activate macrophages and CD8$^+$ cytotoxic T cells to kill intracellular microbes. On the other hand, extracellular parasites, such as helminths, trigger DCs to activate CD4$^+$ Th cells into Th2 cells, secreting proallergic cytokines such as IL-4, IL-5, and IL-13 (Sher *et al.*, 2003b). Th2 cytokines trigger immuno-globulin E (IgE) production, which activates mast cells and eosinophils to eradicate the extracellular microbes. The ability of MDCs and pDCs to direct Th1-versus Th2-type immune responses is determined both by their lineage origin (functional diversity) as well as the environment to which they are exposed (functional plasticity).

1.5.1 MDCs in Adaptive Immune Responses

Myeloid DC subsets have the capacity to produce IL-12 in response to microbial stimuli and thereby to induce protective Th1-type immune responses; however, this capacity may vary with the type of signals delivered to the DC (summarized in Fig. 1.2). Pathogens and/or pathogen-derived products such as LPS from *Escherichia coli* (Pulendran *et al.*, 2001b), peptidoglycan from Gram-positive bacteria (Hilkens *et al.*, 1997), *Mycobacterium tuberculosis* (Stenger and Modlin, 2002), pertussis toxin (Hou *et al.*, 2003), *Toxoplasma gondii* (Sher *et al.*, 2003a), and double-stranded viral RNA (Cella *et al.*, 1999b) may activate myeloid DCs to produce IL-12 and drive Th1 development. By contrast, microbial products such as LPS from *Porphyromonas gingivalis* (Jotwani *et al.*, 2003), *Candida albicans* at the hyphae stage (d'Ostiani *et al.*, 2000), flagellin (Didierlaurent *et al.*, 2004), Der p 1 (house dust mite allergen) (Hammad *et al.*, 2003), *Nippostrongylus brasiliensis* (Balic *et al.*, 2004), and *Schistosoma mansoni* egg extract (Kane *et al.*, 2004) may activate myeloid DCs to induce Th2 cell development, due to their lower capacity to induce IL-12. In addition, prostaglandin E2 (PGE2) (Kalinski *et al.*, 2001), as well as cholera toxin (Lavelle *et al.*, 2004) and histamine (Idzko *et al.*, 2002) induce maturation of Th2-inducing myeloid DCs through active sup-pression of bioactive IL-12 p70. Immunosuppressive cytokines such as IL-10 (Steinbrink *et al.*, 1997) and TGF-β (Strobl and Knapp, 1999), as well

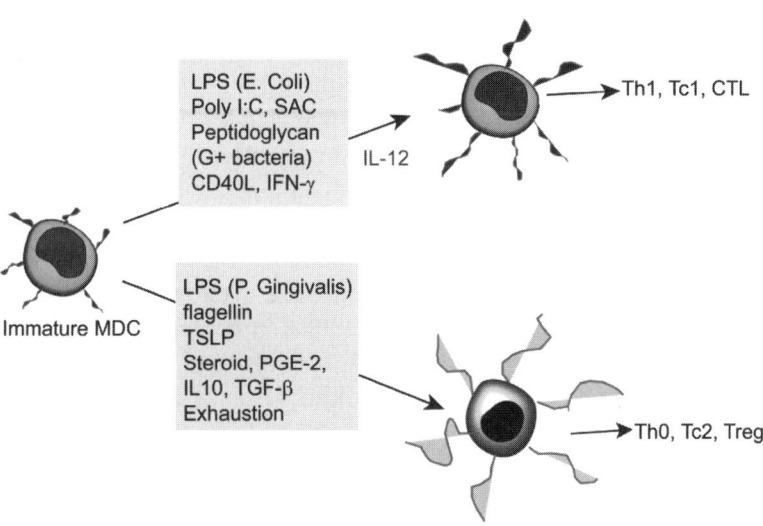

FIGURE 1.2. Functional plasticity of myeloid DCs.

as corticosteroids (de Jong *et al.*, 1999), cyclosporin A (Sauma *et al.*, 2003), and 1α,25-dihydroxyvitamin D (Pedersen *et al.*, 2004), inhibit the maturation of myeloid DCs as well as their IL-12 production. These DCs fail to drive Th1 responses but may rather induce regulatory T cells, similar to immature steady-state DCs (see Section 1.6). Human TSLP, an IL-7-like cytokine, may be the key physiologic mediator that causes allergic inflammation through DCs (Al-Shami *et al.*, 2005; Soumelis *et al.*, 2002; Yoo *et al.*, 2005). TSLP strongly activates human CD11c[+] blood MDCs to upregulate costimulatory molecules and to secrete the Th2 cell–attracting chemokines TARC and MDC but not IL-12 or proinflammatory cytokines. TSLP-activated MDCs induce allogeneic naïve CD4[+] T cells to undergo robust proliferation and subsequently differentiate into Th2 cells capable of secreting large amounts of IL-4, IL-5, IL-13, and TNF-α (Soumelis *et al.*, 2002). These facts, together with the finding of high TSLP expression in keratinocytes from the skin lesions of patients with atopic dermatitis, suggest that TSLP plays a critical role in the initiation of allergic inflammation in humans.

MDCs may also drive Th2 development after prolonged stimulation (a phenomenon called exhaustion). After a burst of IL-12 production, MDCs can no longer produce IL-12 and drive Th1 development, but instead induce Th2 cells and nonpolarized Th0 cells (Langenkamp *et al.*, 2000). Thus, the ability of MDCs to induce Th1 versus Th2 responses seems to be principally regulated through their ability to produce IL-12.

In addition to the microbial stimuli, also T cell–derived signals such as CD40L, IL-4, and IFN-γ can amplify bioactive IL-12 p70 production from MDC (Cella *et al.*, 1996; D'Andrea *et al.*, 1995; Hochrein *et al.*, 2000; Kalinski *et al.*, 2000; Macatonia *et al.*, 1995).

Immature myeloid DCs (MDCs) activated by LPS (lipopolysaccharide from *E. coli*), Poly I:C (double-stranded viral RNA) SAC (*Staphylococcus aureus* Cowan 1), peptidoglycan from Gram-positive bacteria as well as T cell–derived factors including CD40-ligand and IFN-γ produce IL-12 and mature into DCs with the ability to prime naïve $CD4^+$ and $CD8^+$ T cells to produce IFN-γ (Th1 and Tc1 cells) and acquire cytotoxic activity (CTL). By contrast, immature MDCs activated by LPS from *Porphyromonas gingivalis*, flagellin, thymic stromal lymphoietin (TSLP), steroids, prostaglandin (PGL)-2, IL-10, and TGF-β fail to induce significant levels of IL-12 and prime naïve T cells to produce IL-4, IL-5, and IL-13 (Th2), IL-4, IL-5, IL-13, and IFN-γ (Th0) or become IL-10–producing T regulatory cells (Treg).

1.5.2 pDCs in Adaptive Immune Responses

pDCs express low levels of MHC class II and low to undetectable levels of CD80 and CD86 and are incapable of stimulating significant antigen-specific T-cell proliferation (Grouard *et al.*, 1997). *In vitro* studies suggest that pDCs have two pathways to differentiate into mature DCs and acquire the capacity to directly talk to T cells: (a) the IL-3–dependent pathway, in which human pDCs differentiate into DCs in culture with IL-3 or IL-3 plus CD40L; and (b) the IFN-α– and TNF-α–dependent pathway, in which signaling through TLR7 and TLR9 by viruses or by synthetic CpG ODN stimulates pDCs to produce IFN-γ and TNF-β, two cytokines that induce pDCs to differentiate into DCs (Liu, 2005). Whereas pDC-derived mature DCs induced by IL-3 and CD40L preferentially prime naïve $CD4^+$ T cells to produce IL-4, IL-5, and IL-10 (Rissoan *et al.*, 1999b), pDC-derived mature DCs induced by virus preferentially prime naïve $CD4^+$ T cells to produce IFN-γ and IL-10 (Kadowaki *et al.*, 2000) (Fig. 1.3). These studies suggest that, like MDCs, pDCs also display functional plasticity with the ability to drive Th2- or Th1-like responses depending on the type of maturation signals. The biological significance of pDC differentiation into DC in the presence of IL-3 is unknown. Investigators have proposed that IL-3 may be produced by basophils, eosinophils, and mast cells during parasite infection, and pDC-derived DCs may play a role triggering antiparasite adaptive Th2 immune responses. Two recent studies in mice demonstrate that pDCs indeed differentiate into DCs after viral infection *in vivo* (O'Keeffe *et al.*, 2002; Schlecht *et al.*, 2004).

pDCs have two pathways to differentiate into mature DCs: the TLR-dependent pathway, in which signaling through TLR7 and TLR9 by viruses or by synthetic CpG ODN stimulates pDCs to produce IFN-α and TNF-α, two cytokines that induce pDCs to differentiate into DCs; and the IL-3–dependent pathway, in which human pDCs differentiate into DCs in culture with IL-3 or IL-3 plus CD40L. Whereas pDC-derived mature DCs induced by virus preferentially prime naïve CD4⁺ T cells to produce IFN-γ and IL-10 (Th1-like), pDC-derived mature DCs induced by IL-3 and CD40L preferentially prime naïve CD4⁺ T cells to produce IL-4, IL-5, IL-13, and IL-10 (Th2).

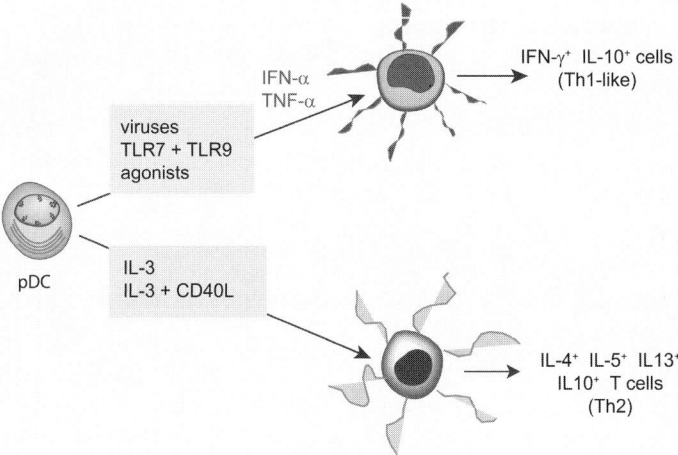

FIGURE 1.3. Functional plasticity of pDC-derived DCs.

These studies indicate that pDCs represent a unique cell lineage within the immune system, which first play a critical role as effector cells in antimicrobial innate immune responses and subsequently differentiate into professional antigen-presenting cells to initiate adaptive immune responses. A puzzling feature of pDCs is their ability to induce IL-10–producing cells regardless of the maturation pathway (IL-3- or TLR-mediated). As discussed in Section 1.6, these pDC-derived DCs may play a key role in peripheral T-cell tolerance by inducing IL-10–producing regulatory cells. However, there is now compelling evidence that pDCs and their activation to produce IFN-α play an important role in inducing productive, adaptive T cell–mediated immune response through the activation of MDCs. This has been suggested in three different types of studies. First, pDCs in systemic lupus erythematosus (SLE) patients appear to be constantly activated through TLR9 by self-chromatin–antichromatin antibody complexes to produce IFN-α (Lovgren et al., 2004). Second, IFN-α within the sera of SLE patients strongly activates monocytes and immature MDCs, which subsequently induce strong Th1-mediated immune responses (Blanco *et al.*, 2001). Third, a recent study showed that human pDCs could induce a bystander maturation of MDCs in response to HIV infection *in vitro* by producing IFN-α and TNF-α. (Fonteneau *et al.*, 2004). The exposure of immature MDCs to IFN-α leads to their activation and enhanced production of IL-12, IL-15, IL-18, and IL-23 (Luft *et al.*, 1998; Nagai *et al.*, 2003; Paquette *et al.*, 1998; Santini *et al.*, 2000). Furthermore, pDCs activated by virus also provide contact-mediated help to bystander MDCs and increase their ability to induce antiviral CTL in a CD2- and CD40-ligand–dependent manner (Yoneyama *et al.*, 2005). In addition to the role in activating bystander MDCs, pDC-derived IFN-γ may also influence directly Th1-polarization of CD4$^+$ T cells (Hibbert *et al.*, 2003), promote expansion of effector CD8$^+$ T cells (Kolumam *et al.*, 2005), participate in the differentiation of B cells into antibody (Ab)-secreting plasma cells (Jego *et al.*, 2003), or directly activate IFN-γ–secreting NK cells, thus further promoting the induction of Th1 inflammation.

1.5.3 Molecular Basis for DC-mediated Th1/ Th2 Polarization

The findings described above suggest that the ability of DC subsets to promote Th1 or Th2 polarization may depend on whether the physiologic conditions are favorable for the induction of IL-12 or IFN-α/β in MDCs or pDCs, respectively. However, this concept was challenged by the obser-vation that IL-12–deficient mice infected by *T. gondii* can still develop Th1 cell responses to some extent (Jankovic *et al.*, 2002) indicating that factors in addition to IL-12 are responsible for the residual Th1 cell induction by MDCs. Two other members of the IL-12 family, IL-23 and IL-27, have been recently identified and may represent additional Th1-inducing factors produced by MDCs (Trinchieri *et al.*, 2003). Also, Th2 differentiation may not be a default mechanism in the absence of Th1-inducing cytokines but may rather require a positive Th2 cell–instructive signal. This is suggested by the finding that IL-12–deficient mice infected with *T. gondii* fail to develop Th2 cell responses (Jankovic *et al.*, 2002). Although IL-4 has been recognized as being an important factor for Th2 cell differentiation, Th2 differentiation can occur in the absence of IL-4 receptor (IL-4R) or IL-4R signaling (Jankovic *et al.*, 2000). Therefore, signals other than IL-4 may instruct Th2 differentiation. Recently, it was shown in the murine system that T-cell polarization by MDCs is regulated by two types of Notch ligands (called Delta and Jagged): whereas LPS-activated myeloid DCs strongly express Delta, which contributes to the

Th1 cell–inducing activity (Amsen *et al.*, 2004; Maekawa *et al.*, 2003), DC activation with PGE2 and cholera toxin preferentially induces the expression of Jagged, which instructs naïve CD4$^+$ T cells to differentiate into Th2 cells independently of IL-4 (Amsen *et al.*, 2004). Other signals than Jagged may also be involved in DC-mediated Th2 responses. Several studies have emphasized the role of OX40-ligand (OX40L), a member of the TNF superfamily, in triggering the development and maintenance of Th2 cells in mice and humans (Akiba *et al.*, 2000; Ito *et al.*, 2004; Jember *et al.*, 2001). *S. mansoni* egg extract induces human MDCs to express OX40L, which contributes to the priming of Th2 cells (de Jong *et al.*, 2002). Furthermore, the ability of TSLP-activated MDCs to drive Th2 polarization is related to their expression of OX40-ligand (Ito and Liu, unpublished data). pDC-derived DCs induced by IL-3 or a virus also express considerable levels of OX40L, and blockade of OX40L significantly inhibits the ability of IL-3-DC to prime naïve CD4$^+$ T cells to produce IL-4, IL-5, and IL-13 (Ito *et al.*, 2004). pDCs activated by a virus induce huge amounts of type I IFN, which conceal the OX40L effect and thus facilitate Th1 cell responses (Ito *et al.*, 2004). A number of other molecules including novel cytokines, TNF-family members, or B7-like molecules may determine the capacity of DC subsets to ultimately drive Th1 or Th2 polarization. The expression of these molecules and their balance in the microenvironment (functional plasticity) as well as the preferential expression of these molecules by distinct DC subsets (functional diversity) may dictate the quality of T-cell responses induced.

1.6 DCs and Tolerance

DCs can induce and maintain immune tolerance, both central and peripheral. Central tolerance may depend on thymic MDCs in the cortico-medullary junction, which are believed to be critical for the deletion of newly generated T cells that have a receptor that recognizes self-components with high affinity (a process called negative selection) (Wu and Shortman, 2005). Recently, it has been shown that Hassall's corpuscles within the thymic medulla express TSLP and that TSLP-activated mature thymic MDCs are able to induce the differentiation of CD4$^+$CD25$^+$ regulatory T cells (Watanabe *et al.*, 2005). Thus, there are two mechanisms of DC-mediated central tolerance: one through the clonal deletion of self-reactive thymocytes, the other by means of converting self-reactive thymocytes into suppressive regulatory T cells (naturally occurring CD4$^+$CD25$^+$ T-regulatory cells).

Central tolerance might not be effective for all antigens. Therefore, there is a need for induction and maintenance of T-cell tolerance in the periphery. Peripheral tolerance occurs in lymphoid organs and is mediated by immature MDCs. In the absence of appropriate activation signals (danger signals delivered by products of microbial invasion), resting immature MDCs may constantly migrate from the peripheral sites into secondary lymphoid organs and present tissue antigens to T cells in the absence of appropriate costimulation, leading to T-cell anergy and to the development of IL-10–secreting inducible regulatory T cells (Steinman and Nussenzweig, 2002). Fusion proteins targeted to immature myeloid DCs lead to the induction of antigen-specific tolerance (Hawiger *et al.*, 2001). By contrast, concomitant activation of these DCs with CD40-specific antibody results in a potent immune response, because the DCs are induced to express a large number of costimulatory molecules (Hawiger *et al.*, 2001).

In addition to having an immunogenic function, pDCs may have tolerogenic functions. Freshly isolated resting pDCs express low levels of MHC class I, class II, and CD86 and no detectable level of CD80 (Grouard *et al.*, 1997). Although resting pDCs do not have the capacity to induce strong T-cell proliferation, they appear to prime $CD4^+$ T cells to differentiate into IL-10–producing Tr1 cells in cultures in both humans (Kuwana *et al.*, 2001) and mice (Bilsborough *et al.*, 2003; Martin *et al.*, 2002). In contrast with MDCs, the ability of pDCs to prime T cells to produce IL-10 is maintained even after differentiation into mature activated DCs. IL-3- and CD40-activated pDC-derived DCs prime naïve $CD4^+$ T cells to Th2 cells secreting high levels of IL-10 (Rissoan *et al.*, 1999b) and naïve $CD8^+$ T cells to IL-10–producing $CD8^+$ T suppressor cells (Gilliet and Liu, 2002). Virus-induced pDC-derived DCs prime naïve T cells into Th1-like T cells producing IL-10 (with suppressive functions) (Kadowaki *et al.*, 2000). In addition, CpG-ODN-induced pDC-derived DCs were shown to induce the generation of T-regulatory cells producing high levels of IL-10 *in vitro* (Moseman *et al.*, 2004). These studies suggest that pDC-derived DCs have an intrinsic ability to prime naïve T cells to produce IL-10, regardless of their maturation stages and activation signals. Results of two recent studies suggest that pDCs may indeed play a critical role in suppressing asthmatic immune responses to inhaled antigens (de Heer *et al.*, 2004) or in suppressing immune responses that mediate *Leishmania major* infection in mice (Baldwin *et al.*, 2004). We hypothesize that pDCs may represent naturally occurring regulatory DCs when directly presenting antigens to T cells, either at a resting stage or at a mature DC stage. pDCs may however trigger productive, T cell–mediated immune responses through activation of MDCs (see Section 1.5). In mice, stimulation of pDCs through the inhibitory ligands CD200-Ig or CTLA4-Ig induces the production of the tryptophan-catabolizing enzyme indoleamine 2,3-dioxygenase, which suppresses T-cell proliferation and survival and could equip DCs with a tolerogenic function for T cells (Fallarino *et al.*, 2004). pDCs within mouse tumor-draining lymph nodes express indoleamine 2,3-dioxygenase, which might contribute to the creation of local immunosuppression against host antitumor T-cell responses (Munn *et al.*, 2004).

1.7 Concluding Remarks

With the progression of our knowledge on DC biology, it appears that DCs are central to the immune system. They are among the first cells to encounter a pathogen, together with epithelial cells, and their intrinsic properties as well as mode of activation have a great impact on the subsequent adaptive immune response. This central role is illustrated by the increasing number of disease situations where MDC function was found to be abnormally increased, such as allergy (Holt and Stumbles, 2000; Lambrecht *et al.*, 2001; Soumelis and Liu, 2004) and autoimmunity (Pulendran *et al.*, 2001c), or decreased, such as neoplasia (Banchereau and Palucka, 2005) or various immune deficiencies, including, but not restricted to, HIV. Although more recently described, pDCs have also been implicated in various pathologic conditions. pDC activation to produce type I IFNs could be a major factor in the pathophysiology of autoimmune diseases such as lupus erythematosus (Blanco *et al.*, 2001; Lovgren *et al.*, 2004; Means *et al.*, 2005) and psoriasis (Nestle *et al.*, 2005). Given their central role in antiviral immunity, pDC deficiency may result in uncontrolled viral infections. Indeed, deficiency of pDCs in a patient with myelodysplasia was associated with unusual chronic herpes simplex infection (Dalloul *et al.*, 2004). Without doubt, the continuing progress of our knowledge of the DC "system" will provide new therapeutic approaches in a variety of immune-related diseases, including HIV.

References

Akiba, H., Miyahira, Y., Atsuta, M., Takeda, K., Nohara, C., Futagawa, T., Matsuda, H., Aoki, T., Yagita, H., and Okumura, K. (2000). Critical contribution of OX40 ligand to T helper cell type 2 differentiation in experimental leishmaniasis. *J. Exp. Med.* 2:375-80.

Akira, S. (2003). Mammalian Toll-like receptors. *Curr. Opin. Immunol.* 1:5-11.

Albert, M.L., Pearce, S.F., Francisco, L.M., Sauter, B., Roy, P., Silverstein, R.L., and Bhardwaj, N. (1998). Immature dendritic cells phagocytose apoptotic cells via alphavbeta5 and CD36, and cross-present antigens to cytotoxic T lymphocytes. *J. Exp. Med.* 7:1359-1368.

Albert, M.L., Kim, J.I., and Birge, R.B. (2000). alphavbeta5 integrin recruits the CrkII-Dock180-rac1 complex for phagocytosis of apoptotic cells. *Nat. Cell Biol.* 12:899-905.

Aliprantis, A.O., Yang, R.B., Mark, M.R., Suggett, S., Devaux, B., Radolf, J.D., Klimpel, G.R., Godowski, P., and Zychlinsky, A. (1999). Cell activation and apoptosis by bacterial lipoproteins through toll-like receptor-2. *Science* 285:736-739.

Al-Shami, A., Spolski, R., Kelly, J., Keane-Myers, A., and Leonard, W.J. (2005). A role for TSLP in the development of inflammation in an asthma model. *J. Exp. Med.* 6:829-839.

Amigorena, S. (2002). Fc gamma receptors and cross-presentation in dendritic cells. *J. Exp. Med.* 1:F1-3.

Amsen, D., Blander, J.M., Lee, G.R., Tanigaki, K., Honjo, T., and Flavell, R.A. (2004). Instruction of distinct CD4 T helper cell fates by different notch ligands on antigen-presenting cells. *Cell* 4:515-526.

Apostolou, I., Sarukhan, A., Klein, L., and von Boehmer, H. (2002). Origin of regulatory T cells with known specificity for antigen. *Nat. Immunol.* 8: 756-763.

Ardavin, C., Wu, L., Li, C.L., and Shortman, K. (1993). Thymic dendritic cells and T cells develop simultaneously in the thymus from a common precursor population. *Nature* 362:761-763.

Arnold-Schild, D., Hanau, D., Spehner, D., Schmid, C., Rammensee, H.G., de la Salle, H., and Schild, H. (1999). Cutting edge: receptor-mediated endocytosis of heat shock proteins by professional antigen-presenting cells. *J. Immunol.* 7:3757-3760.

Asselin-Paturel, C., and Trinchieri, G. (2005). Production of type I interferons: plasmacytoid dendritic cells and beyond. *J. Exp. Med.* 4:461-465.

Austyn, J.M., Kupiec-Weglinski, J.W., and Morris, P.J. (1988). Migration patterns of dendritic cells in the mouse. *Adv. Exp. Med. Biol.* 237:583-589.

Baldwin, T., Henri, S., Curtis, J., O'Keeffe, M., Vremec, D., Shortman, K., and Handman, E. (2004). Dendritic cell populations in Leishmania major-infected skin and draining lymph nodes. *Infect. Immun.* 4:1991-2001.

Balic, A., Harcus, Y., Holland, M.J., and Maizels, R.M. (2004). Selective maturation of dendritic cells by Nippostrongylus brasiliensis-secreted proteins drives Th2 immune responses. *Eur. J. Immunol.* 34:3047-3059.

Banchereau, J., Briere, F., Caux, C., Davoust, J., Lebecque, S., Liu, Y.J., Pulendran, B., and Palucka, K. (2000). Immunobiology of dendritic cells. *Annu. Rev. Immunol.* 18:767-811.

Banchereau, J., and Palucka, A.K. (2005). Dendritic cells as therapeutic vaccines against cancer. *Nat. Rev. Immunol.* 4:296-306.

Banchereau, J., and Steinman, R.M. (1998). Dendritic cells and the control of immunity. *Nature* 392:245-252.

Bangert, C., Friedl, J., Stary, G., Stingl, G., and Kopp, T. (2003). Immunopathologic features of allergic contact dermatitis in humans: participation of plasmacytoid dendritic cells in the pathogenesis of the disease? *J. Invest. Dermatol.* 6:1409-1418.

Banki, Z., Kacani, L., Mullauer, B., Wilflingseder, D., Obermoser, G., Niederegger, H., Schennach, H., Sprinzl, G.M., Sepp, N., Erdei, A., Dierich, M.P., and Stoiber, H. (2003). Cross-linking of CD32 induces maturation of human monocyte-derived dendritic cells via NF-kappa B signaling pathway. *J. Immunol.* 8:3963-3970.

Bave, U., Magnusson, M., Eloranta, M.L., Perers, A., Alm, G.V., and Ronnblom, L. (2003). Fc gamma RIIa is expressed on natural IFN-alpha-producing cells (plasmacytoid dendritic cells) and is required for the IFN-alpha production induced by apoptotic cells combined with lupus IgG. *J. Immunol.* 6: 3296-3302.

Bendriss-Vermare, N., Barthelemy, C., Durand, I., Bruand, C., Dezutter-Dambuyant, C., Moulian, N., Berrih-Aknin, S., Caux, C., Trinchieri, G., and Briere, F. (2001). Human thymus contains IFN-alpha-producing CD11c(-), myeloid CD11c($^+$), and mature interdigitating dendritic cells. *J. Clin. Invest.* 7:835-844.

Bendriss-Vermare, N., Burg, S., Kanzler, H., Chaperot, L., Duhen, T., de Bouteiller, O., D'agostini, M., Bridon, J.M., Durand, I., Sederstrom, J.M., Chen, W., Plumas, J., Jacob, M.C., Liu, Y.J., Garrone, P., Trinchieri, G., Caux, C., and Briere, F. (2005). Virus overrides the propensity of human CD40L-activated plasmacytoid dendritic cells to produce Th2 mediators through synergistic induction of IFN-{gamma} and Th1 chemokine production. *J. Leukoc. Biol.* 78:954-966.

Benlagha, K., Wei, D.G., Veiga, J., Teyton, L., and Bendelac, A. (2005). Characterization of the early stages of thymic NKT cell development. *J. Exp. Med.* 4:485-492.

Bergman, M.P., Engering, A., Smits, H.H., van Vliet, S.J., van Bodegraven, A.A., Wirth, H.P., Kapsenberg, M.L., Vandenbroucke-Grauls, C.M., van Kooyk, Y., and Appelmelk, B.J. (2004). Helicobacter pylori modulates the T helper cell 1/T helper cell 2 balance through phase-variable interaction between lipopolysaccharide and DC-SIGN. *J. Exp. Med.* 8:979-990.

Beutler, B. (2002). Toll-like receptors: how they work and what they do. *Curr. Opin. Hematol.* 1:2-10.

Bilsborough, J., George, T.C., Norment, A., and Viney, J.L. (2003). Mucosal CD8alpha$^+$ DC, with a plasmacytoid phenotype, induce differentiation and support function of T cells with regulatory properties. *Immunology* 4:481-492.

Binder, R.J., Anderson, K.M., Basu, S., and Srivastava, P.K. (2000a). Cutting edge: heat shock protein gp96 induces maturation and migration of CD11c$^+$ cells *in vivo*. *J. Immunol.* 11:6029-6035.

Binder, R.J., Han, D.K., and Srivastava, P.K. (2000b). CD91: a receptor for heat shock protein gp96. *Nat. Immunol.* 2:151-155.

Blanco, P., Palucka, A.K., Gill, M., Pascual, V., and Banchereau, J. (2001). Induction of dendritic cell differentiation by IFN-alpha in systemic lupus erythematosus. *Science* 294:1540-1543.

Blom, B., Ho, S., Antonenko, S., and Liu, Y.J. (2000). Generation of interferon alpha-producing predendritic cell (Pre-DC)2 from human CD34($^+$) hematopoietic stem cells. *J. Exp. Med.* 192:1785-1796.

Bogdan, C. (2000). The function of type I interferons in antimicrobial immunity. *Curr. Opin. Immunol.* 4:419-424.

Bontkes, H.J., Ruizendaal, J.J., Kramer, D., Meijer, C.J., and Hooijberg, E. (2005). Plasmacytoid dendritic cells are present in cervical carcinoma and become activated by human papillomavirus type 16 virus-like particles. *Gynecol. Oncol.* 3:897-901.

Borg, C., Jalil, A., Laderach, D., Maruyama, K., Wakasugi, H., Charrier, S., Ryffel, B., Cambi, A., Figdor, C., Vainchenker, W., Galy, A., Caignard, A., and Zitvogel, L. (2004). NK cell activation by dendritic cells (DCs) requires the formation of a synapse leading to IL-12 polarization in DCs. *Blood* 10:3267-3275.

Borkowski, T.A., Letterio, J.J., Farr, A.G., and Udey, M.C. (1996). A role for endogenous transforming growth factor beta 1 in Langerhans cell biology: the skin of transforming growth factor beta 1 null mice is devoid of epidermal Langerhans cells. *J. Exp. Med.* 6:2417-2422.

Castellino, F., Boucher, P.E., Eichelberg, K., Mayhew, M., Rothman, J.E., Houghton, A.N., and Germain, R.N. (2000). Receptor-mediated uptake of antigen/heat shock protein complexes results in major histocompatibility complex class I antigen presentation via two distinct processing pathways. *J. Exp. Med.* 11:1957-1964.

Caux, C., Massacrier, C., Dubois, B., Valladeau, J., Dezutter-Dambuyant, C., Durand, I., Schmitt, D., and Saeland, S. (1999). Respective involvement of TGF-beta and IL-4 in the development of Langerhans cells and non-Langerhans dendritic cells from CD34$^+$ progenitors. *J. Leukoc. Biol.* 5:781-791.

Caux, C., Massacrier, C., Vanbervliet, B., Dubois, B., Durand, I., Cella, M., Lanzavecchia, A., and Banchereau, J. (1997). CD34$^+$ hematopoietic progenitors from human cord blood differentiate along two independent dendritic cell pathways in response to granulocyte-macrophage colony-stimulating factor plus tumor necrosis factor alpha: II. Functional analysis. *Blood* 4:1458-1470.

Cella, M., Scheidegger, D., Palmer-Lehmann, K., Lane, P., Lanzavecchia, A., and Alber, G. (1996). Ligation of CD40 on dendritic cells triggers production of high levels of interleukin-12 and enhances T cell stimulatory capacity: T-T help via APC activation. *J. Exp. Med.* 184:747-752.

Cella, M., Jarrossay, D., Facchetti, F., Alebardi, O., Nakajima, H., Lanzavecchia, A., and Colonna, M. (1999a). Plasmacytoid monocytes migrate to inflamed lymph nodes and produce large amounts of type I interferon. *Nat. Med.* 8:919-923.

Cella, M., Salio, M., Sakakibara, Y., Langen, H., Julkunen, I., and Lanzavecchia, A. (1999b). Maturation, activation, and protection of dendritic cells induced by double-stranded RNA. *J. Exp. Med.* 5:821-89.

Chen, W., Antonenko, S., Sederstrom, J.M., Liang, X., Chan, A.S., Kanzler, H., Blom, B., Blazar, B.R., and Liu, Y.J. (2004). Thrombopoietin cooperates with

FLT3-ligand in the generation of plasmacytoid dendritic cell precursors from human hematopoietic progenitors. *Blood* 7:2547-2553.

Chicha, L., Jarrossay, D., and Manz, M.G. (2004). Clonal type I interferon-producing and dendritic cell precursors are contained in both human lymphoid and myeloid progenitor populations. *J. Exp. Med.* 11:1519-1524.

Colonna, M., Krug, A., and Cella, M. (2002). Interferon-producing cells: on the front line in immune responses against pathogens. *Curr. Opin. Immunol.* 3:373-379.

Corcoran, L., Ferrero, I., Vremec, D., Lucas, K., Waithman, J., O'Keeffe, M., Wu, L., Wilson, A., and Shortman, K. (2003). The lymphoid past of mouse plasmacytoid cells and thymic dendritic cells. *J. Immunol.* 10:4926-432.

Dalloul, A., Oksenhendler, E., Chosidow, O., Ribaud, P., Carcelain, G., Louvet, S., Massip, P., Lebon, P., and Autran, B. (2004). Severe herpes virus (HSV-2) infection in two patients with myelodysplasia and undetectable NK cells and plasmacytoid dendritic cells in the blood. *J. Clin. Virol.* 4:329-336.

D'Andrea, A., Ma, X., Aste-Amezaga, M., Paganin, C., and Trinchieri, G. (1995). Stimulatory and inhibitory effects of interleukin (IL)-4 and IL-13 on the production of cytokines by human peripheral blood mononuclear cells: priming for IL-12 and tumor necrosis factor alpha production. *J. Exp. Med.* 2:537-546.

de Baey, A., and Lanzavecchia, A. (2000). The role of aquaporins in dendritic cell macropinocytosis. *J. Exp. Med.* 4:743-748.

de Heer, H.J., Hammad, H., Soullie, T., Hijdra, D., Vos, N., Willart, M.A., Hoogsteden, H.C., and Lambrecht, B.N. (2004). Essential role of lung plasmacytoid dendritic cells in preventing asthmatic reactions to harmless inhaled antigen. *J. Exp. Med.* 1:89-98.

de Jong, E.C., Vieira, P.L., Kalinski, P., and Kapsenberg, M.L. (1999). Corticosteroids inhibit the production of inflammatory mediators in immature monocyte-derived DC and induce the development of tolerogenic DC3. *J. Leukoc. Biol.* 2:201-204.

de Jong, E.C., Vieira, P.L., Kalinski, P., Schuitemaker, J.H., Tanaka, Y., Wierenga, E.A., Yazdanbakhsh, M., and Kapsenberg, M.L. (2002). Microbial compounds selectively induce Th1 cell-promoting or Th2 cell-promoting dendritic cells *in vitro* with diverse th cell-polarizing signals. *J. Immunol.* 4:1704-1709.

De Smedt, T., Pajak, B., Muraille, E., Lespagnard, L., Heinen, E., De Baetselier, P., Urbain, J., Leo, O., and Moser, M. (1996). Regulation of dendritic cell numbers and maturation by lipopolysaccharide *in vivo*. *J. Exp. Med.* 4: 1413-1424.

Didierlaurent, A., Ferrero, I., Otten, L.A., Dubois, B., Reinhardt, M., Carlsen, H., Blomhoff, R., Akira, S., Kraehenbuhl, J.P., and Sirard, J.C. (2004). Flagellin promotes myeloid differentiation factor 88-dependent development of Th2-type response. *J. Immunol.* 172:6922-6930.

Diebold, S.S., Kaisho, T., Hemmi, H., Akira, S., and Reis e Sousa, C. (2004). Innate antiviral responses by means of TLR7-mediated recognition of single-stranded RNA. *Science* 303:1529-1531.

Dieu, M.C., Vanbervliet, B., Vicari, A., Bridon, J.M., Oldham, E., Ait-Yahia, S., Briere, F., Zlotnik, A., Lebecque, S., and Caux, C. (1998). Selective recruitment of immature and mature dendritic cells by distinct chemokines expressed in different anatomic sites. *J. Exp. Med.* 2:373-386.

d'Ostiani, C.F., Del Sero, G., Bacci, A., Montagnoli, C., Spreca, A., Mencacci, A., Ricciardi-Castagnoli, P., and Romani, L. (2000). Dendritic cells discriminate between yeasts and hyphae of the fungus Candida albicans. Implications for initiation of T helper cell immunity *in vitro* and *in vivo*. *J. Exp. Med.* 10: 1661-1674.

Duramad, O., Fearon, K.L., Chan, J.H., Kanzler, H., Marshall, J.D., Coffman, R.L., and Barrat, F.J. (2003). IL-10 regulates plasmacytoid dendritic cell response to CpG-containing immunostimulatory sequences. *Blood* 13:4487-4492.

Dzionek, A., Fuchs, A., Schmidt, P., Cremer, S., Zysk, M., Miltenyi, S., Buck, D.W., and Schmitz, J. (2000). BDCA-2, BDCA-3, and BDCA-4: three markers for distinct subsets of dendritic cells in human peripheral blood. *J. Immunol.* 11, 6037-6046.

Dzionek, A., Sohma, Y., Nagafune, J., Cella, M., Colonna, M., Facchetti, F., Gunther, G., Johnston, I., Lanzavecchia, A., Nagasaka, T., Okada, T., Vermi, W., Winkels, G., Yamamoto, T., Zysk, M., Yamaguchi, Y., and Schmitz, J. (2001). BDCA-2, a novel plasmacytoid dendritic cell-specific type II C-type lectin, mediates antigen capture and is a potent inhibitor of interferon alpha/beta induction. *J. Exp. Med.* 12:1823-1834.

Fallarino, F., Asselin-Paturel, C., Vacca, C., Bianchi, R., Gizzi, S., Fioretti, M.C., Trinchieri, G., Grohmann, U., and Puccetti, P. (2004). Murine plasmacytoid dendritic cells initiate the immunosuppressive pathway of tryptophan catabolism in response to CD200 receptor engagement. *J. Immunol.* 6: 3748-3754.

Fanger, N.A., Wardwell, K., Shen, L., Tedder, T.F., and Guyre, P.M. (1996). Type I (CD64) and type II (CD32) Fc gamma receptor-mediated phagocytosis by human blood dendritic cells. *J. Immunol.* 2:541-548.

Farkas, L., Beiske, K., Lund-Johansen, F., Brandtzaeg, P., and Jahnsen, F.L. (2001). Plasmacytoid dendritic cells (natural interferon- alpha/beta-producing cells) accumulate in cutaneous lupus erythematosus lesions. *Am. J. Pathol.* 1:237-243.

Ferlazzo, G., Pack, M., Thomas, D., Paludan, C., Schmid, D., Strowig, T., Bougras, G., Muller, W.A., Moretta, L., and Munz, C. (2004). Distinct roles of IL-12 and IL-15 in human natural killer cell activation by dendritic cells from secondary lymphoid organs. *Proc. Natl. Acad. Sci. U. S. A.* 47:16606-16611.

Fernandez, N.C., Lozier, A., Flament, C., Ricciardi-Castagnoli, P., Bellet, D., Suter, M., Perricaudet, M., Tursz, T., Maraskovsky, E., and Zitvogel, L. (1999). Dendritic cells directly trigger NK cell functions: cross-talk relevant in innate anti-tumor immune responses *in vivo*. *Nat. Med.* 4:405-411.

Figdor, C.G., van Kooyk, Y., and Adema, G.J. (2002). C-type lectin receptors on dendritic cells and Langerhans cells. *Nat. Rev. Immunol.* 2:77-84.

Fonteneau, J.F., Gilliet, M., Larsson, M., Dasilva, I., Munz, C., Liu, Y.J., and Bhardwaj, N. (2003). Activation of influenza virus-specific CD4$^+$ and CD8$^+$ T cells: a new role for plasmacytoid dendritic cells in adaptive immunity. *Blood* 9:3520-3526.

Fonteneau, J.F., Larsson, M., Beignon, A.S., McKenna, K., Dasilva, I., Amara, A., Liu, Y.J., Lifson, J.D., Littman, D.R., and Bhardwaj, N. (2004). Human immunodeficiency virus type 1 activates plasmacytoid dendritic cells and concomitantly induces the bystander maturation of myeloid dendritic cells. *J. Virol.* 10:5223-532.

Foti, M., Granucci, F., and Ricciardi-Castagnoli, P. (2004). A central role for tissue-resident dendritic cells in innate responses. *Trends Immunol.* 12: 650-654.

Garrett, W.S., Chen, L.M., Kroschewski, R., Ebersold, M., Turley, S., Trombetta, S., Galan, J.E., and Mellman, I. (2000). Developmental control of endocytosis in dendritic cells by Cdc42. *Cell* 3:325-334.

Gary-Gouy, H., Lebon, P., and Dalloul, A.H. (2002). Type I interferon production by plasmacytoid dendritic cells and monocytes is triggered by viruses, but the level of production is controlled by distinct cytokines. *J. Interferon Cytokine Res.* 6:653-659.

Gautier, G., Humbert, M., Deauvieau, F., Scuiller, M., Hiscott, J., Bates, E.E., Trinchieri, G., Caux, C., and Garrone, P. (2005). A type I interferon autocrine-paracrine loop is involved in Toll-like receptor-induced interleukin-12p70 secretion by dendritic cells. *J. Exp. Med.* 9:1435-1446.

Geijtenbeek, T.B., Van Vliet, S.J., Koppel, E.A., Sanchez-Hernandez, M., Vandenbroucke-Grauls, C.M., Appelmelk, B., and Van Kooyk, Y. (2003). Mycobacteria target DC-SIGN to suppress dendritic cell function. *J. Exp. Med.* 1:7-17.

Geissmann, F., Prost, C., Monnet, J.P., Dy, M., Brousse, N., and Hermine, O. (1998). Transforming growth factor beta1, in the presence of granulocyte/macrophage colony-stimulating factor and interleukin 4, induces differentiation of human peripheral blood monocytes into dendritic Langerhans cells. *J. Exp. Med.* 6:961-966.

Gerosa, F., Baldani-Guerra, B., Nisii, C., Marchesini, V., Carra, G., and Trinchieri, G. (2002). Reciprocal activating interaction between natural killer cells and dendritic cells. *J. Exp. Med.* 3:327-333.

Gerosa, F., Gobbi, A., Zorzi, P., Burg, S., Briere, F., Carra, G., and Trinchieri, G. (2005). The reciprocal interaction of NK cells with plasmacytoid or myeloid dendritic cells profoundly affects innate resistance functions. *J. Immunol.* 2:727-734.

Gibson, S.J., Lindh, J.M., Riter, T.R., Gleason, R.M., Rogers, L.M., Fuller, A.E., Oesterich, J.L., Gorden, K.B., Qiu, X., McKane, S.W., Noelle, R.J., Miller, R.L., Kedl, R.M., Fitzgerald-Bocarsly, P., Tomai, M.A., and Vasilakos, J.P. (2002). Plasmacytoid dendritic cells produce cytokines and mature in response to the TLR7 agonists, imiquimod and resiquimod. *Cell. Immunol.* 218:74-86.

Gilliet, M., and Liu, Y.J. (2002). Generation of human CD8 T regulatory cells by CD40 ligand-activated plasmacytoid dendritic cells. *J. Exp. Med.* 6:695-704.

Gilliet, M., Boonstra, A., Paturel, C., Antonenko, S., Xu, X.L., Trinchieri, G., O'Garra, A., and Liu, Y.J. (2002). The development of murine Plasmacytoid dendritic cell precursors is differentially regulated by FLT3-ligand and granulocyte/macrophage colony-stimulating factor. *J. Exp. Med.* 7:953-998.

Gilliet, M., Conrad, C., Geiges, M., Cozzio, A., Thurlimann, W., Burg, G., Nestle, F.O., and Dummer, R. (2004). Psoriasis triggered by toll-like receptor 7 agonist imiquimod in the presence of dermal plasmacytoid dendritic cell precursors. *Arch. Dermatol.* 12:1490-1495.

Gordon, S. (2002). Pattern recognition receptors: doubling up for the innate immune response. *Cell* 7:927-930.

Granucci, F., Zanoni, I., Pavelka, N., Van Dommelen, S.L., Andoniou, C.E., Belardelli, F., Degli Esposti, M.A., and Ricciardi-Castagnoli, P. (2004). A contribution of mouse dendritic cell-derived IL-2 for NK cell activation. *J. Exp. Med.* 3:287-295.

Grouard, G., Durand, I., Filgueira, L., Banchereau, J., and Liu, Y.J. (1996). Dendritic cells capable of stimulating T cells in germinal centres. *Nature* 384:364-367.

Grouard, G., Rissoan, M.C., Filgueira, L., Durand, I., Banchereau, J., and Liu, Y.J. (1997). The enigmatic plasmacytoid T cells develop into dendritic cells with interleukin (IL)-3 and CD40-ligand. *J. Exp. Med.* 6:1101-1111.

Guermonprez, P., Valladeau, J., Zitvogel, L., Thery, C., and Amigorena, S. (2002). Antigen presentation and T cell stimulation by dendritic cells. *Annu. Rev. Immunol.* 20:621-667.

Gurney, K.B., Colantonio, A.D., Blom, B., Spits, H., and Uittenbogaart, C.H. (2004). Endogenous IFN-alpha production by plasmacytoid dendritic cells exerts an antiviral effect on thymic HIV-1 infection. *J. Immunol.* 12:7269-7276.

Hammad, H., Smits, H.H., Ratajczak, C., Nithiananthan, A., Wierenga, E.A., Stewart, G.A., Jacquet, A., Tonnel, A.B., and Pestel, J. (2003). Monocyte-derived dendritic cells exposed to Der p 1 allergen enhance the recruitment of Th2 cells: major involvement of the chemokines TARC/CCL17 and MDC/CCL22. *Eur. Cytokine Netw.* 4:219-228.

Hart, D.N., and Fabre, J.W. (1981). Demonstration and characterization of Ia-positive dendritic cells in the interstitial connective tissues of rat heart and other tissues, but not brain. *J. Exp. Med.* 2:347-361.

Hartmann, G., Weiner, G.J., and Krieg, A.M. (1999). CpG DNA: a potent signal for growth, activation, and maturation of human dendritic cells. *Proc. Natl. Acad. Sci. U. S. A.* 16:9305-9310.

Hawiger, D., Inaba, K., Dorsett, Y., Guo, M., Mahnke, K., Rivera, M., Ravetch, J.V., Steinman, R.M., and Nussenzweig, M.C. (2001). Dendritic cells induce peripheral T cell unresponsiveness under steady state conditions *in vivo*. *J. Exp. Med.* 6:769-779.

Heil, F., Hemmi, H., Hochrein, H., Ampenberger, F., Kirschning, C., Akira, S., Lipford, G., Wagner, H., and Bauer, S. (2004). Species-specific recognition of single-stranded RNA via toll-like receptor 7 and 8. *Science* 303:1526-1529.

Hemmi, H., Takeuchi, O., Kawai, T., Kaisho, T., Sato, S., Sanjo, H., Matsumoto, M., Hoshino, K., Wagner, H., Takeda, K., and Akira, S. (2000). A Toll-like receptor recognizes bacterial DNA. *Nature* 408:740-745.

Hemmi, H., Kaisho, T., Takeuchi, O., Sato, S., Sanjo, H., Hoshino, K., Horiuchi, T., Tomizawa, H., Takeda, K., and Akira, S. (2002). Small anti-viral compounds activate immune cells via the TLR7 MyD88-dependent signaling pathway. *Nat. Immunol.* 2:196-200.

Hibbert, L., Pflanz, S., De Waal Malefyt, R., and Kastelein, R.A. (2003). IL-27 and IFN-alpha signal via Stat1 and Stat3 and induce T-Bet and IL-12Rbeta2 in naive T cells. *J. Interferon Cytokine Res.* 9:513-522.

Hilkens, C.M., Kalinski, P., de Boer, M., and Kapsenberg, M.L. (1997). Human dendritic cells require exogenous interleukin-12-inducing factors to direct the development of naive T-helper cells toward the Th1 phenotype. *Blood* 5:1920-196.

Hirschfeld, M., Kirschning, C.J., Schwandner, R., Wesche, H., Weis, J.H., Wooten, R.M., and Weis, J.J. (1999). Cutting edge: inflammatory signaling by Borrelia burgdorferi lipoproteins is mediated by toll-like receptor 2. *J. Immunol.* 5:2382-2386.

Hochrein, H., O'Keeffe, M., Luft, T., Vandenabeele, S., Grumont, R.J., Maraskovsky, E., and Shortman, K. (2000). Interleukin (IL)-4 is a major regulatory cytokine governing bioactive IL-12 production by mouse and human dendritic cells. *J. Exp. Med.* 6:823-833.

Holt, P.G. (2001). Dendritic cell ontogeny as an aetiological factor in respiratory tract diseases in early life. *Thorax* 56:419-20.

Holt, P.G., and Stumbles, P.A. (2000). Regulation of immunologic homeostasis in peripheral tissues by dendritic cells: the respiratory tract as a paradigm. *J. Allergy Clin. Immunol.* 105:421-9.

Hornung, V., Rothenfusser, S., Britsch, S., Krug, A., Jahrsdorfer, B., Giese, T., Endres, S., and Hartmann, G. (2002). Quantitative expression of toll-like receptor 1-10 mRNA in cellular subsets of human peripheral blood mononuclear cells and sensitivity to CpG oligodeoxynucleotides. *J. Immunol.* 9:4531-4537.

Hou, W., Wu, Y., Sun, S., Shi, M., Sun, Y., Yang, C., Pei, G., Gu, Y., Zhong, C., and Sun, B. (2003). Pertussis toxin enhances Th1 responses by stimulation of dendritic cells. *J. Immunol.* 4:1728-136.

Hugot, J.P., Chamaillard, M., Zouali, H., Lesage, S., Cezard, J.P., Belaiche, J., Almer, S., Tysk, C., O'Morain, C.A., Gassull, M., Binder, V., Finkel, Y., Cortot, A., Modigliani, R., Laurent-Puig, P., Gower-Rousseau, C., Macry, J., Colombel, J.F., Sahbatou, M., Thomas, G. (2001). Association of NOD2 leucine-rich repeat variants with susceptibility to Crohn's disease. *Nature* 411:599-603.

Idzko, M., la Sala, A., Ferrari, D., Panther, E., Herouy, Y., Dichmann, S., Mockenhaupt, M., Di Virgilio, F., Girolomoni, G., and Norgauer, J. (2002). Expression and function of histamine receptors in human monocyte-derived dendritic cells. *J. Allergy Clin. Immunol.* 5:839-846.

Ikarashi, Y., Mikami, R., Bendelac, A., Terme, M., Chaput, N., Terada, M., Tursz, T., Angevin, E., Lemonnier, F.A., Wakasugi, H., and Zitvogel, L. (2001). Dendritic cell maturation overrules H-2D-mediated natural killer T (NKT) cell inhibition: critical role for B7 in CD1d-dependent NKT cell interferon gamma production. *J. Exp. Med.* 8:1179-1186.

Ito, T., Inaba, M., Inaba, K., Toki, J., Sogo, S., Iguchi, T., Adachi, Y., Yamaguchi, K., Amakawa, R., Valladeau, J., Saeland, S., Fukuhara, S., Ikehara, S. (1999). A CD1a$^+$/CD11c$^+$ subset of human blood dendritic cells is a direct precursor of Langerhans cells. *J. Immunol.* 163:1409-1419.

Ito, T., Amakawa, R., Inaba, M., Hori, T., Ota, M., Nakamura, K., Takebayashi, M., Miyaji, M., Yoshimura, T., Inaba, K., and Fukuhara, S. (2004). Plasmacytoid dendritic cells regulate Th cell responses through OX40 ligand and type I IFNs. *J. Immunol.* 7:4253-429.

Jahnsen, F.L., Lund-Johansen, F., Dunne, J.F., Farkas, L., Haye, R., and Brandtzaeg, P. (2000). Experimentally induced recruitment of plasmacytoid (CD123high) dendritic cells in human nasal allergy. *J. Immunol.* 7:4062-4068.

Janeway, C.A., Jr., and Medzhitov, R. (2002). Innate immune recognition. *Annu. Rev. Immunol.* 20:197-216.

Jankovic, D., Kullberg, M.C., Noben-Trauth, N., Caspar, P., Paul, W.E., and Sher, A. (2000). Single cell analysis reveals that IL-4 receptor/Stat6 signaling is not required for the *in vivo* or *in vitro* development of CD4$^+$ lymphocytes with a Th2 cytokine profile. *J. Immunol.* 6:3047-3055.

Jankovic, D., Kullberg, M.C., Hieny, S., Caspar, P., Collazo, C.M., and Sher, A. (2002). In the absence of IL-12, CD4($^+$) T cell responses to intracellular pathogens fail to default to a Th2 pattern and are host protective in an IL-10(-/-) setting. *Immunity* 3:429-439.

Jarrossay, D., Napolitani, G., Colonna, M., Sallusto, F., and Lanzavecchia, A. (2001). Specialization and complementarity in microbial molecule recognition by human myeloid and plasmacytoid dendritic cells. *Eur. J. Immunol.* 11:3388-3393.

Jego, G., Palucka, A.K., Blanck, J.P., Chalouni, C., Pascual, V., and Banchereau, J. (2003). Plasmacytoid dendritic cells induce plasma cell differentiation through type I interferon and interleukin 6. *Immunity* 2:225-234.

Jember, A.G., Zuberi, R., Liu, F.T., and Croft, M. (2001). Development of allergic inflammation in a murine model of asthma is dependent on the costimulatory receptor OX40. *J. Exp. Med.* 3:387-392.

Jomantaite, I., Dikopoulos, N., Kroger, A., Leithauser, F., Hauser, H., Schirmbeck, R., and Reimann, J. (2004). Hepatic dendritic cell subsets in the mouse. *Eur. J. Immunol.* 2:355-365.

Jordan, M.S., Boesteanu, A., Reed, A.J., Petrone, A.L., Holenbeck, A.E., Lerman, M.A., Naji, A., and Caton, A.J. (2001). Thymic selection of CD4$^+$CD25$^+$ regulatory T cells induced by an agonist self-peptide. *Nat. Immunol.* 4:301-306.

Jotwani, R., Pulendran, B., Agrawal, S., and Cutler, C.W. (2003). Human dendritic cells respond to Porphyromonas gingivalis LPS by promoting a Th2 effector response *in vitro*. *Eur. J. Immunol.* 11:2980-2986.

Kadowaki, N., Antonenko, S., Lau, J.Y., and Liu, Y.J. (2000). Natural interferon alpha/beta-producing cells link innate and adaptive immunity. *J. Exp. Med.* 2:219-226.

Kadowaki, N., Antonenko, S., Ho, S., Rissoan, M.C., Soumelis, V., Porcelli, S.A., Lanier, L.L., and Liu, Y.J. (2001a). Distinct cytokine profiles of neonatal natural killer T cells after expansion with subsets of dendritic cells. *J. Exp. Med.* 10:1221-1226.

Kadowaki, N., Ho, S., Antonenko, S., Malefyt, R.W., Kastelein, R.A., Bazan, F., and Liu, Y.J. (2001b). Subsets of human dendritic cell precursors express different toll-like receptors and respond to different microbial antigens. *J. Exp. Med.* 194:863-869.

Kaisho, T., and Akira, S. (2001). Dendritic-cell function in Toll-like receptor- and MyD88-knockout mice. *Trends Immunol.* 22:78-83.

Kalergis, A.M., and Ravetch, J.V. (2002). Inducing tumor immunity through the selective engagement of activating Fcgamma receptors on dendritic cells. *J. Exp. Med.* 12:1653-1659.

Kalinski, P., Smits, H.H., Schuitemaker, J.H., Vieira, P.L., van Eijk, M., de Jong, E.C., Wierenga, E.A., and Kapsenberg, M.L. (2000). IL-4 is a mediator of IL-12p70 induction by human Th2 cells: reversal of polarized Th2 phenotype by dendritic cells. *J. Immunol.* 4:1877-1881.

Kalinski, P., Vieira, P.L., Schuitemaker, J.H., de Jong, E.C., and Kapsenberg, M.L. (2001). Prostaglandin E(2) is a selective inducer of interleukin-12p40 (IL-12p40) production and an inhibitor of bioactive IL-12p70 heterodimer. *Blood* 97:3466-3469.

Kane, C.M., Cervi, L., Sun, J., McKee, A.S., Masek, K.S., Shapira, S., Hunter, C.A., and Pearce, E.J. (2004). Helminth antigens modulate TLR-initiated dendritic cell activation. *J. Immunol.* 12:7454-7461.

Keir, M.E., Stoddart, C.A., Linquist-Stepps, V., Moreno, M.E., and McCune, J.M. (2002). IFN-alpha secretion by type 2 predendritic cells up-regulates MHC class I in the HIV-1-infected thymus. *J. Immunol.* 1:325-331.

Kelsall, B.L., and Rescigno, M. (2004). Mucosal dendritic cells in immunity and inflammation. *Nat. Immunol.* 11:1091-1095.

Kolumam, G.A., Thomas, S., Thompson, L.J., Sprent, J., and Murali-Krishna, K. (2005). Type I interferons act directly on CD8 T cells to allow clonal expansion and memory formation in response to viral infection. *J. Exp. Med.* 5:637-650.

Krug, A., Luker, G.D., Barchet, W., Leib, D.A., Akira, S., and Colonna, M. (2004). Herpes simplex virus type 1 activates murine natural interferon-producing cells through toll-like receptor 9. *Blood* 4:1433-1437.

Kupiec-Weglinski, J.W., Austyn, J.M., and Morris, P.J. (1988). Migration patterns of dendritic cells in the mouse. Traffic from the blood, and T cell-dependent and -independent entry to lymphoid tissues. *J. Exp. Med.* 2:632-645.

Kuwana, M., Kaburaki, J., Wright, T.M., Kawakami, Y., and Ikeda, Y. (2001). Induction of antigen-specific human CD4($^+$) T cell anergy by peripheral blood DC2 precursors. *Eur. J. Immunol.* 9:2547-2557.

Lambrecht, B.N. (2001). The dendritic cell in allergic airway diseases: a new player to the game. *Clin. Exp. Allergy* 31:206-218.

Lambrecht, B.N., Hoogsteden, H.C., and Pauwels, R.A. (2001). Dendritic cells as regulators of the immune response to inhaled allergen: recent findings in animal models of asthma. *Int. Arch. Allergy Immunol.* 124:432-446.

Langenkamp, A., Messi, M., Lanzavecchia, A., and Sallusto, F. (2000). Kinetics of dendritic cell activation: impact on priming of TH1, TH2 and nonpolarized T cells. *Nat. Immunol.* 4:311-336.

Lanzavecchia, A., and Sallusto, F. (2001). Regulation of T cell immunity by dendritic cells. *Cell* 106:263-266.

Larsen, C.P., Steinman, R.M., Witmer-Pack, M., Hankins, D.F., Morris, P.J., and Austyn, J.M. (1990). Migration and maturation of Langerhans cells in skin transplants and explants. *J. Exp. Med.* 5:1483-1493.

Lavelle, E.C., Jarnicki, A., McNeela, E., Armstrong, M.E., Higgins, S.C., Leavy, O., and Mills, K.H. (2004). Effects of cholera toxin on innate and adaptive immunity and its application as an immunomodulatory agent. *J. Leukoc. Biol.* 5:756-763.

Liu, Y.J. (2005). IPC: professional type 1 interferon-producing cells and plasmacytoid dendritic cell precursors. *Annu. Rev. Immunol.* 23:275-306.

Liu, Y.J., Kanzler, H., Soumelis, V., and Gilliet, M. (2001). Dendritic cell lineage, plasticity and cross-regulation. *Nat. Immunol.* 2:585-589.

Lovgren, T., Eloranta, M.L., Bave, U., Alm, G.V., and Ronnblom, L. (2004). Induction of interferon-alpha production in plasmacytoid dendritic cells by immune complexes containing nucleic acid released by necrotic or late apoptotic cells and lupus IgG. *Arthritis Rheum.* 6:1861-1872.

Luft, T., Pang, K.C., Thomas, E., Hertzog, P., Hart, D.N., Trapani, J., and Cebon, J. (1998). Type I IFNs enhance the terminal differentiation of dendritic cells. *J. Immunol.* 4:1947-1953.

Lund, J.M., Alexopoulou, L., Sato, A., Karow, M., Adams, N.C., Gale, N.W., Iwasaki, A., and Flavell, R.A. (2004). Recognition of single-stranded RNA viruses by Toll-like receptor 7. *Proc. Natl. Acad. Sci. U. S. A.* 15:5598-5603.

Macatonia, S.E., Hosken, N.A., Litton, M., Vieira, P., Hsieh, C.S., Culpepper, J.A., Wysocka, M., Trinchieri, G., Murphy, K.M., and O'Garra, A. (1995). Dendritic cells produce IL-12 and direct the development of Th1 cells from naive CD4$^+$ T cells. *J. Immunol.* 154:5071-5079.

Maekawa, Y., Tsukumo, S., Chiba, S., Hirai, H., Hayashi, Y., Okada, H., Kishihara, K., and Yasutomo, K. (2003). Delta1-Notch3 interactions bias the functional differentiation of activated CD4$^+$ T cells. *Immunity* 4:549-559.

Manfra, D.J., Chen, S.C., Jensen, K.K., Fine, J.S., Wiekowski, M.T., and Lira, S.A. (2003). Conditional expression of murine Flt3 ligand leads to expansion of multiple dendritic cell subsets in peripheral blood and tissues of transgenic mice. *J. Immunol.* 6:2843-2852.

Martin, P., Del Hoyo, G.M., Anjuere, F., Arias, C.F., Vargas, H.H., Fernandez, L.A., Parrillas, V., and Ardavin, C. (2002). Characterization of a new subpopulation of mouse CD8alpha$^+$ B220$^+$ dendritic cells endowed with type 1 interferon production capacity and tolerogenic potential. *Blood* 2:383-390.

MartIn-Fontecha, A., Sebastiani, S., Hopken, U.E., Uguccioni, M., Lipp, M., Lanzavecchia, A., and Sallusto, F. (2003). Regulation of dendritic cell migration to the draining lymph node: impact on T lymphocyte traffic and priming. *J. Exp. Med.* 4:615-621.

Martin-Fontecha, A., Thomsen, L.L., Brett, S., Gerard, C., Lipp, M., Lanzavecchia, A., and Sallusto, F. (2004). Induced recruitment of NK cells to lymph nodes provides IFN-gamma for T(H)1 priming. *Nat. Immunol.* 12:1260-1265.

Matsumoto, M., Funami, K., Tanabe, M., Oshiumi, H., Shingai, M., Seto, Y., Yamamoto, A., and Seya, T. (2003). Subcellular localization of Toll-like receptor 3 in human dendritic cells. *J. Immunol.* 6:3154-3162.

Mattner, J., Debord, K.L., Ismail, N., Goff, R.D., Cantu, C. 3rd., Zhou, D., Saint-Mezard, P., Wang,V., Gao, Y., Yin, N., Hoebe, K., Schneewind, O., Walker, D., Beutler, B., Teyton, L., Savage, P.B., and Bendelac, A. (2005). Exogenous and endogenous glycolipid antigens activate NKT cells during microbial infections. *Nature* 434:525-529.

Maurer, D., Fiebiger, E., Reininger, B., Ebner, C., Petzelbauer, P., Shi, G.P., Chapman, H.A., and Stingl, G. (1998). Fc epsilon receptor I on dendritic cells delivers IgE-bound multivalent antigens into a cathepsin S-dependent pathway of MHC class II presentation. *J. Immunol.* 6:2731-2739.

Mazzoni, A., Leifer, C.A., Mullen, G.E., Kennedy, M.N., Klinman, D.M., and Segal, D.M. (2003). Cutting edge: histamine inhibits IFN-alpha release from plasmacytoid dendritic cells. *J. Immunol.* 5:2269-2273.

Means, T.K., Latz, E., Hayashi, F., Murali, M.R., Golenbock, D.T., and Luster, A.D. (2005). Human lupus autoantibody-DNA complexes activate DCs through cooperation of CD32 and TLR9. *J. Clin. Invest.* 2:407-417.

Medzhitov, R., and Janeway, C.A., Jr. (1997). Innate immunity: impact on the adaptive immune response. *Curr. Opin. Immunol.* 1:4-9.

Medzhitov, R., and Janeway, C., Jr. (2000). The Toll receptor family and microbial recognition. *Trends Microbiol.* 10:452-456.

Moseman, E.A., Liang, X., Dawson, A.J., Panoskaltsis-Mortari, A., Krieg, A.M., Liu, Y.J., Blazar, B.R., and Chen, W. (2004). Human plasmacytoid dendritic cells activated by CpG oligodeoxynucleotides induce the generation of $CD4^+CD25^+$ regulatory T cells. *J. Immunol.* 7:4433-4442.

Munn, D.H., Sharma, M.D., Hou, D., Baban, B., Lee, J.R., Antonia, S.J., Messina, J.L., Chandler, P., Koni, P.A., and Mellor, A.L. (2004). Expression of indoleamine 2,3-dioxygenase by plasmacytoid dendritic cells in tumor-draining lymph nodes. *J. Clin. Invest.* 2:280-290.

Nagai, T., Devergne, O., Mueller, T.F., Perkins, D.L., van Seventer, J.M., and van Seventer, G.A. (2003). Timing of IFN-beta exposure during human dendritic cell maturation and naive Th cell stimulation has contrasting effects on Th1 subset generation: a role for IFN-beta-mediated regulation of IL-12 family cytokines and IL-18 in naive Th cell differentiation. *J. Immunol.* 10: 5233-5243.

Nestle, F.O., Conrad, C., Tun-Kyi, A., Homey, B., Gombert, M., Boyman, O., Burg, G., Liu, Y.J., and Gilliet, M. (2005). Plasmacytoid predendritic cells initiate psoriasis through interferon-alpha production. *J. Exp. Med.* 1:135-143.

Ngo, V.N., Tang, H.L., and Cyster, J.G. (1998). Epstein-Barr virus-induced molecule 1 ligand chemokine is expressed by dendritic cells in lymphoid tissues and strongly attracts naive T cells and activated B cells. *J. Exp. Med.* 1:181-191.

Nigou, J., Zelle-Rieser, C., Gilleron, M., Thurnher, M., and Puzo, G. (2001). Mannosylated lipoarabinomannans inhibit IL-12 production by human dendritic cells: evidence for a negative signal delivered through the mannose receptor. *J. Immunol.* 12:7477-7485.

O'Doherty, U., Peng, M., Gezelter, S., Swiggard, W.J., Betjes, M., Bhardwaj, N., and Steinman, R.M. (1994). Human blood contains two subsets of dendritic cells, one immunologically mature and the other immature. *Immunology* 3:487-493.

O'Keeffe, M., Hochrein, H., Vremec, D., Caminschi, I., Miller, J.L., Anders, E.M., Wu, L., Lahoud, M.H., Henri, S., Scott, B., Hertzog, P., Tatarczuch, L., and Shortman, K. (2002). Mouse plasmacytoid cells: long-lived cells, heterogeneous in surface phenotype and function, that differentiate into CD8($^+$) dendritic cells only after microbial stimulus. *J. Exp. Med.* 196:1307-1319.

Paquette, R.L., Hsu, N.C., Kiertscher, S.M., Park, A.N., Tran, L., Roth, M.D., and Glaspy, J.A. (1998). Interferon-alpha and granulocyte-macrophage colony-stimulating factor differentiate peripheral blood monocytes into potent antigen-presenting cells. *J. Leukoc. Biol.* 3:358-367.

Pasare, C., and Medzhitov, R. (2003). Toll pathway-dependent blockade of CD4$^+$CD25$^+$ T cell-mediated suppression by dendritic cells. *Science* 299:1033-1036.

Pedersen, A.E., Gad, M., Walter, M.R., and Claesson, M.H. (2004). Induction of regulatory dendritic cells by dexamethasone and 1alpha,25-Dihydroxyvitamin D(3). *Immunol. Lett.* 1:63-69.

Penna, G., Sozzani, S., and Adorini, L. (2001). Cutting edge: selective usage of chemokine receptors by plasmacytoid dendritic cells. *J. Immunol.* 4:1862-1866.

Penna, G., Vulcano, M., Roncari, A., Facchetti, F., Sozzani, S., and Adorini, L. (2002). Cutting edge: differential chemokine production by myeloid and plasmacytoid dendritic cells. *J. Immunol.* 12:6673-6676.

Poltorak, A., He, X., Smirnova, I., Liu, M.Y., Van, Huffel. C., Du, X., Birdwell, D., Alejos, E., Silva, M., Galanos, C., Freudenberg, M., Ricciardi-Castagnoli, P., Layton, B., and Beutler, B. (1998). Defective LPS signaling in C3H/HeJ and C57BL/10ScCr mice: mutations in Tlr4 gene. *Science* 282:2085-2088.

Price, A.A., Cumberbatch, M., Kimber, I., and Ager, A. (1997). Alpha 6 integrins are required for Langerhans cell migration from the epidermis. *J. Exp. Med.* 10:1725-1735.

Pulendran, B., Banchereau, J., Burkeholder, S., Kraus, E., Guinet, E., Chalouni, C., Caron, D., Maliszewski, C., Davoust, J., Fay, J., and Palucka, K. (2000). Flt3-ligand and granulocyte colony-stimulating factor mobilize distinct human dendritic cell subsets *in vivo*. *J. Immunol.* 1:566-572.

Pulendran, B., Banchereau, J., Maraskovsky, E., and Maliszewski, C. (2001a). Modulating the immune response with dendritic cells and their growth factors. *Trends Immunol.* 1:41-47.

Pulendran, B., Kumar, P., Cutler, C.W., Mohamadzadeh, M., Van Dyke, T., and Banchereau, J. (2001b). Lipopolysaccharides from distinct pathogens induce different classes of immune responses *in vivo*. *J. Immunol.* 9:5067-576.

Pulendran, B., Palucka, K., and Banchereau, J. (2001c). Sensing pathogens and tuning immune responses. *Science* 293:253-256.

Randolph, G.J., Angeli, V., and Swartz, M.A. (2005). Dendritic-cell trafficking to lymph nodes through lymphatic vessels. *Nat. Rev. Immunol.* 8:617-628.

Randolph, G.J., Beaulieu, S., Lebecque, S., Steinman, R.M., and Muller, W.A. (1998). Differentiation of monocytes into dendritic cells in a model of transendothelial trafficking. *Science* 282:480-483.

Randolph, G.J., Inaba, K., Robbiani, D.F., Steinman, R.M., and Muller, W.A. (1999). Differentiation of phagocytic monocytes into lymph node dendritic cells *in vivo*. *Immunity* 6:753-761.

Ratzinger, G., Stoitzner, P., Ebner, S., Lutz, M.B., Layton, G.T., Rainer, C., Senior, R.M., Shipley, J.M., Fritsch, P., Schuler, G., and Romani, N. (2002). Matrix metalloproteinases 9 and 2 are necessary for the migration of Langerhans cells and dermal dendritic cells from human and murine skin. *J. Immunol.* 9:4361-4371.

Regnault, A., Lankar, D., Lacabanne, V., Rodriguez, A., Thery, C., Rescigno, M., Saito, T., Verbeek, S., Bonnerot, C., Ricciardi-Castagnoli, P., and Amigorena, S. (1999). Fcgamma receptor-mediated induction of dendritic cell maturation and major histocompatibility complex class I-restricted antigen presentation after immune complex internalization. *J. Exp. Med.* 2:371-380.

Reis e Sousa, C. (2004). Toll-like receptors and dendritic cells: for whom the bug tolls. *Semin. Immunol.* 1:27-34.

Rescigno, M., Urbano, M., Valzasina, B., Francolini, M., Rotta, G., Bonasio, R., Granucci, F., Kraehenbuhl, J.P., and Ricciardi-Castagnoli, P. (2001). Dendritic cells express tight junction proteins and penetrate gut epithelial monolayers to sample bacteria. *Nat. Immunol.* 4:361-367.

Rissoan, M.C., Soumelis, V., Kadowaki, N., Grouard, G., Briere, F., de Waal Malefyt, R., and Liu, Y.J. (1999a). Reciprocal control of T helper cell and dendritic cell differentiation. *Science* 283:1183-1186.

Romani, N., Holzmann, S., Tripp, C.H., Koch, F., and Stoitzner, P. (2003). Langerhans cells - dendritic cells of the epidermis. *APMIS* 111:725-740.

Salio, M., Cella, M., Vermi, W., Facchetti, F., Palmowski, M.J., Smith, C.L., Shepherd, D., Colonna, M., and Cerundolo, V. (2003). Plasmacytoid dendritic cells prime IFN-gamma-secreting melanoma-specific CD8 lymphocytes and are found in primary melanoma lesions. *Eur. J. Immunol.* 4:1052-1062.

Salio, M., Palmowski, M.J., Atzberger, A., Hermans, I.F., and Cerundolo, V. (2004). CpG-matured murine plasmacytoid dendritic cells are capable of *in vivo* priming of functional CD8 T cell responses to endogenous but not exogenous antigens. *J. Exp. Med.* 4:567-579.

Sallusto, F., and Lanzavecchia, A. (1994). Efficient presentation of soluble antigen by cultured human dendritic cells is maintained by granulocyte/macrophage colony-stimulating factor plus interleukin 4 and downregulated by tumor necrosis factor alpha. *J. Exp. Med.* 4:1109-1118.

Santini, S.M., Lapenta, C., Logozzi, M., Parlato, S., Spada, M., Di Pucchio, T., and Belardelli, F. (2000). Type I interferon as a powerful adjuvant for monocyte-derived dendritic cell development and activity *in vitro* and in Hu-PBL-SCID mice. *J. Exp. Med.* 10:1777-1788.

Sauma, D., Fierro, A., Mora, J.R., Lennon-Dumenil, A.M., Bono, M.R., Rosemblatt, M., and Morales, J. (2003). Cyclosporine preconditions dendritic cells during differentiation and reduces IL-2 and IL-12 production following activation: a potential tolerogenic effect. *Transplant. Proc.* 7:2515-2517.

Sauter, B., Albert, M.L., Francisco, L., Larsson, M., Somersan, S., and Bhardwaj, N. (2000). Consequences of cell death: exposure to necrotic tumor cells, but not primary tissue cells or apoptotic cells, induces the maturation of immunostimulatory dendritic cells. *J. Exp. Med.* 3:423-434.

Schlecht, G., Garcia, S., Escriou, N., Freitas, A.A., Leclerc, C., and Dadaglio, G. (2004). Murine plasmacytoid dendritic cells induce effector/memory CD8[+] T-cell responses *in vivo* after viral stimulation. *Blood* 6:1808-1815.

Schotte, R., Nagasawa, M., Weijer, K., Spits, H., and Blom, B. (2004). The ETS transcription factor Spi-B is required for human plasmacytoid dendritic cell development. *J. Exp. Med.* 11:1503-1509.

Schuler, G., Koch, F., Heufler, C., Kampgen, E., Topar, G., and Romani, N. (1993). Murine epidermal Langerhans cells as a model to study tissue dendritic cells. *Adv. Exp. Med. Biol.* 329:243-249.

Sher, A., Collazzo, C., Scanga, C., Jankovic, D., Yap, G., and Aliberti, J. (2003a). Induction and regulation of IL-12-dependent host resistance to Toxoplasma gondii. *Immunol. Res.* 27:521-528.

Sher, A., Pearce, E., and Kaye, P. (2003b). Shaping the immune response to parasites: role of dendritic cells. *Curr. Opin. Immunol.* 4:421-449.

Shortman, K., and Liu, Y.J. (2002). Mouse and human dendritic cell subtypes. *Nat. Rev. Immunol.* 3:151-161.

Siegal, F.P., Kadowaki, N., Shodell, M., Fitzgerald-Bocarsly, P.A., Shah, K., Ho, S., Antonenko, S., and Liu, Y.J. (1999). The nature of the principal type 1 interferon-producing cells in human blood. *Science* 284:1835-1837.

Singh-Jasuja, H., Toes, R.E., Spee, P., Munz, C., Hilf, N., Schoenberger, S.P., Ricciardi-Castagnoli, P., Neefjes, J., Rammensee, H.G., Arnold-Schild, D., and Schild, H. (2000). Cross-presentation of glycoprotein 96-associated antigens on major histocompatibility complex class I molecules requires receptor-mediated endocytosis. *J. Exp. Med.* 11:1965-1974.

Sixt, M., Kanazawa, N., Selg, M., Samson, T., Roos, G., Reinhardt, D.P., Pabst, R., Lutz, M.B., and Sorokin, L. (2005). The conduit system transports soluble antigens from the afferent lymph to resident dendritic cells in the T cell area of the lymph node. *Immunity* 1:19-29.

Smith, A.L., and Fazekas de St Groth, B. (1999). Antigen-pulsed CD8alpha[+] dendritic cells generate an immune response after subcutaneous injection without homing to the draining lymph node. *J. Exp. Med.* 3:593-598.

Soumelis, V., and Liu, Y.J. (2004). Human thymic stromal lymphopoietin: a novel epithelial cell-derived cytokine and a potential key player in the induction of allergic inflammation. *Springer Semin. Immunopathol.* 25:325-333.

Soumelis, V., Reche, P.A., Kanzler, H., Yuan, W., Edward, G., Homey, B., Gilliet, M., Ho, S., Antonenko, S., Lauerma, A., Smith, K., Gorman, D., Zurawski, S., Abrams, J., Menon, S., McClanahan, T., de Waal-Malefyt, Rd. R., Bazan, F., Kastelein, R.A., and Liu, Y.J. (2002). Human epithelial cells trigger dendritic cell mediated allergic inflammation by producing TSLP. *Nat. Immunol.* 3:673-680.

Sozzani, S., Allavena, P., D'Amico, G., Luini, W., Bianchi, G., Kataura, M., Imai, T., Yoshie, O., Bonecchi, R., and Mantovani, A. (1998). Differential regulation of chemokine receptors during dendritic cell maturation: a model for their trafficking properties. *J. Immunol.* 3:1083-1086.

Spits, H., Couwenberg, F., Bakker, A.Q., Weijer, K., and Uittenbogaart, C.H. (2000). Id2 and Id3 inhibit development of CD34([+]) stem cells into predendritic cell (pre-DC)2 but not into pre-DC1. Evidence for a lymphoid origin of pre-DC2. *J. Exp. Med.* 12:1775-1784.

Sporri, R., and Reis e Sousa, C. (2005). Inflammatory mediators are insufficient for full dendritic cell activation and promote expansion of CD4[+] T cell populations lacking helper function. *Nat. Immunol.* 2:163-170.

Steinbrink, K., Wolfl, M., Jonuleit, H., Knop, J., and Enk, A.H. (1997). Induction of tolerance by IL-10-treated dendritic cells. *J. Immunol.* 10:4772-4780.

Steinman, R.M. (1991). The dendritic cell system and its role in immunogenicity. *Annu. Rev. Immunol.* 9:271-296.

Steinman, R.M., and Nussenzweig, M.C. (2002). Avoiding horror autotoxicus: the importance of dendritic cells in peripheral T cell tolerance. *Proc. Natl. Acad. Sci. U.S.A.* 1:351-358.

Steinman, R.M., Hawiger, D., and Nussenzweig, M.C. (2003). Tolerogenic dendritic cells. *Annu. Rev. Immunol.* 21:685-711.

Stenger, S., and Modlin, R.L. (2002). Control of Mycobacterium tuberculosis through mammalian Toll-like receptors. *Curr. Opin. Immunol.* 4:452-457.

Stoitzner, P., Zanella, M., Ortner, U., Lukas, M., Tagwerker, A., Janke, K., Lutz, M.B., Schuler, G., Echtenacher, B., Ryffel, B., Koch, F., and Romani, N. (1999). Migration of langerhans cells and dermal dendritic cells in skin organ cultures: augmentation by TNF-alpha and IL-1beta. *J. Leukoc. Biol.* 3: 462-470.

Stoitzner, P., Ratzinger, G., Koch, F., Janke, K., Scholler, T., Kaser, A., Tilg, H., Cruikshank, W.W., Fritsch, P., and Romani, N. (2001). Interleukin-16 supports the migration of Langerhans cells, partly in a CD4-independent way. *J. Invest. Dermatol.* 5:641-649.

Strobl, H., and Knapp, W. (1999). TGF-beta1 regulation of dendritic cells. *Microbes Infect.* 15:1283-1290.

Strunk, D., Egger, C., Leitner, G., Hanau, D., and Stingl, G. (1997). A skin homing molecule defines the langerhans cell progenitor in human peripheral blood. *J. Exp. Med.* 6:1131-1136.

Tailleux, L., Schwartz, O., Herrmann, J.L., Pivert, E., Jackson, M., Amara, A., Legres, L., Dreher, D., Nicod, L.P., Gluckman, J.C., Lagrange, P.H., Gicquel, B., and Neyrolles, O. (2003). DC-SIGN is the major Mycobacterium tuberculosis receptor on human dendritic cells. *J. Exp. Med.* 197:121-127.

Thery, C., and Amigorena, S. (2001). The cell biology of antigen presentation in dendritic cells. *Curr. Opin. Immunol.* 1:45-51.

Thomson, A.W., and Lu, L. (1999). Are dendritic cells the key to liver transplant tolerance? *Immunol. Today* 1:27-32.

Tilloy, F., Di Santo, J.P., Bendelac, A., and Lantz, O. (1999). Thymic dependence of invariant V alpha 14[+] natural killer-T cell development. *Eur. J. Immunol.* 10:3313-3318.

Treiner, E., Duban, L., Bahram, S., Radosavljevic, M., Wanner, V., Tilloy, F., Affaticati, P., Gilfillan, S., and Lantz, O. (2003). Selection of evolutionarily conserved mucosal-associated invariant T cells by MR1. *Nature* 422:164-169.

Trinchieri, G., Pflanz, S., and Kastelein, R.A. (2003). The IL-12 family of heterodimeric cytokines: new players in the regulation of T cell responses. *Immunity* 5:641-644.

Urosevic, M., Dummer, R., Conrad, C., Beyeler, M., Laine, E., Burg, G., and Gilliet, M. (2005). Disease-independent skin recruitment and activation of plasmacytoid predendritic cells following imiquimod treatment. *J. Natl. Cancer Inst.* 15:1143-1153.

van Gisbergen, K.P., Sanchez-Hernandez, M., Geijtenbeek, T.B., and van Kooyk, Y. (2005). Neutrophils mediate immune modulation of dendritic cells through glycosylation-dependent interactions between Mac-1 and DC-SIGN. *J. Exp. Med.* 8:1281-1292.

van Kooyk, Y., and Geijtenbeek, T.B. (2003). DC-SIGN: escape mechanism for pathogens. *Nat. Rev. Immunol.* 9:697-709.

Vanbervliet, B., Bendriss-Vermare, N., Massacrier, C., Homey, B., de Bouteiller, O., Briere, F., Trinchieri, G., and Caux, C. (2003). The inducible CXCR3 ligands control plasmacytoid dendritic cell responsiveness to the constitutive chemokine stromal cell-derived factor 1 (SDF-1)/CXCL12. *J. Exp. Med.* 5:823-830.

Vandenabeele, S., Hochrein, H., Mavaddat, N., Winkel, K., and Shortman, K. (2001). Human thymus contains 2 distinct dendritic cell populations. *Blood* 6:1733-1741.

Verbovetski, I., Bychkov, H., Trahtemberg, U., Shapira, I., Hareuveni, M., Ben-Tal, O., Kutikov, I., Gill, O., and Mevorach, D. (2002). Opsonization of apoptotic cells by autologous iC3b facilitates clearance by immature dendritic cells, down-regulates DR and CD86, and up-regulates CC chemokine receptor 7. *J. Exp. Med.* 12:1553-1561.

Watanabe, N., Wang, Y.H., Lee, H.K., Ito, T., Cao, W., and Liu, Y.J. (2005). Hassall's corpuscles instruct dendritic cells to induce CD4$^+$CD25$^+$ regulatory T cells in human thymus. *Nature* 436:1181-115.

Watanabe, T., Kitani, A., Murray, P.J., and Strober, W. (2004). NOD2 is a negative regulator of Toll-like receptor 2-mediated T helper type 1 responses. *Nat. Immunol.* 8:800-808.

West, M.A., Prescott, A.R., Eskelinen, E.L., Ridley, A.J., and Watts, C. (2000). Rac is required for constitutive macropinocytosis by dendritic cells but does not control its downregulation. *Curr. Biol.* 14:839-848.

Wollenberg, A., Wagner, M., Gunther, S., Towarowski, A., Tuma, E., Moderer, M., Rothenfusser, S., Wetzel, S., Endres, S., and Hartmann, G. (2002). Plasmacytoid dendritic cells: a new cutaneous dendritic cell subset with distinct role in inflammatory skin diseases. *J. Invest. Dermatol.* 5:1096-1102.

Wu, L., and Shortman, K. (2005). Heterogeneity of thymic dendritic cells. *Semin. Immunol.* 4:304-312.

Yoneyama, H., Matsuno, K., Toda, E., Nishiwaki, T., Matsuo, N., Nakano, A., Narumi, S., Lu, B., Gerard, C., Ishikawa, S., and Matsushima, K. (2005). Plasmacytoid DCs help lymph node DCs to induce anti-HSV CTLs. *J. Exp. Med.* 3:425-435.

Yoneyama, H., Matsuno, K., Zhang, Y., Nishiwaki, T., Kitabatake, M., Ueha, S., Narumi, S., Morikawa, S., Ezaki, T., Lu, B., Gerard, C., Ishikawa, S., and Matsushima, K. (2004). Evidence for recruitment of plasmacytoid dendritic cell precursors to inflamed lymph nodes through high endothelial venules. *Int. Immunol.* 16:915-928.

Yoo, J., Omori, M., Gyarmati, D., Zhou, B., Aye, T., Brewer, A., Comeau, M.R., Campbell, D.J., and Ziegler, S.F. (2005). Spontaneous atopic dermatitis in mice expressing an inducible thymic stromal lymphopoietin transgene specifically in the skin. *J. Exp. Med.* 4:541-549.

Yoshimura, A., Lien, E., Ingalls, R.R., Tuomanen, E., Dziarski, R., and Golenbock, D. (1999). Cutting edge: recognition of Gram-positive bacterial cell wall components by the innate immune system occurs via Toll-like receptor 2. *J. Immunol.* 1:1-5.

Zhou, D., Mattner, J., Cantu, C. 3rd., Schrantz, N., Yin, N., Gao, Y., Sagiv, Y., Hudspeth, K., Wu, Y.P., Yamashita, T., Teneberg, S., Wang, D., Proia, R.L., Levery, S.B., Savage, P.B., Teyton, L., and Bendelac, A. (2004). Lysosomal glycosphingolipid recognition by NKT cells. *Science* 306:1786-1789.

Zuniga, E.I., McGavern, D.B., Pruneda-Paz, J.L., Teng, C., and Oldstone, M.B. (2004). Bone marrow plasmacytoid dendritic cells can differentiate into myeloid dendritic cells upon virus infection. *Nat. Immunol.* 12:1227-1234.

Chapter 2

Dendritic Cells and Their Role in Linking Innate and Adaptive Immune Responses

Mary F. Lipscomb, Julie A. Wilder, and Barbara J. Masten

2.1 Introduction

A novel and relatively infrequent cell that exhibited a striking dendritic shape was first described in murine spleen, lymph nodes, and Peyer's patches by Ralph Steinman and Zanvil Cohn in a seminal paper published in the *Journal of Experimental Medicine* in 1973 (Steinman and Cohn, 1973). They called these cells "dendritic cells" (DCs) and characterized them as nonphagocytic, loosely adherent, and of low buoyant density (Steinman and Cohn, 1974). DCs were found to exist in all lymphoid and most nonlymphoid tissues, constitutively express both MHC class I and class II antigens, and spontaneously cluster T cells via antigen-independent mechanisms [reviewed in Lipscomb and Masten (2002)]. Most important, DCs stimulated the responses of naïve CD4 and CD8 T cells to nominal and alloantigens more effectively than any other previously described antigen-presenting cell (APC). In the past 15 years, DCs have increasingly captured the interest of scientists and physicians, because they are critical adjuvants for vaccines that prevent microbial infections and treat cancer. Because they not only have an enhancing effect on the development of acquired immunity, but can also induce tolerance, they can likely be manipulated to prevent and treat allergic and autoimmune diseases. DCs are known to play an important role in the pathogenesis of HIV infections, and they might also be manipulated to enhance immunization procedures. Thus, understanding more completely the interactions of this deadly virus with the host's DCs will help us to design appropriate preventive and therapeutic strategies. The goal for this chapter is to broadly cover DC ontogeny and biology to set the stage for more detailed discussions that follow.

2.2 DC Origins, Subsets, and Differentiation

DCs are heterogeneous cells that can be divided into a few functionally distinct subsets based on their phenotype, migratory properties, and ability to produce cytokines that modulate both innate and adaptive immune response (Lipscomb and Masten, 2002; MacDonald *et al.*, 2002; Masten, 2004; Masten *et al.*, 2004; Shortman and Liu, 2002). Functions of DCs within a subset are influenced by environmental stimuli that alter their maturation, migration, and/or secretory status (Jarrossay *et al.*, 2001; Kadowaki *et al.*, 2001; Kalinski *et al.*, 2001). All DCs have several features in common. First, DCs arise from $CD34^+$ hematopoietic stem cell (HSC) progenitors in the bone marrow (Steinman *et al.*, 1975). Second, DCs are lineage-negative, major histocompatibility complex (MHC) class I and II positive mononuclear cells found in blood as precursor DCs and as immature DCs in tissues throughout the body, especially at interfaces with the external environment and in lymph nodes. Third, immature DCs are characterized by intermediate surface expression of MHC class II and low levels of costimulatory molecules, including CD80, CD86, and CD40. These immature DCs have a broad innate receptor repertoire, exhibit phagocytic activity, and take up antigens derived from pathogens (e.g., bacteria, viruses, or fungi) and dead or dying cells and respond to indirect inflammatory signals by undergoing an activation and maturation process that includes cytoskeleton reorganization, reduced antigen uptake, and increased antigen processing and expression of MHC class II and costimulatory molecules. Fourth, maturing DCs acquire motility to enter afferent lymphatics and flow into lymph nodes to position themselves for encounters with naïve or central memory T cells. Distinct chemokine receptors regulate the traffic of precursor and immature DCs into peripheral tissue and mature DCs into lymphoid tissue sites (Cyster, 1999; Qu *et al.*, 2004; Randolph *et al.*, 2005; Sallusto and Lanzavecchia, 2000; Sallusto *et al.*, 2000; Zlotnik and Yoshie, 2000). Finally, DCs express T-cell polarizing molecules, either soluble or membrane-bound, that determine the type of immune response that will develop (de Jong *et al.*, 2005; Kalinski *et al.*, 2003; Kapsenberg, 2003).

In humans, DCs are commonly divided into two major subsets based on the expression of the β2-integrin CD11c: $CD11c^+$ myeloid DCs (MDCs) and $CD11c^-$ plasmacytoid DCs (PDCs) (Ito *et al.*, 2005; Liu, 2005; O'Doherty *et al.*, 1994). MDCs express the myeloid markers CD13 and CD33, although (discussed below) their myeloid origin, particularly in mice, is debated (O'Doherty *et al.*, 1994). In this chapter, MDCs will often be referred to generically as DCs. This DC subset is further divided by anatomic location into a) interstitial DCs (which express mannose receptor, coagulation factor XIIa, and DC-SIGN), b) Langerhans DCs (which express CD1a, CD207 or langerin, and E-cadherin and also have a distinct organelle called the Birbeck granule), and c) pre-DC monocytes found in blood (Liu, 2001). Freshly isolated DCs have poorly developed dendrites (Fig. 2.1) and spontaneously acquire a typical DC morphology after a period in culture (O'Doherty *et al.*, 1994; Siegal *et al.*, 1999). Freshly isolated precursor PDCs (Fig. 2.1) have a well-developed rough endoplasmic reticulum and a typical Golgi apparatus (Grouard *et al.*, 1997). PDCs rapidly die during culture without addition of exogenous cytokines. PDCs were named based on the ultrastructural characteristic of resembling Ig-secreting plasma cells with ample rough endoplasmic reticulum and were previously called "plasmacytoid T cells" or "plasmacytoid monocytes" (Facchetti *et al.*, 1999; Grouard *et al.*, 1997; Res *et al.*, 1999; Sorg *et al.*, 1999; Strobl *et al.*, 1998). In contrast with conventional DCs, PDCs do not express CD13 and CD33 nor the markers mentioned above that distinguish interstitial DCs from Langerhans DCs, but do express CD123, the receptor for IL-3, and BDCA-2 (O'Doherty *et al.*, 1994; Rissoan *et al.*, 1999). In human blood, BDCA-2 is specific for PDCs, whereas CD123 is not only present on PDCs but can be on other blood cells, including monocytes (Buelens *et al.*, 2002; Dzionek *et al.*, 2000; Narbutt *et al.*, 2004). BDCA-2 may be downregulated as PDCs mature; thus, this marker is more specific to immature PDCs (Dzionek *et al.*, 2000). The PDC subset is also known as interferon-producing cells (IPCs) (Kadowaki *et al.*, 2000).

Myeloid DC   Plasmacytoid DC

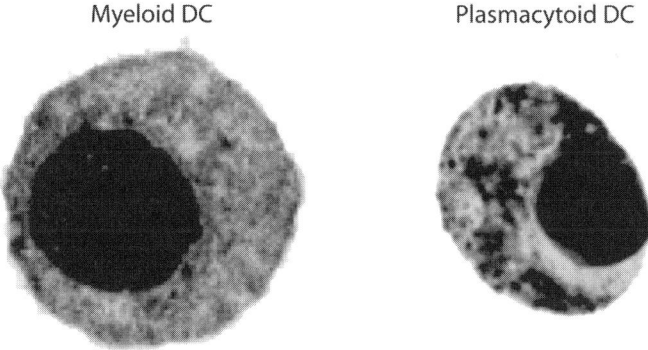

FIGURE 2.1. Human conventional (myeloid) and plasmacytoid DCs. Shown are cells sorted from human peripheral blood mononuclear cells into lineage-negative, CD11c[+], myeloid DC (left) and lineage-negative, CD11c[-], CD123[+] plasmacytoid DC (right). Cytospins of the sorted cells were prepared and stained with Wright-Giemsa stain.

The understanding of human DC subset lineage, phenotype, and differentiation has been largely acquired by studying cytokine-supported cultures of CD34$^+$ HSC progenitors and precursor and immature DCs present in peripheral blood, with limited information derived from isolating DCs from solid tissues. CD34$^+$ HSCs are the earliest progenitors of DCs and develop into CD34$^+$ common lymphoid progenitors (CLPs) and CD34$^+$ common myeloid progenitors (CMPs) (Fig. 2.2) (Ardavin, 2003; Banchereau *et al.*, 2000). The fms-related tyrosine kinase 3 ligand (Flt3L) is critical for PDC development from HSCs; and in combination with stromal cells, substantial numbers of mouse PDCs can be obtained from cultures of HSCs with Flt3L (Pelayo *et al.*, 2005). Flt3L also promotes MDC development from HSCs (Karsunky *et al.*, 2003). Granulocyte-macrophage colony-stimulating factor (GM-CSF) promotes MDC development from CMPs and blocks the growth of PDCs, while interleukin (IL)-7 promotes their development from CLPs (Gilliet *et al.*, 2002; O'Neill and Wilson, 2004). *In vitro*, IL-4 promotes the development of monocyte-derived DCs and maintains them in an immature/semimature state, preserving antigen uptake and processing capacities (O'Neill and Wilson, 2004). Thus, MDCs are often generated *in vitro* by culturing monocytes in

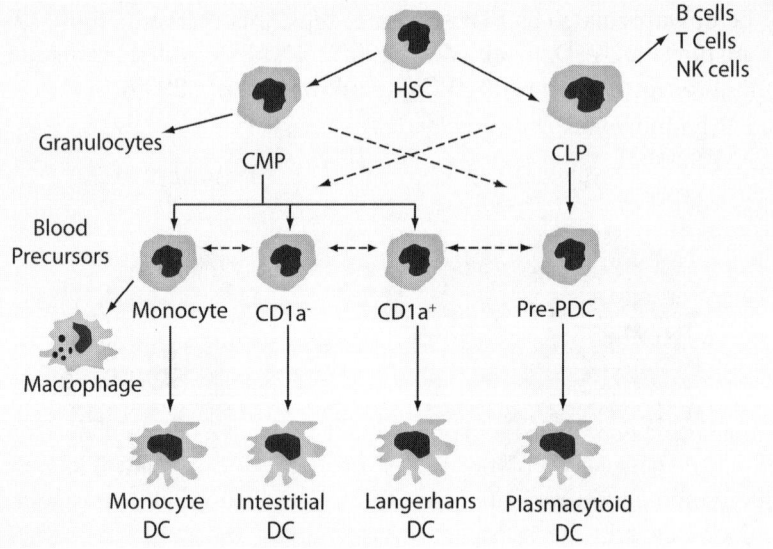

FIGURE 2.2. DC lineage model. Hematopoietic stem cells (HSCs) develop into common myeloid cell precursors (CMPs) or common lymphoid precursors (CLPs). The myeloid precursor generates monocytes, CD1a⁻ precursors of interstitial DCs, and CD1a$^+$ Langerhans DCs. The lymphoid precursor is likely the dominant precursor for plasmacytoid DCs (PDCs). However, the dotted arrows suggest that these cells might also have alternative developmental pathways.

the presence of both GM-CSF and IL-4 (Sallusto and Lanzavecchia, 1994). Tumor necrosis factor-α (TNF-α) also promotes the differentiation of CD34$^+$ HSC and monocytes toward DCs in the presence of stromal cells and blocks the development of PDCs (Chomarat *et al.*, 2003; Gilliet *et al.*, 2002). Transforming growth factor-β (TGF-β) supports Langerhans DC development *in vitro* and is critical *in vivo* (Vandenabeele and Wu, 1999). IL-3 and monocyte-conditioned medium promotes the viability and differentiation of precursor PDCs into immature PDCs (Grouard *et al.*, 1997; O'Doherty *et al.*, 1994). In the presence of IL-3 and CD40L, mature PDCs develop (Grouard *et al.*, 1997; Kadowaki *et al.*, 2000).

A controversy exists over the developmental origin of DC subsets from HSC progenitor cells. One hypothesis is that all DCs arise from a single committed lineage that has phenotypic and functional plasticity (del Hoyo *et al.*, 2002). This hypothesis is supported by studies demonstrating that Flt3L$^+$ cells within either the CLPs or CMPs could differentiate into both MDCs or PDCs in culture and *in vivo* (D'Amico and Wu, 2003; Karsunky *et al.*, 2003). A second hypothesis is that multiple DC lineages exist, each phenotypically and functionally distinct (Corcoran *et al.*, 2003; Grouard *et al.*, 1997; Shortman and Liu, 2002; Wu *et al.*, 2001). This latter notion is supported by the finding that ectopic expression of inhibitor of binding protein 3 (Id3; helix-loop-helix motif containing transcription inhibitor) in CD34$^+$ HSCs blocks the development of T and B lymphocytes and reduces the numbers of PDC-like cells, suggesting most PDCs share an origin with T and B lymphocytes (Spits *et al.*, 2000). Furthermore, the gene encoding CIITA (class II transactivator), a transcription factor crucial for the activation of genes associated with MHC class II antigen presentation, is differentially activated by the myeloid promoter pl in MDCs and the B-cell promoter plll in PDCs (LeibundGut-Landmann *et al.*, 2004). A third hypothesis proposes a synthesis of the other two, that is, DC lineage interconversion can occur (Comeau *et al.*, 2002). This concept is supported by findings that PDCs can differentiate into MDCs and upon doing so lose pre-Tα (pTα) expression and, in the presence of GM-CSF, acquire CD11c (Vandenabeele *et al.*, 2001). Myeloid-like PDCs have been noted as a minor fraction of peripheral blood PDCs in both normal and Flt3L-treated healthy volunteers (Comeau *et al.*, 2002). Furthermore, in a murine model, precursor PDCs in the bone marrow were capable of differentiating into MDCs in a type-1 IFN (IFN-α and IFN-β)-dependent manner in response to a viral infection (Zuniga *et al.*, 2004). Greater understanding of the molecular regulation of DC subset differentiation should help resolve the contradictions arising from the current lineage experimental data.

Murine DCs have been widely employed by researchers investigating the roles of DCs in the generation and regulation of specific immunity (Lambrecht, 2005; Lipscomb and Masten, 2002; Masten and Lipscomb,

1997; Masten *et al.*, 2004; Masten *et al.*, 1997). Like human DCs, murine DCs (a) originate from CD34$^+$ bone marrow stem cells, (b) are found in blood and tissues as precursor and immature DCs, (c) are able to take up and degrade antigen to peptides and express MHC class II molecules complexed with peptide, (e) upregulate the expression of MHC molecules and costimulatory molecules upon maturation, (f) enter afferent lymphatics and traffic to lymph nodes, and (g) express T-cell polarizing molecules that direct antigen-specific immune responses. While in humans CD11c distinguishes MDCs from PDCs, CD11c is expressed on all DC subsets in mice. Murine DCs are commonly divided into two major subsets based on the differential expression of CD45R/B220: CD45R/B220$^-$ MDCs and CD45R/B220$^+$ PDCs (Bjorck, 2001; Bjorck, 2002; Maraskovsky *et al.*, 1996; Nakano *et al.*, 2001; Pulendran *et al.*, 1997; Pulendran *et al.*, 1999). The MDC subset has been further divided into five subsets based on the expression of CD8α, CD4, CD11b, and DEC205 and existence in lymphoid and nonlymphoid tissues (Shortman and Liu, 2002; Vremec *et al.*, 2000). These five MDC subsets consist of three blood-derived subsets (CD4$^+$CD8α^-CD11b$^+$CD205$^-$, CD4$^-$CD8α^+CD11b$^-$CD205$^+$, and CD4$^-$CD8α^-CD11b$^+$CD205$^-$ DCs) and two tissue-derived subsets (Langerhans cells, CD4$^-$CD8$\alpha^{+/-}$CD11b$^+$CD205^{hi}; and interstitial DCs, CD4$^-$CD8α^-CD11b$^{+/-}$CD205$^+$). The expression of CD8α (which correlates with the lack of CD11b) was originally thought to indicate a lymphoid lineage, and lack of CD8α (which correlates with the expression of CD11b) was thought to indicate a myeloid lineage (Ardavin, 1997). However, recent findings suggest that CD8α on DCs does not indicate a lymphoid origin, but rather the maturation or differentiation status of the cell (Merad *et al.*, 2000; Traver *et al.*, 2000). PDCs (Ly6C$^+$CD45R/B220$^+$CD19$^-$120G8$^+$) exist in lymphoid and nonlymphoid tissues and in blood. As in humans, PDCs produce large amounts of type I IFN in response to viral pathogen-associated molecular patterns (PAMPs) (Liu, 2005). PDCs are also described as being Gr-1+ as a result of the Gr-1 monoclonal antibody (mAb) recognizing a cross-reactive epitope detected by both the Ly-6C and Ly6G mAbs (Asselin-Paturel *et al.*, 2001). The 120G8 mAb detects an antigen highly expressed on both resting and activated PDCs (Asselin-Paturel *et al.*, 2001; Asselin-Paturel *et al.*, 2003). PDCs have been noted to express CD8α, especially upon activation, but expression of CD45R/B220 distinguishes them from MDCs (Le Bon *et al.*, 2003). PDCs have been subdivided into CD4$^+$ and CD4$^-$ subsets in peripheral tissues, with the CD4$^-$ PDC subset producing higher relative amounts of type-I IFN, IL-12, and proinflammatory cytokines *in vitro* in response to CpG, and having an increased potential to activate T-cell proliferation (Chang *et al.*, 2005).

The lineage and differentiation relationship of murine DCs with other myeloid and lymphoid cells is not clear, but advances are being made. Committed DC precursors have recently been identified in the blood and bone marrow as a CD11c$^+$ MHC II$^-$ population that can be subdivided into B220$^+$ precursor and B220$^-$ precursor subpopulations, which have distinct functional and proliferative potentials *in vitro* and distinct capacities to generate mature DC populations *in vivo* (del Hoyo *et al.*, 2002; Diao *et al.*, 2004; O'Keeffe *et al.*, 2003). The CD11c$^+$MHCII$^-$B220$^+$CD19$^-$ blood precursor seems to be downstream in development from the CMP and CLP (del Hoyo *et al.*, 2002). The CD11c$^+$MHCII$^-$B220$^-$CD19$^-$ blood precursor appears to be a DC-committed precursor as this population can reconstitute all spleen DC subsets but no other myeloid or lymphoid cell types upon adoptive transfer. The CD11c$^+$MHCII$^-$B220$^+$ and the CD11c$^+$MHCII$^-$B220$^-$ bone marrow precursors display immunophenotye and morphologic resemblance to, respectively, pre-PDCs and pre-MDCs in blood, suggesting these may be equivalent populations. Purified B220$^-$ DC bone marrow precursors were shown to expand and generate mature CD11c$^+$ CD11b$^+$B220$^-$ DCs *in vitro* and after adoptive transfer. Purified B220$^+$ DC bone marrow precursors have a lower regenerative potential than the B220$^-$ DC precursors and yield both CD11b$^-$B220$^+$ and CD11b$^+$B220$^-$ DC populations. Both DC precursor populations can give rise to CD8α^+ and CD8α^- DC subsets (Diao *et al.*, 2004). Analysis of gene expression profiling of precursor DC populations and upstream progenitors should aid in the delineation of factor(s) that drive the differentiation of oligolineage progenitors into DC-restricted precursors and refine these developmental pathways. A recent study in mice suggests that differential expression of interferon regulatory factors (IFRs) may underlie murine (and human) DC development and thus provide a molecular footprint for DC subset development (Tamura *et al.*, 2005). Although differences exist between murine and human DCs, it is evident that the study of murine DCs is relevant to understanding human DCs. However, data from murine DC studies should be interpreted with caution in cases where clear discrepancies exist between murine and human subtype counterparts.

2.3 Plasticity and Trafficking Pathways

Evidence for different functions associated with each DC subset supports the view that DCs are a population of cells that are closely related, but can be functionally distinct. Their ability to respond to their environment and undergo change in their phenotype and function defines them as being highly plastic. Recognition of PAMPs through Toll-like receptors (TLRs) is one very important mechanism by which this plasticity is mediated and allows the induction of a variety of innate immune responses with the subsequent development of the appropriate adaptive immune response against many different pathogens. Pathogen recognition triggers DCs to release mediators of inflammation, including cytokines and chemokines that initiate the recruitment of more phagocytic cells, NK cells, eosinophils, and basophils to the site of infection. Recognition of the infectious agent also triggers signaling pathways in the DC that activate maturation, transforming DCs into efficient APCs for T-cell stimulation (Pasare and Medzhitov, 2004). The pattern of TLR expression is different between MDCs and PDCs and this has been proposed to underlie their unique responsiveness to specific pathogens and determine the type of immune response that is induced by pathogens (Ito *et al.*, 2005; Jarrossay *et al.*, 2001; Kadowaki *et al.*, 2001). Data supporting this concept has mainly been obtained from *in vitro* derived DCs and *ex vivo* isolated DC subsets. A prominent feature of PDCs is the response to TLR7 ligands (antiviral imidazoquinoline is one ligand) and TLR9 ligands (unmethylated CpG DNA from viruses and bacteria) with subsequent maturation and production of high levels of IFN-α, which influences early innate immunity and facilitates the development of Th1 responses and the generation of specific cytotoxic T lymphocytes (Cella *et al.*, 2000; Kadowaki *et al.*, 2000). MDCs can make type-I IFNs, but generally at much lower levels. MDCs respond generally to TLR2 ligands (peptidoglycan from Gram-positive bacteria and lipoteichoic acid from various bacteria), TLR3 (double-stranded RNA and poly I:C), TLR4 ligands (LPS from Gram-negative bacteria), and TLR5 ligands (flagellin) with TNFα, IL-6, and IL-12 secretion (Krug *et al.*, 2001). Human PDCs seem to have less capacity than MDCs to produce IL-12 in response to TLR ligands. The capacity of MDCs to produce IL-12 in response to TLR ligation varies with the type of signals delivered to the DC. For example, lipopolysaccharide (LPS) derived from *Escherichia coli* activates MDCs to produce IL-12 and drive Th1 immune responses, whereas LPS from *Porphyromonas gingivalis* activates MDCs to induce Th2 responses due to a lower capacity of the DC to produce IL-12 (Jotwani *et al.*, 2003; Pulendran *et al.*, 2001). Biological mediators produced as a result of the inflammatory process also influence the ability of DCs to secrete IL-12 and mature. Similar to MDCs, PDCs show functional plasticity to physiologic conditions and inducing T-cell responses. For example, in the presence of IL-3 and CD40L, PDCs mature but produce little to no IL-12 and type I IFNs (Grouard *et al.*, 1997).

DCs are migratory cells that traffic from one site to the next, performing specific functions at each site (Sallusto and Lanzavecchia, 2000; Sallusto *et al.*, 2000; Zlotnik and Yoshie, 2000; Cyster, 1999; Randolph *et al.*, 2005). After leaving the bone marrow, precursor and immature MDCs migrate into nonlymphoid tissues and then into lymph nodes through afferent lymphatics. Precursor MDCs in blood enter inflamed peripheral tissue by interacting with ICAM-2 and P- and E-selectins expressed on activated endothelium and responding to chemokine chemoattractants that interact with their chemokine receptors (Pendl *et al.*, 2002). Chemokines are differentially produced at peripheral tissue sites by endothelial cells, epithelial cells, and leukocytes, including DCs, in response to diverse inflammatory stimuli (Baggiolini, 1998; Cyster, 1999). Chemokines are also constitutively produced by endothelial cells or stromal cells and leukocytes, including DCs, within secondary lymphoid organs to regulate encounters between DCs and T and B cells (Cyster, 1999; Zlotnik and Yoshie, 2000). The anatomical location of inflammatory chemokines within peripheral tissue and constitutive chemokines within lymphoid tissue is thought to regulate the migration of precursor DCs initially to sites of antigen and ultimately to lymphoid tissue to initiate an immune response.

As immature MDCs migrate toward increasing concentrations of inflammatory chemokines, they are also exposed to increasing concentrations of proinflammatory cytokines, such as TNF-α and IL-1, and the pathogen products initiating the inflammatory response, which induce DC maturation. Maturing MDCs downregulate the expression of CCR1, CCR5, and CXCR1 and upregulate the expression of CCR7, which targets DCs to lymphatic vessels and lymph nodes via chemokines CCL19 (Epstein-Barr virus–induced molecule 1 ligand chemokine, ELC) and CCL21 (secondary lymphoid tissue chemokine, SLC; 6Ckine). CCL21 is produced by lymphatic endothelial cells, and both CCL21 and CCL19 are produced by stromal cells and DCs in the T-cell areas of lymphoid organs (Dieu-Nosjean *et al.*, 1999; Foti *et al.*, 1999; Gunn *et al.*, 1998; Lin *et al.*, 1998; Ngo *et al.*, 1998; Sozzani *et al.*, 1998; Vecchi *et al.*, 1999). CCR7 is also selectively expressed on naïve and memory T and B lymphocytes, which allows these cell types to also home to lymphoid tissue (Burgstahler *et al.*, 1995). In addition to chemokines, other molecules are known to modulate migration of DCs to the lymph nodes, such as histamine and cysteinyl leukotrienes (which increase lymph flow) and platelet-activating factor, which decreases lymph flow. Data also suggests that the β2 integrin, lymphocyte function associated antigen-1 (LFA-1), mediates MDC migration to lymph nodes (Ma *et al.*, 1994). Maturation also induces DCs to produce chemokines such as regulated on activation normal T-cell expressed and secreted (RANTES), macrophage inflammatory protein-1α (MIP-1α) and MIP-1β, and thymus and activation-regulated chemokine (TARC), macrophage-derived chemokine (MDC) or interferon-inducible protein-10 (IP-10) (Foti *et al.*, 1999; Sallusto and Lanzavecchia, 2000;

Sallusto *et al.*, 2000). These chemokines together recruit additional monocytes, DCs, and T cells to the sites in which they are secreted.

Expression of CCR7 typically correlates with the upregulation of MHC and costimulatory molecules, but DC migration from the periphery to the lymph node is a process that is distinct from DC maturation. The uptake of apoptotic cells induces CCR7 expression and the ability to respond to CCR7 ligands, but does not induce DC maturation (Verbovetski *et al.*, 2002). A recent study suggests that the signaling adaptor protein DAP12 (DNAX activation protein 12) induces CCR7 expression but incomplete DC maturation (Bouchon *et al.*, 2001). Under homeostatic conditions, where DCs are likely immature and express self-antigens complexed with MHC molecules, lack of DAP12 expression leads to the accumulation of DCs in the periphery (Bakker *et al.*, 2000; Tomasello *et al.*, 2000). The ability of DCs to incompletely express maturation is likely important in their induction of tolerance (see below). Overall, the trafficking of peripheral DCs to lymph nodes, whether under inflammatory, noninflammatory, or homeostatic conditions requires CCR7 expression. However, responsiveness to CCR7 ligands does not depend solely on the presence of CCR7, as CCR7 can be expressed in a biologically "insensitive" state, such that CCR7-expressing DCs fail to migrate in response to CCR7 ligands or require high concentrations of CCR7 ligands to undergo chemotaxis (Robbiani *et al.*, 2000; Scandella *et al.*, 2002). Biological mediators, such as prostaglandin E2 and the ADP-ribosyl cyclase CD38, are found at sites of inflammation and can sensitize CCR7 to its ligands (Kabashima *et al.*, 2003; Partida-Sanchez *et al.*, 2004; Scandella *et al.*, 2002). The mechanism(s) underlying the altered sensitivity of CCR7 to its ligands is not clear but likely involves alterations in signaling cascades that are involved upon exposure to sensitizers (Scandella *et al.*, 2004). In mice, expression of two isoforms of CCL21 may underlie the migratory pattern of precursor DCs into peripheral tissue and peripheral tissue DCs into lymph nodes (Chen *et al.*, 2002; Vassileva *et al.*, 1999). These two isoforms are encoded by separate genes that arose by gene duplication and are modified at position 65 with expression of either a Leu or a Ser. CCL21-Leu is expressed in the periphery by lymphatic vessels, whereas CCL21-Ser is expressed in lymph nodes, including terminal lymphatic vessels that are present in the subcapsular sinus.

Langerhans cells (LCs) express CCR6 in addition to the chemokine receptors expressed by other MDCs (Dieu-Nosjean *et al.*, 2000; Dieu *et al.*, 1998; Sozzani *et al.*, 1999). CCR6 responds to MIP-3β, which is produced constitutively by hepatocytes and lung epithelial cells. Furthermore, MIP-3β is produced by epithelial cells located over inflamed tonsils, in the appendix of humans, and in noninflamed follicle-associated epithelium of murine Peyer's patches (Cook *et al.*, 2000; Dieu-Nosjean *et al.*,

2000; Iwasaki and Kelsall, 2000). MIP-3β and CCR6 represent a chemokine/ receptor pair that has dual function in recruiting Langerhans cells to epidermal and mucosal sites of inflammation and to become sentinels in noninflamed tissues. As LCs mature, CCR6 is downregulated with a con-comitant upregulation of CCR7 to facilitate homing to T-cell areas of lymphoid tissue.

Upon maturation, bacterial and complement receptors on DCs are also differentially regulated. Maturing DCs downregulate their responsiveness and receptor expression for fMLP (N-formyl-methionyl-levcyl-phenylalanine) while maintaining their responsiveness and receptor expression for C5a (Yang *et al.*, 2000). The interaction of C5a with C5aR on mature DCs may participate in guiding mature DCs to B-cell follicles in lymphoid tissue where naïve B cells, a source of C5a, acquire antigens delivered by mature DCs.

The traffic of PDCs contrasts with that of MDCs and is somewhat similar to that of T and B lymphocytes (Butcher and Picker, 1996). After leaving the bone marrow, precursor PDCs migrate into the spleen and via high endothelial venules (HEVs) into lymph nodes and mucosa-associated lymphoid tissue. Precursor PDCs in blood express CD62L, which interacts with L-selectin ligands expressed by HEV and mediates cell adhesion and rolling on HEV endothelium (Girard and Springer, 1995; Gretz *et al.*, 2000; Yang *et al.*, 2005). PDCs express chemokine receptors CXCR4 (which recognize CXCL12) and CXCR3 (which recognize CXCL9), which may mediate homing and migration of these cells into the lymph node in response to inflammatory chemokines (Cella *et al.*, 1999). Human PDCs do not migrate in response to CXCR3 ligands *in vitro*, but exposure to CXCR3 ligands enhances their ability to migrate in response to the constitutive chemokine CXCL12 (Krug *et al.*, 2002; Vanbervliet *et al.*, 2003). As they do for MDCs, CCR7 ligands play a major role in migration of activated PDCs from peripheral tissue into lymphoid organs (Jarrossay *et al.*, 2001).

2.4 DC–T Cell Interactions

As discussed, DCs serve a crucial role in the initiation and regulation of the immune response in that they induce the activation of naïve antigen-specific T cells, influence the differentiation of effector T cells, and induce T-cell tolerance. Furthermore, once T cells have achieved memory or effector function, DCs can activate the function of effector T cells and reactivate memory T cells. MDCs and PDCs play different roles in the immune response, although their plasticity allows considerable overlap as to whether they can drive T cells to become tolerant or differentiate into Th1 versus Th2 cells (Patterson, 2000). Several different factors regulate the activation of naïve T cells including antigen concentration, the type and density of costimulatory molecules on the membranes of the DCs and T cells, the concentration and type of cytokines and other soluble factors that are present in the microenvironment in which the cells are interacting, and the genetic background of the host. CD4 and CD8 T cells respond to peptide antigen displayed on MHC class II and MHC class I molecules, respectively. Accessory molecules on DCs are required to assure that T cells will divide, expand effectively, and differentiate into effector cells. In this section, antigen uptake, antigen presentation, and mechanisms for DC–T cell interactions will be discussed.

2.4.1 Antigen Uptake

DCs take up extracellular antigen via endocytic pathways. The endocytic activity of immature DCs far exceeds that of mature DCs (Steinman and Swanson, 1995). Antigens are taken up by DCs by either receptor-mediated phagocytosis, endocytosis, or by fluid-phase pinocytosis. Antigens are subsequently degraded within endosomal compartments, a later one being rich in MHC class II molecules (MHC Class II Compartment, MIIC) and finally expressed on the cell surface complexed to MHC II molecules. DCs can take up both necrotic and apoptotic cells (Harshyne *et al.*, 2001). Receptors used by DCs for antigen uptake are diverse and include the FcγRs, CD32 and CD64; the high- and low-affinity IgE receptors, FcεRI and FcεRII (CD23); the complement receptors, CD11b and CD11c; DEC205 (CD205), a C-type lectin of mannan binding receptor; and the scavenger receptor pair for apoptotic cells, $\alpha_v\beta_5$ and CD36 [reviewed in Hart (1997)]. The downregulation of these receptors as DCs mature help explain their reduced endocytic activity.

2.4.2 Antigen Processing

2.4.2.1 MHC Class II Presentation

DCs process antigens through both an exogenous endosomal pathway and an endogenous proteosomal pathway (Fig. 2.3). Exogenous antigens enter endosomal compartments where proteases initiate degradation. The resultant peptide fragments then associate with preformed MHC class II molecules within the MIIC. During synthesis of MHC class II α and β peptide chains

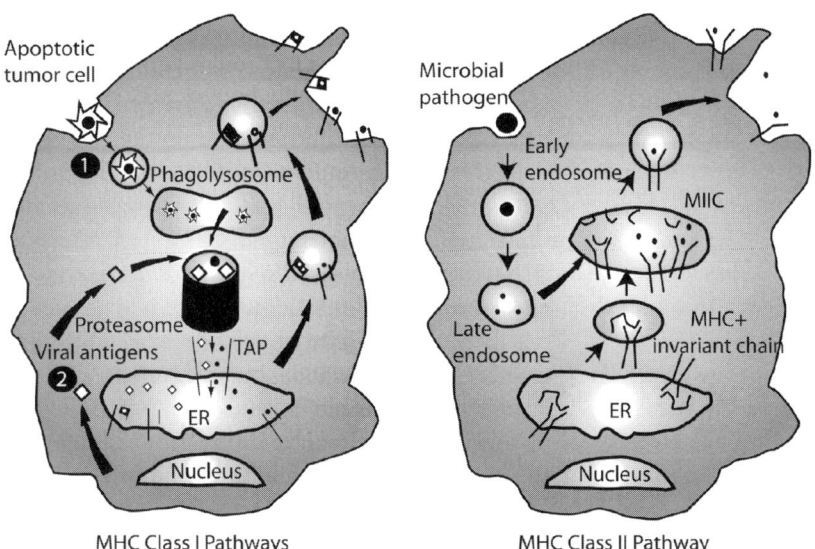

MHC Class I Pathways   MHC Class II Pathway

FIGURE 2.3. Antigen processing by a DC. Shown on the left are the two pathways for DC presentation on MHC class I antigens. For cross-presentation (*1*), an antigen taken up by endocytosis is partially degraded, enters the cytosol, and finds its way into the proteasomal pathway much as endogenous proteins are handled. The endogenous proteins are derived from mutated and viral proteins (*2*) and are degraded via the proteasomal pathway, enter the ER utilizing the transporter of antigenic peptides (TAP), are bound to MHC class I antigen, and are then transported by vesicles to the surface. Not shown is the CD1 pathway. On the right, the MHC class II pathway is shown. Antigen is taken up into the early and late endosomes. After limited proteolysis, the peptides then enter the MHC class II–rich vesicular compartment (MIIC), become bound within the MHC class II peptide-binding groove, and are transported to the cell surface. Also shown is that MHC class II, synthesized in the ER, is protected by invariant chain (Ii) from binding self-peptides, but Ii is degraded into a smaller peptide to ready itself for its replacement by the antigenic peptide in the MIIC.

in the endoplasmic reticulum (ER), a protein called invariant chain (Ii) binds in the peptide-groove, to protect the MHC II molecules from binding self-proteins prior to reaching the endosomal compartment (Castellino and Germain, 1995). Once in the endosome, partial proteolytic cleavage of Ii occurs leaving a small fragment called CLIP (class II–associated invariant-chain peptide) in the peptide-binding groove of the MHC class II molecule. Clip is ultimately exchanged for an antigenic peptide, (Shi *et al.*, 2000) (Morris *et al.*, 1994). Finally, MHC class II, with the new antigenic peptide in its binding groove, is transported for expression on the cell surface. Immunodominant peptides of antigens possess particularly long half-lives on MHC II molecules, which increases their immunogenicity *in vivo*

(Lazarski *et al.*, 2005). It has also been shown that members of the B7 family (i.e. CD80 and CD86) of costimulatory molecules are delivered to the cell surface in association with the MHC class/peptide complexes (Turley *et al.*, 2000). In addition, immature DCs are capable of secreting exosomes, which can transfer functional MHC II/peptide complexes to other DCs (Segura *et al.*, 2005a). Although immature DCs produce more exosomes than mature DCs (Quah and O'Neill H, 2005), exosomes from mature DCs are much more potent at stimulating T-cell activation due to their increased expression of MHC class II, ICAM-1, and the B7-2 (CD86) accessory molecules (Segura et al., 2005b).

2.4.2.2 MHC Class I Presentation

In contrast with CD4 cells, CD8 cells recognize antigen presented by DCs in the context of MHC class I. Association of antigenic peptides with MHC class I occurs after cytoplasmic proteins, often viral antigens, are degraded by proteasomes and the resultant peptides transported via the hetero-dimeric transporter-associated-with-antigen-processing (TAP) into the ER. In the ER, peptide antigen binds to a newly synthesized MHC class Iα chain that is associated with β_2 microglobulin (β_2M). (Pamer and Cresswell, 1998). Finally, MHC class I/β_2M/peptide complexes are transported in exocytic vesicles to the DC plasma membrane by processes similar to those used in MHC class II presentation.

Although it was originally thought that the proteosomal pathway processed only those proteins newly synthesized within APCs, such as viral proteins, it is now recognized that exogenous antigens can escape the endocytic pathways, undergo proteosome-dependent degradation, and subsequently enter the ER via TAP to be presented in the binding groove of MHC class I molecules (Pfeifer *et al.*, 1993; Reis e Sousa and Germain, 1995). This process is termed cross-presentation or cross-priming. The antigens may be particulate or soluble, result from dead or apoptotic cells, or be delivered to the DC on exosomes. Heat-shock proteins (Zheng and Li, 2004) and the TLR adaptor protein MyD88 (Palliser *et al.*, 2004) play important roles in chaperoning the antigens through the endocytic pathway. Many diverse extracellular pathogens, including bacteria (Pfeifer *et al.*, 1993) and fungi (Lin *et al.*, 2005; Ramadan *et al.*, 2005), stimulate CD8 cell activation via cross-priming. Viral antigens and associated live virus taken up from dead or apoptotic cells can also be cross-presented to CD8 T cells via their uptake by DCs (Larsson *et al.*, 2001). In the case of HIV, this facilitates infection of new cells (Maranon *et al.*, 2004). Cross-priming can lead to either productive immunity or tolerance of CD8 T cells, depending on the status of the DC.

2.4.3 Costimulation and Suppression Molecules

Productive interactions between DCs and T cells that lead to effective T-cell activation involve the binding of cell surface molecules on the two cell types in addition to DC MHC class I/II/peptide with the T-cell receptor. Many different molecules, commonly referred to as costimulatory molecules, are involved in the interaction of APCs and T cells. Costimulatory molecules are upregulated on mature DCs and thus enhance their antigen-presenting cell capability. The first costimulatory molecules characterized were CD28 on naïve T cells and the corresponding ligands, CD80 and CD86, on APCs (Coyle and Gutierrez-Ramos, 2001). CD80 and CD86 molecules are members of a larger B7 family. Not all costimulatory molecule interactions, however, lead to an enhanced immune response. CTLA-4, which is homologous to CD28, is upregulated on activated T cells, binds with a higher affinity to CD80 and CD86 than CD28, and its engagement leads to inhibition of T-cell activation (Walunas et al., 1994).

Other B7 family members also have important regulatory roles in immune responses. B7RP-1, expressed predominately on B cells, but also found on macrophages, DCs, and nonlymphoid tissue cells, is the ligand for the inducible immune costimulator (ICOS) protein (Kopf et al., 2000; Swallow et al., 1999; Yoshinaga et al., 1999). ICOS is structurally related to CD28 and, like CTLA-4, is upregulated on activated T cells (Hutloff et al., 1999). ICOS knockout mice fail to form germinal centers or produce antibodies, strongly suggesting that the B7RP-1:ICOS interaction affords positive T-cell stimulation (Dong et al., 2001). It has been suggested that this B7RP-1:ICOS interaction facilitates the downstream CD40L-CD40 interaction, the latter interaction already known as required for T cell–dependent B-cell antibody production (McAdam et al., 2001).

Two other B7 family members, PD-L1 (also called B7-HI) and PD-L2, are unique in that they are constituitively expressed on DCs. They both bind to the programmed cell death 1 (PD-1) receptor on T and B cells [reviewed in Coyle and Gutierrez-Ramos (2001)]. These interactions inhibit T-cell proliferation and cytokine production, with PD-L1 more effectively inhibiting Th1 cell activities and PD-L2 more effectively inhibiting Th2 cells. B7-H3 (also referred to as B7-RP2) is highly expressed on immature DCs and is downregulated on mature DCs, in contrast with the expression of most of the other B7 family molecules (Chapoval et al., 2001). The ligand for B7-H3 is unknown, although it is not CD28, CTLA-4, ICOS, or PD-1. B7-H3, like PD-L1, appears to negatively regulate Th1 rather than Th2 responses (Suh et al., 2003). B7-H4 (also referred to as B7x or B7S1) has been more recently described as a negative regulator of T-cell immune responses (Sica et al., 2003). It is inducibly expressed on a number of hematopoietic cells, including DCs.

It has been proposed that B7-H4 binds to BTLA-4, which is expressed on activated T cells to cause suppression (Watanabe *et al.*, 2003).

Beyond the B7 family of costimulatory molecules, members of the TNF family of ligands and receptors are also important costimulators in DC interactions with T cells. One of the most important of these interactions is that of CD40 on DCs with CD40L (CD154) on T cells. CD40 was originally identified as a critical B-cell molecule that interacted with CD40L on activated T cells. This interaction induced the B cells to undergo antibody isotype switching. However, it was later recognized that during effective APC–T cell interactions, CD40L interaction with CD40 on mature DCs and B cells also triggers IL-12 production, required for Th1 polarization (Caux *et al.*, 1994; Cella *et al.*, 1996). Other TNF family members also play important complementary roles in inducing T-cell proliferation and cytokine production (Weinberg *et al.*, 1998). OX40 is expressed on T cells and OX40L on DCs. OX40 knockout mice demonstrate both reduced CD4 T-cell proliferation and IFN-γ production resulting in reduced protection against an influenza lung infection (Weinberg *et al.*, 1998). Neither OX40 nor OX40L knockout mice demonstrate humoral immune response defects. Still other TNF family members have been shown to provide important costimulatory signals that enhance CD8 T-cell proliferation and IFN-γ production (4-1BBL on DCs and 4-1BB on T cells) and DC cytokine secretion and survival (RANK on DCs and RANK-L on T cells) [reviewed in Banchereau *et al.* (2000)].

A third set of receptor-ligand interactions of importance in T-cell responses to DCs are the integrins and their ligands. Blockade of the adhesive interaction of LFA-1 (CD11a/CD18) with ICAM-1 (CD54) markedly reduces the proliferative response of T cells stimulated by DCs (Bachmann *et al.*, 1997; Masten *et al.*, 1997). These interactions affect not only T-cell proliferation but also their differentiation into cytokine-secreting cells. Thus, the engagement of LFA-1 with ICAM-1 and ICAM-2 seems to favor the generation of Th1 responses (Salomon and Bluestone, 1998). DC-SIGN (which comes from DC-specific, ICAM-3 grabbing nonintegrin) interacts with ICAM-3 on T cells and has been postulated to initiate effective interaction of the MHC class II/peptide complex on DCs with the TCR on T cell (Geijtenbeek *et al.*, 2000).

2.4.4 Events at the DC–T Cell Interface

Interaction with an APC is required for activation of both naïve and effector T cells. Time-lapse microscopy demonstrated the behavior of T cells interacting with planar membranes in which fluorescence-labeled adhesion molecules and MHC-peptide complexes were freely diffusing. The first interactions observed were between LFA-1 with ICAM-1 (within 30 s), which was followed within 20 min by aggregation of the MHC/peptide complexes by the TCR. The area of the membrane at which the clustering of complexes occurs has been referred to as the "immunological synapse" [reviewed in Grakoui et al. (1999)]. In addition to cell-surface molecules, signaling molecules in the T cell such as Lck, Fyn, and ZAP 70 have been observed within the immunological synapse. Recent studies have suggested that LFA-1/ICAM-3 interactions induce the movement of MHC II to the immunological synapse and induce activation of the Vav 1–Rac 1 signaling cascade (de la Fuente et al., 2005). Prolonged binding of T cells to DCs has been shown in some systems to be required not only for optimal antigen stimulation (Banchereau and Steinman, 1998) but also for increased T-cell survival (Feuillet et al., 2005). Others have noted that the time for DC–T cell interaction determines the immunologic outcome with shorter periods of interaction favoring anergy and death; intermediate periods the development of memory, and longer periods resulting in differentiation into effector cells (Lanzavecchia and Sallusto, 2000). The interaction of T cells and DCs in a collagen gel has been shown to induce T-cell calcium influx and activation as measured by increases in activation markers and finally proliferation (Gunzer et al., 2000). In the intact animal, after subcutaneous antigen inoculation, initial interactions occur between DCs and T cells in T-cell areas of the lymph nodes and continue at the interface of the T-cell area with the follicles (Ingulli et al., 1997). Recently, it was shown that T-cell interaction with DCs in vivo was neither dependent on the T-cell activation state nor on the presence of antigen. CD4[+] T cells, instilled into the rat in the absence of exogenously injected antigen, seemed to remain in contact with DCs in T-cell areas of the spleen, lymph nodes, and Peyer's patches for as long as 96 h (Westermann et al., 2005). No differences in this behavior were observed between naïve, memory, or effector CD4[+] cells nor were the interactions dependent on LFA-1. These studies suggest that the architecture of the lymphoid tissue determines that T cells and DCs will interact irrespective of the known molecular interactions required for a productive immune response.

2.5 DCs Link Innate and Acquired Immunity

2.5.1 Influence of the Microenvironment on DC Phenotype and Function

In 1994, Polly Matzinger wrote a review that first popularized the notion that the nonimmune host had the ability to recognize that some antigens were potentially dangerous and then, and only then, to initiate the development of an acquired immune response to focus and amplify innate defenses (Matzinger, 1994). She argued that the older paradigm that T cells were generated in the thymus and further refined in the periphery to primarily distinguish self from not-self was one that had outlived its utility. A number of cells in the body can peruse their microenvironments and respond to danger signals, but circulating leukocytic cells are ideal for the task of recognizing danger because they are also equipped with effective cytotoxic and/or cytostatic machinery. For example, neutrophils can immediately sense the danger of Gram-negative bacilli, then phagocytize and kill them, although the capsule of more virulent organisms are capable of resisting the attack. DCs are also particularly suited to sample their microenvironment because once they perceive danger, they can take it up, process the antigens, and then migrate to secondary lymphoid tissue where naïve T cells abound and/or can be readily recruited through the postcapillary venules to interact with antigen-bearing DCs. In addition, before DCs undertake their journey to draining lymph nodes, they release mediators, such as TNF-α or IL-12, that activate innate immune cells around them, leaving them better able to help reduce the danger in the environment.

DCs respond to both endogenous and exogenous signals, which determine whether they should mature (or become activated) to develop higher levels of MHC class II, more efficiently process and present antigen, and upregulate accessory molecules and secrete cytokines (Fig. 2.4). Activation implies that acquired immunity will be induced in the draining lymph nodes; and that depending on the pathogen, either a Th1 or a Th2 immune response will be initiated. Thus, while IL-12 might be called upon to initiate a Th1 response for many infections, examples exist whereby high levels of antibody and an eosinophilic response, characteristic of a Th2 response, are more appropriate. Such is the case in some parasitic infections.

In order to respond to signals in the environment, DCs express receptors on their surface, within endosomal compartments, and in the cytosol. Cell-surface receptors respond to extracellular signals while endosomal and cytosolic receptors are positioned to respond to degraded microbes, mutated proteins such as occur during oncogenesis, and to intracellular infectious agents. Microbial organisms express molecular structures that are recognized by non-polymorphic receptors on host cells. These molecular structures are referred to as PAMPs and their receptors are pattern-recognition receptors (PRRs). TLRs are among the most important of the PRRs, although other PPRs are continuously being discovered, most recently members of the nucleotide-binding oligomerization domain (Nod) family of intracytoplasmic proteins. DCs respond to autologous cells and their secreted products in the environment and are also well-equipped to

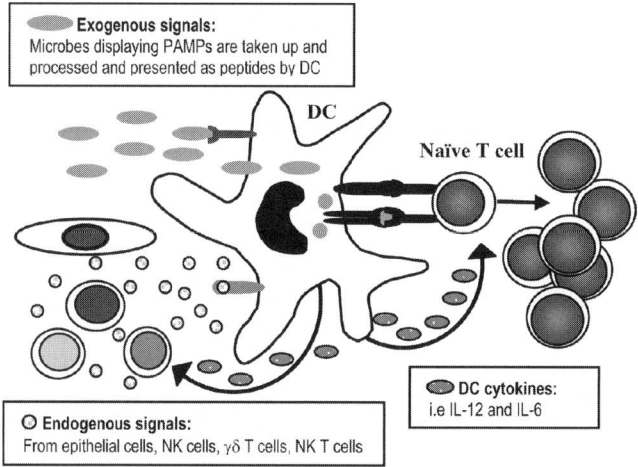

Exogenous signals:
Microbes displaying PAMPs are taken up and
processed and presented as peptides by DC

DC

Naïve T cell

DC cytokines:
i.e IL-12 and IL-6

○ **Endogenous signals:**
From epithelial cells, NK cells, γδ T cells, NK T cells

FIGURE 2.4. DCs and the influence of their microenvironment. A DC can receive
both endogenous and exogenous danger signals. Note the innate lymphocytic cells
are shown as sources of endogenous signals. The signals are integrated through
TLRs and other receptors to initiate either suppression of maturation, partial
maturation, or full maturation. The DC response to full maturation/activation
signals is to (1) upregulate the ability to process antigen and display it on the DC
cell surface as high-density MHC class II-peptide complexes; (2) increase cell-
surface expression of the accessory molecules CD80, CD86, and CD40; (3) shift
from immature to mature chemokine receptors to allow it to respond to lymphatic
and lymph node attracting cytokines; and (4) once it arrives in the lymph node, to
drive a Th1 or Th2 response. Also shown with curved arrows is that activated DCs
can secrete cytokines such as IL-12 and IL-6 that have profound effects on tissue
innate T cells and others that facilitates their lytic activity.

recognize PAMPs. Indeed, the DC response after direct recognition of
PAMPs can be modified by signals from other cells in the microenvironment
of the DC. An increasing literature recognizes the complexity of how the
various signals delivered and the receptor combinations expressed by DCs
fine tune their maturation pathway and finally determine whether tolerance
or adaptive immunity develops (Malissen and Ewbank, 2005).

DCs express most, if not all, of the TLRs, although the expression levels
change depending on their stage of activation. DC TLRs tend to be
downregulated as the DC becomes activated. Further, PDCs differ in their
expression of TLRs compared with MDCs. Other PRRs are used to trigger
endocytosis and initiate antigen processing. These include the type 3
complement receptor (CR3, CD18/CD11b), scavenger receptors (SRs), the
mannose receptor (MR), and other C-type lectin receptors (Gordon, 2002).
The MR, depending on the particular agonist used, was shown to induce

the secretion of several mediators and variably upregulate accessory molecules while inhibiting IL-12 secretion (Chieppa *et al.*, 2003). Nod1 and Nod2 are more recently discovered PPRs and both recognized specific components of microorganisms (Inohara and Nunez, 2003). Nod2 recognizes muramyl dipeptid (MDP) derived from peptidoglycan; binding of Nod2 by MDP stimulates DCs to make IL-6 and IL-12. Nod1 recognizes another muropeptide derived from a peptidoglycan component of a Gram-negative bacterium peptidoglycan (Girardin *et al.*, 2003). Recently, it was confirmed that both Nod1 and Nod2 agonists stimulated human monocytes and DCs; and when LPS was added, synergistic responses were achieved (Fritz *et al.*, 2005).

2.5.1.1 Exogenous Signals

Familiar exogenous danger signals that stimulate through TLRs include the peptidoglycans, lipoteichoic acids, and lipopeptides that simulate through TLR2 and LPS, which stimulates through TLR4 (Lipscomb and Masten, 2002). TLR5 recognizes flagellin and TLR11 recognizes a component of uropathogenic bacteria (Diebold *et al.*, 2004). TLRs on the membranes of endosomes recognize bacterial and/or viral degradation products (O'Neill, 2004). For example, CPG motifs common on bacteria signal through TLR9. dsRNA viruses signal through TLR3 and ssRNA viruses through TLR7 or 8 (Diebold *et al.*, 2004). Unopposed stimulation of DCs through any of these receptors induces maturation. In addition to bacteria and viruses, pathogenic fungi and parasites have been shown to simulate DC maturation, although the molecules and specific receptors are not fully characterized.

2.5.1.2 Endogenous Signals

Endogenous signals are those that are derived from the host environment, generated either by cell-cell contact or soluble chemical mediators. Endogenous danger signals that stimulate partial to full activation of DCs include (a) products of necrotic, but not apoptotic cells; (b) stress-induced purines released from dying cells; (c) intercellular matrix degradation products, such as hyaluronan; (d) endogenous heat-shock proteins, specifically HSP60, HSP70, HSP90, and HSP96; (e) CD40L, generally expressed on cell surfaces; (f) various cytokines, including TNF-α, IL-1β, and type I IFNs; vasointestinal polypeptide (VIP); and aggregated IgG and Fas (Lipscomb and Masten, 2002). Receptors on DCs that recognize these signals are multiple; Toll-like receptors (TLR4 for HSP60) and the low-density lipoprotein (LDL) receptor family (CD91 for HSP70, 90 and 96) to name a few. Cells that die by apoptosis fail to generate signals that result in the maturation of DCs (Basu *et al.*, 2000). Apoptotic death fails to release inflammatory-provoking cellular contents into the microenvironment, which avoids the recruitment and activation of leukocytic cells, including DCs. Still, apoptotic cells can be taken up by DCs; and in the presence of accompanying maturation signals, DC can present the cellular antigens to responder T cells.

Viable cells within the DC microenvironment respond to exogenous or endogenous danger signals to produce many of the activating endogenous signals listed above. For example, mucosal and epidermal epithelial cells deliver signals to DCs that reside within and immediately beneath epithelial surfaces. Keratinocytes, which constitutively express TLR2, TLR3, TLR5, and TLR10, respond to their agonists to secrete the chemokine IL-8 (Kollisch *et al.*, 2005). Furthermore, the Gram-positive coccus, *Staphylococcus aureus*, which stimulates cells via TLR2, stimulated epithelial cells to secrete GM-CSF (Matsubara *et al.*, 2004), a cytokine that enhances the survival and maturation of DCs. The outcome of a two-way interaction of cells of the innate lymphoid system with DCs has been investigated [reviewed in Munz *et al.* (2005)].

Natural killer (NK) cells, $\gamma\,\delta$ T cells, and natural killer T (NKT) cells all recognize unprocessed antigens or antigens displayed on members of the CD1 antigen family, antigens expressed on the surface of many DCs. NK cells express both activating and inhibiting receptors; the balance of their counterbalancing activity determines the affect on DCs. Under certain conditions, NK cells induce DC production of IL-12 and upregulation of costimulatory molecules, thus enhancing their ability to stimulate a Th1 immune response (Raulet, 2004). In addition, the production of IFN-γ by NK cells after their exposure to IL-12 has an enhancing effect on Th1 development. DC-secreted IL-12 activates NK cells and induces their proliferation *in vitro*. Furthermore, in a tumor rejection model, injected DCs were able to enhance NK cytotoxicity against tumor targets. Once activated, NK cells can kill immature DCs, although mature DCs are resistant to attack. Thus, NK cells have both activating and downregulating roles to play in regard to DC function.

Gamma/delta T cells also influence and are influenced by DCs in their microenvironment. These cells can recognize microbial and tumor antigens either directly or in a CD1-restricted fashion (Munz *et al.*, 2005). In mice, the $\gamma\,\delta$ TCR is highly restricted and reflects a tissue distribution, but is more polymorphic and not tissue dependent in humans. Among the antigens directly recognized are phosphorylated isoprenoid precursors and alkylamines, molecules that exist in many species including plants, microbes, and primates. A subset of human $\gamma\,\delta$ T cells recognize antigen on CD1c, an antigen present on human MDCs. In one study, on recognition of the displayed CD1c-associated antigen, γ δ T cells induced DCs to make IL-12 and exhibit enhanced ability to stimulate naïve CD4 T cells (Leslie *et al.*, 2002). The activation of DCs by γ δ T cells was dependent on TNF-α. Very recently, a surprising finding was that human γ δ T cells were found to be capable of serving as fully competent APCs with the ability to process and present antigens to naïve CD4 T cells (Brandes *et al.*, 2005).

NKT cells are the third type of innate lymphocyte that can influence the function of DCs within their microenvironment. These cells express both T cell and NK markers, CD40L, and a restricted TCR with a single Vα chain (Brutkiewicz and Sriram, 2002). When stimulated, NKT cells are capable of secreting IFN-γ or IL-4 and, therefore, play an important role in contributing to the polarization of CD4 T-cell responses. NKT cells recognize glycolipids presented on CD1 molecules on DCs. On responding to DCs displaying the glycolipid antigens on CD1, NKT cells secrete IFN-γ and bind CD40 on the DC, resulting in DC secretion of IL-12 and stimulation of a Th1 response. In turn, DC-secreted IL-12 enhances NKT cell antitumor activity.

2.5.2 DCs in T-Cell Activation and Differentiation

Arrival of DCs bearing antigen in a draining lymph node sets the stage for the initiation of the adaptive immune response. In the cortex, DCs secrete factors to attract naïve T cells to develop an immunological synapse. DCs that express high levels of MHC class II and the accessory molecules CD80, CD86, and CD40 and also secrete IL-12 are likely to drive a Th1 response (Fig. 2.4). However, in the event that they are unable to secrete IL-12, but yet express high levels of antigenic peptide in the context of MHC class II and the requisite accessory molecules, a Th2 response is likely to develop. However, this "default" response requires that appropriate TLR stimulation of the DCs has occurred so that they are truly fully mature. Thus, it has been shown that the simple absence of IL-12 during immune stimulation is not sufficient for a DC to induce a Th2 response (Jankovic *et al.*, 2002). Finally, low expression of MHC class II-peptide complexes, peptide–MHC class II combinations that fail to provide a "tight" fit, low expression of accessory molecules, or the secretion of IL-10 or TGF-β tend to drive the development of tolerance [reviewed in Lipscomb and Masten (2002)].

One of the most exciting areas of research on the biology of DCs relates to determining the molecular mechanisms for how they induce T-cell polarization. Little doubt exists that the ability of DCs to secrete IL-12 directly during an otherwise ideal interaction with naïve T cells leads to a protective Th1 response (Berberich *et al.*, 2003). Effective DC–T cell interaction accompanied by IL-12 secretion results in early tyrosine phosphorylation of the intranuclear T-box protein, T-bet, previously known to be required for CD4$^+$ T cells to produce IFN-γ. T-bet interacts with the transcription factor GATA-2 to interfere with the ability of GATA-2 to bind to its target DNA, thus silencing the transcription of Th2 cytokines (Hwang *et al.*, 2005). A number of stimuli are associated with the production of IL-12 including CD40L, IFN-γ, CCR5 ligands, LPS, CpG motifs, *Staphylococcus aureus*, and the parasites, *Leishmania major*, *L. donovani*, and *Toxoplasma gondii*. Other factors suppress IL-12 secretion by otherwise properly stimulated DCs including PGE2, the chemokines MCP-1 and 4, C5a, the cytokines IL-10 and GM-CSF, corticosteroids, CD47 receptor ligands, vitamin D$_3$, and the measles virus [reviewed in Lipscomb and Masten (2002)].

T-cell tolerance to a given antigen occurs through T-cell death via apoptosis, anergy (paralysis), or by active suppression by regulatory T cells. Central tolerance is mediated in the thymus and requires both thymic epithelial and bone marrow–derived DCs to complete the process of negative and positive selection. Peripheral tolerance in T cells occurs after exit from the thymus. DCs are pivotal in regulating tolerance in the periphery (Steinman *et al.*, 2003). The molecular mechanisms remain somewhat uncertain, but increasing evidence suggests that the combination of ligands that engage DCs at the site where the relevant antigen is taken up, or the lack of sufficient ligand stimulation, influence how DCs induce an immune response. DCs can induce tolerance by failing to deliver costimulatory signals, deleting antigen reactive cells (by making an inhibitory enzyme or by directly delivering an apoptotic signal via Fas/ Fas-ligand interaction), or by generating regulatory T cells (Moser, 2003). Thus, antigens presented to naïve T cells in secondary lymphoid organs by immature DCs lead to the inability of these T cells to respond to a later stimulus (i.e., they become anergic).

DCs that have been activated to express appropriate accessory molecules, but secrete inhibitory cytokines such as TGF-β or IL-10, gene-rate regulatory CD4$^+$ T cells that subsequently downregulate proinflam-matory T cells and other cells in the microenvironment. One explanation for the ability of DCs to induce the development of regulatory T cells is that the DCs are actually "semimature," because their TLRs have not been stimulated appropriately by PAMPs. Therefore, these DCs either fail to express CD40, express inhibitory accessory molecules, or fail to secrete IL-12 on engagement with T cells (Lutz and Schuler, 2002; Sporri and Reis e Sousa, 2005). DCs that secrete relatively high levels of IL-10 tend to cause the development of IL-10 secreting T-regulatory cells, while DCs that secrete TGF-β tend to cause the development of TGF-β–secreting

T cells (Akbari *et al.*, 2001). CD4$^+$ T regulatory cells that secrete high levels of IL-10 and lower levels of TGF-β have been labeled Tr1 cells, while CD4$^+$ T cells that produce greater levels of TGF-β and lower levels of IL-10 have been called Th3 cells (McGuirk and Mills, 2002; Jonuleit and Schmitt, 2003). To be distinguished from the DC-induced tolerogenic cells are natural CD4$^+$CD25$^+$ regulatory T cells, which are generated in the thymus and suppress other T cells by cell contact. The subject of DCs in the induction of tolerance will be treated in greater detail in subsequent chapters.

One important determinant of the ability of DCs to directly suppress a response and induce T-cell death or anergy relates to their ability to secrete indoleamine 2,3-dioxygenase (IDO), an enzyme that degrades the essential amino acid tryptophan (Mellor and Munn, 2004). In the absence of adequate tryptophan levels, T cells cannot proliferate and may become tolerized. IDO expression can be induced in DCs by ligation of the costimulatory molecules, CD80 and CD86, but this activity may be limited to only certain subsets of DCs (Munn *et al.*, 2002; Mellor *et al.*, 2003). Signaling through CTLA-4 can also induce IDO activity, and this mechanism might explain how regulatory T cells expressing CTLA-4 suppress bystander T-cell responses (Fallarino *et al.*, 2003).

2.5.3 Regulation of Pulmonary Immunity by Lung DCs

DCs regulate immunity at all tissue sites. Our laboratory has been particularly interested in the role of lung DCs in the regulation of pulmonary immunity in both health and disease. We and others have found that in both mice and man, both MDCs and PDCs exist (Demedts *et al.*, 2005; Masten *et al.*, 2004; Masten *et al.*, 1997). Lung DCs exist within the epithelium of the airways, in the connective tissue surrounding airways, in alveolar septae, and in the pleura (Lipscomb *et al.*, 1995). Lung DCs participate in the imitation of primary pulmonary immune responses to inhaled and aspirated antigens by processing them and carrying them to native T cells in the draining lymph nodes (Xia *et al.*, 1995), presenting antigens to effector T cells after priming in the lymph node to induce allergic inflammation (van Rijt *et al.*, 2005) and also inducing tolerance in response to antigens delivered to the lungs in a "nondangerous" environment (Akbari *et al.*, 2001).

2.5.4 DCs and B-Cell Function

DCs play a critical indirect role in stimulating B cells in both lymph node T-cell areas and within germinal centers. Well-known is the indirect effect of DCs in that they activate T cells to upregulate CD40L and secrete B-cell helper factors. This upregulation of CD40L by T cells is critical in their ability to directly interact with B cells. However, interdigitating DCs within the paracortical areas of the lymph node can also stimulate directly the proliferation of naïve B cells once the B cell has encountered CD40L on an activating T cell (Dubois *et al.*, 1998; Dubois *et al.*, 1999). DCs also contribute to B-cell differentiation into IgM-secreting plasma cells by virtue of their ability to secrete IL-12 and IL-6. Human tonsillar interdigitiating DCs have been shown to induce CD40-ligated naïve human B cells to secrete high levels of IgM antibody and to switch to both IgG and IgA isotypes (Johansson *et al.*, 2000). In this report, IL-13 was essential for Ig secretion, and DCs were the source.

Follicular DCs (FDCs) are cells unique from MDCs and PDCs. FDCs exist in the germinal centers of the follicles of secondary lymphoid tissue. The origin of these cells has been the subject of some debate, but it now appears that there are two types, one a bone marrow–derived FDC referred to as a germinal center DC (GCDC) and one that is of mesenchymal origin, the classical FDC (Liu *et al.*, 1996). Both FDCs and GCDCs predominately occupy the lighter areas of the germinal center where B cells undergo affinity maturation based on the best fit between of the maturing B cells with its cognate antigen. Antigen is displayed on the elongated processes of the FDCs captured as antigen-antibody or antigen-antibody-complement by complement and Fc receptors. FDCs do not express MHC class II antigens but express CD21 (the complement receptor for C3d), FcγRs, and CD40L. The native antigen within the complexes can also bind the B-cell receptor to reactivate memory B cells and, therefore, FDCs have been thought to be important in maintaining serum antibody long after exposure to microbial agents or vaccination. FcγRIIB on FDCs can act as a decoy for the IgG portion of a complex, thus effectively preventing B cell coligating their inhibitory FcγRIIB when the B-cell receptor engages antigen in the antigen-antibody complex (Qin *et al.*, 2000).

2.6 Concluding Remarks

Since the first description of a novel splenic cell by Ralph Steinmen in 1973, the understanding of DCs and their pivotal role in immune regulation and disease has grown exponentially. They are truly nature's most potent adjuvant for initiating immunity against all microbial agents. At least for the foreseeable future, HIV researchers, including vaccine developers, will continue to seek ways to exploit their biology for relevant medical indications.

Acknowledgments: The authors acknowledge their debt to Spencer Hansen for the photomicrograph shown in Figure 2.1, Michael Grady for preparing Figure 2.3, and Sandy Turner for expert help in formatting the final manuscript. We also gratefully acknowledge funding for our DC research through an NHLBI-funded Specialized Center for Research (SCOR) in Asthma (P50 HL56384; M.F.L., J.A.W., B.J.M.), NIAID U19-AI57234 (M.F.L.), DAIT-BAA-04-18 (M.F.L.), NIAID (PO1 AI56295; M.F.L.), and institutional support from the New Mexico tobacco tax research allocation to the University of New Mexico (M.F.L., J.A.W., B.J.W.).

References

Akbari, O., DeKruyff, R. H., and Umetsu, D. T. (2001). Pulmonary dendritic cells producing IL-10 mediate tolerance induced by respiratory exposure to antigen. *Nat. Immunol.* 2:725-731.

Ardavin, C. (1997). Thymic dendritic cells. *Immunol. Today* 18:350-361.

Ardavin, C. (2003). Origin, precursors and differentiation of mouse dendritic cells. *Nat. Rev. Immunol.* 3:582-590.

Asselin-Paturel, C., Boonstra, A., Dalod, M., Durand, I., Yessaad, N., Dezutter-Dambuyant, C., Vicari, A., O'Garra, A., Biron, C., Briere, F., and Trinchieri, G. (2001). Mouse type I IFN-producing cells are immature APCs with plasmacytoid morphology. *Nat. Immunol.* 2:1144-1150.

Asselin-Paturel, C., Brizard, G., Pin, J. J., Briere, F., and Trinchieri, G. (2003). Mouse strain differences in plasmacytoid dendritic cell frequency and function revealed by a novel monoclonal antibody. *J. Immunol.* 171:6466-6477.

Bachmann, M. F., McKall-Faienza, K., Schmits, R., Bouchard, D., Beach, J., Speiser, D. E., Mak, T. W., and Ohashi, P. S. (1997). Distinct roles for LFA-1 and CD28 during activation of naive T cells: adhesion versus costimulation. *Immunity* 7:549-557.

Baggiolini, M. (1998). Chemokines and leukocyte traffic. *Nature* 392:565-568.

Bakker, A. B., Hoek, R. M., Cerwenka, A., Blom, B., Lucian, L., McNeil, T., Murray, R., Phillips, L. H., Sedgwick, J. D., and Lanier, L. L. (2000). DAP12-deficient mice fail to develop autoimmunity due to impaired antigen priming. *Immunity* 13:345-353.

Banchereau, J., and Steinman, R. M. (1998). Dendritic cells and the control of immunity. *Nature* 392:245-252.

Banchereau, J., Briere, F., Caux, C., Davoust, J., Lebecque, S., Liu, Y. J., Pulendran, B., and Palucka, K. (2000). Immunobiology of dendritic cells. *Annu. Rev. Immunol.* 18:767-811.

Basu, S., Binder, R. J., Suto, R., Anderson, K. M., and Srivastava, P. K. (2000). Necrotic but not apoptotic cell death releases heat shock proteins, which deliver a partial maturation signal to dendritic cells and activate the NF-kappa B pathway. *Int. Immunol.* 12:1539-1546.

Berberich, C., Ramirez-Pineda, J. R., Hambrecht, C., Alber, G., Skeiky, Y. A., and Moll, H. (2003). Dendritic cell (DC)-based protection against an intracellular pathogen is dependent upon DC-derived IL-12 and can be induced by molecularly defined antigens. *J. Immunol.* 170:3171-3179.

Bjorck, P. (2001). Isolation and characterization of plasmacytoid dendritic cells from Flt3 ligand and granulocyte-macrophage colony-stimulating factor-treated mice. *Blood* 98:3520-3526.

Bjorck, P. (2002). The multifaceted murine plasmacytoid dendritic cell. *Hum. Immunol.* 63:1094-1102.

Bouchon, A., Facchetti, F., Weigand, M. A., and Colonna, M. (2001). TREM-1 amplifies inflammation and is a crucial mediator of septic shock. *Nature* 410:1103-1107.

Brandes, M., Willimann, K., and Moser, B. (2005). Professional antigen-presentation function by human gammadelta T Cells. *Science* 309:264-268.

Brutkiewicz, R. R., and Sriram, V. (2002). Natural killer T (NKT) cells and their role in antitumor immunity. *Crit. Rev. Oncol. Hematol.* 41:287-298.

Buelens, C., Bartholome, E. J., Amraoui, Z., Boutriaux, M., Salmon, I., Thielemans, K., Willems, F., and Goldman, M. (2002). Interleukin-3 and interferon beta cooperate to induce differentiation of monocytes into dendritic cells with potent helper T-cell stimulatory properties. *Blood* 99:993-998.

Burgstahler, R., Kempkes, B., Steube, K., and Lipp, M. (1995). Expression of the chemokine receptor BLR2/EBI1 is specifically transactivated by Epstein-Barr virus nuclear antigen 2. *Biochem. Biophys. Res. Commun.* 215:737-743.

Butcher, E. C., and Picker, L. J. (1996). Lymphocyte homing and homeostasis. *Science* 272:60-66.

Castellino, F., and Germain, R. N. (1995). Extensive trafficking of MHC class II-invariant chain complexes in the endocytic pathway and appearance of peptide-loaded class II in multiple compartments. *Immunity* 2:73-88.

Caux, C., Massacrier, C., Vanbervliet, B., Dubois, B., Van Kooten, C., Durand, I., and Banchereau, J. (1994). Activation of human dendritic cells through CD40 cross-linking. *J. Exp. Med.* 180:1263-1272.

Cella, M., Scheidegger, D., Palmer-Lehmann, K., Lane, P., Lanzavecchia, A., and Alber, G. (1996). Ligation of CD40 on dendritic cells triggers production of high levels of interleukin-12 and enhances T cell stimulatory capacity: T-T help via APC activation. *J. Exp. Med.* 184:747-752.

Cella, M., Jarrossay, D., Facchetti, F., Alebardi, O., Nakajima, H., Lanzavecchia, A., and Colonna, M. (1999). Plasmacytoid monocytes migrate to inflamed lymph nodes and produce large amounts of type I interferon. *Nat. Med.* 5: 919-923.

Cella, M., Facchetti, F., Lanzavecchia, A., and Colonna, M. (2000). Plasmacytoid dendritic cells activated by influenza virus and CD40L drive a potent TH1 polarization. *Nat. Immunol.* 1:305-310.

Chang, G. C., Lan, H. C., Juang, S. H., Wu, Y. C., Lee, H. C., Hung, Y. M., Yang, H. Y., Whang-Peng, J., and Liu, K. J. (2005). A pilot clinical trial of vaccination with dendritic cells pulsed with autologous tumor cells derived from malignant pleural effusion in patients with late-stage lung carcinoma. *Cancer* 103:763-771.

Chapoval, A. I., Ni, J., Lau, J. S., Wilcox, R. A., Flies, D. B., Liu, D., Dong, H., Sica, G. L., Zhu, G., Tamada, K., and Chen, L. (2001). B7-H3: a costimulatory molecule for T cell activation and IFN-gamma production. *Nat. Immunol.* 2:269-274.

Chen, S. C., Vassileva, G., Kinsley, D., Holzmann, S., Manfra, D., Wiekowski, M. T., Romani, N., and Lira S. A. (2002). Ectopic expression of the murine chemokines CCL21a and CCL21b induces the formation of lymph node-like structures in pancreas, but not skin, of transgenic mice. *J. Immunol.* 168:1001-1008.

Chiappa, M., Bianchi, G., Doni, A., Del Prete, A., Sironi, M., Laskarin, G., Monti, P., Piemonti, L., Biondi, A., Mantovani, A., Introna, M., and Allavena, P. (2003). Cross-linking of the mannose receptor on monocyte-derived dendritic cells activates an anti-inflammatory immunosuppressive program. *J. Immunol.* 171:4552-4560.

Chomarat, P., Dantin, C., Bennett, L., Banchereau, J., and Palucka, A. K. (2003). TNF skews monocyte differentiation from macrophages to dendritic cells. *J. Immunol.* 171:2262-2269.

Comeau, M. R., Van der Vuurst de Vries, A. R., Maliszewski, C. R., and Galibert, L. (2002). CD123bright plasmacytoid predendritic cells: progenitors undergoing cell fate conversion? *J. Immunol.* 169:75-83.

Cook, D. N., Prosser, D. M., Forster, R., Zhang, J., Kuklin, N. A., Abbondanzo, S. J., Niu, X. D., Chen, S. C., Manfra, D. J., Wiekowski, M. T., Sullivan, L. M., Smith, S. R., Greenberg, H. B., Narula, S. K., Lipp, M., and Lira, S. A. (2000). CCR6 mediates dendritic cell localization, lymphocyte homeostasis, and immune responses in mucosal tissue. *Immunity* 12:495-503.

Corcoran, L., Ferrero, I., Vremec, D., Lucas, K., Waithman, J., O'Keeffe, M., Wu, L., Wilson, A., and Shortman, K. (2003). The lymphoid past of mouse plasmacytoid cells and thymic dendritic cells. *J. Immunol.* 170:4926-4932.

Coyle, A. J., and Gutierrez-Ramos, J. C. (2001). The expanding B7 superfamily: increasing complexity in costimulatory signals regulating T cell function. *Nat. Immunol.* 2:203-209.

Cyster, J. G. (1999). Chemokines and the homing of dendritic cells to the T cell areas of lymphoid organs. *J. Exp. Med.* 189:447-450.

D'Amico, A., and Wu, L. (2003). The early progenitors of mouse dendritic cells and plasmacytoid predendritic cells are within the bone marrow hemopoietic precursors expressing Flt3. *J. Exp. Med.* 198:293-303.

de Jong, E. C., Smits, H. H., and Kapsenberg, M. L. (2005). Dendritic cell-mediated T cell polarization. *Springer Semin Immunopathol.* 26:289-307.

de la Fuente, H., Mittelbrunn, M., Sanchez-Martin, L., Vicente-Manzanares, M., Lamana, A., Pardi, R., Cabanas, C., and Sanchez-Madrid, F. (2005). Synaptic clusters of MHC class II molecules induced on DCs by adhesion molecule-mediated initial T-cell scanning. *Mol. Biol. Cell* 16:3314-3322.

del Hoyo, G. M., Martin, P., Vargas, H. H., Ruiz, S., Arias, C. F., and Ardavin, C. (2002). Characterization of a common precursor population for dendritic cells. *Nature* 415:1043-1047.

Demedts, I. K., Brusselle, G. G., Vermaelen, K. Y., and Pauwels, R. A. (2005). Identification and characterization of human pulmonary dendritic cells. *Am. J. Respir. Cell Mol. Biol.* 32:177-184.

Diao, J., Winter, E., Chen, W., Cantin, C., and Cattral, M. S. (2004). Characterization of distinct conventional and plasmacytoid dendritic cell-committed precursors in murine bone marrow. *J. Immunol.* 173:1826-1833.

Diebold, S. S., Kaisho, T., Hemmi, H., Akira, S., and Reis e Sousa, C. (2004). Innate antiviral responses by means of TLR7-mediated recognition of single-stranded RNA. *Science* 303:1529-1531.

Dieu, M. C., Vanbervliet, B., Vicari, A., Bridon, J. M., Oldham, E., Ait-Yahia, S., Briere, F., Zlotnik, A., Lebecque, S., and Caux, C. (1998). Selective recruitment of immature and mature dendritic cells by distinct chemokines expressed in different anatomic sites. *J. Exp. Med.* 188:373-386.

Dieu-Nosjean, M. C., Vicari, A., Lebecque, S., and Caux, C. (1999). Regulation of dendritic cell trafficking: a process that involves the participation of selective chemokines. *J. Leukoc. Biol.* 66:252-262.

Dieu-Nosjean, M. C., Massacrier, C., Homey, B., Vanbervliet, B., Pin, J. J., Vicari, A., Lebecque, S., Dezutter-Dambuyant, C., Schmitt, D., Zlotnik, A., and Caux, C. (2000). Macrophage inflammatory protein 3alpha is expressed at inflamed epithelial surfaces and is the most potent chemokine known in attracting Langerhans cell precursors. *J. Exp. Med.* 192:705-718.

Dong, C., Juedes, A., Temann, U., Shresta, S., Allison, J., Ruddle, N., and Flavel, R. (2001). ICOS co-stimulatory receptor is essential for T-cell activation and function. *Nature* 409:97-101.

Dubois, B., Massacrier, C., Vanbervliet, B., Fayette, J., Briere, F., Bancbereau, J., and Caux, C. (1998). Critical role of IL-12 in dendritic cell-induced differentiation of naive B lymphocytes. *J. Immunol.* 161:2223-2231.

Dubois, B., Bridon, J. M., Fayette, J., Barthelemy, C., Bancbereau, J., Caux, C., and Briere, F. (1999). Dendritic cells directly modulate B cell growth and differentiation. *J. Leukoc. Biol.* 66:224-230.

Dzionek, A., Fuchs, A., Schmidt, P., Cremer, S., Zysk, M., Miltenyi, S., Buck, D. W., and Schmitz, J. (2000). BDCA-2, BDCA-3, and BDCA-4: three markers for distinct subsets of dendritic cells in human peripheral blood. *J. Immunol.* 165:6037-6046.

Facchetti, F., Candiamo, E., and Vermi, W. (1999). Plasmacytoid monocytes express IL3-receptor alpha and differentiate into dendritic cells. *Histopathology* 35:88-89.

Fallarono, F., Grohmann, U., Hwang, K. W., Orabona, C., Vacca, C., Bianchi, R., Belladonna, M. L., Fioretti, M. C., Alegre, M. L., and Puccetti, P. (2003).

Modulation of tryptophan catabolism by regulatory T cells. *Nat. Immunol.* 4:1206-1212.

Feuillet, V., Lucas, B., Di Santo, J. P., Bismuth, G., and Trautmann, A. (2005). Multiple survival signals are delivered by dendritic cells to naive CD4($^+$) T cells. *Eur. J. Immunol.* 35:2563-2572.

Foti, M., Granucci, F., Aggujaro, D., Liboi, E., Luini, W., Minardi, S., Mantovani, A., Sozzoni, S., and Ricciardi-Castagnoli, P. (1999). Upon dendritic cell (DC) activation chemokines and chemokine receptor expression are rapidly regulated for recruitment and maintenance of DC at the inflammatory site. *Int. Immunol.* 11:979-986.

Fritz, J. H., Girardin, S. E., Fitting, C., Werts, C., Mengin-Lecreulx, D., Caroff, M., Cavaillon, J. M., Philpott, D. J., and Adib-Conquy, M. (2005). Synergistic stimulation of human monocytes and dendritic cells by Toll-like receptor 4 and NOD1- and NOD2-activating agonists. *Eur. J. Immunol.* 35:2459-2470.

Geijtenbeek, T. B., Torensma, R., van Vliet, S. J., van Duijnhoven, G. C., Adema, G. J., van Kooyk, Y., and Figdor, C. G. (2000). Identification of DC-SIGN, a novel dendritic cell-specific ICAM-3 receptor that supports primary immune responses. *Cell* 100:575-585.

Gilliet, M., Boonstra, A., Paturel, C., Antonenko, S., Xu, X. L., Trinchieri, G., O'Garra, A., and Liu, Y. J. (2002). The development of murine plasmacytoid dendritic cell precursors is differentially regulated by FLT3-ligand and granulocyte/macrophage colony-stimulating factor. *J. Exp. Med.* 195:953-958.

Girard, J. P., and Springer, T. A. (1995). High endothelial venules (HEVs): specialized endothelium for lymphocyte migration. *Immunol. Today* 16:449-457.

Girardin, S. E., Boneca, I. G., Carneiro, L. A., Antignac, A., Jehanno, M., Viala, J., Tedin, K., Taha, M. K., Labigne, A., Zahringer, U., Coyle, A. J., Di Stefano, P. S., Bertin, J., Sansonetti, P. J., and Philpott, D. J. (2003). Nod1 detects a unique muropeptide from gram-negative bacterial peptidoglycan. *Science* 300:1584-1587.

Gordon, S. (2002). Pattern recognition receptors: doubling up for the innate immune response. *Cell* 111:927-930.

Grakoui, A., Bromley, S. K., Sumen, C., Davis, M. M., Shaw, A. S., Allen, P. M., and Dustin, M. L. (1999). The immunological synapse: a molecular machine controlling T cell activation. *Science* 285:221-227.

Gretz, J. E., Norbury, C. C., Anderson, A. O., Proudfoot, A. E., and Shaw, S. (2000). Lymph-borne chemokines and other low molecular weight molecules reach high endothelial venules via specialized conduits while a functional barrier limits access to the lymphocyte microenvironments in lymph node cortex. *J. Exp. Med.* 192:1425-1440.

Grouard, G., Rissoan, M. C., Filgueira, L., Durand, I., Banchereau, J., and Liu, Y. J. (1997). The enigmatic plasmacytoid T cells develop into dendritic cells with interleukin (IL)-3 and CD40-ligand. *J. Exp. Med.* 185:1101-1111.

Gunn, M. D., Tangemann, K., Tam, C., Cyster, J. G., Rosen, S. D., and Williams, L. T. (1998). A chemokine expressed in lymphoid high endothelial venules promotes the adhesion and chemotaxis of naive T lymphocytes. *Proc. Natl. Acad. Sci. U.S.A.* 95:258-263.

Gunzer, M., Schafer, A., Borgmann, S., Grabbe, S., Zanker, K. S., Brocker, E. B., Kampgen, E., and Friedl, P. (2000). Antigen presentation in extracellular matrix: interactions of T cells with dendritic cells are dynamic, short lived, and sequential. *Immunity* 13:323-332.

Harshyne, L. A., Watkins, S. C., Gambotto, A., and Barratt-Boyes, S. M. (2001). Dendritic cells acquire antigens from live cells for cross-presentation to CTL. *J. Immunol.* 166:3717-3723.

Hart, D. N. (1997). Dendritic cells: unique leukocyte populations which control the primary immune response. *Blood* 90:3245-3287.

Hutloff, A., Dittrich, A. M., Beier, K. C., Eljaschewitsch, B., Kraft, R., Anagnostopoulos, I., and Kroczek, R. A. (1999). ICOS is an inducible T-cell co-stimulator structurally and functionally related to CD28. *Nature* 397:263-266.

Hwang, E. S., Szabo, S. J., Schwartzberg, P. L., and Glimcher, L. H. (2005). T helper cell fate specified by kinase-mediated interaction of T-bet with GATA-3. *Science* 307:430-433.

Ingulli, E., Mondino, A., Khoruts, A., and Jenkins, M. K. (1997). *In vivo* detection of dendritic cell antigen presentation to CD4($^+$) T cells. *J. Exp. Med.* 185:2133-2141.

Inohara, N., and Nunez, G. (2003). NODs: intracellular proteins involved in inflammation and apoptosis. *Nat. Rev. Immunol.* 3:371-382.

Ito, T., Liu, Y. J., and Kadowaki, N. (2005). Functional diversity and plasticity of human dendritic cell subsets. *Int. J. Hematol.* 81:188-196.

Iwasaki, A., and Kelsall, B. L. (2000). Localization of distinct Peyer's patch dendritic cell subsets and their recruitment by chemokines macrophage inflammatory protein (MIP)-3alpha, MIP-3beta, and secondary lymphoid organ chemokine. *J. Exp. Med.* 191:1381-1394.

Jankovic, D., Kullberg, M. C., Hieny, S., Caspar, P., Collazo, C. M., and Sher, A. (2002). In the absence of IL-12, CD4($^+$) T cell responses to intracellular pathogens fail to default to a Th2 pattern and are host protective in an IL-10(-/-) setting. *Immunity* 16:429-439.

Jarrossay, D., Napolitani, G., Colonna, M., Sallusto, F., and Lanzavecchia, A. (2001). Specialization and complementarity in microbial molecule recognition by human myeloid and plasmacytoid dendritic cells. *Eur. J. Immunol.* 31:3388-3393.

Johansson, B., Ingvarsson, S., Bjorck, P., and Borrebaeck, C. A. (2000). Human interdigitating dendritic cells induce isotype switching and IL-13-dependent IgM production in CD40-activated naive B cells. *J. Immunol.* 164:1847-1854.

Jonuleit, H., and Schmitt, E. (2003). The regulatory T cell family: distinct subsets and their interrelations. *J. Immunol.* 171:6323-6327.

Jotwani, R., Pulendran, B., Agrawal, S., and Cutler, C. W. (2003). Human dendritic cells respond to Porphyromonas gingivalis LPS by promoting a Th2 effector response *in vitro. Eur. J. Immunol.* 33:2980-2986.

Kabashima, K., Sakata, D., Nagamachi, M., Miyachi, Y., Inaba, K., and Narumiya, S. (2003). Prostaglandin E2-EP4 signaling initiates skin immune responses by promoting migration and maturation of Langerhans cells. *Nat. Med.* 9:744-749.

Kadowaki, N., Antonenko, S., Lau, J. Y., and Liu, Y. J. (2000). Natural interferon alpha/beta-producing cells link innate and adaptive immunity. *J. Exp. Med.* 192:219-226.

Kadowaki, N., Ho, S., Antonenko, S., Malefyt, R. W., Kastelein, R. A., Bazan, F., and Liu, Y. J. (2001). Subsets of human dendritic cell precursors express different toll-like receptors and respond to different microbial antigens. *J. Exp. Med.* 194:863-869.

Kalinski, P., Vieira, P. L., Schuitemaker, J. H., de Jong, E. C., and Kapsenberg, M. L. (2001). Prostaglandin E(2) is a selective inducer of interleukin-12 p40 (IL-12p40) production and an inhibitor of bioactive IL-12p70 heterodimer. *Blood* 97:3466-3469.

Kalinski, P., Vieira, P., Schuitemaker, J. H., Cai, Q., and Kapsenberg, M. (2003). Generation of human type 1- and type 2-polarized dendritic cells from peripheral blood. *Methods Mol. Biol.* 215:427-436.

Kapsenberg, M. L. (2003). Dendritic-cell control of pathogen-driven T-cell polarization. *Nat. Rev. Immunol.* 3:984-993.

Karsunky, H., Merad, M., Cozzio, A., Weissman, I. L., and Manz, M. G. (2003). Flt3 ligand regulates dendritic cell development from Flt3$^+$ lymphoid and myeloid-committed progenitors to Flt3$^+$ dendritic cells *in vivo. J. Exp. Med.* 198:305-313.

Kollisch, G., Kalali, B. N., Voelcker, V., Wallich, R., Behrendt, H., Ring, J., Bauer, S., Jakob, T., Mempel, M., and Ollert, M. (2005). Various members of the Toll-like receptor family contribute to the innate immune response of human epidermal keratinocytes. *Immunology* 114:531-541.

Kopf, M., Coyle, A. J., Schmitz, N., Barner, M., Oxenius, A., Gallimore, A., Gutierrez-Ramos, J. C., and Bachmann, M. F. (2000). Inducible costimulator protein (ICOS) controls T helper cell subset polarization after virus and parasite infection. *J. Exp. Med.* 192:53-61.

Krug, A., Towarowski, A., Britsch, S., Rothenfusser, S., Hornung, V., Bals, R., Giese, T., Engelmann, H., Endres, S., Krieg A. M., and Hartmann, G. (2001). Toll-like receptor expression reveals CpG DNA as a unique microbial stimulus for plasmacytoid dendritic cells which synergizes with CD40 ligand to induce high amounts of IL-12. *Eur. J. Immunol.* 31:3026-3037.

Krug, A., Uppaluri, R., Facchetti, F., Dorner, B. G., Sheehan, K. C., Schreiber, R. D., Cella, M., and Colonna, M. (2002). IFN-producing cells respond to CXCR3 ligands in the presence of CXCL12 and secrete inflammatory chemokines upon activation. *J. Immunol.* 169:6079-6083.

Lambrecht, B. N. (2005). Dendritic cells and the regulation of the allergic immune response. *Allergy* 60:271-282.

Lanzavecchia, A., and Sallusto, F. (2000). From synapses to immunological memory: the role of sustained T cell stimulation. Curr. Opin. Immunol. 12: 92-98.

Larsson, M., Fonteneau, J. F., and Bhardwaj, N. (2001). Dendritic cells resurrect antigens from dead cells. *Trends Immunol.* 22:141-148.

Lazarski, C. A., Chaves, F. A., Jenks, S. A., Wu, S., Richards, K. A., Weaver, J. M., and Sant, A. J. (2005). The kinetic stability of MHC class II: peptide complexes is a key parameter that dictates immunodominance. *Immunity* 23:29-40.

Le Bon, A., Etchart, N., Rossmann, C., Ashton, M., Hou, S., Gewert, D., Borrow, P., and Tough, D. F. (2003). Cross-priming of CD8$^+$ T cells stimulated by virus-induced type I interferon. *Nat. Immunol.* 4:1009-1015.

LeibundGut-Landmann, S., Waldburger, J. M., Reis e Sousa, C., Acha-Orbea, H., and Reith, W. (2004). MHC class II expression is differentially regulated in plasmacytoid and conventional dendritic cells. *Nat. Immunol.* 5:899-908.

Leslie, D. S., Vincent, M. S., Spada, F. M., Das, H., Sugita, M., Morita, C. T., and Brenner, M. B. (2002). CD1-mediated gamma/delta T cell maturation of dendritic cells. *J. Exp. Med.* 196:1575-1584.

Lin, C. L., Suri, R. M., Rahdon, R. A., Austyn, J. M., and Roake, J. A. (1998). Dendritic cell chemotaxis and transendothelial migration are induced by distinct chemokines and are regulated on maturation. *Eur. J. Immunol.* 28:4114-4122.

Lin, J. S., Yang, C. W., Wang, D. W., and Wu-Hsieh, B. A. (2005). Dendritic cells cross-present exogenous fungal antigens to stimulate a protective CD8 T cell response in infection by Histoplasma capsulatum. *J. Immunol.* 174:6282-6291.

Lipscomb, M. F., Bice, D. E., Lyons, C. R., Schuyler, M. R., and Wilkes, D. (1995). The regulation of pulmonary immunity. *Adv. Immunol.* 59:369-455.

Lipscomb, M. F., and Masten, B. J. (2002). Dendritic cells: immune regulators in health and disease. *Physiol. Rev.* 82:97-130.

Liu, Y. J. (2001). Dendritic cell subsets and lineages, and their functions in innate and adaptive immunity. *Cell* 106:259-262.

Liu, Y. J. (2005). IPC: Professional type 1 interferon-producing cells and plasmacytoid dendritic cell precursors. *Annu. Rev. Immunol.* 23:275-306.

Liu, Y. J., Grouard, G., de Bouteiller, O., and Banchereau, J. (1996). Follicular dendritic cells and germinal centers. *Int. Rev. Cytol.* 166:139-179.

Lutz, M. B., and Schuler, G. (2002). Immature, semi-mature and fully mature dendritic cells: which signals induce tolerance or immunity? *Trends Immunol.* 23:445-449.

Ma, J., Wang, J. H., Guo, Y. J., Sy, M. S., and Bigby, M. (1994). *In vivo* treatment with anti-ICAM-1 and anti-LFA-1 antibodies inhibits contact sensitization-induced migration of epidermal Langerhans cells to regional lymph nodes. *Cell Immunol.* 158:389-399.

MacDonald, K. P., Munster, D. J., Clark, G. J., Dzionek, A., Schmitz, J., and Hart, D. N. (2002). Characterization of human blood dendritic cell subsets. *Blood* 100:4512-4520.

Malissen, B., and Ewbank, J. J. (2005). "TaiLoRing" the response of dendritic cells to pathogens. *Nat. Immunol.* 6:749-750.

Maranon, C., Desoutter, J. F., Hoeffel, G., Cohen, W., Hanau, D., and Hosmalin, A. (2004). Dendritic cells cross-present HIV antigens from live as well as apoptotic infected CD4$^+$ T lymphocytes. *Proc. Natl. Acad. Sci. U.S.A.* 101: 6092-6097.

Maraskovsky, E., Brasel, K., Teepe, M., Roux, E. R., Lyman, S. D., Shortman, K., and McKenna, H. J. (1996). Dramatic increase in the numbers of functionally mature dendritic cells in Flt3 ligand-treated mice: multiple dendritic cell subpopulations identified. *J. Exp. Med.* 184:1953-1962.

Masten, B. J. (2004). Initiation of lung immunity: the afferent limb and the role of dendritic cells. Semin. *Respir. Crit. Care Med.* 25:11-20.

Masten, B. J., and Lipscomb, M. F. (1997). Methods to isolate and study lung dendritic cells. In *Lung Macrophages and Dendritic Cells* (Edited by Lipscomb M. F. and Russell S. W.), pp. 223-238. Marcel Dekker, Inc., New York.

Masten, B. J., Yates, J. L., Pollard Koga, A. M., and Lipscomb, M. F. (1997). Characterization of accessory molecules in murine lung dendritic cell function: roles for CD80, CD86, CD54, and CD40L. *Am. J. Respir. Cell Mol. Biol.* 16:335-342.

Masten, B. J., Olson, G. K., Kusewitt, D. F., and Lipscomb, M. F. (2004). Flt3 ligand preferentially increases the number of functionally active myeloid dendritic cells in the lungs of mice. *J. Immunol.* 172:4077-4083.

Matsubara, M., Harada, D., Manabe, H., and Hasegawa, K. (2004). Staphylococcus aureus peptidoglycan stimulates granulocyte macrophage colony-stimulating factor production from human epidermal keratinocytes via mitogen-activated protein kinases. *FEBS Lett.* 566:195-200.

Matzinger, P. (1994). Tolerance, danger, and the extended family. *Annu. Rev. Immunol.* 12:991-1045.

McAdam, A., Greenwald, R. J., Levin, M. A., Chernova, T., Malenkovich, N., Ling, V., Freeman, G. J., and Sharpe, A. H. (2001). ICOS is critical for CD40-mediated antibody class switching. *Nature* 409:102-105.

McGuirk, P., and Mills, K. H. (2002). Pathogen-specific regulatory T cells provoke a shift in the Th1/Th2 paradigm in immunity to infectious diseases. *Trends Immunol.* 23:450-455.

Mellor, A. L., and Munn, D. H. (2004). IDO expression by dendritic cells: tolerance and tryptophan catabolism. *Nat. Rev. Immunol.* 4:762-774.

Mellor, A. L., Baban, B., Chandler, P., Marshall, B., Jhaver, K., Hansen, A., Koni, P. A., Iwashima, M., and Munn, D. H. (2003). Cutting edge: induced indoleamine 2,3 dioxygenase expression in dendritic cell subsets suppresses T cell clonal expansion. *J. Immunol.* 171:1652-1655.

Merad, M., Fong, L., Bogenberger, J., and Engleman, E. G. (2000). Differentiation of myeloid dendritic cells into CD8alpha-positive dendritic cells *in vivo*. *Blood* 96:1865-1872.

Morris, P., Shaman, J., Attaya, M., Amaya, M., Goodman, S., Bergman, C., Monaco, J. J., and Mellins, E. (1994). An essential role for HLA-DM in antigen presentation by class II major histocompatibility molecules. *Nature* 368:551-554.

Moser, M. (2003). Dendritic cells in immunity and tolerance-do they display opposite functions? *Immunity* 19:5-8.

Munn, D. H., Sharma, M. D., Lee, J. R., Jhaver, K. G., Johnson, T. S., Keskin, D. B., Marshall, B., Chandler, P., Antonia, S. J., Burgess, R., Slingluff, C. L., Jr., and Mellor, A. L. (2002). Potential regulatory function of human dendritic cells expressing indoleamine 2,3-dioxygenase. *Science* 297:1867-1870.

Munz, C., Steinman, R. M., and Fujii, S. (2005). Dendritic cell maturation by innate lymphocytes: coordinated stimulation of innate and adaptive immunity. *J. Exp. Med.* 202:203-207.

Nakano, H., Yanagita, M., and Gunn, M. D. (2001). CD11c($^+$)B220($^+$)Gr-1($^+$) cells in mouse lymph nodes and spleen display characteristics of plasmacytoid dendritic cells. *J. Exp. Med.* 194:1171-1178.

Narbutt, J., Lesiak, A., Zak-Prelich, M., Wozniacka, A., Sysa-Jedrzejowska, A., Tybura, M., Robak, T., and Smolewski, P. (2004). The distribution of peripheral blood dendritic cells assayed by a new panel of anti-BDCA monoclonal antibodies in healthy representatives of the polish population. *Cell Mol. Biol. Lett.* 9:497-509.

Ngo, V. N., Tang, H. L., and Cyster, J. G. (1998). Epstein-Barr virus-induced molecule 1 ligand chemokine is expressed by dendritic cells in lymphoid tissues and strongly attracts naive T cells and activated B cells. *J. Exp. Med.* 188:181-191.

O'Doherty, U., Peng, M., Gezelter, S., Swiggard, W. J., Betjes, M., Bhardwaj, N., and Steinman, R. M. (1994). Human blood contains two subsets of dendritic cells, one immunologically mature and the other immature. *Immunology* 82:487-493.

O'Keeffe, M., Hochrein, H., Vremec, D., Scott, B., Hertzog, P., Tatarczuch, L., and Shortman, K. (2003). Dendritic cell precursor populations of mouse blood: identification of the murine homologues of human blood plasmacytoid pre-DC2 and CD11c$^+$ DC1 precursors. *Blood* 101:1453-1459.

O'Neill, L. A. (2004). Immunology. After the toll rush. *Science* 303:1481-1482.

O'Neill, H. C., and Wilson, H. L. (2004). Limitations with *in vitro* production of dendritic cells using cytokines. *J. Leukoc. Biol.* 75:600-603.

Palliser, D., Ploegh, H., and Boes, M. (2004). Myeloid differentiation factor 88 is required for cross-priming *in vivo*. *J. Immunol.* 172:3415-3421.

Pamer, E., and Cresswell P. (1998). Mechanisms of MHC class I--restricted antigen processing. *Annu. Rev. Immunol.* 16:323-358.

Partida-Sanchez, S., Goodrich, S., Kusser, K., Oppenheimer, N., Randall, T. D., and Lund, F. E. (2004). Regulation of dendritic cell trafficking by the ADP-ribosyl cyclase CD38: impact on the development of humoral immunity. *Immunity* 20:279-291.

Pasare, C., and Medzhitov, R. (2004). Toll-like receptors: linking innate and adaptive immunity. *Microbes Infect.* 6:1382-1387.

Patterson, S. (2000). Flexibility and cooperation among dendritic cells. *Nat. Immunol.* 1:273-274.

Pelayo, R., Hirose, J., Huang, J., Garrett, K. P., Delogu, A., Busslinger, M., and Kincade, P. W. (2005). Derivation of 2 categories of plasmacytoid dendritic cells in murine bone marrow. *Blood* 105:4407-4415.

Pendl, G. G., Robert, C., Steinert, M., Thanos, R., Eytner, R., Borges, E., Wild, M. K., Lowe, J. B., Fuhlbrigge, R. C., Kupper, T. S., Vestweber, D., and Grabbe, S. (2002). Immature mouse dendritic cells enter inflamed tissue, a process that requires E- and P-selectin, but not P-selectin glycoprotein ligand 1. *Blood* 99:946-956.

Pfeifer, J. D., Wick, M. J., Roberts, R. L., Findlay, K., Normark, S. J., and Harding, C. V. (1993). Phagocytic processing of bacterial antigens for class I MHC presentation to T cells. *Nature* 361:359-362.

Pulendran, B., Lingappa, J., Kennedy, M. K., Smith, J., Teepe, M., Rudensky, A., Maliszewski, C. R., and Maraskovsky, E. (1997). Developmental pathways of dendritic cells *in vivo*: distinct function, phenotype, and localization of dendritic cell subsets in FLT3 ligand- treated mice. *J. Immunol.* 159:2222-2231.

Pulendran, B., Smith, J. L., Caspary, G., Brasel, K., Pettit, D., Maraskovsky, E., and Maliszewski, C. R. (1999). Distinct dendritic cell subsets differentially regulate the class of immune response *in vivo*. *Proc. Natl. Acad. Sci. U.S.A.* 96:1036-1041.

Pulendran, B., Kumar, P., Cutler, C. W., Mohamadzadeh, M., Van Dyke, T., and Banchereau, J. (2001). Lipopolysaccharides from distinct pathogens induce different classes of immune responses *in vivo*. *J. Immunol.* 167:5067-5076.

Qin, D., Wu, J., Vora, K. A., Ravetch, J. V., Szakal, A. K., Manser, T., and Tew, J. G. (2000). Fc gamma receptor IIB on follicular dendritic cells regulates the B cell recall response. *J. Immunol.* 164:6268-6275.

Qu, C., Edwards E. W., Tacke, F., Angeli, V., Llodra, J., Sanchez-Schmitz, G., Garin, A., Haque, N. S., Peters, W., van Rooijen, N., Sanchez-Torres, C., Bromberg, J., Charo, I. F., Jung, S., Lira, S. A., and Randolph G. J. (2004). Role of CCR8 and other chemokine pathways in the migration of monocyte-derived dendritic cells to lymph nodes. *J. Exp. Med.* 200:1231-1241.

Quah, B. J., and O'Neill H. C. (2005). The immunogenicity of dendritic cell-derived exosomes. *Blood Cells Mol. Dis.* 35:94-110.

Ramadan, G., Davies, B., Kurup, V. P., and Keever-Taylor, C. A. (2005). Generation of cytotoxic T cell responses directed to human leucocyte antigen class I restricted epitopes from the Aspergillus f16 allergen. *Clin. Exp. Immunol.* 140:81-91.

Randolph, G. J., Angeli, V., and Swartz, M. A. (2005). Dendritic-cell trafficking to lymph nodes through lymphatic vessels. *Nat. Rev. Immunol.* 5:617-628.

Raulet, D. H. (2004). Interplay of natural killer cells and their receptors with the adaptive immune response. *Nat. Immunol.* 5:996-1002.

Reis e Sousa, C., and Germain, R. N. (1995). Major histocompatibility complex class I presentation of peptides derived from soluble exogenous antigen by a subset of cells engaged in phagocytosis. *J. Exp. Med.* 182:841-851.

Res, P. C., Couwenberg, F., Vyth-Dreese, F. A., and Spits, H. (1999). Expression of pTalpha mRNA in a committed dendritic cell precursor in the human thymus. *Blood* 94:2647-2657.

Rissoan, M. C., Soumelis, V., Kadowaki, N., Grouard, G., Briere, F., de Waal, Malefyt, R., and Liu, Y. J. (1999). Reciprocal control of T helper cell and dendritic cell differentiation. *Science* 283:1183-1186.

Robbiani, D. F., Finch, R. A., Jager, D., Muller, W. A., Sartorelli, A. C., and Randolph, G. J. (2000). The leukotriene C(4) transporter MRP1 regulates CCL19 (MIP-3beta, ELC)-dependent mobilization of dendritic cells to lymph nodes. *Cell* 103:757-768.

Sallusto, F., and Lanzavecchia, A. (1994). Efficient presentation of soluble antigen by cultured human dendritic cells is maintained by granulocyte/macrophage colony-stimulating factor plus interleukin 4 and downregulated by tumor necrosis factor alpha. *J. Exp. Med.* 179:1109-1118.

Sallusto, F., and Lanzavecchia, A. (2000). Understanding dendritic cell and T-lymphocyte traffic through the analysis of chemokine receptor expression. *Immunol. Rev.* 177:134-140.

Sallusto, F., Mackay, C. R., and Lanzavecchia, A. (2000). The role of chemokine receptors in primary, effector, and memory immune responses. *Annu. Rev. Immunol.* 18, 593-620.

Salomon, B., and Bluestone, J. A. (1998). LFA-1 interaction with ICAM-1 and ICAM-2 regulates Th2 cytokine production. *J. Immunol.* 161, 5138-5142.

Scandella, E., Men, Y., Gillessen, S., Forster, R., and Groettrup, M. (2002). Prostaglandin E2 is a key factor for CCR7 surface expression and migration of monocyte-derived dendritic cells. *Blood* 100:1354-1361.

Scandella, E., Men, Y., Legler, D. F., Gillessen, S., Prikler, L., Ludewig, B., and Groettrup, M. (2004). CCL19/CCL21-triggered signal transduction and migration of dendritic cells requires prostaglandin E2. *Blood* 103:1595-1601.

Segura, E., Amigorena, S., and Thery, C. (2005a). Mature dendritic cells secrete exosomes with strong ability to induce antigen-specific effector immune responses. *Blood Cells Mol. Dis.* 35:89-93.

Segura, E., Nicco, C., Lombard, B., Veron, P., Raposo, G., Batteux, F., Amigorena, S., and Thery, C. (2005b). ICAM-1 on exosomes from mature dendritic cells is critical for efficient naive T-cell priming. *Blood* 106:216-223.

Shi, G. P., Bryant, R. A., Riese, R., Verhelst, S., Driessen, C., Li, Z., Bromme, D., Ploegh, H. L., and Chapman, H. A. (2000). Role for cathepsin F in invariant chain processing and major histocompatibility complex class II peptide loading by macrophages. *J. Exp. Med.* 191:1177-1186.

Shortman, K., and Liu, Y. J. (2002). Mouse and human dendritic cell subtypes. *Nat. Rev. Immunol.* 2:151-161.

Sica, G. L., Choi, I. H., Zhu, G., Tamada, K., Wang, S. D., Tamura, H., Chapoval, A. I., Flies, D. B., Bajorath, J., and Chen, L. (2003). B7-H4, a molecule of the B7 family, negatively regulates T cell immunity. *Immunity* 18:849-861.

Siegal, F. P., Kadowaki, N., Shodell, M., Fitzgerald-Bocarsly, P. A., Shah, K., Ho, S., Antonenko, S., and Liu, Y. J. (1999). The nature of the principal type 1 interferon-producing cells in human blood. *Science* 284:1835-1837.

Sorg, R. V., Kogler, G., and Wernet, P. (1999). Identification of cord blood dendritic cells as an immature CD11c- population. *Blood* 93:2302-2307.

Sozzani, S., Allavena, P., D'Amico, G., Luini, W., Bianchi, G., Kataura, M., Imai, T., Yoshie, O., Bonecchi, R., and Mantovani, A. (1998). Differential regulation of chemokine receptors during dendritic cell maturation: a model for their trafficking properties. *J. Immunol.* 161:1083-1086.

Sozzani, S., Allavena, P., Vecchi, A., and Mantovani, A. (1999). The role of chemokines in the regulation of dendritic cell trafficking. *J. Leukoc. Biol.* 66:1-9.

Spits, H., Couwenberg, F., Bakke, A. Q., Weijer, K., and Uittenbogaart, C. H. (2000). Id2 and Id3 inhibit development of CD34($^+$) stem cells into predendritic cell (pre-DC)2 but not into pre-DC1. Evidence for a lymphoid origin of pre-DC2. *J. Exp. Med.* 192:1775-1784.

Sporri, R., and Reis e Sousa, C. (2005). Inflammatory mediators are insufficient for full dendritic cell activation and promote expansion of CD4$^+$ T cell populations lacking helper function. *Nat. Immunol.* 6:163-170.

Steinman, R. M., and Cohn, Z. A. (1973). Identification of a novel cell type in peripheral lymphoid organs of mice. I. Morphology, quantitation, tissue distribution. *J. Exp. Med.* 137:1142-1162.

Steinman, R. M., and Cohn, Z. A. (1974). Identification of a novel cell type in peripheral lymphoid organs of mice. II. Functional properties *in vitro. J. Exp. Med.* 139:380-397.

Steinman, R. M., and Swanson, J. (1995). The endocytic activity of dendritic cells. *J. Exp. Med.* 182:283-288.

Steinman, R. M., Adams, J. C., and Cohn, Z. A. (1975). Identification of a novel cell type in peripheral lymphoid organs of mice. IV. Identification and distribution in mouse spleen. *J. Exp. Med.* 141:804-820.

Steinman R. M., Hawiger, D., and Nussenzweig, M. C. (2003). Tolerogenic dendritic cells. *Annu. Rev. Immunol.* 21:685-711.

Strobl, H., Scheinecker, C., Riedl, E., Csmarits, B., Bello-Fernandez, C., Pickl, W. F., Majdic, O., and Knapp, W. (1998). Identification of CD68$^+$lin-peripheral blood cells with dendritic precursor characteristics. *J. Immunol.* 161:740-748.

Suh, W. K., Gajewska, B. U., Okada, H., Gronski, M. A., Bertram, E. M., Dawicki, W., Duncan, G. S., Bukczynski, J., Plyte, S., Elia, A., Wakeham, A., Itie, A., Chung, S., Da Costa, J., Arya, S., Horan, T., Campbell, P., Gaida, K., Ohashi, P. S., Watts, T. H., Yoshinaga, S. K., Bray, M. R., Jordana, M., and Mak, T. W. (2003). The B7 family member B7-H3 preferentially down-regulates T helper type 1-mediated immune responses. *Nat. Immunol.* 4:899-906.

Swallow, M. M., Wallin, J. J., and Sha, W. C. (1999). B7h, a novel costimulatory homolog of B7.1 and B7.2, is induced by TNFalpha. *Immunity* 11:423-432.

Tamura, T., Tailor, P., Yamaoka, K., Kong, H. J., Tsujimura, H., O'Shea, J. J., Singh, H., and Ozato, K. (2005). IFN regulatory factor-4 and -8 govern dendritic cell subset development and their functional diversity. *J. Immunol.* 174:2573-2581.

Tomasello, E., Desmoulins, P. O., Chemin, K., Guia, S., Cremer, H., Ortaldo, J., Love, P., Kaiserlian, D., and Vivier, E. (2000). Combined natural killer cell and dendritic cell functional deficiency in KARAP/DAP12 loss-of-function mutant mice. *Immunity* 13:355-364.

Traver, D., Akashi, K., Manz, M., Merad, M., Miyamoto, T., Engleman, E. G., and Weissman, I. L. (2000). Development of CD8alpha-positive dendritic cells from a common myeloid progenitor. *Science* 290:2152-2154.

Turley, S. J., Inaba, K., Garrett, W. S., Ebersold, M., Unternaehrer, J., Steinman, R. M., and Mellman, I. (2000). Transport of peptide-MHC class II complexes in developing dendritic cells. *Science* 288:522-527.

van Rijt, L. S., Jung, S., Kleinjan, A., Vos, N., Willart, M., Duez, C., Hoogsteden, H. C., and Lambrecht, B. N. (2005). *In vivo* depletion of lung CD11c$^+$

dendritic cells during allergen challenge abrogates the characteristic features of asthma. *J. Exp. Med.* 201:981-991.

Vanbervliet, B., Bendriss-Vermare, N., Massacrier, C., Homey, B., de Bouteiller, O., Briere, F., Trinchieri, G., and Caux, C. (2003). The inducible CXCR3 ligands control plasmacytoid dendritic cell responsiveness to the constitutive chemokine stromal cell-derived factor 1 (SDF-1)/CXCL12. *J. Exp. Med.* 198:823-830.

Vandenabeele, S., Hochrein, H., Mavaddat, N., Winkel, K., and Shortman, K. (2001). Human thymus contains 2 distinct dendritic cell populations. *Blood* 97:1733-1741.

Vandenabeele, S., and Wu, L. (1999). Dendritic cell origins: puzzles and paradoxes. *Immunol. Cell Biol.* 77:411-419.

Vassileva, G., Soto, H., Zlotnik, A., Nakano, H., Kakiuchi, T., Hedrick, J. A., and Lira, S. A. (1999). The reduced expression of 6Ckine in the plt mouse results from the deletion of one of two 6Ckine genes. *J. Exp. Med.* 190:1183-1188.

Vecchi, A., Massimiliano, L., Ramponi, S., Luini, W., Bernasconi, S., Bonecchi, R., Allavena, P., Parmentier, M., Mantovani, A., and Sozzani, S. (1999). Differential responsiveness to constitutive vs. inducible chemokines of immature and mature mouse dendritic cells. *J. Leukoc. Biol.* 66:489-494.

Verbovetski, I., Bychkov, H., Trahtemberg, U., Shapira, I., Hareuveni, M., Ben-Tal, O., Kutikov, I., Gill, O., and Mevorach, D. (2002). Opsonization of apoptotic cells by autologous iC3b facilitates clearance by immature dendritic cells, down-regulates DR and CD86, and up-regulates CC chemokine receptor 7. *J. Exp. Med.* 196:1553-1561.

Vremec, D., Pooley, J., Hochrein, H., Wu, L., and Shortman, K. (2000). CD4 and CD8 expression by dendritic cell subtypes in mouse thymus and spleen. *J. Immunol.* 164:2978-2986.

Walunas, T. L., Lenschow, D. J., Bakker, C. Y., Linsley, P. S., Freeman, G. J., Green, J. M., Thompson, C. B., and Bluestone, J. A. (1994). CTLA-4 can function as a negative regulator of T cell activation. *Immunity* 1:405-413.

Watanabe, N., Gavrieli, M., Sedy, J. R., Yang, J., Fallarino, F., Loftin, S. K., Hurchla, M. A., Zimmerman, N., Sim, J., Zang, X., Murphy, T. L., Russell, J. H., Allison, J. P., and Murphy, K. M. (2003). BTLA is a lymphocyte inhibitory receptor with similarities to CTLA-4 and PD-1. *Nat. Immunol.* 4:670-679.

Weinberg, A. D., Vella, A. T., and Croft, M. (1998). OX-40: life beyond the effector T cell stage. *Semin. Immunol.* 10:471-480.

Westermann, J., Bode, U., Sahle, A., Speck, U., Karin, N., Bell, E. B., Kalies, K., and Gebert, A. (2005). Naive, effector, and memory T lymphocytes efficiently scan dendritic cells *in vivo*: contact frequency in T cell zones of secondary lymphoid organs does not depend on LFA-1 expression and facilitates survival of effector T cells. *J. Immunol.* 174:2517-2524.

Wu, L., Vandenabeele, S., and Georgopoulos, K. (2001). Derivation of dendritic cells from myeloid and lymphoid precursors. *Int. Rev. Immunol.* 20:117-135.

Xia, W., Pinto, C. E., and Kradin, R. L. (1995). The antigen-presenting activities of Ia[+] dendritic cells shift dynamically from lung to lymph node after an airway challenge with soluble antigen. *J. Exp. Med.* 181:1275-1283.

Yang, D., Chen, Q., Stoll, S., Chen, X., Howard, O. M., and Oppenheim, J. J. (2000). Differential regulation of responsiveness to fMLP and C5a upon dendritic cell maturation: correlation with receptor expression. *J. Immunol.* 165:2694-2702.

Yang, G. X., Lian, Z. X., Kikuchi, K., Liu, Y. J., Ansari, A. A., Ikehara, S., and Gershwin, M. E. (2005). CD4-plasmacytoid dendritic cells (pDCs) migrate in lymph nodes by CpG inoculation and represent a potent functional subset of pDCs. *J. Immunol.* 174:3197-3203.

Yoshinaga, S. K., Whoriskey, J. S., Khare, S. D., Sarmiento, U., Guo, J., Horan, T., Shih, G., Zhang, M., Coccia, M. A., Kohno, T., Tafuri-Bladt, A., Brankow, D., Campbell, P., Chang, D., Chiu, L., Dai, T., Duncan, G., Elliott, G. S., Hui, A., McCabe, S. M., Scully, S., Shahinian, A., Shaklee, C. L., Van, G., Mak, T. W., and Senaldi, G. (1999). T-cell co-stimulation through B7RP-1 and ICOS. *Nature* 402:827-832.

Zheng, H., and Li, Z. (2004). Cutting edge: cross-presentation of cell-associated antigens to MHC class I molecule is regulated by a major transcription factor for heat shock proteins. *J. Immunol.* 173:5929-5933.

Zlotnik, A., and Yoshie, O. (2000). Chemokines: a new classification system and their role in immunity. *Immunity* 12:121-127.

Zuniga, E. I., McGavern, D. B., Pruneda-Paz, J. L., Teng, C., and Oldstone, M. B. (2004). Bone marrow plasmacytoid dendritic cells can differentiate into myeloid dendritic cells upon virus infection. *Nat. Immunol.* 5:1227-1234.

Chapter 3

Dendritic Cell and Pathogen Interactions in the Subversion of Protective Immunity

John E. Connolly, Damien Chaussabel, and Jacques Banchereau

3.1 Introduction

The immune response to viral challenge involves the coordinated action of both innate and adaptive effectors. In their immature state, dendritic cells (DCs) constitutively internalize and survey cellular debris through the action of a variety of endocytic and innate pattern-recognition receptors. Upon detection of microbial incursion, DCs recruit innate effector cells through the coordinated expression of specific proinflammatory cytokines and chemokines. Signals resulting from microbial recognition initiate a program of DC maturation, which is highlighted by a functional transition to a cell specialized for antigen presentation and adaptive immune priming. As the most potent antigen-presenting cells (APCs), DCs serve multiple functions in the initiation and control of adaptive immune responses. Given their unique and central role in the initiation and maintenance of antiviral immunity, DCs represent a preferred target for viruses seeking to evade immune surveillance. The life of a DC from progenitor to a full mature differentiated APC is marked by a number of highly regulated control points, namely (1) DC subset ontogeny; (2) antigen uptake, processing, and presentation; (3) maturation; (4) death. Viruses have evolved a startling variety of approaches that target each of these critical points in DC development. This chapter summarizes our current understanding of the central role performed by DCs in immune regulation and how viruses have evolved to interfere with this process to successfully evade host immunity.

3.2 Dendritic Cell Biology

3.2.1 Dendritic Cells Are Composed of Subsets

Classically, two main DC differentiation pathways are recognized (Banchereau *et al.*, 2000; Shortman and Liu, 2002). A myeloid pathway generates both Langerhans cells (LCs), which are found in stratified epithelia such as the skin, and interstitial (int) DCs, which are found in all other tissues (Caux *et al.*, 1996). Another pathway generates plasmacytoid DCs (pDCs) (Liu, 2005), which secrete large amounts of IFN-α after viral infection (Asselin-Paturel *et al.*, 2005; Cella *et al.*, 1999; Siegal *et al.*, 1999). These two primary differentiation pathways serve important, nonredundant roles in the establishment of antiviral immunity.

3.2.1.1 Distinct DC Subsets Are Endowed with Distinct Functional Properties

DC subsets are endowed with common as well as unique biological functions. This functional diversity may permit the body to deal with the immunologic challenges brought by the myriad of evolving microbes.

Myeloid Dendritic Cells: *In vivo* human myeloid DCs (mDCs) exist in at least three subset varieties; circulating blood mDCs and tissue resident intDCs and LCs (Banchereau *et al.*, 2000). Tissue resident mDC subsets are thought to encounter microbes most commonly at a mucosal surface or a point of damage in the skin. Maturation by either microbial products and/or proinflammatory agents leads to a shift in DC chemokine receptor expression and migration to the lymph node via afferent lymphatics. Although it is thought that all three subsets arise from a common myeloid progenitor, the relationship between blood mDCs and tissue resident DCs is unclear. Circulating blood mDCs may represent a homeostatic reservoir of immature mDCs that migrate into the tissue to replenish depleted mDC populations. Alternatively, this population may be present to detect and initiate immune responses against blood-borne pathogens. Recent studies indicate that subsets of DCs may be able to divide under the influence of stromal-derived factors in lymphoid organs (Falcon *et al.*, 2005; Geiss *et al.*, 2000; Kabashima *et al.*, 2005; Zhang *et al.*, 2004). *In vitro* experiments have demonstrated that LCs and intDCs generated from cultures of CD34[+] hematopoietic progenitors differ in their capacity to activate lymphocytes: intDCs induce the differentiation of naïve B cells into immunoglobulin-secreting plasma cells (Caux *et al.*, 1996; Caux *et al.*, 1997), whereas LCs seem to be particularly efficient activators of cytotoxic CD8[+] T cells

(J.E. Connolly, unpublished observation). They also differ in the cytokines that they secrete (only intDCs produce IL-10) and in their enzymatic activity (Caux et al., 1996; Caux et al., 1997), which may be important for the generation of peptides that will be presented to T cells. Indeed, different enzymes are likely to degrade a given antigen into different peptide repertoires, as recently shown for HIV nef protein (Seifert et al., 2003). This will lead to different sets of peptide-MHC complexes being presented and to distinct antigen-specific T-cell repertoires. Importantly, DC subsets express specific receptors for sensing pathogens, a reflection of both location and specificity for specific microbes. In humans, mDCs express the Toll-like receptors (TLRs) 1, 2, 3, 4, 5, 6, and 8.

Plasmacytoid Dendritic Cells: The maturation state of a pDC is distinguished by two functionally distinct stages, that is, (i) ability of precursor pDCs to secrete large amounts of IFN-α after viral exposure (Asselin-Paturel et al., 2005; Cella et al., 1999; Siegal et al., 1999); (ii) ability of mature pDCs to activate and modulate T-cell responses (Kadowaki et al., 2000). In humans, pDCs express TLR6, 7, 9, and 10. Although all nucleated cells can produce type I IFNs, pDCs are recognized as a main source of type I IFN produced in response to viral (Siegal et al., 1999) or CpG (Hartmann et al., 1999) triggering. An activated pDC is capable of producing up to 10 pg/cell of IFN-α (Barchet et al., 2005). Recent studies demonstrated that pDC IFN production is fundamentally different from nonimmune cells. Interferon regulatory factor 7 (IRF-7) is critical for pDC IFN-α secretion in response to both RNA and CpG stimuli. In humans, pDCs but not mDCs can direct CpG to endosomal compartments thereby permitting MyD88/ IRF-7 activation and IFN-α secretion (Honda et al., 2005). In nonimmune cells, the activation of the RNA helicase RIG-I appears to be the primary pathway for the production of type I IFNs in response to double-stranded RNA recognition. This pathway appears to be dispensable in pDCs (Kato et al., 2005). Recently, the existence of pDC subsets has been demonstrated. Thus, in the mouse, expression of lymphoid-related genes (RAG1 and Ig rearrangement products) (Kadowaki et al., 2000) or proteins (CD4) (Yang et al., 2005) distinguishes between two subsets of pDCs. Our own results in humans demonstrate that CD2 expression distinguishes pDCs with common, IFN-α secretion, and unique, cytotoxic activity against K562 cells, functions (J.E. Connolly, unpublished observation). The existence of these pDC subsets may indicate an attempt by host to circumvent viral immune evasion strategies.

3.2.2 Dendritic Cell Ontogeny

3.2.2.1 DC Progenitors and Precursors

DC progenitors reside within CD34$^+$ hematopoietic progenitor cells (HPCs) (Caux *et al.*, 1992). Both lymphoid and common myeloid progenitors yield, at the clonal level, mDCs and pDCs (Chicha *et al.*, 2004). Interestingly, the progenitors of pDCs and mDCs can be found within FLT3$^+$ HPCs (D'Amico and Wu, 2003; Karsunky *et al.*, 2003). This is consistent with the well-established role of FLT3 ligand (FLT3-L) in DC differentiation/mobilization *in vivo* in both humans and mice (Maraskovsky *et al.*, 1996; Maraskovsky *et al.*, 2000; McKenna *et al.*, 2000; Pulendran *et al.*, 1997; Pulendran *et al.*, 2000). Accordingly, FLT3-L is essential in the generation of pDCs and mDCs (Blom *et al.*, 2000; Chen *et al.*, 2005; Spits *et al.*, 2000). In contrast with granulocyte-macrophage colony-stimulating factor (GM-CSF) knockouts, FLT3-deficient mice demonstrate decreased DC numbers in the steady state. Thus, FLT3-L appears as a major factor governing DC homeostasis in the steady state. Given the role of GM-CSF in DC generation (Caux *et al.*, 1992; Markowicz and Engleman, 1990; Santiago-Schwarz *et al.*, 1992), activation, and survival (Witmer-Pack *et al.*, 1987), it is tempting to postulate that GM-CSF is actually a major factor governing DC differentiation upon infection and inflammation. In this context, GM-CSF preferentially expands the myeloid DC subset *in vivo* (Pulendran *et al.*, 1999).

Recent studies indicate that cells not traditionally thought of as APCs, such as γ δ T cells, may acquire DC-like phenotype and function (Brandes *et al.*, 2005). Furthermore, proinflammatory cytokines can endow human NK cells with ability to acquire antigen and to stimulate T cells (Hanna *et al.*, 2004). The existence of multiple subsets, possibly with redundant antigen-presenting function, may represent an attempt by the host to avoid pathogen immune evasion strategies. The question to resolve is whether all DCs are equal and to define *bone fide* DCs (e.g., LCs) from cells that can acquire DC function under the environmental pressure (e.g., γ δ T cell–derived DCs). A possible parameter for such distinction could be the extent of their professionalism measured by their capacity to prime naïve T cells.

3.2.3 Antigen Capture, Processing, and Presentation

Immature DCs are extremely efficient in antigen capture, serving a sentinel role in the immune system, while mature DCs are specialized in antigen presentation and are therefore efficient in the initiation of immune responses.

The process of antigen capture, processing, and presentation represents a critical checkpoint in DC biology.

DCs are well equipped to acquire microbial antigens: DCs utilize a host of receptor-mediated and bulk-phase uptake mechanisms to capture antigen including (1) macropinocytosis; (2) receptor-mediated endocytosis via C-type lectins (e.g., mannose receptor, DEC-205, DC-SIGN) (Cambi *et al.*, 2005; Engering *et al.*, 1997; Figdor *et al.*, 2002; Jiang *et al.*, 1995; Mommaas *et al.*, 1999; Reis e Sousa *et al.*, 1993; Sallusto *et al.*, 1995; Tan *et al.*, 1997) or Fcγ receptors type I (CD64) and type II (CD32) (uptake of immune complexes or opsonized particles) (Fanger *et al.*, 1996); (3) phagocytosis of apoptotic and necrotic cells (Albert *et al.*, 1998b; Albert *et al.*, 1998a; Rubartelli *et al.*, 1997), viruses, bacteria including mycobacteria (Inaba *et al.*, 1993; Regnault *et al.*, 1999), as well as intracellular parasites such as *Leishmania major*, and (4) internalization of heat-shock proteins, hsp70 or gp96-peptide complexes, through multiple receptors including LOX-1 (Delneste *et al.*, 2002) and TLR2/4 (Asea *et al.*, 2002). (Fig. 3.1).

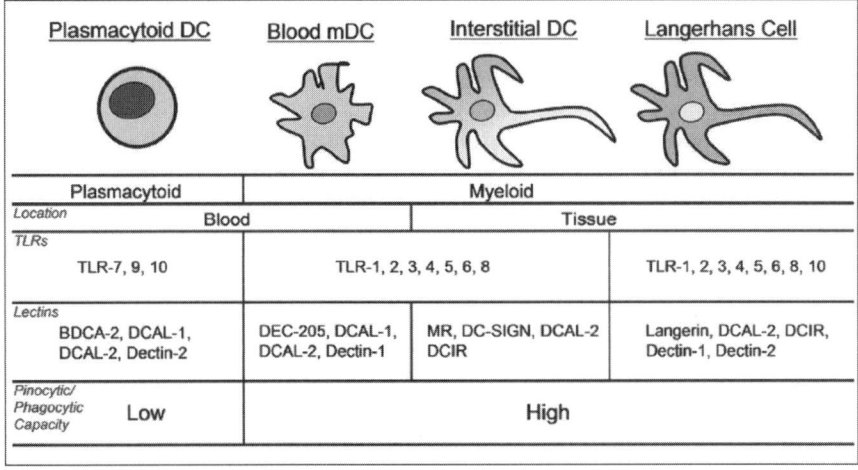

Plasmacytoid DC	Blood mDC	Interstitial DC	Langerhans Cell
Plasmacytoid	Myeloid		
Location Blood	Tissue		
TLRs TLR-7, 9, 10	TLR-1, 2, 3, 4, 5, 6, 8		TLR-1, 2, 3, 4, 5, 6, 8, 10
Lectins BDCA-2, DCAL-1, DCAL-2, Dectin-2	DEC-205, DCAL-1, DCAL-2, Dectin-1	MR, DC-SIGN, DCAL-2, DCIR	Langerin, DCAL-2, DCIR, Dectin-1, Dectin-2
Pinocytic/ Phagocytic Capacity Low	High		

FIGURE 3.1. Human dendritic cell subsets are specialized for microbial uptake. Two main DC differentiation groups are recognized: a myeloid and plasmacytoid group. Myeloid and plasmacytoid cell subsets reside in different locations throughout the body. This subset specialization is further reflected in the expression of specific endocytic receptors such as C-type lectins and pathogen-associated pattern-recognition receptors such as TLRs.

The process of maturation regulates all modes of DC antigen capture. Immediately after microbial recognition, bulk-phase internalization actually increases, thereby enhancing the amount of internalized pathogen-associated antigen. The maturing DC then downregulates both macropinocytic uptake and many of the receptors associated with antigen internalization.

Furthermore, the ability of the maturing DC to process and present antigen in the context of MHC class II is downregulated.

3.2.3.1 Presentation via MHC Class I

DCs are the principal initiators of $CD8^+$ T-cell immunity through the presentation of antigens in the context of MHC class I (Trombetta and Mellman, 2005). This involves the classical presentation of endogenous peptides, originating from cellular and viral proteins, as well as the presentation of exogenous antigens via cross-priming/presentation. In fact, cross-priming/presentation may be the main pathway through which immunity to microbes that do not infect DCs directly is generated (Albert *et al.*, 1998a; Carbone and Heath, 2003; Sigal *et al.*, 1999). It remains to be defined how DCs are able to put exogenous antigens for presentation on MHC class I. Recent studies from Yewdell and colleagues suggest that cross-priming may be based on the transfer of proteasome substrates rather than peptides (Norbury *et al.*, 2004). Furthermore, the exact nature of the loading compartment remains unknown, that is, it remains to be determined whether it is the endoplasmic reticulum (ER) or a mixed phagosome-ER compartment.

Nonclassical HLA class I molecules are also targeted by microbial pathogens. Recent work on nonclassical HLA molecules indicates that they may serve an inhibitory role in DC interactions with both T cells and NK cells. HLA-G and HLA-E are members of the MHC class IB family, which, unlike the classic MHC class IA molecules, displays a very low degree of polymorphism and limited peptide repertoire (Heinrichs and Orr, 1990; Kovats *et al.*, 1990). HLA-E is expressed on a variety of cells in all tissues. Constitutive expression of both soluble and membrane bound forms of HLA-G have been characterized at the maternal fetal interface (Le Bouteiller *et al.*, 2003). Both HLA-E and HLA-G have been demonstrated to interact with inhibitory, immunoreceptor tyrosine-based inhibition motifs (ITIMs) containing receptors on T and NK cells.

Expression of HLA-G by APCs appears to have profound inhibitory effects on immune function. DCs, monocytes, and macrophages have been characterized in the maternal dicidua implicating them as possible mediators of HLA-G immunomodulation (Hunt *et al.*, 2000; Sukhikh and Vanko, 1999). $CD34^+$HPCs derived DCs express low levels of HLA-G (Le Friec *et al.*, 2004). HLA-G is also found on lung DCs and macrophages during inflammation (Pangault *et al.*, 2002). DCs treated with soluble HLA-G fail to activate T cells in an allogeneic proliferation assay (Le Friec *et al.*, 2003). Transfection of U937 cells with HLA-G1 also strongly inhibited allogeneic proliferation of purified CD4 T cells. Interestingly,

T cells sensitized with HLA-G transfected APCs themselves could inhibit T-cell proliferation when added to a third-party allo assay, indicating that these cells may have regulatory function (Lemaoult *et al.*, 2004).

3.2.3.2 Presentation via MHC Class II

Captured antigens are presented by MHC class II molecules (Inaba *et al.*, 1998), which upon DC maturation are transported from lysosomal compart-ments to the cell membrane (Inaba *et al.*, 2000; Turley *et al.*, 2000). In fact, this translocation of peptide-MHC (pMHC) class II complexes from intracellular compartments to cell membrane represents a hallmark of DC maturation. These pMHC complexes are very stable on the cell membrane of mature DCs thereby facilitating TCR recognition. Furthermore, as opposed to macrophages that favor antigen degradation, DCs show lower levels of lysosomal protease activity thereby permitting low rate of antigen degradation (Delamarre *et al.*, 2005). This in turn permits antigen retention in lymphoid organs *in vivo* for extended periods that might favor antigen presentation (Delamarre *et al.*, 2005). Thus, the prolonged availability of antigen for generation of pMHC complexes and prolonged presentation of such complexes on cell surface might both explain a unique efficiency of DCs in triggering naïve T-cell differentiation.

MHC class II molecules are under the control of a transcriptional coacti-vator, MHC class II transactivator (CIITA) (Ting and Trowsdale, 2002). The expression of CIITA is regulated by three independent promoters whose activity quantitatively determines MHC class II expression (Muhlethaler-Mottet *et al.*, 1997). Recent studies demonstrate that distinct subsets of a APC utilize different promoters, that is, plasmacytoid DCs (pDCs) and B cells rely on promoter pIII while mDCs and macrophages use pI (LeibundGut-Landmann *et al.*, 2004). These differences may have fundamental impact on the antigen presentation on MHC class II by these cell types and on ensuing immune responses.

3.2.3.3 Presentation via CD1 Family of MHC Molecules

CD1 proteins present lipid antigens to effector T cells. This family consists of five members CD1a–c (group 1 molecules), CD1d (group 2 molecule), and CD1e each with distinct expression patterns (Beckman *et al.*, 1994; Hunger *et al.*, 2004). Different CD1 molecules display distinct intracellular trafficking patterns likely resulting in antigen delivery into distinct compartments (Beckman *et al.*, 1994; Hunger *et al.*, 2004). Differential expression of CD1 molecules and its impact on T-cell immunity can be illustrated by CD1a and CD1d. CD1a expression *in vivo* is restricted to LCs and thymocytes. LCs have been shown to use langerin (a unique lectin expressed by LCs) and CD1a to capture and present nonpeptide antigens of *Mycobacterium leprae* to T-cell clones derived from a leprosy patient (Hunger *et al.*, 2004). Using CD1a restricted TCR α β T cells, Zajonc and colleagues demonstrated that the mechanism of lipopeptide antigen presentation by CD1a involves the anchoring of antigens in the hydrophobic binding groove, resulting in exposure of the peptide moiety for TCR contact (Zajonc *et al.*, 2005). CD1d presents to natural killer (NK)T cells, a unique subset of T cells expressing limited T cell receptor (TCR) repertoire consisting mainly of $V\alpha 24V\beta 11$ (Bendelac *et al.*, 1997), which contribute to immune response to infection and malignancy. Bacterial glycosylceramides are antigens for presentation by CD1d to NKT cells (Mattner *et al.*, 2005). Interestingly, CD1d ligation on monocytes triggers translocation of NFκB and IL-12 secretion (Yue *et al.*, 2005), possibly providing a mechanism through which CD1d-restricted NKT cells activate DCs (Munz *et al.*, 2005). The role of intracellular CD1e has recently been elucidated in DC antigen presentation. In a manner reminiscent of HLA-DM, CD1e has been shown to serve an accessory function in the selection and loading of group1 CD1 molecules. The action of CD1e therefore helps expand the repertoire of available glycolipidic T-cell antigens (de la *et al.*, 2005). The next challenge will be to understand which pathways of antigen presentation are being preferentially utilized by distinct DC subsets and the consequences of such differential presentation on the outcomes of immune responses.

3.2.4 Dendritic Cell Maturation

3.2.4.1 Maturation Signals

The multifaceted life of the DC is marked by the maturational transition between high-capacity antigen capture and a highly specialized antigen presentation abilities. This event is a critical control point in DC biology as it represents a shift between two functionally distinct states. Immature DCs are responsible for sampling the peripheral microenvironment and are capable of internalizing and processing large amounts of antigen through receptor-mediated, macropinocytic, and phagocytic mechanisms. Upon maturation by microbial products or proinflammatory cytokines, DCs transform into highly efficient antigen presenting and T lymphocyte stimulatory cells.

DCs can receive maturation signals through (1) pathogen-associated molecular patterns (PAMPs) engaging DC surface molecules involved in microbial recognition including TLRs and C-type lectins [reviewed in

Klechevsky *et al.*, (2005)]; (2) host-derived soluble factors such as proinflammatory cytokines including IL-1β, TNF, IL-6, and PGE2 (Jonuleit *et al.*, 1997), or an apparently more potent combination of IL-1β and TNF with type I (IFN-β) and II (IFN-γ) IFN (Mailliard *et al.*, 2004); and (3) cells, including T cells, (Caux *et al.*, 1994); as well as NK, NKT cells, and α β T-cells [reviewed in Munz *et al.*, (2005)]. Most likely at any given time point, the DCs will be exposed to a combination of these signals, which will influence the net result of T-cell activation. Interestingly, TLR receptors are differentially expressed by distinct DC subsets. For example, TLR9 (a receptor for unmethylated DNA) is expressed only by pDCs, whereas mDCs preferentially express TLR2 and TLR4 (receptors for bacterial products such as peptidoglycan and lipopolysaccharide, respectively) (Kadowaki *et al.*, 2001). Similarly, distinct DC subsets express unique lectins (Figdor *et al.*, 2002), which can display immunostimulatory (ITAM) or inhibitory (ITIM) motifs. Such differential expression may confer distinct maturation signals yielding distinct type of immune responses (Boonstra *et al.*, 2003). Beside a direct triggering of TLRs and C-type lectins on DCs, PAMPs can lead to the activation of DCs in trans. Indeed, TLR-mediated signaling of stromal cells will trigger expression of a specific combination of chemokines/cytokines and adhesion molecules, which in turn will modulate DC maturation (Colonna, 2004). Furthermore soluble markers of cell stress or disruption such as uric acid, free ATP, and heat-shock proteins have been reported to lead to the activation and subsequent maturation of DCs. The receptors for these markers of cellular stress have recently been identified and may play a role in inflammatory responses to pathogens (Kanneganti *et al.*, 2006).

Maturation Phenotype: Once a maturation signal is delivered, a number of changes occur in the phenotype and function of the DC. These changes again represent important control points in the development of antiviral immune responses and are therefore attractive targets for viral exploitation. The DC maturation is a continuous process that is associated with several coordinated events such as (1) loss of endocytic/phagocytic receptors; (2) upregulation of costimulatory molecules CD80, CD86, and several members of TNF/TNF receptor family including CD40, CD70 (ligand for CD27), 4-1BB-L, and OX40-L, all of which can have costimulatory effects on T cells (Watts, 2005); (3) changes in morphology, which include a loss of adhesive structures, cytoskeleton organization, and the acquisition of high cellular motility (Winzler *et al.*, 1997); (4) shift in lysosomal compartments with downregulation of CD68 and upregulation of DC-LAMP (Saint-Vis *et al.*, 1998); (5) change in class II MHC compartments as discussed above; and (6) secretion of cytokines including IL-12 and IL-23, which are important for the type 1 polarization of T-cell immunity.

This basic process of DC maturation can be modulated by pathogens via interaction with TLRs expressed on DCs. For example, TLR receptors ligands together with a T cell-like signal delivered through CD40 may enhance the phenotypic maturation of DCs (Reis e Sousa, 2001). Indeed, TLR-mediated signals are involved in the control of $CD4^+$ T-cell activation (Pasare and Medzhitov, 2004) and, for example, DCs loaded with a heart-specific self-peptide induce $CD4^+$ T cell–mediated myocarditis in mice if activated through both CD40 and TLRs (Eriksson *et al.*, 2003). Pathogens may also contain several TLR agonists that could engage several TLRs on the same DCs or on two distinct DC subsets (Palucka and Banchereau, 2002). Napolitani and colleagues. showed recently that in both human and mouse DCs, TLR3 and TLR4 acted in synergy with TLR7, TLR8, and TLR9, which led to increased production of IL-12 and IL-23 and increased the ratio of Delta-4/Jagged-1 that are dictating the type of T-cell immunity elicited by DCs (Napolitani *et al.*, 2005; Amsen *et al.*, 2004). As expected, this led to sustained Th type 1 polarizing capacity of exposed DCs (Napolitani *et al.*, 2005). These results suggest that TLR signaling might polarize DC maturation toward Th1 or Th2 inducing cells by modulating Notch ligands on DCs.

DC maturation phenotype can also be modulated by C-type lectins. Thus, Dectin-1, a yeast binding C-type lectin, synergizes with TLR2 to induce TNF-α and IL-12 (Rogers *et al.*, 2005). Yet, Dectin-1 can also promote synthesis of IL-2 and IL-10 through recruitment of Syk kinase. Accordingly, syk DCs do not make IL-10 or IL-2 upon yeast stimulation but produce IL-12, indicating that the Dectin-1/Syk and Dectin-1/TLR2 pathways can operate independently (Rogers *et al.*, 2005). These results demonstrate that pathogens utilize several surface molecules to modulate DC function.

3.2.5 DC Migration

The physical location of DCs at the time of antigen capture and at the time of antigen presentation is important for ensuing T-cell immunity. For example, upon subcutaneous immunization with fluorescent antigen, two waves of DC-mediated antigen presentation occur (Itano *et al.*, 2003). The first one is mediated by skin-derived DCs that acquired the antigen while in the draining lymph nodes, since these were the first cells to display pMHC II complexes of the red fluorescent protein (Itano *et al.*, 2003). DCs migrating from the actual antigen injection site and presenting the captured antigen arrived several hours later in the second wave (Itano *et al.*, 2003). The first wave of DCs triggered T-cell activation and proliferation while the second wave was necessary for development of delayed type hyper-sensitivity (Itano *et al.*, 2003).

Furthermore, the tissue origin of DCs determine the homing of elicited T cells (Mora *et al.*, 2003; Mullins *et al.*, 2003). Thus, whereas DCs from Peyer's patches, peripheral lymph nodes, and spleen induced equivalent activation markers and effector activity in CD8$^+$ T cells, only Peyer's patch DCs induced CD8$^+$ T cells with the ability to home to the small intestine (Mora *et al.*, 2003; Mullins *et al.*, 2003). Similarly, when *ex vivo* generated DCs were injected into the mice bearing melanoma, both intravenous and subcutaneous injection induced specific memory T cells in spleen and permitted control of lung metastasis. However, whereas subcutaneous immunization also induced memory T cells in the lymph nodes allowing subsequent protection against subcutaneously growing tumors, intravenous immunization failed to do so (Mullins *et al.*, 2003). Thus, DCs that have migrated to different tissues can prime T cells with different homing capa-cities. Finally, inflammation will enhance DC migration as demonstrated in mice by conditioning the site of *ex vivo* generated DC injection with TNF-α. This in turn significantly increased DC migration to the draining lymph nodes and the magnitude of the CD4$^+$ T-cell response (MartIn-Fontecha *et al.*, 2003).

DC migration within lymphatics and to specific areas within lymphoid organs represents another important parameter for T-cell immunity. We have already discussed the strategic localization of distinct DCs within specific areas of secondary lymphoid organs as analyzed on tissue sections. Furthermore, it appears that DCs recirculate within lymphoid organs possibly to deliver help to each other as shown by cytotoxic T cells for pDCs delivering help for mDCs to prime herpes simplex virus (HSV) specific cytotoxic T cells (CTLs) (Yoneyama *et al.*, 2005). This important aspect of DC biology can be now appreciated *in vivo* thanks to the development of intravital imaging technologies (Huang *et al.*, 2004). This will allow us to dissect *in vivo* the DC/lymphocyte interaction in steady state (Lindquist *et al.*, 2004a) and upon launching of protective immunity. For example, it appears that both tolerance and immunity are proceeded by stable and lasting several hours interaction of DCs with T cells (Lindquist *et al.*, 2004a).

3.2.6 Dendritic Cells and Tolerance

DCs are now thought to play a pivotal role in the control of both central and peripheral tolerance (Heath and Carbone, 2001; Finkelman *et al.*, 1996; Steinman *et al.*, 2003b; Adler *et al.*, 1998).

3.2.6.1 Central Tolerance

The thymus steadily produces thymocytes expressing newly assembled TCR, some of which may be reactive with components of self. High-affinity autoreactive thymocytes are eliminated upon encountering self-MHC peptide (Sprent and Kishimoto, 2002; Starr *et al.*, 2003; Marrack and Kappler, 1997). There is evidence that both thymic epithelial cells as well as mature DCs in the thymus may be involved in this process (Brocker, 1999; Fujimoto *et al.*, 2002). However, autoreactive T cells that are not deleted in the thymus need to be controlled in the periphery to prevent immune responses to self. An important crossroad of central and peripheral tolerance can be found in the thymic structures called the Hassall's corpuscles (HCs). In these structures, HC resident mDCs stimulated by thymic stromal lymphopoietin (TSLP) drive the positive selection of self-reactive $CD4^+CD25^+$ regulatory T cells (Watanabe *et al.*, 2005). These cells are thought to be critical for the maintenance of self tolerance in the periphery.

3.2.6.2 Peripheral Tolerance Through DCs

There is now evidence that immature/steady-state DCs control peripheral tolerance [reviewed in Steinman *et al.* (2003b)]. In the absence of inflammation, immature DCs capture dying cells and migrate at slow rate through the afferent lymphatics to the draining lymph nodes. These DCs then present tissue antigens to T cells in the absence of appropriate costimulation, leading to T-cell anergy or deletion (Steinman *et al.*, 2003b) or the development of IL-10–producing, regulatory T cells (Jonuleit *et al.*, 2000; Dhodapkar *et al.*, 2001). By suppressing a mandatory T-cell help, DCs may also avoid a self-Ag specific and T-dependent B-cell activation. The molecular mechanisms underlying the tolerogenic properties of peripheral DCs might involve (a) lack of and/or inappropriate costimulation; (b) cell death induction by expression of indoleamine 2,3-dioxygenase (IDO), which induces the catabolism of tryptophan, or by Fas/Fas-L interaction; (c) secretion of IL-10/TGF-β; and (d) inhibitory receptors.

3.2.6.3 Inhibitory Receptors

The effector function of immune cells is regulated by positive and negative signals provided through class I recognition receptors (Pende *et al.*, 2001; Shiroishi *et al.*, 2003; Cantoni *et al.*, 1999; Cerwenka and Lanier, 2001; Pende *et al.*, 2002; Perez-Villar *et al.*, 1997). In humans, these are provided by KIR (killer cell immunoglobulin receptor) and KLR (killer cell lectin-like receptor) family members that interact specifically with certain HLA allotypes and class I related genes (Middleton *et al.*, 2002). A third class of receptors with both activating and inhibitory functions includes immunoglobulin-like molecules (LILR, for leukocyte immunoglobulin-like receptors, also referred to as ILT, for immunoglobulin-like transcript) (Fig. 3.2). ILT family members are expressed on a wide variety of cells of both lymphoid and myeloid origin and can be divided into three groups according to function (Allan *et al.*, 2000): (1) the inhibitory receptors with cytoplasmic ITIM motifs; (2) the activating receptors that associate with the ITAM containing FcR-gamma chain; and (3) a single-member ILT6 that does not contain a transmembrane region and may be secreted.

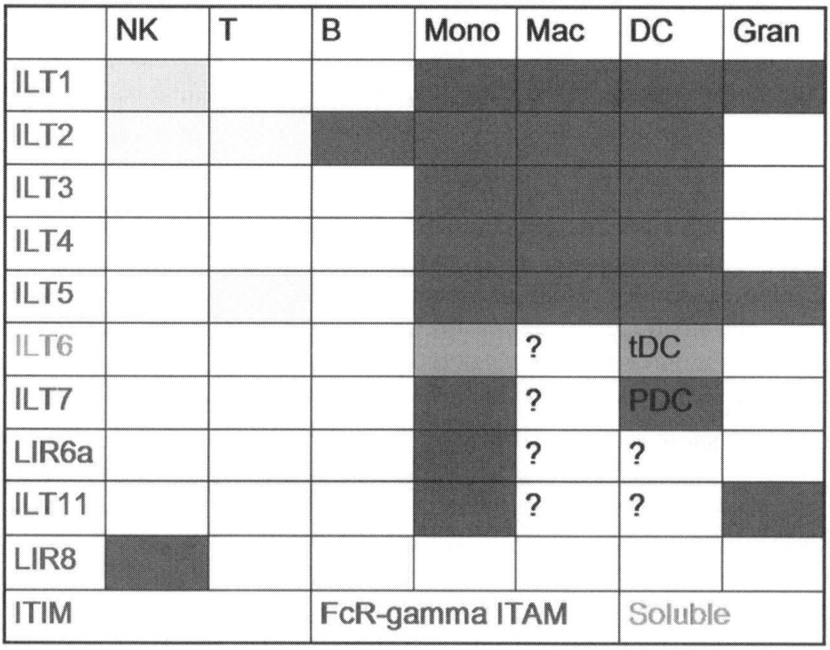

	NK	T	B	Mono	Mac	DC	Gran
ILT1							
ILT2							
ILT3							
ILT4							
ILT5							
ILT6					?	tDC	
ILT7					?	PDC	
LIR6a					?	?	
ILT11					?	?	
LIR8							
ITIM			FcR-gamma ITAM			Soluble	

FIGURE 3.2. Cell type–specific ILT receptor expression. ILT receptors regulate the activation state of a wide variety of cell types. Inhibitory ITIM-containing recaptors are highlighted in blue, activating ITAM-containing receptors are highlighted in red; the only know soluble receptor is highlighted in green. (See Plate 1).

The two most extensively studied members are the inhibitory receptors ILT2 and ILT4. ILT2 is broadly expressed on all monocytes, DCs, most B cells and subsets of T and NK cells (Allan *et al.*, 2000). ILT4 is restricted to myeloid cells and expressed on all monocytes and some DC populations. Both ILT2 and ILT4 interact with multiple class I alleles including HLA-G (Colonna *et al.*, 1999). Signaling through these molecules may be at least in part responsible for the immunosuppressive effects of HLA-G on antigen-presenting cell function. Ligation of ILT4 on DCs by tetrameric HLA-G attenuates maturation in response to CD40L and reduces DC alloproliferative capacity (Liang and Horuzsko, 2003). Expression of inhibitory ILT receptors appears to be a general feature of tolerogenic DCs (Manavalan *et al.*, 2003). DCs treated with IL-10 specifically upregulate inhibitory ILT receptors (Manavalan *et al.*, 2003) while blocking inhibitory ILT receptors leads to a restoration of the alloproliferative capacity in spite of reduced costimulatory molecule expression (Beinhauer *et al.*, 2004).

3.2.7 Dendritic Cell Death

After maturation and lymph node migration, DCs are thought to die in the lymph node. The life span of a DC is influenced by T-cell engagement and pathogen-associated TLR signaling. TLR signaling and T-cell costimulatory molecules both trigger Bcl-XL production and promote DC survival. However, unlike T-cell interaction, TLR engagement increases expression of Bim and triggers cell death by a pathway that is blocked by Bcl-2. The relative activity of these proteins serves as a molecular "timer" that fixes the life span of DCs after pathogen engagement (Hou and Van Parijs, 2004). The life span of a DC is also dependent both on its subtype and tissue location. Studies from Kamath and colleagues demonstrated that in the spleen, CD8$^+$ DC subtype displays the most rapid turnover, with a uniformly short (3 day) life span but demonstrates kinetically distinct short-lived and longer-lived subpopulations in thymus. Epidermal-derived Langerhans cells displayed a longer life span than the dermal-derived DCs. However, once they arrived in lymphoid organs, all DCs display a rapid turnover. This rapid turnover is significantly increased after microbial stimulation (Kamath *et al.*, 2002). Although at first glance the increase death of DCs after reaching the lymph nodes may seem counter to the generation of productive immunity, recent data indicates that this may not be the case. *In vivo* studies using intra vita multiphoton imaging indicate that antigen-loaded DCs migrating from the periphery enter large clusters containing resident DCs (Lindquist *et al.*, 2004b). Death of the recently migrated DC within this cluster has been proposed as a means of antigen distribution within the lymph node, effectively increasing the number of antigen-loaded DCs. Proinflammatory cytokines produced by the mature migrant DCs may further mature lymph node–resident DCs permitting antigen-specific T-cell priming.

3.3 Subversion of Dendritic Cell Function by Pathogens

3.3.1 Pathogens Evade Dendritic Cell Antigen Processing

3.3.1.1 Evasion of DC Uptake

Pathogens can subvert DC function by evading capture. Extracellular pathogens such as *Yersinia pestis* and *Pseudomonas aeruginosa* utilize type III cytotoxins to block phagocytic uptake by interfering with actin cytoskeletal rearrangement (Baldwin and Barbieri, 2005). Soluble HIV Tat protein has been shown to inhibit phagocytic uptake by DCs thereby limiting downstream antigen availability (Poggi *et al.*, 2002).

Internalized antigens will then be processed and loaded onto MHC molecules. A wide range of pathogens have developed means to interfere with antigen processing: intracellular pathogens such as *Toxoplasma gondii*, *Salmonella typhimurium*, and *Mycobacterium tuberculosis* modulate phagosome acidification and trafficking in order to avoid processing (Sibley *et al.*, 1985; Alvarez-Dominguez *et al.*, 1996; Ferrari *et al.*, 1999). HIV escapes endosomal processing by exploiting the differential intracellular trafficking patterns of DC lectins. HIV binds to the C-type lectin DC-SIGN on the surface of the DC. Internalization of HIV through DC-SIGN leads to sequestration of the virus within an endosomal compartment where it is protected from antigen processing. Upon DC T-cell interaction, components of this compartment, including virus, are directed to the immunologic synapse where active infection of the interacting T cell can occur (Geijtenbeek *et al.*, 2000). Naturally occurring variants of DC-SIGN in the human population are associated with both protection against and susceptibility to HIV infection (Liu *et al.*, 2004; Martin *et al.*, 2004). Other viruses use DC-SIGN as an entry receptor for subverting DC function. Human herpes virus-8 (HHV-8) utilizes DC-SIGN to selectively target and enter DCs, inhibiting their CD8[+] T-cell stimulatory capacity (Rappocciolo *et al.*, 2006). Dengue virus appears to utilize DC-SIGN as an attachment receptor, and natural variants are associated with sever disease in humans (Lozach *et al.*, 2005; Sakuntabhai *et al.*, 2005). Targeting of DC-SIGN by pathogens as a means of immune subversion has also been observed in nonviral pathogens. *Mycobacterium tuberculosis* binds to DC-SIGN on DC and inhibits subsequent TLR signaling (Geijtenbeek *et al.*, 2003).

Once captured, antigens are processed in distinct intracellular compartments and loaded onto DC antigen-presenting molecules [reviewed in Trombetta and Mellman (2005)]. Protein antigens are presented by classical MHC class I and class II molecules while lipid antigens are presented through nonclassical CD1 antigen-presenting molecules (Banchereau *et al.*, 2000).

3.3.1.2 Inhibition of Classical Class I Processing and Presentation

CD8$^+$ T-cell immunity represents a critical aspect of antiviral response, and it comes as no surprise that a number of viruses have evolved strategies to interfere with MHC class I processing and presentation (Fig. 3.3). Human cytomegalovirus (HCMV) dedicates a significant portion of its genome to the inhibition of MHC class I expression at multiple stages. Inhibition by HCMV occurs through the coordinated expression of US, or "unique short," sequences. US6 is an integral membrane ER protein that binds to the luminal face of the TAP transporter complex and prevents peptide transport into the ER (Ahn *et al.*, 1997; Hewitt *et al.*, 2001; Halenius *et al.*, 2005). A similar strategy has been utilized by herpes virus through the expression of ICP47, which also inhibits TAP function (Galocha *et al.*, 1997). The ER resident US10 and US3 glycoproteins of HCMV bind to MHC class I and delay its transport to the cell surface (Furman *et al.*, 2002; Jones *et al.*, 1996). Furthermore, the US3 protein binds to tapasin and inhibits class I loading (Park *et al.*, 2004). This approach has been exploited by adenovirus, which expresses a structurally unrelated protein E19 to delay class I ER egress and inhibit peptide loading (Bennett *et al.*, 1999). US2 and US11 are type I ER integral membrane products of HCMV that bind directly to the class I heavy chain and direct its retrograde transport out of the ER for subsequent proteosomal degradation (Story *et al.*, 1999; van der Wal *et al.*, 2002). This is accomplished by interacting with cellular machinery normally used for the ER-associated protein degradation (ERAD) stress response (van der Wal *et al.*, 2002). Other viruses such as HIV have developed strategies to limit class I surface expression by increasing the rate of endocytosis of surface MHC class I. HIV Nef protein binds to the cytoplasmic tail of class I in an allele-specific manner and mediates its internalization (Le Gall *et al.*, 2000; Piguet *et al.*,

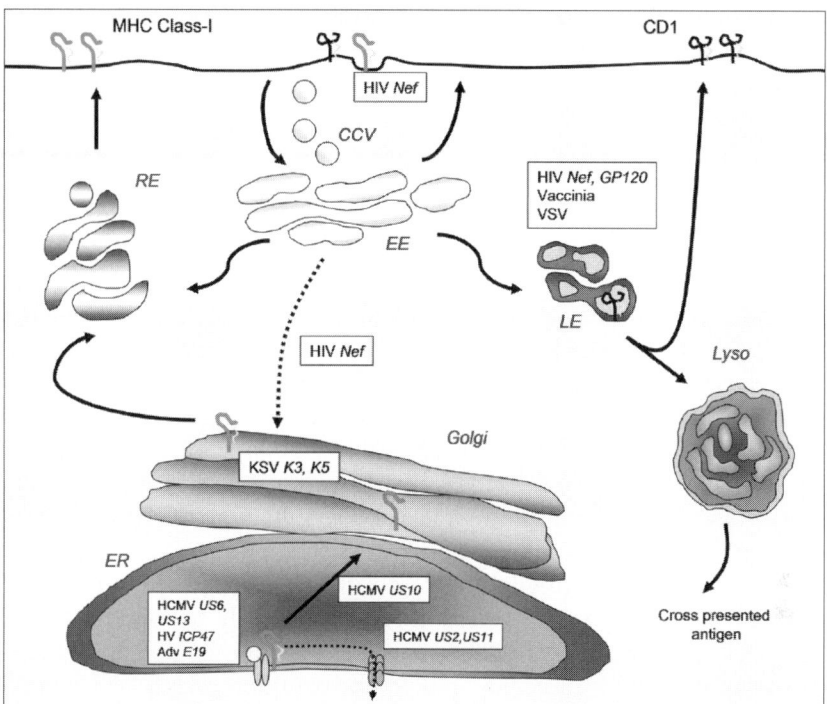

FIGURE 3.3. Pathogen inhibition of class I and CD1 antigen presentation. MHC class I presentation begins with peptide loading in the endoplasmic reticulum (ER) in association with the TAP transporter. Class I peptide complex is transported through the Golgi and endosomal networks to the surface. Surface class I is internalized in clathrin-coated vesicle (CCV) into the early endosomal (EE) network where it can then be either recycled back to the surface through recycling endosomes (RE) or targeted for lysosomal (Lyso) degradation. In contrast, CD1 molecules are internalized from the surface and loaded in acidic late endosomes (LE) and lysosomes. Once loaded, CD1 molecules are trafficked to the surface. Pathogens inhibit both MHC class I and CD1 antigen presentation at multiple stages.

1999). Nef internalizes class I, is directed away from its normal recycling pathway, and is sequestered in the trans-Golgi network (Le Gall *et al.*, 2000). Kaposi's sarcoma–associated herpes virus (KSHV) proteins K3 and K5 downregulate surface expression of MHC class I in a clathrin-dependent manner by ubiquitination of the class I heavy chain during its maturation in post-ER compartments (Ishido *et al.*, 2000; Lorenzo *et al.*, 2002; Furman and Ploegh, 2002). Upon reaching the surface, this ubiqitin modified class I is rapidly internalized and sorted to late endosomes and lysosomes for degradation (Lorenzo *et al.*, 2002; Furman and Ploegh, 2002).

3.3.1.3 Upregulation of Inhibitory Nonclassical Class I Molecules

Viruses appear to have evolved mechanisms to take advantage of HLA-G–mediated immunosuppression. As described above, HCMV utilizes multiple strategies to downregulate MHC class I (Park *et al.*, 2004). The cytoplasmic tail of HLA-G appears to confer resistance to HCMV-mediated downregulation (Barel *et al.*, 2003). In addition, the HCMV-derived cytokine mimetic IL-10 gene selectively upregulates HLA-G (Spencer *et al.*, 2002). Monocytes infected with HCMV strongly upregulate HLA-G even in the context of HLA-A downregulation (Onno *et al.*, 2000; Pizzato *et al.*, 2004). Selective expression of HLA-G also occurs during HIV infection. The HIV Nef protein causes the selective internalization of HLA class I from the surface of infected cells without affecting HLA-G expression (Pizzato *et al.*, 2004). All monocytes and 30% of T cells from HIV-infected individuals express surface HLA-G, and during the course of anti-retroviral therapy, ablation of HLA-G levels correlated with absence of disease progression (Lorenzo *et al.*, 2002).

HLA-E expression, although far less restricted, appears to be similarly modulated by many types of viral infection. HLA-E is resistant to HIV Nef-mediated surface downregulation of class I, and its increased surface expression after viral infection leads to an inhibition of NK cell function (Nattermann *et al.*, 2005). Like HLA-G, HLA-E expression is not subjected to HCMV-mediated inhibition by US2 or US11 (Llano *et al.*, 2003). Inhibition of NK cell activation by viral HLA-E expression may be an important mechanism of disrupting NK cell–mediated DC maturation. Such a mechanism may play an important role in the context of chronic viral infections such as HIV, HCMV, and hepatitis C virus (HCV).

3.3.1.4 Inhibition of Class II Processing and Presentation

The development of CD4$^+$ T-cell immunity is an important part of host defense and long-term antiviral protection. Central to this is the processing and presentation of viral antigen in the context of MHC class II. A number of viruses have targeted this important control point in order to subvert DC function (Fig. 3.4). The HCMV proteins US2 and US3 disrupt class II processing at multiple stages (Johnson and Hegde, 2002; Hegde *et al.*, 2002; Tomazin *et al.*, 1999). Similar to its effect on class I processing, US2 induces the retrograde transport of HLA-DR out of the ER forsubsequent proteosomal degradation (Tomazin *et al.*, 1999). US3 protein binds directly to HLA-DR α and β chains, disrupting their interaction with the class II invariant chain (Hegde *et al.*, 2002). The subsequent DR/US3 complex

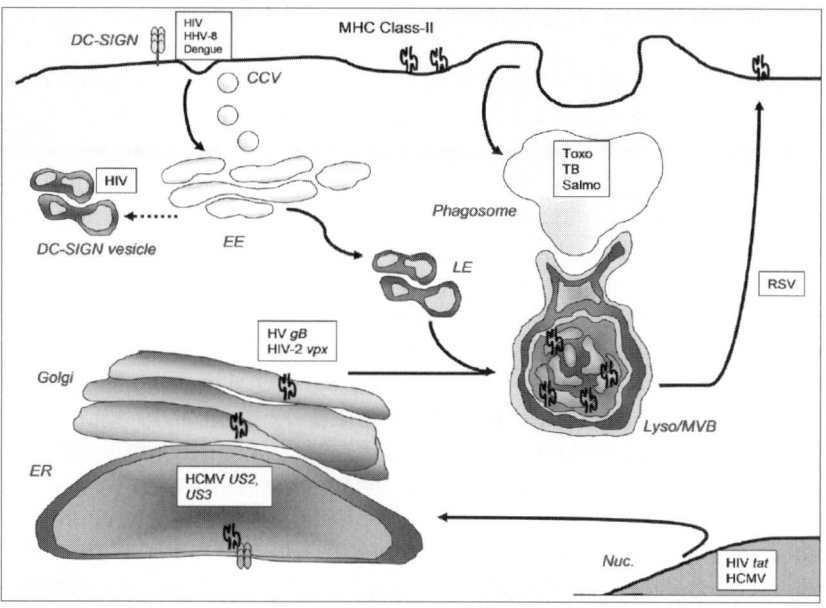

FIGURE 3.4. Pathogen antigen uptake and MHC class II presentation. Immature DCs are specialized for antigen uptake. Antigen is primarily internalized "specifically" via receptor-mediated [both clathrin-dependent (CCV) and -independent mechanisms through early endosomes (EE)] or "nonspecifically" through bulk-phase mechanisms such as pinocytosis or phagocytosis. In both cases, antigen is trafficked through acidified edosomes (LE) and into the lysosome (Lyso) for processing and presentation on MHC class II. In DCs, lysosomal acidification is a tightly regulated event linked to maturation signaling. Acidification leads to antigen processing, loading of presynthesized MHC class II, and rapid surface translocation of the peptide-MHC complex. Pathogens inhibit both antigen uptake and MHC class II presentation at multiple stages.

is then targeted away from sites of class II loading. Both the gB protein of herpes virus and Vpx protein of HIV-2 disrupt class II processing by binding directly to invariant chain leading to its degradation (Sievers *et al.*, 2002; Neumann *et al.*, 2003; Pancio *et al.*, 2000). We have observed inhibition of HLA-DR surface translocation after DC exposure to respiratory syncytial virus (RSV), even in the context of otherwise normal mDC maturation (J.E. Connolly, unpublished observations). In addition to disrupting localization and function of class II, viruses also target the transcriptional activation of class II genes. Inhibition CIITA activation is one of the early events in HCMV viral infection (Miller *et al.*, 1998; Le Roy *et al.*, 1999). HIV tat protein competes with CIITA for binding to cyclin T1 leading to inhibition of class II transcription (Tosi *et al.*, 2000; Mudhasani and Fontes, 2002).

3.3.1.5 Inhibition of CD1 Processing and Presentation

NKT cell activation and cytokine production after CD1 recognition leads to the activation of a broad spectrum of innate and adaptive antiviral effectors, both directly and through polarized DC maturation. HIV employs multiple proteins to subvert CD1 antigen presentation. Monocytes from HIV-infected patients show significantly lower CD1d expression when compared with uninfected individuals (Hage *et al.*, 2005). Similar to its effect on MHC class I, the HIV Nef protein binds directly to CD1d and directs its rapid internalization. Nef expression in DC leads to the lysosomal targeting and degradation of surface CD1a molecules (Shinya *et al.*, 2004). HIV gp120 protein has also been shown to be sufficient for inhibition of CD1d expression, although the mechanism remains unclear (Hage *et al.*, 2005). Other viruses target the CD1 antigen presentation pathway in an effort to avoid immune recognition. Lin and colleagues (Lin *et al.*, 2005) demonstrated that murine splenic DCs show decreased CD1d expression after vaccinia virus or vesicular stomatitis virus (VSV) challenge. Furthermore, these viruses selectively inhibited the expression of CD1d on thymocytes within 1 h of viral exposure, leading to impaired NKT cell recognition (Renukaradhya *et al.*, 2005).

In conclusion, antigen capture, processing, and presentation are essential steps for the initiation of adaptive immune responses by DCs. As such, it is not surprising to find that pathogens have developed strategies to disrupt this sequence of events at multiple levels. As we progress with our understating of these antigen-presentation pathways, we will certainly uncover more points for immune subversion by pathogens. Studying the pathogens and how they interact with DC machinery may in fact be the most efficient means to understand the critical mechanisms behind antigen processing.

3.3.2 Selective Targeting of DC Subsets by Pathogens

Pathogens have developed specific mechanisms to target the molecular pathways permitting the differential responses of pDCs and mDCs to microbes. This is best illustrated by studies on mechanisms regulating type I IFN secretion. Given the importance of type I IFN production in regulating the antiviral response in all cell types, it is not surprising that a variety of viruses have converged to evolve strategies limiting type I IFN production. Kaposi's Sarcoma-associated herpesvirus (KSHV) immediate-early protein ORF45 inhibits the phosphorylation and subsequent nuclear translocation of IRF-7, resulting in an inhibition of type I IFN production after viral infection (Zhu *et al.*, 2002). Similarly, the vaccinia protein E3L inhibits the phosphorylation of IRF-7 and IRF-3 leading to suppressed IFN production (Smith *et al.*, 2001). In addition, the vaccinia protein A46R inhibits multiple TLR adaptor proteins including Myd88 leading to an inhibition of type I IFN response (Stack *et al.*, 2005). The NS1 protein of influenza virus inhibits type I IFN production at least in part by blocking the nuclear translocation of IRF-3 (Donelan *et al.*, 2004). A similar IFN inhibitory role has been attributed to the NS1 protein of RSV (Spann *et al.*, 2004). Knock down of this single viral product in the context of infection leads to productive vaccination (Zhang *et al.*, 2005b). The RIG-I adaptor protein Cardif is the target of NS3-4A, a serine protease from HCV (Meylan *et al.*, 2005). The action of this protein may limit the production of IFN from non-pDCs, thereby making them more susceptible to viral infection.

Myeloid DC subsets are similarly targeted by viruses as a means of immune evasion. Human papillomavirus-16 has been demonstrated to target LCs in the epithelium by binding to the LC-specific lectin langerin. Internalization was rapid and did not lead to activation of the LC (Bousarghin *et al.*, 2005). It has recently been demonstrated that HIV productively infects only a subset of mDCs in the human blood. This mDC subset expresses the HIV receptors CD4, CCR5, and CXCR4 and is distinguished by the expression of BDCA1 (Granelli-Piperno *et al.*, 2006). Infected mDCs appear be of an immature phenotype and are no longer able to drive T-cell allogeneic proliferation (Granelli-Piperno *et al.*, 2006; Patterson *et al.*, 2005). The targeted inhibition of a specific subset of DCs may indicate that these cells are particularly adept at viral antigen presentation and their neutralization would be advantageous to the virus. Indeed, Granelli-Piperno demonstrated that uninfected BDCA-1 positive mDCs are strong stimulators of T-cell proliferation and IL-12 p70 production (Granelli-Piperno *et al.*, 2006).

3.3.3 Pathogens Target DC Precursors

DCs in the blood and periphery are continually replenished from their precursor populations. As described above, these populations include monocytes populations pre-pDCs and CD34[+] hematopoetic precursors. The antigen processing and presentation capabilities of all DC subsets far surpass that of their precursor populations. As described above, the differentiation of a DC is a well coordinated and tightly regulated event. Differentiation of DCs therefore represents a regulatory control point that can be manipulated by pathogens. Epstein-Barr virus (EBV) also inhibits mDC development, this time by promoting apoptosis in monocytes precursor populations under conditions that normally favor DC differen-tiation (Li *et al.*, 2002). Evidence also exists for both measles virus and lymphocoriomeningitis virus variant 13 targeting DC precursors in a manner dependent on type I IFN signaling (Hahm *et al.*, 2005). Similarly, the number of both myeloid and plasmacytoid DC progenitors is severely reduced in patients with chronic HCV infection (Kanto *et al.*, 2004; Szabo and Dolganiuc, 2005). Furthermore, DC precursor populations appear to be the site of latent infection for a diverse variety of viruses. HCMV, dengue, and HIV all specifically target DC precursors (Kwan *et al.*, 2005; Patterson *et al.*, 2005; Reeves *et al.*, 2005). Aside from transporting virus for subsequent dissemination, as has been described for HIV, viral infection may have an impact on subsequent DC differentiation and function.

3.3.4 Inhibition of DC Maturation by Pathogens

3.3.4.1 Inhibition of DC Activation

A variety of pathogens target the maturation control point in DC biology in an effort to subvert antimicrobial immunity (Fig. 3.5). Perhaps the most direct method, and one used by a variety of pathogens, is the inhibition of detection by blocking of the TLR signaling pathways.   HCV NS3/4A protein is a serine protease that cleaves the TLR3 adaptor protein TRIF, rendering the DC incapable of activating maturation pathways in response to viral RNA ligands (Li *et al.*, 2005; Patterson *et al.*, 2005). Vaccinia virus protein A52R blocks NFκB activation by multiple TLRs, including TLR3. A52R binds directly to, and disrupts the function of, interleukin 1 receptor–associated kinase 2 (IRAK2) and tumor necrosis factor receptor–associated factor 6 (TRAF6) proteins important in TLR signaling (Harte *et al.*, 2003). In addition, vaccinia virus also expresses the A46R protein, which inhibits multiple intracellular signaling pathways by a range of TLRs. Similar to all TLR and IL-1 family receptors, A46R contains a TIR domain allowing it to function as a competitive inhibitor of TLR signaling (Stack *et al.*, 2005). The poxvirus protein N1L shares structural similarity with vaccinia A52R and inhibits TLR-mediated NFκB and IRF-3 activation by binding directly to the upstream I-kappa B kinase (IKK) complex (DiPerna *et al.*, 2004). The fact that bacteria have evolved a convergent strategy to inhibit TLR signaling is an indication of their importance in initiating immune response. The lethal factor protein of anthrax has been demonstrated to specifically cleave activators of the MAP kinase pathways downstream of TLR receptors and to block subsequent DC maturation (Agrawal *et al.*, 2003). By inhibiting signaling downstream of TLR pattern recognition receptors, pathogens may block this critical

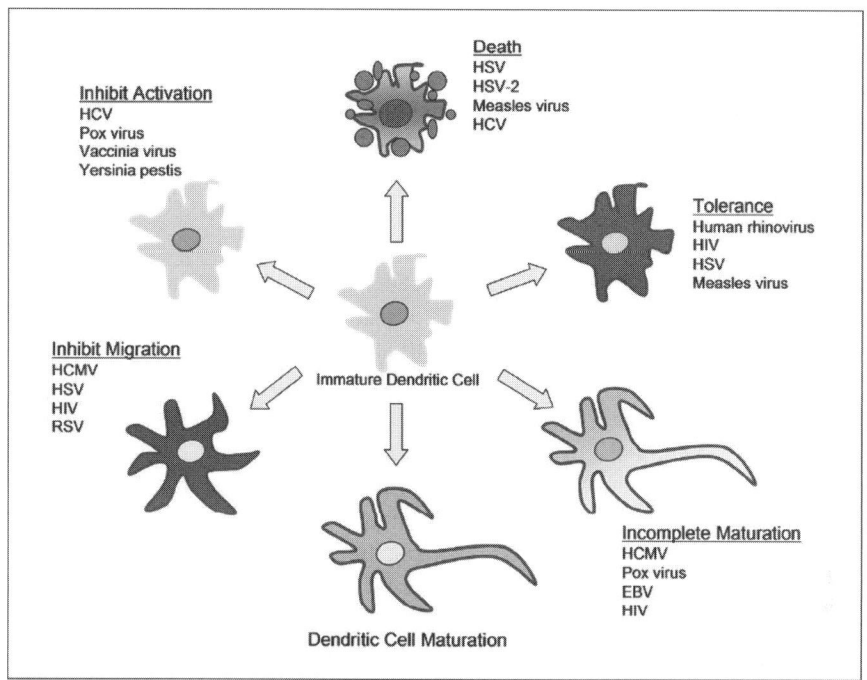

FIGURE 3.5. Pathogens modulate dendritic cell maturation. Activation of DC by a variety of mechanisms initiates a tightly regulated program of maturation. Disruption of this process may lead to profound differences in DC phenotype and downstream immunologic outcome. A wide variety of pathogens have developed strategies that target dendritic cells during the maturation process.

control point in DC development and effectively evade immune detection. Not all viruses target the TLR pathway as a means of inhibition DC maturation. HIV-infected CD4$^+$ T-cells downregulate expression of CD40 ligand and are therefore incapable of providing an important secondary activation signal to autologus DCs (Zhang *et al.*, 2005a). HCV core protein and NS3 specifically inhibit maturation downstream of TNF receptor and the TNF family member CD40, while leaving TLR-mediated maturation pathway intact (Sarobe *et al.*, 2003). This strategy may allow the virus to evade inflammation-induced antigen presentation during the course of chronic infection.

3.3.4.2 Pathogen Modulation of DC Maturation

Many viruses modulate specific aspects of DC maturation and thereby skew downstream immunity. Perturbations in both magnitude and profile of cytokine response to stimulus is a common strategy used by a wide range of viruses. In addition to inhibiting many features of TLR signaling as discussed above, vaccinia virus protein A52R activates p38 mitogen-activated protein kinase driving IL-10 production from the DC after stimulation (Maloney *et al.*, 2005). This results in a DC that gives immunosuppressive cytokine signals to surrounding immune cells after viral activation. HCMV infection of DCs does not induce their maturation and instead efficiently downregulated the expression of surface MHC class I, CD40, and CD80 molecules. Furthermore, the infected DCs are refractory to subsequent TLR stimulation as measured by an inhibition of MHC class I and II, CD83, CD86, CD40, and cytokine upregulation. These HCMV-infected cell are unable to generate virus-specific cytotoxic T cells (Moutaftsi *et al.*, 2002). HCMV virus expresses a homologue of human IL-10, UL111A protein, which may be responsible for many of these inhibitory actions on downstream DC immunity (Chang *et al.*, 2004). Both the EBV and Orf poxvirus encode viral IL-10 homologues that act in a similar manner to perturb antiviral immune function by influencing DCs during this critical maturation event (Takayama *et al.*, 1998; Takayama *et al.*, 1998; Lateef *et al.*, 2003). The HIV Vpr protein inhibits the production of IL-12 and upregulates IL-10 production in DCs, leading to significant reduction in CD8[+] memory T-cell responses (Majumder *et al.*, 2005).

3.3.5 Pathogen-mediated Inhibition of DC Migration

A host of viruses have exploited the critical role of DC migration in the establishment of antiviral immunity. Viruses selectively modulate chemokine and chemokine receptor expression as well as adhesion molecules in an effort to disrupt DC migration. For instance, the viral IL-10 homologue of EBV prevents *in vivo* DC chemotaxis to the lymph node (LN) by modulating chemokine receptor expression. Indeed, DCs treated with EBV vIL-10 show a pronounced upregulation of the peripheral homing CCR1 and CCR5 receptors but downregulate CCR7 lymph node homing chemokine receptor (Takayama *et al.*, 2001). HCMV blocks the migration of infected DCs in response to the lymph node chemotactic chemokines CCL19 and CCL21. Expression of CCR7 is markedly downregulated in infected cells even in response to subsequent stimulation by TLR agonists (Moutaftsi *et al.*, 2004). A similar profound inhibition of CCR7 expression and function can be observed after HSV-1 infection of DCs (Prechtel *et al.*, 2005). HIV infection of DCs leads to the selective expression of the CXCR3 ligands CXCL10 and CXCL11. The recruitment of CXCR3/CCR5 expressing CD4[+] T cells, known to be susceptible to HIV infection, to infected DCs promotes viral dissemination (Foley *et al.*, 2005). Many viral genomes contain molecular mimics of mammalian chemokines, chemokine receptors, and receptor antagonists as has recently been reviewed elsewhere (Rosenkilde, 2005). The effect of many of these viral agents on the migration of DCs and the subsequent establishment of antiviral immunity remains to be determined.

3.3.6 Viral Modulation of DC Tolerance

This recently described DC immunoregulatory function may be a particularly effective method for pathogens to evade immunity, and viruses have developed systems to actively induce tolerogenic DCs. Measles virus infected LCs block naïve allogeneic CD4$^+$ T-cell proliferation in response to uninfected LCs (Steineur *et al.*, 1998). This active inhibitory effect requires cell contact and is associated with both phenotypic and functional DC maturation (Dubois *et al.*, 2001). Stimulation of pDCs through CD40 or CpG has been associated with the induction of both CD4 and CD8$^+$ regulatory T-cell populations (Moseman *et al.*, 2004; Gilliet and Liu, 2002). HSV stimulation of pDCs has recently been shown to lead to the generation of cytotoxic CD4$^+$ regulatory T cells, which may be involved in the transition to a site of persistent viral infection (Kawamura *et al.*, 2006). Human rhinoviruses (HRV) exposure of DCs generates tolerogenic DCs capable of inhibiting T-cell proliferation. Paradoxically, viral-exposed DCs demonstrated high levels of MHC and costimulatory molecules. Induction of T-cell anergy is dependent on viral-induced B7-H1 and sialic acid binding Ig-like lectin 1 (SIGLEC-1) (Kirchberger *et al.*, 2005). DCs isolated from the lymph nodes of HIV-infected patients have recently been reported to demonstrate tolerogenic function DCs. Treatment of patients with highly active antiretroviral therapy correlated with the disappearance of these cells and restoration of T-cell stimulatory capacity (Krathwohl *et al.*, 2006). It will be interesting to examine situations in which repeated viral exposure offers little or no protection for similar mechanisms of immune subversion.

3.3.7 Pathogen-mediated Dendritic Cell Death

Perhaps the most direct route to subverting DC function is to kill DCs before they are able to activate antiviral immunity. Measles virus induces Fas expression on infected DCs. Infected cells are then susceptible to lysis from Fas ligand–expressing T cells (Servet-Delprat *et al.*, 2000). HSV infection of immature DCs leads to their rapid apoptosis in a manner dependent on tumor necrosis factor-related apoptosis-inducing ligand (TRAIL). Infection leads to a downregulation of the important anti-apoptotic protein c-FLIP making DCs susceptible to TRAIL killing (Muller *et al.*, 2004). HSV-2–infected bone marrow–derived DCs show a similar rapid apoptosis (Jones *et al.*, 2003). DCs express the poliovirus receptor and are highly susceptible to infection by multiple strains of poliovirus. Infection of DCs leads to cell death and lysis within 24 h postinfection (Wahid *et al.*, 2005). In contrast with many viruses, HCV induces apoptosis in mature DCs. The HCV core, NS3, NS5A, or NS5B redundantly induce apoptosis in mature DCs after infection (Siavoshian *et al.*, 2005).

3.4 Concluding Remarks

DCs play a unique and central role in directing the immune response to pathogens. As such, pathogens have developed mechanisms to inhibit DC function at all levels. Specifically targeting the molecular mechanisms by which pathogens disrupt DC function is already leading to the development of new vaccines and to improved immunogenicity of existing stocks (Falcon *et al.*, 2005; Geiss *et al.*, 2000; Zhang *et al.*, 2005b).

As the most potent antigen-presenting cell, DCs are the focus of the next generation of rationally designed vaccines (Falcon *et al.*, 2005; Geiss *et al.*, 2000; Lori *et al.*, 2004; Pope, 2003; Steinman *et al.*, 2003a). The complex life of a DC is distinguished by diverse states of differentiation, activation, and maturation. Each of these states varies greatly in phenotype, function, and downstream immunologic outcome. The complexity of the DC system necessitates more research to be performed before we can fully realize DC potential as vaccines. We will gain a deeper understanding of DC biology and the critical mechanisms involved by looking at which proteins pathogens have chosen to disrupt in order to subvert DC function.

References

Adler, A.J., Marsh, D.W., Yochum, G.S., Guzzo, J.L., Nigam, A., Nelson, W.G., and Pardoll, D.M. (1998). CD4+ T cell tolerance to parenchymal self-antigens requires presentation by bone marrow-derived antigen-presenting cells. J. Exp. Med. 187, 1555-1564.

Agrawal, A., Lingappa, J., Leppla, S.H., Agrawal, S., Jabbar, A., Quinn, C., and Pulendran, B. (2003). Impairment of dendritic cells and adaptive immunity by anthrax lethal toxin. Nature 424, 329-334.

Ahn, K., Gruhler, A., Galocha, B., Jones, T.R., Wiertz, E.J., Ploegh, H.L., Peterson, P.A., Yang, Y., and Fruh, K. (1997). The ER-luminal domain of the HCMV glycoprotein US6 inhibits peptide translocation by TAP. Immunity 6, 613-621.

Albert, M.L., Pearce, S.F., Francisco, L.M., Sauter, B., Roy, P., Silverstein, R.L., and Bhardwaj, N. (1998a). Immature dendritic cells phagocytose apoptotic cells via alphavbeta5 and CD36, and cross-present antigens to cytotoxic T lymphocytes. J. Exp. Med. 188, 1359-1368.

Albert, M.L., Sauter, B., and Bhardwaj, N. (1998b). Dendritic cells acquire antigen from apoptotic cells and induce class I-restricted CTLs. Nature 392, 86-89.

Allan, D.S., McMichael, A.J., and Braud, V.M. (2000). The ILT family of leukocyte receptors. Immunobiology 202, 34-41.

Alvarez-Dominguez, C., Barbieri, A.M., Beron, W., Wandinger-Ness, A., and Stahl, P.D. (1996). Phagocytosed live Listeria monocytogenes influences Rab5-regulated *in vitro* phagosome-endosome fusion. J. Biol. Chem. 271, 13834-13843.

Amsen, D., Blander, J.M., Lee, G.R., Tanigaki, K., Honjo, T., and Flavell, R.A. (2004). Instruction of distinct CD4 T helper cell fates by different notch ligands on antigen-presenting cells. Cell 117, 515-526.

Asea, A., Rehli, M., Kabingu, E., Boch, J.A., Bare, O., Auron, P.E., Stevenson, M.A., and Calderwood, S.K. (2002). Novel signal transduction pathway utilized by extracellular HSP70: role of toll-like receptor (TLR) 2 and TLR4. J. Biol. Chem. 277, 15028-15034.

Asselin-Paturel, C., Brizard, G., Chemin, K., Boonstra, A., O'Garra, A., Vicari, A., and Trinchieri, G. (2005). Type I interferon dependence of plasmacytoid dendritic cell activation and migration. J. Exp. Med. 201, 1157-1167.

Baldwin, M.R. and Barbieri, J.T. (2005). The type III cytotoxins of Yersinia and Pseudomonas aeruginosa that modulate the actin cytoskeleton. Curr. Top. Microbiol. Immunol. 291, 147-166.

Banchereau, J., Briere, F., Caux, C., Davoust, J., Lebecque, S., Liu, Y.J., Pulendran, B., and Palucka, K. (2000). Immunobiology of dendritic cells. Annu. Rev. Immunol. 18, 767-811.

Barchet, W., Cella, M., and Colonna, M. (2005). Plasmacytoid dendritic cells-- virus experts of innate immunity. Semin. Immunol. 17, 253-261.

Barel, M.T., Ressing, M., Pizzato, N., van Leeuwen, D., Le Bouteiller, P., Lenfant, F., and Wiertz, E.J. (2003). Human cytomegalovirus-encoded US2 differentially affects surface expression of MHC class I locus products and targets membrane-bound, but not soluble HLA-G1 for degradation. J. Immunol. 171, 6757-6765.

Beckman, E.M., Porcelli, S.A., Morita, C.T., Behar, S.M., Furlong, S.T., and Brenner, M.B. (1994). Recognition of a lipid antigen by CD1-restricted alpha beta[+] T cells. Nature 372, 691-694.

Beinhauer, B.G., McBride, J.M., Graf, P., Pursch, E., Bongers, M., Rogy, M., Korthauer, U., de Vries, J.E., Aversa, G., and Jung, T. (2004). Interleukin 10 regulates cell surface and soluble LIR-2 (CD85d) expression on dendritic cells resulting in T cell hyporesponsiveness *in vitro*. Eur. J. Immunol. 34, 74-80.

Bendelac, A., Rivera, M.N., Park, S.H., and Roark, J.H. (1997). Mouse CD1-specific NK1 T cells: development, specificity, and function. Annu. Rev. Immunol. 15, 535-562.

Bennett, E.M., Bennink, J.R., Yewdell, J.W., and Brodsky, F.M. (1999). Cutting edge: adenovirus E19 has two mechanisms for affecting class I MHC expression. J. Immunol. 162, 5049-5052.

Blom, B., Ho, S., Antonenko, S., and Liu, Y.J. (2000). Generation of interferon alpha-producing predendritic cell (Pre-DC)2 from human CD34($^+$) hematopoietic stem cells. J. Exp. Med. 192, 1785-1796.

Boonstra, A., Asselin-Paturel, C., Gilliet, M., Crain, C., Trinchieri, G., Liu, Y.J., and O'Garra, A. (2003). Flexibility of mouse classical and plasmacytoid-derived dendritic cells in directing T helper type 1 and 2 cell development: dependency on antigen dose and differential toll-like receptor ligation. J. Exp. Med. 197, 101-109.

Bousarghin, L., Hubert, P., Franzen, E., Jacobs, N., Boniver, J., and Delvenne, P. (2005). Human papillomavirus 16 virus-like particles use heparan sulfates to bind dendritic cells and colocalize with langerin in Langerhans cells. J. Gen. Virol. 86, 1297-1305.

Brandes, M., Willimann, K., and Moser, B. (2005). Professional antigen-presentation function by human gammadelta T Cells. Science 309, 264-268.

Brocker, T. (1999). The role of dendritic cells in T cell selection and survival. J. Leukoc. Biol. 66, 331-335.

Cambi, A., Koopman, M., and Figdor, C.G. (2005). How C-type lectins detect pathogens. Cell Microbiol. 7, 481-488.

Cantoni, C., Falco, M., Pessino, A., Moretta, A., Moretta, L., and Biassoni, R. (1999). P49, a putative HLA-G1 specific inhibitory NK receptor belonging to the immunoglobulin Superfamily. J. Reprod. Immunol. 43, 157-165.

Carbone, F.R. and Heath, W.R. (2003). The role of dendritic cell subsets in immunity to viruses. Curr. Opin. Immunol. 15, 416-420.

Caux, C., Dezutter-Dambuyant, C., Schmitt, D., and Banchereau, J. (1992). GM-CSF and TNF-alpha cooperate in the generation of dendritic Langerhans cells. Nature 360, 258-261.

Caux, C., Massacrier, C., Vanbervliet, B., Dubois, B., Durand, I., Cella, M., Lanzavecchia, A., and Banchereau, J. (1997). CD34$^+$ hematopoietic progenitors from human cord blood differentiate along two independent dendritic cell pathways in response to granulocyte-macrophage colony-stimulating factor plus tumor necrosis factor alpha: II. Functional analysis. Blood 90, 1458-1470.

Caux, C., Massacrier, C., Vanbervliet, B., Dubois, B., Van Kooten, C., Durand, I., and Banchereau, J. (1994). Activation of human dendritic cells through CD40 cross-linking. J. Exp. Med. 180, 1263-1272.

Caux, C., Vanbervliet, B., Massacrier, C., Dezutter-Dambuyant, C., Saint-Vis, B., Jacquet, C., Yoneda, K., Imamura, S., Schmitt, D., and Banchereau, J. (1996). CD34$^+$ hematopoietic progenitors from human cord blood differentiate along two independent dendritic cell pathways in response to GM-CSF$^+$TNF alpha. J. Exp. Med. 184, 695-706.

Cella, M., Jarrossay, D., Facchetti, F., Alebardi, O., Nakajima, H., Lanzavecchia, A., and Colonna, M. (1999). Plasmacytoid monocytes migrate to inflamed lymph nodes and produce large amounts of type I interferon. Nat. Med. 5, 919-923.

Cerwenka, A. and Lanier, L.L. (2001). Ligands for natural killer cell receptors: redundancy or specificity. Immunol. Rev. 181, 158-169.

Chang, W.L., Baumgarth, N., Yu, D., and Barry, P.A. (2004). Human cytomegalovirus-encoded interleukin-10 homolog inhibits maturation of dendritic cells and alters their functionality. J. Virol. 78, 8720-8731.

Chen, W., Chan, A.S., Dawson, A.J., Liang, X., Blazar, B.R., and Miller, J.S. (2005). FLT3 ligand administration after hematopoietic cell transplantation increases circulating dendritic cell precursors that can be activated by CpG oligodeoxynucleotides to enhance T-cell and natural killer cell function. Biol. Blood Marrow Transplant. 11, 23-34.

Chicha, L., Jarrossay, D., and Manz, M.G. (2004). Clonal type I interferon-producing and dendritic cell precursors are contained in both human lymphoid and myeloid progenitor populations. J. Exp. Med. 200, 1519-1524.

Colonna, M. (2004). Alerting dendritic cells to pathogens: the importance of Toll-like receptor signaling of stromal cells. Proc. Natl. Acad. Sci. U. S. A. 101, 16083-16084.

Colonna, M., Nakajima, H., Navarro, F., and Lopez-Botet, M. (1999). A novel family of Ig-like receptors for HLA class I molecules that modulate function of lymphoid and myeloid cells. J. Leukoc. Biol. 66, 375-381.

D'Amico, A. and Wu, L. (2003). The early progenitors of mouse dendritic cells and plasmacytoid predendritic cells are within the bone marrow hemopoietic precursors expressing Flt3. J. Exp. Med. 198, 293-303.

de la, S.H., Mariotti, S., Angenieux, C., Gilleron, M., Garcia-Alles, L.F., Malm, D., Berg, T., Paoletti, S., Maitre, B., Mourey, L., Salamero, J., Cazenave, J.P., Hanau, D., Mori, L., Puzo, G., and De Libero, G. (2005). Assistance of microbial glycolipid antigen processing by CD1e. Science 310, 1321-1324.

Delamarre, L., Pack, M., Chang, H., Mellman, I., and Trombetta, E.S. (2005). Differential lysosomal proteolysis in antigen-presenting cells determines antigen fate. Science 307, 1630-1634.

Delneste, Y., Magistrelli, G., Gauchat, J., Haeuw, J., Aubry, J., Nakamura, K., Kawakami-Honda, N., Goetsch, L., Sawamura, T., Bonnefoy, J., and Jeannin, P. (2002). Involvement of LOX-1 in dendritic cell-mediated antigen cross-presentation. Immunity 17, 353-362.

Dhodapkar, M.V., Steinman, R.M., Krasovsky, J., Munz, C., and Bhardwaj, N. (2001). Antigen-specific inhibition of effector T cell function in humans after injection of immature dendritic cells. J. Exp. Med. 193, 233-238.

DiPerna, G., Stack, J., Bowie, A.G., Boyd, A., Kotwal, G., Zhang, Z., Arvikar, S., Latz, E., Fitzgerald, K.A., and Marshall, W.L. (2004). Poxvirus protein N1L targets the I-kappaB kinase complex, inhibits signaling to NF-kappaB by the tumor necrosis factor superfamily of receptors, and inhibits NF-kappaB and IRF3 signaling by toll-like receptors. J. Biol. Chem. 279, 36570-36578.

Donelan, N.R., Dauber, B., Wang, X., Basler, C.F., Wolff, T., and Garcia-Sastre, A. (2004). The N- and C-terminal domains of the NS1 protein of influenza B virus can independently inhibit IRF-3 and beta interferon promoter activation. J. Virol. 78, 11574-11582.

Dubois, B., Lamy, P.J., Chemin, K., Lachaux, A., and Kaiserlian, D. (2001). Measles virus exploits dendritic cells to suppress CD4$^+$ T-cell proliferation via expression of surface viral glycoproteins independently of T-cell trans-infection. Cell Immunol. 214, 173-183.

Engering, A.J., Cella, M., Fluitsma, D., Brockhaus, M., Hoefsmit, E.C., Lanzavecchia, A., and Pieters, J. (1997). The mannose receptor functions as a high capacity and broad specificity antigen receptor in human dendritic cells. Eur. J. Immunol. 27, 2417-2425.

Eriksson, U., Ricci, R., Hunziker, L., Kurrer, M.O., Oudit, G.Y., Watts, T.H., Sonderegger, I., Bachmaier, K., Kopf, M., and Penninger, J.M. (2003). Dendritic cell-induced autoimmune heart failure requires cooperation between adaptive and innate immunity. Nat. Med. 9, 1484-1490.

Falcon, A.M., Fernandez-Sesma, A., Nakaya, Y., Moran, T.M., Ortin, J., and Garcia-Sastre, A. (2005). Attenuation and immunogenicity in mice of temperature-sensitive influenza viruses expressing truncated NS1 proteins. J. Gen. Virol. 86, 2817-2821.

Fanger, N.A., Wardwell, K., Shen, L., Tedder, T.F., and Guyre, P.M. (1996). Type I (CD64) and type II (CD32) Fc gamma receptor-mediated phagocytosis by human blood dendritic cells. J. Immunol. 157, 541-548.

Ferrari, G., Langen, H., Naito, M., and Pieters, J. (1999). A coat protein on phagosomes involved in the intracellular survival of mycobacteria. Cell 97, 435-447.

Figdor, C.G., van Kooyk, Y., and Adema, G.J. (2002). C-type lectin receptors on dendritic cells and Langerhans cells. Nat. Rev. Immunol. 2, 77-84.

Finkelman, F.D., Lees, A., Birnbaum, R., Gause, W.C., and Morris, S.C. (1996). Dendritic cells can present antigen *in vivo* in a tolerogenic or immunogenic fashion. J. Immunol. 157, 1406-1414.

Foley, J.F., Yu, C.R., Solow, R., Yacobucci, M., Peden, K.W., and Farber, J.M. (2005). Roles for CXC chemokine ligands 10 and 11 in recruiting CD4$^+$ T cells to HIV-1-infected monocyte-derived macrophages, dendritic cells, and lymph nodes. J. Immunol. 174, 4892-4900.

Fujimoto, Y., Tu, L., Miller, A.S., Bock, C., Fujimoto, M., Doyle, C., Steeber, D.A., and Tedder, T.F. (2002). CD83 expression influences CD4$^+$ T cell development in the thymus. Cell 108, 755-767.

Furman, M.H. and Ploegh, H.L. (2002). Lessons from viral manipulation of protein disposal pathways. J. Clin. Invest 110, 875-879.

Furman, M.H., Dey, N., Tortorella, D., and Ploegh, H.L. (2002). The human cytomegalovirus US10 gene product delays trafficking of major histocompatibility complex class I molecules. J. Virol. 76, 11753-11756.

Galocha, B., Hill, A., Barnett, B.C., Dolan, A., Raimondi, A., Cook, R.F., Brunner, J., McGeoch, D.J., and Ploegh, H.L. (1997). The active site of ICP47, a herpes simplex virus-encoded inhibitor of the major histo-compatibility complex (MHC)-encoded peptide transporter associated with antigen processing (TAP), maps to the NH2-terminal 35 residues. J. Exp. Med. 185, 1565-1572.

Geijtenbeek, T.B., Kwon, D.S., Torensma, R., van Vliet, S.J., van Duijnhoven, G.C., Middel, J., Cornelissen, I.L., Nottet, H.S., KewalRamani, V.N., Littman, D.R., Figdor, C.G., and van Kooyk, Y. (2000). DC-SIGN, a dendritic cell-specific HIV-1-binding protein that enhances trans-infection of T cells. Cell 100, 587-597.

Geijtenbeek, T.B., van Vliet, S.J., Koppel, E.A., Sanchez-Hernandez, M., Vandenbroucke-Grauls, C.M., Appelmelk, B., and van Kooyk, Y. (2003). Mycobacteria target DC-SIGN to suppress dendritic cell function. J. Exp. Med. 197, 7-17.

Geiss, B.J., Smith, T.J., Leib, D.A., and Morrison, L.A. (2000). Disruption of virion host shutoff activity improves the immunogenicity and protective capacity of a replication-incompetent herpes simplex virus type 1 vaccine strain. J. Virol. 74, 11137-11144.

Gilliet, M. and Liu, Y.J. (2002). Generation of human CD8 T regulatory cells by CD40 ligand-activated plasmacytoid dendritic cells. J. Exp. Med. 195, 695-704.

Granelli-Piperno, A., Shimeliovich, I., Pack, M., Trumpfheller, C., and Steinman, R.M. (2006). HIV-1 selectively infects a subset of nonmaturing BDCA1-positive dendritic cells in human blood. J. Immunol. 176, 991-998.

Hage, C.A., Kohli, L.L., Cho, S., Brutkiewicz, R.R., Twigg, H.L., III, and Knox, K.S. (2005). Human immunodeficiency virus gp120 downregulates CD1d cell surface expression. Immunol. Lett. 98, 131-135.

Hahm, B., Trifilo, M.J., Zuniga, E.I., and Oldstone, M.B. (2005). Viruses evade the immune system through type I interferon-mediated STAT2-dependent, but STAT1-independent, signaling. Immunity 22, 247-257.

Halenius, A., Momburg, F., Reinhard, H., Bauer, D., Lobigs, M., and Hengel, H. (2005). Physical and functional interactions of the cytomegalovirus US6 glycoprotein with the transporter associated with antigen processing (TAP). J. Biol. Chem. 281(9):5383-90.

Hanna, J., Gonen-Gross, T., Fitchett, J., Rowe, T., Daniels, M., Arnon, T.I., Gazit, R., Joseph, A., Schjetne, K.W., Steinle, A., Porgador, A., Mevorach, D., Goldman-Wohl, D., Yagel, S., LaBarre, M.J., Buckner, J.H., and Mandelboim, O. (2004). Novel APC-like properties of human NK cells directly regulate T cell activation. J. Clin. Invest 114, 1612-1623.

Harte, M.T., Haga, I.R., Maloney, G., Gray, P., Reading, P.C., Bartlett, N.W., Smith, G.L., Bowie, A., and O'Neill, L.A. (2003). The poxvirus protein A52R targets Toll-like receptor signaling complexes to suppress host defense. J. Exp. Med. 197, 343-351.

Hartmann, G., Weiner, G.J., and Krieg, A.M. (1999). CpG DNA: a potent signal for growth, activation, and maturation of human dendritic cells. Proc. Natl. Acad. Sci. U. S. A. 96, 9305-9310.

Heath, W.R. and Carbone, F.R. (2001). Cross-presentation, dendritic cells, tolerance and immunity. Annu. Rev. Immunol. 19, 47-64.

Hegde, N.R., Tomazin, R.A., Wisner, T.W., Dunn, C., Boname, J.M., Lewinsohn, D.M., and Johnson, D.C. (2002). Inhibition of HLA-DR assembly, transport, and loading by human cytomegalovirus glycoprotein US3: a novel mechanism for evading major histocompatibility complex class II antigen presentation. J. Virol. 76, 10929-10941.

Heinrichs, H. and Orr, H.T. (1990). HLA non-A, B, C class I genes: their structure and expression. Immunol. Res. 9, 265-274.

Hewitt, E.W., Gupta, S.S., and Lehner, P.J. (2001). The human cytomegalovirus gene product US6 inhibits ATP binding by TAP. EMBO J. 20, 387-396.

Honda, K., Yanai, H., Negishi, H., Asagiri, M., Sato, M., Mizutani, T., Shimada, N., Ohba, Y., Takaoka, A., Yoshida, N., and Taniguchi, T. (2005). IRF-7 is the master regulator of type-I interferon-dependent immune responses. Nature 434, 772-777.

Hou, W.S. and Van Parijs, L. (2004). A Bcl-2-dependent molecular timer regulates the lifespan and immunogenicity of dendritic cells. Nat. Immunol. 5, 583-589.

Huang, A.Y., Qi, H., and Germain, R.N. (2004). Illuminating the landscape of *in vivo* immunity: insights from dynamic in situ imaging of secondary lymphoid tissues. Immunity 21, 331-339.

Hunger, R.E., Sieling, P.A., Ochoa, M.T., Sugaya, M., Burdick, A.E., Rea, T.H., Brennan, P.J., Belisle, J.T., Blauvelt, A., Porcelli, S.A., and Modlin, R.L. (2004). Langerhans cells utilize CD1a and langerin to efficiently present nonpeptide antigens to T cells. J. Clin. Invest. 113, 701-708.

Hunt, J.S., Petroff, M.G., Morales, P., Sedlmayr, P., Geraghty, D.E., and Ober, C. (2000). HLA-G in reproduction: studies on the maternal-fetal interface. Hum. Immunol. 61, 1113-1117.

Inaba, K., Inaba, M., Naito, M., and Steinman, R.M. (1993). Dendritic cell progenitors phagocytose particulates, including bacillus Calmette-Guerin organisms, and sensitize mice to mycobacterial antigens *in vivo*. J. Exp. Med. 178, 479-488.

Inaba, K., Turley, S., Yamaide, F., Iyoda, T., Mahnke, K., Inaba, M., Pack, M., Subklewe, M., Sauter, B., Sheff, D., Albert, M., Bhardwaj, N., Mellman, I., and Steinman, R.M. (1998). Efficient presentation of phagocytosed cellular fragments on the major histocompatibility complex class II products of dendritic cells. J. Exp. Med. 188, 2163-2173.

Inaba, K., Turley, S., Iyoda, T., Yamaide, F., Shimoyama, S., Reis e Sousa, Germain, R.N., Mellman, I., and Steinman, R.M. (2000). The formation of immunogenic major histocompatibility complex class II-peptide ligands in lysosomal compartments of dendritic cells is regulated by inflammatory stimuli. J. Exp. Med. 191, 927-936.

Ishido, S., Wang, C., Lee, B.S., Cohen, G.B., and Jung, J.U. (2000). Downregulation of major histocompatibility complex class I molecules by Kaposi's sarcoma-associated herpesvirus K3 and K5 proteins. J. Virol. 74, 5300-5309.

Itano, A.A., McSorley, S.J., Reinhardt, R.L., Ehst, B.D., Ingulli, E., Rudensky, A.Y., and Jenkins, M.K. (2003). Distinct dendritic cell populations sequentially present antigen to CD4 T cells and stimulate different aspects of cell-mediated immunity. Immunity 19, 47-57.

Jiang, W., Swiggard, W.J., Heufler, C., Peng, M., Mirza, A., Steinman, R.M., and Nussenzweig, M.C. (1995). The receptor DEC-205 expressed by dendritic cells and thymic epithelial cells is involved in antigen processing. Nature 375, 151-155.

Johnson, D.C. and Hegde, N.R. (2002). Inhibition of the MHC class II antigen presentation pathway by human cytomegalovirus. Curr. Top. Microbiol. Immunol. 269, 101-115.

Jones, T.R., Wiertz, E.J., Sun, L., Fish, K.N., Nelson, J.A., and Ploegh, H.L. (1996). Human cytomegalovirus US3 impairs transport and maturation of major histocompatibility complex class I heavy chains. Proc. Natl. Acad. Sci. U. S. A. 93, 11327-11333.

Jones, C.A., Fernandez, M., Herc, K., Bosnjak, L., Miranda-Saksena, M., Boadle, R.A., and Cunningham, A. (2003). Herpes simplex virus type 2 induces rapid cell death and functional impairment of murine dendritic cells *in vitro*. J. Virol. 77, 11139-11149.

Jonuleit, H., Wiedemann, K., Muller, G., Degwert, J., Hoppe, U., Knop, J., and Enk, A.H. (1997). Induction of IL-15 messenger RNA and protein in human blood-derived dendritic cells: a role for IL-15 in attraction of T cells. J. Immunol. 158, 2610-2615.

Jonuleit, H., Schmitt, E., Schuler, G., Knop, J., and Enk, A.H. (2000). Induction of interleukin 10-producing, nonproliferating CD4($^+$) T cells with regulatory properties by repetitive stimulation with allogeneic immature human dendritic cells. J. Exp. Med. 192, 1213-1222.

Kabashima, K., Banks, T.A., Ansel, K.M., Lu, T.T., Ware, C.F., and Cyster, J.G. (2005). Intrinsic lymphotoxin-beta receptor requirement for homeostasis of lymphoid tissue dendritic cells. Immunity 22, 439-450.

Kadowaki, N., Antonenko, S., Lau, J.Y., and Liu, Y.J. (2000). Natural interferon alpha/beta-producing cells link innate and adaptive immunity. J. Exp. Med. 192, 219-226.

Kadowaki, N., Ho, S., Antonenko, S., Malefyt, R.W., Kastelein, R.A., Bazan, F., and Liu, Y.J. (2001). Subsets of human dendritic cell precursors express different toll-like receptors and respond to different microbial antigens. J. Exp. Med. 194, 863-869.

Kamath, A.T., Henri, S., Battye, F., Tough, D.F., and Shortman, K. (2002). Developmental kinetics and lifespan of dendritic cells in mouse lymphoid organs. Blood 100, 1734-1741.

Kanneganti, T.D., Ozoren, N., Body-Malapel, M., Amer, A., Park, J.H., Franchi, L., Whitfield, J., Barchet, W., Colonna, M., Vandenabeele, P., Bertin, J., Coyle, A., Grant, E.P., Akira, S., and Nunez, G. (2006). Bacterial RNA and small antiviral compounds activate caspase-1 through cryopyrin/Nalp3. Nature 440(7081):233-6.

Kanto, T., Inoue, M., Miyatake, H., Sato, A., Sakakibara, M., Yakushijin, T., Oki, C., Itose, I., Hiramatsu, N., Takehara, T., Kasahara, A., and Hayashi, N. (2004). Reduced numbers and impaired ability of myeloid and plasmacytoid dendritic cells to polarize T helper cells in chronic hepatitis C virus infection. J. Infect. Dis. 190, 1919-1926.

Karsunky, H., Merad, M., Cozzio, A., Weissman, I.L., and Manz, M.G. (2003). Flt3 ligand regulates dendritic cell development from Flt3$^+$ lymphoid and myeloid-committed progenitors to Flt3$^+$ dendritic cells *in vivo*. J. Exp. Med. 198, 305-313.

Kato, H., Sato, S., Yoneyama, M., Yamamoto, M., Uematsu, S., Matsui, K., Tsujimura, T., Takeda, K., Fujita, T., Takeuchi, O., and Akira, S. (2005). Cell type-specific involvement of RIG-I in antiviral response. Immunity 23, 19-28.

Kawamura, K., Kadowaki, N., Kitawaki, T., and Uchiyama, T. (2006). Virus-stimulated plasmacytoid dendritic cells induce CD4$^+$ cytotoxic regulatory T cells. Blood 107, 1031-1038.

Kirchberger, S., Majdic, O., Steinberger, P., Bluml, S., Pfistershammer, K., Zlabinger, G., Deszcz, L., Kuechler, E., Knapp, W., and Stockl, J. (2005). Human rhinoviruses inhibit the accessory function of dendritic cells by inducing sialoadhesin and B7-H1 expression. J. Immunol. 175, 1145-1152.

Klechevsky, E., Kato, H., and Sponaas, A.M. (2005). Dendritic cells star in Vancouver. J. Exp. Med. 202, 5-10.

Kovats, S., Main, E.K., Librach, C., Stubblebine, M., Fisher, S.J., and DeMars, R. (1990). A class I antigen, HLA-G, expressed in human trophoblasts. Science 248, 220-223.

Krathwohl, M.D., Schacker, T.W., and Anderson, J.L. (2006). Abnormal presence of semimature dendritic cells that induce regulatory T cells in HIV-infected subjects. J. Infect. Dis. 193, 494-504.

Kwan, W.H., Helt, A.M., Maranon, C., Barbaroux, J.B., Hosmalin, A., Harris, E., Fridman, W.H., and Mueller, C.G. (2005). Dendritic cell precursors are permissive to dengue virus and human immunodeficiency virus infection. J. Virol. 79, 7291-7299.

Lateef, Z., Fleming, S., Halliday, G., Faulkner, L., Mercer, A., and Baird, M. (2003). Orf virus-encoded interleukin-10 inhibits maturation, antigen presentation and migration of murine dendritic cells. J. Gen. Virol. 84, 1101-1109.

Le Bouteiller, P., Legrand-Abravanel, F., and Solier, C. (2003). Soluble HLA-G1 at the materno-foetal interface—a review. Placenta 24 Suppl A, S10-S15.

Le Friec, G., Laupeze, B., Fardel, O., Sebti, Y., Pangault, C., Guilloux, V., Beauplet, A., Fauchet, R., and Amiot, L. (2003). Soluble HLA-G inhibits human dendritic cell-triggered allogeneic T-cell proliferation without altering dendritic differentiation and maturation processes. Hum. Immunol. 64, 752-761.

Le Friec, G., Gros, F., Sebti, Y., Guilloux, V., Pangault, C., Fauchet, R., and Amiot, L. (2004). Capacity of myeloid and plasmacytoid dendritic cells especially at mature stage to express and secrete HLA-G molecules. J. Leukoc. Biol. 76, 1125-1133.

Le Gall, S., Buseyne, F., Trocha, A., Walker, B.D., Heard, J.M., and Schwartz, O. (2000). Distinct trafficking pathways mediate Nef-induced and clathrin-dependent major histocompatibility complex class I down-regulation. J. Virol. 74, 9256-9266.

Le Roy, E., Muhlethaler-Mottet, A., Davrinche, C., Mach, B., and Davignon, J.L. (1999). Escape of human cytomegalovirus from HLA-DR-restricted CD4($^+$) T-cell response is mediated by repression of gamma interferon-induced class II transactivator expression. J. Virol. 73, 6582-6589.

LeibundGut-Landmann, S., Waldburger, J.M., Reis e Sousa, Acha-Orbea, H., and Reith, W. (2004). MHC class II expression is differentially regulated in plasmacytoid and conventional dendritic cells. Nat. Immunol. 5, 899-908.

Lemaoult, J., Krawice-Radanne, I., Dausset, J., and Carosella, E.D. (2004). HLA-G1-expressing antigen-presenting cells induce immunosuppressive CD4$^+$ T cells. Proc. Natl. Acad. Sci. U. S. A. 101, 7064-7069.

Li, L., Liu, D., Hutt-Fletcher, L., Morgan, A., Masucci, M.G., and Levitsky, V. (2002). Epstein-Barr virus inhibits the development of dendritic cells by promoting apoptosis of their monocyte precursors in the presence of granulocyte macrophage-colony-stimulating factor and interleukin-4. Blood 99, 3725-3734.

Li, K., Foy, E., Ferreon, J.C., Nakamura, M., Ferreon, A.C., Ikeda, M., Ray, S.C., Gale, M., Jr., and Lemon, S.M. (2005). Immune evasion by hepatitis C virus NS3/4A protease-mediated cleavage of the Toll-like receptor 3 adaptor protein TRIF. Proc. Natl. Acad. Sci. U. S. A. 102, 2992-2997.

Liang, S. and Horuzsko, A. (2003). Mobilizing dendritic cells for tolerance by engagement of immune inhibitory receptors for HLA-G. Hum. Immunol. 64, 1025-1032.

Lin, Y., Roberts, T.J., Wang, C.R., Cho, S., and Brutkiewicz, R.R. (2005). Long-term loss of canonical NKT cells following an acute virus infection. Eur. J. Immunol. 35, 879-889.

Lindquist, R.L., Shakhar, G., Dudziak, D., Wardemann, H., Eisenreich, T., Dustin, M.L., and Nussenzweig, M.C. (2004a). Visualizing dendritic cell networks *in vivo*. Nat. Immunol. 5, 1243-1250.

Lindquist, R.L., Shakhar, G., Dudziak, D., Wardemann, H., Eisenreich, T., Dustin, M.L., and Nussenzweig, M.C. (2004b). Visualizing dendritic cell networks *in vivo*. Nat. Immunol. 5, 1243-1250.

Liu, H., Hwangbo, Y., Holte, S., Lee, J., Wang, C., Kaupp, N., Zhu, H., Celum, C., Corey, L., McElrath, M.J., and Zhu, T. (2004). Analysis of genetic polymorphisms in CCR5, CCR2, stromal cell-derived factor-1, RANTES, and dendritic cell-specific intercellular adhesion molecule-3-grabbing nonintegrin in seronegative individuals repeatedly exposed to HIV-1. J. Infect. Dis. 190, 1055-1058.

Liu, Y.J. (2005). IPC: professional type 1 interferon-producing cells and plasmacytoid dendritic cell precursors. Annu. Rev. Immunol. 23, 275-306.

Llano, M., Guma, M., Ortega, M., Angulo, A., and Lopez-Botet, M. (2003). Differential effects of US2, US6 and US11 human cytomegalovirus proteins on HLA class Ia and HLA-E expression: impact on target susceptibility to NK cell subsets. Eur. J. Immunol. 33, 2744-2754.

Lorenzo, M.E., Jung, J.U., and Ploegh, H.L. (2002). Kaposi's sarcoma-associated herpesvirus K3 utilizes the ubiquitin-proteasome system in routing class major histocompatibility complexes to late endocytic compartments. J. Virol. 76, 5522-5531.

Lori, F., Kelly, L.M., and Lisziewicz, J. (2004). APC-targeted immunization for the treatment of HIV-1. Expert. Rev. Vaccines 3, S189-S198.

Lozach, P.Y., Burleigh, L., Staropoli, I., Navarro-Sanchez, E., Harriague, J., Virelizier, J.L., Rey, F.A., Despres, P., Arenzana-Seisdedos, F., and Amara, A.

(2005). Dendritic cell-specific intercellular adhesion molecule 3-grabbing non-integrin (DC-SIGN)-mediated enhancement of dengue virus infection is independent of DC-SIGN internalization signals. J. Biol. Chem. 280, 23698-23708.

Mailliard, R.B., Wankowicz-Kalinska, A., Cai, Q., Wesa, A., Hilkens, C.M., Kapsenberg, M.L., Kirkwood, J.M., Storkus, W.J., and Kalinski, P. (2004). alpha-type-1 polarized dendritic cells: a novel immunization tool with optimized CTL-inducing activity. Cancer Res. 64, 5934-5937.

Majumder, B., Janket, M.L., Schafer, E.A., Schaubert, K., Huang, X.L., Kan-Mitchell, J., Rinaldo, C.R., Jr., and Ayyavoo, V. (2005). Human immunodeficiency virus type 1 Vpr impairs dendritic cell maturation and T-cell activation: implications for viral immune escape. J. Virol. 79, 7990-8003.

Maloney, G., Schroder, M., and Bowie, A.G. (2005). Vaccinia virus protein A52R activates p38 mitogen-activated protein kinase and potentiates lipo-polysaccharide-induced interleukin-10. J. Biol. Chem. 280, 30838-30844.

Manavalan, J.S., Rossi, P.C., Vlad, G., Piazza, F., Yarilina, A., Cortesini, R., Mancini, D., and Suciu-Foca, N. (2003). High expression of ILT3 and ILT4 is a general feature of tolerogenic dendritic cells. Transpl. Immunol. 11, 245-258.

Maraskovsky, E., Brasel, K., Teepe, M., Roux, E.R., Lyman, S.D., Shortman, K., and McKenna, H.J. (1996). Dramatic increase in the numbers of functionally mature dendritic cells in Flt3 ligand-treated mice: multiple dendritic cell subpopulations identified. J. Exp. Med. 184, 1953-1962.

Maraskovsky, E., Daro, E., Roux, E., Teepe, M., Maliszewski, C.R., Hoek, J., Caron, D., Lebsack, M.E., and McKenna, H.J. (2000). *In vivo* generation of human dendritic cell subsets by Flt3 ligand. Blood 96, 878-884.

Markowicz, S. and Engleman, E.G. (1990). Granulocyte-macrophage colony-stimulating factor promotes differentiation and survival of human peripheral blood dendritic cells *in vitro*. J. Clin. Invest. 85, 955-961.

Marrack, P. and Kappler, J. (1997). Positive selection of thymocytes bearing alpha beta T cell receptors. Curr. Opin. Immunol. 9, 250-255.

Martin, M.P., Lederman, M.M., Hutcheson, H.B., Goedert, J.J., Nelson, G.W., van Kooyk, Y., Detels, R., Buchbinder, S., Hoots, K., Vlahov, D., O'Brien, S.J., and Carrington, M. (2004). Association of DC-SIGN promoter polymorphism with increased risk for parenteral, but not mucosal, acquisition of human immunodeficiency virus type 1 infection. J. Virol. 78, 14053-14056.

MartIn-Fontecha, A., Sebastiani, S., Hopken, U.E., Uguccioni, M., Lipp, M., Lanzavecchia, A., and Sallusto, F. (2003). Regulation of dendritic cell migration to the draining lymph node: impact on T lymphocyte traffic and priming. J. Exp. Med. 198, 615-621.

Mattner, J., Debord, K.L., Ismail, N., Goff, R.D., Cantu, C., III, Zhou, D., Saint-Mezard, P., Wang, V., Gao, Y., Yin, N., Hoebe, K., Schneewind, O., Walker, D., Beutler, B., Teyton, L., Savage, P.B., and Bendelac, A. (2005). Exogenous and endogenous glycolipid antigens activate NKT cells during microbial infections. Nature 434, 525-529.

McKenna, H.J., Stocking, K.L., Miller, R.E., Brasel, K., De Smedt, T., Maraskovsky, E., Maliszewski, C.R., Lynch, D.H., Smith, J., Pulendran, B., Roux, E.R., Teepe, M., Lyman, S.D., and Peschon, J.J. (2000). Mice lacking flt3 ligand have deficient hematopoiesis affecting hematopoietic progenitor cells, dendritic cells, and natural killer cells. Blood 95, 3489-3497.

Meylan, E., Curran, J., Hofmann, K., Moradpour, D., Binder, M., Bartenschlager, R., and Tschopp, J. (2005). Cardif is an adaptor protein in the RIG-I antiviral pathway and is targeted by hepatitis C virus. Nature 437, 1167-1172.

Middleton, D., Curran, M., and Maxwell, L. (2002). Natural killer cells and their receptors. Transpl. Immunol. 10, 147-164.

Miller, D.M., Rahill, B.M., Boss, J.M., Lairmore, M.D., Durbin, J.E., Waldman, J.W., and Sedmak, D.D. (1998). Human cytomegalovirus inhibits major histocompatibility complex class II expression by disruption of the Jak/Stat pathway. J. Exp. Med. 187, 675-683.

Mommaas, A.M., Mulder, A.A., Jordens, R., Out, C., Tan, M.C., Cresswell, P., Kluin, P.M., and Koning, F. (1999). Human epidermal Langerhans cells lack functional mannose receptors and a fully developed endosomal/lysosomal compartment for loading of HLA class II molecules. Eur. J. Immunol. 29, 571-580.

Mora, J.R., Bono, M.R., Manjunath, N., Weninger, W., Cavanagh, L.L., Rosemblatt, M., and Von Andrian, U.H. (2003). Selective imprinting of gut-homing T cells by Peyer's patch dendritic cells. Nature 424, 88-93.

Moseman, E.A., Liang, X., Dawson, A.J., Panoskaltsis-Mortari, A., Krieg, A.M., Liu, Y.J., Blazar, B.R., and Chen, W. (2004). Human plasmacytoid dendritic cells activated by CpG oligodeoxynucleotides induce the generation of CD4$^+$CD25$^+$ regulatory T cells. J. Immunol. 173, 4433-4442.

Moutaftsi, M., Mehl, A.M., Borysiewicz, L.K., and Tabi, Z. (2002). Human cytomegalovirus inhibits maturation and impairs function of monocyte-derived dendritic cells. Blood 99, 2913-2921.

Moutaftsi, M., Brennan, P., Spector, S.A., and Tabi, Z. (2004). Impaired lymphoid chemokine-mediated migration due to a block on the chemokine receptor switch in human cytomegalovirus-infected dendritic cells. J. Virol. 78, 3046-3054.

Mudhasani, R. and Fontes, J.D. (2002). The class II transactivator requires brahma-related gene 1 to activate transcription of major histocompatibility complex class II genes. Mol. Cell Biol. 22, 5019-5026.

Muhlethaler-Mottet, A., Otten, L.A., Steimle, V., and Mach, B. (1997). Expression of MHC class II molecules in different cellular and functional compartments is controlled by differential usage of multiple promoters of the transactivator CIITA. EMBO J. 16, 2851-2860.

Muller, D.B., Raftery, M.J., Kather, A., Giese, T., and SchOnrich, G. (2004). Frontline: Induction of apoptosis and modulation of c-FLIPL and p53 in immature dendritic cells infected with herpes simplex virus. Eur. J. Immunol. 34, 941-951.

Mullins, D.W., Sheasley, S.L., Ream, R.M., Bullock, T.N., Fu, Y.X., and Engelhard, V.H. (2003). Route of immunization with peptide-pulsed dendritic cells controls the distribution of memory and effector T cells in lymphoid tissues and determines the pattern of regional tumor control. J. Exp. Med. 198, 1023-1034.

Munz, C., Steinman, R.M., and Fujii, S. (2005). Dendritic cell maturation by innate lymphocytes: coordinated stimulation of innate and adaptive immunity. J. Exp. Med. 202, 203-207.

Napolitani, G., Rinaldi, A., Bertoni, F., Sallusto, F., and Lanzavecchia, A. (2005). Selected Toll-like receptor agonist combinations synergistically trigger a T helper type 1-polarizing program in dendritic cells. Nat. Immunol. 6, 769-776.

Nattermann, J., Nischalke, H.D., Hofmeister, V., Kupfer, B., Ahlenstiel, G., Feldmann, G., Rockstroh, J., Weiss, E.H., Sauerbruch, T., and Spengler, U. (2005). HIV-1 infection leads to increased HLA-E expression resulting in impaired function of natural killer cells. Antivir. Ther. 10, 95-107.

Neumann, J., Eis-Hubinger, A.M., and Koch, N. (2003). Herpes simplex virus type 1 targets the MHC class II processing pathway for immune evasion. J. Immunol. 171, 3075-3083.

Norbury, C.C., Basta, S., Donohue, K.B., Tscharke, D.C., Princiotta, M.F., Berglund, P., Gibbs, J., Bennink, J.R., and Yewdell, J.W. (2004). CD8$^+$ T cell cross-priming via transfer of proteasome substrates. Science 304, 1318-1321.

Onno, M., Le Friec, G., Pangault, C., Amiot, L., Guilloux, V., Drenou, B., Caulet-Maugendre, S., Andre, P., and Fauchet, R. (2000). Modulation of HLA-G antigens expression in myelomonocytic cells. Hum. Immunol. 61, 1086-1094.

Palucka, K. and Banchereau, J. (2002). How dendritic cells and microbes interact to elicit or subvert protective immune responses. Curr. Opin. Immunol. 14, 420-431.

Pancio, H.A., Vander, H.N., Kosuri, K., Cresswell, P., and Ratner, L. (2000). Interaction of human immunodeficiency virus type 2 Vpx and invariant chain. J. Virol. 74, 6168-6172.

Pangault, C., Le Friec, G., Caulet-Maugendre, S., Lena, H., Amiot, L., Guilloux, V., Onno, M., and Fauchet, R. (2002). Lung macrophages and dendritic cells express HLA-G molecules in pulmonary diseases. Hum. Immunol. 63, 83-90.

Park, B., Kim, Y., Shin, J., Lee, S., Cho, K., Fruh, K., Lee, S., and Ahn, K. (2004). Human cytomegalovirus inhibits tapasin-dependent peptide loading and optimization of the MHC class I peptide cargo for immune evasion. Immunity 20, 71-85.

Pasare, C. and Medzhitov, R. (2004). Toll-dependent control mechanisms of CD4 T cell activation. Immunity 21, 733-741.

Patterson, S., Donaghy, H., Amjadi, P., Gazzard, B., Gotch, F., and Kelleher, P. (2005). Human BDCA-1-positive blood dendritic cells differentiate into phenotypically distinct immature and mature populations in the absence of exogenous maturational stimuli: differentiation failure in HIV infection. J. Immunol. 174, 8200-8209.

Pende, D., Cantoni, C., Rivera, P., Vitale, M., Castriconi, R., Marcenaro, S., Nanni, M., Biassoni, R., Bottino, C., Moretta, A., and Moretta, L. (2001). Role of NKG2D in tumor cell lysis mediated by human NK cells: cooperation with natural cytotoxicity receptors and capability of recognizing tumors of nonepithelial origin. Eur. J. Immunol. 31, 1076-1086.

Pende, D., Rivera, P., Marcenaro, S., Chang, C.C., Biassoni, R., Conte, R., Kubin, M., Cosman, D., Ferrone, S., Moretta, L., and Moretta, A. (2002). Major histocompatibility complex class I-related chain A and UL16-binding protein expression on tumor cell lines of different histotypes: analysis of tumor susceptibility to NKG2D-dependent natural killer cell cytotoxicity. Cancer Res. 62, 6178-6186.

Perez-Villar, J.J., Melero, I., Navarro, F., Carretero, M., Bellon, T., Llano, M., Colonna, M., Geraghty, D.E., and Lopez-Botet, M. (1997). The CD94/NKG2-A inhibitory receptor complex is involved in natural killer cell-mediated recognition of cells expressing HLA-G1. J. Immunol. 158, 5736-5743.

Piguet, V., Schwartz, O., Le Gall, S., and Trono, D. (1999). The downregulation of CD4 and MHC-I by primate lentiviruses: a paradigm for the modulation of cell surface receptors. Immunol. Rev. 168, 51-63.

Pizzato, N., Derrien, M., and Lenfant, F. (2004). The short cytoplasmic tail of HLA-G determines its resistance to HIV-1 Nef-mediated cell surface downregulation. Hum. Immunol. 65, 1389-1396.

Poggi, A., Carosio, R., Rubartelli, A., and Zocchi, M.R. (2002). Beta(3)-mediated engulfment of apoptotic tumor cells by dendritic cells is dependent on CAMKII: inhibition by HIV-1 Tat. J. Leukoc. Biol. 71, 531-537.

Pope, M. (2003). Dendritic cells as a conduit to improve HIV vaccines. Curr. Mol. Med. 3, 229-242.

Prechtel, A.T., Turza, N.M., Kobelt, D.J., Eisemann, J.I., Coffin, R.S., McGrath, Y., Hacker, C., Ju, X., Zenke, M., and Steinkasserer, A. (2005). Infection of mature dendritic cells with herpes simplex virus type 1 dramatically reduces lymphoid chemokine-mediated migration. J. Gen. Virol. 86, 1645-1657.

Pulendran, B., Lingappa, J., Kennedy, M.K., Smith, J., Teepe, M., Rudensky, A., Maliszewski, C.R., and Maraskovsky, E. (1997). Developmental pathways of dendritic cells in vivo: distinct function, phenotype, and localization of dendritic cell subsets in FLT3 ligand-treated mice. J. Immunol. 159, 2222-2231.

Pulendran, B., Smith, J.L., Caspary, G., Brasel, K., Pettit, D., Maraskovsky, E., and Maliszewski, C.R. (1999). Distinct dendritic cell subsets differentially regulate the class of immune response in vivo. Proc. Natl. Acad. Sci. U. S. A. 96, 1036-1041.

Pulendran, B., Banchereau, J., Burkeholder, S., Kraus, E., Guinet, E., Chalouni, C., Caron, D., Maliszewski, C., Davoust, J., Fay, J., and Palucka, K. (2000). Flt3-ligand and granulocyte colony-stimulating factor mobilize distinct human dendritic cell subsets in vivo. J. Immunol. 165, 566-572.

Rappocciolo, G., Jenkins, F.J., Hensler, H.R., Piazza, P., Jais, M., Borowski, L., Watkins, S.C., and Rinaldo, C.R., Jr. (2006). DC-SIGN Is a Receptor for Human Herpesvirus 8 on Dendritic Cells and Macrophages. J. Immunol. 176, 1741-1749.

Reeves, M.B., MacAry, P.A., Lehner, P.J., Sissons, J.G., and Sinclair, J.H. (2005). Latency, chromatin remodeling, and reactivation of human cytomegalovirus in the dendritic cells of healthy carriers. Proc. Natl. Acad. Sci. U. S. A. 102, 4140-4145.

Regnault, A., Lankar, D., Lacabanne, V., Rodriguez, A., Thery, C., Rescigno, M., Saito, T., Verbeek, S., Bonnerot, C., Ricciardi-Castagnoli, P., and Amigorena, S. (1999). Fcgamma receptor-mediated induction of dendritic cell maturation and major histocompatibility complex class I-restricted antigen presentation after immune complex internalization. J. Exp. Med. 189, 371-380.

Reis e Sousa (2001). Dendritic cells as sensors of infection. Immunity 14, 495-498.

Reis e Sousa, Stahl, P.D., and Austyn, J.M. (1993). Phagocytosis of antigens by Langerhans cells *in vitro*. J. Exp. Med. 178, 509-519.

Renukaradhya, G.J., Webb, T.J., Khan, M.A., Lin, Y.L., Du, W., Gervay-Hague, J., and Brutkiewicz, R.R. (2005). Virus-induced inhibition of CD1d1-mediated antigen presentation: reciprocal regulation by p38 and ERK. J. Immunol. 175, 4301-4308.

Rogers, N.C., Slack, E.C., Edwards, A.D., Nolte, M.A., Schulz, O., Schweighoffer, E., Williams, D.L., Gordon, S., Tybulewicz, V.L., Brown, G.D., and Reis e Sousa (2005). Syk-dependent cytokine induction by Dectin-1 reveals a novel pattern recognition pathway for C type lectins. Immunity 22, 507-517.

Rosenkilde, M.M. (2005). Virus-encoded chemokine receptors—putative novel antiviral drug targets. Neuropharmacology 48, 1-13.

Rubartelli, A., Poggi, A., and Zocchi, M.R. (1997). The selective engulfment of apoptotic bodies by dendritic cells is mediated by the alpha(v)beta3 integrin and requires intracellular and extracellular calcium. Eur. J. Immunol. 27, 1893-1900.

Saint-Vis, B., Vincent, J., Vandenabeele, S., Vanbervliet, B., Pin, J.J., Ait-Yahia, S., Patel, S., Mattei, M.G., Banchereau, J., Zurawski, S., Davoust, J., Caux, C., and Lebecque, S. (1998). A novel lysosome-associated membrane glycoprotein, DC-LAMP, induced upon DC maturation, is transiently expressed in MHC class II compartment. Immunity 9, 325-336.

Sakuntabhai, A., Turbpaiboon, C., Casademont , I., Chuansumrit, A., Lowhnoo, T., Kajaste-Rudnitski, A., Kalayanarooj, S.M., Tangnararatchakit, K., Tangthawornchaikul, N., Vasanawathana, S., Chaiyaratana, W., Yenchitsomanus, P.T., Suriyaphol, P., Avirutnan, P., Chokephaibulkit, K., Matsuda, F., Yoksan, S., Jacob, Y., Lathrop, G.M., Malasit, P., Despres, P., and Julier, C. (2005). A variant in the CD209 promoter is associated with severity of dengue disease. Nat. Genet. 37, 507-513.

Sallusto, F., Cella, M., Danieli, C., and Lanzavecchia, A. (1995). Dendritic cells use macropinocytosis and the mannose receptor to concentrate macromolecules in the major histocompatibility complex class II compartment: downregulation by cytokines and bacterial products. J. Exp. Med. 182, 389-400.

Santiago-Schwarz, F., Belilos, E., Diamond, B., and Carsons, S.E. (1992). TNF in combination with GM-CSF enhances the differentiation of neonatal cord blood stem cells into dendritic cells and macrophages. J. Leukoc. Biol. 52, 274-281.

Sarobe, P., Lasarte, J.J., Zabaleta, A., Arribillaga, L., Arina, A., Melero, I., Borras-Cuesta, F., and Prieto, J. (2003). Hepatitis C virus structural proteins impair dendritic cell maturation and inhibit *in vivo* induction of cellular immune responses. J. Virol. 77, 10862-10871.

Seifert, U., Maranon, C., Shmueli, A., Desoutter, J.F., Wesoloski, L., Janek, K., Henklein, P., Diescher, S., Andrieu, M., de la, S.H., Weinschenk, T., Schild, H., Laderach, D., Galy, A., Haas, G., Kloetzel, P.M., Reiss, Y., and Hosmalin, A. (2003). An essential role for tripeptidyl peptidase in the generation of an MHC class I epitope. Nat. Immunol. 4, 375-379.

Servet-Delprat, C., Vidalain, P.O., Azocar, O., Le Deist, F., Fischer, A., and Rabourdin-Combe, C. (2000). Consequences of Fas-mediated human dendritic cell apoptosis induced by measles virus. J. Virol. 74, 4387-4393.

Shinya, E., Owaki, A., Shimizu, M., Takeuchi, J., Kawashima, T., Hidaka, C., Satomi, M., Watari, E., Sugita, M., and Takahashi, H. (2004). Endogenously expressed HIV-1 nef down-regulates antigen-presenting molecules, not only class I MHC but also CD1a, in immature dendritic cells. Virology 326, 79-89.

Shiroishi, M., Tsumoto, K., Amano, K., Shirakihara, Y., Colonna, M., Braud, V.M., Allan, D.S., Makadzange, A., Rowland-Jones, S., Willcox, B., Jones, E.Y., van der Merwe, P.A., Kumagai, I., and Maenaka, K. (2003). Human inhibitory receptors Ig-like transcript 2 (ILT2) and ILT4 compete with CD8 for MHC class I binding and bind preferentially to HLA-G. Proc. Natl. Acad. Sci. U.S.A. 100, 8856-8861.

Shortman, K. and Liu, Y.J. (2002). Mouse and human dendritic cell subtypes. Nat. Rev. Immunol. 2, 151-161.

Siavoshian, S., Abraham, J.D., Thumann, C., Kieny, M.P., and Schuster, C. (2005). Hepatitis C virus core, NS3, NS5A, NS5B proteins induce apoptosis in mature dendritic cells. J. Med. Virol. 75, 402-411.

Sibley, L.D., Weidner, E., and Krahenbuhl, J.L. (1985). Phagosome acidification blocked by intracellular Toxoplasma gondii. Nature 315, 416-419.

Siegal, F.P., Kadowaki, N., Shodell, M., Fitzgerald-Bocarsly, P.A., Shah, K., Ho, S., Antonenko, S., and Liu, Y.J. (1999). The nature of the principal type 1 interferon-producing cells in human blood. Science 284, 1835-1837.

Sievers, E., Neumann, J., Raftery, M., SchOnrich, G., Eis-Hubinger, A.M., and Koch, N. (2002). Glycoprotein B from strain 17 of herpes simplex virus type I contains an invariant chain homologous sequence that binds to MHC class II molecules. Immunology 107, 129-135.

Sigal, L.J., Crotty, S., Andino, R., and Rock, K.L. (1999). Cytotoxic T-cell immunity to virus-infected non-haematopoietic cells requires presentation of exogenous antigen. Nature 398, 77-80.

Smith, E.J., Marie, I., Prakash, A., Garcia-Sastre, A., and Levy, D.E. (2001). IRF3 and IRF7 phosphorylation in virus-infected cells does not require double-stranded RNA-dependent protein kinase R or Ikappa B kinase but is blocked by Vaccinia virus E3L protein. J. Biol. Chem. 276, 8951-8957.

Spann, K.M., Tran, K.C., Chi, B., Rabin, R.L., and Collins, P.L. (2004). Suppression of the induction of alpha, beta, and lambda interferons by the NS1 and NS2 proteins of human respiratory syncytial virus in human epithelial cells and macrophages [corrected]. J. Virol. 78, 4363-4369.

Spencer, J.V., Lockridge, K.M., Barry, P.A., Lin, G., Tsang, M., Penfold, M.E., and Schall, T.J. (2002). Potent immunosuppressive activities of cytomegalovirus-encoded interleukin-10. J. Virol. 76, 1285-1292.

Spits, H., Couwenberg, F., Bakker, A.Q., Weijer, K., and Uittenbogaart, C.H. (2000). Id2 and Id3 inhibit development of CD34($^+$) stem cells into predendritic cell (pre-DC)2 but not into pre-DC1. Evidence for a lymphoid origin of pre-DC2. J. Exp. Med. 192, 1775-1784.

Sprent, J. and Kishimoto, H. (2002). The thymus and negative selection. Immunol. Rev. 185, 126-135.

Stack, J., Haga, I.R., Schroder, M., Bartlett, N.W., Maloney, G., Reading, P.C., Fitzgerald, K.A., Smith, G.L., and Bowie, A.G. (2005). Vaccinia virus protein A46R targets multiple Toll-like-interleukin-1 receptor adaptors and contributes to virulence. J. Exp. Med. 201, 1007-1018.

Starr, T.K., Daniels, M.A., Lucido, M.M., Jameson, S.C., and Hogquist, K.A. (2003). Thymocyte sensitivity and supramolecular activation cluster formation are developmentally regulated: a partial role for sialylation. J. Immunol. 171, 4512-4520.

Steineur, M.P., Grosjean, I., Bella, C., and Kaiserlian, D. (1998). Langerhans cells are susceptible to measles virus infection and actively suppress T cell proliferation. Eur. J. Dermatol. 8, 413-420.

Steinman, R.M., Granelli-Piperno, A., Pope, M., Trumpfheller, C., Ignatius, R., Arrode, G., Racz, P., and Tenner-Racz, K. (2003a). The interaction of immunodeficiency viruses with dendritic cells. Curr. Top. Microbiol. Immunol. 276, 1-30.

Steinman, R.M., Hawiger, D., and Nussenzweig, M.C. (2003b). Tolerogenic dendritic cells. Annu. Rev. Immunol. 21, 685-711.

Story, C.M., Furman, M.H., and Ploegh, H.L. (1999). The cytosolic tail of class I MHC heavy chain is required for its dislocation by the human cytomegalovirus US2 and US11 gene products. Proc. Natl. Acad. Sci. U. S. A. 96, 8516-8521.

Sukhikh, G.T. and Vanko, L.V. (1999). Interrelationships between immune and reproductive systems in human. Russ. J. Immunol. 4, 312-314.

Szabo, G. and Dolganiuc, A. (2005). Subversion of plasmacytoid and myeloid dendritic cell functions in chronic HCV infection. Immunobiology 210, 237-247.

Takayama, T., Morelli, A.E., Onai, N., Hirao, M., Matsushima, K., Tahara, H., and Thomson, A.W. (2001). Mammalian and viral IL-10 enhance C-C chemokine receptor 5 but down-regulate C-C chemokine receptor 7 expression by myeloid dendritic cells: impact on chemotactic responses and *in vivo* homing ability. J. Immunol. 166, 7136-7143.

Takayama, T., Nishioka, Y., Lu, L., Lotze, M.T., Tahara, H., and Thomson, A.W. (1998). Retroviral delivery of viral interleukin-10 into myeloid dendritic cells markedly inhibits their allostimulatory activity and promotes the induction of T-cell hyporesponsiveness. Transplantation 66, 1567-1574.

Tan, M.C., Mommaas, A.M., Drijfhout, J.W., Jordens, R., Onderwater, J.J., Verwoerd, D., Mulder, A.A., van der Heiden, A.N., Scheidegger, D., Oomen, L.C., Ottenhoff, T.H., Tulp, A., Neefjes, J.J., and Koning, F. (1997). Mannose receptor-mediated uptake of antigens strongly enhances HLA class II-restricted antigen presentation by cultured dendritic cells. Eur. J. Immunol. 27, 2426-2435.

Ting, J.P. and Trowsdale, J. (2002). Genetic control of MHC class II expression. Cell 109 Suppl, S21-S33.

Tomazin, R., Boname, J., Hegde, N.R., Lewinsohn, D.M., Altschuler, Y., Jones, T.R., Cresswell, P., Nelson, J.A., Riddell, S.R., and Johnson, D.C. (1999). Cytomegalovirus US2 destroys two components of the MHC class II pathway, preventing recognition by CD4[+] T cells. Nat. Med. 5, 1039-1043.

Tosi, G., Meazza, R., De Lerma, B.A., D'Agostino, A., Mazza, S., Corradin, G., Albini, A., Noonan, D.M., Ferrini, S., and Accolla, R.S. (2000). Highly stable oligomerization forms of HIV-1 Tat detected by monoclonal antibodies and requirement of monomeric forms for the transactivating function on the HIV-1 LTR. Eur. J. Immunol. 30, 1120-1126.

Trombetta, E.S. and Mellman, I. (2005). Cell biology of antigen processing *in vitro* and *in vivo*. Annu. Rev. Immunol. 23, 975-1028.

Turley, S.J., Inaba, K., Garrett, W.S., Ebersold, M., Unternaehrer, J., Steinman, R.M., and Mellman, I. (2000). Transport of peptide-MHC class II complexes in developing dendritic cells. Science 288, 522-527.

van der Wal, F.J., Kikkert, M., and Wiertz, E. (2002). The HCMV gene products US2 and US11 target MHC class I molecules for degradation in the cytosol. Curr. Top. Microbiol. Immunol. 269, 37-55.

Wahid, R., Cannon, M.J., and Chow, M. (2005). Dendritic cells and macrophages are productively infected by poliovirus. J. Virol. 79, 401-409.

Watanabe, N., Wang, Y.H., Lee, H.K., Ito, T., Wang, Y.H., Cao, W., and Liu, Y.J. (2005). Hassall's corpuscles instruct dendritic cells to induce CD4[+]CD25[+] regulatory T cells in human thymus. Nature 436, 1181-1185.

Watts, T.H. (2005). TNF/TNFR family members in costimulation of T cell responses. Annu. Rev. Immunol. 23, 23-68.

Winzler, C., Rovere, P., Rescigno, M., Granucci, F., Penna, G., Adorini, L., Zimmermann, V.S., Davoust, J., and Ricciardi-Castagnoli, P. (1997). Maturation stages of mouse dendritic cells in growth factor-dependent long-term cultures. J. Exp. Med. 185, 317-328.

Witmer-Pack, M.D., Olivier, W., Valinsky, J., Schuler, G., and Steinman, R.M. (1987). Granulocyte/macrophage colony-stimulating factor is essential for the viability and function of cultured murine epidermal Langerhans cells. J. Exp. Med. 166, 1484-1498.

Yang, G.X., Lian, Z.X., Kikuchi, K., Moritoki, Y., Ansari, A.A., Liu, Y.J., Ikehara, S., and Gershwin, M.E. (2005). Plasmacytoid dendritic cells of different origins have distinct characteristics and function: studies of lymphoid progenitors versus myeloid progenitors. J. Immunol. 175,7281-7287.

Yoneyama, H., Matsuno, K., Toda, E., Nishiwaki, T., Matsuo, N., Nakano, A., Narumi, S., Lu, B., Gerard, C., Ishikawa, S., and Matsushima, K. (2005). Plasmacytoid DCs help lymph node DCs to induce anti-HSV CTLs. J. Exp. Med. 202, 425-435.

Yue, S.C., Shaulov, A., Wang, R., Balk, S.P., and Exley, M.A. (2005). CD1d ligation on human monocytes directly signals rapid NF-kappaB activation and production of bioactive IL-12. Proc. Natl. Acad. Sci. U. S. A. 102, 11811-11816.

Zajonc, D.M., Crispin, M.D., Bowden, T.A., Young, D.C., Cheng, T.Y., Hu, J., Costello, C.E., Rudd, P.M., Dwek, R.A., Miller, M.J., Brenner, M.B., Moody, D.B., and Wilson, I.A. (2005). Molecular mechanism of lipopeptide presentation by CD1a. Immunity 22, 209-219.

Zhang, M., Tang, H., Guo, Z., An, H., Zhu, X., Song, W., Guo, J., Huang, X., Chen, T., Wang, J., and Cao, X. (2004). Splenic stroma drives mature dendritic cells to differentiate into regulatory dendritic cells. Nat. Immunol. 5, 1124-1133.

Zhang, R., Lifson, J.D., and Chougnet, C.A. (2005a). Failure of HIV-exposed CD4[+] T cells to activate dendritic cells is reversed by restoration of CD40/CD154 interactions. Blood 107(5):1989-95.

Zhang, W., Yang, H., Kong, X., Mohapatra, S., Juan-Vergara, H., Hellermann, G., Behera, S., Singam, R., Lockey, R.F., and Mohapatra, S.S. (2005b). Inhibition of respiratory syncytial virus infection with intranasal siRNA nanoparticles targeting the viral NS1 gene. Nat. Med. 11, 56-62.

Zhu, F.X., King, S.M., Smith, E.J., Levy, D.E., and Yuan, Y. (2002). A Kaposi's sarcoma-associated herpesviral protein inhibits virus-mediated induction of type I interferon by blocking IRF-7 phosphorylation and nuclear accumulation. Proc. Natl. Acad. Sci. U.S.A. 99, 5573-5578.

Chapter 4

Dendritic Cells as Keepers of Peripheral Tolerance

Sabine Ring, Alexander H. Enk, and Karsten Mahnke

4.1 Introduction

At first glance, the induction of peripheral tolerance seems a redundant feature, because the adaptive immune system has developed mechanisms that teach developing T cells to discriminate between "self" and "non-self." During the random rearrangement process of T-cell-receptor generation, it is likely that autoreactive T cells are generated; but these potentially dangerous T cells are deleted in the thymus upon interaction with dendritic cells (DCs) via an educational process that has been classically ascribed to clonal deletion of self-reactive T cells. However, this so-called central tolerance is incomplete, because not all self-antigens gain access to the thymus and several self-reactive lymphocytes escape central deletion (Nossal, 2001). Accordingly, autoreactive T cells that are present in the periphery have the ability to respond to autoantigens expressed by different tissues and hence produce an immune reaction against "self." Therefore, additional mechanisms ensuring tolerance in the periphery must be present.

For a long time, DCs have been regarded as the key inducers of immunity. Indeed, many features discovered in the early years of DC biology support their function as "nature's adjuvant" (Bancherau and Steinman, 1998). DCs have been characterized by their excessive expression of MHC molecules and T cell costimulatory molecules that enable them to generate an immune response *de novo*. In agreement with this function, DCs are located in virtually all peripheral tissues where they display a first barrier for invading pathogens and tissue damage. Here they are the guardians for the integrity of the body tissues and after antigen encounter, the DCs leave the periphery and migrate through the lymphatics to secondary lymphoid organs.

However, even in the absence of pathogens or tissue disruption, DCs are distributed throughout the body in virtually all tissues and come in contact with a full repertoire of self-antigens. These self-antigens originate from cellular proteins or cell-derived detritus and thus enable the DCs to constantly sample their environment. Therefore, DCs are an effective source of "self-peptide–MHC" complexes, and presentation of these complexes can be used to tolerize potentially self-reactive T cells (Inaba *et al.*, 1997). Although the migratory DCs arriving from the periphery are a good source for self-peptide and may provide a basis to maintain peripheral tolerance, another dilemma arises: now the immune system has to distinguish between potentially dangerous foreign antigens and harmless self-proteins, which are both presented by DCs. The two types of antigens require opposite T-cell reactions. For example, presentation of microorganism-derived antigens will result in T-cell activation, whereas presentation of "self" should lead to T-cell silencing.

These two functions of DCs (i.e., maintaining tolerance to self and inducing immunity to pathogens) are not in conflict, because they are elicited under different circumstances. When tissue disruption occurs or bacteria enter the body, it is likely that these events influence the resident DC population in certain ways. For instance, bacterial proteins may modify DCs in a way that only T-cell activating properties prevail, or a specialized "immunizing" subset of DCs might be recruited. Finally, a spatial separation of "tolerizing" and "activating" DCs may exist in the periphery. However, so far a unique mechanism has not been determined, but there is evidence that several mechanisms operate to guide DCs in their decision making of whether tolerance or immunity is induced (Mahnke *et al.*, 2002).

4.2 The Concept of "Steady State" versus "Activated" DC

In the course of their individual development, DCs pass through different developmental stages that are marked by changes in their phenotype and their physiologic behavior (Pierre *et al.*, 1997; Turley *et al.*, 2000). Regardless of their origin or their tissue distribution, it is obvious that DCs undergo maturation steps. Accordingly, so-called immature DCs can be distinguished from mature DCs by various means.

From initial observations, Steinman and Schuler concluded that freshly isolated and hence immature DCs need proper stimuli to mature to fully mature, T-cell stimulatory cells (Schuler and Steinman, 1985). In these early experiments, DCs were purified from spleens or lymph nodes (LN) from mice and subjected to mixed lymphocytes reactions to test their T-cell stimulatory capacity. Here it was shown that freshly isolated DCs do not stimulate T-cell proliferation better than conventional antigen-presenting cells, whereas after 2 days of culture in medium, the very same DCs acquired their unsurpassed T-cell stimulatory capacity.

This maturation process, as initially marked by augmented T-cell stimulatory capacity, is accompanied by up regulation of the T-cell costimulatory molecules CD80, CD86, ICAM-1, and the appearance of MHC class II and MHC class I on the cell surface. In addition, Th1 T-cell promoting cytokines such as IL-12 are secreted. Therefore, a fully mature DC is very well equipped to stimulate T cells and to induce T-cell responses (Inaba et al., 2000; Mellman and Steinman, 2001).

This scheme of DC development is adapted to the physiologic tasks that are fulfilled by DCs during protection and initiation of an immune response. Frequently, DCs are regarded as sentinels in the peripheral tissues, where they are able to endocytose and to process foreign and potentially dangerous antigens. Once these pathogens are captured, DCs have to meet with T cells in the T-cell areas of LN and to present the "foreign" proteins to T cells in order to stimulate an immune response. According to this scenario, the different functional requirements for DCs are reflected by the maturation process: in the periphery, immature DCs have to be able to endocytose potentially dangerous antigens and to load peptide fragments onto MHC class I and MHC class II molecules; once the DCs are loaded, they start their migration toward secondary lymphoid organs and establish contact with T cells for the induction of an immune response. Here DCs are required to express molecules that facilitate the interaction with T cells. In parallel, the termination of antigen uptake prevents the loss of potentially important antigens originating from the periphery. However, in this consequential scenario, DCs possess only one function (i.e., the induction of an immune response). This led to the assumption that immature DCs are just a precursor of what would later be a "real" and fully developed DC (Mellman and Steinman, 2001).

However, recently a novel concept of immature or steady-state DCs emerged, assigning the induction of tolerance to immature DCs. This concept takes into account that DCs function as sentinels in the periphery of the body and here they may also encounter various self-antigens or harmless foreign proteins such as apoptotic cell detritus and/or antigens derived from commensal microbes, which are present on body surfaces like lung and gut epithelia.

When these antigens would be presented in an immunostimulatory way, autoimmunity and/or overboarding immune responses against otherwise harmless and ubiquitous antigens would be generated. Therefore, it is reasonable to predict that not all DCs that encounter antigens in the periphery present these antigens in an immunostimulatory fashion, which results in T-cell activation.

Indeed, there is evidence that peripheral DCs also take up particles that are harmless. For instance, it has been shown that DCs take up LATEX particles or colloidal carbon in the periphery or transport melanin from the skin to draining lymph nodes. Peripheral tissue DCs also take up molecules by the internalization of plasma membrane, shed microvesicles or exosomes (Harshyne *et al.*, 2001; Thery *et al.*, 2002) and can pick up parts of epithelial cells lining the gut and lung (Henri *et al.*, 2001; Holt *et al.*, 1990; Huang *et al.*, 2000). These antigens are further transported to draining LN without driving any immune response toward "self."

Altogether, these data indicate that substantial amounts of self-antigens are captured and transported to secondary lymphoid organs. However, formal proof that these antigens are also actively presented on MHC class II molecules derives from investigations using the Y-ae antibody. This antibody recognizes MHC class II molecules of the IAb haplotype that are loaded with a peptide that is derived from IE (IEα52-68). Using this antibody, Inaba *et al.* (1997) could show that murine DCs in the LN express huge amounts of self-peptide in the steady state, however, autoimmunity against IEα did not occur under these circumstances.

These data clearly state a dilemma, because some DCs are undoubtedly loaded with self-antigens picked up in periphery, but others may carry potentially "dangerous" pathogen-derived antigens. Thus, how can DCs "decide" whether an immune reaction or tolerance has to be established? This would imply that DCs, by whatever means, can realize whether antigens are "harmless" self or "dangerous" foreign (Matzinger, 1994).

This discrepancy can be addressed by bearing in mind that antigen uptake takes place during different environmental contexts. That is, self-antigens such as apoptotic cells normally do not disrupt tissue or deliver other DC activation signals. Therefore, DCs remain in the steady state during antigen uptake and antigen delivery to T cells. On the contrary, when tissue damage and pathogens are present, a different set of signals is delivered. Here DCs become activated either by Toll-like receptors (TLRs) and/or other pattern-recognition receptors (PRRs) (Medzhitov, 2001). Accordingly, under these circumstances DCs become activated and antigens presented to T cells favor the development of immunity rather then tolerance. Thus, environmental circumstances, that is, inflammation (or danger) versus steady state (or harmless), affects the outcome of an immune response (Fig. 4.1.).

This concept is backed by several *in vivo* studies analyzing the effects of activated DCs on the outcome of immune response. For instance, tissue disruption and thus a "danger-signal" leads to expression of T-cell costimulatory molecules CD80, CD86, MHC-peptide complexes, and chemokine receptors, all of which are clear indicators of DC activation (Caux *et al.*, 1994a; Caux *et al.*, 1994b; Sallusto *et al.*, 1998). Vice versa, presence of steady-state DC seems to be crucial for maintenance of tolerance, as

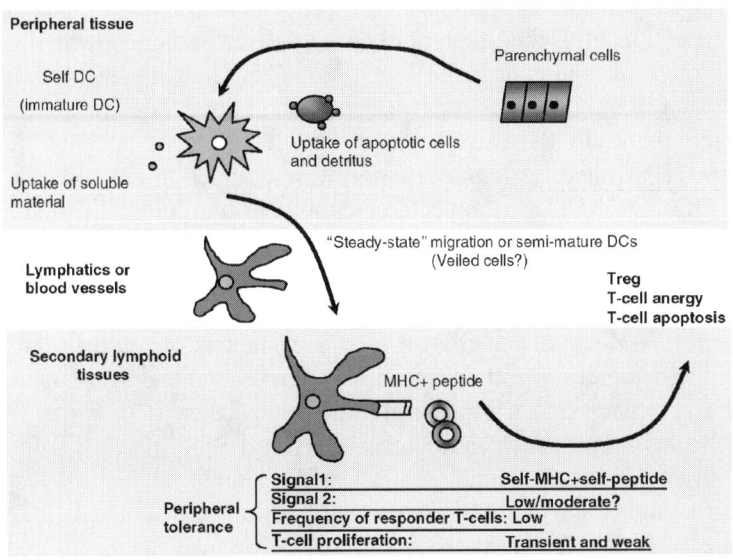

FIGURE 4.1. The concept of "steady state" versus "activated" DC. In non-inflamed tissues, DCs are normally not activated and remain in the steady state. They display an immature phenotype and constantly sample their environment. They mainly take up self-antigens by pinocytosis and endocytosis and remain in the steady state as long as no further activation signals reach the tissue. From here, DCs continuously migrate to secondary lymphoid organs where they eventually meet T cells. In the lymph nodes, the antigens picked up in the periphery and thus mainly self-antigens are presented to T cells. Because mainly self-antigens are presented, it is mandatory that tolerance ensues. For the full activation of T cells, MHC-peptide complexes and concomitant costimulatory signals are necessary, otherwise anergy will ensue. Because "steady-state" DCs express only low numbers of costimulatory signals, the presentation of surface MHC-peptide complexes to T cells may indeed induce anergy. Furthermore, as discussed in the text, T cells may undergo apoptosis or regulatory T cells may be induced by those steady-state DCs. Although the means by which DCs accomplish these changes are not entirely clear yet, all effects depicted here lead to suppression of T-cell responses and thus ensure tolerance to self-antigen presented by DCs.

Mehling *et al.* have shown that CD40-transduced and therefore constantly activated DCs induced severe autoimmunity in the skin (Mehling *et al.*, 2001). Moreover, it has been shown that in humans suffering from the autoimmune disease systemic lupus erythematosus (SLE), the DCs display a chronically activated phenotype induced by IFN-α (Blanco *et al.*, 2001).

However, the further analysis of the events during antigen presentation in the steady state was hampered by the fact that DCs had to be isolated and to be loaded with antigens in order to study the subsequent events of antigen presentation. Although these experimental protocols avoided external DC-activating stimuli like lipopolysaccharide (LPS) contamination, the isolation and purification from the tissue is likely to deliver enough "danger" signals that DCs mature spontaneously over time. This shortcoming was avoided when *in vivo* antigen targeting was applied to load DCs *in situ* without disturbing them. In these experiments, model antigens such as hen egg lysozyme (HEL) and ovalbumin (OVA) were coupled to antibodies specific for the DEC-205 (CD205) molecule (Mahnke *et al.*, 2003c). The DEC-205 receptor is a potent antigen receptor that is solely expressed by DCs and not by macrophages, as the MMR (Macrophage Mannose Receptor) is, or by B cells, as the Fc-receptor is. Its DC-specific expression pattern opened the possibility to selectively load DCs *in situ* by means of antibody targeting. This antigen targeting will mimic the *in vivo* situation, in which antigen uptake takes place under steady-state conditions, that is, in the absence of DC-activating stimuli such as microbial-derived LPS or tissue disruption. In these investigations, a model OVA or HEL antigen, respectively, was covalently linked to anti-DEC-205 antibodies and injected into mice (Bonifaz *et al.*, 2002; Hawiger *et al.*, 2001; Mahnke *et al.*, 2003c). This resulted in antigen loading of the DCs in the LN and to presentation of those antigens to OVA- and HEL-specific T cells, respectively. The analysis of the induced immune response after targeting DCs in the steady state revealed that tolerance was induced and two possible mechanisms were obvious: Hawiger *et al.* (2001) demonstrated the disappearance of HEL-specific T cells and speculated on T-cell deletion. In contrast, our own results point toward induction of regulatory T cells (Treg) (Mahnke *et al.*, 2003c). Both mechanisms are not mutually exclusive, as induction of Treg was observed approximately 8 days after loading the DCs *in vivo* and T-cell deletion in the "HEL system" was recorded 3 weeks after the original challenge of the DCs. Possibly both pathways (i.e., induction of Treg as well as deletion) may interact with each other. The induction of Treg might be a very early response to dampen activation of T cells immediately and locally contained, followed by the deletion as a final means to extinguish the T cells and to ensure long-lasting tolerance toward that respective antigen. These *in vitro* studies have formally proved that steady-state DCs preferentially induce tolerance and that activated DCs are required for the induction of immunity.

From a conceptional point of view, this mechanism provides a means by which peripheral tolerance is constantly under examination. During normal tissue turnover, the apoptosis of various body cells releases a comprehensive spectrum of self-proteins, and this material displays an ideal source for DCs to sample the repertoire of peripheral proteins. In this regard, immature DCs seem especially equipped to take up these self-proteins, as they express a broad spectrum of scavenger receptors, such as the integrin $\alpha v \beta 3$, $\alpha v \beta 5$, and the thrombospondin receptor. The mechanism by which these receptors bind apoptotic material is not clear yet, but it is conceivable that apoptotic and necrotic material differs in many ways. That is, during apoptosis, as opposed to necrosis, cells modify their surface specifically by exposing phosphatidylserine and altered carbohydrates. These components of the inner membrane leaflet serve as specific ligands for αvβ3 and/or αvβ5 integrin receptors expressed by DCs (Albert et al., 1998; Urban et al., 2001).

The importance of these mechanisms for the maintenance of tolerance is underlined by the fact that engagement of the scavenger receptor CD36 is able to block subsequent DC maturation (Urban et al., 2001) and that uptake of apoptotic cells via the CD11b/CD18 integrin inhibits production of proinflammatory cytokines (Verbovetski et al., 2002). Thus, engagement of these "apoptotic cell receptors" blocks DC maturation effectively, acting as a safeguard mechanism that prevents immunity and ensures induction of tolerance.

Even the uptake of foreign, bacteria-derived proteins does not automatically trigger DC activation. Instead, there is evidence that DCs react differentially to uniformly Gram-positive bacteria. For instance, it has been shown that only contact to the pathogenic strain of *Streptococcus pyogenes* resulted in a fully mature DC phenotype, whereas the non-pathogenic *Lactobacillus rhamnosus* induced semimature DCs with rather low T-cell stimulatory capacity (Veckman et al., 2002); and these interactions may further explain why certain probiotic bacteria exert beneficial effects on a number of inflammatory diseases, including atopic dermatitis and Crohn's disease. Smits and collegues (Smits et al., 2005) have further shown that two common strains of lactobacilli (*L. casei* and *L. reuteri*) primed DCs for the development of Tregs and not for the induction of effector T cells, although these bacteria are foreign to the host DCs and indeed also expressed ligands that potentially engage TLR or other PRR. Here it was shown that the binding to specialized "downregulatory" receptors, such as DC-SIGN in this case, was able to shift the balance between DC-activating and DC-inhibiting signals toward the induction of tolerance. Therefore, the final result of an immune response may be determined by the integration of different sets of signals received by antigen receptors, which deliver either stimulating or deactivating signals to the DCs.

This system of activation-controlled T-cell stimulation by DCs does not seem to be restricted to MHC class II and CD4$^+$ T cells, as there is evidence that exogenous antigens can also enter the MHC class I pathway (Larsson et al., 2001). Normally, MHC class I binding peptides derive from endogenously synthesized proteins. However, evidence for cross-presentation, that is, the presentation of exogenous material in context of MHC class I, has been described earlier. Although originally defined in the context of T-cell priming, recent evidence shows that cross-presentation may also contribute to T-cell tolerance (Kurts, 2000; Kurts et al., 1997; Kurts et al., 2001; Larsson et al., 2001). In a series of experiments, Liu et al. (2002) showed that OVA, which has been encapsulated into apoptotic cells, was efficiently phagocytosed and presented to OVA-specific T cells upon injection into mice. On the contrary, free OVA or OVA encapsulated into necrotic cells were not able to drive T-cell proliferation. In these experiments, a subset of CD8[+] DCs was the main cell population that selectively took up apoptotic bodies and drove initial T-cell proliferation. Notably, the observed T-cell proliferation was short-lived and within weeks the OVA-specific T cells were deleted from the body rendering the animals tolerant. Similar observations had been made earlier in transgenic RIP-OVA mice that express OVA selectively in the pancreas. Again, upon transfer of OVA specific T cells, initial proliferation was observed but eventually deletion of these T cells occurred. Thus, these results suggest that similar to CD4[+] T cells, mechanisms of tolerance induction by steady-state DCs also operate on CD8[+] T cells, that is, cross-presentation allows cross-tolerizing of potentially autoreactive CD8[+] T lymphocytes under physiologic conditions.

However, the distinction between immature and mature DCs is sophisticated, and DCs undergo several intermediate stages between these two end points. Therefore, under certain circumstances two opposite functions, that is, tolerance in the CD4 compartment and CD8-driven immunity, can be executed from the very same DC. Using the model of autoimmune encephalomyelitis (EAE), it has been shown that immature DCs, which have simultaneously been pulsed with EAE-derived MHC class I and MHC class II peptides, did indeed induce tolerance of the CD4[+] T cells but in parallel triggered cytolytic activity of respective CD8[+] T cells (Kleindienst *et al.*, 2005). Thus, these data implicate that not only DCs decide about tolerance and/or immunity but also the type of T cells and their requirements for activation are involved in balancing tolerance versus immunity.

4.3 Factors That Affect Tolerogenic DCs

4.3.1 TNF-α and Semimature DCs

The "two-stage concept" of immature DCs (tolerance) and mature DCs (immunity) has its merits and serves well to explain fundamental phenomena observed in immunology. However, the maturation process itself is not a binary switch. DCs undergo a development from the immature state to the mature state. Depending on the measured values, DCs can be mature as judged by some parameters but still maintain their tolerogenic properties. Even in the *in vivo* situation, so-called immature and thus tolerogenic DCs have to migrate from peripheral tissues to the LN in order to fulfill their tolerizing duties. By definition, DCs need an impetus to start migration from the periphery to lymphoid organ, therefore even steady-state DCs are activated to a certain degree.

Most definitions of immature versus mature DCs rely on measurement of surface markers such as CD80, CD86, CD40, and MHC class II. These markers are indeed the most powerful characteristics for mature DCs, however, cytokine production seems also to be important and provides another criterion to further subdivide the DC maturation status. Evaluation of cytokine production during maturation revealed that not all mature DCs, as judged by surface marker expression, are fully mature in terms of cytokine production. Only some DCs among the population of mature DCs secrete substantial amounts of interleukin (IL)-12, IL-6, TNF-α, and IL-1β; and because this cytokine profile was not released by all DCs uniformly, the term semimature DCs (i.e., DCs that express mature surface molecules but miss cytokine production) was coined by Lutz and co-workers (Lutz and Schuler, 2002). These semimature DCs, as characterized by their limited cytokine production, are prone to induce $CD4^+/IL-10^+$ Treg cells that exert suppressive effect in a variety of allergic and autoimmune diseases (Fig. 4.2.).

In the course of development of semimature DCs, TNF-α might be critically involved (Menges et al., 2002). It has been shown that exposure of immature DCs to TNF-α resulted in a mature phenotype by the means of CD80, CD86, and MHC class II surface expression but they lacked secretion of IL-12, IL-1, and IL-6.

Upon infusion into mice, these cells prevented the development of EAE (the murine EAE serves as model for human multiple sclerosis). By contrast, DCs matured with LPS and anti-CD40 did display the same surface marker profile as compared with their TNF-α exposed counterpart but in addition secreted substantial amounts of proinflammatory cytokines and were thus unable to affect EAE development.

However, these results were obtained in a murine system where DCs were cultivated in vitro and transfer of the results into the human system seems difficult. Here it has been shown that after blockade of TNF-α in vivo and in vitro, human DCs secrete reduced amounts of CCLS. Hence, in this study neutralization of TNF-α and not presence of it did induce a semimature phenotype (van Lieshout et al., 2005). However, whether TNF-α is able to stably induce a semimature phenotype awaits further proof. But the obvious differences in cytokine production between mature and nonmature DCs open further means by which DCs can influence the outcome of an immune response.

Recently, Sporri and Sousa (Sporri and Reis e Sousa, 2005) have developed a more detailed model of DC development. This model relies

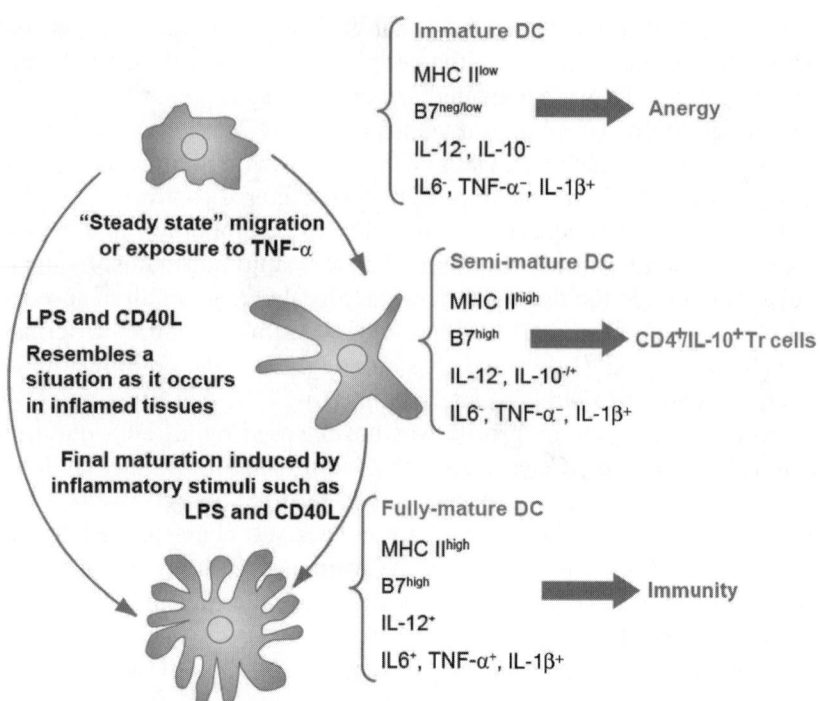

Immature DC

MHC II^{low}

B7$^{neg/low}$ ➡ Anergy

IL-12$^-$, IL-10$^-$

IL6$^-$, TNF-α$^-$, IL-1β$^+$

"Steady state" migration
or exposure to TNF-α

LPS and CD40L

Resembles a
situation as it occurs
in inflamed tissues

Semi-mature DC

MHC II^{high}

B7^{high} ➡ CD4$^+$/IL-10$^+$Tr cells

IL-12$^-$, IL-10$^{-/+}$

IL6$^-$, TNF-α$^-$, IL-1β$^+$

Final maturation induced by
inflammatory stimuli such as
LPS and CD40L

Fully-mature DC

MHC II^{high}

B7^{high} ➡ Immunity

IL-12$^+$

IL6$^+$, TNF-α$^+$, IL-1β$^+$

FIGURE 4.2. Factors that affect tolerogenic DCs. During maturation, DCs undergo several changes and pass through different stages. Immature DCs are present in the tissue and remain in the steady state without further stimulation. However, a first impetus for maturation can be provided by TNF-α. Here, semimature DCs develop, which are well equipped with costimulatory molecules but lack cytokine secretion. Despite their surface molecule expression, the semimature DCs are prone to induce tolerance, by induction of regulatory T cells. Finally, the mature DC phenotype develops either directly from immature or from semimature DCs after induction by inflammatory stimuli such as LPS or CD40L. These DCs are fully equipped for the induction of T cell immunity.

not only on intrinsic development of DC maturation but also takes into account that external factors may trigger DC maturation in certain ways. In this model, DCs become "licensed" to induce immune responses. Vice versa, "unlicensed" DCs may predominately induce tolerance. In this scenario, two different stages of mature DCs can be distinguished. Chimeric mice were generated that express two different sets of DCs. One subset was able to respond to TLR ligands and thus become activated. But this subset expresses only mismatched MHC regarding the corresponding T cells. Therefore, an immune response could not be generated. The other subset did express the right MHC-haplotype but it was engineered not to respond to TLR ligands anymore. In this system, DC development into a mature phenotype, that is, B7 and MHC class II expression occurred in both subsets, but only those DCs that could react directly to stimulating agents such as LPS or CpG induced $CD4^+$ effector T cell. Thus, only direct activation through pathogen-derived stimuli is able to "license" DCs for induction of immunity. The mere activation via bystander cytokines was not sufficient to mount an immune response.

More evidence about how TLR pathways are involved in DC licensing derived from studies in knockout mice. Here it has been shown that only properly activated and hence "licensed" DCs are able to overcome Treg-mediated suppression of immune responses. In these experiments, it has become clear that IL-6 is critically involved in that process and only after stimulation with TLR ligands, DCs produced sufficient amounts of IL-6 to overcome suppression by Treg. Interestingly, other hallmark cytokines for DC maturation such as IL-12 or TNF-α were not involved.

Although those studies did not directly address the function of "unlicensed" or semimature DCs, it seemed obvious that not fully activated DCs are inferior in mounting an immune response. Henceforth, from a conceptually point of view the default action of DCs would be tolerance induction unless the DCs are properly licensed, or matured by all means (Hamilton-Williams *et al.*, 2005; Napolitani *et al.*, 2005; Turnbull *et al.*, 2005).

4.3.2 Vitamin D₃ Affects DC Maturation

1α,25-Dihydroxyvitamin D_3 (1,25(OH)2D_3), the biologically active meta-bolite of vitamin D_3, is a secosteroid hormone with pleiotropic actions. 1,25(OH)2D_3 not only regulates bone and calcium/phosphate homeostasis but also exerts a number of other biological activities, including regulation of growth and differentiation of cancer cells and modulation of the immune response via specific receptors expressed in antigen-presenting cells (APCs) and activated T cells.

The metabolism of vitamin D to its active form is well-known. Vitamin D is transported in the blood by the vitamin D binding protein (DBP) to the liver where it is hydroxylated at C-25, resulting in the formation of 25-hydroxyvitamin D_3 (25(OH)D_3), which is transported by the DBP to the kidney. In the proximal convoluted and straight tubules of the kidney, 25(OH)D_3 is hydroxylated at the 1α position by the mitochondrial cytochrome P450 enzyme 25-hydroxyvitamin D_3-1α-hydroxylase (1-α-(OH)ase), resulting in the formation of the active form of vitamin D [reviewed in Christakos *et al.* (2003)].

1α,25-(OH)2D_3 binds to a nuclear receptor (vitamin D_3 receptor; VDR), a member of the superfamily of nuclear receptors for steroid hormones, thyroid hormone, and retinoic acid, all of which have important effects on the immune system. The constitutive presence of VDR in most cell types of the immune system led to the assumption that VDR ligands operate as immunomodulatory agents, which are able to control the immune response.

The immunosuppressive functions of 1,25(OH)2D$_3$ and its analogues have been demonstrated in several studies using different models of autoimmune diseases and experimental organ transplantations, including experimental allergic encephalomyelitis (the murine model for multiple sclerosis) (Lemire and Archer, 1991), experimental lupus erythematosus in lpr/lpr mice (Abe *et al.*, 1990), collagen-induced arthritis, and autoimmune thyroiditis (Fournier *et al.*, 1990). 1,25(OH)2D$_3$ has also been reported to prevent autoimmune diabetes in non-obese diabetic (NOD) mice (Mathieu *et al.*, 1994) and prolongs the survival of heart (Lemire *et al.*, 1992) and small bowel allografts (Johnsson *et al.*, 1994). Analogues of 1,25(OH)2D$_3$ are clinically relevant in the treatment of psoriasis [reviewed in Fogh *et al.* (2004)], a Th1-mediated autoimmune disease of the skin, where they show efficacy comparable with topical steroids. Type 1 diabetes in different populations is associated with polymorphisms of the vitamin D receptor gene (Pani *et al.*, 2000; Chang *et al.*, 2000) and in addition, epidemiologic studies have shown a higher incidence of type 1 diabetes in northern than in southern regions (Green *et al.*, 1992), suggesting a possible involvement of a 1,25(OH)2D$_3$ deficiency in the pathogenesis of type 1 diabetes. A large population-based case-control study also showed that the intake of vitamin D$_3$ decreased the risk of type 1 diabetes development significantly (The EURODIAB Substudy 2 Study Group., 1999).

1,25(OH)2D$_3$ modulates the immune response via the specific receptor VDR expressed in APCs and activated T cells. In addition to direct effects on T-cell activation, the numerous immune effects of 1,25(OH)2D$_3$ are mainly mediated through its action on DCs, as it affects all major stages of the DC life cycle: differentiation, maturation, activation, and survival. *In vitro* and *in vivo* experiments have shown that VDR ligands induce DCs to acquire tolerogenic properties that favor the induction of regulatory rather than effector T cells. The fact that the suppressive effect of 1,25(OH)2D$_3$ is mediated by DCs was shown by bidirectional human mixed leukocyte reactions (MLRs). Proliferation as well as IFN-γ production of alloreactive T cells was inhibited by incubation with 1,25(OH)2D$_3$ (Penna and Adorini, 2000). This inhibitory effect in T-cell responses appeared to be indirect because T-cell activation by plate-bound anti-CD3 with or without costimulation by anti-CD28 was scarcely affected by 1,25(OH)2D$_3$ as determined by cell proliferation or IFN-γ secretion. This indicated that 1,25(OH)2D$_3$ inhibits the ability of DCs to induce alloreactive T-cell activation rather than directly inhibiting T cells. Several studies performed on either monocyte-derived DCs from human peripheral blood or on bone marrow–derived mouse DCs have clearly demonstrated that *in vitro* treatment of DCs with VDR ligands leads to interference of differentiation of monocytes into DCs (Piemonti *et al.*, 2000; Griffin *et al.*, 2000). DCs can be generated from monocytes by culture for 1 week under plastic-adherent conditions in the presence of granulocyte-macrophage colony-stimulating factor (GM-CSF) and IL-4. The resulting immature DCs express high levels of CD1a and intracellular MHC class II, low levels of CD83, CD40, CD80, and CD86, and are negative or low positive for CD14 and CD16. Addition of 1,25(OH)2D$_3$ at day 0 during *in vitro* generation completely inhibited DC differentiation from monocytes and the generation of CD1a$^+$ cells from progenitor cells is prevented while numbers of CD14$^+$ cells are increased. Expression of costimulatory molecules such as CD40 and CD86 were slightly decreased or non-significantly affected (CD80 and MHC class II). Antigen uptake molecules such as CD32 and the mannose receptor were significantly increased. The phenotype of DCs cultured in presence of 1,25(OH)2D3 resembled that of original monocytes, and functional tests demonstrated reduced capacity to stimulate CD4$^+$ T-cell proliferation in a primary MLR assay. Thus, 1,25(OH)2D3 inhibits phenotypically and functionally the differentiation of peripheral blood monocytes into mature DCs (Canning *et al.*, 2001).

DCs obtained after 7-day culture with GM-CSF and IL-4 could be further differentiated *in vitro* into fully mature DCs by exposure to LPS or CD40L. Slight downregulation of CD1a, decreased expression of mannose receptor, induction of the maturation marker CD83, and further upre-gulation of HLA-DR, CD40, and CD86 accompany maturation. Addition of 1,25(OH)2D$_3$ prevents this LPS-induced maturation of immature DCs, maintaining DCs in an immature stage characterized by high mannose receptor and low CD83 expression. In fact, after exposure to LPS or CD40L, 1,25(OH)2D$_3$-DCs were unable to upregulate CD83 as well as the molecules involved in Ag presentation (MHC I, MHC II, CD80, CD86, and CD40) and to downregulate the Ag uptake molecules (CD32 and mannose receptor) (Piemonti *et al.*, 2000). 1,25(OH)2D$_3$ inhibited the expression of MHC class II, CD58, CD40, CD80, and CD86 molecules by about 50%. This is associated with a reduced capacity of DCs to activate alloreactive T cells, as determined by decreased proliferation and IFN-γ secretion in MLR. The modulation of phenotype and function of DCs matured in the presence of 1,25(OH)2D$_3$ induces cocultured alloreactive CD4$^+$ cells to secrete less IFN-γ upon restimulation, to upregulate CD152, and to downregulate CD154 molecules (Berer *et al.*, 2000).

Besides the effects on cell surface marker expression and on maturation, the addition of 1,25(OH)2D$_3$ during LPS-induced maturation of DCs also blocked the secretion of IL-12p75, whereas the secretion of IL-10 was sevenfold higher compared with DCs cultured under conventional conditions. TGF-β1 levels, however, did not change significantly (Penna and Adorini, 2000). The abrogation of IL-12 production and the strongly enhanced production of IL-10 highlight the important functional effects of 1,25(OH)2D$_3$ on DCs and are, at least in part, responsible for the induction of DCs with tolerogenic properties.

The prevention of DC differentiation and maturation as well as the modulation of their activation and survival leads to a DC phenotype with tolerogenic properties and may explain the immunosuppressive activity of 1,25(OH)2D$_3$.

These effects are not limited to *in vitro* activity: 1,25(OH)2D$_3$ can also induce DCs with tolerogenic properties *in vivo*. This has been demonstrated in models of allograft rejection by direct oral administration or by adoptive transfer of *in vitro*–treated DCs. Tolerogenic DCs induced by a short treatment with 1,25(OH)2D$_3$ are probably responsible for the capacity of this hormone to induce CD4$^+$CD25$^+$ regulatory T cells that are able to mediate transplantation tolerance (Adorini *et al.*, 2003).

The observations on the effects of 1,25(OH)2D$_3$ on *in vitro*–derived DCs suggested a physiologic role for 1,25(OH)2D$_3$/VDR in DC homeostasis *in vivo*. Analysis of LN and spleens of wild-type (WT) and VDR KO mice showed that LN from VDR-deficient mice, which are unresponsiveness to 1,25(OH)2D$_3$, were larger than those of WT animals associated with an increased number of mature DCs. Thus one can speculate about physiologically relevant inhibition of DC differentiation by 1,25(OH)2D$_3$ *in vivo* (Griffin and Kumar, 2003).

In vivo studies with naïve or 1,25(OH)2D$_3$ treated DCs demonstrated that 1,25(OH)2D$_3$ also induces a state of dendritic cell immaturity in phenotype and function *in vivo*, which results in tolerogenic immune responses. In recent studies, myeloid DCs were described for the first time as a possible extrarenal source of 1,25(OH)2D$_3$. LPS-induced different-tiation and maturation of human DCs is associated with significant 1,25(OH)2D$_3$ synthesis, which is caused by transcriptional upregulation of 25(OH)D$_3$-1α-hydroxylase, an enzyme that catalyzes the conversion of 25(OH)D$_3$ (Fritsche *et al.*, 2003). In contrast, the nuclear receptor VDR is significantly downregulated upon differentiation of monocytes to DCs, suggesting a potential autocrine/paracrine loop in the regulation of DC differentiation and function (Hewison *et al.*, 2003). Here, end-stage DCs are able to block further maturation of incoming immature DCs, without being affected by the 1,25(OH)2D$_3$ themselves. These results identify the synthesis of 1,25(OH)2D$_3$ by DCs as an inhibitory factor of undesirable immune reactivity.

4.3.3 IL-10 Modulates DCs for Tolerance Induction

The absence of activation stimuli is one crucial factor that determines the development of tolerogenic DCs, however more active mechanisms that directly interact with DCs may be provided by immunomodulatory cytokines such as IL-10. IL-10 has originally been described as cytokine-synthesis-inhibiting factor (CSIF) with regard to its effects exerted on IFN-γ production by Th1 cells. Meanwhile, it has been found to exert suppressive effects on a wide range of different populations of lymphocytes. When human or murine DCs are exposed to IL-10 in *in vitro* culture systems, the cells display reduced surface expression of MHC class I and MHC class II molecules and reduced expression of T-cell costimulatory molecules of the B7 family. In addition, the release of proinflammatory cytokines, that is, IL-1β, IL-6, TNF-α and most markedly IL-12, is abolished after IL-10 treatment (Buelens *et al.*, 1997; De Smedt *et al.*, 1997; Koch *et al.*, 1996). However, all of these effects could only be recorded when immature DCs were exposed to IL-10. In contrast, mature DCs are insensitive to IL-10 and display a stable phenotype in the presence of IL-10 once they have matured (Steinbrink *et al.*, 1997; Steinbrink *et al.*, 1999; Steinbrink *et al.*, 2002).

According to their reduced MHC and B7 expression, the IL-10–treated DCs are inferior in T-cell stimulation as opposed to their fully activated counterparts, but IL-10 does not merely keep DCs in an immature state, instead there is evidence that IL-10 modulates DC maturation enabling DCs to induce T cells with regulatory properties. For example, freshly isolated Langerhans cells inhibit proliferation of Th1 cells after exposure to IL-10 but had no effect on Th2 cells (Chang *et al.*, 1995). Moreover, it has been shown that IL-10–modulated DCs from peripheral blood induce alloantigen-specific anergy or anergy in melanoma-specific CD4$^+$ and CD8$^+$ T cells (Sato *et al.*, 2003a). Further analysis of these anergic T cells revealed reduced IL-2 and IFN-γ production and, in contrast with genuine Treg, reduced expression of the IL-2 receptor α-chain CD25. However, in addition to these anergic T cells, some authors have also observed the emergence of genuine Treg after injection of IL-10 as indicated by CD25$^+$ upregulation and cell-cell contact requirement for their suppressive activity. The therapeutic use of these IL-10–modulated DCs is under investigation because injection of *in vitro*–generated, IL-10–modified DCs is able to prevent autoimmunity in the EAE murine model and prolonged graft survival significantly in a murine GVHD model (Muller *et al.*, 2002; Sato *et al.*, 2003a; Sato *et al.*, 2003b). Although most of these protocols involved *in vitro* exposure of DCs to IL-10, there is recent evidence that IL-10–driven DC modulation may also play a role in generation of regulatory T cells *in vivo*. For instance, Wakkach *et al.* (Wakkach *et al.*, 2003) could not only confirm previous *in vitro* results showing that addition of IL-10 to *in vitro* cultures differentiated DCs to a CD45^{high} tolerogenic phenotype but could also demonstrate that this tolerogenic phenotype, along with regulatory Tr1 cells, is significantly enriched in spleens of IL-10 transgenic mice. Thus, these data show that IL-10 plays an important role in rendering DCs not merely immature but also modifies their ability to induce regulatory T cells *in vivo*.

However, IL-10 may play a dual role, because it acts simultaneously on DCs as well as on T cells. For example, Misra *et al.* (2004) have shown that DCs cocultured with Treg remain in an immature state as judged by surface marker expression. These "Treg-exposed" DCs were inferior in induction of T-cell proliferation and produced significant amounts of IL-10. In another murine cardiac transplantation model, increased numbers of splenic CD4$^+$CD25$^+$ Treg and immature DCs were observed after treatment of the recipients with a commonly used antirejection drug. As expected, these immature DCs purified from tolerant recipients induced the generation of CD4$^+$CD25$^+$ Treg when incubated with T cells. Surprisingly, when these Treg isolated from tolerant recipients were incubated with DC progenitors, generation of DCs with tolerogenic properties, that is, inferior T-cell stimulatory capacity and IL-10 production, was observed. In conclusion, these results support the notion that IL-10 is a critical factor in a self-maintaining feedback loop, that is, IL-10 derived from regulatory T cells has been shown to play a role in locking immature DCs in a tolerogenic state, which in turn induces further regulatory T cells that may contribute to IL-10 production (Min *et al.*, 2003). However, this positive feedback loop can ensure prolonged immunosuppression and does not only rely on the cell-cell contact required by genuine Treg.

4.4 Molecular Mechanisms Involved in Tolerance Induction

4.4.1 Costimulatory Molecules

Current results indicate that resting or steady-state DCs are prone to induce tolerance, however the molecular means by which the DCs accomplish silencing of T cells is not clear yet. Partial explanation can be deduced by considering how T cells are normally stimulated. Here it has long been obvious that T cells need signal 1 (provided by MHC-peptide complex) and signal 2 (costimulatory molecules) in order to be activated properly. If one signal is missing, T-cell anergy results; and indeed this situation may occur when T cells encounter steady-state or immature DCs (Inaba *et al.*, 1994). By definition, immature DCs express only modest amounts of MHC-peptide complexes and are devoid of T-cell costimulatory molecules. Accordingly, migratory T cells might be anergized when entering the tissue because they here encounter mostly immature DCs that engage the TCR via MHC class II–peptide complexes without concomitant costimulation. Also, most tissue cells do not possess the function of costimulatory molecules but do express mediocre amounts of MHC class I–peptide complexes, facilitating the tolerization of CD8$^+$ T cells.

Similar to the situation in peripheral tissues, the resident DCs in lymphoid organs also display an immature phenotype, thus, in the steady state most T cells passing through the LN and other secondary organs may be anergized by immature DCs via engagement of the TCR by MHC-peptide complexes and concomitant absence of costimulation. However, this scenario complies with a rather passive role of DCs during T-cell tolerization but active mechanisms such as secretion of immune-modulatory cytokines or T-cell-suppressive surface molecules would provide further means by which T-cell responses can be curbed.

Among those negative regulatory molecules are ligands for the "programmed cell death" (PD) molecule family (Greenwald *et al.*, 2005; Prasad *et al.*, 2004). These B7 relatives confer negative regulation to T cells, that is, engagement of these receptors prevents T-cell activation (Rodig *et al.*, 2003; Strome *et al.*, 2003; Suh *et al.*, 2003). Some ligands for "PD" molecules are characterized so far: PD-L1, PD-L2, and PD-L3, all of them are expressed by DCs. However, *in vivo* data is limited so far, but it seems conceivable that these molecules may contribute to T-cell anergy, as expression of these ligands on DCs converts them into T-cell-suppressing DCs (Schreiner *et al.*, 2004; Selenko-Gebauer *et al.*, 2003).

In the human system, more data is available about the potentially tolerogenic effects of the receptors ILT3 and ILT4 (Ju *et al.*, 2004; Penna *et al.*, 2005). The "immunoglobulin-like transcripts" (ILT) belong to a family of inhibitory receptors that are expressed by human monocytes and DCs.

These receptors are structurally related to KIR receptors and contain an ITIM motive as it is known from inhibitory FcγRIIb (Ravetch and Bolland, 2001). Upon engagement of these receptors by classical and nonclassical MHC class I molecules, the calcium flux as well as phosphorylation pattern is modified in a way that these cells cannot become activated anymore (Shiroishi *et al.*, 2003). The ILT3 and ILT4 receptors are expressed by DCs, and after binding to respective ligands, the DCs acquire tolerizing properties and are no longer able to mount an immune response (Cella *et al.*, 1997; Manavalan *et al.*, 2003).

Interestingly, these receptors are also upregulated by DCs after contact to CD8⁺/CD28⁻ suppressor T cells (Ts). Because these Ts are immuno-suppressive by themselves, this mechanism provides another means by which suppressive activity is spread among immune cells in an immune-tolerance stimulating feedback loop. First, Ts are able to suppress T-cell activation by direct T cell–T cell interaction(s) and second, surrounding DCs are converted into a tolerizing phenotype, which may block CD4⁺ T-cell activation. And indeed it has been shown that DCs of successfully trans-planted patients, which have high numbers of Ts, expressed significantly higher amounts of ILT3 und ILT4 as compared with controls or patients, which had recently rejected a transplant. Thus, upregulation of ILT on DCs is strongly associated with the absence of acute rejection (Chang *et al.*, 2002).

4.4.2 RelB Translocation Is Crucial for DC Maturation

Many different exogenous stimuli that confer DC activation (IL-1, LPS, TNF-α, etc.) are funneled into the NFκB pathway. NFκB comprises a family of transcription factors one of which is RelB. Upon activation of the DCs, the transcription factors translocate from the cytoplasm into the nucleus of the cells (Thompson *et al.*, 2002). Among the five members of the NFκB family, RelB, p50, and c-Rel are the major players in DC maturation. They are involved in induction of IL-12 secretion, CD40 and MHC class II expression, and in the transcription of anti-apoptotic factors. The most crucial effect of RelB is marked by the upregulation of CD40. Blockade or absence of CD40 has been shown to convert B cells as well as DCs into tolerance-inducing APCs (Dong *et al.*, 2003; O'Sullivan *et al.*, 2000; Pettit *et al.*, 1997). Moreover, CD40 is directly involved in a stimulating feedback loop with RelB transition. Here, CD40 expression is upregulated by RelB and in parallel engagement of CD40 molecules further augments RelB transition.

The interaction between CD40 molecules and RelB may also be of importance in the regulation of self-tolerance. For instance, in the skin the maturation of Langerhans cells (LCs) is critically controlled by TGF-β. This strong connection successfully blocks maturation of LCs even in the presence of TNF-α, IL-1, and/or LPS. Therefore, epidermal LCs are not sufficiently activated solely by microbial products. Only after engagement of CD40 via CD40-L expressing T cells and thus activation of RelB, LCs can be fully activated (Geissmann *et al.*, 1999). This mechanism may provide another safeguard mechanism to avoid overboarding immune reactions. Here, microbial stimuli are not enough to fully activate DCs. Instead, DCs only then turn into T-cell activators after they have made contact with a matching T-cell partner. That way, an effective DC–T cell interaction is ensured and unspecific activation of DCs at the site of antigen contact is avoided.

On a molecular basis, a strong correlation between immature, that is, tolerogenic DCs and absence of nuclear RelB exists. In the periphery, steady-state DCs have been shown to be negative for active (nuclear) RelB and agents such as IL-10 and TGF-β, suppressing DC activation, do also suppress the cytoplasm-nuclear transition of RelB. Vice versa, blocking of these RelB "suppressors" results in autoimmunity and exacerbated inflammation in various murine models (Fig. 4.3.).

Most of the pharmaceuticals that drive DCs into a more tolerogenic phenotype interact with the RelB pathway. There is evidence that mycophenolate mofetil, glucocorticoids, and vitamin D$_3$ all downregulate NFκB, and when incubated with DCs their function is indeed modulated in a way

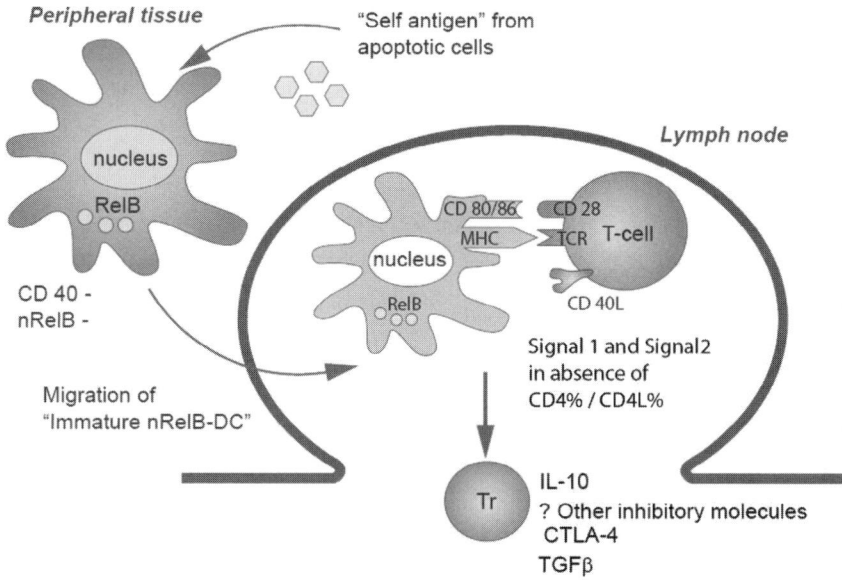

FIGURE 4.3. RelB translocation is crucial for DC activation. Immature DCs in the periphery do not express RelB in the nucleus and CD40 is not expressed either. When proper DC activating stimuli are missing, these nuclear RelB-negative, CD40-negative DCs migrate to the lymph node. Here, the mere contact with T cells is not enough to induce nuclear RelB expression. Only after proper stimulation via CD40/CD40L nuclear translocation occurs, and several DCs activating pathways will be induced. Without this stimulation, DCs remain negative for nuclear RelB and are prone to induce tolerance via several mechanisms.

that induction of regulatory T cells is promoted (Gregori *et al.*, 2001; Mehling *et al.*, 2001; Penna and Adorini, 2001). This system can further be utilized for therapeutic efforts on a molecular level. For instance, when DCs were transfected with the dominant negative form of IκBα, these DCs induce T-cell anergy (Calder *et al.*, 2003; Yoshimura *et al.*, 2001). Also, the introduction of oligo-desoxynucleotides with decoy binding sites for NFκB converted already mature DCs into tolerogenic DCs that induce tolerance as indicated by hyporesponsiveness and prolongation of allograft survival (Tomasoni *et al.*, 2005). In aggregate, nuclear translocation of RelB in DCs is a reliable marker for DC activation, and application of pharmaceuticals preventing or delaying nuclear RelB expression *in vivo* can "lock" DCs in an immature state and may provide a tool by which regulatory T cell are induced via immature DCs.

4.4.3 The Role of Indoleamine 2,3-dioxygenase (IDO) in Tolerance Induction

The enzyme indoleamine 2,3-dioxygenase (IDO) catalyzes the oxidative cleavage of the indole ring of several important regulatory molecules, including tryptophan. Tryptophan is the fewest of the 20 amino acids in proteins to be found and therefore the most important essential amino acid for the biosynthesis of proteins and the growth of viruses, bacteria, and parasites. When infectious agents infiltrate tissues, leukocytes and lymphocytes accumulate at the inflamed site and release interferons into the local tissue microenvironment. Interferon then acts in an autocrine or paracrine fashion to activate IDO expression transcriptional. Tryptophan is degraded and depleted by IDO, which results in several metabolites like kynurenine, which is the main metabolite of tryptophan catabolism, 3-hydroxykynurenine, 3-hydroxyanthranilic acid, and quinolinic acid (Taylor and Feng, 1991).

IDO is expressed particularly in human and animal APCs of lymphoid organs and in the placenta. During infection or inflammation, the IDO production significantly increases. One of the first studies demonstrating the essential role of IDO in controlling immune responses in mice showed that IDO expression in the placenta is critically involved in preventing immunological rejection of allogeneic fetuses (Munn *et al.*, 1998). Here, the enzyme suppresses maternal T cells that otherwise would make their way through the placenta and attack the fetus. Inhibition of IDO with 1-methy-DL-tryptophan (1-MT), a well-known inhibitor of IDO, results in rejection of the fetus in pregnant mice, indicating that IDO expressed by placental trophpblasts and macrophages normally contributes to the maintenance of maternal T-cell tolerance (Mellor *et al.*, 2001).

Also in adults, the expression of IDO in APCs mediates the ability to suppress unwanted cells, particularly autoreactive T cells and henceforth contributes to the maintenance of peripheral tolerance against self-antigens. In humans and mice, distinct subpopulations of APCs have been identified to express IDO. In mice, the functional enzyme is particularly expressed by the IFN-γ–treated CD8α$^+$ DC subset residing in the spleen (Fallarino et al., 2002a). Likewise in humans, there is one IDO$^+$ subset of human monocyte-derived DCs described that is characterized by the coexpression of CD123 and CCR6 (Munn et al., 2002). These are minor subsets of CD11c$^+$ DCs, which exert suppressive function on T-cell responses and promote tolerance to specific antigens. Expression of IDO in APCs, especially in DCs, is dependent of their status of activation. Thus, there is only a negligible amount of IDO produced by inactive DCs, whereas activation of cultured monocyte-derived human DCs with CD40L/IFN-γ or CD40L/LPS/IFN-γ induces IDO mRNA significantly (Hwu et al., 2000). The translated enzyme is functionally active and capable of metabolizing tryptophan to kynurenine. Also, OKT3-activated T cells are able to induce IDO production in DCs, and presumably IFN-γ plays a major role, as neutralizing antibodies against IFN-γ can inhibit IDO induction. Production of IDO by DCs results in degradation of tryptophan and inhibition of T-cell proliferation, which can be prevented by adding the IDO inhibitor 1-MT. Accordingly, the presence of increasing concentrations of 1-MT results in enhanced T-cell proliferation (Hwu et al., 2000). These results suggest that activation of DCs induces the production of functional IDO, which causes depletion of tryptophan and subsequent inhibition of T-cell responses. This may represent a potential mechanism for DCs to regulate the immune responses (Fig. 4.4.).

One regulatory molecule that is involved in the in vivo tryptophan catabolism in DCs is CTLA-4-Ig, which was shown to be involved in the induction of tolerance to islet allografts (Grohmann et al., 2002). Activation of the B7 molecules through the binding of CTLA-4-Ig leads to intracellular signals in the DCs mediating the production of IFN-γ and subsequently the production of IDO. This therapeutically effect of CTLA-4-Ig could also be completely reversed by inhibiting IDO with 1-MT. DCs isolated from IFN-γ-KO mice did not express IDO upon exposure to CTLA-4-Ig, showing again the essential role of IFN-γ in the upregulation of IDO after B7 ligation.

IL-10, a major cytokine secreted by Tr cells, is another possible candidate, which influences the expression and function of IDO. Indeed it has been shown that stable IDO expression is potentiated in DCs by IL-10 (Munn et al., 2002) but is suppressed in human monocytes by presence of IL-4 (Musso et al., 1994). In vitro, TNF-α also enhances synergistically

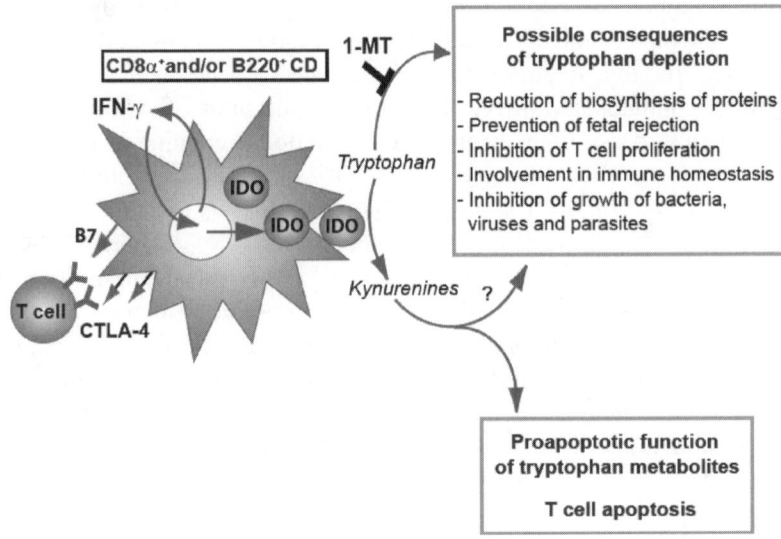

FIGURE 4.4. Possible mechanism of tolerance induction by DCs producing indoleamine 2,3-dioxygenase (IDO). Activation of the B7 molecules on DCs through the binding of CTLA-4, expressed on T cells and other cell types (e.g., placental fibroblasts), leads to intracellular signals in the DCs mediating the production of IFN-γ, which acts in an autocrine or paracrine fashion to activate IDO expression transcriptionally. In addition to the immunosuppressive effects resulting from the IDO-mediated deletion of tryptophan, pro-apoptotic tryptophan metabolites also acts on T cells to control their proliferation and/or survival *in vivo*.

the IDO activity induced by IFN-γ (Robinson *et al.*, 2005). However, the exact mechanisms of IDO induction and activation in DCs and the roles of several cytokines are not fully understood yet.

A recent study by Braun *et al.* (Braun *et al.*, 2005) characterized a new mechanism by which IDO is upregulated in monocyte-derived DCs. It is shown that treatment with PGE2 induces mRNA expression of IDO in DCs, which is not functionally active at first. Only when combined with a second signal via TNF-R or a TLR tryptophan metabolites were measured, demonstrating the active IDO enzyme. This finding suggests a link between DC migration and the acquisition of tolerogenic functions. As monocyte-derived DCs exhibit only a poor migration capability to CCR7 ligand unless matured in the presence of PGE2, it is conceivable that only migratory (PGE2-exposed) DCs are susceptible to IDO induction.

Some controversial results regarding the mechanisms involved in the inhibition of T-cell activation by IDO exist. There are two possible mechanisms discussed: either IDO depletes tryptophan, the essential amino acid required for cell proliferation (Mellor and Munn, 1999), or the appearance of metabolites resulting from tryptophan breakdown has immunomodulatory effects on lymphocytes (Terness *et al.*, 2002).

For the tolerogenic function of DCs, however, various mechanisms, including induction of T-cell anergy, immune deviation, Treg cell activity, and promotion of activated T-cell apoptosis are discussed. IFN-γ–activated, IDO-expressing CD8α^+ DCs can affect apoptosis of Th1 cell clones, which can be reversed by IDO inhibitor (Grohmann *et al.*, 2001). Thereby it was shown that the IDO-induced tryptophan metabolites in the kynurenine pathway, such as 3-hydroxykynurenine and 3-hydroxyanthranilic acid, induce the selective apoptosis of murine thymocytes and of Th1 but not Th2 cells. Both suppressive effects have been shown to be additive. The suppressed T-cell proliferation cannot be reversed by restimulation, further demonstrating that the mechanism of suppression is cell death. This T-cell apoptosis is independent of Fas/FasL interaction and is associated with activation of caspases-8 and release of cytochrome c from mitochondria (Fallarino *et al.*, 2002b).

Not only T cells, but also B and NK cells are sensitive to tryptophan metabolite cytotoxicity, whereas DCs, the cells that produce IDO, are resistant (Terness *et al.*, 2002). Thus, apoptosis of autoreactive lymphocytes, initiated by IDO expressing DCs, could represent a crucial means of maintaining peripheral tolerance.

DCs are the most potent activators of naïve T cells. But on the other hand, upon activation DCs produce an enzyme that inhibits T-cell proliferation. This may be part of a negative feedback loop whereby DCs may regulate immune responses in the presence of a large number of activated T cells expressing CD40L and IFN-γ. Another possible explanation is that IDO is produced by a subset of inhibitory DCs within the bulk population, as evidence exists for the presence of inhibitory or tolerogenic DCs (Mellor *et al.*, 2003; Mellor and Munn, 2004; Munn *et al.*, 2005).

4.5 Are There Specialized Subsets of Tolerogenic DCs?

4.5.1 Surface Marker Expression

DCs are clearly not a single homogeneous population. The relationship between different DC subpopulations is still better understood in mice compared with humans, as lymphoid tissues are available in great numbers and more established surface markers for murine DCs are known.

Recent evidence suggests that the diverse functions of DCs in immunoregulation depend on the diversity of DC subsets and lineages as well as on the functional plasticity of DCs in the immature stage. Currently, two possible mechanisms by which DCs might maintain peripheral tolerance are discussed. The first proposes that tolerogenic DCs represent a specialized lineage, whereas an alternative explanation would be that the same DC type may be responsible for inducing either tolerance or immunity depending on the context in which DCs are matured and/or activated.

The original concept was that immature DCs (iDCs) induce tolerance, whereas mature DCs are responsible for immunity induction (Mahnke *et al.*, 2003a). Thereby, there are at least three possible mechanisms that may be involved in the tolerogenic activity of iDCs: antigenic presentation in absence of costimulation, deletion of antigen-reactive cells by apoptosis (Hawiger *et al.*, 2001), and the induction of regulatory T cells (Mahnke *et al.*, 2003c). However, the idea has been revised to suggest that different DC lineages regulate immunity versus tolerance (Shortman and Heath, 2001; Vremec and Shortman, 1997).

Experiments, performed particularly in different lymphoid and non-lymphoid organs of the mouse have defined two main DC subtypes characterized on the basis of their origin, phenotypic profile, and physiologic properties. Lymphoid DCs, such as thymic DCs or $CD8\alpha^+$ splenic DCs, are defined as $CD8\alpha^+$, DEC-205^{high}, Mac-1^{low}, whereas myeloid DC, such as $CD8\alpha^-$ splenic DCs, are $CD8\alpha^-$, DEC-205^{low}, Mac-1^{high}.

$CD8\alpha^+$ and $CD8\alpha^-$ subsets represent approximately 60% and 40% of total splenic DCs, respectively. $CD8\alpha^+$ splenic DCs have the same phenotype as thymic DCs, that is, they are $CD8\alpha^+$, express the endocytic receptor DEC-205 recognized by the MoAb NLDC-145, express low levels of the myeloid marker Mac-1, and have high levels of LFA-1, FcR, B7-2, and CD40 on their surface. On the other hand, $CD8\alpha^-$ splenic DCs are DEC-205^{low} Mac-1^{high}, display high levels of LFA-1 and FcR, and express the costimulatory molecules B7-2 and CD40 at lower levels than $CD8\alpha^+$ splenic DCs (Kronin *et al.*, 1997b; Kronin *et al.*, 1997a; Vremec and Shortman, 1997; Wu *et al.*, 2001).

It was shown that splenic $CD8\alpha^-$ DCs induced a potent proliferative response in $CD4^+$ T cells, whereas $CD8\alpha^+$ DCs expressed high levels of CD95L and induced a lesser response that was associated with Fas-dependent T-cell apoptosis *in vitro* (Suss and Shortman, 1996). Notably, the same subtype of CD95L-expressing DCs in mice also expresses the DEC-205 receptor, and recently a DEC-205$^+$ subset of liver DCs has been shown to exert rather tolerogenic functions (Lu *et al.*, 2001). Therefore, it might be conceivable that the DEC-205 receptor, beyond its function as an antigen receptor, may serve as an indicator for a tolerogenic subtype of DCs (Kronin *et al.*, 1997a; Kronin *et al.*, 1997b).

More recently, the same subset, which was involved in cross-priming, was shown to induce peripheral self-tolerance to tissue-associated antigens, a phenomenon referred to as cross-tolerance (Belz *et al.*, 2002; Liu *et al.*, 2002). In addition, functional differences between CD8α$^+$ and CD8α$^-$ splenic DCs concerning their T-cell stimulation potential, phagocytic activity, IL-12 secretion capacity, and localization within the spleen have been reported. The potential role of CD8α$^+$ DCs in peripheral tolerance is, however, challenged by several reports showing that this DC subpopulation is a potent inducer of Th1-biased cytokine responses in reactive CD4$^+$ T cells associated with a high production of IL-12, whereas CD8α$^-$ DCs tend to a Th2 biased response (Kronin *et al.*, 1996; Kronin *et al.*, 2001; Pooley *et al.*, 2001). However, Th2 cells are still activated and immunologically active T cells, but in contrast with their Th1 counterparts, they secrete IL-10, stimulate only few cytotoxic effects, and mediate overall a less fulminant immune response. Therefore, Th2 is not a hallmark of tolerance induction, but because these cells mount a more "tolerable" immune response, they are frequently attributed to tolerogenic phenomena.

4.5.2 Cytokine Expression

The dogma of an irrevocably tolerogenic phenotype of DCs is further challenged by results showing that the same DC type may be responsible for inducing either tolerance or immunity depending on the context of maturation or activation and/or cytokine production. Thus, it was shown that pathogen-derived molecules subvert the immune response by inducing tolerogenic DC phenotypes, as demonstrated in studies with the virulence factor of *Bordetella pertussis*. This filamentous hemagglutinin interacts directly with DCs to induce IL-10 thus preventing an effective immune reaction (McGuirk *et al.*, 2002). Another study demonstrated that mature pulmonary DCs from mice exposed to respiratory antigen transiently produce IL-10 and mediate tolerance (Akbari *et al.*, 2001). Additionally, the DC production of IL-10 or other immunosuppressive cytokines may be critical for the differentiation of suppressor cells. In human, lymphoid DCs polarize naïve T cells toward IL-10–producing T cells, and different DC subtypes (depending on their anatomical location) seem to suppress immune responses by the development of Th2 cells or regulatory T cells. Similar results could be obtained in mice, showing that DC subpopulations residing in the respiratory tract, the Peyer's patch, or the liver induce Th2 cells and therefore prevent the induction of Th1-driven inflammatory T-cell responses. Although Th2 cells per se are not tolerogenic themselves, the Th2-derived IL-10 can then in turn further augment tolerance, as exposure of DCs to IL-10 seems to convert them into a tolerogenic phenotype. In summary, IL-10 may be involved in two different ways: (1) IL-10 may be required to differentiate regulatory T cells, or (2) IL-10 can exert direct effects on DCs, leading to decreased stimulatory function and induction of tolerogenic function(s). However, these experiments rely on *in vitro* experiments using cultured DCs and it remains to be determined whether DCs in the steady state *in vivo* produce IL-10 (Levings *et al.*, 2001; Steinbrink *et al.*, 1997; Steinbrink *et al.*, 1999).

4.6 "Designer DC": Tailored for Tolerance Induction

Emerging evidence about molecules and mechanisms involved in tolerance induction by DCs initiated the search for molecular means by which DCs can irrevocably be driven into a tolerogenic phenotype.

As outlined before, different cytokines as well as pharmaceuticals can induce a relatively stable phenotype of tolerogenic DCs. In these trials, the DC development under influence of the drugs were tested in *in vitro* cultures; however, once these immature DCs are injected, the further control and monitoring of the maturation process is impossible. Therefore, recently many different attempts were made to manipulate the DC phenotype by molecular and/or pharmaceutical means, which would provide a promising tool to constantly keep DCs in a tolerogenic status.

These attempts include the transfection of DCs with Fas ligand (FasL), CTLA-4, suppressive cytokines such as IL-10 and TGF-β, NFκB blocking factors, and T-cell receptor mimic peptides.

As outlined before, IL-10 has the ability to prevent DC maturation (Steinbrink *et al.*, 1997). Likewise, TGF-β is one of the major immunosuppressive cytokines (Yamaguchi *et al.*, 1997) and accordingly several attempts have been made to transfect DCs *ex vivo* with cDNA coding for these immunoregulatory cytokines. These experiments aimed to convert DCs into IL-10– and/or TGF-β–producing DCs with tolerogenic properties (Gorczynski *et al.*, 1999). As expected, IL-10–transfected DCs exhibited reduced allogenic T-cell stimulatory capacity and induced T-cell anergy *in vitro* and *in vivo* (Coates *et al.*, 2001a; Coates *et al.*, 2001b). Moreover, in an *in vivo* approach the injection of IL-10–transduced DCs did indeed lead to prolonged allograft survival in a murine kidney transplantation model (Takayama *et al.*, 1998; Takayama *et al.*, 2000); and in sheep the induction of a alloantigen-specific Th2 response was observed (Coates *et al.*, 2001b).

As mentioned before, Th2 cells are only indirectly involved in tolerance. These cells secrete IL-10 and mount a somewhat narrowed immune response. Similar to IL-10, IL-4 is also a "Th2 promoting" cytokine and thus is able to block development of Th1 responses. Therefore, transfection of DCs with IL-4 subsequently leads to the suppression of immune responses in a murine model of adjuvant-induced arthritis and had beneficial effect on the course of diabetes in a NOD mouse model (Kim *et al.*, 2001). As for IL-10, similar approaches with TGF-β producing DCs were performed. Here the data also show that in rhesus monkeys, CD4$^+$ as well as CD8$^+$ T-cell responses were inhibited by TGF-β transfected DCs (Asiedu *et al.*, 2002).

In aggregate, these results have demonstrated that the production of immunoregulatory cytokines in an autocrine fashion is able to modify DC function during tolerance induction. However, this secretion may facilitate induction of tolerance beyond the direct effects on DC maturation: first IL-10 acts in an autocrine fashion preventing the maturation of DCs and thus locks them in an immature and thus tolerogenic status. Second, cytokines also act in a paracrine fashion and therefore also affect the induction of T-cell responses and/or antigen presentation by other APCs directly. Therefore, the secretion of regulatory factors may be effective in some clinical settings, but safety concerns may arise from the fact that their suppressive effects may spread out as third-party cells are affected.

A different set of approaches, however, takes advantage of DC specific expression of immunomodulatory molecules on the surface of DCs. Among them are apoptosis-inducing molecules such as the Fas ligand (FasL) and T-cell negative regulatory molecules such as CTLA-4 and B7H's. These molecules directly interfere at the site of DC–T cell interaction by different means (Fig. 4.5.).

The transfection of DCs with CTLA-4Ig takes advantage of the fact that T cells express regulatory molecules on their surface, which would in case of engagement suppress T-cell activation. In this regard, the CTLA-4 molecule is involved in counterregulation of T-cell activation. Therefore, when DCs express anti-CTLA-4 single chain antibodies, those chimeric molecules will bind to the T cells and prevent T-cell activation upon contact (Takayama *et al.*, 2000). This strategy has successfully been tested as those CTLA-4Ig–transduced DCs blocked T-cell activation in an allogenic system *in vitro* (O'Rourke *et al.*, 2000).

However, recently several novel candidates of negative regulatory molecules expressed by T cells have been characterized. For instance, the programmed cell death ligand 1 (PD-L1 or B7-H1) is an enhancer for T-cell anergy, whereas the blockade of this receptor results in enhanced IL-2 production and hyperproliferative T cells. Interestingly, B7-H1 expression

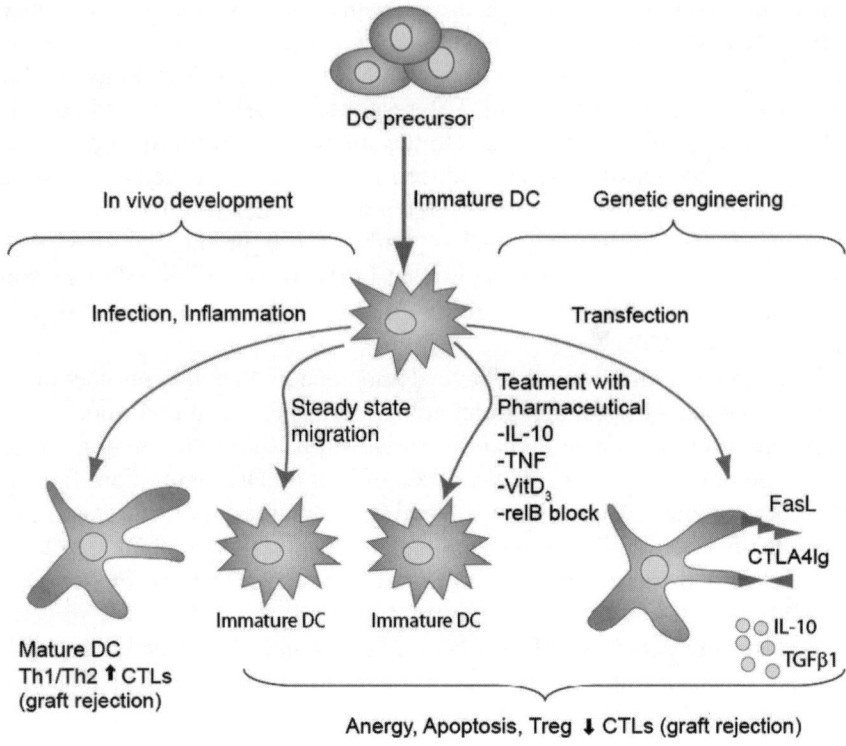

FIGURE 4.5. Development of tolerogenic DCs. Either ex vivo manipulation with cytokines or transfection with immunomodulatory molecules can induce a tolerogenic phenotype of DCs.

is enhanced/upregulated on immature DCs and downregulated during maturation. Therefore, transfection of DCs with cDNA coding for B7-H1 may provide us with another tool to genetically "rejuvenate" maturing DCs and engineer tolerogenic phenotype.

Even more effective than silencing or anergizing T cells would be their removal from the body, that is, the "killing" of them. In this regard, several apoptosis-inducing molecules have been tried for the generation of so-called killer DCs (Mahnke *et al.*, 2003b; Matsue *et al.*, 1999). These DCs are transfected to carry the FasL and therefore are able to provoke suicide of the T cells via engagement of the Fas (CD95) molecule. This FasL-Fas interaction induces apoptosis, and in several animal models it has been shown that injection of these killer DCs did indeed protect animals from EAE, CHS, and other T cell–mediated diseases (Kusuhara *et al.*, 2002; Matsue *et al.*, 2001). However, the therapeutic benefits of this treatment have to be carefully revisited.

Novel emerging molecules involved in apoptosis may further revitalize this field of DC engineering. For instance, the novel lectin-like molecule galectin 1 has implications on Treg development and is able to induce apoptosis similar to FasL. Although not used for DC transfection yet, the recombinant protein has been shown to prevent collagen-induced arthritis in *in vivo* models (Rabinovich *et al.*, 1999).

However, clinical use of "killer DCs" to delete potentially autoreactive T cells may bear the risk that accidentally other cell types that express Fas also undergo apoptosis. Therefore, means that specifically block T cell by interaction with the TCR are preferable.

An approach along these lines has been tested using a T-cell receptor (TCR) mimic peptide (Manolios *et al.*, 1997). This peptide consists of the transmembrane sequence of the α-chain of the TCR. Normally upon engagement of the TCR, the CD3 molecule interacts with the transmembrane domain of the α-chain of the TCR, eventually leading to activation of the intracellular activation cascade. The TCR mimic peptide consists of the very same transmembrane sequence of the α-chain and when added to T cells, this peptide is integrated into the T-cell membrane and blocks CD3 molecules. As a consequence, the CD3 molecules are no longer available to engage with the TCR and therefore T-cell activation is abrogated completely. This peptide is highly specific for T cells as it only interacts with the TCR. However, systemic application leads to an unwanted complete abrogation of T-cell activation (Gollner *et al.*, 2000). To utilize this system for *in vivo* therapy, experiments were performed to transduce DCs with a secreted form of the TCR mimic peptide in order to make them "tolerogenic". It was shown that TCR mimic peptide–transduced DCs were able to inhibit allergic and autoimmune responses in different animal models (Mahnke *et al.*, 2003b). Interestingly, this effect was antigen specific, as only those T cells were inhibited that recognized the respective antigen presented by the TCR mimic peptide–transduced DCs.

This effect might be explained by the fact that the DC–T cell contact is rather short for T cells that do not recognize the antigens presented by the respective DC. On the contrary, DC– T cell contacts are prolonged after matching TCR and MHC-peptide molecules have met. Therefore, only those T cells pick up substantial amounts of peptide that recognize their specific antigen and, hence, this system allows induction of antigen-specific tolerance.

4.7 Concluding Remarks

Dendritic cells are traditionally characterized by their immunostimulatory capacity; however, several lines of evidence indicate that DCs are also able to induce tolerance. DCs are often called the sentinels of the immune system and indeed they are favorably located in nearly all peripheral tissues. Here they can take up either self-antigens derived from apoptotic cells or detritus from tissue remodeling, or they are the first barrier for invading microorganisms.

Depending on which antigen is taken up, immunity (against "foreign") or tolerance (toward "self") has to be induced. How can a DC "decide" which reaction to induce? This dilemma is solved by the impact of the environment on DC development. During noninflammatory conditions, that is, when no foreign pathogens are present, DCs remain nonactivated in the steady state. This state is characterized by low numbers of T-cell stimulatory molecules and the secretion of putative immunosuppressive cytokines. Because these nonactivated DCs mainly carry self-antigens from the periphery, subsequent interaction with T cells will lead to tolerance induction.

On the contrary, the peripheral environment changes dramatically after infections occur. Now pathogens and inflammatory mediators are present, which induce a fully mature phenotype of DCs. Because these DCs originate from a diseased area, their antigen-load is mainly derived from foreign organisms (bacteria, virus, etc.). Therefore, these activated DCs are now prone to avoid tolerance and to induce immunity.

Thus, this concept explains how a single cell type is able to govern two opposite reactions, and it allows the body to utilize its pivotal sentinel cells in peripheral tissues for induction of tolerance and immunity.

Although this concept has its merits, there may be other safeguard mechanisms in place, ensuring tolerance beyond the "steady-state DCs." For example, it is still conceivable that specialized tolerogenic DC subsets exist. These DCs, which may intrinsically be programmed to tolerize T cells, are located at body surfaces where foreign, but otherwise harmless commensals are present. Here the specialized DCs continuously dampen the immune response, whereas otherwise migratory DCs monitor the tissue for intruders and mount an immune response in case pathogens are detected.

References

Abe, J., Nakamura, K., Takita, Y., Nakano, T., Irie, H., and Nishii, Y. (1990). Prevention of immunological disorders in MRL/l mice by a new synthetic analogue of vitamin D_3: 22-oxa-1 alpha,25-dihydroxyvitamin D_3. J. Nutr. Sci. Vitaminol. (Tokyo) 36:21-31.

Adorini, L., Penna, G., Giarratana, N., and Uskokovic, M. (2003). Tolerogenic dendritic cells induced by vitamin D receptor ligands enhance regulatory T cells inhibiting allograft rejection and autoimmune diseases. J. Cell. Biochem. 88:227-233.

Akbari, O., DeKruyff, R. H., and Umetsu, D. T. (2001). Pulmonary dendritic cells producing IL-10 mediate tolerance induced by respiratory exposure to antigen. Nat. Immunol. 2:725-731.

Albert, M. L., Pearce, S. F., Francisco, L. M., Sauter, B., Roy, P., Silverstein, R. L., and Bhardwaj, N. (1998). Immature dendritic cells phagocytose apoptotic cells via alphavbeta5 and CD36, and cross-present antigens to cytotoxic T lymphocytes. J. Exp. Med. 188:1359-1368.

Asiedu, C., Dong, S. S., Pereboev, A., Wang, W., Navarro, J., Curiel, D. T., and Thomas, J. M. (2002). Rhesus monocyte-derived dendritic cells modified to over-express TGF-beta1 exhibit potent veto activity. Transplantation 74:629-637.

Banchereau, J., and Steinman, R. M. (1998). Dendritic cells and the control of immunity. Nature 392:245-252.

Belz, G. T., Behrens, G. M., Smith, C. M., Miller, J. F., Jones, C., Lejon, K., Fathman, C. G., Mueller, S. N., Shortman, K., Carbone, F. R., and Heath, W. R. (2002). The CD8alpha($^+$) dendritic cell is responsible for inducing peripheral self-tolerance to tissue-associated antigens. J. Exp. Med. 196:1099-1104.

Berer, A., Stockl, J., Majdic, O., Wagner, T., Kollars, M., Lechner, K., Geissler, K., and Oehler, L. (2000). 1,25-Dihydroxyvitamin D(3) inhibits dendritic cell differentiation and maturation *in vitro*. Exp. Hematol. 28:575-583.

Blanco, P., Palucka, A. K., Gill, M., Pascual, V., and Banchereau, J. (2001). Induction of dendritic cell differentiation by IFN-alpha in systemic lupus erythematosus. Science 294:1540-1543.

Bonifaz, L., Bonnyay, D., Mahnke, K., Rivera, M., Nussenzweig, M. C., and Steinman, R. M. (2002). Efficient targeting of protein antigen to the dendritic cell receptor DEC-205 in the steady state leads to antigen presentation on major histocompatibility complex class I products and peripheral CD8$^+$ T cell tolerance. J. Exp. Med. 196:1627-1638.

Braun, D., Longman, R. S., and Albert, M. L. (2005). A two-step induction of indoleamine 2,3 dioxygenase (IDO) activity during dendritic-cell maturation. Blood 106:2375-2381.

Buelens, C., Verhasselt, V., De Groote. D., Thielemans, K., Goldman, M., and Willems, F. (1997). Human dendritic cell responses to lipopolysaccharide and CD40 ligation are differentially regulated by interleukin-10. Eur. J. Immunol. 27:1848-1852.

Calder, V. L., Bondeson, J., Brennan, F. M., Foxwell, B. M., and Feldmann, M. (2003). Antigen-specific T-cell downregulation by human dendritic cells following blockade of NF-kappaB. Scand. J. Immunol. 57:261-270.

Caux, C., Massacrier, C., Vanbervliet, B., Dubois, B., Van Kooten, C., Durand, I., and Banchereau, J. (1994a). Activation of human dendritic cells through CD40 cross-linking. J. Exp. Med. 180:1263-1272.

Caux, C., Vanbervliet, B., Massacrier, C., Azuma, M., Okumura, K., Lanier, L. L., and Banchereau, J. (1994b). B70/B7-2 is identical to CD86 and is the major functional ligand for CD28 expressed on human dendritic cells. J. Exp. Med. 180:1841-1847.

Cella, M., Dohring, C., Samaridis, J., Dessing, M., Brockhaus, M., Lanzavecchia, A., and Colonna, M. (1997). A novel inhibitory receptor (ILT3) expressed on monocytes, macrophages, and dendritic cells involved in antigen processing. J. Exp. Med. 185:1743-1751.

Canning, M. O., Grotenhuis, K., de Wit, H., Ruwhof, C., and Drexhage, H. A. (2001). 1-alpha,25-Dihydroxyvitamin D_3 (1,25(OH)(2)D(3)) hampers the maturation of fully active immature dendritic cells from monocytes. Eur. J. Endocrinol. 145:351-357.

Chang, C. H., Furue, M., and Tamaki, K. (1995). B7-1 expression of Langerhans cells is up-regulated by proinflammatory cytokines, and is down-regulated by interferon-gamma or by interleukin-10. Eur. J. Immunol. 25:394-398.

Chang, T. J., Lei, H. H., Yeh, J. I., Chiu, K. C., Lee, K. C., Chen, M. C., Tai, T. Y., and Chuang, L. M. (2000). Vitamin D receptor gene polymorphisms influence susceptibility to type 1 diabetes mellitus in the Taiwanese population. Clin. Endocrinol. 52:575-580.

Chang, C. C., Ciubotariu, R., Manavalan, J. S., Yuan, J., Colovai, A. I., Piazza, F., Lederman, S., Colonna, M., Cortesini, R., Dalla-Favera, R., and Suciu-Foca, N. (2002). Tolerization of dendritic cells by T(S) cells: the crucial role of inhibitory receptors ILT3 and ILT4. Nat. Immunol. 3:237-243.

Christakos, S., Barletta, F., Huening, M., Dhawan, P., Liu, Y., Porta, A., and Peng, X. (2003). Vitamin D target proteins: function and regulation. J. Cell. Biochem. 88:238-244.

Coates, P. T., Krishnan, R., Kireta, S., Johnston, J., and Russ, G. R. (2001a) Human myeloid dendritic cells transduced with an adenoviral interleukin-10 gene construct inhibit human skin graft rejection in humanized NOD-scid chimeric mice. Gene Ther. 8:1224-1233.

Coates, P. T., Krishnan, R., Chew, G., Kireta, S., Johnston, J., Kanachanabat, B., Russell, C. H., Siddins, M., and Russ, G. R. (2001b) Dendritic cell TH2 cytokine gene therapy in sheep. Transplant. Proc. 33:180-181.

De Smedt, T., Van Mechelen, M., De Becker, G., Urbain, J., Leo, O., and Moser, M. (1997). Effect of interleukin-10 on dendritic cell maturation and function. Eur. J. Immunol. 27:1229-1235.

Dong, X., Craig, T., Xing, N., Bachman, L. A., Paya, C. V., Weih, F., McKean, D. J., Kumar, R., and Griffin, M. D. (2003). Direct transcriptional regulation of RelB by 1alpha,25-dihydroxyvitamin D_3 and its analogs: physiologic and therapeutic implications for dendritic cell function. J. Biol. Chem. 278:49378-49385.

Fallarino, F., Grohmann, U., Hwang, K. W., Orabona, C., Vacca, C., Bianchi, R., Belladonna, M. L., Fioretti, M. C., Alegra, M. L., and Puccetti, P. (2003). Modulation of tryptophan catabolism by regulatory T cells. Nat. Immunol. 4:1206-1212.

Fallarino, F., Vacca, C., Orabona, C., Belladonna, M. L., Bianchi, R., Marshall, B., Keskin, D. B., Mellor, A. L., Fioretti, M. C., Grohmann, U., and Puccetti, P. (2002a). Functional expression of indoleamine 2,3-dioxygenase by murine CD8 alpha($^+$) dendritic cells. Int. Immunol. 14:65-68.

Fallarino, F., Grohmann, U., Vacca, C., Bianchi, R., Orabona, C., Spreca, A., Fioretti, M. C., and Puccetti, P. (2002b). T cell apoptosis by tryptophan catabolism. Cell Death Differ. 9:1069-1077.

Fogh, K., and Kragballe, K. (2004). New vitamin D analogs in psoriasis. Curr Drug Targets Inflamm. Allergy 3:199-204.

Fournier, C., Gepner, P., Sadouk, M., and Charreire, J. (1990). *In vivo* beneficial effects of cyclosporin A and 1,25-dihydroxyvitamin D_3 on the induction of experimental autoimmune thyroiditis. Clin. Immunol. Immunopathol. 54:53-63.

Fritsche, J., Mondal, K., Ehrnsperger, A., Andreesen, R., and Kreutz, M. (2003). Regulation of 25-hydroxyvitamin D3-1 alpha-hydroxylase and production of 1 alpha,25-dihydroxyvitamin D_3 by human dendritic cells. Blood 102:3314-3316.

Geissmann, F., Revy, P., Regnault, A., Lepelletier, Y., Dy, M., Brousse, N., Amigorena, S., Hermine, O., and Durandy, A. (1999). TGF-beta 1 prevents the noncognate maturation of human dendritic Langerhans cells. J. Immunol. 162:4567-4575.

Gollner, G. P., Muller, G., Alt, R., Knop, J., and Enk, A. H. (2000) Therapeutic application of T cell receptor mimic peptides or cDNA in the treatment of T cell-mediated skin diseases. Gene Ther. 7:1000-1004.

Gorczynski, L., Chen, Z., Hu, J., Kai, Y., Lei, J., Ramakrishna, V., and Gorczynski, R. M. (1999) Evidence that an OX-2-positive cell can inhibit the stimulation of type 1 cytokine production by bone marrow-derived B7-1 (and B7-2)-positive dendritic cells. J. Immunol. 162:774-781.

Green, A., Gale, E. A., and Patterson, C. C. (1992). Incidence of childhood-onset insulin-dependent diabetes mellitus: the EURODIAB ACE Study. Lancet 339:905-909.

Greenwald, R. J., Freeman, G. J., and Sharpe, A. H. (2005). The B7 family revisited. Annu. Rev. Immunol. 23:515-548.

Gregori, S., Casorati, M., Amuchastegui, S., Smiroldo, S., Davalli, A. M., and Adorini, L. (2001). Regulatory T cells induced by 1 alpha, 25-dihydroxyvitamin D_3 and mycophenolate mofetil treatment mediate transplantation tolerance. J. Immunol. 167:1945-1953.

Griffin, M. D., and Kumar, R. (2003). Effects of 1alpha,25(OH)2D3 and its analogs on dendritic cell function. J. Cell. Biochem. 88:323-326.

Griffin, M. D., and Kumar, R. (2005). Multiple potential clinical benefits for 1alpha,25-dihydroxyvitamin D(3) analogs in kidney transplant recipients. J. Steroid Biochem. Mol. Biol. 97:213-218.

Griffin, M. D., Lutz, W. H., Phan, V. A., Bachman, L. A., McKean, D. J., and Kumar, R. (2000). Potent inhibition of dendritic cell differentiation and maturation by vitamin D analogs. Biochem. Biophys. Res. Commun. 270:701-708.

Griffin, M. D., Lutz, W., Phan, V. A., Bachman, L. A., McKean, D. J., and Kumar, R. (2001). Dendritic cell modulation by 1alpha,25 dihydroxyvitamin D_3 and its analogs: a vitamin D receptor-dependent pathway that promotes a persistent state of immaturity *in vitro* and *in vivo*. Proc. Natl. Acad. Sci. U. S. A. 98:6800-6805.

Grohmann, U., Fallarino, F., Bianchi, R., Belladonna, M. L., Vacca, C., Orabona, C., Uyttenhove, C., Fioretti, M. C., and Puccetti, P. (2001). IL-6 inhibits the tolerogenic function of CD8 alpha[+] dendritic cells expressing indoleamine 2,3-dioxygenase. J. Immunol. 167:708-714.

Grohmann, U., Orabona, C., Fallarino, F., Vacca, C., Calcinaro, F., Falorni, A., Candeloro, P., Belladonna, M. L., Bianchi, R., Fioretti, M. C., and Puccetti, P. (2002). CTLA-4-Ig regulates tryptophan catabolism *in vivo*. Nat. Immunol. 3:1097-1101.

Grohmann, U., Fallarino, F., and Puccetti, P. (2003). Tolerance, DCs and tryptophan: much ado about IDO. Trends Immunol. 24:242-248.

Hamilton-Williams, E. E., Lang, A., Benke, D., Davey, G. M., Wiesmuller, K. H., and Kurts, C. (2005). Cutting edge: TLR ligands are not sufficient to break cross-tolerance to self-antigens. J. Immunol. 174:1159-1163.

Harshyne, L. A., Watkins, S. C., Gambotto, A., and Barratt-Boyes, S. M. (2001). Dendritic cells acquire antigens from live cells for cross-presentation to CTL. J. Immunol. 166:3717-3723.

Hawiger, D., Inaba, K., Dorsett, Y., Guo, M., Mahnke, K., Rivera, M., Ravetch, J. V., Steinman, R. M., and Nussenzweig, M. C. (2001). Dendritic cells induce peripheral T cell unresponsiveness under steady state conditions *in vivo*. J. Exp. Med. 194:769-779.

Henri, S., Vremec, D., Kamath, A., Waithman, J., Williams, S., Benoist, C., Burnham, K., Saeland, S., Handman, E., and Shortman, K. (2001). The dendritic cell populations of mouse lymph nodes. J. Immunol. 167:741-748.

Hewison, M., Freeman, L., Hughes, S. V., Evans, K. N., Bland, R., Eliopoulos, A. G., Kilby, M. D., Moss, P. A., and Chakraverty, R. (2003). Differential regulation of vitamin D receptor and its ligand in human monocyte-derived dendritic cells. J. Immunol. 170:5382-5390.

Holt, P. G., Schon-Hegrad, M. A., Oliver, J., Holt, B. J., and McMenamin, P. G. (1990). A contiguous network of dendritic antigen-presenting cells within the respiratory epithelium. Int. Arch. Allergy Appl. Immunol. 91:155-159.

Huang, F. P., Platt, N., Wykes, M., Major, J. R., Powell, T. J., Jenkins, C. D., and MacPherson, G. G. (2000). A discrete subpopulation of dendritic cells transports apoptotic intestinal epithelial cells to T cell areas of mesenteric lymph nodes. J. Exp. Med. 191:435-444.

Hwu, P., Du, M. X., Lapointe, R., Do, M., Taylor, M. W., and Young, H. A. (2000). Indoleamine 2,3-dioxygenase production by human dendritic cells results in the inhibition of T cell proliferation. J. Immunol. 164:3596-3599.

Inaba, K., Witmer-Pack, M., Inaba, M., Hathcock, K. S., Sakuta, H., Azuma, M., Yagita, H., Okumura, K., Linsley, P. S., Ikehara, S., Muramatsu, S., Hodes, R. J., and Steinman R. M. (1994). The tissue distribution of the B7-2 costimulator in mice: abundant expression on dendritic cells in situ and during maturation *in vitro*. J. Exp. Med. 180:1849-1860.

Inaba, K., Pack, M., Inaba, M., Sakuta, H., Isdell, F., and Steinman, R. M. (1997). High levels of a major histocompatibility complex II-self peptide complex on dendritic cells from the T cell areas of lymph nodes. J. Exp. Med. 186:665-672.

Inaba, K., Turley, S., Iyoda, T., Yamaide, F., Shimoyama, S., Reis e Sousa, C., Germain, R. N., Mellman, I., and Steinman, R. M. (2000). The formation of immunogenic major histocompatibility complex class II-peptide ligands in lysosomal compartments of dendritic cells is regulated by inflammatory stimuli. J. Exp. Med. 191:927-936.

Johnsson, C., and Tufveson, G. (1994). MC 1288—a vitamin D analogue with immunosuppressive effects on heart and small bowel grafts. Transpl. Int. 7:392-397.

Ju, X. S., Hacker, C., Scherer, B., Redecke, V., Berger, T., Schuler, G., Wagner, H., Lipford, G. B., and Zenke, M. (2004). Immunoglobulin-like transcripts ILT2, ILT3 and ILT7 are expressed by human dendritic cells and down-regulated following activation. Gene 331:159-164.

Kim, S. H., Kim, S., Evans, C. H., Ghivizzani, S. C., Oligino, T., and Robbins, P. D. (2001). Effective treatment of established murine collagen-induced arthritis by systemic administration of dendritic cells genetically modified to express IL-4. J. Immunol. 166:3499-3505.

Kleindienst, P., Wiethe, C., Lutz, M. B., and Brocker, T. (2005). Simultaneous induction of CD4 T cell tolerance and CD8 T cell immunity by semimature dendritic cells. J. Immunol. 174:3941-3947.

Koch, F., Stanzl, U., Jennewein, P., Janke, K., Heufler, C., Kampgen, E., Romani, N., and Schuler, G. (1996) High level IL-12 production by murine dendritic cells: upregulation via MHC class II and CD40 molecules and downregulation by IL-4 and IL-10. J. Exp. Med. 184:741-746.

Kronin, V., Winkel, K., Suss, G., Kelso, A., Heath, W., Kirberg, A., von Boehmer, H., and Shortman, K. (1996). A subclass of dendritic cells regulates the response of naive CD8 T cells by limiting their IL-2 production. J. Immunol. 157:3819-3827.

Kronin, V., Suss, G., Winkel, K., and Shortman, K. (1997a). The regulation of T cell responses by a subpopulation of $CD8^+DEC205^+$ murine dendritic cells. Adv. Exp. Med. Biol. 417:239-248.

Kronin, V., Vremec, D., Winkel, K., Classon, B. J., Miller, R. G., Mak, T. W., Shortman, K., and Suss, G. (1997b). Are $CD8^+$ dendritic cells (DC) veto cells? The role of CD8 on DC in DC development and in the regulation of CD4 and CD8 T cell responses. Int. Immunol. 9:1061-1064.

Kronin, V., Fitzmaurice, C. J., Caminschi, I., Shortman, K., Jackson, D. C., and Brown, L. E. (2001). Differential effect of $CD8^{(+)}$ and CD8(-) dendritic cells in the stimulation of secondary $CD4^{(+)}$ T cells. Int. Immunol.13:465-473.

Kurts, C. (2000). Cross-presentation: inducing CD8 T cell immunity and tolerance. J. Mol. Med. 78:326-332.

Kurts, C., Kosaka, H., Carbone, F. R., Miller, J. F., and Heath, W. R. (1997). Class I-restricted cross-presentation of exogenous self-antigens leads to deletion of autoreactive CD8($^+$) T cells. J. Exp. Med. 186:239-245.

Kurts, C., Cannarile, M., Klebba, I., and Brocker, T. (2001). Dendritic cells are sufficient to cross-present self-antigens to CD8 T cells *in vivo*. J. Immunol. 166:1439-1442.

Kusuhara, M., Matsue, K., Edelbaum, D., Loftus, J., Takashima, A., and Matsue, H. (2002). Killing of naive T cells by CD95L-transfected dendritic cells (DC): *in vivo* study using killer DC-DC hybrids and CD4($^+$) T cells from DO11.10 mice. Eur. J. Immunol. 32:1035-1043.

Larsson, M., Fonteneau, J. F., Somersan, S., Sanders, C., Bickham, K., Thomas, E. K., Mahnke, K., and Bhardwaj N. (2001). Efficiency of cross presentation of vaccinia virus-derived antigens by human dendritic cells. Eur. J. Immunol. 31:3432-3442.

Lemire, J. M., and Archer, D. C. (1991). 1,25-dihydroxyvitamin D_3 prevents the *in vivo* induction of murine experimental autoimmune encephalomyelitis. J. Clin. Invest. 87:1103-1107.

Lemire, J. M., Archer, D. C., Khulkarni, A., Ince, A., Uskokovic, M. R., and Stepkowski, S. (1992). Prolongation of the survival of murine cardiac allografts by the vitamin D_3 analogue 1,25-dihydroxy-delta 16-cholecalciferol. Transplantation 54:762-763.

Levings, M. K., Sangregorio, R., Galbiati, F., Squadrone, S., de Waal Malefyt, R., and Roncarolo, M. G. (2001). IFN-alpha and IL-10 induce the differentiation of human type 1 T regulatory cells. J. Immunol. 166:5530-5539.

Liu, K., Iyoda, T., Saternus, M., Kimura, Y., Inaba, K., and Steinman, R. M. (2002). Immune tolerance after delivery of dying cells to dendritic cells in situ. J. Exp. Med. 196:1091-1097.

Lu, L., Bonham, C. A., Liang, X., Chen, Z., Li, W., Wang, L., Watkins, S. C., Nalesnik, M. a., Schlissel, M. S., Demestris, A. J., Fung, J. J., and Qian, S. (2001). Liver-derived DEC205$^+$B220$^+$CD19- dendritic cells regulate T cell responses. J. Immunol. 166:7042-7052.

Lutz, M. B., and Schuler, G. (2002). Immature, semi-mature and fully mature dendritic cells: which signals induce tolerance or immunity? Trends Immunol. 23:445-449.

Mahnke, K., Schmitt, E., Bonifaz, L., Enk, A. H., and Jonuleit, H. (2002). Immature, but not inactive: the tolerogenic function of immature dendritic cells. Immunol. Cell Biol. 80:477-483.

Mahnke, K., Knop, J., and Enk, A. H. (2003a). Induction of tolerogenic DCs: 'you are what you eat.' Trends Immunol. 24:646-651.

Mahnke, K., Qian, Y., Knop, J., and Enk, A. H. (2003b). Dendritic cells, engineered to secrete a T-cell receptor mimic peptide, induce antigen-specific immunosuppression *in vivo*. Nat. Biotechnol. 21:903-908.

Mahnke, K., Qian, Y., Knop, J., and Enk, A. H. (2003c). Induction of CD4$^+$/CD25$^+$ regulatory T cells by targeting of antigens to immature dendritic cells. Blood 101:4862-4869.

Manavalan, J. S., Rossi, P. C., Vlad, G., Piazza, F., Yarilina, A., Cortesini, R., Mancini, D., and Suciu-Foca, N. (2003). High expression of ILT3 and ILT4 is a general feature of tolerogenic dendritic cells. Transpl. Immunol. 11:245-258.

Manolios, N., Collier, S., Taylor, J., Pollard, J., Harrison, L. C., and Bender, V. (1997). T-cell antigen receptor transmembrane peptides modulate T-cell function and T cell-mediated disease. Nat. Med. 3:84-88.

Matsue, H., Matsue, K., Kusuhara, M., Kumamoto, T., Okumura, K., Yagita, H., and Takashima, A. (2001). Immunosuppressive properties of CD95L-transduced "killer" hybrids created by fusing donor- and recipient-derived dendritic cells. Blood 98:3465-3472.

Mathieu, C., Waer, M., Laureys, J., Rutgeerts, O., and Bouillon, R. (1994). Prevention of autoimmune diabetes in NOD mice by 1,25 dihydroxyvitamin D_3. Diabetologia 37:552-558.

Matzinger, P. (1994). Tolerance, danger, and the extended family. Annu. Rev. Immunol. 12:991-1045.

McGuirk, P., McCann, C., and Mills, K. H. (2002). Pathogen-specific T regulatory 1 cells induced in the respiratory tract by a bacterial molecule that stimulates interleukin 10 production by dendritic cells: a novel strategy for evasion of protective T helper type 1 responses by Bordetella pertussis. J. Exp. Med. 195:221-231.

Medzhitov, R. (2001) Toll-like receptors and innate immunity. Nat. Rev. Immunol. 1:135-145.

Mehling, A., Loser, K., Varga, G., Metze, D., Luger, T. A., Schwarz, T., Grabbe, S., and Beissert, S. (2001). Overexpression of CD40 ligand in murine epidermis results in chronic skin inflammation and systemic autoimmunity. J. Exp. Med.194:615-628.

Mellman, I., and Steinman, R. M. (2001). Dendritic cells: specialized and regulated antigen processing machines. Cell 106:255-258.

Mellor, A. L., and Munn, D. H. (2004). IDO expression by dendritic cells: tolerance and tryptophan catabolism. Nat. Rev. Immunol. 4:762-774.

Mellor, A. L., and Munn, D. H. (1999). Tryptophan catabolism and T-cell tolerance: immunosuppression by starvation? Immunol. Today 20:469-473.

Mellor, A. L., Sivakumar, J., Chandler, P., Smith, K., Molina, H., Mao, D., and Munn, D. H. (2001). Prevention of T cell-driven complement activation and inflammation by tryptophan catabolism during pregnancy. Nat. Immunol. 2:64-68.

Mellor, A. L., Baban, B., Chandler, P., Marshall, B., Jhaver, K., Hansen, A., Koni, P. A., Iwashima, M., and Munn, D. H. (2003). Cutting edge: induced indoleamine 2,3 dioxygenase expression in dendritic cell subsets suppresses T cell clonal expansion. J. Immunol. 171:1652-1655.

Menges, M., Rossner, S., Voigtlander, C., Schindler, H., Kukutsch, N. A., Bogdan, C., Erb, K., Schuler, G., and Lutz M.B. (2002). Repetitive injections of dendritic cells matured with tumor necrosis factor alpha induce antigen-specific protection of mice from autoimmunity. J. Exp. Med. 195:15-21.

Min, W. P., Zhou, D., Ichim, T. E., Strejan, G. H., Xia, X., Yang, J., Huang, X., Garcia, B., White, D., Dutartre, P., Jevnikar, A. M., and Zhong, R. (2003).

Inhibitory feedback loop between tolerogenic dendritic cells and regulatory T cells in transplant tolerance. J. Immunol. 170:1304-1312.

Misra, N., Bayry, J., Lacroix-Desmazes, S., Kazatchkine, M. D., and Kaveri, S. V. (2004). Cutting edge: human CD4$^+$CD25$^+$ T cells restrain the maturation and antigen-presenting function of dendritic cells. J. Immunol. 172:4676-4680.

Muller, G., Muller, A., Tuting, T., Steinbrink, K., Saloga, J., Szalma, C., Knop, J., and Enk, A. H. (2002). Interleukin-10-treated dendritic cells modulate immune responses of naive and sensitized T cells *in vivo*. J. Invest. Dermatol. 119:836-841.

Munn, D. H., Zhou, M., Attwood, J. T., Bondarev, I., Conway, S. J., Marshall, B., Brown, C., and Mellor, A. L. (1998). Prevention of allogeneic fetal rejection by tryptophan catabolism. Science 281:1191-1193.

Munn, D. H., Sharma, M. D., Lee, J. R., Jhaver, K. G., Johnson, T. S., Keskin, D. B., Marshall, B., Chandler, P., Antonia, S. J., Burgess, R., Slingluff, C. L. Jr, and Mellor, A. L. (2002). Potential regulatory function of human dendritic cells expressing indoleamine 2,3-dioxygenase. Science 297:1867-1870.

Munn, D. H., Mellor, A. L., Rossi, M., and Young, J. W. (2005). Dendritic cells have the option to express IDO-mediated suppression or not. Blood 105:2618.

Musso, T., Gusella, G. L., Brooks, A., Longo, D. L., and Varesio, L. (1994). Interleukin-4 inhibits indoleamine 2,3-dioxygenase expression in human monocytes. Blood 83:1408-1411.

Napolitani, G., Rinaldi, A., Bertoni, F., Sallusto, F., and Lanzavecchia, A. (2005). Selected Toll-like receptor agonist combinations synergistically trigger a T helper type 1-polarizing program in dendritic cells. Nat. Immunol. 6:769-776.

Nossal, G. J. (2001). A purgative mastery. Nature 412:685-686.

O'Rourke, R. W., Kang, S. M., Lower, J. A., Feng, S., Ascher, N. L., Baekkeskov, S., and Stock, P. G. (2000). A dendritic cell line genetically modified to express CTLA4-IG as a means to prolong isl*et al*lograft survival. Transplantation 69:1440-1446.

O'Sullivan, B. J., MacDonald, K. P., Pettit, A. R., and Thomas, R. (2000). RelB nuclear translocation regulates B cell MHC molecule, CD40 expression, and antigen-presenting cell function. Proc. Natl. Acad. Sci. U. S. A. 97:11421-11426.

Pani, M. A., Knapp, M., Donner, H., Braun, J., Baur, M. P., Usadel, K. H., Badenhoop, K. (2000). Vitamin D receptor allele combinations influence genetic susceptibility to type 1 diabetes in Germans. Diabetes 49:504-507.

Penna, G., and Adorini, L. (2000). 1 Alpha,25-dihydroxyvitamin D$_3$ inhibits differentiation, maturation, activation, and survival of dendritic cells leading to impaired alloreactive T cell activation. J. Immunol. 164:2405-2411.

Penna, G., and Adorini, L. (2001). Inhibition of costimulatory pathways for T-cell activation by 1,25-dihydroxyvitamin D(3). Transplant. Proc. 33:2083-2084.

Penna, G., Roncari, A., Amuchastegui, S., Daniel, K. C., Berti, E., Colonna, M., and Adorini, L. (2005). Expression of the inhibitory receptor ILT3 on dendritic cells is dispensable for induction of CD4$^+$Foxp3$^+$ regulatory T cells by 1,25-dihydroxyvitamin D$_3$. Blood 106:3490-3497.

Pettit, A. R., Quinn, C., MacDonald, K. P., Cavanagh, L. L., Thomas, G., Townsend, W., Handel, M., and Thomas, R. (1997). Nuclear localization of RelB is associated with effective antigen-presenting cell function. J. Immunol. 159:3681-3691.

Piemonti, L., Monti, P., Sironi, M., Fraticelli, P., Leone, B. E., Dal Cin, E., Allavena, P., and Di Carlo, V. (2000). Vitamin D_3 affects differentiation, maturation, and function of human monocyte-derived dendritic cells. J. Immunol. 164:4443-4451.

Pierre, P., Turley, S. J., Gatti, E., Hull, M., Meltzer, J., Mirza, A., Inaba, K., Steinman, R. M., and Mellman, I. (1997). Developmental regulation of MHC class II transport in mouse dendritic cells. Nature 388:787-792.

Pooley, J. L., Heath, W. R., and Shortman, K. (2001). Cutting edge: intravenous soluble antigen is presented to CD4 T cells by CD8- dendritic cells, but cross-presented to CD8 T cells by $CD8^+$ dendritic cells. J. Immunol. 166:5327-5330.

Prasad, D. V., Nguyen, T., Li, Z., Yang, Y., Duong, J., Wang, Y., and Dong, C. (2004). Murine B7-H3 is a negative regulator of T cells. J. Immunol. 173:2500-2506.

Rabinovich, G. A., Daly, G., Dreja, H., Tailor, H., Riera, C. M., Hirabayashi, J., and Chernajovsky, Y. (1999). Recombinant galectin-1 and its genetic delivery suppress collagen-induced arthritis via T cell apoptosis. J. Exp. Med. 190:385-398.

Ravetch, J. V., and Bolland, S. (2001). IgG Fc receptors. Annu. Rev. Immunol. 19:275-290.

Robinson, C. M., Shirey, K. A., and Carlin, J. M. (2003). Synergistic transcript-tional activation of indoleamine dioxygenase by IFN-gamma and tumor necrosis factor-alpha. J. Interferon Cytokine Res. 23:413-421.

Rodig, N., Ryan, T., Allen, J. A., Pang, H., Grabie, N., Chernova, T., Greenfield, E. A., Liang, S. C., Sharpe, A. H., Lichtman, A. H., and Freeman, G. J. (2003). Endothelial expression of PD-L1 and PD-L2 down-regulates $CD8^+$ T cell activation and cytolysis. Eur. J. Immunol. 33:3117-3126.

Sallusto, F., Schaerli, P., Loetscher, P., Schaniel, C., Lenig, D., Mackay, C. R., Qin, S., and Lanzavecchia, A. (1998). Rapid and coordinated switch in chemokine receptor expression during dendritic cell maturation. Eur. J. Immunol. 28:2760-2769.

Sato, K., Yamashita, N., Baba, M., and Matsuyama, T. (2003a). Modified myeloid dendritic cells act as regulatory dendritic cells to induce anergic and regulatory T cells. Blood 101:3581-3589.

Sato, K., Yamashita, N., Yamashita, N., Baba, M., and Matsuyama, T. (2003b). Regulatory dendritic cells protect mice from murine acute graft-versus-host disease and leukemia relapse. Immunity 18:367-379.

Schreiner, B., Mitsdoerffer, M., Kieseier, B. C., Chen, L., Hartung, H. P., Weller, M., and Wiendl, H. (2004). Interferon-beta enhances monocyte and dendritic cell expression of B7-H1 (PD-L1), a strong inhibitor of autologous T-cell activation: relevance for the immune modulatory effect in multiple sclerosis. J. Neuroimmunol. 155:172-182.

Schuler, G., and Steinman, R. M. (1985). Murine epidermal Langerhans cells mature into potent immunostimulatory dendritic cells *in vitro*. J. Exp. Med. 161:526-546.

Selenko-Gebauer, N., Majdic, O., Szekeres, A., Hofler, G., Guthann, E., Korthauer, U., Zlabinger, G., Steinberger, P., Pickl, W. F., Stockinger, H., Knapp, W., and Stockl, J. (2003). B7-H1 (programmed death-1 ligand) on dendritic cells is involved in the induction and maintenance of T cell anergy. J. Immunol. 170:3637-3644.

Shiroishi, M., Tsumoto, K., Amano, K., Shirakihara, Y., Colonna, M., Braud, V. M., Allan, D. S., Makadzange, A., Rowland-Jones, S., Willcox, B., Jones, E. Y., van der Merwe, P. A., Kumagai, I., and Maenaka, K. (2003). Human inhibitory receptors Ig-like transcript 2 (ILT2) and ILT4 compete with CD8 for MHC class I binding and bind preferentially to HLA-G. Proc. Natl. Acad. Sci. U. S. A. 100:8856-8861.

Shortman, K., and Heath, W. R. (2001). Immunity or tolerance? That is the question for dendritic cells. Nat. Immunol. 2:988-989.

Smits, H. H., Engering, A., van der Kleiy, D., de Jong, E. C., Schipper, K., van Capel, T. M., Zaat, B. A., Yazdanbakhsh, M., Wierenga, E. A., van Kooyk, Y., and Kapsenberg, M. L. (2005). Selective probiotic bacteria induce IL-10-producing regulatory T cells *in vitro* by modulating dendritic cell function through dendritic cell-specific intercellular adhesion molecule 3-grabbing nonintegrin. J. Allergy Clin. Immunol. 115:1260-1267.

Sporri, R., and Reis e Sousa, C. (2005). Inflammatory mediators are insufficient for full dendritic cell activation and promote expansion of $CD4^+$ T cell populations lacking helper function. Nat. Immunol. 6:163-170.

Steinbrink, K., Wolfl, M., Jonuleit, H., Knop, J., and Enk, A. H. (1997). Induction of tolerance by IL-10-treated dendritic cells. J. Immunol. 159:4772-4780.

Steinbrink, K., Jonuleit, H., Muller, G., Schuler, G., Knop, J., and Enk, A. H. (1999). Interleukin-10-treated human dendritic cells induce a melanoma-antigen-specific anergy in $CD8(^+)$ T cells resulting in a failure to lyse tumor cells. Blood 93:1634-1642.

Steinbrink, K., Graulich, E., Kubsch, S., Knop, J., and Enk, A. H. (2002). $CD4(^+)$ and $CD8(^+)$ anergic T cells induced by interleukin-10-treated human dendritic cells display antigen-specific suppressor activity. Blood 99:2468-2476.

Strome, S. E., Dong, H., Tamura, H., Voss, S. G., Flies, D. B., Tamada, K., Salomao, D., Cheville, J., Hirano, F., Lin, W., Kasperbauer, J. L., Ballman, K. V., and Chen, L. (2003). B7-H1 blockade augments adoptive T-cell immunotherapy for squamous cell carcinoma. Cancer Res. 63:6501-6505.

Suh, W. K., Gajewska, B. U., Okada, H., Gronski, M. A., Bertram, E. M., Dawicki, W., Duncan, G. S., Bukczynski, J., Plyte, S., Elia, A., Wakeham, A., Itie, A., Chung, S., Da Costa, J., Arya, S., Horan, T., Campbell, P., Gaida, K., Ohashi, P. S., Watts, T. H., Yoshinaga, S. K., Bray, M. R., Jordana, M., and Mak, T. W. (2003). The B7 family member B7-H3 preferentially down-regulates T helper type 1-mediated immune responses. Nat. Immunol. 4:899-906.

Suss, G., and Shortman, K. (1996). A subclass of dendritic cells kills CD4 T cells via Fas/Fas-ligand-induced apoptosis. J. Exp. Med. 183:1789-1796.

Takayama, T., Nishioka, Y., Lu, L., Lotze, M. T., Tahara, H., and Thomson, A. W. (1998). Retroviral delivery of viral interleukin-10 into myeloid dendritic cells markedly inhibits their allostimulatory activity and promotes the induction of T-cell hyporesponsiveness. Transplantation 66:1567-1574.

Takayama, T., Morelli, A. E., Robbins, P. D., Tahara, H., and Thomson, A. W. (2000). Feasibility of CTLA4Ig gene delivery and expression *in vivo* using retrovirally transduced myeloid dendritic cells that induce alloantigen-specific T cell anergy *in vitro*. Gene Ther. 7:1265-1273.

Taylor, M. W., and Feng, G. S. (1991). Relationship between interferon-gamma, indoleamine 2,3-dioxygenase, and tryptophan catabolisma. FASEB J. 5:2516-2522.

Terness, P., Bauer, T. M., Rose, L., Dufter, C., Watzlik, A., Simon, H., and Opelz, G. (2002). Inhibition of allogeneic T cell proliferation by indoleamine 2,3-dioxygenase-expressing dendritic cells: mediation of suppression by tryptophan metabolites. J. Exp. Med. 196:447-457.

The EURODIAB Substudy 2 Study Group. (1999) Vitamin D supplement in early childhood and risk for Type I (insulin-dependent) diabetes mellitus. Diabetologia 42:51-54.

Thery, C., Duban, L., Segura, E., Veron, P., Lantz, O., and Amigorena, S. (2002). Indirect activation of naive CD4$^+$ T cells by dendritic cell-derived exosomes. Nat. Immunol. 3:1156-1162.

Thompson, A. G., Pettit, A. R., Padmanabha, J., Mansfield, H., Frazer, I. H., Strutton, G. M., and Thomas, R. (2002). Nuclear RelB$^+$ cells are found in normal lymphoid organs and in peripheral tissue in the context of inflammation, but not under normal resting conditions. Immunol. Cell. Biol. 80:164-169.

Tomasoni, S., Aiello, S., Cassis, L., Noris, M., Longaretti, L., Cavinato, R. A., Azzollini, N., Pezzotta, A., Remuzzi, G., and Benigni, A. (2005). Dendritic cells genetically engineered with adenoviral vector encoding dnIKK2 induce the formation of potent CD4$^+$ T-regulatory cells. Transplantation 79:1056-1061.

Turley, S. J., Inaba, K., Garrett, W. S., Ebersold, M., Unternaehrer, J., Steinman, R. M. *et al.* (2000). Transport of peptide-MHC class II complexes in developing dendritic cells. Science 288:522-527.

Turnbull, E. L., Yrlid, U., Jenkins, C. D., and MacPherson, G. G. (2005). Intestinal dendritic cell subsets: differential effects of systemic TLR4 stimulation on migratory fate and activation *in vivo*. J. Immunol. 174:1374-1384.

Urban, B. C., Willcox, N., and Roberts, D. J. (2001). A role for CD36 in the regulation of dendritic cell function. Proc. Natl. Acad. Sci. U. S. A. 98:8750-8755.

van Lieshout, A. W., Barrera, P., Smeets, R. L., Pesman, G. J., van Riel, P. L., van den Berg, W. B., and Radstake, T. R. (2005). Inhibition of TNF alpha during maturation of dendritic cells results in the development of semi-mature cells: a potential mechanism for the beneficial effects of TNF alpha blockade in rheumatoid arthritis. Ann. Rheum. Dis. 64:408-414.

Veckman, V., Miettinen, M., Pirhonen, J., Siren, J., Matikainen, S., and Julkunen, I. (2002). Streptococcus pyogenes and Lactobacillus rhamnosus differentially induce maturation and production of Th1-type cytokines and chemokines in human monocyte-derived dendritic cells. J. Leukoc. Biol. 75:764-771.

Verbovetski, I., Bychkov, H., Trahtemberg, U., Shapira, I., Hareuveni, M., Ben-Tal, O., Kutikov, I., Gill, O., and Mevorach, D. (2002). Opsonization of apoptotic cells by autologous iC3b facilitates clearance by immature dendritic cells, down-regulates DR and CD86, and up-regulates CC chemokine receptor 7. J. Exp. Med. 196:1553-1561.

Vremec, D., and Shortman, K. (1997). Dendritic cell subtypes in mouse lymphoid organs: cross-correlation of surface markers, changes with incubation, and differences among thymus, spleen, and lymph nodes. J. Immunol. 159:565-573.

Wakkach, A., Fournier, N., Brun, V., Breittmayer, J. P., Cottrez, F., and Groux, H. (2003). Characterization of dendritic cells that induce tolerance and T regulatory 1 cell differentiation *in vivo*. Immunity 18:605-617.

Wu, L., D'Amico, A., Hochrein, H., O'Keeffe, M., Shortman, K., and Lucas, K. (2001). Development of thymic and splenic dendritic cell populations from different hemopoietic precursors. Blood 98:3376-3382.

Yamaguchi, Y., Tsumura, H., Miwa, M., and Inaba, K. (1997). Contrasting effects of TGF-beta 1 and TNF-alpha on the development of dendritic cells from progenitors in mouse bone marrow. Stem Cells 15:144-153.

Yoshimura, S., Bondeson, J., Brennan, F. M., Foxwell, B. M., and Feldmann, M. (2001). Role of NFkappaB in antigen presentation and development of regulatory T cells elucidated by treatment of dendritic cells with the proteasome inhibitor PSI. Eur. J. Immunol. 31:1883-1893.

Chapter 5

Adjuvants, Dendritic Cells, and Cytokines: Strategies for Enhancing Vaccine Efficacy

Paola Rizza, Imerio Capone, and Filippo Belardelli

5.1 Introduction

Adjuvants are functionally defined as components added to vaccine formulations that are capable of enhancing the *in vivo* immunogenicity of defined antigens. For a long time, the studies pursued on adjuvants were based on phenomenological observations, and their mechanisms of action remained obscure for years. Therefore, it is not surprising that as late as in 1989, Janeway referred to adjuvants as "the immunologist's little dirty secret" (Janeway, 1989). This view has profoundly changed with the great advances in immunology and the technological revolution that has taken over in recent years that have provided both the knowledge and powerful tools for elucidating the action mechanisms of many traditional adjuvants. Hence, not only has this progress facilitated the comprehension of the roles performed by known adjuvants, but also it has allowed the rational design of new adjuvants particularly suited for modern vaccines. Much of the recent progress on research of vaccine adjuvants is due to the recently accumulated knowledge on the biology of a special type of professional antigen-presenting cells (APCs), namely dendritic cells (DCs), which play a fundamental role in linking innate and adaptive immunity (Banchereau and Steinman, 1998; Banchereau *et al.*, 2000).

DCs originate from pluripotent stem cells in the bone marrow, entering the bloodstream and localizing into almost all organs (Hart, 1997). Circulating immature DCs typically enter tissues in response to inflammatory chemoattractant cytokines and chemokines. After having ingested and processed incoming pathogens, DCs switch their chemokine receptor set and migrate to regional lymph nodes in response to lymphoid chemokines, which also direct their position within lymphoid tissues. This occurs so that they can efficiently present processed antigens to lymphocytes, priming them for specific immune response (Forster *et al.*, 1999; Sozzani *et al.*, 1999). Some types of DCs and DC precursors have recently drawn the attention of immunologists (Liu, 2001), and their main characteristics are extensively described in other chapters of this volume. The recent progress in immunology has revealed that DCs are the main cell players for the generation of a protective immune response against both infectious diseases and tumors (Banchereau *et al.*, 2000; Banchereau *et al.*, 2001). Today, understanding how to manipulate DCs could be considered as strictly instrumental in the progress of vaccine research. In fact, the development of vaccines has often been hampered by the poor immunogenicity of the relevant antigenic components, which require potent adjuvants in order to promote efficiently a protective immune response. Hence, critical issues in vaccinology are how to enhance the immune response to defined antigens, as well as how to shape qualitatively this response toward the correct correlates of protection; addressing these issues means understanding how to exploit the recent knowledge on the interactions between DCs and lymphocytes for the identification of more effective and safe adjuvants for modern vaccines. In this chapter, we first summarize the main milestones of research progress on adjuvants, providing the reader with an updated classification according to the characteristics and mechanisms of action of the different molecules/ formulations. We then focus on those adjuvants, which specifically act on DCs and show promise as potential candidates for enhancing the immuno-genicity of modern vaccines.

5.2 Brief Historical Background on Adjuvants

The concept of adjuvant (from the Latin *adjuvare*, "to help") goes back to early studies of vaccination, performed at the beginning of the last century. Figure 5.1 describes the key discoveries involved in adjuvant research. In 1916, Le Moignic and Pinoy reported the findings that mineral oil emulsions coadministered with inactivated *Salmonella typhimurium* increased the immune response to the antigen (Le Moignic and Pinoy, 1916). In the 1920s, Gaston Ramon observed that horses inoculated with the diphtheria toxoid generated vigorous antitoxin sera if they also developed an abscess at the inoculation site (Ramon, 1925). Hence, he demonstrated that the increased responses were indeed due to unrelated substances coinjected with the toxoid and causing local inflammation (Ramon, 1926). Ramon coined the term "adjuvant" to describe "substances used in combination with a specific antigen that produce a more robust

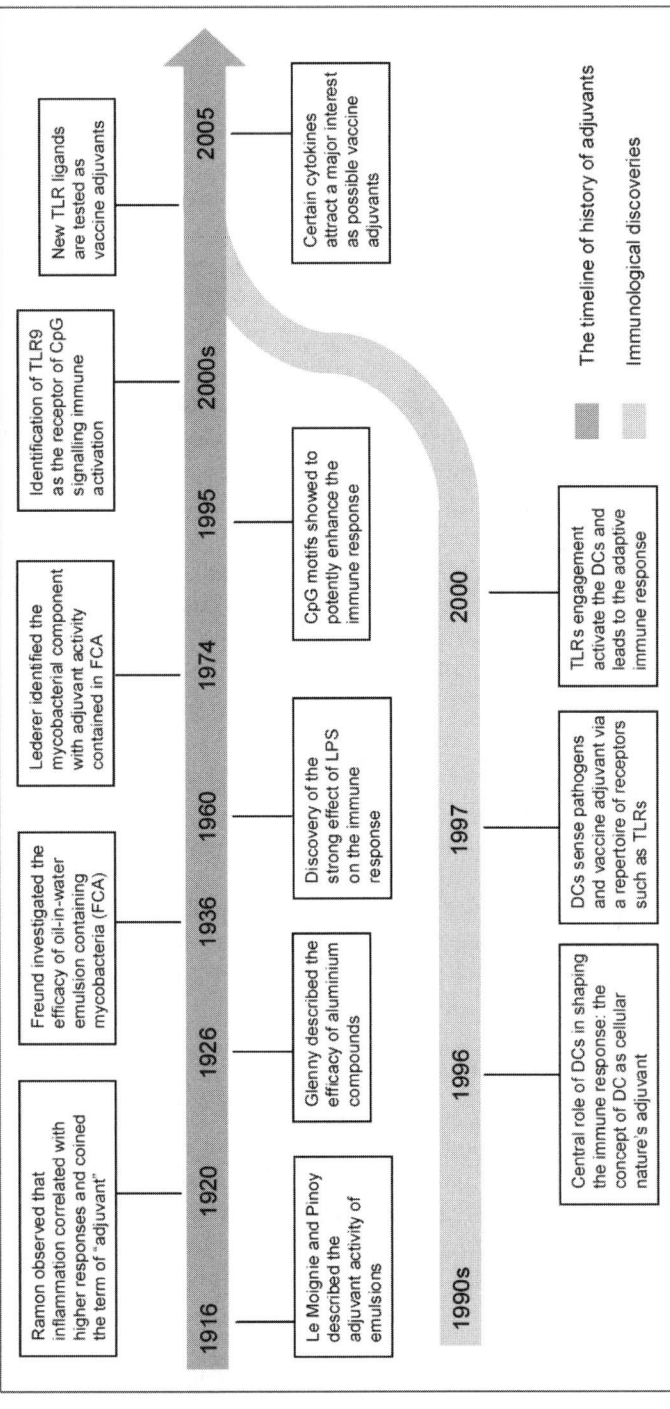

FIGURE 5.1. The historical road map of adjuvants. The history of adjuvants can be traced back to the early 1900s. The most important milestones are depicted in this figure. In the 1990s, important breakthroughs in immunology have provided new clues for explaining both the mechanism of action of many adjuvants and designing novel and effective molecules endowed with adjuvant activity.

Abbreviations: FCA, Freund complete adjuvant; LPS, lipopolysaccharide; TLR, Toll-like receptor; DC, dendritic cell.

immune response than the antigen alone." In the following years, Glenny and co-workers described the potent activity exerted by aluminum compounds on the development of the immune response (Glenny *et al.*, 1926). In particular, they found that the injection of laboratory animals with diphtheria toxoid adsorbed with aluminum salts led to a significant increase of the specific immune response, mainly due to a delayed clearing of the alum-precipitated toxoid from the injection sites, in comparison with toxoid alone. In the mid-1930s, Freund investigated the efficacy of oil-in-water emulsions containing killed mycobacteria in enhancing the antibody response to tubercle bacilli (Freund *et al.*, 1937). Since then, this formulation was named Freund's complete adjuvant (FCA), one of the most potent adjuvants ever developed. Nevertheless, the use of FCA had been limited by its severe toxicity. Freund's incomplete adjuvants (FIAs), the oil-in-water emulsions without added mycobacteria, showed a decreased toxicity and were used to some extent in human vaccine studies (Stuart-Harris, 1969). The early findings that the immune response to antibacterial vaccines were elevated when Gram-negative bacteria were included in the vaccine preparations inspired studies to isolate the responsible factor (Greenberg and Fleming, 1947). These studies identified the lipopolysaccharide (LPS), an important component of the outer wall of the Gram-negative bacteria, as the factor enhancing the antibody response to protein antigens (Johnson *et al.*, 1956). However, the elevated toxicity associated with the adjuvant activity of this compound had severely limited its use in human vaccines. In later studies, attempts to detoxify the molecule without destroying adjuvanticity had been pursued and led to the identification of monophosphoryl lipid A (MPL) as the active component (Qureshi *et al.*, 1982; Ribi, 1984). This substance further proved to be effective in enhancing the antibody response to a reference antigen in mouse models (Schneerson *et al.*, 1991). In 1974, Lederer and collaborators identified the component of the mycobacteria contained in the FCA that was responsible for its strong adjuvant activity, namely the muramyldipeptide (MDP) (Ellouz *et al.*, 1974). Safety concerns have often restricted the possible use in humans of molecules shown to exhibit potent vaccine adjuvant activity in animal models. To date, aluminum salts remain the only compounds approved for human use worldwide, with the only exception of MF59, a microfluidized oil/water emulsion, which in recent years has been licensed for elderly people in association with influenza vaccine (Podda and Del Giudice, 2003).

The mechanism of action of most adjuvants has remained poorly understood until recently, when the molecular interactions involved in the triggering of the immune response have been better clarified. In particular, the finding that cytokines produced on pathogen exposure are the major mediators of the immune and inflammatory response has provided the key in the explanation of how most adjuvants function (Belardelli and

Ferrantini, 2002; Villinger, 2003). Of particular note, in several studies, some cytokines, such as granulocyte-monocyte colony-stimulating factor (GM-CSF), interferon (IFN)-γ, and IL-12, have been used as adjuvants of experimental vaccines with variable success (Rizza *et al.*, 2002); these studies will be reviewed in a subsequent part of this chapter.

Most of the adjuvants studied so far are bacterial derivative products and have been considered as important signals, whose principle natural function is to alert the immune system to respond rapidly against infections. These pathogen components have been called pathogen-associated molecular patterns (PAMPs) (Janeway, 1989; Janeway, 1992), and their strong effects in improving the immune response to coinjected antigens have not until recently been fully understood, when complex receptors, called pattern-recognition receptors (PRRs), crucially involved in the innate immune responses to PAMPs have been identified (Medzhitov *et al.*, 1997). PRRs, such as the TLRs (Takeda *et al.*, 2003), are differentially expressed on immune cells such as macrophages and DCs (Akira and Takeda, 2004). The engagement of TLRs by their ligand counterparts activates a cascade of signaling, which leads to the secretion of inflammatory cytokines and chemokines, which in turn influence cell activation, maturation, and migration, finally resulting in the shaping of the adaptive immune response (Akira and Takeda, 2004). Hence, the understanding of the types of immune response triggered on interactions of PAMPs with TLRs, which represent an important natural mechanism for pathogen-structure recognition by DCs, may hold the key for the discovery of selectively targeted immune potentiators and the development of new vaccine adjuvants.

5.3 The Need of New Adjuvants for Vaccine Development and Crucial Importance of Adjuvant–Dendritic Cell Interactions for the Polarization of Immune Response

Vaccines are widely considered among the most successful medical interventions against pathogens. Most of the research on vaccine development is aimed at finding new protective antigens. In recent years, however, it has become apparent that the critical issue in the development of modern vaccines, which are based on purified pathogen components, is the identification of safe and effective adjuvants (O'Hagan and Valiante, 2003; Singh and O'Hagan, 1999). Notably, early vaccines, mainly based on live replicating, attenuated pathogens, exhibited a strong capability of stimulating the innate immune system, by virtue of their intrinsic properties of mimicking their natural counterpart (Fig. 5.2). However, safety concerns have drastically reduced their usage in humans. The killed vaccines offer a great advantage over the live vaccines, but they are less effective in inducing a strong immunity. New-generation vaccines, mainly based on recombinant subunit preparations, are more rationally defined. In fact, these are generally based on selected antigens, in order for the desired immune response to be induced. However, they exhibit poor immunogenicity. As such, highly purified recombinant antigens are less efficient than a "dirty" vaccine. Even if the intrinsic nature of the antigen itself may positively affect the quality of immune response, the antigen *per se* lacks a strong capability of eliciting an immune response: in order to be an effective immunogen, any antigen requires to be combined with a suitable adjuvant. In this respect, the search for new adjuvants reflects the need of restoring in the newly designed vaccines those immune-potentiating properties that have been lost during the development of more selective and pure vaccines.

Today, the search for new and effective adjuvants has assumed much more importance. This reflects the need for a systematic approach to adjuvant development and selection that can respond to the specific requirements of each vaccine. It has become increasingly clear that one of the crucial prerequisites to be accomplished by a specific vaccine is to stimulate optimally the innate immune system, in particular the DCs, to trigger downstream the establishment of a strong and durable adaptive immune response (Hoebe *et al.*, 2004; Pashine *et al.*, 2005). In this scenario, the role of adjuvants is to deliver the antigen in the correct way to the correct cells. This enhances good cross-talk between innate and adaptive immune system and ensures that an optimal immune response is evoked against the vaccine antigen components (Fig. 5.2).

The protective immunity against a pathogen is achieved through a complex integration of functions fulfilled by the two branches of the immune system: the innate and the adaptive immune response. The innate response is quickly activated after pathogen exposure by a sensitive and nonspecific detection system present on APCs, most importantly the TLRs (Takeda *et al.*, 2003), and represents an immediate strategy to respond broadly to the invading pathogens (Fig. 5.2). This natural mechanism has evolved not only for ensuring a first line of defense but also for providing the host with the inflammatory milieu sufficient to initiate the antigen-specific (namely adaptive) immune response. The delicate balance of factors (cytokines and chemokines) produced upon innate immune stimulation, together with the instructions generated after antigen processing and presentation, subsequently leads to the clonal selection and expansion of antigen-specific $CD4^+$ and $CD8^+$ T cells and B cells. The

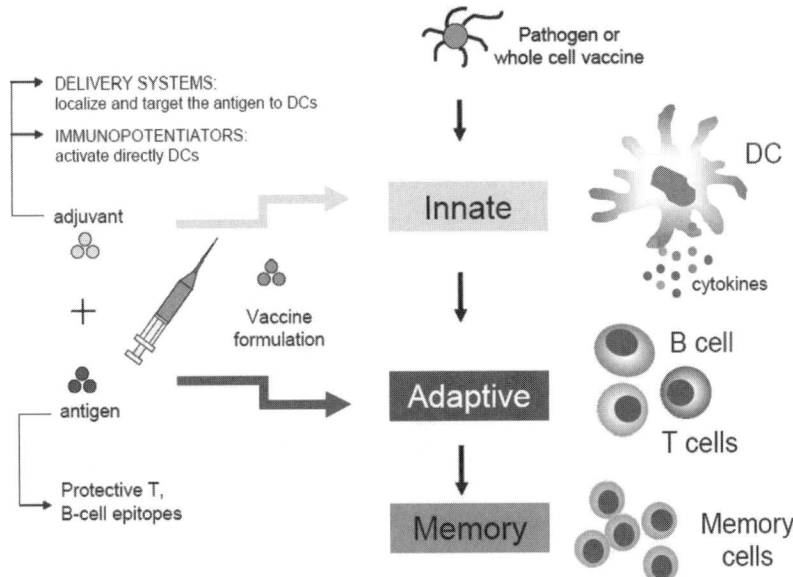

FIGURE 5.2. Vaccine components and their effects on the main cell players of the immune response. The adjuvant and the antigen display different and complementary roles. The antigen provides the immune system with the instructions on how to shape the specific immune response. In order to be effective, the antigen requires the aid of an adjuvant. Adjuvants may pertain to delivery systems or immunopotentiators. Delivery systems aid the antigen to be efficiently targeted to DCs, whereas immunopotentiators act directly on receptors harbored by DCs, strongly increasing the competence of these cells in triggering the immune response. Both the antigen and the adjuvant are essential to achieve long-term immunological memory. Whole-cell vaccines or pathogens are constitutively able to stimulate both the innate and the adaptive immune response.

activation of these effector cells will ultimately lead to the elimination of the specific pathogen and the establishment of a long-term immunologic memory (Fig. 5.2). Thus, vaccines cannot induce protective immunity without properly stimulating upstream the innate immune response. As the dependency of the adaptive immunity on the innate immune response has become more appreciated, a better-defined role for adjuvants has emerged: to target the innate immune response (Pashine *et al.*, 2005). Consistent with this concept, most of the effective adjuvants, such as microbial derivatives, are strong activators of the innate immune response through DCs; the biochemical pathways involved in signaling these innate immune mechanisms have just started to be understood. Interestingly, certain cytokines produced at the early phase of the initiation of the immune response and involved in DC activation have been shown to have a role in boosting immunity. These cytokines differently influence the type of immune response. The dichotomy of the type of immune response, Th1 versus Th2, occurring on certain types of infection or vaccination strategies has raised the question of whether the current adjuvants used in human vaccines against well-defined pathogens are effective in inducing the desired correlates of protection. It is accepted that the Th-polarization of the immune response to a pathogen depends on the binding of the pathogen-derived molecules to PRRs (e.g., TLRs) of DCs in their immature state, resulting in selective programming of these DCs during their maturation (De Jong *et al.*, 2005). Basically, DCs exposed to intracellular pathogens promote Th1 responses, characterized by the induction of IL-12, IL-2, and IFN-γ secretion, this being the last factor instrumental in the induction of cell-mediated immunity (De Jong *et al.*, 2002; Kalinski *et al.*, 1999). This type of response would be particularly desirable for protection against viral infections (Seder and Hill, 2000). In contrast, certain helminth infections promote DCs to drive the development of Th2 responses, which are predominantly accompanied by IL-4, IL-5, and IL-10 release, and favor the production of circulating or secretory antibodies (De Jong *et al.*, 2005). It is clear that type 1 and type 2 responses involve opposite effector functions, and the upregulation of one type of response normally results in downregulation of the alternative differentiation pathway. Adjuvants may elicit a Th1 or Th2 response and preferentially enhance a cell-mediated immunity (Th1) or a humoral response (Th2). For example, aluminum salts, the adjuvant currently used in licensed human vaccines, mainly induce a Th2-type of immune response (Gupta, 1998; Singh and O'Hagan, 2002), which may not be optimal for stimulating CTL responses against intracellular pathogens. The view that the alum activity was only based on "the depot effect" (helping antigen persistence at the injection sites) has been recently changed. Alum has been shown to promote IL-4 release, thus further supporting the concept of its role as inducer of Th2-type immunity (Ulanova *et al.*, 2001; Lindblad, 2004). Notably, the Th2 polarization of the immune response could be deleterious to the host when a type-1 response would be protective and vice versa; therefore, successful vaccination critically depends on the ability of the adjuvant to induce the desired type of response selectively and reliably (Moingeon *et al.*, 2001). One main problem related to the choice of the adjuvant fitting with this requirement is that the correlates of protection remain to be determined for some pathogens representing major challenges to vaccinologists. In addition, although the number of natural and synthetic compounds showing promising activity as Th1-adjuvants in animal models has increased, only clinical studies will verify their possible effectiveness as adjuvants of human vaccines. Some of these substances have attracted much interest. For instance, CpG oligonucleotides (Krieg *et al.*, 1995), as well as MPL (Qureshi *et al.*, 1982), have been shown to strongly bias the immune response to coinjected antigen toward Th1-type effector functions, which include the promotion of CTL generation (Baldridge *et al.*, 2000; Chu *et al.*, 1997; Klinmann *et al.*, 2004). It is worth noting that the usage of some cytokines, which have been shown to play a major role in the development of Th1-type immunity, could offer a certain advantage in redirecting the polarization of the adaptive immune response. In this respect, it has been shown that certain strains of mice may benefit from the administration of IL-12 as adjuvant in the vaccination against *Leishmania major* (Afonso *et al.*, 1994). In this model, the experimental infection with *Leishmania major* induces a Th2-type of immune response and is ineffective in eliminating the parasite. Conversely, the coadministration of IL-12 with a vaccine containing *Leishmania* antigens stimulates a Th1 response, which protects the mice against the subsequent challenge with the parasite (Afonso *et al.*, 1994). Therefore, selected adjuvants may affect the rate of success of a given vaccine, molding the polarization of the adaptive immune response.

5.4 New Classification of Vaccine Adjuvants

As previously mentioned, adjuvants have traditionally been developed by empirical methods, and their mechanisms of action have been elusive for decades. The enormous amount of data acquired over the past few years in the field of vaccinology and molecular immunology have provided scientific explanations on how the many traditional adjuvants function. Thus, adjuvants are now being redesigned in a more rational way as substances capable of profoundly affecting the magnitude and the type of the adaptive immune response and novel criteria are now being established to define both the traditional and newly developed adjuvants (O'Hagan and Valiante, 2003; Van der Laan, 2005). With our new understanding of their action mechanism, adjuvants have been divided into two main categories: (i) delivery systems; (ii) immunopotentiators (Table 5.1).

One crucial issue in vaccination is to efficiently deliver the antigen to DCs in order to induce a protective immune response (Fig. 5.3) (Reis and Sousa, 2004; Schijns, 2003). Conventional adjuvants, such as aluminum salts or mineral oil emulsions, act by trapping the antigen, thus allowing antigen stabilization and slow release (Zinkernagel, 2000). A number of delivery systems have been developed to deliver more selectively the antigen to the APCs and to promote antigen uptake for an efficient initiation of the immune response. These systems act principally indirectly, by accessing specific endosomal uptake functions of APCs and include

TABLE 5.1. Main types of adjuvants and delivery systems.

Adjuvant category or strategy	Representative example	Mechanism of action
Mineral salts	Aluminum and calcium salts	Antigen stabilization and slow release, leading to sustained immune activation
Emulsions and surfactant-based formulations	FCA and IFA MF59 Montanide ISA-51, ISA-720 AS02 QS21	Depot effects enabling prolonged antigen presentation
Particulate delivery vehicles	Microparticles ISCOMs Liposomes Virosomes VLPs	Efficient antigen uptake by APCs
Receptor-mediated DC targeting	Antibody to DEC-205 receptor conjugated to OVA	Efficient endocytosis of antigen by DC via DEC-205 receptor
Microbials derivatives (natural and synthetic)	Monophosphoryl lipid A lipoprotein CT and LT from *Escherichia coli* CpG ODNs	Direct activation of APCs through specific receptors
Cytokines	GM-CSF IL-12 IL-2 IL-15 IFN-γ IFN-α	Immune modulation: activation and recruitment of APCs, upregulation of costimulatory molecules, induction of Th-1 response
Cells	Antigen-loaded dendritic cells	Efficient triggering of the immune response

The main categories of vaccine adjuvants and delivery systems are shown. Some of these are not strictly a substance or a formulation but rather a strategy to better stimulate the immune response, such as using the antibody to DEC-205 as a vehicle to target the antigen onto DCs or the use of the DC as cellular adjuvants.

emulsions, liposomes, immunostimulatory complexes (ISCOMs), virus-like particles (VLPs), and microparticles (Kenney and Edelman, 2003; O'Hagan and Valiante, 2003). Particulate delivery systems, such as microparticles, ISCOMs, liposomes, virosomes, and virus-like particles adsorb the antigen onto particles, which display sizes comparable with those

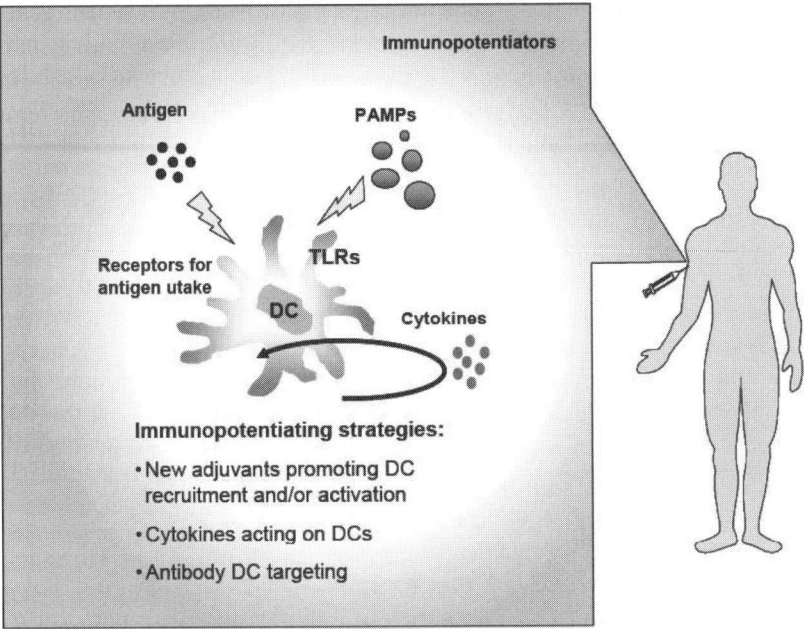

FIGURE 5.3. DCs as the main targets of new adjuvants. DCs can be targeted *in vivo* by immunopotentiating agents (TLRs ligands, cytokines, antibodies, etc.), which may enhance their ability to take-up and process antigens, to promote their activation, or both, and migration to lymphoid organs in order to efficiently prime immune responses.

exhibited by the pathogens against which the response is desired. Therefore, these structures, by mimicking the original pathogen, are very efficiently delivered to DCs. In other delivery systems, including emulsions and surfactant formulations, the antigen is complexed into particles, which not only facilitate antigen uptake by DCs but also serve to display the antigen in a repetitive fashion, thus facilitating the direct recognition by B-cell antigen receptor repertoire (O'Hagan and Valiante, 2003).

While delivery systems act mainly on the targeting of the antigen to professional APCs (mainly DCs), immune potentiators exert their effects directly on these cells through specific receptors (Fig. 5.3) (Degen *et al.*, 2003; Marciani, 2003). For example, PAMPs act as strong immune activators, by virtue of their natural role of activating the DCs through PRRs; furthermore, some cytokines, particularly those linking innate and adaptive immunity, can directly stimulate and amplify DC activity (Belardelli and Ferrantini, 2002; Moingeon *et al.*, 2001; Villinger, 2003). It is clear that both immune potentiators and delivery systems represent essential components of successful vaccines, and often a combination is required to elicit an optimal immune response.

An innovative strategy that has attracted major interest in recent years is the *in vivo* targeting of the antigen to DCs through endocytic receptors (Table 5.1). This approach represents a particularly selective and potentially effective strategy of *in vivo* delivery system (Table 5.1). In this regard, Steinman and co-workers have exploited the property exhibited by the DEC-205 molecule, a member of the C-type lectin-like receptor family, which is specifically expressed by DCs within the T-cell area of lymphoid tissues and is particularly abundant on lymphoid DCs (Witmer-Pack *et al.*, 1995). It has been reported that DEC-205 binding to microbial antigens may result in an efficient transfer into the endocytic compartment for antigen processing and loading on MHC class II molecules, thus leading to CD4$^+$ T-cell proliferation (Engering *et al.*, 2002). More interestingly, DEC-205 engagement may also induce CD8$^+$ T-cell responses, implying that cross-presentation mechanisms are ensured by recognition of antigen through this receptor (Bonifaz *et al.*, 2002). Therefore, DEC-205 provides an efficient receptor-based mechanism for DCs to process proteins for MHC class I presentation *in vivo*.

To assess the potential of this receptor to act as a specific key for introducing antigen into DCs, a series of experiments were performed on mice by injecting a reference antigen fused to an antibody to DEC-205 (Bonifaz *et al.*, 2002). The first experiments showed that this treatment, in the absence of additional stimuli for DC maturation, induced transient antigen-specific T-cell activation followed by T-cell deletion and unresponsiveness (Bonifaz *et al.*, 2002). These results demonstrated that DCs play a crucial role in the maintenance of peripheral tolerance and that the state of maturation/activation of DCs does influence the balance of tolerance versus immunity. Consistent with these results, further experiments demonstrated that strong CD4$^+$ and CD8$^+$ T-cell immunity is induced by the antibody-mediated antigen targeting via the DEC-205 receptor, as long as a DC maturation stimulus (agonistic anti-CD40 antibody) was also administered (Bonifaz *et al.*, 2004). Interestingly, this approach was found to generate much stronger immunity than the standard use of the soluble antigen injected with Freund's complete adjuvant. Furthermore, the antigen-coupled antibody rapidly reached distal sites, allowing a systemic delivery of the antigen to a large number of DCs. A prolonged presentation of the antigen in most lymphoid tissues was another feature observed in this model; this implies that antigens gaining access to DCs via this receptor may persist in subcellular compartments for long periods. These results could have major implications in the designing of vaccines. They represent important evidence of the efficacy of antigen targeting *in vivo* to DCs, if maturation *stimuli* are also supplied, in inducing T cell–mediated protective immunity.

The following sections will provide the reader with an overview of substances that have shown promise as potentially novel adjuvants, including molecules recognized as strong DC activators and certain cytokines that exhibit a special role in linking innate and adaptive immunity.

5.5 Ligands for TLRs and Their Potential Role as Adjuvants

5.5.1 PAMP-dependent Targeting of DCs

An important impetus in the search for new immune-potentiating adjuvants has been provided by the recent discovery of the PAMPs, a family of evolutionary conserved structural elements shared by microbial pathogens, such as LPS and CpG motifs, which can act as powerful activators of the immune system (Janeway, 1989; Janeway, 1992). These compounds represent signatures of potentially noxious substances. They are recognized by PRRs (Medzhitov et al., 1997). Importantly, these receptors are expressed by DCs that constantly screen the environment by using their PRRs as "sensors" for pathogens and process this information by eliciting distinct functional responses. An increasing number of PRRs have been identified, including the 10 recently discovered members of the human TLR family, and some receptors of the C-type lectin family (Figdor et al., 2002; Takeda et al., 2003). It is worth noting that the TLR members attract a significant interest among immunologists by virtue of their role as initiators of the entire program of host defense (O'Neill, 2004). Since their discovery, advances have been made in assigning natural ligands for TLRs, as well as designing synthetic homologues mimicking their natural counterparts (Takeda et al., 2003). Hence, it has been found that LPS from *Escherichia coli* interacts with TLR4, double-stranded RNA (poly I:C) is recognized by TLR3, TLR7 can be triggered by the synthetic compounds imidazoquinolines, whereas CpG-rich DNA motifs are natural ligands for TLR9 (Pulendran, 2004a). The recognition of the microbial *stimulus* through the specific receptor expressed by DCs is translated into a biochemical signal inside the cell. This in turn leads to the transcriptional activation of proinflammatory cytokines and chemokines and upregulation of costimulatory molecules, finally resulting in the activation of the adaptive immune response.

The TLRs are differentially expressed by distinct DC subsets (and in distinct cellular compartments), and the trigger of the different TLRs may mediate distinct Th-polarized responses, or induce T-regulatory pathways, CTL responses, or antibody production. Thus, it is possible to target selectively functionally distinct DCs through the selection of specific TLR ligands. For example, there is evidence showing that the engagement of TLR7 and TLR9, selectively expressed on human plasmacytoid DCs (PDC), induces strong production of IL-12 (p70) and IFN-γ, which subsequently stimulates Th1-type immune responses (Pulendran, 2004b). Myeloid human DCs have been found to express all TLRs except TLR7 and TLR9 and produce high levels of IL-12 upon TLR2- or TLR4-mediated activation (Pulendran, 2004b). Notably, only some TLRs, particularly TLR3, TLR4, TLR7, and TLR9, can induce type I IFNs, which play an important role in antiviral defense by multiple mechanisms, including autocrine DC activation (Lebon *et al.*, 2001; Santini *et al.*, 2002), and for TLR2 controversial results have been found on whether this receptor can induce Th2 responses. To date, several studies on the signaling triggered by certain TRL2 ligands support a role of this receptor in mediating Th2-biased or T-regulatory responses (Pulendran, 2004b). The involvement of distinct TLRs in the induction of CTL responses has been demonstrated. Interestingly, TLR3, TLR7, and TLR9 engagement by their relative ligands may induce type I IFN production, which is well-known to enhance cross-presentation by DCs and favor CTL activation [reviewed by Tough (2004)]. This effect is of particular relevance for vaccines facing viral infections, for which CTL responses represent a fundamental correlate of protection (Seder and Hill, 2000). It is clear that the understanding a of the signaling pathways triggered by microbial stimuli engaged by TLRs on distinct subsets of DCs is fundamental in identifying new candidates to be used as adjuvants for DC targeting (Akira and Takeda, 2004).

5.5.2 CpG Oligonucleotides

Among the novel adjuvants showing strong immunostimulatory effects on DC activation, the synthetic oligodeoxynucleotides (ODNs) containing unmethylated CpG motifs represent an attractive candidate. CpG motifs are sequences typical of bacterial DNA, 20 times more common than in mammalian DNA (Krieg *et al.*, 1995; Klinman *et al.*, 2004). The release of bacterial DNA upon infection functions as a "danger signal," alerting the innate immune system to react against the microbe. In mice and humans, CpG motifs bind to TLR9 expressed by immune cells, B cells and plasmacytoid DCs being the predominant human cell types expressing TLR9 (Klinman *et al.*, 2004). It has been reported that, unlike most TLRs, TLR9 is not expressed at the cell surface. On exposure, CpG is rapidly internalized by immune cells, thus binding to TLR9 inside endocytic vesicles.

Since the first report demonstrating immunostimulatory effects on immune cells such as NK cells, a large number of preclinical studies has been performed with CpG motifs in animal models to determine whether CpG ODNs could boost the immune response elicited by vaccines. An ensemble of early experiments showed that the administration of these compounds in combination with or cross-linked to a reference antigen resulted in an antibody immune response enhanced by several orders of magnitude. The requirement for this to occur was the close spatial and temporal proximity of injection of both antigen and adjuvant. The antibody isotype profiling correlated with the induction of a Th1-type immune response; this was further supported in a variety of experiments, which confirmed that CpG ODNs are potent inducers of Th1-type responses (Chu et al., 1997). In addition, CpG ODNs have been shown to increase, in a dose-dependent manner, the in vivo IFN-γ secretion, consistently with a Th1-biased immune response. Subsequent studies established that CpG ODNs induce activation of DCs, implying improved antigen presentation of soluble proteins to class I–restricted T cells and stimulation of CTL responses (Askew et al., 2000; Shirota et al., 2001).

A large number of reports has presented data obtained in animal models by using CpG ODNs as adjuvant of conventional antiviral vaccines. The coadministration of CpG ODNs and vaccines against influenza, measles, lymphocytic choriomeningitis viruses, and hepatitis B surface antigen or tetanus toxoid was shown to induce an increase of the antibody titers up to threefold with respect to vaccine alone, an antibody profile consistent with IFN-γ production and the development of CTL responses. Of note, protection from virus challenge for some viral models was also observed. Studies in non-human primates for evaluating the potential of CpG ODNs to improve immune responses to reference antigens and experimental and conventional vaccines have also been performed (Verthelyi et al., 2002). The ensemble of preclinical data obtained in these animal models confirmed the expectations that selected compounds could boost immunity in vivo to otherwise ineffective vaccines. However, these studies have shown that different types of CpG ODNs may have optimal adjuvant activity for different vaccines. Therefore, although much data are indicating how to design these compounds in a more rational way, only the results from clinical studies will provide conclusive evidence on the effectiveness and advantages of using specific CpG ODNs in conjunction with human vaccines.

Recently, some CpG ODNs have been developed for clinical use, and clinical trials with these compounds have been started. These studies consisted of phase I clinical trials aimed at evaluating the safety and the immunomodulatory activity of CpG ODNs administered alone or in combination with licensed vaccines, antibodies, or allergens, and phase II studies in the field of cancer immunotherapy and of allergy (Klinman *et al.*, 2004). The results of a study in which CpG ODNs have been used as adjuvant of hepatitis B vaccine have shown an earlier development of antibody responses and a significant increase of geometric mean titers in recipients receiving vaccine plus ODNs (Halperin *et al.*, 2003). In another study where CpG was administered as adjuvant of the influenza vaccine (Klinman *et al.*, 2000), an increase in antibody response was detected only among subjects with preexisting anti-flu antibodies, and PBMCs from these subjects proved to be capable of secreting large amounts of IFN-γ on *in vitro* restimulation. Even though this ensemble of clinical data represents proof of the concept, further clinical investigations and optimization of the use of CpG ODNs are necessary to validate their use as vaccine adjuvants in humans.

5.5.3 PAMP-independent Activation of DCs

The mostly accepted model for DC activation assumes that the innate immune system has evolved mechanisms to discriminate self versus non-self molecules through a set of PRRs expressed by DCs. A related, though distinct concept of DC activation is the "danger theory" proposed by Matzinger and co-workers (Gallucci *et al.*, 1999). This group has emphasized that DCs are capable of sensing not just microbial *stimuli* but also any form of cellular stress produced by damaged or infected cells, which can then represent major DC activation *stimuli* driving the immune response (Gallucci *et al.*, 1999). This paradigm implies that immune system can discriminate dangerous from harmless rather than self from non-self and fits with the notion that the magnitude of the immune response is proportional to the tissue damage evoked by the adjuvant. Notably, the ability to cause danger by a local reaction at the injection site (Ramon, 1926; Freund *et al.*, 1937) seems to be a characteristic shared by most of the widely used adjuvants.

It has been suggested that the endogenous signals derived by wound cells may mimic PAMPs and act as ligands for some PRRs. Consistent with this notion, heat-shock proteins (HSPs) released by cells undergoing necrosis have been found to activate DCs via TLR2 and TLR4 in a PAMP-similar fashion (Srivastava and Amato, 2001). Although some controversy exists on this matter, the documented ability of HSPs to activate DCs through TLRs as well as to bind peptides and promote a MHC class I presentation pathway makes these molecules attractive candidates for peptides-based vaccines against tumors and intracellular pathogens (Srivastava and Amato, 2001). Further support for the "danger" model has been recently provided by studies showing that uric acid contained in dead cells can act as a danger signal for the immune system, and compelling evidence on the capability of these compounds to activate DC has been shown (Pulendran, 2004b; Shi *et al.*, 2003).

5.5.4 Small-Molecule Immune Potentiators: Imidazoquinolines

The study of PAMPs has represented an important impetus for the selection of products to be used as natural adjuvants for boosting DC-driven immunity; many laboratories are also focusing their efforts in establishing platforms to develop new small-molecule immune potentiator (SMIP) approaches. This is mainly because of the need of relying on more potent and less toxic molecules. A family of compounds, which may hold promise for the future, is represented by the imidazoquinolines (Hemmi *et al.*, 2002). These molecules have been shown to enhance the antigen-specific immune response in mouse models, by binding to TLR7 and TLR8 expressed on DCs (Jurk *et al.*, 2002). It is clear that SMIP-based adjuvant development offers several advantages such as low-cost manufacturing, high purity, and more standardized chemical profiles. In the future, an intensive investigation is expected on these small compounds for their use as possible vaccine adjuvants in humans.

5.6 Interaction of Cytokines with DCs: Importance for Vaccine Development

Cytokines are proteins produced by different cells of the immune system, which can play a major role in the regulation of cell differentiation and immune response. A generally accepted functional definition of cytokines distinguishes between type-1 and type-2 cytokines. Type-1 cytokines (IL-2, IL-12, IL-15, IFN-α, IFN-β, and IFN-γ) are involved in Th1 immune responses and induce mainly a cell-mediated immunity. By contrast, type-2 cytokines (IL-4, IL-5, IL-6, IL-10, and IL-13) are involved in Th2 immune response; they tend to promote humoral immunity and, under some conditions, immune deviation to a nonprotective response (tolerance). While the number of cytokines is continuously increasing with the identification of new members endowed with multiple and complex roles in the regulation of the immune response, some cytokines have gained great attention for their major role in linking innate ad adaptive immunity (Belardelli and Ferrantini, 2002). In general, these cytokines are produced in the local microenvironment by cells of the innate immunity (macrophages, NK cells, and DCs) and are capable of acting on DCs, other immune cells, or both, affecting their phenotype and function, thus profoundly shaping the type of the resulting immune response. In fact, DC functions are not predetermined; on the contrary, they are extremely adaptable to perturbations of the microenvironment as a consequence of their response to the locally produced cytokines. Initial commitment of DCs to promote Th1/Th2 or T-regulatory pathways may be shaped by the cytokines to which the DCs are exposed. Figure 5.4 illustrates the main cytokines exerting an effect on DCs, including enhancement of differentiation or activation as well as induction of migration capability and maturation. Of special note, some cytokines, such as IL-10 can exert an inhibitory effect of DC activation, thus contributing to the generation of DCs endowed with tolerogenic activity (Steinbrink *et al.*, 1997; Steinbrink *et al.*, 1999).

Figure 5.5 illustrates the production of different cytokines by DCs in response to various stimuli and summarizes the type of preferential response elicited by these cytokines. As an example, viruses are known to stimulate the production of IFN-α by human precursors of plasmacytoid DCs and induce their differentiation into DCs that elicit IFN-γ– and IL-10– producing T cells (Kadowaki *et al.*, 2000); however, the IL-3 exposure induces their differentiation into Th2-inducing pDCs (Rissoan *et al.*, 1999). The ability of different DC subsets to elicit different Th-type of responses is the result of integration between the genetically determined potential of a given subset and the diverse signals from the environment. The cytokines released in the microenvironment by activated immune cells may be determinant in tuning the type of response elicited by DCs (Pulendran *et al.*, 2001). In mice for example, two subsets of DCs, freshly isolated from the spleen, namely the CD8α^+ and CD8α^- DCs, are functionally divergent in that they induce Th1 and Th2 responses, respectively (Pulendran *et al.*, 1999). Consistent with this, differential skewing cytokines, which differentially expand these DC subsets *in vivo*, may have a role in promoting different types of immune responses. As an example, GM-CSF has been shown to expand preferentially CD8α^- DCs populations and elicits Th2 responses (Pulendran *et al.*, 1999). *In vitro* studies performed with human DCs suggest that the exposure to IL-10 or TGF-β can switch the initial DC commitment to induce Th1 responses toward Th2 responses (Kalinski *et al.*, 1999; Steinbrink *et al.*, 1997; Tacheuchi *et al.*, 1997). Conversely, the IFN-γ exposure can instruct the DCs to acquire a Th1-inducing capacity, while monocyte-derived human DCs cultured with IFN-α have been shown to be potent inducers of Th1 responses (Santini *et al.*, 2000).

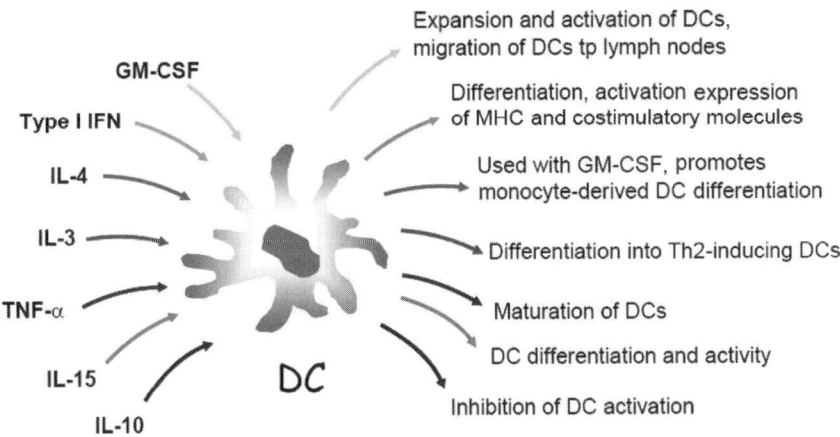

FIGURE 5.4. Cytokines showing an activity on DC differentiation/maturation. This figure illustrates the different effects of the main cytokines on DC functions. Each cytokine and the respective effect are depicted in the same color.

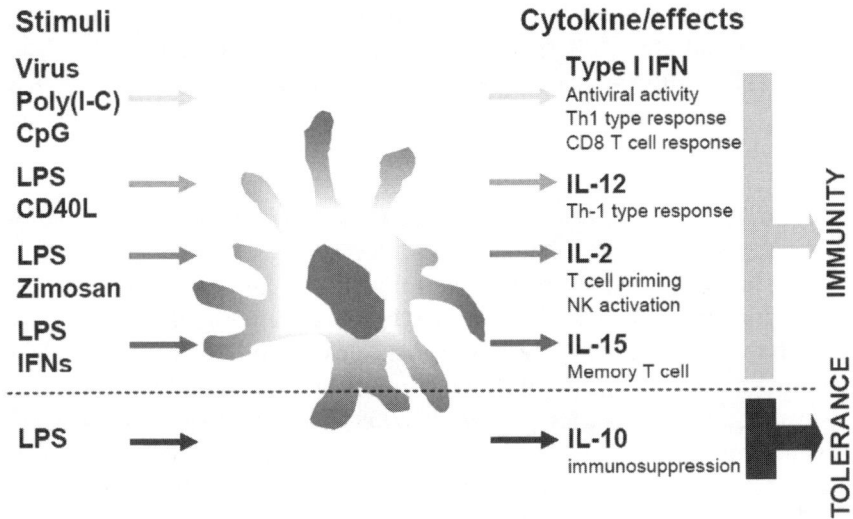

FIGURE 5.5. Cytokines produced by DCs on stimulation. This figure illustrates the production of different cytokines by DCs in response to various stimuli and summarizes the type of preferential response elicited by these cytokines.

During the past decade, a large amount of evidence has been supporting the concept that adjuvants generally act through the induction of cytokines, which are the key mediators of the immune response. This fact has raised the question of whether the direct use of certain selected cytokines as natural adjuvants could be of some advantage with respect to the conventional adjuvants (Lin *et al.*, 1995; Rizza *et al.*, 2002; Villinger, 2003). It is worth noting that the induction of a wide spectrum of proinflammatory cytokines and chemokines by adjuvants may not always be beneficial for the development of the protective immune response. First, this mechanism acts nonspecifically on a broad range of immune cells and does not ensure that the key APCs are selected for stimulation. Furthermore, it does not ensure that the desired cytokine milieu is evoked. Actually, aluminum salts, the currently used adjuvants in human vaccines, have been shown to be potent inducers of Th2-promoting cytokine profile, which does not always mirror the immune correlates of protection for many infectious agents (Gupta, 1998). In addition, failure of efficient antigen presentation or suboptimal DC stimulation could impair the triggering of the immune response. The administration of cytokines as adjuvants in experimental vaccination strategies has recently been pursued [reviewed by Rizza *et al.* (2002)]. Here, we will summarize the most relevant preclinical investigation on cytokines potentially valuable as vaccine adjuvants as well as some remarkable clinical results. We will focus on cytokines known to play an important role in linking innate and adaptive immunity, with special attention given to IL-2, GM-CSF, and type I IFN and to their action as vaccine adjuvants. It should be noted, however, that some of the effects induced by these cytokines may be due to the induction of other cytokines. For example, part of the IFN-α–induced effects on differentiation/activation of DCs and expansion/survival of memory $CD8^+$ T cells may be mediated by IL-15, which is produced by DCs in response to IFN-α (Santini *et al.*, 2000). Thus, complex interactions between host cells and cytokines determine the modalities and mechanisms involved in linking innate and adaptive immunity and regulate quality and intensity of the immune response. An ensemble of preclinical data, including the effects on DC differentiation (Mattei *et al.*, 2001) and on the survival of $CD8^+$ T cells (Berard *et al.*, 2003), supports the concept that IL-15 can represent a valuable vaccine adjuvant. However, because clinical studies have not yet been performed to address this issue, the possible role of IL-15 in enhancing the immunogenicity of human vaccines will not be further discussed. Finally, we will not review the biological effects of recently characterized cytokines, such as IL-23 and IL-27. These cytokines are produced by DCs and play an important role in the regulation of the Th-1 immune response (Matsui *et al.*, 2004). Currently, however, insufficient information is available on the possible vaccine adjuvant activity of these cytokines.

IL-2: Typically known as a cytokine of the adaptive immune response, IL-2 has been extensively evaluated in a large number of clinical trials designed to boost T and NK cell immunity in therapeutic setting against cancer and chronic viral infections (Atkins, 2002). High-dose IL-2 therapy, even if showing significant side effects, still remains a first line of effective treatment against a number of tumors (Atkins *et al.*, 2000; Fisher *et al.*, 2000; Keilholz *et al.*, 1998). In HIV infection, IL-2 therapy has been applied, with variable rate of success, as a strategy for restoring immune competence in patients undergoing antiviral chemotherapy (Emery *et al.*, 2000). However, the improvement in immune function has been shown to be effective only in patients showing more than 200 CD4$^+$ T cells/μl in the blood before treatment and required an extended period of treatment (Paredes *et al.*, 2002). It has been found that spleen and bone marrow–derived mouse DCs are able to produce IL-2 on activation with different bacterial stimuli (Granucci *et al.*, 2001). The secretion of IL-2 is tightly regulated and is induced in a narrow window after microbial encounter. This previously unknown production of IL-2 by DCs has provided a new clue to be pursued for exploring the mechanisms by which DCs control innate and adaptive immunity. Hence, it has been suggested that the early production of IL-2 may confer to DCs the unique capability of T-cell priming. This assumption is supported by the finding that IL-2–deficient, but not wild-type DCs are severely impaired in inducing allogeneic CD8$^+$ and CD4$^+$ T-cell proliferation (Granucci *et al.*, 2001). Furthermore, the persistent infection by an immunosuppressive virus, such MCMV, determines the block of IL-2 production by DCs and of their ability to activate T cells (Andrews *et al.*, 2001). In addition to the role of T-cell priming, DC-derived IL-2 may be involved in sustaining NK response, as this cytokine has been described as a NK cell growth factor (Granucci *et al.*, 2004). In the human system, it has been recently demonstrated that under special conditions, DCs may be a source of IL-2 production. In particular, monocyte-derived DCs (MoDCs) generated in the presence of IL-15 express intracellular IL-2 only if CD40 stimulation is also provided (Feau *et al.*, 2005). Thus, a T cell–derived signal, such as CD40L, is required for IL-2 production by human MoDCs. It has been hypothesized that this mechanism can boost and sustain CD8$^+$ T-cell activation when T-cell contact is established. In contrast, peripheral blood–derived DCs have been shown to start IL-2 transcription by following distinct kinetics under viral or CpG stimulation, whereas CD40L exposure has not been found to stimulate IL-2 mRNA expression, unless the cells were previously activated with CpG or viral challenge. On the whole, the finding that IL-2 can be produced, under special conditions, by DCs in response to certain stimuli emphasizes the interest of performing further studies aimed at the evaluation of the possible adjuvant activity of this cytokine for human vaccines. However, some concerns on the use of IL-2 have recently been raised because of the finding that, under some conditions, this cytokine can enhance the proliferation of $CD25^+/CD4^+$ regulatory T cells, which might be expanded in certain categories of patients and can produce immune-suppressive effects (Malek and Bayer, 2004).

GM-CSF: The immunomodulatory effects of GM-CSF (Mellstedt *et al.*, 1999; Pulendran *et al.*, 2000) have represented the rationale of its use as adjuvant, especially with regard to attempts to develop therapeutic cancer vaccines (Borrello and Pardoll, 2002; Weber *et al.*, 2003). Some studies have shown that the use of soluble GM-CSF for vaccination therapy can enhance the induction of a humoral and a type 1 T-cell response in cancer patients. In patients with advanced colon carcinoma, the combination of GM-CSF with an anti-colon carcinoma monoclonal antibody resulted in a significant response rate. GM-CSF may augment the antibody antitumor effect by enhancing ADCC and by amplifying the induction of a humoral and cellular idiotypic network response. However, GM-CSF may also induce an impairment of the immune response by virtue of its ability to recruit myeloid suppressor cells (Serafini *et al.*, 2004). Activation versus suppression of the immune response by GM-CSF may depend on the dose, with high doses promoting the release of immunosuppressive factors by macrophages (Mellstedt *et al.*, 1999). GM-CSF has been shown to promote a protective and therapeutic immunity in a variety of mouse tumor models, by using different strategies such as fusion protein vaccines, gene transfection into tumor cells or DCs, and DNA immunization (Warren and Weiner, 2000). The encouraging results of the preclinical studies have prompted the testing of GM-CSF gene-transduced tumor vaccines in phase I and phase II clinical trials for immunotherapy of some malignancies (Borrello and Pardoll, 2002). GM-CSF has also been used as adjuvant of viral vaccines with little or poor success so far (Looney *et al.*, 2001). Further studies are needed to better define the possible optimal modalities for using this cytokine as adjuvant of human vaccines.

IL-12: IL-12 is mostly produced by macrophages and DCs in response to certain pathogens and activation factors and has been considered as a key cytokine in linking innate and adaptive immunity by the induction of a Th1-type of immune response (Colombo and Trinchieri, 2002). However, in spite of the promising results obtained in animal tumor models, the clinical use of IL-12 has been restricted because of its severe toxicity. In order to minimize the toxic effects associated with systemic administration of IL-12 and to exploit fully the immunomodulatory activities of this cytokine, a large number of studies in preclinical models of cancer have focused on gene therapy approaches (Melero *et al.*, 2001). Overall, the results obtained in a variety of mouse tumor models indicate that IL-12 gene transfer leads to significant antitumor activity and to the development of a potent cellular immune response, with CD8$^+$ T cells playing a major role in most models.

Type I IFN: Originally described for its antiviral activities, type I IFN has recently been shown to exert important effects on the immune system (Belardelli *et al.*, 2002; Tough, 2004), including promotion of cellular and humoral responses by virtue of its adjuvant effects on APCs (Santini *et al.*, 2002). Type I IFN (especially IFN-α) is the most widely used cytokine in patients with cancer and certain viral infections, such as hepatitis C. Recent studies have pointed out the potential interest of using type I IFN as an adjuvant of vaccines against cancer and infectious diseases (Belardelli *et al.*, 2002; Proietti *et al.*, 2002). In fact, these cytokines potently enhance both T cell and antibody responses to a soluble protein and promote immunologic memory by acting on DCs (Le Bon *et al.*, 2001). Furthermore, recent studies in mice (Bracci *et al.*, 2005; Proietti *et al.*, 2002) have shown that (i) endogenous type I IFN is indispensable for the action of several Th1-promoting adjuvants; (ii) administration of this cytokine as an adjuvant of the human influenza vaccine results in a remarkable enhancement of vaccine immunogenicity, comparable or even superior to that obtained with the most powerful adjuvants. Recently, studies have shown that IFN-α is a potent inducer of the differentiation and activation of both mouse and human DCs (Parlato *et al.*, 2001; Santini *et al.*, 2003; Santodonato *et al.*, 2003). Of note, studies in murine models have revealed that IFN-α/β is a potent enhancer of the cross-priming of CD8$^+$ T cells against exogenous antigens and needs to interact with DCs in order to exert this action (Le Bon *et al.*, 2003). Furthermore, in chimeric models susceptible to HIV infection (SCID mice reconstituted with PBLs), protection from virus challenge has been demonstrated after vaccination with inactivated virus-pulsed DCs generated in the presence of IFN-α (Lapenta *et al.*, 2003). The challenge will now be to evaluate whether the remarkable vaccine adjuvant activity observed in mouse models can also be demonstrated in humans. It should be noted, however, that a recent study (Mennechet and Uzè, 2006) has shown that culture of monocytes with IFN-λ, a type I IFN molecule whose production is also induced in response to viral infection, can result in the generation of DCs endowed with a tolerogenic phenotype. This intriguing finding suggests that different IFN subtypes can render DCs capable of inducing either immunity (IFN-α and IFN-β) or tolerance (IFN-λ), further emphasizing the complex roles of the IFN system in the regulation of the immune response.

5.7 Concluding Remarks

The discovery of DCs and the understanding of their central role in shaping the initiation and regulation of the immune response have represented major milestones in the history of the research on adjuvants, opening new fields of investigation and novel perspectives for vaccine development (Fig. 5.1). Today, DCs are considered as the important targets of modern vaccines. Any kind of vaccine has, in fact, to reach host DCs in a suitable manner in order to induce a protective immunity. With this general concept, DCs have been defined as "nature's cellular adjuvant" (Steinman, 1996). We have recently understood the heterogenicity and the complexity of these cells, and some key factors capable of shaping the DC response toward either immunity or tolerance have been identified. Likewise, only recently, have we begun to understand the action mechanisms of conventional adjuvants and the importance of DCs and of certain cytokines in mediating their biological activities. We can now design new and more selective vaccine adjuvants, taking into consideration the key interactions between pathogens or pathogen-induced danger signals (including certain cytokine) and DCs, which are essential for the generation of a protective immune response. The recent progress in immunology and biotechnologies has led to the development of new strategies for the *in vivo* targeting of the relevant antigens to DCs. This has opened new perspectives for vaccine development and has emphasized the importance of DCs as tools for more effective therapeutic vaccines to be used in patients with cancer or severe chronic infections (Cerundolo *et al.*, 2004; Figdor *et al.*, 2004; Hart, 2001; Steinman and Dhodapkar, 2001; Svane *et al.*, 2003). The results of clinical studies on DC-based vaccines are encouraging, in terms of the extent of immune response induced in patients in the absence of toxicity, but there is a major need for the implementation of methods for the generation and antigen loading of DCs as well as for the clinical testing of this special type of cell vaccines. The overview of the state-of-the-art research on the development of DC-based vaccines is the topic of two other chapters of this book. Here, we only point out that the progress of research on new adjuvants has remarkably benefited from the ensemble of studies aimed at implementing the contemporary methods of generating human DCs from monocytes or CD34[+] progenitor cells. In fact, the use of DC cultures for testing the effects of different natural or synthetic molecules has become a current approach, complementary to the use of animal models, for the identification of new potentially effective vaccine adjuvants. Although the progress of the research on adjuvants has been relatively slow for many years, we predict that now, as a consequence of the emerging knowledge of DC biology, a major advance in the identification of new and more effective adjuvants will rapidly occur, thus resulting in novel perspectives for vaccine development.

Acknowledgments: We are grateful to Cinzia Gasparrini for her precious technical assistance.

References

Afonso, L.C., Scharton, T.M., Vieira, L.Q., Wysocka, M., Trinchieri, G., and Scott, P. (1994). The adjuvant effect of interleukin-12 in a vaccine against Leishmania major. Science 263:235-237.

Akira, S., and Takeda, K. (2004). Toll-like receptor signalling. Nat. Rev. Immunol. 4:499-511.

Andrews, D.M., Andoniou, C.E., Granucci, F., Ricciardi-Castagnoli, P., and Degli-Esposti, M.A. (2001). Infection of dendritic cells by murine cytomegalovirus induces functional paralysis. Nat. Immunol. 2:1077-1084.

Askew, D., Chu, R.S., Krieg, A.M., and Harding, C.V. (2000). CpG DNA induces maturation of dendritic cells with distinct effects on nascent and recycling MHC-II antigen-processing mechanisms. J. Immunol. 165:6889-6895.

Atkins, M.B. (2002). Interleukin-2: clinical applications. Semin. Oncol. 29:12-17.

Atkins, M.B., Kunkel, L., Sznol, M., and Rosenberg, S.A. (2000). High-dose recombinant interleukin-2 therapy in patients with metastatic melanoma: long-term survival update. Cancer J. Sci. Am. 6 Suppl 1:S11-4.

Baldridge, J.R., Yorgensen, Y., Ward, J.R., and Ulrich, J.T. (2000). Monophosphoryl lipid A enhances mucosal and systemic immunity to vaccine antigens following intranasal administration. Vaccine 18:2416-2425.

Banchereau, J., and Steinman, R.M. (1998). Dendritic cells and the control of immunity. Nature 392:245-252.

Banchereau, J., Briere, F., Caux, C., Davoust, J., Lebecque, S., Liu, Y.J., Pulendran, B., and Palucka, K. (2000). Immunobiology of dendritic cells. Annu. Rev. Immunol. 18:767-811.

Banchereau, J., Palucka, A.K., Dhodapkar, M., Burkeholder, S., Taquet, N., Rolland, A., Taquet, S., Coquery, S., Wittkowski, K.M., Bhardwaj, N., Pineiro, L., Steinman, R., and Fay, J. (2001). Immune and clinical responses in patients with metastatic melanoma to CD34($^+$) progenitor-derived dendritic cell vaccine. Cancer Res. 61:6451-6458.

Belardelli, F., and Ferrantini, M. (2002). Cytokines as a link between innate and adaptive antitumor immunity. Trends Immunol. 23:201-208.

Belardelli, F., Ferrantini, M., Proietti, E., and Kirkwood, J.M. (2002). Interferon-alpha in tumor immunity and immunotheraphy. Cytokine Growth Factor Rev. 13:119-134.

Berard, M., Brandt, K., Bulfone-Paus, S., and Tough, D.F. (2003). IL-15 promotes the survival of naive and memory phenotype CD8$^+$ T cells.

J. Immunol. 170:5018-26. [Erratum in: J. Immunol. 2003; 171(4):following 2169.]

Bonifaz, L., Bonnyay, D., Mahnke, K., Rivera, M., Nussenzweig, M.C., and Steinman, R.M. (2002). Efficient targeting of protein antigen to the dendritic cell receptor DEC-205 in the steady state leads to antigen presentation on major histocompatibility complex class I products and peripheral CD8+ T cell tolerance. J. Exp. Med. 196:1627-1638.

Bonifaz, L.C., Bonnyay, D.P., Charalambous, A., Darguste, D.I., Fujii, S., Soares, H., Brimnes, M.K., Moltedo, B., Moran, T.M., and Steinman, R.M. (2004). *In vivo* targeting of antigens to maturing dendritic cells via the DEC-205 receptor improves T cell vaccination. J. Exp. Med. 199:815-824.

Borrello, I., and Pardoll, D. (2002). GM-CSF-based cellular vaccines: a review of the clinical experience. Cytokine Growth Factor Rev. 13:185-193.

Bracci, L., Canini, I., Puzelli, S., Sestili, P., Venditti, M., Spada, M., Donatelli, I., Belardelli, F., and Proietti, E. (2005). Type I IFN is a powerful mucosal adjuvant for a selective intranasal vaccination against influenza virus in mice and affects antigen entrapment at mucosal level. Vaccine 23:2994-3004.

Cerundolo, V., Hermans, I.F., and Salio, M. (2004). Dendritic cells: a journey from laboratory to clinic. Nat. Immunol. 5:7-10.

Chu, R.S., Targoni, O.S., Krieg, A.M., Lehmann, P.V., and Harding, C.V. (1997). CpG oligodeoxynucleotides act as adjuvants that switch on T helper 1 (Th1) immunity. J. Exp. Med. 186:1623-1631.

Colombo, M.P., and Trinchieri, G. (2002). Interleukin-12 in anti-tumor immunity and immunotherapy. Cytokine Growth Factor Rev. 13:155-168.

de Jong, E.C., Vieira, P.L., Kalinski, P., Schuitemaker, J.H., Tanaka, Y., Wierenga, E.A., Yazdanbakhsh, M., and Kapsenberg, M.L. (2002). Microbial compounds selectively induce Th1 cell-promoting or Th2 cell-promoting dendritic cells *in vitro* with diverse th cell-polarizing signals. J. Immunol. 168:1704-1709.

de Jong, E.C., Smits, H.H., and Kapsenberg, M.L. (2005). Dendritic cell-mediated T cell polarization. Springer Semin. Immunopathology 26:289-307.

Degen, W.G., Jansen, T., and Schijns, V.E. (2003). Vaccine adjuvant technology: from mechanistic concepts to practical applications. Expert Rev. Vaccines 2:327-235.

Ellouz, F., Adam, A., Ciobaru, R., and Lederer, E. (1974). Minimal structural requirements for adjuvant activity of bacterial peptido glycan derivates. Biochem. Biophys. Res. Commun. 59: 1317-1325.

Emery, S., Capra, W.B., Cooper, D.A., Mitsuyasu, R.T., Kovacs, J.A., Vig, P., Smolskis, M., Saravolatz, L.D., Lane, H.C., Fyfe, G.A., and Curtin, P.T. (2000). Pooled analysis of 3 randomized, controlled trials of interleukin-2 therapy in adult human immunodeficiency virus type 1 disease. J. Infect. Dis. 182:428-434.

Engering, A., Geijtenbeek, T.B., and van Kooyk, Y. (2002). Immune escape through C-type lectins on dendritic cells. Trends Immunol. 23:480-485.

Feau, S., Facchinetti, V., Granucci, F., Citterio, S., Jarrossay, D., Seresini, S., Protti, M.P., Lanzavecchia, A., and Ricciardi-Castagnoli, P. (2005). Dendritic cell-derived IL-2 production is regulated by IL-15 in humans and in mice. Blood 105:697-702.

Figdor, C.G., van Kooyk, Y., and Adema, G.J. (2002). C-type lectin receptors on dendritic cells and Langerhans cells. Nat. Rev. Immunol. 2:77-84.

Figdor, C.G., de Vries, I.J., Lesterhuis, W.J., and Melief, C.J. (2004). Dendritic cell immunotherapy: mapping the way. Nat. Med. 10:475-480.

Fisher, R.I., Rosenberg, S.A., and Fyfe, G. (2000). Long-term survival update for high-dose recombinant interleukin-2 in patients with renal cell carcinoma. Cancer J. Sci. Am. 6 Suppl 1:S55-7

Forster, R., Schubel, A., Breitfeld, D., Kremmer, E., Renner-Muller, I., Wolf, E., and Lipp, M. (1999). CCR7 coordinates the primary immune response by establishing function in secondary lymphoid organs. Cell 99:23-33.

Freund, J., Casals, J., and Hosmer, E.P. (1937). Sensitization and antibody formation after injection of tubercle bacili and parafin oil. Proc. Soc. Exp. Biol. Med. 37:50913.

Gallucci, S., Lolkema, M., and Matzinger, P. (1999). Natural adjuvants: endogenous activators of dendritic cells. Nat. Med. 5:1249-1255.

Glenny, A.T., Pope, C.G., Waddington, H., and Wallace, V. (1926). The antigenic value of toxoid precipitated by potassium-alum. J. Path. Bacteriol. 29:3845.

Granucci, F., Vizzardelli, C., Pavelka, N., Feau, S., Persico, M., Virzi, E., Rescigno, M., Moro, G., and Ricciardi-Castagnoli, P. (2001). Inducible IL-2 production by dendritic cells revealed by global gene expression analysis. Nat. Immunol. 2:882-888.

Granucci, F., Zanoni, I., Pavelka, N., Van Dommelen, S.L., Andoniou, C.E., Belardelli, F., Degli Esposti, M.A., and Ricciardi-Castagnoli, P. (2004). A contribution of mouse dendritic cell-derived IL-2 for NK cell activation. J. Exp. Med. 200:287-295.

Greenberg, L., and Fleming, D.S. (1947). Increased efficiency of diphteria toxoid when combined with pertussis vaccine. Can. J. Public. Health 38:279-286

Gupta, R.K. (1998). Aluminum compounds as vaccine adjuvants. Adv. Drug. Deliv. Rev. 32:155-172.

Halperin, S.A., Van Nest, G., Smith, B., Abtahi, S., Whiley, H., and Eiden, J.J. (2003). A phase I study of the safety and immunogenicity of recombinant hepatitis B surface antigen co-administered with an immunostimulatory phosphorothioate oligonucleotide adjuvant. Vaccine 21:2461-2467.

Hart, D.N. (1997). Dendritic cells: unique leukocyte populations which control the primary immune response. Blood 90:3245-3287.

Hart, D.N. (2001). Dendritic cells and their emerging clinical applications. Pathology 33:479-492.

Hemmi, H., Kaisho, T., Takeuchi, O., Sato, S., Sanjo, H., Hoshino, K., Horiuchi, T., Tomizawa, H., Takeda, K., and Akira, S. (2002). Small anti-viral compounds activate immune cells via the TLR7 MyD88-dependent signaling pathway. Nat. Immunol. 3:196-200.

Hoebe, K., Janssen, E., and Beutler, B. (2004). The interface between innate and adaptive immunity. Nat. Immunol. 5:971-974.

Janeway, C.A. Jr. (1989). Approaching the asymptote? Evolution and revolution in immunology. Cold Spring Harb. Symp. Quant. Biol. 54 Pt 1:1-13.

Janeway, C.A. Jr. (1992). The immune system evolved to discriminate infectious nonself from noninfectious self. Immunol. Today. 13:11-16.

Johnson, A.G., Gaines, S., and Landy, M. (1956). Studies on the O-antigen of Salmonella typhosa V. Enhancement of antibody response to protein antigens by the purified lipopolysaccharide. J. Exp. Med. 103:225-246.

Jurk, M., Heil, F., Vollmer, J., Schetter, C., Krieg, A.M., Wagner, H., Lipford, G., and Bauer, S. (2002). Human TLR7 or TLR8 independently confer responsiveness to the antiviral compound R-848. Nat. Immunol. 3:499.

Kadowaki, N., Antonenko, S., Lau, J.Y., and Liu, Y.J. (2000). Natural interferon alpha/beta-producing cells link innate and adaptive immunity. J. Exp. Med. 192:219-226.

Kalinski, P., Hilkens, C.M., Wierenga, E.A., and Kapsenberg, M.L. (1999). T-cell priming by type-1 and type-2 polarized dendritic cells: the concept of a third signal. Immunol. Today 20:561-567.

Keilholz, U., Conradt, C., Legha, S.S., Khayat, D., Scheibenbogen, C., Thatcher, N., Goey, S.H., Gore, M., Dorval, T., Hancock, B., Punt, C.J., Dummer, R., Avril, M.F., Brocker, E.B., Benhammouda, A., Eggermont, A.M., and Pritsch, M. (1998). Results of interleukin-2-based treatment in advanced melanoma: a case record-based analysis of 631 patients. J. Clin. Oncol. 16:2921-2929.

Kenney, R.T. and Edelman, R. (2003). Survey of human-use adjuvants. Expert Rev. Vaccines 2:167-188.

Klinman, D.M., Ishii, K.J., Gursel, M., Gursel, I., Takeshita, S., and Takeshita, F. (2000). Immunotherapeutic applications of CpG-containing oligodeoxy-nucleotides. Drug News Perspect. 13:289-296.

Klinman, D.M., Currie, D., Gursel, I., and Verthelyi, D. (2004). Use of CpG oligodeoxynucleotides as immune adjuvants. Immunol. Rev. 199:201-216.

Krieg, A.M., Yi, A.K., Matson, S., Waldschmidt, T.J., Bishop, G.A., Teasdale, R., Koretzky, G.A., and Klinman, D.M. (1995). CpG motifs in bacterial DNA trigger direct B-cell activation. Nature 374:546-549.

Lapenta, C., Santini, S.M., Logozzi, M., Spada, M., Andreotti, M., Di Pucchio, T., Parlato, S., and Belardelli, F. (2003). Potent immune response against HIV-1 and protection from virus challenge in hu-PBL-SCID mice immunized with inactivated-virus-pulsed dendritic cells generated in the presence of IFN-α. J. Exp. Med. 198:361-367.

Le Bon, A., Schiavoni, G., D'Agostino, G., Gresser, I., Belardelli, F., and Tough, D.F. (2001). Type I interferons potently enhance humoral immunity and can promote isotype switching by stimulating dendritic cells *in vivo*. Immunity 14:461-470.

Le Bon, A., Etchart, N., Rossmann, C., Ashton, M., Hou, S., Gewert, D., Borrow, P., and Tough, D.F. (2003). Cross-priming of CD8$^+$ T cells stimulated by virus-induced type I interferon. Nat. Immunol. 4:1009-1015.

Le Moignic and Pinoy. (1916). Les vaccines en emulsion dans le corps gras ou 'lipovaccins.' Cmptes redus de la societe de biologie 79:201-203.

Lin, R., Tarr, P.E., and Jones, T.C. (1995). Present status of the use of cytokines as adjuvants with vaccines to protect against infectious diseases. Clin. Infect. Dis. 21:1439-1449.

Lindblad, E.B. (2004). Aluminium adjuvants—in retrospect and prospect. Vaccine 22:3658-68.

Liu, Y.J. (2001). Dendritic cell subsets and lineages, and their functions in innate and adaptive immunity. Cell 106:259-262.

Looney, R.J., Hasan, M.S., Coffin, D., Campbell, D., Falsey, A.R., Kolassa, J., Agosti, J.M., Abraham, G.N., and Evans, T.G. (2001). Hepatitis B immunization of healthy elderly adults: relationship between naive CD4$^+$ T cells and primary immune response and evaluation of GM-CSF as an adjuvant. J. Clin. Immunol. 21:30-36.

Malek, T.R., and Bayer, A.L. (2004). Tolerance, not immunity, crucially depends on IL-2. Nat. Rev. Immunol. 4:665-674.

Marciani, D.J. (2003). Vaccine adjuvants: role and mechanisms of action in vaccine immunogenicity. Drug Discov. Today. 8:934-943.

Matsui, M.O. Moriya, M.L. Belladonna, S. Kamiya, F.A. Lemonnier, T. Yoshimoto, and T. Akatsuka. (2004). Adjuvant activities of novel cytokines, interleukin-23 (IL-23) and IL-27, for induction of hepatitis C virus-specific cytotoxic T lymphocytes in HLA-A*0201 transgenic mice. J. Virol. 78:9093-9104.

Mattei, F., Schiavoni, G., Belardelli, F., and Tough, D.F. (2001). IL-15 is expressed by dendritic cells in response to type I IFN, double-stranded RNA, or lipopolysaccharide and promotes dendritic cell activation. J. Immunol. 167:1179-1187.

Medzhitov, R., Preston-Hurlburt, P., and Janeway, C.A. Jr. (1997). A human homologue of the Drosophila Toll protein signals activation of adaptive immunity. Nature 388:323-324.

Melero, I., Mazzolini, G., Narvaiza, I., Qian, C., Chen, L., and Prieto, J. (2001). IL-12 gene therapy for cancer: in synergy with other immunotherapies. Trends Immunol. 22:113-115.

Mellstedt, H., Fagerberg, J., Frodin, J.E., Henriksson, L., Hjelm-Skoog, A.L., Liljefors, M., Ragnhammar, P., Shetye, J., and Osterborg, A. (1999). Augmentation of the immune response with granulocyte-macrophage colony-stimulating factor and other hematopoietic growth factors. Curr. Opin. Hematol. 6:169-175.

Mennechet, F.J., and Uze, G. (2006). Interferon {lambda}-treated dendritic cells specifically induce proliferation of FOXP3-expressing suppressor T cells. Blood (in press).

Moingeon, P., Haensler, J., Lindberg, A. (2001). Towards the rational design of Th1 adjuvants. Vaccine 19:4363-4372.

O'Hagan, D.T., and Valiante, N.M. (2003). Recent advances in the discovery and delivery of vaccine adjuvants. Nat. Rev. Drug. Discov. 2:727-735.

O'Neill, L.A. (2004). TLRs: Professor Mechnikov, sit on your hat. Trends Immunol. 25:687-693.

Paredes, R., Lopez Benaldo de Quiros, J.C., Fernandez-Cruz, E., Clotet, B., and Lane, H.C. (2002). The potential role of interleukin-2 in patients with HIV infection. AIDS Rev. 4:36-40.

Parlato, S., Santini, S.M., Lapenta, C., Di Pucchio, T., Logozzi, M., Spada, M., Giammarioli, A.M., Malorni, W., Fais, S., and Belardelli, F. (2001).

Expression of CCR-7, MIP-3beta, and Th-1 chemokines in type I IFN-induced monocyte-derived dendritic cells: importance for the rapid acquisition of potent migratory and functional activities. Blood 98:3022-3029.

Pashine, A., Valiante, N.M., and Ulmer, J.B. (2005). Targeting the innate immune response with improved vaccine adjuvants. Nat. Med. 11:S63-68.

Podda, A., and Del Giudice, G. (2003). MF59-adjuvanted vaccines: increased immunogenicity with an optimal safety profile. Expert Rev. Vaccines 2: 197-203.

Proietti, E., Bracci, L., Puzelli, S., Di Pucchio, T., Sestili, P., De Vincenti, E., Venditti, M., Capone, I., Seif, I., De Maeyer, E., Tough, D., Donatelli, I., and Belardelli, F. (2002). Type I interferon as a natural adjuvant for a protective immune response: lessons from the influenza vaccine model. J. Immunol. 169:375-383.

Pulendran, B. (2004a). Modulating vaccine responses with dendritic cells and Toll-like receptors. Immunol. Rev. 199:227-250.

Pulendran, B. (2004b). Immune activation: death, danger and dendritic cells. Curr. Biol. 14:R30-32.

Pulendran, B., Smith, J.L., Caspary, G., Brasel, K., Pettit, D., Maraskovsky, E., and Maliszewski, C.R. (1999). Distinct dendritic cell subsets differentially regulate the class of immune response *in vivo*. Proc. Natl. Acad. Sci. U. S. A. 96:1036-1041.

Pulendran, B., Banchereau, J., Burkeholder, S., Kraus, E., Guinet, E., Chalouni, C., Caron, D., Maliszewski, C., Davoust, J., Fay, J., and Palucka, K. (2000). Flt3-ligand and granulocyte colony-stimulating factor mobilize distinct human dendritic cell subsets *in vivo*. J. Immunol. 165:566-572.

Pulendran, B., Palucka, K., and Banchereau, J. (2001). Sensing pathogens and tuning immune responses. Science 293:253-256.

Qureshi, N., Takayama, K., and Ribi, E. (1982). Purification and structural determination of nontoxic lipid A obtained from the lipopolysaccharide of Salmonella typhimurium. J. Biol. Chem. 257:11808-11815.

Ramon, G. (1925). Sur l'augmentation anormale de l'antitoxine chez les chevaux producteurs de serum antidiphterique. Bull. Soc. Centr. Med. Vet. 101: 227-234.

Ramon, G. (1926). Procedes pour accroître la production des antitoxins. Ann. Inst. Pasteur 40:110.

Reis, E., and Sousa, C. (2004). Activation of dendritic cells: translating innate into adaptive immunity. Curr. Opin. Immunol. 16:21-25.

Ribi, E. (1984). Beneficial modification of the endotoxin molecule. J. Biol. Response Mod. 3:19.

Rissoan, M.C., Soumelis, V., Kadowaki, N., Grouard, G., Briere, F., de Waal Malefyt, R., and Liu, Y.J. (1999). Reciprocal control of T helper cell and dendritic cell differentiation. Science 283:1183-1186.

Rizza, P., Ferrantini, M., Capone, I., and Belardelli, F. (2002). Cytokines as natural adjuvants for vaccine: where are we now? Trends Immunol. 23: 381-383.

Santini, S.M., Lapenta, C., Logozzi, M., Parlato, S., Spada, M., Di Pucchio, T., and Belardelli, F. (2000). Type I interferon as a powerful adjuvant for monocyte-derived dendritic cell development and activity *in vitro* and in Hu-PBL-SCID mice. J. Exp. Med. 191:1777-1788.

Santini, S.M., Di Pucchio, T., Lapenta, C., Logozzi, M., Parlato, S., and Belardelli, F. (2002). The natural alliance between type I IFN and dendritic cells and its role in linking innate and adaptive immunity. J. Interferon Cytokine Res. 11:1071-1080.

Santini, S.M., Di Pucchio, T., Lapenta, C., Parlato, S., Logozzi, M., and Belardelli, F. (2003) A new type I IFN-mediated pathway for the rapid differentiation of monocytes into highly active dendritic cells. Stem Cells 21:357-362.

Santodonato, L., D'Agostino, G., Nisini, R., Mariotti, S., Monque, D.M., Spada, M., Lattanzi, L., Perrone, M.P., Andreotti, M., Belardelli, F., and Ferrantini, M. (2003). Monocyte-derived dendritic cells generated after a short-term culture with IFN-a and granulocyte-macrophage colony-stimulating factor stimulate a potent Epstein-Barr virus-specific CD8$^+$ T cell response. J. Immunol. 170:5195-5202.

Schijns, V.E. (2003). Mechanisms of vaccine adjuvant activity: initiation and regulation of immune responses by vaccine adjuvants. Vaccine 21:829-831.

Schneerson, R., Fattom, A., Szu, S.C., Bryla, D., Ulrich, J.T., Rudbach, J.A., Schiffman, G., and Robbins, J.B. (1991). Evaluation of monophosphoryl lipid A (MPL) as an adjuvant. Enhancement of the serum antibody response in mice to polysaccharide-protein conjugates by concurrent injection with MPL. J. Immunol. 147:2136-2140.

Seder, R.A., and Hill, A.V. (2000). Vaccines against intracellular infections requiring cellular immunity. Nature 406:793-798.

Serafini, P., Carbley, R., Noonan, K.A., Tan, G., Bronte, V., and Borrello, I. (2004). High-dose granulocyte-macrophage colony-stimulating factor-producing vaccines impair the immune response through the recruitment of myeloid suppressor cells. Cancer Res. 64:6337-6343.

Shi, Y., Evans, J.E., and Rock, K.L. (2003). Molecular identification of a danger signal that alerts the immune system to dying cells. Nature 425:516-521.

Shirota, H., Sano, K., Hirasawa, N., Terui, T., Ohuchi, K., Hattori, T., Shirato, K., and Tamura, G. (2001). Novel roles of CpG oligodeoxynucleotides as a leader for the sampling and presentation of CpG-tagged antigen by dendritic cells. J. Immunol. 167:66-74.

Singh, M., and O'Hagan, D. (1999). Advances in vaccine adjuvants. Nat. Biotechnol. 17:1075-1081.

Singh, M., and O'Hagan, D.T. (2002). Recent advances in vaccine adjuvants. Pharm. Res. 19:715-728.

Sozzani, S., Allavena, P., Vecchi, A., and Mantovani, A. (1999). The role of chemokines in the regulation of dendritic cell trafficking. J. Leukoc. Biol. 66:1-9.

Srivastava, P.K., and Amato, R.J. (2001). Heat shock proteins: the 'Swiss Army Knife' vaccines against cancers and infectious agents. Vaccine 19:2590-2597.

Steinbrink, K., Wolfl, M., Jonuleit, H., Knop, J., and Enk, A.H. (1997). Induction of tolerance by IL-10-treated dendritic cells. J. Immunol. 159:4772-4780.

Steinbrink, K., Jonuleit, H., Muller, G., Schuler, G., Knop, J., and Enk, A.H. (1999). Interleukin-10-treated human dendritic cells induce a melanoma-antigen-specificanergy in CD8($^+$) T cells resulting in a failure to lyse tumor cells. Blood 93:1634-1642.

Steinman, R.M. (1996). Dendritic cells and immune-based therapies. Exp. Hematol. 24:859-862.

Steinman, R.M., and Dhodapkar, M. (2001). Active immunization against cancer with dendritic cells: the near future. Int. J. Cancer 94:459-473.

Stuart-Harris, C.H. (1969). Adjuvant influenza vaccines. Bull. WHO 41:617-621.

Svane, I.M., Soot, M.L., Buus, S., and Johnsen, H.E. (2003). Clinical application of dendritic cells in cancer vaccination therapy. APMIS 111:818-834.

Takeda, K., Kaisho, T., and Akira, S. (2003). Toll-like receptors. Annu. Rev. Immunol. 21:335-376.

Takeuchi, M., Kosiewicz, M.M., Alard, P., and Streilein, J.W. (1997). On the mechanisms by which transforming growth factor-beta 2 alters antigen-presenting abilities of macrophages on T cell activation. Eur. J. Immunol. 27:1648-1656.

Tough, D.F. (2004). Type I interferon as a link between innate and adaptive immunity through dendritic cell stimulation. Leuk. Lymphoma 45:257-264.

Ulanova, M., Tarkowski, A., Hahn-Zoric, M., and Hanson, L.A. (2001). The Common vaccine adjuvant aluminum hydroxide up-regulates accessory properties of human monocytes via an interleukin-4-dependent mechanism. Infect. Immun. 69:1151-1159.

van der Laan, J.W. (2005) Adjuvants enhancing an integral immune response to antigens. Expert Rev. Vaccine 4:15-18.

Verthelyi, D., Kenney, R.T., Seder, R.A., Gam, A.A., Friedag, B., and Klinman, D.M. (2002). CpG oligodeoxynucleotides as vaccine adjuvants in primates. J. Immunol. 168:1659-1663.

Villinger, F. (2003). Cytokines as clinical adjuvants: how far are we? Expert Rev. Vaccines 2:317-326.

Warren, T.L., and Weiner, G.J. (2000). Uses of granulocyte-macrophage colony-stimulating factor in vaccine development. Curr. Opin. Hematol. 7:168-173.

Weber, J., Sondak, V.K., Scotland, R., Phillip, R., Wang, F., Rubio, V., Stuge, T.B., Groshen, S.G., Gee, C., Jeffery, G.G., Sian, S., and Lee, P.P. (2003). Granulocyte-macrophage-colony-stimulating factor added to a multipeptide vaccine for resected Stage II melanoma. Cancer 97:186-200.

Witmer-Pack, M.D., Swiggard, W.J., Mirza, A., Inaba, K., and Steinman, R.M. (1995). Tissue distribution of the DEC-205 protein that is detected by the monoclonal antibody NLDC-145. II. Expression in situ in lymphoid and nonlymphoid tissues. Cell. Immunol. 163:157-162.

Zinkernagel, R.M. (2000). Localization dose and time of antigens determine immune reactivity. Semin. Immunol. 12:163-171.

Chapter 6

Ex Vivo–Generated Dendritic Cells for Clinical Trials versus *In Vivo* Targeting to Dendritic Cells: Critical Issues

Joannes F.M. Jacobs, Cândida F. Pereira, Paul J. Tacken, I. Jolanda M. de Vries, Cornelus J.A. Punt, Gosse J. Adema, and Carl G. Figdor

6.1 Introduction

Dendritic cells (DCs) are antigen-presenting cells with the unique ability to take up and process antigens in the peripheral blood and tissues. They subsequently migrate to draining lymph nodes, where they present antigen to resting lymphocytes (Banchereau and Steinman, 1998). Depending on the activation state of DCs, they can either be immunostimulatory or tolerogenic. Immunostimulatory DCs stimulate immune responses by activating T and B cells. Tolerogenic DCs inhibit immune responses by activating a specific T-cell subset, the regulatory T cells. Literature from anticancer immunotherapy studies shows that specific T-cell responses against an immunogenic target can be induced or enhanced by vaccination with *ex vivo*–generated DCs loaded with the immunogenic target.

Our increased understanding of DC biology and the possibility to obtain large numbers of DCs *in vitro* has boosted the use of DCs in tumor immunotherapy (Banchereau *et al.*, 2001; Steinman and Dhodapkar, 2001). New insights broaden the application of DCs for the treatment of autoimmune diseases and the prevention of transplant rejection and therapy-resistant infections such as HIV infection (Fig. 6.1).

DCs have the capacity to take up HIV particles at the site of infection through cell-surface receptors such as CCR5/CD4 and DC-SIGN and induce an HIV-specific immune response. Vigorous HIV-specific CTL and CD4$^+$ Th1-cell responses have been associated with control of viremia and long-term nonprogression in infected individuals. However, the natural immune

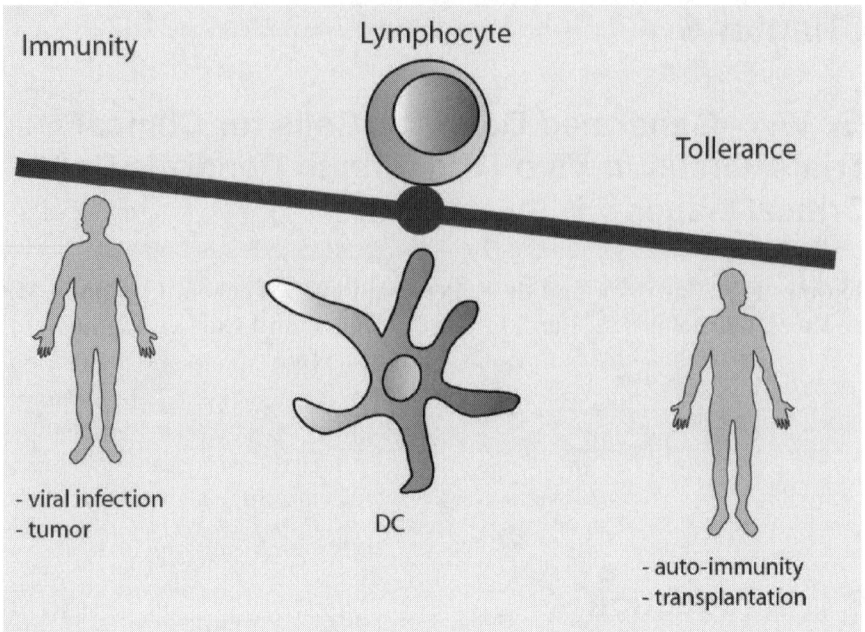

FIGURE 6.1. Immune balance. Depending on the DC subtype and activation state, DCs can induce either immunity or tolerance. CD8$^+$ cytotoxic T cells, CD4$^+$ T helper cells, and B cells are the effector cells for an effective adaptive immune response. Regulatory T cells are involved in the induction of tolerance.

response to HIV is not effective in eradicating the virus. Cellular immune responses decline along the course of the infection, suggesting a progressive loss of function of antigen-presenting cells such as DCs (Pitcher *et al.*, 1999; Kalams *et al.*, 1999). These observations led to the design of new vaccination strategies. These strategies are extensively reviewed in the last chapter of this book.

In this chapter, we review the requirements for an optimal and reproducible clinical-grade DC vaccine. *Ex vivo* generation of DCs, choice and loading of antigen target, vaccine administration, and immunomonitoring are important aspects that will be discussed. Because *ex vivo* generation of DCs is time consuming and a costly process, it would be advantageous to directly activate and target DCs *in vivo*. Therefore, we will discuss the most recent developments of *in vivo* targeting of antigen to DCs.

6.2 DC Culture, from Bench to Clinical-Grade Product

6.2.1 Introduction

Because DCs mainly reside in the peripheral tissues and the lymphoid organs, only a small amount of these cells can be isolated from the peripheral blood. To overcome this problem, DCs can be generated from their blood precursors. DCs originate from $CD34^+$ bone marrow stem cells differentiating into either myeloid or lymphoid precursors. These precursors subsequently differentiate into Langerhans cells, interstitial DCs, or plasmacytoid DCs. All subtypes have their own phenotype and specialized function. For the eradication of HIV-infected cells, both a strong cellular and humoral response seems to be needed. Several studies have shown that human monocyte-derived DCs induce both specific CTL and B-cell responses (Hilkens *et al.*, 1997). So far, most studies have been carried out with monocyte-derived DCs. Figure 6.2 demonstrates a widely used vaccination protocol with monocyte-derived DCs loaded with selected antigens. Each DC vaccination study must be carefully designed to fit its specific purpose. The large majority of clinical studies of DC-based vaccination have been performed in patients with certain types of cancer. Figure 6.3 represents a map with an overview of the different decisions to make. A prerequisite of DC vaccination is to develop a good manufacturing practice (GMP)-compatible procedure for clinical applications that is of high quality and that is cost-effective and reproducible. In this section, the basic principles of DC culture from monocyte precursors are outlined. However, clinical DC vaccination studies have also been conducted with DCs derived from $CD34^+$ progenitor cells and DCs directly isolated from peripheral blood (Banchereau and Palucka, 2005).

6.2.2 Precursor Isolation

It is relatively easy to isolate large numbers of monocytes for clinical use. They can be purified from peripheral blood by a variety of methods, including immunoselection based on CD14 expression, counterflow elutriation based on specific size and weight, and adherence based on the capacity of monocytes to adhere to plastic. In general, sufficient DCs for a

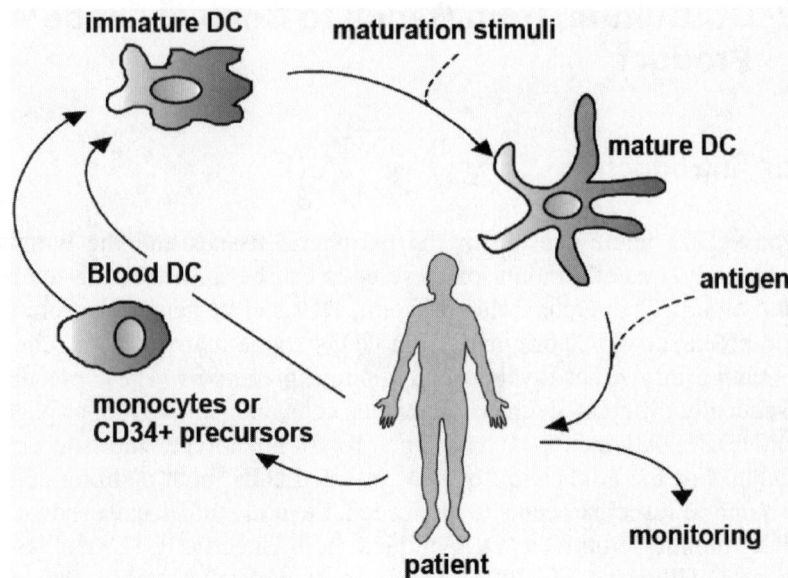

FIGURE 6.2. Schematic representation of DC-vaccination treatment protocol. DCs cultured directly from blood, monocytes, or CD34$^+$ progenitor cells can be loaded *ex vivo* with antigen and administered to patients after culture in the presence of maturation stimuli such as proinflammatory cytokines.

vaccination trial can be cultured from 500 ml of peripheral blood. However, additional DCs are needed for multiple vaccinations, monitoring purposes, and different routes of administration. Therefore, autologous DCs for vaccination trials are usually cultured from leukapheresis product.

6.2.3 Differentiation into DC Phenotype

In 1994, it was discovered that GM-CSF and IL-4 promote differentiation of monocytes into immature DCs over a period of 3 to 5 days (Romani *et al.*, 1994; Sallusto and Lanzavecchia, 1994). This allowed the generation of large numbers of DCs and boosted understanding of DC biology. Pioneering clinical studies with DCs quickly followed, initially focusing on immunotherapy against various cancers. New insights have broadened the application of DCs for the treatment of autoimmune diseases and the prevention of transplant rejection and therapy-resistant infections such as HIV infection (Colino and Snapper, 2003; Thompson and Thomas, 2002). Although HIV infection can adversely affect DC function, it is important to note that monocyte-derived DCs isolated from patients with HIV infection are mostly uninfected and functionally intact (Chougnet *et al.*, 1999; Sapp *et al.*, 1999).

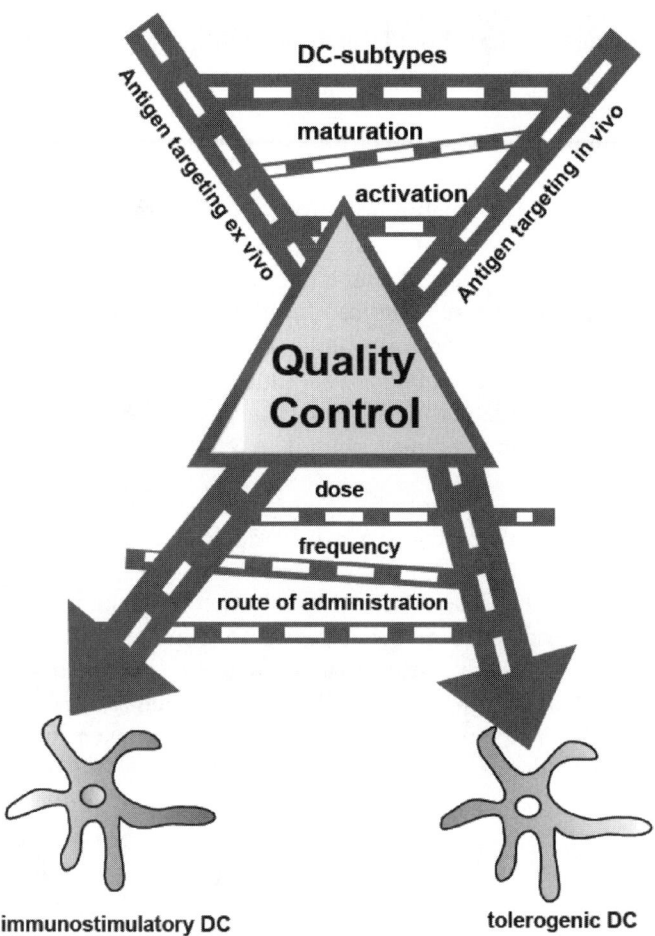

FIGURE 6.3. Dendritic cell vaccination: Which way to go? Because of the many variables in DC vaccination, there is no standard vaccination protocol. Each protocol has to choose its optimal DC subtype, culture conditions, antigen, and route of administration.

6.2.4 Maturation

6.2.4.1 Maturation Signals

Immature DCs have the unique ability to take up and process antigens in peripheral blood and tissues. Mature DCs have the capacity to migrate to draining lymph nodes, present antigen to resting lymphocytes, and induce

T-cell activation. *In vivo*, DC maturation occurs in response to micro-environmental signals that allow DCs to switch their functional phenotype. In the presence of "danger" signals, DCs can undergo this maturation process (Matzinger, 2002). Danger signals can be generated by tissue damage, inflammatory mediators, or directly by microbial products such as lipopolysaccharide and peptidoglycans associated with bacterial cell walls (Reis e Sousa, 2004). The maturation process can be mimicked *in vitro* by a 1- to 2-day culture protocol using various exogenous stimuli, such as LPS, dsRNA, apoptotic cells, immune complexes, CpG DNA, proinflammatory cytokines and prostaglandins. Jonuleit and colleagues first standardized the maturation procedure under GMP conditions. Their efforts led to the current "gold standard" method of DC maturation with a cocktail of proinflammatory cytokines (IL-1β, TNF-α, IL-6, and prostaglandin E2), now known as the "Jonuleit cocktail" (Jonuleit *et al.*, 1997).

6.2.4.2 Effect of Maturation on the Function of DCs

During maturation, DCs lose their ability to efficiently take up and process antigens and, instead, acquire the capacity to migrate toward lymphoid organs. In the T cell–rich parafollicular areas of lymph nodes, several chemokines, such as CCL19 and CCL21, are produced (Kellermann *et al.*, 1999). Mature DCs respond to these chemokines as a consequence of upregulation of chemokine receptors such as CCR7 (Caux *et al.*, 2000; Cyster, 1999).

After arrival of the mature DC in the lymph node, its main function is to activate lymphocytes. In order to achieve this, the mature DCs express antigenic peptides on their MHC complexes, upregulate expression of costimulatory molecules, and secrete cytokines. The principal costimulatory molecules CD80 and/or CD86, intracellular adhesion molecule 1 (ICAM-1) and/or ICAM-2, and OX40 ligand interact with their respective T-cell counterreceptors, CD28, LFA-1, and OX40. The activation signals received by a DC determine its activation program, including which cytokines are produced. These cytokines determine the type of immune response that is induced.

It is thought that for an optimal immune response against HIV, both CD8[+] cytotoxic T cells and CD4[+] Th1 cells are needed (Schoenberger *et al.*, 1998). Although the exact nature of the signals that induce maturation are not completely understood, there is evidence that DCs matured with TNF-α and IL-1β mainly result in CTL-polarizing DCs, and DCs matured with IFN-γ result in Th1-polarizing DCs (Kalinski *et al.*, 1999). Other studies report that the most effective maturation signals must come from microbial and viral products themselves (pathogen-associated molecular patterns). They are directly recognized by pattern-recognition receptors such as members of the Toll-like receptor family. Pattern-recongnition receptors control the expression of genes that directly signal for DC maturation (Sporri and Reis e Sousa, 2005).

6.2.5 The Immune Target

Both $CD8^+$ T cells and $CD4^+$ T cells recognize antigens presented as small peptides in the groove of human leukocyte antigen (HLA; the human analogue of the major histocompatibility complex). $CD8^+$ cells recognize small peptides derived from intracellular cytoplasmic proteins, digested in proteasomes and presented on cell-surface class I molecules. In contrast, $CD4^+$ cells recognize engulfed extracellular proteins, digested to peptides in intracellular endosomes and presented on cell-surface class II HLA molecules. The antigen source and the method of antigen loading can direct an immune response toward a $CD8^+$, $CD4^+$, or combined immune response. In this section, we discuss the rationale for selecting the antigen source and the methods to deliver defined and undefined antigens to DCs. An important question in vaccinology concerns the best antigen source for induction of a specific immune response. Inactivated whole target vaccines are highly immunogenic and usually induce both cellular and humoral immune responses. However, the use of live inactivated targets such as HIV in vaccines raises safety concerns because of potential mutations and incomplete inactivation. Therefore, an attractive alternative is the administration of a single recombinant protein, peptide, or DNA encoding a defined antigen. Unfortunately, this approach induces only partial immune responses. Therefore, new strategies include cocktails of conserved viral proteins or peptides, DNA vaccines composed of genes that encode for several HIV subunit proteins, recombinant vectors and prime-boost strategies, which use two different vaccine strategies to administer the same antigen.

6.2.5.1 MHC Class I and Class II Loading

The major route for presentation of exogenous antigens that enter the DC via endocytosis or phagocytosis is via MHC class II. After uptake, soluble and particulate antigens are directed to the MHC class II compartments (MIIC), where they are degraded into peptide fragments and loaded onto MHC class II molecules. In general, strategies involving pulsing DCs with protein antigens or whole targets will induce class II presentation. The class I pathway is mainly involved in presentation of endogenous self- and viral antigens that are present in the cytosol. These cytosolic proteins are degraded into peptide fragments by the proteasome. Subsequently, the peptides are transported into the endoplasmatic reticulum (ER) by transporters associated with antigen processing (TAP), where they are loaded onto MHC class I molecules. This implies that strategies delivering antigens directly into the cytosol of DCs, such as loading DCs with RNA or DNA encoding antigens, will efficiently induce class I presentation.

In addition, DCs have the capacity to present exogenous antigens on MHC class I via the process of cross-presentation. This process provides internalized proteins access to the cytosolic proteasome and their derived peptides access to the ER-based class I processing machinery (Ackerman and Cresswell, 2004). Particulate antigens taken up by phagocytosis have access to this machinery because phagosomes fuse with the ER soon after or during their formation (Guermonprez *et al.*, 2003). Soluble antigens are less efficiently cross-presented than particulate antigens (Carbone and Bevan, 1990; Reis e Sousa and Germain, 1995). However, internalized soluble proteins can escape proteolysis and gain access to the lumen of the ER, from where they might be transported to the cytosol and processed by the proteasome (Ackerman *et al.*, 2005).

6.2.5.2 Antigen Source

6.2.5.2.1 Loading DCs with Peptides

Peptides derived from pathogens and tumors have been extensively used for loading of DCs to induce an immune response against those targets. The main advantage of using a defined peptide is to generate an immune response that is very specific for that epitope. This approach minimizes the risk of autoimmunity and unwanted tissue damage. Peptide vaccination is often associated with epitope spreading, which is the generation of an immune response against distinct antigens released from the target cells eliminated by the vaccine (el-Shami *et al.*, 1999). Furthermore, loading DCs with peptides circumvents the use of dangerous pathogens or infected cells and facilitates quality control. In addition, the immune response can be accurately monitored because the immune target is well defined. These properties render peptides very attractive for clinical vaccination trials.

However, the use of peptides has several disadvantages that may limit the capacity for clinical application. First, the target epitope must be defined. The use of peptides is therefore limited to those targets for which rejection antigens are identified. Second, the induction of a $CD4^+$ and $CD8^+$ mediated immune response requires a combination of both MHC class I and class II peptides (Wilson and Villadangos, 2005). Third, because of the MHC restriction of the peptides, they are only applicable to MHC-matched patients.

Therefore, large clinical studies using peptides are usually restricted to common HLA-types such as HLA-A1 or HLA-A2. Fourth, a single mutation can cause pathogen escape from immune recognition (Mosier, 2005). The selection of a broader spectrum of peptides may prevent the escape mechanism by loss of epitope variants. Indeed, previous studies indicate that vaccines that include a combination of several HIV peptides can induce a strong cellular immune response against virus challenge (Abdel-Motal *et al.*, 2001; Nehete *et al.*, 2005).

6.2.5.2.2 Loading DCs with Proteins

When a pathogenic antigen has been defined, but the peptide epitope has not yet been identified, the DC loading with the whole protein is a possibility. This does not only circumvent the need for identifying peptide epitopes but also expands the clinical application to patients who are excluded due to the MHC restriction associated with peptide loading. As mentioned previously, when given alone, recombinant viral proteins are poor immunogens. Therefore, an attractive alternative is to combine several proteins into one vaccine. For instance, Huang *et al.*, showed that loading of immature DCs with liposome-complexed HIV structural proteins (p24 gag, gp160 env, or p55 gag) followed by maturation of these DCs can prime CD8$^+$ T cells to recognize HIV *in vitro* (Huang *et al.*, 2003). These results indicate that DCs loaded with whole HIV protein in liposomes may be an effective way for HIV vaccine protocols. However, some HIV proteins may induce adverse effects when used in a whole-protein-based vaccine. Several studies suggest that the HIV envelope glycoprotein gp120 can directly contribute to the immunopathogenesis of HIV by influencing various cell populations of the immune system, including hematopoietic progenitors, T and B lymphocytes, monocytes/macrophages, and DCs (Fantuzzi *et al.*, 2004; Kawamura *et al.*, 2003). The HIV regulatory protein Tat can produce, directly or indirectly, damaging effects in different organs and host systems, such as myocardium, kidney, liver, and central nervous system. In addition, extracellular Tat also has important effects on immunoregulatory functions, such as maturation of DCs (Caputo *et al.*, 2004), DC activation toward a Th-1–inducing phenotype (Fanales-Belasio *et al.*, 2002a; Fanales-Belasio *et al.*, 2002b), access to the MHC class I pathway of presentation (Kim *et al.*, 1997), and modulation of the proteasome catalytic subunit composition, modifying the hierarchy of the CTL epitopes presented in favor of subdominant and cryptic epitopes (Gavioli *et al.*, 2004). Although previous studies have shown that immunizations with active or inactive Tat (Pauza *et al.*, 2000; Silvera *et al.*, 2002) and with gp120 (Pitisuttithum *et al.*, 2003) are safe, the observations described above suggest caution in the use of these full-length HIV proteins for vaccination purposes.

6.2.5.2.3 Loading DCs with Whole Target

When immunogenic antigens have not yet been identified for a certain target or when a more potent T-cell immune response is needed, DCs can be loaded with the whole target. This may reduce the possibility of target escape by loss of epitope variants, as discussed above. If the target is processed and presented through both MHC class I and II, a more potent immune response is generated because $CD4^+$ Th cells play a critical role in inducing and maintaining effective CTL responses (Kalams and Walker, 1998). Especially for tumor therapy, the main disadvantage of using whole target is that the generated immune response could be less specific and may cause more concern for autoimmune diseases. Specific immune monitoring is also more difficult because the epitopes involved are not known. An additional disadvantage of using a whole target is the potential of pathogens to regain infectivity. To overcome this safety issue, pathogens can be treated by irradiation, heat, formalin, or aldrithiol-2 (AT-2), which renders them nonreplicative before *in vivo* administration. AT-2 inactivates virus infectivity by covalently modifying the nucleocapsid zinc finger motifs (Arthur *et al.*, 1998) while preserving the conformational and functional integrity of the virion surface proteins (Rossio *et al.*, 1998).

Although inactivated virus is still considered by many as a potentially dangerous form of HIV, two human clinical studies have shown that this form of vaccination is safe and well tolerated when administered to HIV-infected patients. In these studies, autologous DCs were loaded *ex vivo* with whole heat-inactivated (Garcia *et al.*, 2005) or AT-2–inactivated (Lu *et al.*, 2004) autologous HIV, followed by administration to the patients. Although these vaccines did not induce a humoral immune response, they elicited a cellular immune response that was associated with a partial and transient control of viral replication. The results of these trials are reviewed more extensively in Chapter 14.

6.2.5.2.4 Loading DCs with RNA or DNA

Using RNA isolated from tumor cells, pathogens, or infected cells has some unique advantages. First, sufficient RNA can be generated from a small amount of tissue by amplification of the RNA. Second, target-restricted RNA can be enriched before DC loading by subtractive hybridization with RNA from normal tissues. Pathogen-specific immune responses are thereby augmented and the risk of autoimmunity is reduced. Major drawback for the use of RNA is the instability of RNA products and the greater technical demands (Mitchell and Nair, 2000).

In addition to RNA, the use of DNA that encodes for target antigen has the advantage of expressing the antigen within the cell. Therefore, the antigen can be processed and presented in a way that closely resembles the processing of endogenous proteins. Because DNA is less prone to degradation than RNA, antigen presentation may last longer. However, the generation of immunocompetent DCs with target DNA remains difficult because of the limitations of the DNA delivery systems (Arthur *et al.*, 1997).

6.2.5.3 Methods for Loading DCs with Antigen

Although several methods for loading DCs with antigen have been described (Zhou *et al.*, 2002), the loading method is often restricted by the choice of antigen. Synthetic peptides are effective when pulsed onto mature DCs. In contrast, proteins and whole targets are usually loaded into immature DCs. If a whole target is used, immature DCs can be incubated with apoptotic cells, necrotic cells, cell lysates, or inactivated pathogens. Afterwards, the immature DCs must be fully matured before readministration into patients.

The most common method for introducing RNA into the DC is electroporation (Gilboa and Vieweg, 2004). Genetically modified recombinant vectors are known as the most suitable vehicle for DNA delivery into DCs (gene therapy). Recombinant vector vaccines use attenuated virus or bacteria as carriers of modified target antigens into the host. The target genetic information is incorporated into the vector genome and target proteins are produced. It has previously been shown that DCs loaded *ex vivo* with recombinant vectors, such as replication-defective herpes simplex virus type 1 (HSV-1) amplicons that expressed HIV gp120 (HSV gp120MN/LAI) (Gorantla *et al.*, 2005), replication-defective adenovirus serotype 5 vectors expressing SIV Gag antigens p17 and p45 (Brown *et al.*, 2003), and recombinant *Lactococcus lactis* vector expressing the V2-V4 loops of HIV Env on its cell surface (IL1403-pHIV) (Xin *et al.*, 2003), induce HIV-specific immune responses *in vivo*. Brown and colleagues (Brown *et al.*, 2003) also reported that repeated vaccination with an adenoviral vector did not induce T-cell responses to the vector and did not prevent antigen-expressing DC injected under the capsule of the lymph node from migrating to the paracortex and interposing between T cells. However, the authors described that boost injections of adenovirus-loaded DC generated weak vector-specific immune responses, in particular neutralizing antibody responses that may limit the effectiveness of repeated boosting injections. Therefore, the use of these vectors in clinical vaccination trials is still limited (Kappes and Wu, 2001). Of interest, AT-2–inactivated SIV/HIV has been shown to enter the DCs through a receptor-mediated mechanism and to elicit a potent HLA-I–restricted CTL response (Frank *et al.*, 2002). One of the more experimental methods is targeting antigen to specific receptors on the DC. Several receptors expressed on the surface of DCs are capable of delivering antigens into the DC for MHC-mediated processing and presentation. They include Fc gamma-receptors, receptors for mannose, heat-shock proteins, C-type lectins, and DEC-205. Because these receptors can also be used for *in vivo* targeting of DCs, the mechanism of receptor-mediated loading will be discussed in the section about *in vivo* targeting.

6.2.5.4 Loading of DCs with Antigens: Summary

In summary, the goal of DC-based vaccines is to induce strong immune responses against an antigen. The choice of antigen source and loading method is a crucial step for the development of DC-based vaccines that (1) induce both humoral and cellular immune responses; (2) target a wide range of antigens; (3) avoid unwanted autoimmune responses; (4) are applicable to patients with different HLA types; and (5) meet the safety requirements.

6.2.6 Storage

Culturing DCs is a time-consuming process. To prevent multiple culture steps, the effect of freezing DCs has been investigated (de Vries *et al.*, 2002; Feuerstein *et al.*, 2000). It is demonstrated that DCs can be stored for long time periods in liquid nitrogen. Both phenotype and function are unaffected; however, up to 20% of the DCs can be lost due to the freezing procedure. Based on these results, it is common practice to culture all DCs in one procedure, divide them into lots, and freeze the DCs for administration at later time points.

6.3 DC Quality Check

6.3.1 Introduction

Poor-quality vaccines are not only a health hazard, but they also bias research results and are therefore a waste of research funds. Good manufacturing practice (GMP) is a system to ensure that products are consistently produced and controlled according to quality standards. It is designed to minimize the risks involved in any pharmaceutical production process. WHO has established detailed guidelines for good manufacturing practice; these are available online at http://www.who.int/medicines /organization/

qsm/activities/qualityassurance/gmp. Before administration to patients, microbiological tests must rule out the possibility of any bacterial or fungal contamination.

A prerequisite for DCs used in clinical vaccination studies is that they are cultured under GMP conditions; however, it is also important that the DCs retain their functional properties. Therefore, *ex vivo*–generated DCs for clinical use must meet several quality criteria. Morphologic and functional criteria and guidelines for DC quality controls have been published (Figdor *et al.*, 2004; Whiteside and Odoux, 2004).

6.3.2 DC Phenotype and Purity

Flow cytometry with antibodies directed against cell surface markers demonstrates whether all necessary molecules are expressed. Receptors for antigen recognition and uptake: Fc receptors (CD32 and CD64), complement receptor C3bi (CD11b), which increase efficiency of immune complex endocytosis. C-type lectin receptors, such as the macrophage mannose receptor, DEC-205 and DC-SIGN, which bind pathogenic carbohydrates. Antigen presenting molecules: CD1a, MHC I, and MHC II. Accessory/costimulatory molecules: LFA-3 (CD59), B7-1 (CD80), B7-2 (CD86), CD40, and ICAM-1 (CD54). Monoclonal antibodies against DC-specific markers can measure the purity of the DC product. The presence of low numbers of contaminating lymphocytes and monocytes in a final batch is common but not desirable.

6.3.3 Function

Proper functioning of DCs is evaluated by measuring cytokine release using multiplex technologies. IL-12 production has always been of specific interest as it is considered the most important cytokine for CTL and Th1 induction. The IL-12 production can be significantly impaired if monocytes are even transiently exposed to endotoxin during culture (Karp *et al.*, 1998). To test the proper functioning of DCs, several assays are available that mainly focus on their ability to induce T-cell proliferation (e.g., in a mixed lymphocyte reaction) and their ability to migrate using *in vitro* DC-migration models (de Vries *et al.*, 2003a).

6.4 Vaccine Administration

The ability of DCs to migrate to appropriate regions of lymphatic tissues is critical for the success of DC-based vaccines. DCs administered intravenously or intradermally can end up in the spleen and draining lymph nodes, respectively. However, using radiolabeled DCs, it has been determined that only small percentages reach these lymphoid organs (de Vries *et al.*, 2003a; de Vries *et al.*, 2003b). Some clinical trials try to overcome this by directly injecting DCs into lymph vessels or, under ultrasound guidance, into lymph nodes. Recent data suggest that the location of the primary immune response can determine the distribution of effector cells at different sites of the body (Mullins *et al.*, 2003). DCs injected intravenously home to the spleen and induce visceral immunity. DCs administered subcutaneously or directly in peripheral lymph nodes lead to the control of cutaneous lesions. Animal experiments have not provided a clear-cut set of guidelines for DC administration; therefore, the different routes are often combined.

6.5 Immunomonitoring

Once the DC-based vaccine is delivered to the patient, its therapeutic and immunologic effects must be measured. The immunomonitoring is a crucial issue of clinical studies with cancer vaccines. Enhancement of the frequency of antigen-specific $CD8^+$ T cells in the peripheral blood is generally considered as an important end-point of the patient's response to the injected vaccine. Therapeutic end points in HIV infection can be defined as plasma viral load and $CD4^+$ T-cell counts. For insight in immunologic effects, serum-neutralizing antibody titers can be determined with commercial ELISA kits. HIV-specific $CD4^+$ and $CD8^+$ T cells circulating in peripheral blood in response to the DC vaccination can be quantified with IFN-γ release measured in an ELISPOT assay and with tetramer analysis to defined epitopes (Clay *et al.*, 2001) (Fig. 6.4). After isolation of these specific T cells, their function can be determined by intracellular cytokine production or in cytotoxicity assays such as the chromium release assay. The delayed type hypersensitivity assay is another functional monitoring assay in which antigen, alone or loaded on DCs, is injected intradermally. If functional-specific T cells are present, a local immune response can be observed. The exact nature of this immune response can be analyzed by immunohistochemistry and by analyzing

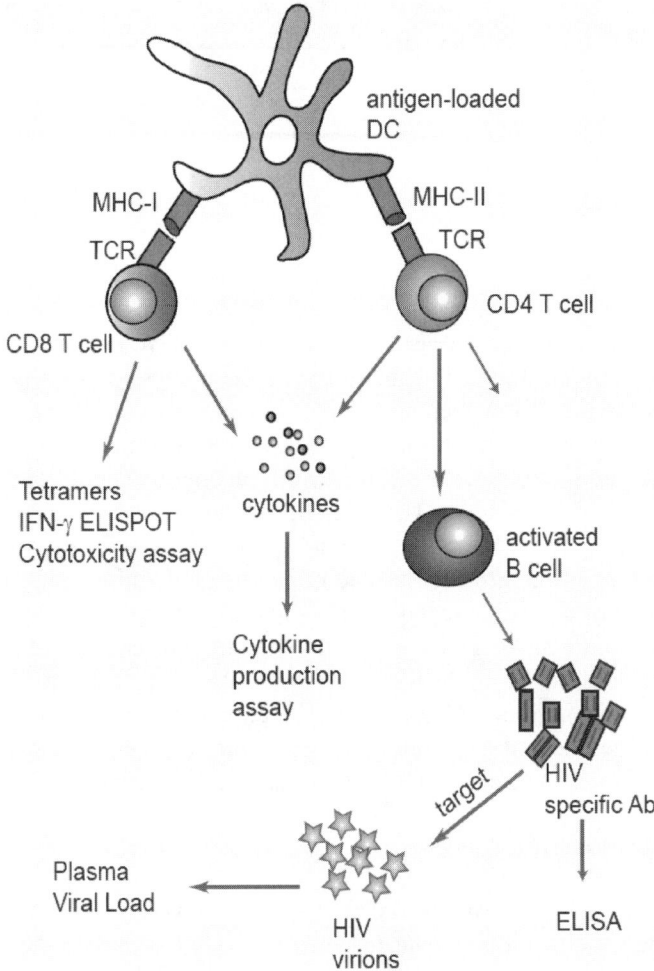

FIGURE 6.4. Monitoring tools. The design of the immunomonitoring is a crucial aspect for all clinical studies involving the use of DC-based vaccines. Much information is available on immunomonitoring strategies and platforms with regard to the use of DC-based cancer vaccines. The figure anticipates the main features of the monitoring for clinical studies with DC-based vaccines in HIV-infected patients, including the evaluation of T-cell responses, antibodies, cytokines, and viral load. The little information so far available on the use of DC-based vaccines in HIV infection is reviewed in the last chapter of this book.

freshly isolated lymphocytes from the lesion (de Vries *et al.*, 2005). Whereas a lot of information is available on the immunomonitoring in clinical studies with DC-based cancer vaccines, only very few studies have reported the immunomonitoring in HIV-infected patients vaccinated with autologous DC-based vaccines (Garcia *et al.*, 2005; Lu *et al.*, 2004). The results of these studies are extensively reviewed in Chapter 14.

6.6 Targeting DCs *In Vivo*

6.6.1 Introduction

Studies involving *ex vivo* antigen loading of DCs have provided, and will continue to provide, much insight in the prerequisites for successful DC-based vaccination. However, the *ex vivo* approach is expensive and elaborate and is not suitable for large-scale immunization programs. The next step in the development of DC-based vaccination involves targeting of antigens to DCs *in vivo*. This can be accomplished by targeting the antigens to specific receptors on the surface of the DCs. There are several ways to deliver antigen to DC surface receptors. The most direct way is to attach a receptor ligand or receptor-specific antibody to the antigen. Alternatively, the antigen, or DNA encoding the antigen, can be incorporated into a more complex drug delivery system containing a receptor ligand or receptor-specific antibody (Table 6.1). The choice of drug delivery system will determine the antigen route of entry into the cell and affects the efficiency of presentation via the MHC class I and II pathway.

TABLE 6.1. Drug delivery systems.

Group	Delivery system	Delivers
Live vectors	Live attenuated virus	DNA
	Live (attenuated) bacteria	DNA
Microparticles	Liposomes	DNA/protein/peptide
	Polymer microparticles	DNA/protein/peptide
	Bacterial ghosts	DNA/protein/peptide
	Virus-like particles	Protein/peptide
	Virosomes	DNA/protein/peptide
Receptor ligands	Bacterial toxins	Protein/peptide
	Heat shock proteins	Protein/peptide
	Sugar residues	Protein/peptide
	Antibodies	Protein/peptide

Drug delivery systems can be divided into three groups: live vectors, microparticles, and receptor ligands. Whereas some systems are suitable to deliver either DNA or protein/peptide antigens to DCs, others deliver both.

6.6.2 Maturing DCs *In Vivo*

Strategies aimed at inducing immunity will require more than strictly targeting of antigen to DCs. The DCs will have to mature and migrate to the lymph nodes to present the antigen to T cells. This is illustrated by studies in mice, where targeting of antigen to the DC surface receptor DEC-205 leads to tolerance, and coadministration of a DC maturation stimulus is required to induce immunity (Bonifaz *et al.*, 2002; Bonifaz *et al.*, 2004; Hawiger *et al.*, 2001). These findings are consistent with DC-based vaccination studies in humans, showing that DC maturation is a prerequisite for induction of immunity (de Vries *et al.*, 2003b). As will be discussed in the following paragraphs, some DC targeting constructs have an inherent capacity to stimulate DC maturation. Targeting constructs that lack this capacity should be administered together with DC maturation factors. Agents that show promise for DC therapy and have been shown to induce maturation *in vivo* include anti-CD40 antibody (Hawiger *et al.*, 2001; Bonifaz *et al.*, 2004), interferon (IFN)-γ (van Broekhoven *et al.*, 2004), α-galactosylceramide (Fujii *et al.*, 2003), and the Toll-like receptor ligands LPS (De Smedt *et al.*, 1996) and CpG oligonucleotides (Jakob *et al.*, 1998).

6.6.3 Targeting Antigens to the MHC Class I and II Pathway

Efficient vaccination strategies for HIV will induce both humoral and cellular responses. This requires that antigens need to be presented via both the MHC class I and class II pathway. In general, targeting strategies delivering antigens directly into the cytosol, or transfecting the DC with DNA encoding antigens, will efficiently induce class I presentation. Strategies involving targeting protein antigen to DC surface receptors will mainly result in presentation of antigen via MHC class II, although the antigens can gain access to the MHC class I loading machinery via cross-presentation (Fig. 6.5). Although endocytosed soluble antigens are less efficiently cross-presented than particulate antigens taken up via phagocytosis (Carbone and Bevan, 1990; Reis e Sousa and Germain, 1995), it is possible to target soluble and particulate antigens to DC surface receptors for class I presentation. However, it remains to be established whether the efficiency of cross-presentation, especially of endocytosed antigens, is sufficient for induction of clinical responses.

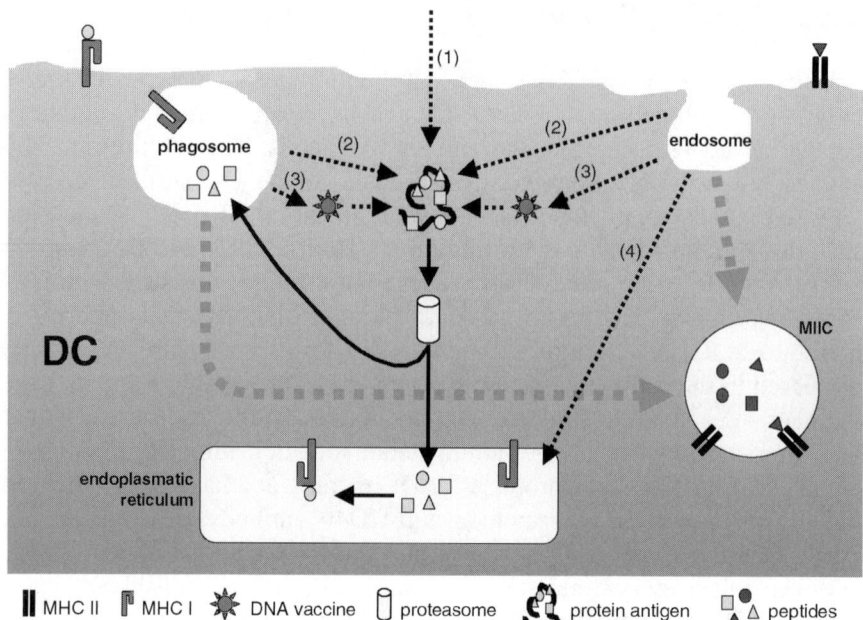

FIGURE 6.5. Targeting antigens to MHC class I and II pathways. Endocytotic and phagocytotic vesicles fuse with the MIIC compartment to deliver exogenous antigens into the MHC class II pathway (grey dotted arrows). There are various vaccination strategies to target the MHC class I pathway (black dotted arrows). Antigens can be targeted directly across the cytoplasmic membrane (*1*). Antigens targeted to the phagosomal and endosomal compartment can escape into the cytosol (*2*). DNA vaccines encoding antigen can be targeted to the endosomal or phagosomal compartment, from where they escape, resulting in expression of the antigen in the cytosol (*3*). Finally, certain bacterial toxins target incorporated antigens via the endosome and Golgi to the ER via the retrograde pathway. In addtion, proteins that are endocytosed can find their way from the endosome to the ER, from where they could be transported to the cytosol (*4*).

6.6.4 Drug Delivery Systems

Drug delivery systems that can be used to target DCs can be subdivided into live vectors, microparticles, and receptor ligands (Table 6.1). Live infectious vectors and certain microparticles are particularly well suited to deliver DNA to DCs. For delivery of protein and peptide antigens to DCs, both microparticles and receptor ligands are used.

6.6.4.1 Live Vectors to Deliver DNA to DCs

Live attenuated viruses and bacteria have been shown to potently induce cellular immune responses in HIV vaccination therapy in animal models (Rayevskaya and Frankel, 2001; Rose *et al.*, 2001). The live vectors drive expression of antigen encoded by DNA inside the cells they invade, which enters the classical MHC class I pathway.

Especially viruses show promise as vehicles for targeted gene delivery. Incorporation of DC surface receptor ligands, mostly antibodies, into viruses allows them to use these receptors for cell entry and specifically infect DCs (Belousova *et al.*, 2003; Korokhov *et al.*, 2005a; Korokhov *et al.*, 2005b). Unfortunately, not all viruses infect DCs, while others decrease DC viability or decrease their immunostimulatory capacity by arresting maturation or downregulation of costimulatory molecules (Jenne *et al.*, 2001). Adenoviruses are promising for DC therapy, as they do not show adverse effects on DC viability and phenotype (Jenne *et al.*, 2001). Targeting of adenovirus to DCs *ex vivo* has been shown to potently induce cellular immunity (Tillman *et al.*, 1999).

Several issues remain to be addressed in the use of live vectors as vaccine. These are safety issues, as reversion of attenuated live vectors to virulent strains by mutation or genetic recombination cannot be excluded. Furthermore, preexisting immunity against the parental vector, or immunity induced by prior treatment with the attenuated vector, might reduce vaccine efficacy (Moron *et al.*, 2004).

6.6.4.2 Microparticles to Deliver DNA, Protein, or Peptides to DCs

Microparticles include lipid particles such as liposomes, polymer micro-particles, bacterial ghosts, virus-like particles (VLPs), and virosomes. Particulate delivery of antigens often increases their stability and bioavail-ability. Furthermore, microparticles are efficiently taken up by APCs through phagocytosis, resulting in antigen presentation. However, the efficacy of microparticles might be hampered by immune responses directed against the carrier-particle itself.

Liposomes are phospholipid vesicles with a bilayer structure. Most liposomes contain an internal space, which allows the incorporation of hydrophilic antigens apart from bilayer association of more amphiphilic antigens. Antigen-containing liposomes in general show modest to low adjuvant effects. Targeting of liposomes containing peptide antigen or DNA to professional APCs effectively enhances their capacity to induce CTL responses *in vivo* (Fukasawa *et al.*, 1998; Hattori *et al.*, 2004). Liposomes can be targeted to DCs by introduction of receptor ligands into the bilayer (Fukasawa *et al.*, 1998), by metal-chelating linkage of his-tagged antibodies to the bilayer (van Broekhoven *et al.*, 2004), or by introduction of protein A into the bilayer, and attaching receptor-specific antibodies to the protein A (Gieseler *et al.*, 2004).

Polymer microparticles allow protection of antigenic proteins, peptides, and DNA from immediate degradation. Entrapment of HIV protein antigen in poly(lactide-*co*-glycolide) microparticles results in an effective vaccine that induces specific class I– and class II–mediated responses in mice (Moore *et al.*, 1995). Targeting of polymer microparticles to specific cells is achieved by introduction of receptor ligands, such as mannose, into the particle. Topical application of particles composed of polyethylenimine mannose and plasmid DNA encoding simian immunodeficiency virus (SIV) proteins to mice results in transfection of APCs, most likely Langerhans cells, present in the skin. Subsequently, the SIV protein expressing APCs migrate to the inguinal lymph nodes, where the antigens can be presented to T cells. Topical application of this mannosylated vaccine to naïve rhesus macaques results in SIV-specific $CD4^+$ and $CD8^+$ T cells in peripheral blood and in SIV-specific DTH responses (Lisziewicz *et al.*, 2005).

Bacterial ghosts are empty bacterial envelopes generated by conditional expression of a lethality gene. They retain all morphologic, structural, and antigenic features of the bacterial cell wall. Bacterial ghosts can be used to transfer DNA or proteins to cells, resulting in induction of cellular and humoral responses (Eko *et al.*, 2003; Ebensen *et al.*, 2004). A major advantage of using bacterial ghosts for vaccination lies in their intrinsic capacity to induce DC maturation, which is likely due to recognition of pathogen-associated molecular patterns on the bacterial envelope by specific DC receptors (Ebensen *et al.*, 2004).

VLPs are generated in insect and mammalian cell expression systems by ectopic expression of subsets of viral proteins. These viral proteins assemble into noninfectious particles, with the ability to enclose specific antigens into their structure. VLPs are capable of inducing both CD4 and CD8 responses *in vivo* (Griffiths *et al.*, 1991; Layton *et al.*, 1993). The mechanisms used by VLPs to deliver antigens into the class I pathway are poorly understood. Certain types of VLP are processed within the endocytic vesicles, and their derived peptides are bound by preexisting class I molecules (Schirmbeck and Reimann, 1996), whereas others reach the cytosol resulting in binding of peptides to neosynthesized class I molecules in the ER (Moron *et al.*, 2003). Similar to bacterial ghosts, VLPs have been shown to induce DC maturation (Rudolf *et al.*, 2001).

Virosomes are semisynthetic complexes derived from nucleic acid–free viral particles. They retain their fusogenic activity, which allows them to deliver the incorporated antigen or DNA into the cytosol of their target cell. Consequently, virosomes are exceptionally well suited for induction of class I responses (Arkema *et al.*, 2000; Bungener *et al.*, 2002), while their capacity to load peptides on MHC class II is comparable with those of other microparticles that enter the cell via endocytosis (Bungener *et al.*, 2002). Strategies involving specific targeting of bacterial ghosts, VLPs, and virosomes to DCs are likely to increase their capacity to elicit immune responses. However, in contrast with DC targeting strategies involving liposomes and polymer microparticles, these have not yet been explored.

6.6.4.3 Receptor Ligands to Deliver Protein or Peptides to DCs

In addition to conjugation of receptor ligands to the drug delivery systems described above, they can be attached directly to protein and peptide antigens for targeting purposes. There are two categories of receptor ligands that target antigen to DC receptors: (i) "natural" receptor ligands, including bacterial toxins, heat-shock proteins (HSPs), and sugar residues; and (ii) receptor-specific antibodies.

Several bacterial toxins have been used for delivery of MHC class I restricted antigen. These toxins bind to cell-surface receptors and are translocated either directly across the plasma membrane into the cytosol, across the endosomal membrane into the cytosol, or to the ER via a retrograde endosome-Golgi-ER pathway (Moron *et al.*, 2004). In this way, antigens coupled to toxins that target to the cytosol and ER gain direct access to the classical MHC class I pathway. Two bacterial toxins have been described to bind to DC surface receptors and induce CTL responses *in vivo* when conjugated to antigens (see Sections 6.6.5.1 and 6.6.5.2).

HSPs are intracellular chaperones expressed by most animal cells and microorganisms. Intact proteins or protein fragments chaperoned by HSPs that are released upon cell lysis can be taken up by APCs via specific receptors, resulting in cross-priming (Binder and Srivastava, 2005). Viral antigens can be bound to HSPs either by covalent coupling, *in vitro* mixing of HSPs and antigen, or by isolation of HSPs from virus-infected cells. Injection of HSP–viral antigen complexes in mice induces protective humoral and cellular responses (Suzue and Young, 1996), resulting in reduced viral titers (Ciupitu *et al.*, 1998). There are several advantages to the use of HSPs as targeting vectors. First, HSP-antigen complexes can be isolated from viral-infected cells, without the need for determining immunogenic epitopes. Second, the interaction of HSPs with APCs leads to production of inflamematory cytokines and chemokines and induces maturation of DCs (Srivastava, 2002), which will enhance vaccination efficacy.

Attachment of sugar residues to antigens allows them to be recognized by C-type lectin receptors on the DC and enter the MHC class I and II pathway. The choice of sugar residue dictates which receptors are targeted, thereby determining the immunologic outcome. For example, antigens coupled to oxidized mannan induce mostly cellular responses, whereas antigens coupled to reduced mannan induce humoral responses (Apostolopoulos et al., 1995; Apostolopoulos et al., 2000). A disadvantage of using sugars for targeting purposes is that most sugars are recognized by multiple receptors, which makes this targeting method rather aspecific. However, the development of synthetic oligolysine-based oligosaccharide clusters has generated high-affinity ligands for specific C-type lectins that could be useful for targeting purposes (Frison et al., 2003).

Antibodies recognize their ligands with remarkable specificity, which makes them a powerful tool for targeting purposes. Recent developments in antibody engineering provide the tools for the production of humanized and human antibodies. These techniques have improved their clinical efficiency and led to the approval of many antibody-based therapies for use in the clinic (Hudson and Souriau, 2003). For DC-based therapy, protein and peptide antigens can be fused to a receptor specific antibody or a single chain Fv of the antibody. Many antibodies targeted to cell-surface receptors induce receptor-mediated endocytosis. Although endocytosis is a relatively inefficient process for induction of cross-presentation, antibody-mediated targeting of protein and peptide antigens to DC surface receptors induces strong CTL responses in vivo (Bonifaz et al., 2004; Hawiger et al., 2001). Moreover, antibody-mediated targeting of antigens has been shown to be more potent than immunization with DCs pulsed with antigen ex vivo (Bonifaz et al., 2004).

6.6.5 Targeting DC Surface Receptors

Receptors expressed by DCs that have been reported to enhance antigen presentation include Mac-1 (CD11b/CD18), globotriaosylceramide (Gb3/CD77), CD40, Fc-receptors (FcRs), and C-type lectin receptors. Many surface receptors activate intracellular signaling pathways upon engagement of ligand, which can affect the behavior of the cell. In general, qualities that allow receptors to serve as vaccine targets include main expression on DCs, high expression levels, endocytic receptor, ability to induce antigen presentation via class I and II, and ability to trigger immunostimulatory signaling pathways.

6.6.5.1 Mac-1

Mac-1 is expressed on PMN, macrophages, myeloid DCs, and NK cells (Arnaout, 1990; Bell *et al.*, 1999). The adenylate cyclase toxin derived from *Bordetella pertussis*, CyaA, is a ligand for Mac-1 (Guermonprez *et al.*, 2001). Multiple studies have shown induction of CTL responses in mice vaccinated with viral epitopes introduced into the catalytic domain of genetically detoxified CyaA (Fayolle *et al.*, 1996; Saron *et al.*, 1997; Sebo *et al.*, 1999; Fayolle *et al.*, 2001). Remarkably, CyaA-mediated delivery of antigens into the class I pathway does not require endocytosis, in contrast with delivery into the class II pathway (Guermonprez *et al.*, 1999; Schlecht *et al.*, 2004). This might be explained by the capacity of CyaA to insert its catalytic domain directly across the plasma membrane into the cytosol (Ladant and Ullmann, 1999; Moron *et al.*, 2004).

6.6.5.2 Gb3

Epithelial, endothelial, and B cells, monocytes, and DCs express the glycosphingolipid Gb3 (Jacewicz *et al.*, 1995; Kiyokawa *et al.*, 1998; LaCasse *et al.*, 1996; Lee *et al.*, 1998). Gb3 is the receptor for the B-subunit of the *Shigella*-derived Shiga toxin (Lindberg *et al.*, 1987). Following binding to Gb3, the B-subunit of Shiga toxin is internalized and transported via endosomes and Golgi to the ER (Smith *et al.*, 2004). Conjugation of antigens to the B-subunit of Shiga toxin results in presentation of the antigens in class I and induces CTL responses *in vivo* (Haicheur *et al.*, 2000; Lee *et al.*, 1998).

6.6.5.3 CD40

CD40 is expressed on B cells, macrophages, DCs, endothelial cells, keratinocytes, fibroblasts, $CD34^+$ hematopoietic cell progenitors, and thymic epithelial cells (Grousson and Peguet-Navarro, 1998). The efficacy of CD40 targeting has been studied using fusion proteins between antigens and CD40 ligand as well as specific targeting of virus to CD40. Adenovirus-driven expression of a fusion protein between soluble CD40 ligand and the tumor-associated antigen MUC-1 induces tumor protection and reversal of T-cell anergy *in vivo* (Zhang *et al.*, 2003). Targeting of adenovirus to CD40 by incorporating soluble CD40 ligand in the virus, or by bispecific antibodies, results in efficient transfection of DCs in both *in vitro* and *ex vivo* models (Belousova *et al.*, 2003; Korokhov *et al.*, 2005b; Tillman *et al.*, 1999). Targeting CD40 has the advantage that it induces maturation of the DC without the need for additional maturation stimuli

(Tillman *et al.*, 1999). Whether this outweighs the disadvantage of rather aspecific targeting due to the broad expression pattern of CD40 remains to be determined.

6.6.5.4 Fc Receptors

Fc receptors (FcRs) comprise a family of membrane proteins that interact with the constant region of immunoglobulins (Ig). They are important mediators of immune responses, such as phagocytosis, antibody-dependent cell-mediated cytotoxicity (ADCC), regulation of lymphocyte proliferation and antigen presentation (Ortiz-Stern and Rosales 2003). DCs express receptors for IgG (FcγR), IgA (FcαR), and IgE (FcεR). The human IgG receptor family consists of several activating receptors, FcγRI (CD64), FcγRIIa (CD32a), and FcγRIIIa/b (CD16a/b), and the inhibitory receptor FcγRIIb (CD32b). FcγRI and FcγRIIIa signal via noncovalent association with a dimeric FcRγ-chain that is shared with FcεRI and FcαRI. FcγRIIa does not associate with the FcRγ-chain but signals via an immunoreceptor tyrosine-based activation motif (ITAM) in its cytoplasmic tail (Van den Herik-Oudijk *et al.*, 1995). In contrast with the activating receptors, FcγRIIb mediates its inhibitory effect via the immunoreceptor tyrosine-based inhibitory motif (ITIM) in its cytoplasmic tail (Daeron, 1997). Immune complexes of IgG and antigen can induce DC maturation and priming of peptide-specific CD8[+] T cells *in vivo* (Regnault *et al.*, 1999; Schuurhuis *et al.*, 2002). However, the balance between activating and inhibitory signaling after engagement of the various FcRs on the DC dictates whether uptake of the immune complexes results in efficient naïve T-cell activation and protective immunity (Kalergis and Ravetch, 2002). This can be overcome by targeting vectors containing antibodies recognizing a specific activating FcR. These vectors should be carefully designed, as FcRs need to be cross-linked in order to induce internalization. Therefore, single-chain Fv constructs are not suitable for FcR targeting, unless these are cross-linked themselves. Moreover, single-chain Fv constructs might inhibit immune responses, as continuous stimulation of FcαRI with single-chain Fv has been described to inhibit activating responses of heterologous FcγRI and FcεRI (Pasquier *et al.*, 2005).

FcγRI is the most studied receptor with regard to FcγR targeting. FcγRI is expressed on monocytes, macrophages, DCs, and activated neutrophils (Deo *et al.*, 1997). In contrast with FcγRII and FcγRIII, FcγRI is a high-affinity receptor for IgG with a high affinity for monomeric IgG (Daeron, 1997). Targeting antigens specifically to FcγRI results in antigen presentation via the class I and class II pathway *in vitro* (Gosselin *et al.*, 1992; Wallace *et al.*, 2001). In FcγRI transgenic mice, targeting antibodies directed against FcγRI induce potent humoral responses against the immunoglobulin-idiotype (Id) of the antibody (Heijnen *et al.*, 1996; Keler *et al.*, 2000). Furthermore, a bispecific antibody against Id on the B cell, and FcγRI on the effector cells, targets lymphoma cells to FcγRI in human FcγRI transgenic mice and effectively clears the tumor cells (Honeychurch *et al.*, 2000). Upon rechallenge, mice treated with the bispecific antibody are protected against the tumor. In this model, deletion of either the CD4 or CD8 cells abrogates tumor protection, suggesting FcγRI targeting leads to class I and II presentation *in vivo* (Honeychurch *et al.*, 2000).

FcαRI (CD89) is a receptor for monomeric IgA and is exclusively expressed on myeloid cells, including monocytes, macrophages, DCs, Kupffer cells, neutrophils, and eosinophils (Monteiro and Van De Winkel, 2003). So far, FcαRI-targeting has mainly been studied in relation to ADCC (Deo *et al.*, 1998; Sundarapandiyan *et al.*, 2001) and phagocytosis of pathogens (van Spriel *et al.*, 1999; Tacken *et al.*, 2004). However, targeting FcαRI has been described to enhance antigen presentation via MHC class II (Shen *et al.*, 2001).

6.6.5.5 C-type Lectin Receptors

The C-type lectins represent a family of lectins that share primary structural homology in their carbohydrate-recognition domain (CRD), which binds to sugar residues in a calcium-dependent manner. C-type lectin receptors expressed by DCs are implicated in immunoregulatory processes, such as antigen capture, DC trafficking, and DC–T cell interactions (Figdor *et al.*, 2002). Based on the location of the amino (N) terminus, two types of membrane-bound C-type lectin receptors can be distinguished on DCs. Type I C-type lectins have their N terminus located outside, whereas type II C type lectins have their N terminus located inside the cell. Several studies have been conducted on antigen targeting to C-type lectin receptors for vaccination purposes, mainly focusing on the mannose receptor (MR), DEC-205, lectin-like oxidized low-density lipoprotein receptor-1 (LOX-1), and DC-SIGN.

The MR is a type I C-type lectin expressed on a variety of cell types, including macrophages, immature DCs, subsets of endothelial cells, retinal pigment epithelium, kidney mesangial cells, and tracheal smooth muscle cells (Keler *et al.*, 2004). Care should be taken when choosing MR ligands for targeting purposes, as certain MR ligands and antibodies induce an immunosuppressive program in DCs (Chieppa *et al.*, 2003; Nigou *et al.*, 2001). Conjugation of "natural" MR ligands, such as mannose and mannan, to antigen has been widely applied to enhance antigen-specific immunity. The results from a clinical study involving the tumor-associated

MUC-I antigen conjugated to mannan in patients with advanced carcinoma of the breast, colon, stomach, or rectum has shown humoral and cellular responses in a number of patients, although no objective clinical responses were observed (Karanikas *et al.*, 1997). However, the sugar residues used in these studies lack specificity for the MR and may target the antigen to a variety of other lectins with overlapping binding specificities. Recent studies involving monoclonal antibodies specifically targeting antigen to MR conclusively demonstrate enhanced antigen uptake and presentation via both the MHC class I and class II pathway (Keler *et al.*, 2004).

DEC-205 is a type I C-type lectin receptor that is abundantly expressed on thymic epithelial cells, mature DCs, and weakly expressed on certain leukocyte subpopulations that remain to be identified (Guo *et al.*, 2000). DEC-205–specific antibodies fused to protein antigen, peptide antigen, or protein antigen–loaded liposomes induce antigen presentation via MHC class I and II. Moreover, they induce immunity in mice, provided DC maturation stimuli, such as IFN-γ, LPS, or anti-CD40 antibodies, are coadministered (Bonifaz *et al.*, 2004; Hawiger *et al.*, 2001; van Broekhoven *et al.*, 2004).

LOX-1 is a type II C-type lectin receptor expressed on macrophages, fibroblasts, smooth muscle cells, endothelial cells, and peripheral blood myeloid DCs (Delneste *et al.*, 2002; Moriwaki *et al.*, 1998; Sawamura *et al.*, 1997). LOX-1 is a receptor for Hsp70, and targeting DCs with protein antigen conjugated to Hsp70 results in cross-presentation *in vitro*. Moreover, antibody-mediated targeting of OVA protein to LOX-1 induces protective antitumor CD8$^+$ T-cell responses in mice challenged with OVA-expressing tumor cells (Delneste *et al.*, 2002).

DC-SIGN represents a member of the type II C-type lectins. A major advantage of targeting DC-SIGN over other C-type lectin receptors is its expression pattern. In humans, DC-SIGN expression is restricted to professional APCs. Human DC-SIGN is abundantly expressed on DCs residing in lymphoid tissues and at mucosal surfaces, dermal DCs, and specialized macrophages in placenta and lung (Geijtenbeek *et al.*, 2000; Soilleux *et al.*, 2002). The expression pattern of mouse DC-SIGN homologues is very different from that of human DC-SIGN (Park *et al.*, 2001; Geijtenbeek *et al.*, 2002). Because the mouse represents the most favorite animal model for *in vivo* targeting studies, this might explain the lack of *in vivo* studies on DC-SIGN targeting. Although it remains to be determined whether DC-SIGN directly triggers intracellular signaling events, binding of the DC-SIGN ligand ManLAM has been described to inhibit LPS-induced DC maturation (Geijtenbeek *et al.*, 2003). *In vitro* studies show efficient targeting of liposomes (Gieseler *et al.*, 2004), adenovirus (Korokhov *et al.*, 2005a), and protein antigens (Engering *et al.*,

2002; Tacken *et al.*, 2005) to DCs via DC-SIGN. Moreover, antibody-mediated targeting of protein antigens to DC-SIGN induces both naïve and recall T-cell responses via MHC class II, and possibly class I, mediated antigen presentation (Engering *et al.*, 2002; Tacken *et al.*, 2005).

6.7 Concluding Remarks

There is great potential in optimizing antigen delivery to DCs *in vivo*, although many questions remain to be answered. Only few studies have directly compared different DC surface receptors in targeting studies, making it difficult to predict what strategy will elicit the most potent immune response. Differences between receptors in expression patterns and levels, route of intracellular trafficking, and association with specific signaling pathways affect their suitability to serve as targets for antigen delivery to DCs. Apart from the receptor target, the targeting vehicle itself will determine the efficiency of antigen delivery to the DC. Especially DNA targeting vectors and vectors delivering proteins directly into the cytosol were believed to effectively drive class I presentation of antigens. The finding that endocytosis of antigens conjugated to receptor-specific antibodies effectively activates naïve $CD8^+$ T cells *in vivo* (Bonifaz *et al.*, 2004; Delneste *et al.*, 2002; Hawiger *et al.*, 2001) was somewhat surprising, as soluble antigens are poorly cross-presented. As monoclonal antibodies are regarded as safe, and more than 20 drugs containing monoclonal antibodies as their main active ingredient have already been approved by the U.S. Food and Drug Administration (http://www.accessdata.fda.gov/scripts/cder/drugsatfda/), this might facilitate the introduction of *in vivo* DC targeting vaccination strategies in the clinic.

References

Abdel-Motal, U. M., Friedline, R., Poligone, B., Pogue-Caley, R. R., Frelinger, J. A., and Tisch, R. (2001). Dendritic cell vaccination induces cross-reactive cytotoxic T lymphocytes specific for wild-type and natural variant human immunodeficiency virus type 1 epitopes in HLA-A*0201/Kb transgenic mice. Clin. Immunol. 101:51-58.

Ackerman, A. L., and Cresswell, P. (2004). Cellular mechanisms governing cross-presentation of exogenous antigens. Nat. Immunol. 5:678-684.

Ackerman, A. L., Kyritsis, C., Tampe, R., and Cresswell, P. (2005). Access of soluble antigens to the endoplasmic reticulum can explain cross-presentation by dendritic cells. Nat. Immunol. 6:107-113.

Apostolopoulos, V., Pietersz, G. A., Loveland, B. E., Sandrin, M. S., and McKenzie, I. F. (1995). Oxidative/reductive conjugation of mannan to antigen selects for T1 or T2 immune responses. Proc. Natl. Acad. Sci. U. S. A. 92:10128-10132.

Apostolopoulos, V., Pietersz, G. A., Gordon, S., Martinez-Pomares, L., and McKenzie, I. F. (2000). Aldehyde-mannan antigen complexes target the MHC class I antigen-presentation pathway. Eur. J. Immunol. 30:1714-1723.

Arkema, A., Huckriede, A., Schoen, P., Wilschut, J., and Daemen, T. (2000). Induction of cytotoxic T lymphocyte activity by fusion-active peptide-containing virosomes. Vaccine 18:1327-1333.

Arnaout, M. A. (1990). Structure and function of the leukocyte adhesion molecules CD11/CD18. Blood 75:1037-1050.

Arthur, J. F., Butterfield, L. H., Roth, M. D., Bui, L. A., Kiertscher, S. M., Lau, R., Dubinett, S., Glaspy, J., McBride, W. H., and Economou, J. S. (1997). A comparison of gene transfer methods in human dendritic cells. Cancer Gene Ther. 4:17-25.

Arthur, L. O., Bess, J. W., Jr., Chertova, E. N., Rossio, J. L., Esser, M. T., Benveniste, R. E., Henderson, L. E., and Lifson, J. D. (1998). Chemical inactivation of retroviral infectivity by targeting nucleocapsid protein zinc fingers: a candidate SIV vaccine. AIDS Res. Hum. Retroviruses 14 Suppl. 3:S311-S319.

Banchereau, J., and Palucka, A. K. (2005). Dendritic cells as therapeutic vaccines against cancer. Nat. Rev. Immunol. 5:296-306.

Banchereau, J., and Steinman, R. M. (1998). Dendritic cells and the control of immunity. Nature 392:245-252.

Banchereau, J., Schuler-Thurner, B., Palucka, A. K., and Schuler, G. (2001). Dendritic cells as vectors for therapy. Cell 106:271-274.

Bell, D., Young, J. W., and Banchereau, J. (1999). Dendritic cells. Adv. Immunol. 72:255-324.

Belousova, N., Korokhov, N., Krendelshchikova, V., Simonenko, V., Mikheeva, G., Triozzi, P. L., Aldrich, W. A., Banerjee, P. T., Gillies, S. D., Curiel, D. T., and Krasnykh, V. (2003). Genetically targeted adenovirus vector directed to CD40-expressing cells. J. Virol. 77:11367-11377.

Binder, R. J., and Srivastava, P. K. (2005). Peptides chaperoned by heat-shock proteins are a necessary and sufficient source of antigen in the cross-priming of CD8$^+$ T cells. Nat. Immunol. 6:593-599.

Bonifaz, L., Bonnyay, D., Mahnke, K., Rivera, M., Nussenzweig, M. C., and Steinman, R. M. (2002). Efficient targeting of protein antigen to the dendritic cell receptor DEC-205 in the steady state leads to antigen presentation on major histocompatibility complex class I products and peripheral CD8$^+$ T cell tolerance. J. Exp. Med. 196:1627-1638.

Bonifaz, L. C., Bonnyay, D. P., Charalambous, A., Darguste, D. I., Fujii, S., Soares, H., Brimnes, M. K., Moltedo, B., Moran, T. M., and Steinman, R. M. (2004). *In vivo* targeting of antigens to maturing dendritic cells via the DEC-205 receptor improves T cell vaccination. J. Exp. Med. 199:815-824.

Brown, K., Gao, W., Alber, S., Trichel, A., Murphey-Corb, M., Watkins, S. C., Gambotto, A., and Barratt-Boyes, S. M. (2003). Adenovirus-transduced dendritic cells injected into skin or lymph node prime potent simian immunodeficiency virus-specific T cell immunity in monkeys. J. Immunol. 171:6875-6882.

Bungener, L., Serre, K., Bijl, L., Leserman, L., Wilschut, J., Daemen, T., and Machy, P. (2002). Virosome-mediated delivery of protein antigens to dendritic cells. Vaccine 20:2287-2295.

Caputo, A., Gavioli, R., and Ensoli, B. (2004). Recent advances in the development of HIV-1 Tat-based vaccines. Curr. HIV Res. 2:357-376.

Carbone, F. R., and Bevan, M. J. (1990). Class I-restricted processing and presentation of exogenous cell-associated antigen *in vivo*. J. Exp. Med. 171:377-387.

Caux, C., Ait-Yahia, S., Chemin, K., de Bouteiller, O., Dieu-Nosjean, M. C., Homey, B., Massacrier, C., Vanbervliet, B., Zlotnik, A., and Vicari, A. (2000). Dendritic cell biology and regulation of dendritic cell trafficking by chemokines. Springer Semin. Immunopathol. 22:345-369.

Chieppa, M., Bianchi, G., Doni, A., Del Prete, A., Sironi, M., Laskarin, G., Monti, P., Piemonti, L., Biondi, A., Mantovani, A., Introna, M., and Allavena, P. (2003). Cross-linking of the mannose receptor on monocyte-derived dendritic cells activates an anti-inflammatory immunosuppressive program. J. Immunol. 171:4552-4560.

Chougnet, C., Cohen, S. S., Kawamura, T., Landay, A. L., Kessler, H. A., Thomas, E., Blauvelt, A., and Shearer, G. M. (1999). Normal immune function of monocyte-derived dendritic cells from HIV-infected individuals: implications for immunotherapy. J. Immunol. 163:1666-1673.

Ciupitu, A. M., Petersson, M., O'Donnell, C. L., Williams, K., Jindal, S., Kiessling, R., and Welsh, R. M. (1998). Immunization with a lymphocytic choriomeningitis virus peptide mixed with heat shock protein 70á results in protective antiviral immunity and specific cytotoxic T lymphocytes. J. Exp. Med. 187:685-691.

Clay, T. M., Hobeika, A. C., Mosca, P. J., Lyerly, H. K., and Morse, M. A. (2001). Assays for monitoring cellular immune responses to active immunotherapy of cancer. Clin. Cancer Res. 7:1127-1135.

Colino, J., and Snapper, C. M. (2003). Dendritic cells, new tools for vaccination. Microbes Infect. 5:311-319.

Cyster, J. G. (1999). Chemokines and the homing of dendritic cells to the T cell areas of lymphoid organs. J. Exp. Med. 189:447-450.

Daeron, M. (1997). Fc receptor biology. Annu. Rev. Immunol. 15:203-234.

De Smedt, T., Pajak, B., Muraille, E., Lespagnard, L., Heinen, E., De Baetselier, P., Urbain, J., Leo, O., and Moser, M. (1996). Regulation of dendritic cell numbers and maturation by lipopolysaccharide *in vivo*. J. Exp. Med. 184:1413-1424.

de Vries, I. J., Eggert, A. A., Scharenborg, N. M., Vissers, J. L., Lesterhuis, W. J., Boerman, O. C., Punt, C. J., Adema, G. J., and Figdor, C. G. (2002). Phenotypical and functional characterization of clinical grade dendritic cells. J. Immunother. 25:429-438.

de Vries, I. J., Krooshoop, D. J., Scharenborg, N. M., Lesterhuis, W. J., Diepstra, J. H., Van Muijen, G. N., Strijk, S. P., Ruers, T. J., Boerman, O. C., Oyen, W. J., Adema, G. J., Punt, C. J., and Figdor, C. G. (2003a). Effective migration of antigen-pulsed dendritic cells to lymph nodes in melanoma patients is determined by their maturation state. Cancer Res. 63:12-17.

de Vries, I. J., Lesterhuis, W. J., Scharenborg, N. M., Engelen, L. P., Ruiter, D. J., Gerritsen, M. J., Croockewit, S., Britten, C. M., Torensma, R., Adema, G. J., Figdor, C. G., and Punt, C. J. (2003b). Maturation of dendritic cells is a prerequisite for inducing immune responses in advanced melanoma patients. Clin. Cancer Res. 9:5091-5100.

de Vries, I. J., Bernsen, M. R., Lesterhuis, W. J., Scharenborg, N. M., Strijk, S. P., Gerritsen, M. J., Ruiter, D. J., Figdor, C. G., Punt, C. J., and Adema, G. J. (2005). Immunomonitoring tumor-specific T cells in delayed-type hypersensitivity skin biopsies after dendritic cell vaccination correlates with clinical outcome. J. Clin. Oncol. 23:5779-5787.

Delneste, Y., Magistrelli, G., Gauchat, J., Haeuw, J., Aubry, J., Nakamura, K., Kawakami-Honda, N., Goetsch, L., Sawamura, T., Bonnefoy, J., and Jeannin, P. (2002). Involvement of LOX-1 in dendritic cell-mediated antigen crosspresentation. Immunity 17:353-362.

Deo, Y. M., Graziano, R. F., Repp, R., and van De Winkel, J. G. (1997). Clinical significance of IgG Fc receptors and Fc gamma R-directed immunotherapies. Immunol. Today 18:127-135.

Deo, Y. M., Sundarapandiyan, K., Keler, T., Wallace, P. K., and Graziano, R. F. (1998). Bispecific molecules directed to the Fc receptor for IgA (Fc{alpha}RI, CD89) and tumor antigens efficiently promote cell-mediated cytotoxicity of tumor targets in whole blood. J. Immunol. 160:1677-1686.

Ebensen, T., Paukner, S., Link, C., Kudela, P., de Domenico, C., Lubitz, W., and Guzman, C. A. (2004). Bacterial ghosts are an efficient delivery system for DNA vaccines. J. Immunol. 172:6858-6865.

Eko, F. O., Lubitz, W., McMillan, L., Ramey, K., Moore, T. T., Ananaba, G. A., Lyn, D., Black, C. M., and Igietseme, J. U. (2003). Recombinant Vibrio cholerae ghosts as a delivery vehicle for vaccinating against Chlamydia trachomatis. Vaccine 21:1694-1703.

el-Shami, K., Tirosh, B., Bar-Haim, E., Carmon, L., Vadai, E., Fridkin, M., Feldman, M., and Eisenbach, L. (1999). MHC class I-restricted epitope spreading in the context of tumor rejection following vaccination with a single immunodominant CTL epitope. Eur. J. Immunol. 29:3295-3301.

Engering, A., Geijtenbeek, T. B. H., van Vliet, S. J., Wijers, M., van Liempt, E., Demaurex, N., Lanzavecchia, A., Fransen, J., Figdor, C. G., Piguet, V., and van Kooyk, Y. (2002). The dendritic cell-specific adhesion receptor DC-SIGN internalizes antigen for presentation to T cells. J. Immunol. 168:2118-2126.

Fanales-Belasio, E., Cafaro, A., Cara, A., Negri, D. R., Fiorelli, V., Butto, S., Moretti, S., Maggiorella, M. T., Baroncelli, S., Michelini, Z., Tripiciano, A., Sernicola, L., Scoglio, A., Borsetti, A., Ridolfi, B., Bona, R., Ten, H. P., Macchia, I., Leone, P., Pavone-Cossut, M. R., Nappi, F., Vardas, E., Magnani, M., Laguardia, E., Caputo, A., Titti, F., and Ensoli, B. (2002a). HIV-1

Tat-based vaccines: from basic science to clinical trials. DNA Cell Biol. 21:599-610.

Fanales-Belasio, E., Moretti, S., Nappi, F., Barillari, G., Micheletti, F., Cafaro, A., and Ensoli, B. (2002b). Native HIV-1 Tat protein targets monocyte-derived dendritic cells and enhances their maturation, function, and antigen-specific T cell responses. J. Immunol. 168:197-206.

Fantuzzi, L., Purificato, C., Donato, K., Belardelli, F., and Gessani, S. (2004). Human immunodeficiency virus type 1 gp120 induces abnormal maturation and functional alterations of dendritic cells: a novel mechanism for AIDS pathogenesis. J. Virol. 78:9763-9772.

Fayolle, C., Sebo, P., Ladant, D., Ullmann, A., and Leclerc, C. (1996). *In vivo* induction of CTL responses by recombinant adenylate cyclase of Bordetella pertussis carrying viral CD8$^+$ T cell epitopes. J. Immunol. 156:4697-4706.

Fayolle, C., Osickova, A., Osicka, R., Henry, T., Rojas, M. J., Saron, M. F., Sebo, P., and Leclerc, C. (2001). Delivery of multiple epitopes by recombinant detoxified adenylate cyclase of Bordetella pertussis induces protective antiviral immunity. J. Virol. 75:7330-7338.

Feuerstein, B., Berger, T. G., Maczek, C., Roder, C., Schreiner, D., Hirsch, U., Haendle, I., Leisgang, W., Glaser, A., Kuss, O., Diepgen, T. L., Schuler, G., and Schuler-Thurner, B. (2000). A method for the production of cryopreserved aliquots of antigen-preloaded, mature dendritic cells ready for clinical use. J. Immunol. Methods 245:15-29.

Figdor, C. G., van Kooyk, Y., and Adema, G. J. (2002). C-type lectin receptors on dendritic cells and Langerhans cells. Nat. Rev. Immunol. 2:77-84.

Figdor, C. G., de Vries, I. J., Lesterhuis, W. J., and Melief, C. J. (2004). Dendritic cell immunotherapy: mapping the way. Nat. Med. 10:475-480.

Frank, I., Piatak, M., Jr., Stoessel, H., Romani, N., Bonnyay, D., Lifson, J. D., and Pope, M. (2002). Infectious and whole inactivated simian immunodeficiency viruses interact similarly with primate dendritic cells (DCs): differential intracellular fate of virions in mature and immature DCs. J. Virol. 76:2936-2951.

Frison, N., Taylor, M. E., Soilleux, E., Bousser, M. T., Mayer, R., Monsigny, M., Drickamer, K., and Roche, A. C. (2003). Oligolysine-based oligosaccharide clusters: selective recognition and endocytosis by the mannose receptor and dendritic cell-specific intercellular adhesion molecule 3 (ICAM-3)-grabbing nonintegrin. J. Biol. Chem. 278:23922-23929.

Fujii, S., Shimizu, K., Smith, C., Bonifaz, L., and Steinman, R. M. (2003). Activation of natural killer T cells by a-galactosylceramide rapidly induces the full maturation of dendritic cells *in vivo* and thereby acts as an adjuvant for combined CD4 and CD8 T cell immunity to a coadministered protein. J. Exp. Med. 198:267-279.

Fukasawa, M., Shimizu, Y., Shikata, K., Nakata, M., Sakakibara, R., Yamamoto, N., Hatanaka, M., and Mizuochi, T. (1998). Liposome oligomannose-coated with neoglycolipid, a new candidate for a safe adjuvant for induction of CD8$^+$ cytotoxic T lymphocytes. FEBS Lett. 441:353-356.

Garcia, F., Lejeune, M., Climent, N., Gil, C., Alcami, J., Morente, V., Alos, L., Ruiz, A., Setoain, J., Fumero, E., Castro, P., Lopez, A., Cruceta, A., Piera, C.,

Florence, E., Pereira, A., Libois, A., Gonzalez, N., Guila, M., Caballero, M., Lomena, F., Joseph, J., Miro, J. M., Pumarola, T., Plana, M., Gatell, J. M., and Gallart, T. (2005). Therapeutic immunization with dendritic cells loaded with heat-inactivated autologous HIV-1 in patients with chronic HIV-1 infection. J. Infect. Dis. 191:1680-1685.

Gavioli, R., Gallerani, E., Fortini, C., Fabris, M., Bottoni, A., Canella, A., Bonaccorsi, A., Marastoni, M., Micheletti, F., Cafaro, A., Rimessi, P., Caputo, A., and Ensoli, B. (2004). HIV-1 tat protein modulates the generation of cytotoxic T cell epitopes by modifying proteasome composition and enzymatic activity. J. Immunol. 173:3838-3843.

Geijtenbeek, T. B. H., Torensma, R., van Vliet, S. J., van Duijnhoven, G. C. F., Adema, G. J., van Kooyk, Y., and Figdor, C. G. (2000). Identification of DC-SIGN, a novel dendritic cell-specific ICAM-3 receptor that supports primary immune responses. Cell 100:575-585.

Geijtenbeek, T. B. H., Groot, P. C., Nolte, M. A., van Vliet, S. J., Gangaram-Panday, S. T., van Duijnhoven, G. C. F., Kraal, G., van Oosterhout, A. J. M., and van Kooyk, Y. (2002). Marginal zone macrophages express a murine homologue of DC-SIGN that captures blood-borne antigens *in vivo*. Blood 100:2908-2916.

Geijtenbeek, T. B. H., van Vliet, S. J., Koppel, E. A., Sanchez-Hernandez, M., Vandenbroucke-Grauls, C. M. J. E., Appelmelk, B., and van Kooyk, Y. (2003). Mycobacteria target DC-SIGN to suppress dendritic cell function. J. Exp. Med. 197:7-17.

Gieseler, R. K., Marquitan, G., Hahn, M. J., Perdon, L. A., Driessen, W. H., Sullivan, S. M., and Scolaro, M. J. (2004). DC-SIGN-specific liposomal targeting and selective intracellular compound delivery to human myeloid dendritic cells: implications for HIV disease. Scand. J. Immunol. 59:415-424.

Gilboa, E., and Vieweg, J. (2004). Cancer immunotherapy with mRNA-transfected dendritic cells. Immunol. Rev. 199:251-263.

Gorantla, S., Santos, K., Meyer, V., Dewhurst, S., Bowers, W. J., Federoff, H. J., Gendelman, H. E., and Poluektova, L. (2005). Human dendritic cells transduced with herpes simplex virus amplicons encoding human immuno-deficiency virus type 1 (HIV-1) gp120 elicit adaptive immune responses from human cells engrafted into NOD/SCID mice and confer partial protection against HIV-1 challenge. J. Virol. 79:2124-2132.

Gosselin, E. J., Wardwell, K., Gosselin, D. R., Alter, N., Fisher, J. L., and Guyre, P. M. (1992). Enhanced antigen presentation using human Fc gamma receptor (monocyte/macrophage)-specific immunogens. J. Immunol. 149:3477-3481.

Griffiths, J. C., Berrie, E. L., Holdsworth, L. N., Moore, J. P., Harris, S. J., Senior, J. M., Kingsman, S. M., Kingsman, A. J., and Adams, S. E. (1991). Induction of high-titer neutralizing antibodies, using hybrid human immunodeficiency virus V3-Ty viruslike particles in a clinically relevant adjuvant. J. Virol. 65:450-456.

Grousson, J., and Peguet-Navarro, J. (1998). Functional relevance of CD40 expression in human epidermis. Cell Biol. Toxicol. 14:345-350.

Guermonprez, P., Ladant, D., Karimova, G., Ullmann, A., and Leclerc, C. (1999). Direct delivery of the Bordetella pertussis adenylate cyclase toxin to the MHC class I antigen presentation pathway. J. Immunol. 162:1910-1916.

Guermonprez, P., Khelef, N., Blouin, E., Rieu, P., Ricciardi-Castagnoli, P., Guiso, N., Ladant, D., and Leclerc, C. (2001). The adenylate cyclase toxin of Bordetella pertussis binds to target cells via the {{alpha}}M{beta}2 integrin (CD11b/CD18). J. Exp. Med. 193:1035-1044.

Guermonprez, P., Saveanu, L., Kleijmeer, M., Davoust, J., van Endert, P., and Amigorena, S. (2003). ER-phagosome fusion defines an MHC class I cross-presentation compartment in dendritic cells. Nature 425:397-402.

Guo, M., Gong, S., Maric, S., Misulovin, Z., Pack, M., Mahnke, K., Nussenzweig, M. C., and Steinman, R. M. (2000). A monoclonal antibody to the DEC-205 endocytosis receptor on human dendritic cells. Hum. Immunol. 61:729-738.

Haicheur, N., Bismuth, E., Bosset, S., Adotevi, O., Warnier, G., Lacabanne, V., Regnault, A., Desaymard, C., Amigorena, S., Ricciardi-Castagnoli, P., Goud, B., Fridman, W. H., Johannes, L., and Tartour, E. (2000). The B Subunit of Shiga toxin fused to a tumor antigen elicits CTL and targets dendritic cells to allow MHC class I-restricted presentation of peptides derived from exogenous antigens. J. Immunol. 165:3301-3308.

Hattori, Y., Kawakami, S., Suzuki, S., Yamashita, F., and Hashida, M. (2004). Enhancement of immune responses by DNA vaccination through targeted gene delivery using mannosylated cationic liposome formulations following intravenous administration in mice. Biochem. Biophys. Res. Commun. 317:992-999.

Hawiger, D., Inaba, K., Dorsett, Y., Guo, M., Mahnke, K., Rivera, M., Ravetch, J. V., Steinman, R. M., and Nussenzweig, M. C. (2001). Dendritic cells induce peripheral T cell unresponsiveness under steady state conditions *in vivo*. J. Exp. Med. 194:769-779.

Heijnen, I. A., Vugt, M. J., Fanger, N. A., Graziano, R. F., Wit, T. P., Hofhuis, F. M., Guyre, P. M., Capel, P. J., Verbeek, J. S., and Winkel, J. G. (1996). Antigen targeting to myeloid-specific human Fcgamma RI/CD64 triggers enhanced antibody responses in transgenic mice. J. Clin. Invest. 97:331-338.

Hilkens, C. M., Kalinski, P., de Boer, M., and Kapsenberg, M. L. (1997). Human dendritic cells require exogenous interleukin-12-inducing factors to direct the development of naive T-helper cells toward the Th1 phenotype. Blood 90:1920-1926.

Honeychurch, J., Tutt, A. L., Valerius, T., Heijnen, I. A. F. M., van de Winkel, J. G. J., and Glennie, M. J. (2000). Therapeutic efficacy of Fcgamma RI/CD64-directed bispecific antibodies in B-cell lymphoma. Blood 96:3544-3552.

Huang, X. L., Fan, Z., Zheng, L., Borowski, L., Li, H., Thomas, E. K., Hildebrand, W. H., Zhao, X. Q., and Rinaldo, C. R., Jr. (2003). Priming of human immunodeficiency virus type 1 (HIV-1)-specific $CD8^+$ T cell responses by dendritic cells loaded with HIV-1 proteins. J. Infect. Dis. 187:315-319.

Hudson, P. J., and Souriau, C. (2003). Engineered antibodies. Nat. Med. 9:129-134.

Jacewicz, M. S., Acheson, D. W., Mobassaleh, M., Donohue-Rolfe, A., Balasubramanian, K. A., and Keusch, G. T. (1995). Maturational regulation of globotriaosylceramide, the Shiga-like toxin 1 receptor, in cultured human gut epithelial cells. J. Clin. Invest. 96:1328-1335.

Jakob, T., Walker, P. S., Krieg, A. M., Udey, M. C., and Vogel, J. C. (1998). Activation of cutaneous dendritic cells by CpG-containing oligodeoxy-nucleotides: a role for dendritic cells in the augmentation of Th1 responses by immunostimulatory DNA. J. Immunol. 161:3042-3049.

Jenne, L., Schuler, G., and Steinkasserer, A. (2001). Viral vectors for dendritic cell-based immunotherapy. Trends Immunol. 22:102-107.

Jonuleit, H., Kuhn, U., Muller, G., Steinbrink, K., Paragnik, L., Schmitt, E., Knop, J., and Enk, A. H. (1997). Pro-inflammatory cytokines and prostaglandins induce maturation of potent immunostimulatory dendritic cells under fetal calf serum-free conditions. Eur. J. Immunol. 27:3135-3142.

Kalams, S. A., and Walker, B. D. (1998). The critical need for CD4 help in maintaining effective cytotoxic T lymphocyte responses. J. Exp. Med. 188:2199-2204.

Kalams, S. A., Goulder, P. J., Shea, A. K., Jones, N. G., Trocha, A. K., Ogg, G. S., and Walker, B. D. (1999). Levels of human immunodeficiency virus type 1-specific cytotoxic T-lymphocyte effector and memory responses decline after suppression of viremia with highly active antiretroviral therapy. J. Virol. 73:6721-6728.

Kalergis, A. M., and Ravetch, J. V. (2002). Inducing tumor immunity through the selective engagement of activating Fc{gamma} receptors on dendritic cells. J. Exp. Med. 195:1653-1659.

Kalinski, P., Hilkens, C. M., Wierenga, E. A., and Kapsenberg, M. L. (1999). T-cell priming by type-1 and type-2 polarized dendritic cells: the concept of a third signal. Immunol. Today 20:561-567.

Kappes, J. C., and Wu, X. (2001). Safety considerations in vector development. Somat. Cell Mol. Genet. 26:147-158.

Karanikas, V., Hwang, L. A., Pearson, J., Ong, C. S., Apostolopoulos, V., Vaughan, H., Xing, P. X., Jamieson, G., Pietersz, G., Tait, B., Broadbent, R., Thynne, G., and McKenzie, I. F. (1997). Antibody and T cell responses of patients with adenocarcinoma immunized with mannan-MUC1 fusion protein. J. Clin. Invest. 100:2783-2792.

Karp, C. L., Wysocka, M., Ma, X., Marovich, M., Factor, R. E., Nutman, T., Armant, M., Wahl, L., Cuomo, P., and Trinchieri, G. (1998). Potent suppression of IL-12 production from monocytes and dendritic cells during endotoxin tolerance. Eur. J. Immunol. 28:3128-3136.

Kawamura, T., Gatanaga, H., Borris, D. L., Connors, M., Mitsuya, H., and Blauvelt, A. (2003). Decreased stimulation of CD4$^+$ T cell proliferation and IL-2 production by highly enriched populations of HIV-infected dendritic cells. J. Immunol. 170:4260-4266.

Keler, T., Guyre, P. M., Vitale, L. A., Sundarapandiyan, K., van De Winkel, J. G., Deo, Y. M., and Graziano, R. F. (2000). Targeting weak antigens to CD64 elicits potent humoral responses in human CD64 transgenic mice. J. Immunol. 165:6738-6742.

Keler, T., Ramakrishna, V., and Fanger, M. W. (2004). Mannose receptor-targeted vaccines. Expert Opin. Biol. Ther. 4:1953-1962.

Kellermann, S. A., Hudak, S., Oldham, E. R., Liu, Y. J., and McEvoy, L. M. (1999). The CC chemokine receptor-7 ligands 6Ckine and macrophage inflammatory protein-3 beta are potent chemoattractants for *in vitro-* and *in vivo*-derived dendritic cells. J. Immunol. 162:3859-3864.

Kim, D. T., Mitchell, D. J., Brockstedt, D. G., Fong, L., Nolan, G. P., Fathman, C. G., Engleman, E. G., and Rothbard, J. B. (1997). Introduction of soluble proteins into the MHC class I pathway by conjugation to an HIV tat peptide. J. Immunol. 159:1666-1668.

Kiyokawa, N., Taguchi, T., Mori, T., Uchida, H., Sato, N., Takeda, T., and Fujimoto, J. (1998). Induction of apoptosis in normal human renal tubular epithelial cells by Escherichia coli Shiga toxins 1 and 2. J. Infect. Dis. 178:178-184.

Korokhov, N., de Gruijl, T. D., Aldrich, W. A., Triozzi, P. L., Banerjee, P. T., Gillies, S. D., Curiel, T. J., Douglas, J. T., Scheper, R. J., and Curiel, D. T. (2005a). High efficiency transduction of dendritic cells by adenoviral vectors targeted to DC-SIGN. Cancer Biol. Ther. 4:289-94.

Korokhov, N., Noureddini, S. C., Curiel, D. T., Santegoets, S. J., Scheper, R. J., and de Gruijl, T. D. (2005b). A single-component CD40-targeted adenovirus vector displays highly efficient transduction and activation of dendritic cells in a human skin substrate system. Mol. Pharm. 2:218-223.

LaCasse, E. C., Saleh, M. T., Patterson, B., Minden, M. D., and Gariepy, J. (1996). Shiga-like toxin purges human lymphoma from bone marrow of severe combined immunodeficient mice. Blood 88:1561-1567.

Ladant, D., and Ullmann, A. (1999). Bordatella pertussis adenylate cyclase: a toxin with multiple talents. Trends Microbiol. 7:172-176.

Layton, G. T., Harris, S. J., Gearing, A. J., Hill-Perkins, M., Cole, J. S., Griffiths, J. C., Burns, N. R., Kingsman, A. J., and Adams, S. E. (1993). Induction of HIV-specific cytotoxic T lymphocytes *in vivo* with hybrid HIV-1 V3:Ty-virus-like particles. J. Immunol. 151:1097-1107.

Lee, R. S., Tartour, E., van der, B. P., Vantomme, V., Joyeux, I., Goud, B., Fridman, W. H., and Johannes, L. (1998). Major histocompatibility complex class I presentation of exogenous soluble tumor antigen fused to the B-fragment of Shiga toxin. Eur. J. Immunol. 28:2726-2737.

Lindberg, A. A., Brown, J. E., Stromberg, N., Westling-Ryd, M., Schultz, J. E., and Karlsson, K. A. (1987). Identification of the carbohydrate receptor for Shiga toxin produced by Shigella dysenteriae type 1. J. Biol. Chem. 262:1779-1785.

Lisziewicz, J., Trocio, J., Whitman, L., Varga, G., Xu, J., Bakare, N., Erbacher, P., Fox, C., Woodward, R., Markham, P., Arya, S., Behr, J. P., and Lori, F. (2005). DermaVir: a novel topical vaccine for HIV/AIDS. J. Invest. Dermatol. 124:160-169.

Lu, W., Arraes, L. C., Ferreira, W. T., and Andrieu, J. M. (2004). Therapeutic dendritic-cell vaccine for chronic HIV-1 infection. Nat. Med. 10:1359-1365.

Matzinger, P. (2002). The danger model: a renewed sense of self. Science 296:301-305.

Mitchell, D. A., and Nair, S. K. (2000). RNA-transfected dendritic cells in cancer immunotherapy. J. Clin. Invest. 106:1065-1069.

Moore, A., McGuirk, P., Adams, S., Jones, W. C., McGee, J. P., O'Hagan, D. T., and Mills, K. H. (1995). Immunization with a soluble recombinant HIV protein entrapped in biodegradable microparticles induces HIV-specific CD8[+] cytotoxic T lymphocytes and CD4[+] Th1 cells. Vaccine 13:1741-1749.

Moriwaki, H., Kume, N., Sawamura, T., Aoyama, T., Hoshikawa, H., Ochi, H., Nishi, E., Masaki, T., and Kita, T. (1998). Ligand specificity of LOX-1, a novel endothelial receptor for oxidized low density lipoprotein. Arterioscler. Thromb. Vasc. Biol. 18:1541-1547.

Moron, V. G., Rueda, P., Sedlik, C., and Leclerc, C. (2003). *In vivo*, dendritic cells can cross-present virus-like particles using an endosome-to-cytosol pathway. J. Immunol. 171:2242-2250.

Moron, G., Dadaglio, G., and Leclerc, C. (2004). New tools for antigen delivery to the MHC class I pathway. Trends Immunol. 25:92-97.

Mosier, D. E. (2005). HIV-1 envelope evolution and vaccine efficacy. Curr. Drug Targets Infect. Disord. 5:171-177.

Mullins, D. W., Sheasley, S. L., Ream, R. M., Bullock, T. N., Fu, Y. X., and Engelhard, V. H. (2003). Route of immunization with peptide-pulsed dendritic cells controls the distribution of memory and effector T cells in lymphoid tissues and determines the pattern of regional tumor control. J. Exp. Med. 198:1023-1034.

Nehete, P. N., Nehete, B. P., Manuri, P., Hill, L., Palmer, J. L., and Sastry, K. J. (2005). Protection by dendritic cells-based HIV synthetic peptide cocktail vaccine: preclinical studies in the SHIV-rhesus model. Vaccine 23:2154-2159.

Nigou, J., Zelle-Rieser, C., Gilleron, M., Thurnher, M., and Puzo, G. (2001). Mannosylated lipoarabinomannans inhibit IL-12 production by human dendritic cells: evidence for a negative signal delivered through the mannose receptor. J. Immunol. 166:7477-7485.

Ortiz-Stern, A., and Rosales, C. (2003). Cross-talk between Fc receptors and integrins. Immunol. Lett. 90:137-143.

Park, C. G., Takahara, K., Umemoto, E., Yashima, Y., Matsubara, K., Matsuda, Y., Clausen, B. E., Inaba, K., and Steinman, R. M. (2001). Five mouse homologues of the human dendritic cell C-type lectin, DC-SIGN. Int. Immunol. 13:1283-1290.

Pasquier, B., Launay, P., Kanamaru, Y., Moura, I. C., Pfirsch, S., Ruffie, C., Henin, D., Benhamou, M., Pretolani, M., Blank, U., and Monteiro, R. C. (2005). Identification of FcalphaRI as an inhibitory receptor that controls inflammation: dual role of FcRgamma ITAM. Immunity 22:31-42.

Pauza, C. D., Trivedi, P., Wallace, M., Ruckwardt, T. J., Le, B. H., Lu, W., Bizzini, B., Burny, A., Zagury, D., and Gallo, R. C. (2000). Vaccination with tat toxoid attenuates disease in simian/HIV-challenged macaques. Proc. Natl. Acad. Sci. U. S. A. 97:3515-3519.

Pitcher, C. J., Quittner, C., Peterson, D. M., Connors, M., Koup, R. A., Maino, V. C., and Picker, L. J. (1999). HIV-1-specific CD4[+] T cells are detectable in most individuals with active HIV-1 infection, but decline with prolonged viral suppression. Nat. Med. 5:518-525.

Pitisuttithum, P., Nitayaphan, S., Thongcharoen, P., Khamboonruang, C., Kim, J., de, S. M., Chuenchitra, T., Garner, R. P., Thapinta, D., Polonis, V., Ratto-Kim, S., Chanbancherd, P., Chiu, J., Birx, D. L., Duliege, A. M., McNeil, J. G., and Brown, A. E. (2003). Safety and immunogenicity of combinations of recombinant subtype E and B human immunodeficiency virus type 1 envelope glycoprotein 120 vaccines in healthy Thai adults. J. Infect. Dis. 188:219-227.

Rayevskaya, M. V., and Frankel, F. R. (2001). Systemic immunity and mucosal immunity are induced against human immunodeficiency virus Gag protein in mice by a new hyperattenuated strain of Listeria monocytogenes. J. Virol. 75:2786-2791.

Regnault, A., Lankar, D., Lacabanne, V., Rodriguez, A., Thery, C., Rescigno, M., Saito, T., Verbeek, S., Bonnerot, C., Ricciardi-Castagnoli, P., and Amigorena, S. (1999). Fcgamma receptor-mediated induction of dendritic cell maturation and major histocompatibility complex class I-restricted antigen presentation after immune complex internalization. J. Exp. Med. 189:371-380.

Reis e Sousa. (2004). Activation of dendritic cells: translating innate into adaptive immunity. Curr. Opin. Immunol. 16:21-25.

Reis e Sousa, and Germain, R. N. (1995). Major histocompatibility complex class I presentation of peptides derived from soluble exogenous antigen by a subset of cells engaged in phagocytosis. J. Exp. Med. 182:841-851.

Romani, N., Gruner, S., Brang, D., Kampgen, E., Lenz, A., Trockenbacher, B., Konwalinka, G., Fritsch, P. O., Steinman, R. M., and Schuler, G. (1994). Proliferating dendritic cell progenitors in human blood. J. Exp. Med. 180:83-93.

Rose, N. F., Marx, P. A., Luckay, A., Nixon, D. F., Moretto, W. J., Donahoe, S. M., Montefiori, D., Roberts, A., Buonocore, L., and Rose, J. K. (2001). An effective AIDS vaccine based on live attenuated vesicular stomatitis virus recombinants. Cell 106:539-549.

Rossio, J. L., Esser, M. T., Suryanarayana, K., Schneider, D. K., Bess, J. W., Jr., Vasquez, G. M., Wiltrout, T. A., Chertova, E., Grimes, M. K., Sattentau, Q., Arthur, L. O., Henderson, L. E., and Lifson, J. D. (1998). Inactivation of human immunodeficiency virus type 1 infectivity with preservation of conformational and functional integrity of virion surface proteins. J. Virol. 72:7992-8001.

Rudolf, M. P., Fausch, S. C., Da Silva, D. M., and Kast, W. M. (2001). Human dendritic cells are activated by chimeric human papillomavirus type-16 virus-like particles and induce epitope-specific human T cell responses in vitro. J. Immunol. 166:5917-5924.

Sallusto, F., and Lanzavecchia, A. (1994). Efficient presentation of soluble antigen by cultured human dendritic cells is maintained by granulocyte/macrophage colony-stimulating factor plus interleukin 4 and downregulated by tumor necrosis factor alpha. J. Exp. Med. 179:1109-1118.

Sapp, M., Engelmayer, J., Larsson, M., Granelli-Piperno, A., Steinman, R., and Bhardwaj, N. (1999). Dendritic cells generated from blood monocytes of HIV-1 patients are not infected and act as competent antigen presenting cells eliciting potent T-cell responses. Immunol. Lett. 66:121-128.

Saron, M. F., Fayolle, C., Sebo, P., Ladant, D., Ullmann, A., and Leclerc, C. (1997). Anti-viral protection conferred by recombinant adenylate cyclase

toxins from Bordetella pertussis carrying a CD8⁺ T cell epitope from lymphocytic choriomeningitisávirus. Proc. Natl. Acad. Sci. U. S. A. 94:3314-3319.

Sawamura, T., Kume, N., Aoyama, T., Moriwaki, H., Hoshikawa, H., Aiba, Y.,
Tanaka, T., Miwa, S., Katsura, Y., Kita, T., and Masaki, T. (1997). An
endothelial receptor for oxidized low-density lipoprotein. Nature 386:73-77.

Schirmbeck, R., and Reimann, J. (1996). 'Empty' Ld molecules capture peptides
from endocytosed hepatitis B surface antigen particles for major histocompatibility complex class I-restricted presentation. Eur. J. Immunol. 26:2812-2822.

Schlecht, G., Loucka, J., Najar, H., Sebo, P., and Leclerc, C. (2004). Antigen
targeting to CD11b allows efficient presentation of CD4⁺ and CD8⁺ T cell epitopes and *in vivo* Th1-polarized T cell priming. J. Immunol. 173:6089-6097.

Schoenberger, S. P., Toes, R. E., van der Voort, E. I., Offringa, R., and Melief,
C. J. (1998). T-cell help for cytotoxic T lymphocytes is mediated by CD40-
CD40L interactions. Nature 393:480-483.

Schuurhuis, D. H., Ioan-Facsinay, A., Nagelkerken, B., van Schip, J. J., Sedlik, C.,
Melief, C. J. M., Verbeek, J. S., and Ossendorp, F. (2002). Antigen-antibody
immune complexes empower dendritic cells to efficiently prime specific
CD8⁺ CTL responses *in vivo*. J. Immunol. 168:2240-2246.

Sebo, P., Moukrim, Z., Kalhous, M., Schaft, N., Dadaglio, G., Sheshko, V.,
Fayolle, C., and Leclerc, C. (1999). *In vivo* induction of CTL responses by
recombinant adenylate cyclase of Bordetella pertussis carrying multiple
copies of a viral CD8(⁺) T-cell epitope. FEMS Immunol. Med. Microbiol.
26:167-173.

Shen, L., van Egmond, M., Siemasko, K., Gao, H., Wade, T., Lang, M. L., Clark,
M., van de Winkel, J. G. J., and Wade, W. F. (2001). Presentation of
ovalbumin internalized via the immunoglobulin-A Fc receptor is enhanced
through Fc receptor {gamma}-chain signaling. Blood 97:205-213.

Silvera, P., Richardson, M. W., Greenhouse, J., Yalley-Ogunro, J., Shaw, N.,
Mirchandani, J., Khalili, K., Zagury, J. F., Lewis, M. G., and Rappaport, J.
(2002). Outcome of simian-human immunodeficiency virus strain 89.6p
challenge following vaccination of rhesus macaques with human
immunodeficiency virus Tat protein. J. Virol. 76:3800-3809.

Smith, D. C., Lord, J. M., Roberts, L. M., and Johannes, L. (2004). Glycosphin-
golipids as toxin receptors. Semin. Cell Dev. Biol. 15:397-408.

Soilleux, E. J., Morris, L. S., Leslie, G., Chehimi, J., Luo, Q., Levroney, E.,
Trowsdale, J., Montaner, L. J., Doms, R. W., Weissman, D., Coleman, N., and
Lee, B. (2002). Constitutive and induced expression of DC-SIGN on dendritic
cell and macrophage subpopulations in situ and *in vitro*. J. Leukoc. Biol.
71:445-457.

Sporri, R., and Reis e Sousa. (2005). Inflammatory mediators are insufficient for
full dendritic cell activation and promote expansion of CD4⁺ T cell
populations lacking helper function. Nat. Immunol. 6:163-170.

Srivastava, P. (2002). Roles of heat-shock proteins in innate and adaptive
immunity. Nat. Rev. Immunol. 2:185-194.

Steinman, R. M., and Dhodapkar, M. (2001). Active immunization against cancer
with dendritic cells: the near future. Int. J. Cancer 94:459-473.

Sundarapandiyan, K., Keler, T., Behnke, D., Engert, A., Barth, S., Matthey, B., Deo, Y. M., and Graziano, R. F. (2001). Bispecific antibody-mediated destruction of Hodgkin's lymphoma cells. J. Immunol. Methods 248:113-123.

Suzue, K., and Young, R. A. (1996). Adjuvant-free hsp70 fusion protein system elicits humoral and cellular immune responses to HIV-1 p24. J. Immunol. 156:873-879.

Tacken, P. J., Hartshorn, K. L., White, M. R., van Kooten, C., van de Winkel, J. G. J., Reid, K. B. M., and Batenburg, J. J. (2004). Effective targeting of pathogens to neutrophils via chimeric surfactant protein D/anti-CD89 protein. J. Immunol. 172:4934-4940.

Tacken, P. J., de Vries, I. J., Gijzen, K., Joosten, B., Wu, D., Rother, R. P., Faas, S. J., Punt, C. J., Torensma, R., Adema, G. J., and Figdor, C. G. (2005). Effective induction of naive and recall T cell responses by targeting antigen to human dendritic cells via a humanized anti-DC-SIGN antibody. Blood 106:1278-1285.

Thompson, A. G., and Thomas, R. (2002). Induction of immune tolerance by dendritic cells: implications for preventative and therapeutic immunotherapy of autoimmune disease. Immunol. Cell Biol. 80:509-519.

Tillman, B. W., Gruijl, T. D., Bakker, S. A. L., Scheper, R. J., Pinedo, H. M., Curiel, T. J., Gerritsen, W. R., and Curiel, D. T. (1999). Maturation of dendritic cells accompanies high-efficiency gene transfer by a CD40-targeted adenoviral vector. J. Immunol. 162:6378-6383.

van Broekhoven, C. L., Parish, C. R., Demangel, C., Britton, W. J., and Altin, J. G. (2004). Targeting dendritic cells with antigen-containing liposomes: a highly effective procedure for induction of antitumor immunity and for tumor immunotherapy. Cancer Res. 64:4357-4365.

Van den Herik-Oudijk, I., Capel, P. J., van der Bruggen, T., and van De Winkel, J. G. (1995). Identification of signaling motifs within human Fc gamma RIIa and Fc gamma RIIb isoforms. Blood 85:2202-2211.

van Spriel, A. B., van den Herik-Oudijk IE, van Sorge, N. M., Vile, H. A., van Strijp, J. A., and van De Winkel, J. G. (1999). Effective phagocytosis and killing of Candida albicans via targeting FcgammaRI (CD64) or FcalphaRI (CD89) on neutrophils. J. Infect. Dis. 179:661-669.

Wallace, P. K., Tsang, K. Y., Goldstein, J., Correale, P., Jarry, T. M., Schlom, J., Guyre, P. M., Ernstoff, M. S., and Fanger, M. W. (2001). Exogenous antigen targeted to FcgammaRI on myeloid cells is presented in association with MHC class I. J. Immunol. Methods 248:183-194.

Whiteside, T. L., and Odoux, C. (2004). Dendritic cell biology and cancer therapy. Cancer Immunol. Immunother. 53:240-248.

Wilson, N. S., and Villadangos, J. A. (2005). Regulation of antigen presentation and cross-presentation in the dendritic cell network: facts, hypothesis, and immunological implications. Adv. Immunol. 86:241-305.

Xin, K. Q., Hoshino, Y., Toda, Y., Igimi, S., Kojima, Y., Jounai, N., Ohba, K., Kushiro, A., Kiwaki, M., Hamajima, K., Klinman, D., and Okuda, K. (2003). Immunogenicity and protective efficacy of orally administered recombinant Lactococcus lactis expressing surface-bound HIV Env. Blood 102:223-228.

Zhang, L., Tang, Y., Akbulut, H., Zelterman, D., Linton, P. J., and Deisseroth, A. B. (2003). An adenoviral vector cancer vaccine that delivers a tumor-associated antigen/CD40-ligand fusion protein to dendritic cells. Proc. Natl. Acad. Sci. U. S. A. 100:15101-15106.

Zhou, Y., Bosch, M. L., and Salgaller, M. L. (2002). Current methods for loading dendritic cells with tumor antigen for the induction of antitumor immunity. J. Immunother. 25:289-303.

Part II

**General Aspects of the Pathogenesis
and Immune Response to HIV-1 Infection**

Chapter 7

Immunopathogenesis of HIV Infection

Elisa Vicenzi, Massimo Alfano, Silvia Ghezzi, and Guido Poli

7.1 Introduction

HIV is the first pathogenic human retrovirus responsible for a lethal disease of pandemic proportions. Its ability to integrate in the host genome renders its infection life-long, whereas anti-retroviral agents can limit and control its extent but not lead to viral eradication. Its selectivity for $CD4^+$ cells (T lymphocytes, mononuclear phagocytes, and dendritic cells) is biologically sustained by the interaction of its gp120 Env molecule with CD4 itself (primary receptor), which must be followed by binding to a chemokine coreceptor (CCR5 and/or CXCR4) leading to a conformational change of gp41 Env, which causes the fusion between the virion and the cell membrane. A number of intracellular factors have been recently highlighted that control and influence the efficiency of HIV infection in target cells, among which APOBEC3G appears currently as the most relevant or simply best characterized. The virus counteracts these intracellular mechanisms with an array of regulatory and accessory proteins (such as Vif, which inactivates APOBEC3G). After viral integration by means of the reverse transcriptase (RT), several viral (Tat, Rev) and host factors (cytokines, chemokines) control the efficiency of proviral DNA transcription and post-transcriptional steps leading to assembly and release of new progeny virions. Virus replication and pathogenesis occurs mostly in lymphoid organs such as gut-associated lymphoid tissue (GALT) and the central nervous system (CNS), whereas the peripheral blood represents a useful accessible compartment reflective to some extent of organ and tissue events. HIV infection causes a profound depletion of $CD4^+$ T lymphocytes and immunodeficiency that becomes lethal in the stage defined as acquired immunodeficiency syndrome

(AIDS). Whether these events are caused directly by viral replication or whether HIV-induced cytopathicity is mostly dependent upon a host immunologic response remains a controversial issue.

7.2 The Life Cycle of HIV-1 in CD4⁺ T Cells and Mononuclear Phagocytes

HIV infection targets $CD4^+$ cells, mostly represented by a subset of T lymphocytes and mononuclear phagocytes, including monocyte-derived dendritic cells (DCs). Infection occurs mostly in secondary lymphoid organs, such as the GALT, where selective or privileged infection of memory T cells occurs (Veazey *et al.*, 1998), but also in immune-privileged sites such as the CNS and testis. Although T-cell activation plays a fundamental role in virus propagation, also resting T cells can be productively infected at least in the genital mucosa (Zhang *et al.*, 2004). In addition to CD4 and a chemokine receptor (either CCR5 or CXCR4), DC can internalize virions through dendritic cell–specific intercellular adhesion molecule 3-grabbing nonintegrin (DC-SIGN). DC-SIGN and other mannose-binding receptors bind virus particles and transmit them to $CD4^+$ T cells that are simultaneously activated by interacting with HIV-carrying/infected DCs through an infectious synapse, thereby generating an amplification system for viral replication.

HIV entry is mediated by viral envelope glycoproteins (gp), namely gp120 Env, which is noncovalently linked to the transmembrane gp41 Env. Both molecules are derived from a common precursor, gp160 Env, which is processed by a cellular protease in the two mature products. Functional gp120 Env is an oligomer composed of three identical, highly glycosylated units that include both constant and variable domains (V1 to V4) representing important determinants in terms of virion antigenicity. Functional gp41 Env is also an oligomer of three identical subunits encompassing a hydrophobic N-terminus domain mediating fusion of the viral Env with the host plasma membrane. gp120 Env must bind first to CD4 via conserved conformational domains; this interaction exposes cryptic gp120 Env domains that enables the molecule to bind CCR5 or CXCR4 (entry coreceptors) (Berger *et al.*, 1999). gp120 Env binding to CD4 and chemokine coreceptors induces a conformational change in gp41 Env leading to a coiled-coil structure exposing the hydrophobic fusion domain, therefore allowing the fusion between the virion and the plasma membranes (Berger *et al.*, 1999) (Fig. 7.1). The dispersion of the virion membrane into cellular membrane allows the injection of the virion core into the cytoplasm of the target cell and the formation of the so-called preintegration complex.

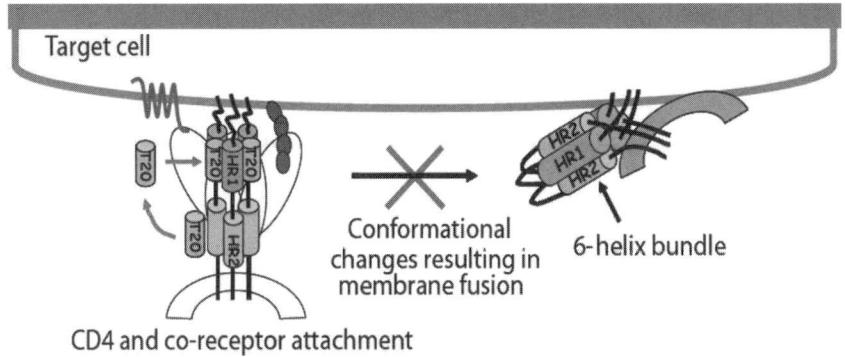

Target cell

Conformational changes resulting in membrane fusion

6-helix bundle

CD4 and co-receptor attachment

FIGURE 7.1. HIV-1 chemokine coreceptor interaction. Courtesy of Ben Bezkhout.

The complex includes the double-stranded RNA genome bound to the viral reverse transcriptase (RT), the very enzyme that characterizes the family of retroviridiae (Fig. 7.2). The RT process leading to the synthesis of a double-stranded DNA version of the viral genome starts in the cytoplasm, and it is completed in the nucleus. In this regard, the viral preintegration complex enters the nucleus via an active uptake process that depends on multiple viral nuclear localization signals present in different viral proteins, that is, Vpr, integrase (IN), Gag-derived p17 matrix (MA), and by a DNA flap (central polypurine tract) located at the viral DNA center (Sherman and Greene, 2002). The ability to transverse the nuclear pore allows HIV, unlike oncoretroviruses, to infect nondividing cells such as tissue macrophages. The synthesis of linear viral DNA is completed in the nucleus and it is followed by integration into the host DNA. To accomplish its task, RT literally jumps both intrastrand and interstrands, in order to synthesize identically repeated structures at the 5′ and 3′ ends of the viral genome (defined as long terminal repeats; LTRs). LTRs are conventionally subdivided into U3, R, and U5 regions and contain the promoter/enhancer elements controlling HIV transcription (Cullen, 1995). Inefficient integration leads to the accumulation of nonfunctional viral DNA circular forms (with either one or two LTRs). It has been proposed that the inner nuclear membrane facilitates the interactions between the HIV DNA and the chromatin, thereby preventing the inactivation of viral cDNA by recombinases and ligases within the nucleus (Montes de Oca et al., 2005). Integration is not sequence specific and, therefore, all chromosomal sites could potentially serve as integration events. However, in vivo integrated HIV DNA (defined as provirus) is preferentially detected in "hot spots" of chromatin characterized by an open structure, a hallmark of actively transcribed genes (Schroder et al., 2002).

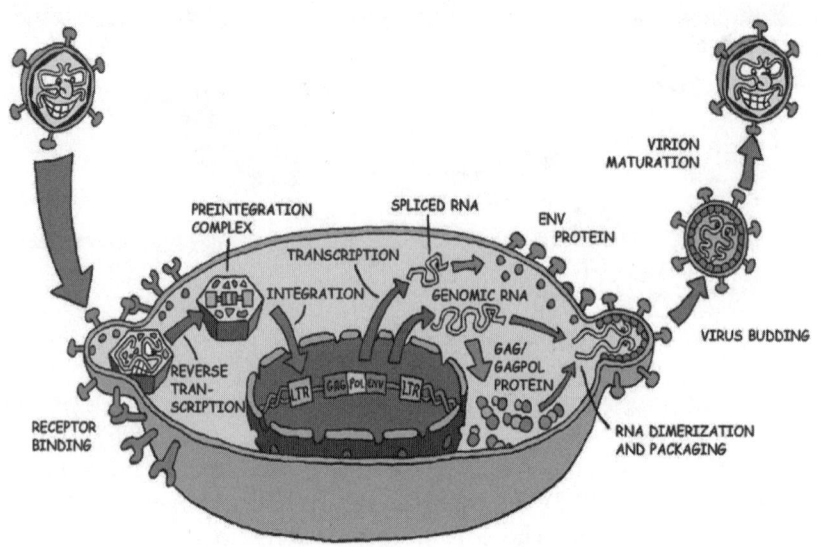

FIGURE 7.2. The HIV life cycle. Courtesy of Ben Bezkhout.

Once integrated, the HIV genome is transcribed into viral RNA by the cellular RNA polymerase II complex. The 5′ LTR serves as starter of HIV transcription whereas the LTR in 3′ functions as terminator of transcription. The U3 region of the 5′ LTR includes a core promoter (where the RNA polymerase II is recruited) and enhancer elements containing the binding sites for cellular transcription factors such as NFAT and NF-kB (Cullen, 1995), thereby enabling HIV to be transcriptionally triggered by cellular activation. HIV transcription driven by cellular stimuli would be, however, quite inefficient in the absence of the major viral transactivator (i.e., Tat).

The neosynthesized viral RNA undergoes splicing. In this regard, fully spliced 2-kb mRNAs leave the nucleus and are translated into three nonstructural proteins: Tat, Nef, and Rev. Early synthesis of Rev, a crucial shuttling viral protein, must occur in order to overcome the restriction on nuclear export of partially spliced (4.5 kb) and unspliced (9.2 kb) transcripts (Fig. 7.3). The Rev protein contains an arginine-rich RNA binding motif that binds to a stem-loop structure, known as the Rev responsive element (RRE), located in the *env* gene (Malim *et al.*, 1989). The same arginine-rich motif in Rev also acts as a nuclear localization signal (NLS), which is required for the transport of Rev from the cytoplasm to the nucleus. The NLS promotes the direct binding of Rev to the nuclear import factor Importin, which targets the resultant protein complex to the nucleus (Truant and Cullen, 1999). Rev binding to RRE leads to its multimerization until it will exit the nucleus together with its

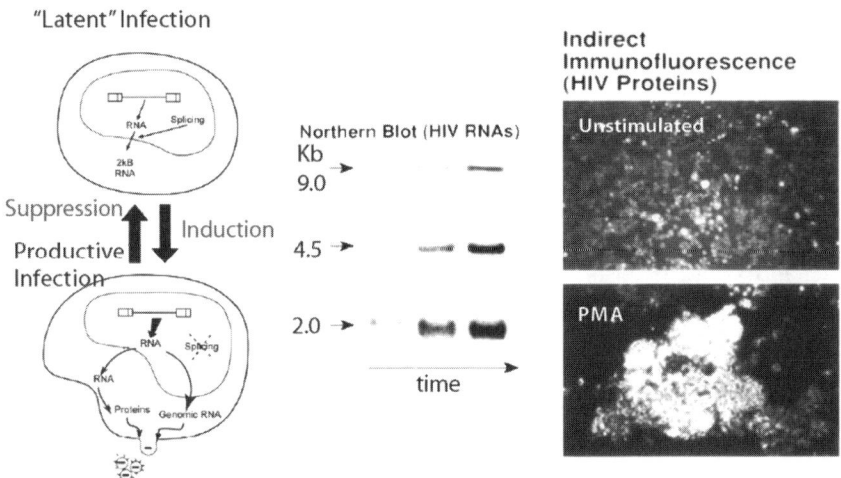

FIGURE 7.3. Chronically infected cell lines as models for dissecting out host and viral factors influencing HIV transcription and post-transcriptional events leading to new progeny virion production. The figure is based on results utilizing the promonocytic U1 cell line (Folks *et al.*, 1987; Folks *et al.*, 1988a).

protein cargo (Malim *et al.*, 1990). In this regard, the nuclear export of Rev-RRE complexes requires a nuclear export signal (NES) that mediates the interaction of Rev with nuclear export factors. The nuclear export of Rev-RRE complexes involves Crm1, a conserved protein that acts as a cellular receptor for NES-containing proteins. The association of Crm1 with NES-containing proteins such as Rev is regulated by Ran, a cellular guanosine triphosphate (GTP)ase, which forms a complex with Rev and its associated RNA allowing their export to the cytoplasm (Neville *et al.*, 1997).

The Rev-independent synthesis of Tat, another viral protein capable of reentering the nucleus via an NLS leads to a self-propagating increase of viral transcription that, together with the action of Rev, ensures high levels of expression of structural (Gag, Pol, and Env), regulatory (Tat, Rev), and accessory (VpR, VpU, Vif, and Nef) viral proteins (Strebel, 2003). In the absence of Tat, transcription from the HIV-1 LTR produces predominately short, non-polyadenylated RNA that include the trans-activation response region (TAR) stem-loop structure binding Tat. The expression of Tat results in the production of longer, polyadenylated RNA and in increased gene expression (Zhou and Sharp, 1995), therefore functioning as an anti-terminator of transcription. Tat binding to nascent viral RNA allows its elongation through the recruitment of elongation-competent RNA polymerase II complexes capable of processive transcription (Wei *et al.*, 1998). Recently, a novel function of the Tat protein has been demonstrated, that is to suppress RNA interference (RNAi). In this regard, in order to generate a productive infection, HIV has to counteract a viral small interfering RNA that elicits RNAi-based restriction in human cells. Tat abrogates RNA silencing by affecting the ability of Dicer to process double stranded RNA into siRNA (Bennasser *et al.*, 2005).

Once viral proteins are optimally synthesized, they need to assemble to give rise to a new progeny of virions to perpetuate the infection in novel target cells. Virion assembly at the plasma membrane involves essentially Gag and Gag-Pol polyproteins, the Envelope gp, and the viral genomic RNA, but it is the Gag polyprotein to promote virion budding and release (virus-like particles can be formed by expression of only Gag) (Freed, 2003). The Gag-Pol polyprotein is produced by a ribosomal frameshift between the gag and pol reading frames. It consists of the Gag polyprotein fused to a large C-terminal extension containing the pol-encoded enzymes, that is, protease (P), RT, and integrase (IN). Gag is myristylated at the amino-terminus leading to its accumulation at the inner side of plasma membrane where it multimerizes in order to form the virion macromolecular complex (Freed, 2003; Gomez and Hope, 2005) (Fig. 7.4A). The Envelope gp is an integral membrane protein expressed in the secretory pathway; it is cleaved by a cellular protease in the Golgi to produce a trans-membrane component (TM), which spans the virion membrane, and a surface component (SU), and it is noncovalently attached to TM, outside the virion (Freed, 2002; Gomez and Hope, 2005).

The viral genomic RNA is incorporated into virions via interaction with zinc-finger regions located in the nucleocapsid (NC) domain of the Gag polyprotein. The final step of HIV virion assembly is determined by the activation of viral protease. When virions exit the cells, the Gag and Gag-Pol polyproteins are intact and the virion has an "immature morphology". Activation of the viral protease lead, to cleavage of the two polyproteins conferring a "mature" morphology to the virus particle (Fig. 7.4B). Of interest is the fact that HIV infection of macrophages leads to budding and accumulation of infectious virions in intracellular compartments known as multivescicular bodies (MVBs), as identified by the CD63 marker, and belonging to the hexosomal secretory pathway (Pelchen-Matthews *et al.*, 2003). This diversity in the morphogenetic strategy likely bears importance for HIV pathogenesis in that macrophages can hide copious amounts of infectious virus without proper recognition by the host immune system (so-called "Trojan horse" hypothesis) (Koenig *et al.*, 1986; Gendelman *et al.*, 1988).

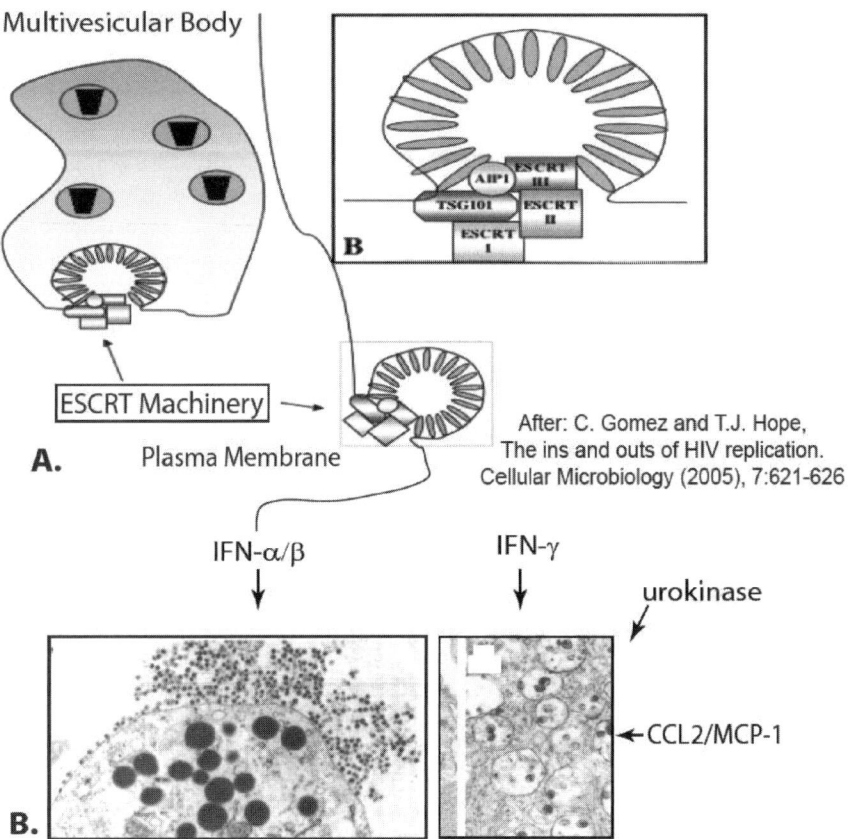

FIGURE 7.4. Multiple signals influence HIV-1 morphogenesis at either the plasma membrane (A) or in intracytoplasmic vacuoles of likely Golgi origin (B) in mature mononuclear phagocytes. The picture were obtained from U1 cells stimulated and differentiated by phorbol-myristate acetate (PMA) in the presence of either IFN-α/β (left panel) or IFN-γ (right panel); references in the text. After Gomez and Hope (2005) (A) and Poli *et al.* (1989), Biswas *et al.* (1992), Alfano et al. (2002), and Fantuzzi et al. (2003) (B).

Figure 7.4(A) Reproduced from Gomez, C., *et al.*, *The Ins and Outs of HIV Replication*, Vol 7(5), pgs 621-626, 2005, with permission form Blackwell Publishing.

Figure 7.4(B) Reproduced with permission from Poli, *et al.*, SCIENCE 244:1989, AAAS, and Biswas et al., J. Exp. Med. 176:739-750, copyright permission of The Rockefeller University Press.

7.3 CCR5- versus CXCR4-dependent HIV-1 Infections

Soon after the discovery that the CD4 molecule (expressed by the very subset of T lymphocytes undergoing progressive depletion as a consequence of infection and by mononuclear phagocytes) served as entry receptor for HIV (Dalgleish *et al.*, 1984; McDougal *et al.*, 1986) small animals were elegantly engineered to provide convenient tools for studies of pathogenesis, drug and vaccine discovery. Transgenic animal cells correctly expressed human CD4 at levels comparable with those of human immune cells and bound HIV with affinity and kinetic properties comparable with those of human cells. However, these transgenic cells were not susceptible to HIV infection, and when analyzed by immunofluorescence or electron microscopy they appeared as nicely decorated by viral proteins (Lores *et al.*, 1992). Virions perfectly bound these cells but were incompetent for fusion and penetration. Something was missing . . . a novel receptor or a coreceptor required for HIV fusion and/or entry. It took 10 years of research, and several false alarms, to solve the mistery when the first HIV entry coreceptor, named fusin, was identified (Feng *et al.*, 1996). Fusin was a G-protein coupled, 7-transmembrane domains receptor already known under different names to researchers investigating the migratory capacity of bone marrow–derived multipotent CD34$^+$ hematopoietic cells. Its natural ligand, CXCL12/stromal cell–derived factor-1α (SDF-1α), was chemotactic for these cells and was soon demonstrated to be an inhibitor of HIV entry (Bleul *et al.*, 1996; Oberlin *et al.*, 1996). However, only those HIV classified as "syncytia-inducing (SI)," that is, giving rise to a cytopathic infection in the MT-2 cell line, used fusin as entry coreceptor, whereas the remainder of the viruses did not infect these cells and were therefore classified as "non-SI (NSI)" (Table 7.1) (Koot *et al.*, 1992). Frequently, these latter viruses could replicate efficiently, unlike SI viruses, in monocyte-derived macrophages (MDMs), establishing a paradigm that NSI viruses essentially encompassed macrophage-tropic viruses. NSI viruses were typically isolated from individuals early after infection, whereas SI viruses emerged in approximately 50% of people infected by subtype B viruses before the onset of AIDS (Koot *et al.*, 1993; Koot *et al.*, 1999). The discovery of fusin, renamed CXCR4 once established that it was the receptor for the chemokine CXCL12/SDF-1α, provided a scientific explanation to the SI/NSI empirical classification: the MT-2 cell line expressed abundantly fusin/CXCR4 and was, therefore, permissive for those viruses utilizing this entry coreceptor in addition to CD4, the primary

TABLE 7.1. Old and new classifications of HIV based on infection/fusion in MT-2 cells and chemokine coreceptor use.

MT-2 phenotype (Koot et al., 1992; Koot et al., 1993)	Coreceptor use	New nomenclature (Berger et al., 1998)	Cell tropism	Notes
Syncytia-inducing (SI)	CXCR4	X4	Primary T cells, cell lines	Infection of MDM reported by some authors (Verani et al., 1998; Yi et al., 1998), rarely observed in late stage patients infected with subtype B HIV-1.
Non-SI (NSI)	CCR5	R5	Primary T cells, macrophages, immature DC	Predominately transmitted either via blood transfusion, sexually, or mother to child. All HIV-1 subtypes.
SI	CCR5$^+$ CXCR4$^+$	R5X4	Primary T cells, macrophages, immature DC, cell lines	Emerging in 50% individuals infected with subtype B HIV-1 (Koot et al., 1999). They represent either mixtures of R5 and X4 viruses or viruses with gp120 Env capable of interacting with both coreceptors (Connor et al., 1997; Scarlatti et al., 1997; Ghezzi et al., 2001).
SI	CCR5$^+$ CXCR4$^+$ other chemokine receptors (CCR2, CCR3, etc.)	R2R3R5X4 (etc.)	Primary T cells, macrophages, immature DC, cell lines	Rarely observed in late stage patients infected with subtype B HIV-1 (Doranz et al., 1996).

receptor (Table 7.1). Half the mistery of how HIV could infect human CD4$^+$ cells was solved! But, what about the other half?

Concomitantly with the discovery of CXCR4, another chemokine receptor was discovered to be deeply involved with HIV infection: CCR5. Molecular epidemiologic studies demonstrated that individuals homozygous for a 32-base-pair deletion in CCR5 (resulting in the lack of expression of this chemokine receptor at the cell surface) were virtually resistant to HIV infection, thereby providing a scientific explanation for the so-called highly-HIV exposed seronegative individuals (ESN) (Liu *et al.*, 1996; Paxton *et al.*, 1996). In addition, individuals heterozygous for CCR5-Δ32 were infectable by HIV but frequently show a slower disease progression (individuals defined as "long-term nonprogressors"; LTNP) (Huang *et al.*, 1996). Only a minority of CCR5-Δ32 homozygotes turned infected and, in these cases, the infecting virus was invariably using CXCR4 as entry coreceptor (Biti *et al.*, 1997).

A few months before the discovery of fusin/CXCR4 and of the existence of HIV-resistant individuals, an independent line of research provided a first, important step to the final solution. Searching for a CD8$^+$ T-cell non-lytic soluble inhibitor of HIV replication, three molecules unknown to the HIV research community were purified from culture supernatant: regulated upon activation normal T-cell expressed and secreted (RANTES), macrophage-inflammatory protein-1α (MIP-1α), and MIP-1β (now CCL5, CCL3 and CCL4, respectively). They were three chemokines know to immunologists because of their ability to induce the migration of T cells and monocytes upon their expression in inflammatory conditions. Addition of these chemokines before or during HIV infection resulted in the inhibition of HIV replication, whereas neutralization of all three chemokines was required to obtain a significant antiviral effect (Cocchi *et al.*, 1995). The following discovery of the chemokine receptor CXCR4 as entry receptor for certain HIV strains suggested that also these chemokines may inhibit HIV infection with a similar mechanism, whereas the epidemiologic observation of ESN carrying the CCR5-Δ32 homozygotic mutation pointed toward this receptor as a likely obligatory determinant for infection by HIV. Within a few months, CCR5 was indeed recognized as the receptor for the NSI viruses and for the three inhibitory chemokines earlier identified as CD8-derived non-lytic antiviral factors (Alkhatib *et al.*, 1996). These chemokines indeed blocked the entry of NSI viruses in T cells and macrophages (Fig. 7.5). The MT-2 cell line utilized for classifying viruses therefore expressed one of the two essential entry coreceptors, CXCR4, but not the other Thus, the SI/NSI phenotypic classification indeed accurately

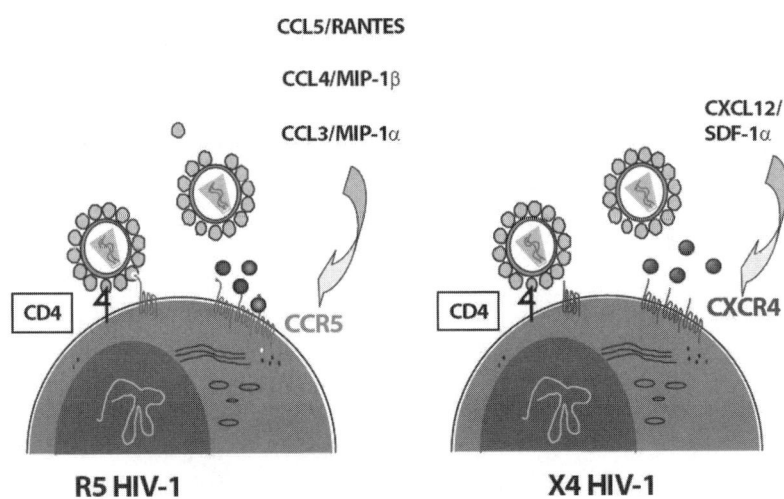

FIGURE 7.5. HIV-1 entry into CD4$^+$ cells. Host and viral components. (Reproduced from P. Lusso, Chemokines and Viruses: The Dearest Enemies, Virology, 273:2, pgs. 228-240, 2000, with permission from Elsevier.

reflected the biology of HIV entry in to CD4$^+$ cells according to either CXCR4 or CCR5 use (Table 7.1).

The discovery of chemokine receptors as HIV entry coreceptors in CD4$^+$ cell has allowed, on the one hand, a better comprehension of the entry/fusion process (as already discussed in the first paragraph of this chapter), and, on the other hand, a new appreciation of the dynamics of HIV infection both in a single individual and on a worldwide scale. Most, if not all, NSI viruses utilize CCR5 as sole coreceptor for entry into CD4$^+$ cells, whereas only a minor fraction of SI viruses uses exclusively CXCR4 in that most SI viruses can also interact with CCR5 (Table 7.1). They are defined as "dualtropic" viruses. Other chemokine receptors, including CCR2 and CCR3, have been described as facultative entry coreceptors, although almost invariably in association with CCR5 and/or CXCR4. These viruses, quite rare, are frequently named "multitropic." A novel classification, based on the use of these coreceptors, has been adopted according to which CCR5- and CXCR4- dependent viruses are defined as "R5" and "X4," respectively, whereas dualtropic viruses are "R5X4" and the occasional use of additional coreceptors indicated accordingly (e.g., a virus utilizing CCR2 and CCR3 in addition to CCR5 and CXCR4, such as the original 89.6 strain, is classified as a multitropic R5R2R3X4 virus) (Table 7.1) (Berger *et al.*, 1998).

CCR5 is clear-cut the most relevant entry coreceptor for HIV infection, both at the individual and population levels. The world pandemics, particularly in sub-Saharan Africa and Southeast Asia, is almost entirely carried on by R5 viruses (Fig. 7.6) in spite of their high propensity to generate

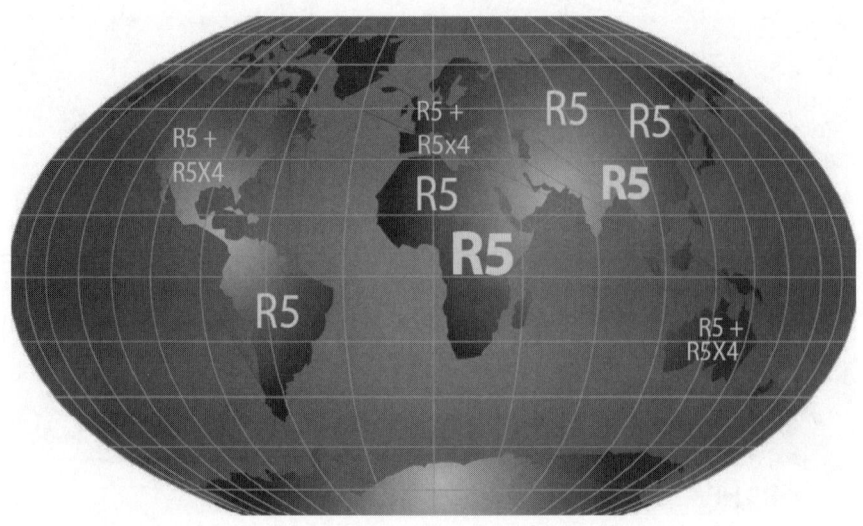

FIGURE 7.6. R5 HIV-1 dominant role in the HIV-1 pandemics.

circulating recombinant forms reassorting other viral genes. CXCR4 infection is confined to the HIV-1 subtype B, dominant in North America, Europe, and Australia and New Zeland for unclear reasons. It is suggestive, however, that these high-income countries also experience the highest prevalence of allergic diseases, a condition associated with increased expression of interleukin-4 (IL-4) and IL-13, known to upregulate the otherwise constitutive expression of CXCR4 (Galli *et al.*, 1998; Jourdan *et al.*, 1998; Valentin *et al.*, 1998; Wang *et al.*, 1998). Even in the context of subtype B infection, almost invariably viral transmission is followed by the expansion of monotropic R5 HIV (including in those cases where the viral transmitter harbors a CXCR4-using virus as major species). Usage of CXCR4 occurs later in approximately 50% of individuals infected with subtype B (as correctly predicted by the SI/NSI classification) and represents frequently an addition to, rather than a substitution of, CCR5 use. In other words, most SI viruses emerging in late-stage disease are indeed dualtropic R5X4 viruses, whereas solitary CXCR4 use is a quite rare event (Table 7.1). In a collection of more than 600 primary HIV isolates, monotropic X4 viruses represented <5% of the SI viruses (our unpublished observation). Emergence of CXCR4-using viruses, either dualtropic or monotropic, nonetheless represents an important event in the natural history of HIV infection in that coincides with an accelerate disease progression toward AIDS. Several antagonists of viral entry, most of which directed against CCR5, have already demonstrated good tolerability and partial efficacy in reducing plasma viremia in phase I/II studies and are currently being tested for clinical efficacy in combination with the preexisting classes of antiretroviral agents such RT and protease inhibitors (Barber, 2004).

More recently, the importance of a segmental duplication encompassing the gene for CC chemokine ligand 3-like protein (CCL3L1), which identifies a genetic variant of MIP-α also known as LD78β, known to be the natural chemokine with the most potent anti-HIV activity (Menten et al., 1999; Nibbs et al., 1999), has been identified in that a low CCL3L1 copy number is associated with an increased susceptibility to AIDS progression (Gonzalez et al., 2005). No major polymorphisms have been discovered in CXCR4, the coreceptor of X4 viruses, however a polymorphism in the 3′ untranslated region of its ligand SDF-1α has controversial effects in either HIV transmission or disease progression. Additional coreceptors used at lower frequency compared with CCR5 and CXCR4 is CCR2b. The CCR2-64I genetic variant [a G to A substitution resulting in a valine (V) to isoleucine (I) change at position 64] is in strong linkage disequilibrium with a mutation within the CCR5 regulatory region (CCR5-59653T). Individuals with two CCR2-64I alleles are not resistant to sexual transmission of HIV-1 but are frequently LTNP (Michael et al., 1997; Rizzardi et al., 1998).

At the cellular levels, several aspects regarding CCR5 versus CXCR4 use remain to be fully understood. CXCR4 is a chemokine receptor constitutively and abundantly expressed at high levels on the surface of different cell types. It is expressed particularly at higher levels on naïve T cells and on Th0/Th2 cells (as mentioned above, the Th2 cytokines IL-4 and IL-13 upregulate its expression). CXCR4 is also expressed by mononuclear phagocytes, although several investigators, including us, have failed to demonstrate—unlike others (Verani et al., 1998)— productive infection of these cells by pure X4 monotropic viruses. CCR5 is usually expressed at much lower levels than CXCR4 on different cell types, but it is induced in the context of inflammation and immune response. It is also preferentially expressed by memory T cells, monocytes, macrophages, as well as by iDe cells. Pathologic conditions, such as coinfections or helmintic infestation, leading to increased levels of CCR5 expression also resulted in higher levels of HIV replication in vivo (Clerici et al., 2000; Kalinkovich et al., 2001). Conversely, HTLV-II infection (a condition not clearly associated with a human disease) results in the upregulation of CCR5-binding chemokines, and particularly CCL3/ MIP-1α, which counteract R5 HIV infection (Casoli et al., 2000 and Casoli et al., Blood 2006, in press). Therefore, an increased availability of cell-surface CCR5 would results in higher chances of infection and/or efficiency of HIV entry. Alternatively, or in addition, increased expression of CCR5 may reflect a pro-inflammatory microenvironment favoring HIV replication by acting at post-entry levels, such as NFκB-dependent viral transcription. In this regard, when we investigated the susceptibility to R5, X4, and R5X4 infection of cord blood–derived primary T cells and T-cell clones either polarized or not toward Th1 or Th2 (with the former expressing higher levels of CCR5 than the latter) and maintained in condition of suboptimal activation, we observed that R5 HIVs replicated efficiently, whereas X4 viruses did not, in both cell types and regardless of the levels of CCR5 or CXCR4 expression at the cell surface. R5 superior capacity of spreading in these cells was not accounted for by a more efficient entry versus X4 viruses but rather by an increased capacity of propagating after infection that, unlike X4 virus infection, was insensitive to anti-CD3 stimulation (Vicenzi *et al.*, 1999). When R5X4 dualtropic viruses were tested in this model system, they efficiently replicated in Th0 and Th2 cells and clones, but they were restricted in Th1 cells, again by a post-entry mechanism (Vicenzi *et al.*, 2002). These experimental observations suggest that HIV utilization of CCR5 and CXCR4 implies functional consequences for the infected cells beyond their mechanical role as entry coreceptors. In this regard, several studies have demonstrated that trimeric gp120 R5 Env has differential and superior signaling capacity than its X4 counterpart in both T cells and macrophages (Arthos *et al.*, 2000; Liu *et al.*, 2000b; Weissman *et al.*, 1997). It is conceivable that these functional differences contribute to the superior capacity of R5 HIV to spread worldwide.

Anti-CCR5 antibodies (Abs) have been developed as potential anti-HIV agents, either as drugs or microbicides (Trkola *et al.*, 2001). In addition, anti-CCR5 Abs have been described in a cohort of highly exposed HIV-uninfected individuals but not in control uninfected subjects or infected individuals with progressive disease (Lopalco *et al.*, 2000), and, more recently, in some HIV-infected LTNP (L. Lopalco *et al.*, Blood, in press). The mechanism of generation of these (auto?) Abs and their potential relevance in the prevention of infection or disease progression remains speculative.

In addition to interference with HIV entry by CCR5 antagonists, noncompetitive inhibition of CCR5 has been demonstrated in the case of pertussis toxin B-oligomer (PTX-B) and the bacterial-derived chemotactic peptide formyl-leucyl-phenylalanine (FMLP). These molecules were shown to prevent entry of R5 HIV in primary T cells, MDMs, and cell lines (Alfano *et al.*, 1999; Alfano *et al.*, 2000; Alfano *et al.*, 2001; Li *et al.*, 2001). In the case of PTX-B, also post-entry inhibitory effects against both R5 and X4 viruses were demonstrated (Alfano *et al.*, 2001). This molecule, endowed with immunostimulating capacity and adjuvant properties has shown anti-HIV effects in both lymphoid histocultures (Alfano *et al.*, 2005), and SCID-hu mice reconstituted with human PBMCs (Lapenta *et al.*, 2005).

7.4 Intrinsic Resistance to HIV Infection

Productive HIV infection is the result of the interaction between the host and the virus. Recently, intracellular factors playing a key role in determining the fate of such an interaction have been discovered. Intracellular antiviral responses represent a first line of natural defense in preventing persistent infection as proved by the fact that HIV-1 has evolved efficient counteracting measures. APOBEC3G is the best example of the importance of host cell restriction factors (Sheehy et al., 2002). APOBEC3G is expressed by primary human CD4$^+$ T cells and by some transformed T-cell lines, whereas its action is counteracted by the HIV-1 accessory protein Vif. APOBEC3G is a member of a family of enzymes that edit RNA/DNA by deaminating cytosine to uracile. In the absence of Vif, APOBEC is incorporated in virions; as described above, upon infection of a new target cell, a minus-strand copy of DNA is copied from the viral RNA by the RT enzyme. This nascent single-stranded DNA is the exquisite target of APOBEC3G that induces G to A hypermutations potentially generating stop codons and amino acid changes in regions critical for the viral proteins function(s) (Mangeat et al., 2003; Navarro and Landau, 2004; Vartanian et al., 2003). The G to A mutations generally occur with a graded frequency being most detectable toward the 3' end of HIV genome. An in vitro model suggests that the length of time that each nucleotide remains single-stranded during reverse transcription determines the frequency of APOBEC3G-induced mutations (Yu et al., 2004a). The accessory gene product Vif is a relatively small protein made by 192 amino acids (Strebel et al., 1987). Its in vitro function is evident only in "nonpermissive" cells that do not support replication of a Vif-minus HIV (Sova and Volsky, 1993). Nonpermissive cells are converted to become permissive for HIV replication by expression of APOBEC3G (Sheehy et al., 2002). Vif can indeed bind to APOBEC3G and induce its ubiquitination followed by proteosomal degradation, thereby preventing its incorporation into new progeny virions (Sheehy et al., 2003). In addition to virion-associated APOBEC3G, intracellular APOBEC3G can also mediate antiviral functions. However, most of this protein seems to be inactivated by a high-MW complex bound to an endogenous RNA and polysomes (Kozak et al., J. Biol. Chem. 281:29105-29119, 2006) in activated lymphocytes or macrophages. Conversely, resting T cells and monocytes express APOBEC3G as a low-MW complex that efficiently inactivates incoming virions (Chiu et al., 2005).

Two other components of the APOBEC family, APOBEC3F and APOBEC3B, exhibit anti-HIV activities (Yu et al., 2004b; Zheng et al., 2004). However, APOBEC3F appears less potent than APOBEC3G and is partially resistant to HIV-1 Vif. Indeed, it has also been shown that APOBEC3G C-terminal mutants that cannot edit the DNA are still antiviral suggesting an alternative antiviral mechanism independent of cytidine deamination (Newman *et al.*, 2005). Both APOBEC3G and 3F are expressed in cell populations susceptible to HIV-1 infection, whereas APOBEC3B is not normally expressed in the lymphoid cells. Thus, activation of the endogenous APOBEC3B gene in primary human lymphoid cells could form a novel and effective strategy for inhibition of HIV-1 replication *in vivo* (Doehle *et al.*, 2005).

Recently, a second restriction factor governs species-specific infection in retroviruses and lentiviruses and it is responsible for the inefficiency of HIV-1 to replicate in non-human primate cells (Stremlau *et al.*, 2004). TRIM5α belongs to a family of proteins defined by a tripartite motif consisting of RING, B-box 2, and coiled-coil domains. The TRIM5α contains a C-terminal variable domain that differs depending upon species of origin. When only three amino acids within the first variable region of the human TRIM5α are altered to resemble monkey TRIM5α, the human chimera TRIM5α, which is more than 98% identical to the human protein, potently suppresses HIV-1 infection (Stremlau *et al.*, 2005). This factor, previously identified as Ref1 in humans (Hatziioannou *et al.*, 2004) and Lv1 in rhesus macaque (Keckesova *et al.*, 2004), interacts with HIV-1 capsid protein to interfere with virion uncoating, therefore blocking a step in the HIV-life cycle before reverse transcription but after entry into the target cells. HIV-1 overcomes the TRIM5α restriction by binding cellular cyclophilin A (CypA) to capsid (Sayah *et al.*, 2004). Cyclophilin A is a peptidyl-prolcyl isomerase first identified through its ability to interact with the HIV capsid. Studies in which cyclophilin A was inhibited by RNAi or was knocked out by gene deletion revealed that cyclophilin is packaged into virions through its interaction with the viral capsid (Braaten and Luban, 2001; Colgan *et al.*, 2005).

Other factors, including yet unidentified restriction factors in human cells, impedes HIV-1 particle assembly and release. These factor(s) are counteracted by the HIV-1 accessory protein VpU (Varthakavi *et al.*, 2003), MURR1 (Ganesh *et al.*, 2003), and nuclear membrane–associated molecules (Montes de Oca *et al.*, 2005) and this may play important roles in HIV infection and integration. In conclusion, the currently known restriction factors and their mechanism of action may represent only the tip of the iceberg of mechanisms evolved to protect eukaryotic cells from viral infections to preserve the integrity of the host cell and of its genome.

7.5 HIV Cytopathicity

Whether HIV should be considered a "cytopathic" virus or not remains a highly unresolved issue. According to the dogma, cytopathic viruses (such as influenza virus, cold viruses, etc.) damage directly their target cells and tissue and cause little immune response, whereas the cell/tissue damage observed upon infection by non-cytopathic viruses (such as hepatitis B virus; HBV) is essentially caused by the immune response to infected cells, for example by cytotoxic T lymphocytes (CTLs) (Zinkernagel and Hengartner, 1994). A confounding variable in this dispute regards whether cytopathicity *in vitro* may or may not accurately reflect cytopathicity *in vivo*. In this regard, X4 HIV, particularly after laboratory adaptation by multiple passages, is highly cytopathic for several T-cell lines (such as Jurkat, SupT1, CEM, etc.), an effect that may or may not be associated with giant cell formation in form of multinucleated syncytia (in some cases clustering hundreds of nuclei). Indeed, most cell lines, also not belonging to T lymphocytes, such as myelomonocytic HL-60, U937, THP1, and MonoMac, express abundant levels of CD4 and CXCR4 and, therefore, are susceptible to X4 virus infection, not necessarily in the presence of syncytia, but eventually undergoing single cell death (Vicenzi and Poli, 2005). HTLV-I transformed cell lines (such as MT-2 and MT-4), expressing variable levels of the regulatory protein Tax with trans-activating effects on the HIV-1 LTR, are highly susceptible to cytopathic HIV-1 infection (i.e., most cells rapidly form syncytia and die in culture upon infection) to the point that it is extremely difficult to obtain chronically infected survivor cell lines.

With the possible exception of HTLV-transformed cell lines, survival of cells after acute *in vitro* infection is almost the rule for most cell lines. Some of these survivor cells have been cloned by limiting dilution (such as the promonocytic U1 and U33, the myelomonocytic OM10.1, the T lymphocytic ACH-2 and 8E5 cell lines) and further characterized as surrogate models of *in vitro* latency, as discussed further. Their viability and growth features are grossly comparable with that of their parental uninfected cell lines (although an increased fragility, for example, to cell centrifugation is often observed). These cells downregulate CD4 from their cell surface, likely as a consequence of multiple interactions with gp120 (Folks *et al.*, 1985; Folks *et al.*, 1986a; Folks *et al.*, 1986b) whereas it is currently unclear whether CXCR4 expression is affected by the infection. They are characterized by possessing a low (1–2) copy number of integrated proviruses, mostly defective (although ACH-2 cells can generate infectious virions after repair of a defective provirus).

The biological correlates associated with the different fate of immortalized cells (i.e., death or survival as a consequence of HIV infection) remain speculative and obscure. Early studies have indicated that either the number of intracellular CD4-gp120 Env complexes or the drift in protein synthesis imposed by the infection over the physiologic levels of a given cell type could explain HIV-induced cytotoxicity (Hoxie *et al.*, 1986). Induction of robust virion production (to an extent comparable with that of acutely infected cells) in chronically infected U1 and ACH-2 cell lines *per se* does not lead to substantial cytopathicity (Biswas *et al.*, 1994) (with the caveat that the reduced expression of CD4 at the cell surface may impede or reduce the formation of intracellular complexes with gp120 Env and subsequent internalization). Infection of primary T lymphocytes may or may not result in profound cytopathicity as a result of both the stimulatory conditions (mitogens such as PHA cause a more rapid blast transformation than stimulation with only IL-2), the type of virus (X4 HIV is more cytopathic *in vitro* than R5 virus), and its infectious titer (in some case, high levels of the infectious titer result in an abortive infection due to acute cytopathicity). Like cell lines, persistently infected lymphocytes arise from primary cell infection *in vitro*, although they are usually $CD8^+$ rather than $CD4^+$. $CD8^+$ T lymphocytes become indeed infected *in vitro* in a manner strictly dependent upon interaction with $CD4^+$ T cells and not by free virions (De Maria *et al.*, 1991). It is conceivable that gp120 Env activation in *trans* by surface CD4 (expressed by $CD4^+$ lymphocytes or mononuclear phagocytes) results in the capacity of HIV to interact with CXCR4 on the surface of $CD8^+$ T cells and, therefore, infect them in a strictly cell-cell contact dependent manner. Of interest, this infection was not obviously cytopathic raising the question of whether there is something peculiar of $CD4^+$ T cells that makes them susceptible not only to infection but also to virus-induced cytotoxic effect. A similar mechanism has been invoked for $CD8^+$ T cell–dependent killing of infected macrophages (Herbein *et al.*, 1998). Because no substantial depletion of $CD8^+$ T cells or monocyte-macrophages occurs in most individuals, also in the AIDS phases, it is legitimate to wonder whether these mechanisms are truly reflective of pathologic events in infected individuals or are "artifactual" events occurring in the test tube.

In vivo, efficient depletion of $CD4^+$ T lymphocytes is observed upon X4 infection of lymphoid histocultures (established from tonsils, lymph nodes, thymus, or spleen), whereas R5 infection has milder effects (Glushakova *et al.*, 1995; Eckstein *et al.*, 2001). This has been correlated with a differential quantitative distribution of the two related chemokines receptors (namely, CXCR4 and CCR5) in the lymphocyte populations present in these lymphoid tissues (Grivel and Margolis, 1999; Grivel *et al.*, 2000).

Plate 1

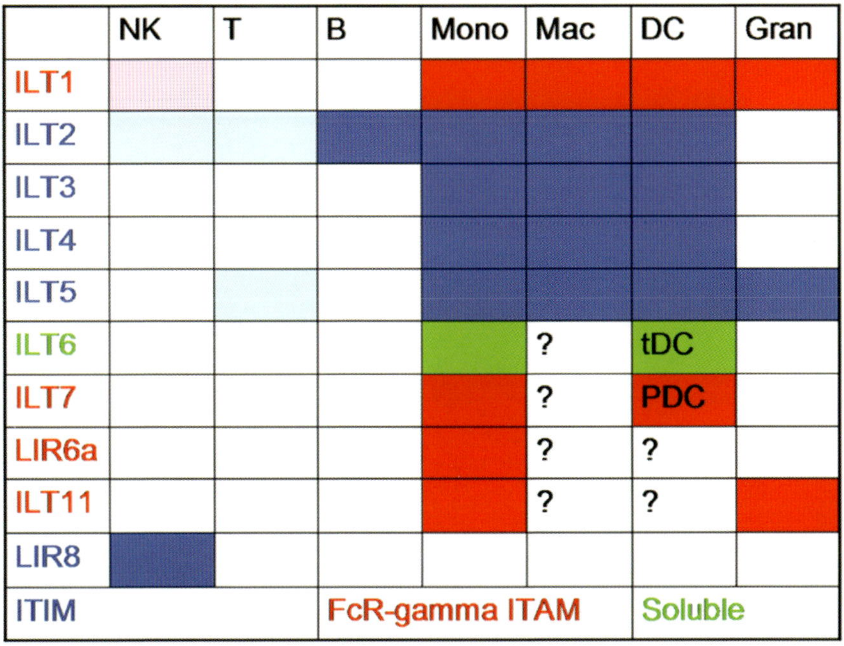

	NK	T	B	Mono	Mac	DC	Gran
ILT1							
ILT2							
ILT3							
ILT4							
ILT5							
ILT6					?	tDC	
ILT7					?	PDC	
LIR6a					?	?	
ILT11					?	?	
LIR8							
ITIM			FcR-gamma ITAM			Soluble	

FIGURE 3.2. Cell type–specific ILT receptor expression. ILT receptors regulate the activation state of a wide variety of cell types. Inhibitory ITIM-containing receptors are highlighted in blue, activating ITAM-containing receptors are highlighted in red; the only know soluble receptor is highlighted in green.

Plate 2

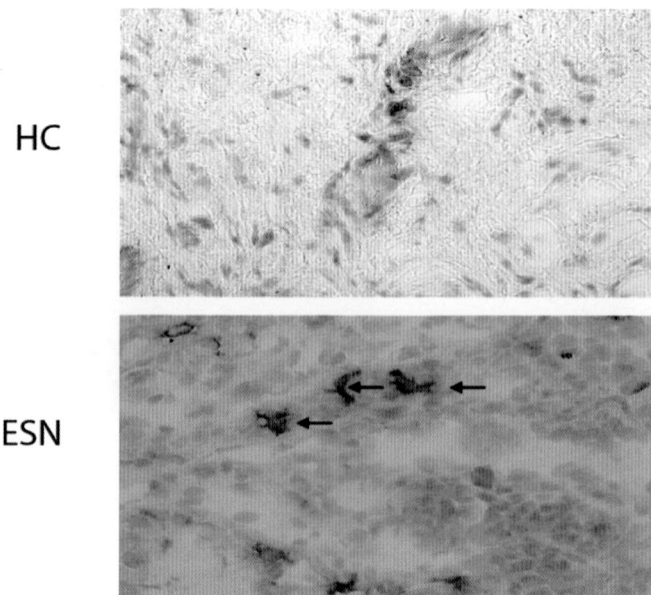

HC

ESN

FIGURE 9.3. α-Defensin expression in HIV-infected individuals. α-Defensin detection by immunohistochemistry staining in cervical biopsies of a representative ESN and a representative healthy control (HC). The arrows indicate positive cells.

Plate 3

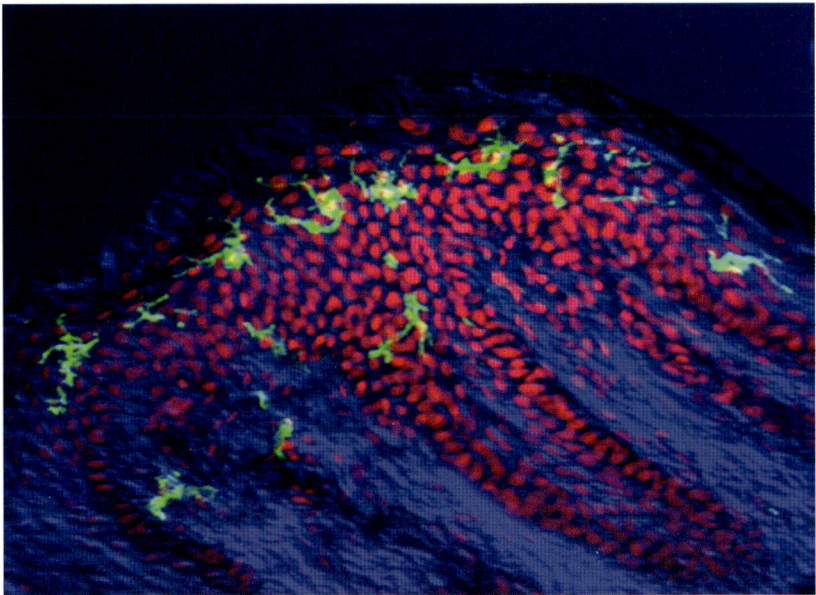

FIGURE 10.2. Langerhan's cells in human foreskin with processes extending to the tissue surface (image kindly provided by Scott McCoombe).

Plate 4

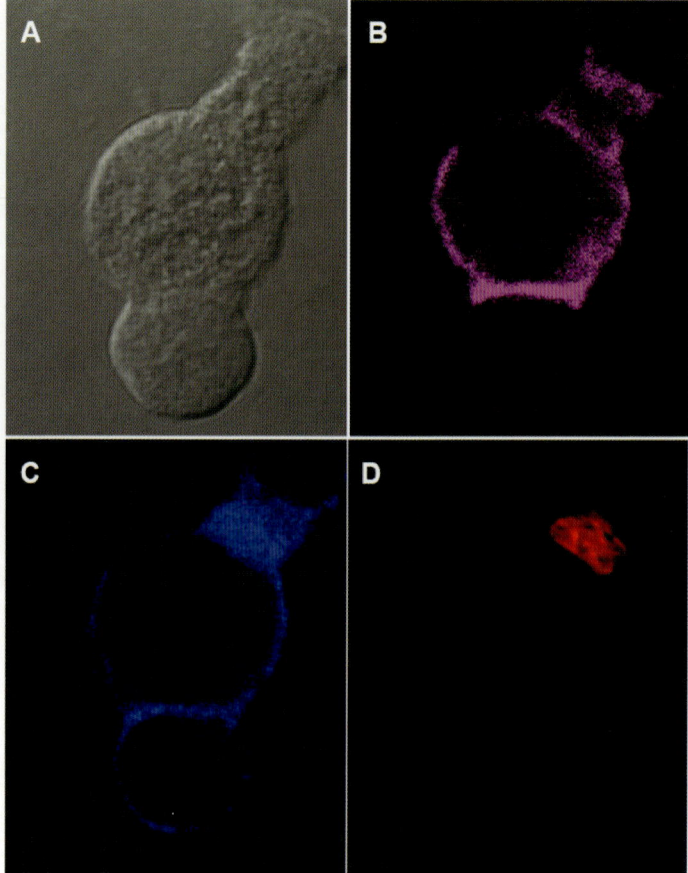

FIGURE 13.2. DCs acquire molecules from live as well as apoptotic infected CD4[+] T lymphocytes. DCs and HIV-infected CD4[+] T cells interact closely when cultured together, as shown by confocal transmission microscopy (A). DCs are labeled in magenta by immunofluorescence using anti-CD11c antibodies. (B) DCs acquire CD3 molecules (in blue; C) from the T-cell receptor complex. T cells (smaller, round, labeled in blue and not in magenta) have been irradiated. Some are apoptotic, as shown by TUNEL labeling in red (D), others are alive, but both are in close contact with DCs. After this contact and material transfer, DCs cross-present the HIV antigens from the infected CD4[+] T lymphocytes to effector CD8[+] T lymphocytes.

Concerning the phenomenon of giant cell formation, it is likely a magnification of HIV-induced cytopathicity *in vitro*, in that syncytia have not been observed in HIV-infected individuals (e.g., in their lymphoid organs). *In vitro*, expression of the integrin LFA-1 (CD11a/CD18) as well as of its ligands ICAM-1, -2, and –3 has been clearly demonstrated to play a crucial role in HIV-induced cell fusion and giant cell formation (Hildreth and Orentas, 1989). Lymphocytes of individuals genetically deficient for LFA-1 expression (patients with leukocyte adhesion deficiency; LAD) do not form syncytia upon infection with X4 viruses (Pantaleo *et al.*, 1991a). Similarly, interfering with monoclonal Abs (mAbs) against either LFA-1 or its ligands together result in a complete inhibition of cell-cell fusion (Pantaleo *et al.*, 1991b). Yet, both LAD and normal lymphocytes treated with these Abs show only partially (50%) reduced levels of virus replication, likely resulting from infection by free virions in the absence of cell-cell spreading infection.

Unlike T lymphocytes, macrophages and DC do not die significantly upon productive infection with HIV-1. Although syncytia formation has been described during *in vitro* infection of MDMs, this does not result in a substantial depletion of cell in culture. Productively infected MDM cultures can be maintained literally for months without evidence of obvious cell proliferation or death. As mentioned previously, macrophage infection both *in vitro* and *in vivo* is associated with the budding and accumulation of virions in intracytoplasmic vacuoles characterized by expression of CD63, that is, MVB, belonging to the hexosomal pathway (Freed, 2002; Gould *et al.*, 2003; Pelchen-Matthews *et al.*, 2003). Also, CD34$^+$-derived chronically infected cell lines assume these intracellular budding features over time, likely as a result of differentiation of these cells into macrophages (Folks *et al.*, 1988b).

Surrogate animal models, such as severe combined immunodeficiency (SCID) mice reconstituted with human tissue, have provided conflicting results in terms of HIV-induced cytopathicity. Although the paradigm of an increased cellular toxicity being associated with X4 infection is usually observed, at least one study has highlighted that R5 viruses are more cytopathic in these models than X4 HIVs (Picchio *et al.*, 1997; Fais *et al.*, 1999; Scoggins *et al.*, 2000). As mentioned above, experimentally infected macaques undergo rapid T-cell depletion with X4 SIV, whereas natural SIV infection in Sooty Mangebeys (the natural host) results in very high levels of virus replication in the absence of immunodeficiency and/or cell depletion (Silvestri *et al.*, 2003). Similarly, experimental infection of chimpanzees with HIV-1 leads to high levels of viral replication without inducing significant T-cell depletion or disease (Saksela *et al.*, 1993). Because this feature is also associated with a lack of immune response to the infection, these results can be interpreted under the light that, as for HBV, also HIV-induced cytopathicity results mainly from a host response rather than from intrinsic properties of the virus.

The dispute over HIV direct versus indirect cytopathic effects clearly goes beyond an academic arena. In the first case, an effective anti-retroviral therapy should essentially block HIV-induced CD4$^+$ T-cell depletion and allow the progressive restoration of a normal number of functional lymphocytes through the bone marrow–thymus axis. Although substantial immune recovery can be obtained with most highly aggressive antiretroviral therapy (HAART) protocols, frequently this remains incomplete (both in terms of CD4$^+$ T-cell counts and function). The emergence of resistant viruses and/or compliance problems related to the assumption of these regimens has not allowed a thorough analysis of whether HAART regimens prolonged for several years can lead indeed to a complete immunologic restoration, although some studies support this view. Combination of HAART with intermittent IL-2 administration usually results in more robust and rapid restoration of a normal or near normal number of peripheral naïve and memory CD4$^+$ T cells (Kovacs *et al.*, 1995; Kovacs *et al.*, 1996), at least some of which with a restored functionality in terms of antigen recognition and immune response (Carcelain *et al.*, 2003; Levy *et al.*, 2003).

Compelling evidence in favor of a direct cytopathic effect of HIV against CCR5$^+$ memory CD4$^+$ T lymphocytes has been recently empha-sized. After the original observation that the main site of virus replication and cytopathicity in experimentally infected macaques is the (GALT) (Veazey *et al.*, 1998; Mattapallil *et al.*, 2005), studies in humans undergoing primary HIV infection (PHI) have essentially confirmed this finding (Brenchley *et al.*, 2004). According to these studies, up to 60% of CD4$^+$ memory T cells are wiped out by the virus in the GALT in the first days of virus replication, whereas the remainder of HIV-induced cytopathicity is a smoldering erosion of T cells counterbalanced in part by the *ex novo* capacity of the T-cell regenerative axis until the breakdown coinciding with the AIDS immunodeficient phase of the disease.

This "virocentric" view—resembling the previous view based on mathematical modeling on plasma viremia perturbation upon introduction of antiretroviral agents (Ho, 1995; Ho *et al.*, 1995; Wei *et al.*, 1995)—is counterbalanced by the observation that a substantial fraction of apoptotic CD4$^+$ T-cell death occurs in uninfected cells bystanding infected cells or cells displaying attached virions on their surface (Finkel *et al.*, 1995). According to this second model, death of CD4$^+$ T cells is mostly or at least in part dependent upon Fas-Fas ligand interaction and not virus infection-driven (Finkel *et al.*, 1995; Li *et al.*, 2005). In this regard, the observation that PHI patients receiving cyclosporin A (CsA) together with HAART show a very rapid increase of circulating CD4$^+$ T cells in the face of comparable levels of decrease in plasma viremia supports the view that CD4[+] T cell depletion during PHI is largely immune-mediated rather than virus driven (Rizzardi *et al.*, 2002). However, a number of caveats should be underscored in that this peculiar study was based on a limited number of patients that were matched *a posteriori* with a group of similar individuals experiencing PHI. In addition, CsA is known to affect HIV-1 replication via interference with the Gag polyprotein (Braaten and Luban, 2001; Luban *et al.*, 1993) at least *in vitro*, and the lack of effect on plasma viremia may mask a direct antiviral effect in certain tissue and reservoir.

The dispute remains unresolved at present, and HIV remains suspended between the categories of cytopathic and noncytopathic viruses.

7.6 HIV Replication in Lymphoid Organs and Central Nervous System

Although plasma viremia (reflecting HIV replication) and CD4[+] T-cell count (a main parameter of immune deterioration) are routinely evaluated in peripheral blood samples, crucial events of HIV pathogenesis occur in lymphoid tissue (Pantaleo *et al.*, 1998) and CNS (Kramer-Hammerle *et al.*, 2005). As previously discussed, recent studies have emphasized the GALT as the primary site of infection and CD4[+] T-cells depletion during PHI (Guadalupe *et al.*, 2003). In more general terms, lymphoid tissues represent the main reservoir of HIV infection, with the virus carried by DCs into the draining lymph nodes (Moll, 2003), a process initiated by responsiveness to CCR7 binding chemokines (CCL19/ELC and CCL21/SLC) expressed by high endothelial cells in the lymph nodes (LNs) afferent venules and by other LN cells, respectively (Baekkevold *et al.*, 2001). HIV dissemination via the lymphatic system and viral replication into LNs occurs very soon after infection and before immune responses develop, with the viral reservoir during chronic infection maintained by follicular dendritic cells (FDCs) (Pantaleo *et al.*, 1998). In this regard, virus-bearing FDCs have also been proposed to be responsible of the second-phase of the biphasic plasma viral decay (Hlavacek *et al.*, 2000). Moreover, destruction of LN architecture seems to be proportional with the level of viremia (Perrin *et al.*, 1998) and gp120 Env–induced apoptosis of CD4[+] T cells (Sunila *et al.*, 1997), as supported by the observation that in LTNP and individuals under efficient HAART the LN architecture is preserved or restored (Bart *et al.*, 2000), respectively.

A peculiar lymphoid organ targeted by HIV is thymus. HIV-infected thymocytes have been found *in vivo* and thymopoiesis is interrupted by direct infection (Su *et al.*, 1995). However, also indirect cytopathic mechanisms, mediated by multiple cytokines (Table 7.2), in HIV$^+$ individuals have been shown to mediate disruption of stromal architecture and thymocyte depletion (Gaulton *et al.*, 1997) as consequence of apoptosis of uninfected cells (Su *et al.*, 1995). Therefore, the impairment of thymic function results in a diminished capacity to replenish the lymphocyte pool, therefore contributing substantially to the overall CD4$^+$ T-cells depletion observed during the disease progression (Hazenberg *et al.*, 2000). In this regard, the immune restoration observed in HIV$^+$ children under HAART (Gibb *et al.*, 2000) or HIV$^+$ adults under HAART plus IL-2 (Pido-Lopez *et al.*, 2003) has been described to be either thymus dependent or independent, respectively. However, the observation of an increased number of naïve T cells emerging after several months of HAART plus IL-2 therapy (Carcelain *et al.*, 2003; Marchetti *et al.*, 2004) suggests the possibility that thymic function can be stimulated also in adults.

TABLE 7.2. Upregulation of cytokines expression in lymphoid tissues during *in vivo* infection.

Cytokine	HIV$^+$ individuals or SIV-infected animals
IFN-γ	Increased expression in germinal centers of HIV-infected individuals (Emilie *et al.*, 1990; Graziosi *et al.*, 1996), as well as in HIV-1 –infected chimpanzees (Villinger *et al.*, 1997) and during primary SIV infection of macaques (Cheret *et al.*, 1999).
IL-1α, IL-1β, IL-2, IL-6, IL-12	Overexpression in tonsils and LN of HIV$^+$ individuals, and during primary SIV infection of macaques (Devergne *et al.*, 1991; Khatissian *et al.*, 1996; Andersson *et al.*, 1998).
TNF-α	Correlation between TNF-α levels in LN and SIV replication (Khatissian *et al.*, 1996).

Investigating the pathogenesis of HIV infection in lymphoid tissues can be accomplished in *ex vivo* explants, derived either from human LN, thymus, or tonsils (Glushakova *et al.*, 1995). In these explants, the *in vivo*– like cytoarchitecture and cellular repertoire (Glushakova *et al.*, 1998), including the network of FDCs, B and T lymphocytes (Glushakova *et al.*, 1999), are retained in physiologic numerical proportion and three-dimensional interrelationship (Grivel and Margolis, 1999). *Ex vivo*– cultured human lymphoid tissue supports productive infection of both R5 and X4 HIV-1 and is suitable for investigating the effects of antiviral drugs (Bonyhadi *et al.*, 1995). Finally, *in vivo* study of the cytopathic effects of HIV infection in lymphoid tissue has greatly improved by the development of SCID mice reconstituted with either fetal liver, thymus, LN, or PBMCs (Meissner *et al.*, 2003; Mosier *et al.*, 1991; Namikawa *et al.*, 1988).

HIV enters the CNS very soon after systemic infection (Bell, 2004), either as free virus or by infected cells. HIV virions can diffuse from blood through the blood-brain barrier (BBB) into the CNS by migrating between brain microvascular endothelial cells (BMVECs) or by transcytosis (in the case that virions are into vacuoles and are taken up and released by BMVECs) (Bobardt *et al.*, 2004). In addition, it has also been proposed that BMVECs may be able to support productive HIV replication (Wiley and Nelson, 1988), although contrasting results have been also reported (Takahashi *et al.*, 1996). A third possible modality for HIV to enter the CNS is via infiltration of HIV-infected leukocytes (the so-called Trojan horse hypothesis) throughout the BBB (Albright *et al.*, 2003). In the CNS, only brain macrophages and microglial cells are able to efficiently replicate HIV, whereas astrocytes, although found infected *in vivo* and infectable *in vitro,* are not efficient in virion expression (Kramer-Hammerle *et al.*, 2005), but can produce abundant amounts of Nef (Brack-Werner, 1999). In spite of the fact that HIV spreading in the CNS occurs very early after systemic infection, HIV replication in the CNS becomes evident at the onset of AIDS (Gray *et al.*, 1996) in coincidence with increased levels of CD8[+] T cells infiltrating the brain (Albright *et al.*, 2003).

HIV infection of the CNS gives rise to a variety of neurologic symptoms such as motor disturbances, cognitive impairments, behavioral changes, and dementia (Cinque *et al.*, 1997; McArthur *et al.*, 2003). Neuro-pathologic abnormalities in the CNS of HIV[+] individuals, such as axonal damage and brain atrophy (Bell, 2004), are commonly observed in advanced patients (Kaul *et al.*, 2001). Intracellular expression of Tat has been shown to activate expression of inducible nitric oxide synthase (iNOS), NO (Liu *et al.*, 2002), inflammatory cytokines, and adhesion molecules (Huigen *et al.*, 2004) in infected CNS. In addition, soluble and/or virion-associated viral proteins such as gp120 Env, Tat, Nef, and VpR can trigger the expression of pro-inflammatory and neurotoxic factors inducing apoptosis of uninfected neurons (Kramer-Hammerle *et al.*, 2005). Introduction of HAART has proved to be efficacious in the reduction of the incidence of HIV-associated dementia (HAD) (Kandanearatchi *et al.*, 2003), although resurgence of HIV-associated encephalitis and HIV leukoencephalopathy has been reported (Langford *et al.*, 2003). In this regard, HIV[+] individuals under HAART survive longer and, in some instances, a poor diffusion of some agents through the BBB may result in persistent levels of viral replication in the CNS even if viremia is efficiently suppressed. Among other inflammatory markers, increased levels of CCL2/monocyte chemotactic protein-1 (MCP-1) are highly associated with HIV encephalitis (and cytomegalovirus-induced encephalitis), likely recruiting monocytes (infected?) from peripheral blood to the CNS (Cinque *et al.*, 1998; Kelder *et al.*, 1998). Productive HIV infection of MDMs (Mengozzi *et al.*, 1999) as well as stimulation of MDMs or astrocytes with either Tat (Conant *et al.*, 1998; Mengozzi *et al.*, 1999) or gp120 Env interaction with MDMs (Fantuzzi *et al.*, 2001) resulted in increased production of CCL2/MCP-1 *in vitro*. Therefore, it is likely that HIV-induced CCL2/MCP-1 expression plays a major role in the pathogenesis of HIV infection in the CNS.

7.7 Host Determinants of HIV Propagation: Cytokines and Chemokines

Cytokines are soluble factors produced by immune and nonimmune cells, such as leukocytes and endothelial cells, characterized by pleiotropy (one molecule with multiple cellular targets) and redundancy (different cytokines exert the same biological activity on the same cell) (Paul and Seder, 1994). Cytokines control both innate and specific immune responses, including the on-off state of inflammatory processes (Slifka and Whitton, 2000), and specific immune responses. They can also exert mitogenic activity or promote growth arrest, cell differentiation, and, in some case, trigger apoptotic pathways (Refaeli *et al.*, 1998). Cytokines are conventionally subdivided into pro- and anti-inflammatory molecules, whereas some of them exert predominant immunoregulatory effects, for example in terms of polarization toward Th1 (cellular, phagocyte-dependent) or Th2 (humoral, phagocyte-independent) immune responses (O'Garra and Murphy, 1996). The recognition of microbial agents by pattern recognition receptors such as Toll-like receptors (TLRs) (Moll, 2003), in addition to maturation of DCs (mature DCs; mDCs), leads to increased secretion of pro-inflammatory cytokines, including IL-1β, IL-12, IL-18, and tumor necrosis factor-α (TNF-α), and chemokines such as CCL3/MIP-1α, CCL4/MIP-1β, and CXCL1/MIP-2 (Granucci *et al.*, 2003) (Table 7.3).

In this regard, CCR5 ligands can stimulate the replication of X4 viruses in activated primary T cells (Dolei *et al.*, 1998; Kinter *et al.*, 1998). In addition, it is unclear whether physiologic levels of these molecules may ultimately exert a positive preventive role in natural infection or whether

TABLE 7.3. Bidirectional regulation of HIV replication and cytokine/chemokine expression.

Cytokines	Effect on HIV replication	Effect of HIV or its proteins on cytokine expression
Proinflammatory cytokines		
IL-1α/β	Enhancement of HIV replication in PBMC, MDM, and U1 cells (Poli *et al.*, 1994; Kinter *et al.*, 1995) (Granowitz *et al.*, 1995)	
IL-2	Synergy with IL-4 in the induction of HIV replication (Hays *et al.*, 1992, Chun *et al.*, 1998; Chun *et al.*, 1999) and, indirectly, via upregulation of other cytokines (Kinter *et al.*, 1995a; Kinter *et al.*, 1995b). Enhancement of HIV replication in the absence of HAART in individuals with <200 CD4$^+$ T cell counts/μl; no increase in viremia HAART in individuals with >200 CD4$^+$ T cell counts/μl (Kovacs *et al.*, 2000).	Decreased serum levels and expression from PBMC of HIV$^+$ individuals (Klein *et al.*, 1997).
IL-6	Enhancement of HIV expression in U1 cells and MDM (Poli *et al.*, 1990). Induction of HIV replication from resting memory T cells in the presence of other proinflammatory cytokines (Chun *et al.*, 1998).	
IL-7	Enhances HIV replication in CD8-depleted PBMC (Smithgall *et al.*, 1996), viral expression in chronically infected human cells (Scripture-Adams *et al.*, 2002) and viral spreading in CD4$^+$ thymocytes (Chene *et al.*, 1999). Induction of replication from resting memory CD4$^+$ T cells (Ducrey-Rundquist *et al.*, 2002; Wang *et al.*, 2005).	Elevated levels in HIV$^+$ individuals (Chiappini *et al.*, 2003) and in uninfected children born from HIV$^+$ mothers (Clerici *et al.*, 2000).

IL-12	Both enhancing (Al-Harthi *et al.*, 1998, Kinter *et al.*, 1995) and inhibitory (Wang *et al.*, 1999) effects have been reported.	IL-12 induction by p17 MA in human NK cells (Vitale *et al.*, 2003), Nef and p55 Gag (Quaranta *et al.*, 2003). Inhibition by HIV infection *in vitro* (Marshall *et al.*, 1999) and *in vivo* (Marshall *et al.*, 1999).
IL-15	Enhancement of R5 HIV replication (Al-Harthi *et al.*, 1998).	Induction by p17 MA in human NK cells (Vitale *et al.*, 2003).
TNF-α	Enhancement of HIV transcription (Nabel and Baltimore, 1987; Griffin *et al.*, 1989) and spreading in the CNS (Obregon *et al.*, 1999; Cota *et al.*, 2000). Inhibition of HIV entry via induction of CD4 downmodulation (Herbein *et al.*, 1996).	Enhancement of expression by HIV infection or gp 120 Env stimulation of macrophages, T and B cells (Rieckmann *et al.*, 1991; Khanna *et al.*, 2000).
Anti-inflammatory cytokines		
IL-4	Enhancement of HIV replication in monocytes, but post-transcriptional inhibition in MDM (Kazazi *et al.*, 1992; Schuitemaker *et al.*, 1992; Naif *et al.*, 1994). Enhancement of virus expression in U1 cells costimulated with IL-1 and IL-10 (Weissman *et al.*, 1995).	
IL-10	Inhibition HIV replication in macrophages (Borghi *et al.*, 1995) by blocking TNF-α and IL-6 expression (Weissman *et al.*, 1994), but enhancement of monocyte susceptibility to both R5 (Sozzani *et al.*, 1998) and X4 (Ancuta *et al.*, 2001) infection. Enhancement of HIV transcription in U1 cells (Weissman *et al.*, 1995).	Increased expression in LN of HIV$^+$ individuals (Falciola *et al.*, 1997) HIV-1 gp 120 induced IL-10 expression in monocytes/macrophages infected *in vitro* (Borghi *et al.*, 1995).

they would ultimately favor viral spreading by recruiting T cells at the site of virus infection (e.g., in genital mucosa). In this regard, it is worthy of note that the accessory protein Nef, which plays an important role in promoting *in vivo* pathogenesis of retroviral infection, upregulates two out of three CCR5 ligands (CCL3/MIP1α and CCL4/MIP1β) upon infection of macro–phages and DCs (Swingler *et al.*, 1999), while HIV-1 gp120 Env enhances chemokine production independently of CD4 (Fantuzzi *et al.*, 2001) via activation of Pyk2 and mitogen-activated protein kinases (Del Cornò *et al.*, 2001). In addition, other chemokines of both the CC (such as CCL2/MCP1) (Kinter *et al.*, 1998; Vicenzi *et al.*, 2000) and CXC (e.g., CXCL8/IL-8 and CXCL1/Gro-α) families can stimulate HIV replication in both T cells and MDMs (Lane *et al.*, 2001a; Lane *et al.*, 2001b).

Pro-inflammatory cytokines, including TNF-α, IL-1β, and IL-6, have been associated with increased levels of HIV replication *in vivo* and *in vitro*. At the molecular level, both NFκB and AP-1 can mediate transcriptional activation of HIV expression either by a direct interaction with DNA binding sites in the HIV LTR or, in the case of AP-1, via binding to an intragenic enhancer (Van Lint *et al.*, 1994). However, additional post-transcriptional control of virus replication is likely to play an important role in cytokine-mediated upregulation of virus replication, as discussed in the case of IL-4, IL-6, IL-13, and CXC chemokines (Fig. 7.7) (Poli *et al.*, 1990; Montaner *et al.*, 1997; Lane *et al.*, 2001a; Lane *et al.*, 2001b).

A peculiar example of post-transcriptional and post-translational control of the HIV life cycle is provided by interferons (IFNs). Class I IFNs (IFN-α and IFN-β) can affect multiple pre- and postintegration steps of the virus life cycle, including exerting a postbudding effect on the release of otherwise mature virions (Gessani *et al.*, 1994; Pitha, 1994). In contrast, IFN-γ can mediate either suppressive (Bovolenta *et al.*, 1999) or upregulatory effects on HIV replication as a function of whether cells are acutely or chronically infected by HIV and whether other stimuli are present. For example, IFN-γ strongly synergizes with TNF-α in inducing HIV transcription from integrated provirus, but also in terms of cytokine-mediated cytotoxicity (Biswas *et al.*, 1994). Unlike IFN-α/-β, IFN-γ stimulation of infected monocytic cells may lead to the redirection of the primary site of virion budding from the plasma membrane to Golgi-derived intracytoplasmic vacuoles (Biswas *et al.*, 1992), today recognized as belonging to MVB, as discussed above, and resembling morphologic features typical of brain macrophages of individuals with HIV encephalitis (Koenig *et al.*, 1986; Orenstein and Jannotta, 1988). Of interest is the fact that signaling generated

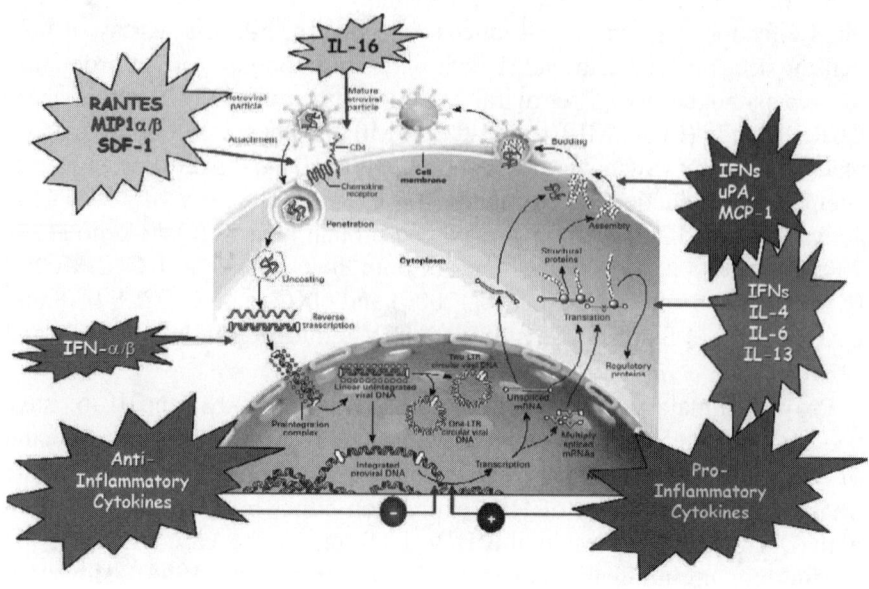

FIGURE 7.7. Cytokine control of the HIV life cycle. Cytokines and chemokines influence virtually all the crucial steps in the HIV life cyle in CD4[+] T cells and macrophages. Modified after Furtado *et al.*, N.E.J.M., 340:1614-1622, 1999.

by IFN-γ–unrelated molecules, including CCL2/MCP-1 (Fantuzzi *et al.*, 2003), and urokinase/urokinase receptor (Alfano *et al.*, 2002), may induce similar features in infected and differentiated monocytic cells (Fig. 7.4B). These observations suggest the existence of a common target of diverse signaling pathways that can influence the "choice" of the virion assembly machinery to target either the plasma membrane or endosomal membranes (Pelchen-Matthews *et al.*, 2003). Of interest, this seems to be restricted to mononuclear phagocytes, although inactivation of VpU results in partially similar features in T lymphocytes (Klimkait *et al.*, 1990).

In addition to class I IFNs, very few molecules have been reported to exert clear-cut inhibitory effects on HIV replication. IL-4, IL-10, IL-13, and TGF-β (known to exert anti-inflammatory activities) and leukemia inhibitory factor (LIF) (Patterson *et al.*, 2001) have been associated with inhibition of viral replication. However, these molecules have been reported to activate HIV replication in different experimental conditions (Broor *et al.*, 1994; Patterson *et al.*, 2001) making it difficult to assign them a precise role in the balance between virus replication and latency.

In vivo, administration of intermittent IL-2 has been demonstrated to induce "blips" of virus replication without inducing an increase in steady-state viremia, even in the absence of a full suppression of HIV replication by HAART (Kovacs *et al.*, 1995; Kovacs *et al.*, 1996). It is likely that the net effect of cytokines and chemokines in *in vivo* infection will result not only from a direct effect on the infected cells but also, and perhaps even more importantly, from the indirect activation of other effectors, such as CD8[+] T and Natural Killer cells (Liu *et al.*, 2000a) which may contribute and even take advantage of transiently increased expression of viral proteins to eliminate infected cells. These concepts are currently explored in experimental protocols aimed at eradicating infection from latent reservoirs, including resting memory T cells (Chun *et al.*, 1997; Finzi *et al.*, 1997) and monocytes (Fulcher *et al.*, 2004; Zhu *et al.*, 2002).

7.8 Concluding Remarks

HIV has evolved an efficient way to infect and persist in the human host by taking advantage of his cell-surface receptors (CD4, chemokine receptors) and transcriptional machinery. Conversely, a number of natural antiviral weapons (intracellular proteins such as APOBEC3G and TRIM-5α and extracellular cytokines, chemokines, and IFNs) counteract the ability of the virus to spread in the infected individual. Although not discussed in detail in this chapter, the most peculiar feature of HIV as a retrovirus is its ability to integrate in the host genome to give rise to a permanent infection that may remain in a state of relative or absolute latency or continue a retroviral life cycle of cell-to-cell and free virion spreading. Although the potent HAART regimens can curtail HIV spreading to minimal undetectable levels, they cannot to date decrease the pool of latently infected cells. As soon as the pharmacological control of HIV replication is suspended, viral replication restarts as vigorously as before therapy and the march toward a severe immunodeficient state goes along with it. Therefore, only by unraveling the intimate connection between the virus and its host cells may we hope to identify key molecules and pathways for future pharmacological development. The possibility of combining immunologic approaches (such as IL-2 administration) to conventional anti-retroviral regimens may also influence the HIV/host relationship in favor of the latter.

References

Al-Harthi, L., Roebuck, K. A. and Landay, A. (1998). Induction of HIV-1 replication by type 1-like cytokines, interleukin (IL)-12 and IL-15: effect on viral transcriptional activation, cellular proliferation, and endogenous cytokine production. *J. Clin. Immunol.* 18:124-131.

Albright, A. V., Soldan, S. S. and Gonzalez-Scarano, F. (2003). Pathogenesis of human immunodeficiency virus-induced neurological disease. *J. Neurovirol.* 9:222-227.

Alfano, M., Schmidtmayerova, H., Amella, C. A., Pushkarsky, T. and Bukrinsky, M. (1999). The B-oligomer of pertussis toxin deactivates CC chemokine receptor 5 and blocks entry of M-tropic HIV-1 strains. *J. Exp. Med.* 190: 597-606.

Alfano, M., Pushkarsky, T., Poli, G. and Bukrinsky, M. (2000). The B-oligomer of pertussis toxin inhibits human immunodeficiency virus type 1 replication at multiple stages. *J. Virol.* 74:8767-8770.

Alfano, M., Vallanti, G., Biswas, P., Bovolenta, C., Vicenzi, E., Mantelli, B., Pushkarsky, T., Rappuoli, R., Lazzarin, A., Bukrinsky, M. and Poli, G. (2001). The binding subunit of pertussis toxin inhibits HIV replication in human macrophages and virus expression in chronically infected promonocytic U1 cells. *J. Immunol.* 166:1863-1870.

Alfano, M., Sidenius, N., Panzeri, B., Blasi, F. and Poli, G. (2002). Urokinase-urokinase receptor interaction mediates an inhibitory signal for HIV-1 replication. *Proc. Natl. Acad. Sci. U. S. A.* 99:8862-8867.

Alfano, M., Grivel, J. C., Ghezzi, S., Corti, D., Trimarchi, M., Poli, G. and Margolis, L. (2005). Pertussis toxin B-oligomer dissociates T cell activation and HIV replication in CD4 T cells released from infected lymphoid tissue. *AIDS* 19:1007-1014.

Alkhatib, G., Combadiere, C., Broder, C. C., Feng, Y., Kennedy, P. E., Murphy, P. M. and Berger, E. A. (1996). CC CKR5: a RANTES, MIP-1alpha, MIP-1beta receptor as a fusion cofactor for macrophage-tropic HIV-1. *Science* 272: 1955-1958.

Alonso, K., Pontiggia, P., Medenica, R. and Rizzo, S. (1997). Cytokine patterns in adults with AIDS. *Immunol. Invest.* 26:341-350.

Ancuta, P., Bakri, Y., Chomont, N., Hocini, H., Gabuzda, D. and Haeffner-Cavaillon, N. (2001). Opposite effects of IL-10 on the ability of dendritic cells and macrophages to replicate primary CXCR4-dependent HIV-1 strains. *J. Immunol.* 166:4244-4253.

Andersson, J., Fehniger, T. E., Patterson, B. K., Pottage, J., Agnoli, M., Jones, P., Behbahani, H. and Landay, A. (1998). Early reduction of immune activation in lymphoid tissue following highly active HIV therapy. *AIDS* 12:F123-9.

Arthos, J., Rubbert, A., Rabin, R. L., Cicala, C., Machado, E., Wildt, K., Hanbach, M., Steenbeke, T. D., Swofford, R., Farber, J. M. and Fauci, A. S. (2000). CCR5 signal transduction in macrophages by human immunodeficiency virus and simian immunodeficiency virus envelopes. *J. Virol.* 74:6418-6424.

Baca-Regen, L., Heinzinger, N., Stevenson, M. and Gendelman, H. E. (1994). Alpha interferon-induced antiretroviral activities: restriction of viral nucleic acid synthesis and progeny virion production in human immunodeficiency virus type 1-infected monocytes. *J. Virol.* 68:7559-7565.

Baekkevold, E. S., Yamanaka, T., Palframan, R. T., Carlsen, H. S., Reinholt, F. P., von Andrian, U. H., Brandtzaeg, P. and Haraldsen, G. (2001). The CCR7 ligand elc (CCL19) is transcytosed in high endothelial venules and mediates T cell recruitment. *J. Exp. Med.* 193:1105-1112.

Bailer, R. T., Holloway, A., Sun, J., Margolick, J. B., Martin, M., Kostman, J. and Montaner, L. J. (1999). IL-13 and IFN-gamma secretion by activated T cells in HIV-1 infection associated with viral suppression and a lack of disease progression. *J. Immunol.* 162:7534-7542.

Bailer, R. T., Lee, B. and Montaner, L. J. (2000). IL-13 and TNF-alpha inhibit dual-tropic HIV-1 in primary macrophages by reduction of surface expression of CD4, chemokine receptors CCR5, CXCR4 and post-entry viral gene expression. *Eur. J. Immunol.* 30:1340-1349.

Barber, C. G. (2004). CCR5 antagonists for the treatment of HIV. *Curr. Opin. Invest. Drugs* 5:851-861.

Bart, P. A., Rizzardi, G. P., Tambussi, G., Chave, J. P., Chapuis, A. G., Graziosi, C., Corpataux, J. M., Halkic, N., Meuwly, J. Y., Munoz, M., Meylan, P., Spreen, W., McDade, H., Yerly, S., Perrin, L., Lazzarin, A. and Pantaleo, G. (2000). Immunological and virological responses in HIV-1-infected adults at early stage of established infection treated with highly active antiretroviral therapy. *AIDS* 14:1887-1897.

Bell, J. E. (2004). An update on the neuropathology of HIV in the HAART era. *Histopathology* 45:549-559.

Bennasser, Y., Le, S. Y., Benkirane, M. and Jeang, K. T. (2005). Evidence that HIV-1 encodes an siRNA and a suppressor of RNA silencing. *Immunity* 22:607-619.

Berger, E. A., Doms, R. W., Fenyo, E. M., Korber, B. T., Littman, D. R., Moore, J. P., Sattentau, Q. J., Schuitemaker, H., Sodroski, J. and Weiss, R. A. (1998). A new classification for HIV-1 [letter]. *Nature* 391:240.

Berger, E. A., Murphy, P. M. and Farber, J. M. (1999). Chemokine receptors as HIV-1 coreceptors: roles in viral entry, tropism, and disease. *Annu. Rev. Immunol.* 17:657-700.

Biswas, P., Poli, G., Kinter, A. L., Justement, J. S., Stanley, S. K., Maury, W. J., Bressler, P., Orenstein, J. M. and Fauci, A. S. (1992). Interferon gamma induces the expression of human immunodeficiency virus in persistently infected promonocytic cells (U1) and redirects the production of virions to intracytoplasmic vacuoles in phorbol myristate acetate-differentiated U1 cells. *J. Exp. Med.* 176:739-750.

Biswas, P., Poli, G., Orenstein, J. M. and Fauci, A. S. (1994). Cytokine-mediated induction of human immunodeficiency virus (HIV) expression and cell death in chronically infected U1 cells: do tumor necrosis factor alpha and gamma interferon selectively kill HIV- infected cells? *J. Virol.* 68:2598-2604.

Biti, R., Ffrench, R., Young, J., Bennetts, B., Stewart, G. and Liang, T. (1997). HIV-1 infection in an individual homozygous for the CCR5 deletion allele. *Nature* Med. 3:252-253.

Bleul, C. C., Farzan, M., Choe, H., Parolin, C., Clark-Lewis, I., Sodroski, J. and Springer, T. A. (1996). The lymphocyte chemoattractant SDF-1 is a ligand for LESTR/fusin and blocks HIV-1 entry. *Nature* 382:829-833.

Bobardt, M. D., Salmon, P., Wang, L., Esko, J. D., Gabuzda, D., Fiala, M., Trono, D., Van der Schueren, B., David, G. and Gallay, P. A. (2004). Contribution of proteoglycans to human immunodeficiency virus type 1 brain invasion. *J. Virol.* 78:6567-6584.

Bonyhadi, M. L., Su, L., Auten, J., McCune, J. M. and Kaneshima, H. (1995). Development of a human thymic organ culture model for the study of HIV pathogenesis. *AIDS Res. Hum. Retroviruses* 11:1073-1080.

Borghi, P., Fantuzzi, L., Varano, B., Gessani, S., Puddu, P., Conti, L., Capobianchi, M. R., Ameglio, F. and Belardelli, F. (1995). Induction of interleukin-10 by human immunodeficiency virus type 1 and its gp120 protein in human monocytes/macrophages. *J. Virol.* 69:1284-1287.

Bovolenta, C., Lorini, A. L., Mantelli, B., Camorali, L., Novelli, F., Biswas, P. and Poli, G. (1999). A selective defect of IFN-gamma- but not of IFN-alpha-induced JAK/STAT pathway in a subset of U937 clones prevents the antiretroviral effect of IFN-gamma against HIV-1. *J. Immunol.* 162:323-330.

Braaten, D. and Luban, J. (2001). Cyclophilin A regulates HIV-1 infectivity, as demonstrated by gene targeting in human T cells. *EMBO J.* 20:1300-1309.

Brack-Werner, R. (1999). Astrocytes: HIV cellular reservoirs and important participants in neuropathogenesis [editorial]. *AIDS* 13:1-22.

Brenchley, J. M., Schacker, T. W., Ruff, L. E., Price, D. A., Taylor, J. H., Beilman, G. J., Nguyen, P. L., Khoruts, A., Larson, M., Haase, A. T. and Douek, D. C. (2004). CD4[+] T cell depletion during all stages of HIV disease occurs predominantly in the gastrointestinal tract. *J. Exp. Med.* 200:749-759.

Broor, S., Kusari, A. B., Zhang, B., Seth, P., Richman, D. D., Carson, D. A., Wachsman, W. and Lotz, M. (1994). Stimulation of HIV replication in mononuclear phagocytes by leukemia inhibitory factor. *J. Acquir. Immune Defic. Syndr.* 7:647-654.

Buhl, R., Jaffe, H. A., Holroyd, K. J., Borok, Z., Roum, J. H., Mastrangeli, A., Wells, F. B., Kirby, M., Saltini, C. and Crystal, R. G. (1993). Activation of alveolar macrophages in asymptomatic HIV-infected individuals. *J. Immunol.* 150:1019-1028.

Capobianchi, M. R., Mattana, P., Mercuri, F., Conciatori, G., Ameglio, F., Ankel, H. and Dianzani, F. (1992). Acid lability is not an intrinsic property of interferon-alpha induced by HIV-infected cells. *J. Interferon Res.* 12:431-438.

Capobianchi, M. R., Barresi, C., Borghi, P., Gessani, S., Fantuzzi, L., Ameglio, F., Belardelli, F., Papadia, S. and Dianzani, F. (1997). Human immunodeficiency virus type 1 gp120 stimulates cytomegalovirus replication in monocytes: possible role of endogenous interleukin-8. *J. Virol.* 71:1591-1597.

Carcelain, G., Saint-Mezard, P., Altes, H. K., Tubiana, R., Grenot, P., Rabian, C., de Boer, R., Costagliola, D., Katlama, C., Debre, P. and Autran, B. (2003).

IL-2 therapy and thymic production of naive CD4 T cells in HIV-infected patients with severe CD4 lymphopenia. *AIDS* 17:841-850.

Casoli, C., Vicenzi, E., Cimarelli, A., Magnani, G., Ciancianaini, P., Cattaneo, E., Dall'Aglio, P., Poli, G. and Bertazzoni, U. (2000). HTLV-II down-regulates HIV-1 replication in IL-2-stimulated primary PBMC of coinfected individuals through expression of MIP-1alpha. *Blood* 95:2760-2769.

Chene, L., Nugeyre, M. T., Barre-Sinoussi, F. and Israel, N. (1999). High-level replication of human immunodeficiency virus in thymocytes requires NF-kappaB activation through interaction with thymic epithelial cells. *J. Virol.* 73:2064-2073.

Cheret, A., Le Grand, R., Caufour, P., Neildez, O., Matheux, F., Theodoro, F., Vaslin, B. and Dormont, D. (1999). RANTES, IFN-gamma, CCR1, and CCR5 mRNA expression in peripheral blood, lymph node, and bronchoalveolar lavage mononuclear cells during primary simian immunodeficiency virus infection of macaques. *Virology* 255:285-293.

Chiappini, E., Galli, L., Azzari, C. and de Martino, M. (2003). Interleukin-7 and immunologic failure despite treatment with highly active antiretroviral therapy in children perinatally infected with HIV-1. *J. Acquir. Immune Defic. Syndr.* 33:601-604.

Chiu, Y. L., Soros, V. B., Kreisberg, J. F., Stopak, K., Yonemoto, W. and Greene, W. C. (2005). Cellular APOBEC3G restricts HIV-1 infection in resting CD4[+] T cells. *Nature* 435:108-114.

Chun, T. W., Carruth, L., Finzi, D., Shen, X., DiGiuseppe, J. A., Taylor, H., Hermankova, M., Chadwick, K., Margolick, J., Quinn, T. C., Kuo, Y. H., Brookmeyer, R., Zeiger, M. A., Barditch-Crovo, P. and Siliciano, R. F. (1997). Quantification of latent tissue reservoirs and total body viral load in HIV-1 infection [see comments]. *Nature* 387:183-188.

Chun, T. W., Engel, D., Mizell, S. B., Ehler, L. A. and Fauci, A. S. (1998). Induction of HIV-1 replication in latently infected CD4[+] T cells using a combination of cytokines. *J. Exp. Med.* 188:83-91.

Chun, T. W., Engel, D., Mizell, S. B., Hallahan, C. W., Fischette, M., Park, S., Davey, R. T., Jr., Dybul, M., Kovacs, J. A., Metcalf, J. A., Mican, J. M., Berrey, M. M., Corey, L., Lane, H. C. and Fauci, A. S. (1999). Effect of interleukin-2 on the pool of latently infected, resting CD4[+] T cells in HIV-1-infected patients receiving highly active anti- retroviral therapy. *Nat. Med.* 5:651-655.

Cinque, P., Scarpellini, P., Vago, L., Linde, A. and Lazzarin, A. (1997). Diagnosis of central nervous system complications in HIV-infected patients – cerebrospinal fluid analysis by the polymerase chain reaction. *AIDS* 11:1-17.

Cinque, P., Vago, L., Mengozzi, M., Torri, V., Ceresa, D., Vicenzi, E., Transidico, P., Vagani, A., Sozzani, S., Mantovani, A., Lazzarin, A. and Poli, G. (1998). Elevated cerebrospinal fluid levels of monocyte chemotactic protein-1 correlate with HIV-1 encephalitis and local viral replication. *AIDS* 12:1327-1332.

Clerici, M., Butto, S., Lukwiya, M., Saresella, M., Declich, S., Trabattoni, D., Pastori, C., Piconi, S., Fracasso, C., Fabiani, M., Ferrante, P., Rizzardini, G. and Lopalco, L. (2000). Immune activation in africa is environmentally-driven and is associated with upregulation of CCR5. Italian-Ugandan AIDS Project. *AIDS* 14:2083-2092.

Cocchi, F., DeVico, A. L., Garzino-Demo, A., Arya, S. K., Gallo, R. C. and Lusso, P. (1995). Identification of RANTES, MIP-1 alpha, and MIP-1 beta as the major HIV- suppressive factors produced by $CD8^+$ T cells [see comments]. *Science* 270:1811-1815.

Colgan, J., Asmal, M., Yu, B. and Luban, J. (2005). Cyclophilin A-deficient mice are resistant to immunosuppression by cyclosporine. *J. Immunol.* 174: 6030-6038.

Conant, K., Garzino-Demo, A., Nath, A., McArthur, J. C., Halliday, W., Power, C., Gallo, R. C. and Major, E. O. (1998). Induction of monocyte chemoattractant protein-1 in HIV-1 Tat-stimulated astrocytes and elevation in AIDS dementia. *Proc. Natl. Acad. Sci. U. S. A.* 95:3117-3121.

Connor, R. I., Sheridan, K. E., Ceradini, D., Choe, S. and Landau, N. R. (1997). Change in coreceptor use coreceptor use correlates with disease progression in HIV-1–infected individuals. *J. Exp. Med.* 185:621-628.

Cota, M., Mengozzi, M., Vicenzi, E., Panina-Bordignon, P., Sinigaglia, F., Transidico, P., Sozzani, S., Mantovani, A. and Poli, G. (2000). Selective inhibition of HIV replication in primary macrophages but not T lymphocytes by macrophage-derived chemokine. *Proc. Natl. Acad. Sci. U. S. A.* 97:9162-9167.

Cullen, B. R. (1995). Regulation of HIV gene expression. *Aids* 9 Suppl A:S19-32.

Dalgleish, A. G., Beverley, P. C., Clapham, P. R., Crawford, D. H., Greaves, M. F. and Weiss, R. A. (1984). The CD4 (T4) antigen is an essential component of the receptor for the AIDS retrovirus. *Nature* 312:763-767.

De Maria, A., Pantaleo, G., Schnittman, S. M., Greenhouse, J. J., Baseler, M., Orenstein, J. M. and Fauci, A. S. (1991). Infection of $CD8^+$ T lymphocytes with HIV. Requirement for interaction with infected $CD4^+$ cells and induction of infectious virus from chronically infected $CD8^+$ cells. *J. Immunol.* 146:2220-2226.

de Paulis, A., Florio, G., Prevete, N., Triggiani, M., Fiorentino, I., Genovese, A. and Marone, G. (2002). HIV-1 envelope gp41 peptides promote migration of human Fc epsilon RI^+ cells and inhibit IL-13 synthesis through interaction with formyl peptide receptors. *J. Immunol.* 169:4559-4567.

Del Corno, M., Liu, Q. H., Schols, D., de Clercq, E., Gessani, S., Freedman, B. D. and Collman, R. G. (2001). HIV-1 gp120 and chemokine activation of Pyk2 and mitogen-activated protein kinases in primary macrophages mediated by calcium-dependent, pertussis toxin-insensitive chemokine receptor signaling. *Blood* 98:2909-2916.

Devergne, O., Peuchmaur, M., Humbert, M., Navratil, E., Leger-Ravet, M. B., Crevon, M. C., Petit, M. A., Galanaud, P. and Emilie, D. (1991). *In vivo* expression of IL-1 beta and IL-6 genes during viral infections in human. *Eur. Cytokine Netw.* 2:183-194.

Doehle, B. P., Schafer, A. and Cullen, B. R. (2005). Human APOBEC3B is a potent inhibitor of HIV-1 infectivity and is resistant to HIV-1 *Vif. Virology* 339:281-288.

Dolei, A., Biolchini, A., Serra, C., Curreli, S., Gomes, E. and Dianzani, F. (1998). Increased replication of T-cell-tropic HIV strains and CXC-chemokine receptor-4 induction in T cells treated with macrophage inflammatory protein (MIP)-1alpha, MIP-1beta and RANTES beta-chemokines. *AIDS* 12:183-190.

Doranz, B. J., Rucker, J., Yi, Y., Smyth, R. J., Samson, M., Peiper, S. C., Parmentier, M., Collman, R. G. and Doms, R. W. (1996). A dual-tropic primary HIV-1 isolate that uses fusin and the beta- chemokine receptors CKR-5, CKR-3, and CKR-2b as fusion cofactors. *Cell* 85:1149-1158.

Ducrey-Rundquist, O., Guyader, M. and Trono, D. (2002). Modalities of interleukin-7-induced human immunodeficiency virus permissiveness in quiescent T lymphocytes. *J. Virol.* 76:9103-9111.

Eckstein, D. A., Penn, M. L., Korin, Y. D., Scripture-Adams, D. D., Zack, J. A., Kreisberg, J. F., Roederer, M., Sherman, M. P., Chin, P. S. and Goldsmith, M. A. (2001). HIV-1 actively replicates in naive CD4($^+$) T cells residing within human lymphoid tissues. *Immunity* 15:671-682.

Emilie, D., Peuchmaur, M., Maillot, M. C., Crevon, M. C., Brousse, N., Delfraissy, J. F., Dormont, J. and Galanaud, P. (1990). Production of interleukins in human immunodeficiency virus-1- replicating lymph nodes. *J. Clin. Invest.* 86:148-159.

Fais, S., Lapenta, C., Santini, S. M., Spada, M., Parlato, S., Logozzi, M., Rizza, P. and Belardelli, F. (1999). Human immunodeficiency virus type 1 strains R5 and X4 induce different pathogenic effects in hu-PBL-SCID mice, depending on the state of activation/differentiation of human target cells at the time of primary infection. *J. Virol.* 73:6453-6459.

Falciola, L., Spada, F., Calogero, S., Langst, G., Voit, R., Grummt, I. and Bianchi, M. E. (1997). High mobility group 1 protein is not stably associated with the chromosomes of somatic cells. *J. Cell. Biol.* 137:19-26.

Fantuzzi, L., Canini, I., Belardelli, F. and Gessani, S. (2001). HIV-1 gp120 stimulates the production of beta-chemokines in human peripheral blood monocytes through a CD4-independent mechanism. *J. Immunol.* 166: 5381-5387.

Fantuzzi, L., Spadaro, F., Vallanti, G., Canini, I., Ramoni, C., Vicenzi, E., Belardelli, F., Poli, G. and Gessani, S. (2003). Endogenous CCL2 (monocyte chemotactic protein-1) modulates human immunodeficiency virus type-1 replication and affects cytoskeleton organization in human monocyte-derived macrophages. *Blood* 102:2334-2337.

Feng, Y., Broder, C. C., Kennedy, P. E. and Berger, E. A. (1996). HIV-1 entry cofactor: functional cDNA cloning of a seven-transmembrane, G protein-coupled receptor [see comments]. *Science* 272:872-877.

Finkel, T. H., Tudor-Williams, G., Banda, N. K., Cotton, M. F., Curiel, T., Monks, C., Baba, T. W., Ruprecht, R. M. and Kupfer, A. (1995). Apoptosis occurs predominantly in bystander cells and not in productively infected cells of HIV- and SIV-infected lymph nodes. *Nat. Med.* 1:129-134.

Finzi, D., Hermankova, M., Pierson, T., Carruth, L. M., Buck, C., Chaisson, R. E., Quinn, T. C., Chadwick, K., Margolick, J., Brookmeyer, R., Gallant, J., Markowitz, M., Ho, D. D., Richman, D. D. and Siliciano, R. F. (1997).

Identification of a reservoir for HIV-1 in patients on highly active antiretroviral therapy [see comments]. *Science* 278:1295-1300.

Folks, T., Benn, S., Rabson, A., Theodore, T., Hoggan, M. D., Martin, M., Lightfoote, M. and Sell, K. (1985). Characterization of a continuous T-cell line susceptible to the cytopathic effects of the acquired immunodeficiency syndrome (AIDS)-associated retrovirus. *Proc. Natl. Acad. Sci. U. S. A.* 82:4539-4543.

Folks, T., Powell, D. M., Lightfoote, M. M., Benn, S., Martin, M. A. and Fauci, A. S. (1986a). Induction of HTLV-III/LAV from a nonvirus-producing T-cell line: implications for latency. *Science* 231:600-602.

Folks, T. M., Powell, D., Lightfoote, M., Koenig, S., Fauci, A. S., Benn, S., Rabson, A., Daugherty, D., Gendelman, H. E., Hoggan, M. D., Venkatesan, S., and Martin M.A. (1986b). Biological and biochemical characterization of a cloned Leu-3- cell surviving infection with the acquired immune deficiency syndrome retrovirus. *J. Exp. Med.* 164:280-290.

Folks, T. M., Justement, J., Kinter, A., Dinarello, C. A. and Fauci, A. S. (1987). Cytokine-induced expression of HIV-1 in a chronically infected promonocyte cell line. *Science* 238:800-802.

Folks, T. M., Justement, J., Kinter, A., Schnittman, S., Orenstein, J., Poli, G. and Fauci, A. S. (1988a). Characterization of a promonocyte clone chronically infected with HIV and inducible by 13-phorbol-12-myristate acetate. *J. Immunol.* 140:1117-1122.

Folks, T. M., Kessler, S. W., Orenstein, J. M., Justement, J. S., Jaffe, E. S. and Fauci, A. S. (1988b). Infection and replication of HIV-1 in purified progenitor cells of normal human bone marrow. *Science* 242:919-922.

Freed, E. O. (2002). Viral late domains. *J. Virol.* 76:4679-4687.

Freed, E. O. (2003). The HIV-TSG101 interface: recent advances in a budding field. *Trends Microbiol.* 11:56-59.

Fuchs, D., Hansen, A., Reibnegger, G., Werner, E. R., Dierich, M. P. and Watcher, H. (1988). Neopterin as a marker for activated cell-mediated immunity: application in HIV infection. *Immunol. Today* 9:150-155.

Fulcher, J. A., Hwangbo, Y., Zioni, R., Nickle, D., Lin, X., Heath, L., Mullins, J. I., Corey, L. and Zhu, T. (2004). Compartmentalization of human immunodeficiency virus type 1 between blood monocytes and CD4[+] T cells during infection. *J. Virol.* 78:7883-7893.

Galli, G., Annunziato, F., Mavilia, C., Romagnani, P., Cosmi, L., Manetti, R., Pupilli, C., Maggi, E. and Romagnani, S. (1998). Enhanced HIV expression during Th2-oriented responses explained by the opposite regulatory effect of IL-4 and IFN-gamma of fusin/CXCR4. *Eur. J. Immunol.* 28:3280-3290.

Ganesh, L., Burstein, E., Guha-Niyogi, A., Louder, M. K., Mascola, J. R., Klomp, L. W., Wijmenga, C., Duckett, C. S. and Nabel, G. J. (2003). The gene product Murr1 restricts HIV-1 replication in resting CD4[+] lymphocytes. *Nature* 426:853-857.

Garba, M. L., Pilcher, C. D., Bingham, A. L., Eron, J. and Frelinger, J. A. (2002). HIV antigens can induce TGF-beta(1)-producing immunoregulatory CD8[+] T cells. *J. Immunol.* 168:2247-2254.

Gaulton, G. N., Scobie, J. V. and Rosenzweig, M. (1997). HIV-1 and the thymus. *AIDS* 11:403-414.

Gendelman, H. E., Orenstein, J. M., Martin, M. A., Ferrua, C., Mitra, R., Phipps, T., Wahl, L. A., Lane, H. C., Fauci, A. S., Burke, D. S., Skillman D., and Meltzer M.S. (1988). Efficient isolation and propagation of human immunodeficiency virus on recombinant colony-stimulating factor 1-treated monocytes. *J. Exp. Med.* 167:1428-1441.

Gessani, S., Puddu, P., Varano, B., Borghi, P., Conti, L., Fantuzzi, L. and Belardelli, F. (1994). Induction of beta interferon by human immunodeficiency virus type 1 and its gp120 protein in human monocytes-macrophages: role of beta interferon in restriction of virus replication. *J. Virol.* 68:1983-1986.

Ghezzi, S., Noonan, D. M., Aluigi, M. G., Vallanti, G., Cota, M., Benelli, R., Morini, M., Reeves, J. D., Vicenzi, E., Poli, G. and Albini, A. (2000). Inhibition of CXCR4-dependent HIV-1 infection by extracellular HIV-1 Tat. *Biochem. Biophys. Res. Commun.* 270:992-996.

Ghezzi, S., Menzo, S., Brambilla, A., Bordignon, P. P., Lorini, A. L., Clementi, M., Poli, G. and Vicenzi, E. (2001). Inhibition of R5X4 dualtropic HIV-1 primary isolates by single chemokine co-receptor ligands. *Virology* 280: 253-261.

Gibb, D. M., Newberry, A., Klein, N., de Rossi, A., Grosch-Woerner, I. and Babiker, A. (2000). Immune repopulation after HAART in previously untreated HIV-1-infected children. Paediatric European Network for Treatment of AIDS (PENTA) Steering Committee. *Lancet* 355:1331-1332.

Glushakova, S., Baibakov, B., Margolis, L. B. and Zimmerberg, J. (1995). Infection of human tonsil histocultures: a model for HIV pathogenesis. *Nat. Med.* 1:1320-1322.

Glushakova, S., Grivel, J. C., Fitzgerald, W., Sylwester, A., Zimmerberg, J. and Margolis, L. B. (1998). Evidence for the HIV-1 phenotype switch as a causal factor in acquired immunodeficiency. *Nat. Med.* 4:346-349.

Glushakova, S., Yi, Y., Grivel, J. C., Singh, A., Schols, D., De Clercq, E., Collman, R. G. and Margolis, L. (1999). Preferential coreceptor utilization and cytopathicity by dual-tropic HIV-1 in human lymphoid tissue *ex vivo* [see comments]. *J. Clin. Invest.* 104:R7-R11.

Gomez, C. and Hope, T. J. (2005). The ins and outs of HIV replication. *Cell. Microbiol.* 7:621-626.

Gonzalez, E., Kulkarni, H., Bolivar, H., Mangano, A., Sanchez, R., Catano, G., Nibbs, R. J., Freedman, B. I., Quinones, M. P., Bamshad, M. J., Murthy, K. K., Rovin, B. H., Bradley, W., Clark, R. A., Anderson, S. A., O'Connell. R. J., Agan, B. K., Ahuja, S. S., Bologna, R., Sen, L., Dolan, M. J. and Ahuja, S. K. (2005). The influence of CCL3L1 Gene-containing segmental duplications on HIV-1/AIDS susceptibility. *Science* 307:1434-1440.

Gould, S. J., Booth, A. M. and Hildreth, J. E. (2003). The Trojan exosome hypothesis. *Proc. Natl. Acad. Sci. U. S. A.* 100:10592-10597.

Granowitz, E. V., Saget, B. M., Wang, M. Z., Dinarello, C. A. and Skolnik, P. R. (1995). Interleukin 1 induces HIV-1 expression in chronically infected U1 cells: blockade by interleukin 1 receptor antagonist and tumor necrosis factor binding protein type 1. *Mol. Med.* 1:667-677.

Granucci, F., Zanoni, I., Feau, S. and Ricciardi-Castagnoli, P. (2003). Dendritic cell regulation of immune responses: a new role for interleukin 2 at the intersection of innate and adaptive immunity. *EMBO J.* 22:2546-2451.

Gray, F., Scaravilli, F., Everall, I., Chretien, F., An, S., Boche, D., Adle-Biassette, H., Wingertsmann, L., Durigon, M., Hurtrel, B., Chiodi, F., Bell, J. and Lantos, P. (1996). Neuropathology of early HIV-1 infection. *Brain. Pathol.* 6:1-15.

Graziosi, C., Gantt, K. R., Vaccarezza, M., Demarest, J. F., Daucher, M., Saag, M. S., Shaw, G. M., Quinn, T. C., Cohen, O. J., Welbon, C. C., Pantaleo, G. and Fauci, A. S. (1996). Kinetics of cytokine expression during primary human immunodeficiency virus type 1 infection. *Proc. Natl. Acad. Sci. U. S. A.* 93:4386-4391.

Griffin, G. E., Leung, K., Folks, T. M., Kunkel, S. and Nabel, G. J. (1989). Activation of HIV gene expression during monocyte differentiation by induction of NF-kappa B. *Nature* 339:70-73.

Grivel, J. C. and Margolis, L. B. (1999). CCR5- and CXCR4-tropic HIV-1 are equally cytopathic for their T-cell targets in human lymphoid tissue. *Nat. Med.* 5:344-346.

Grivel, J. C., Penn, M. L., Eckstein, D. A., Schramm, B., Speck, R. F., Abbey, N. W., Herndier, B., Margolis, L. and Goldsmith, M. A. (2000). Human immunodeficiency virus type 1 coreceptor preferences determine target T-cell depletion and cellular tropism in human lymphoid tissue. *J. Virol.* 74: 5347-5351.

Guadalupe, M., Reay, E., Sankaran, S., Prindiville, T., Flamm, J., McNeil, A. and Dandekar, S. (2003). Severe CD4$^+$ T-cell depletion in gut lymphoid tissue during primary human immunodeficiency virus type 1 infection and substantial delay in restoration following highly active antiretroviral therapy. *J. Virol.* 77:11708-11717.

Haas, D. W., Lavelle, J., Nadler, J. P., Greenberg, S. B., Frame, P., Mustafa, N., St. Clair, M., McKinnis, R., Dix, L., Elkins, M. and Rooney, J. (2000). A randomized trial of interferon alpha therapy for HIV type 1 infection. *AIDS Res. Hum. Retroviruses* 16:183-190.

Hatziioannou, T., Perez-Caballero, D., Yang, A., Cowan, S. and Bieniasz, P. D. (2004). Retrovirus resistance factors Ref1 and Lv1 are species-specific variants of TRIM5alpha. *Proc. Natl. Acad. Sci. U. S. A.* 101:10774-10779.

Hays, E. F., Uittenbogaart, C. H., Brewer, J. C., Vollger, L. W. and Zack, J. A. (1992). *In vitro* studies of HIV-1 expression in thymocytes from infants and children. *AIDS* 6:265-272.

Hazenberg, M. D., Otto, S. A., Cohen Stuart, J. W., Verschuren, M. C., Borleffs, J. C., Boucher, C. A., Coutinho, R. A., Lange, J. M., Rinke de Wit, T. F., Tsegaye, A., van Dongen, J. J., Hamann, D., de Boer, R. J. and Miedema, F. (2000). Increased cell division but not thymic dysfunction rapidly affects the T-cell receptor excision circle content of the naive T cell population in HIV-1 infection. *Nat. Med.* 6:1036-1042.

Herbein, G., Montaner, L. J. and Gordon, S. (1996). Tumor necrosis factor alpha inhibits entry of human immunodeficiency virus type 1 into primary human macrophages: a selective role for the 75-kilodalton receptor [published erratum appears in J Virol 1997 Mar;71(3):1581]. *J. Virol.* 70:7388-7397.

Herbein, G., Mahlknecht, U., Batliwalla, F., Gregersen, P., Pappas, T., Butler, J., O'Brien, W. A. and Verdin, E. (1998). Apoptosis of CD8[+] T cells is mediated by macrophages through interaction of HIV gp120 with chemokine receptor CXCR4 [see comments]. *Nature* 395:189-194.

Hildreth, J. E. and Orentas, R. J. (1989). Involvement of a leukocyte adhesion receptor (LFA-1) in HIV-induced syncytium formation. *Science* 244: 1075-1078.

Hlavacek, W. S., Stilianakis, N. I. and Perelson, A. S. (2000). Influence of follicular dendritic cells on HIV dynamics. *Philos. Trans. R. Soc. Lond. B Biol. Sci.* 355:1051-1058.

Ho, D. D. (1995). Time to hit HIV, early and hard. *N. Engl. J. Med.* 333:450-451.

Ho, D. D., Neumann, A. U., Perelson, A. S., Chen, W., Leonard, J. M. and Markowitz, M. (1995). Rapid turnover of plasma virions and CD4 lymphocytes in HIV-1 infection. *Nature* 373:123-126.

Hoxie, J. A., Alpers, J. D., Rackowski, J. L., Huebner, K., Haggarty, B. S., Cedarbaum, A. J. and Reed, J. C. (1986). Alterations in T4 (CD4) protein and mRNA synthesis in cells infected with HIV. *Science* 234:1123-1127.

Hu, R., Oyaizu, N., Than, S., Kalyanaraman, V. S., Wang, X. P. and Pahwa, S. (1996). HIV-1 gp160 induces transforming growth factor-beta production in human PBMC. *Clin. Immunol. Immunopathol.* 80:283-289.

Huang, Y., Paxton, W. A., Wolinsky, S. M., Neumann, A. U., Zhang, L., He, T., Kang, S., Ceradini, D., Jin, Z., Yazdanbakhsh, K., Kunstman, K., Erickson, D., Dragon, E., Landau, N. R., Phair, J., Ho, D. D. and Koup, R. A. (1996). The role of a mutant CCR5 allele in HIV-1 transmission and disease progression [see comments]. *Nat. Med.* 2:1240-1243.

Huigen, M. C., Kamp, W. and Nottet, H. S. (2004). Multiple effects of HIV-1 trans-activator protein on the pathogenesis of HIV-1 infection. *Eur. J. Clin. Invest.* 34:57-66.

Johnson, M. D. and Gold, L. I. (1996). Distribution of transforming growth factor-beta isoforms in human immunodeficiency virus-1 encephalitis. *Hum. Pathol.* 27:643-649.

Jourdan, P., Abbal, C., Nora, N., Hori, T., Uchiyama, T., Vendrell, J.-P., Bousquet, J., Taylor, N., Pène, J. and Yssel, H. (1998). Cutting edge: IL-4 induces functional cell-surface expression of CXCR4 on human T cells. *J. Immunol.* 160:4153-4157.

Kalinkovich, A., Borkow, G., Weisman, Z., Tsimanis, A., Stein, M. and Bentwich, Z. (2001). Increased CCR5 and CXCR4 expression in Ethiopians living in Israel: environmental and constitutive factors. *Clin. Immunol.* 100:107-117.

Kandanearatchi, A., Williams, B. and Everall, I. P. (2003). Assessing the efficacy of highly active antiretroviral therapy in the brain. *Brain. Pathol.* 13:104-110.

Kaul, M., Garden, G. A. and Lipton, S. A. (2001). Pathways to neuronal injury and apoptosis in HIV-associated dementia. *Nature* 410:988-994.

Kazazi, F., Mathijs, J. M., Chang, J., Malafiej, P., Lopez, A., Dowton, D., Sorrell, T. C., Vadas, M. A. and Cunningham, A. L. (1992). Recombinant interleukin 4 stimulates human immunodeficiency virus production by infected monocytes and macrophages. *J. Gen. Virol.* 73:941-949.

Keckesova, Z., Ylinen, L. M. and Towers, G. J. (2004). The human and African green monkey TRIM5alpha genes encode Ref1 and Lv1 retroviral restriction factor activities. *Proc. Natl. Acad. Sci. U. S. A.* 101:10780-10785.

Kekow, J., Wachsman, W., McCutchan, J. A., Cronin, M., Carson, D. A. and Lotz, M. (1990). Transforming growth factor beta and noncytopathic mechanisms of immunodeficiency in human immunodeficiency virus infection. *Proc. Natl. Acad. Sci. U. S. A.* 87:8321-8325.

Kekow, J., Wachsman, W., McCutchan, J. A., Gross, W. L., Zachariah, M., Carson, D. A. and Lotz, M. (1991). Transforming growth factor-beta and suppression of humoral immune responses in HIV infection. *J. Clin. Invest.* 87:1010-1016.

Kelder, W., McArthur, J. C., Nance-Sproson, T., McClernon, D. and Griffin, D. E. (1998). Beta-chemokines MCP-1 and RANTES are selectively increased in cerebrospinal fluid of patients with human immunodeficiency virus-associated dementia. *Ann. Neurol.* 44:831-835.

Khanna, K. V., Yu, X. F., Ford, D. H., Ratner, L., Hildreth, J. K. and Markham, R. B. (2000). Differences among HIV-1 variants in their ability to elicit secretion of TNF-alpha. *J. Immunol.* 164:1408-1415.

Khatissian, E., Chakrabarti, L. and Hurtrel, B. (1996). Cytokine patterns and viral load in lymph nodes during the early stages of SIV infection. *Res. Virol.* 147:181-189.

Kinter, A. L., Bende, S. M., Hardy, E. C., Jackson, R. and Fauci, A. S. (1995a). Interleukin 2 induces CD8[+] T cell-mediated suppression of human immunodeficiency virus replication in CD4[+] T cells and this effect overrides its ability to stimulate virus expression. *Proc. Natl. Acad. Sci. U. S. A.* 92:10985-10989.

Kinter, A. L., Poli, G., Fox, L., Hardy, E. and Fauci, A. S. (1995b). HIV replication in IL-2-stimulated peripheral blood mononuclear cells is driven in an autocrine/paracrine manner by endogenous cytokines. *J. Immunol.* 154:2448-2459.

Kinter, A. L., Ostrowski, M., Goletti, D., Oliva, A., Weissman, D., Gantt, K., Hardy, E., Jackson, R., Ehler, L. and Fauci, A. S. (1996). HIV replication in CD4[+] T cells of HIV-infected individuals is regulated by a balance between the viral suppressive effects of endogenous beta-chemokines and the viral inductive effects of other endogenous cytokines. *Proc. Natl. Acad. Sci. U. S. A.* 93:14076-14081.

Kinter, A., Catanzaro, A., Monaco, J., Ruiz, M., Justement, J., Moir, S., Arthos, J., Oliva, A., Ehler, L., Mizell, S., Jackson, R., Ostrowski, M., Hoxie, J., Offord, R. and Fauci, A. S. (1998). CC-chemokines enhance the replication of T-tropic strains of HIV-1 in CD4([+]) T cells: role of signal transduction. *Proc. Natl. Acad. Sci. U. S. A.* 95:11880-11885.

Klein, S. A., Dobmeyer, J. M., Dobmeyer, T. S., Pape, M., Ottmann, O. G., Helm, E. B., Hoelzer, D. and Rossol, R. (1997). Demonstration of the Th1 to Th2 cytokine shift during the course of HIV-1 infection using cytoplasmic cytokine detection on single cell level by flow cytometry. *AIDS* 11: 1111-1118.

Klimkait, T., Strebel, K., Hoggan, M. D., Martin, M. A. and Orenstein, J. M. (1990). The human immunodeficiency virus type 1-specific protein vpu is required for efficient virus maturation and release. *J. Virol.* 64:621-629.

Koenig, S., Gendelman, H. E., Orenstein, J. M., Dal Canto, M. C., Pezeshkpour, G. H., Yungbluth, M., Janotta, F., Aksamit, A., Martin, M. A. and Fauci, A. S. (1986). Detection of AIDS virus in macrophages in brain tissue from AIDS patients with encephalopathy. *Science* 223:1089-1093.

Koot, M., Vos, A. H., Keet, R. P., de Goede, R. E., Dercksen, M. W., Terpstra, F. G., Coutinho, R. A., Miedema, F. and Tersmette, M. (1992). HIV-1 biological phenotype in long-term infected individuals evaluated with an MT-2 cocultivation assay. *AIDS* 6:49-54.

Koot, M., Keet, I. P., Vos, A. H., de Goede, R. E., Roos, M. T., Coutinho, R. A., Miedema, F., Schellekens, P. T. and Tersmette, M. (1993). Prognostic value of HIV-1 syncytium-inducing phenotype for rate of CD4$^+$ cell depletion and progression to AIDS [see comments]. *Ann. Intern. Med.* 118:681-688.

Koot, M., van Leeuwen, R., de Goede, R. E., Keet, I. P., Danner, S., Eeftinck Schattenkerk, J. K., Reiss, P., Tersmette, M., Lange, J. M. and Schuitemaker, H. (1999). Conversion rate towards a syncytium-inducing (SI) phenotype during different stages of human immunodeficiency virus type 1 infection and prognostic value of SI phenotype for survival after AIDS diagnosis. *J. Infect. Dis.* 179:254-258.

Kovacs, J. A., Baseler, M., Dewar, R. J., Vogel, S., Davey, R. T., Jr., Falloon, J., Polis, M. A., Walker, R. E., Stevens, R. and Salzman, N. P. (1995). Increases in CD4 T lymphocytes with intermittent courses of interleukin-2 in patients with human immunodeficiency virus infection. A preliminary study. *N. Engl. J. Med.* 332:567-575.

Kovacs, J. A., Vogel, S., Albert, J. M., Falloon, J., Davey, R. T., Jr., Walker, R. E., Polis, M. A., Spooner, K., Metcalf, J. A., Baseler, M., Fyfe, G. and Lane, H. C. (1996). Controlled trial of interleukin-2 infusions in patients infected with the human immunodeficiency virus [see comments]. *N. Engl. J. Med.* 335:1350-1356.

Kovacs, J. A., Imamichi, H., Vogel, S., Metcalf, J. A., Dewar, R. L., Baseler, M., Stevens, R., Adelsberger, J., Lambert, L., Davey, R. T., Jr., Walker, R. E., Falloon, J., Polis, M. A., Masur, H. and Lane, H. C. (2000). Effects of intermittent interleukin-2 therapy on plasma and tissue human immunodeficiency virus levels and quasi-species expression. *J. Infect. Dis.* 182:1063-1069.

Koyanagi, Y., O'Brien, W. A., Zhao, J. Q., Golde, D. W., Gasson, J. C. and Chen, I. S. (1988). Cytokines alter production of HIV-1 from primary mononuclear phagocytes. *Science* 241:1673-1675.

Kramer-Hammerle, S., Rothenaigner, I., Wolff, H., Bell, J. E. and Brack-Werner, R. (2005). Cells of the central nervous system as targets and reservoirs of the human immunodeficiency virus. *Virus Res.* 111:194-213.

Kutsch, O., Oh, J., Nath, A. and Benveniste, E. N. (2000). Induction of the chemokines interleukin-8 and IP-10 by human immunodeficiency virus type 1 tat in astrocytes. *J. Virol.* 74:9214-9221.

Lane, H. C., Kovacs, J. A., Feinberg, J., Herpin, B., Davey, V., Walker, R., Deyton, L., Metcalf, J. A., Baseler, M., Salzman, N. and *et al.* (1988). Anti-retroviral effects of interferon-alpha in AIDS-associated Kaposi's sarcoma. *Lancet* 2:1218-1222.

Lane, B. R., Lore, K., Bock, P. J., Andersson, J., Coffey, M. J., Strieter, R. M. and Markovitz, D. M. (2001a). Interleukin-8 stimulates human immunodeficiency virus type 1 replication and is a potential new target for antiretroviral therapy. *J. Virol.* 75:8195-8202.

Lane, B. R., Strieter, R. M., Coffey, M. J. and Markovitz, D. M. (2001b). Human immunodeficiency virus type 1 (HIV-1)-induced GRO-alpha production stimulates HIV-1 replication in macrophages and T lymphocytes. *J. Virol.* 75:5812-5822.

Langford, T. D., Letendre, S. L., Larrea, G. J. and Masliah, E. (2003). Changing patterns in the neuropathogenesis of HIV during the HAART era. *Brain. Pathol.* 13:195-210.

Lapenta, C., Spada, M., Santini, S. M., Racca, S., Dorigatti, F., Poli, G., Belardelli, F. and Alfano, M. (2005). Pertussis toxin B-oligomer inhibits HIV infection and replication in hu-PBL-SCID mice. *Int. Immunol.* 17:469-475.

Lazdins, J. K., Klimkait, T., Woods-Cook, K., Walker, M., Alteri, E., Cox, D., Cerletti, N., Shipman, R., Bilbe, G. and McMaster, G. (1991). *In vitro* effect of transforming growth factor-β on progression of HIV-1 infection in primary mononuclear phagocytes. *J. Immunol.* 147:1201-1207.

Levy, Y., Durier, C., Krzysiek, R., Rabian, C., Capitant, C., Lascaux, A. S., Michon, C., Oksenhendler, E., Weiss, L., Gastaut, J. A., Goujard, C., Rouzioux, C., Maral, J., Delfraissy, J. F., Emilie, D. and Aboulker, J. P. (2003). Effects of interleukin-2 therapy combined with highly active antiretroviral therapy on immune restoration in HIV-1 infection: a randomized controlled trial. *AIDS* 17:343-351.

Li, B. Q., Wetzel, M. A., Mikovits, J. A., Henderson, E. E., Rogers, T. J., Gong, W., Le, Y., Ruscetti, F. W. and Wang, J. M. (2001). The synthetic peptide WKYMVm attenuates the function of the chemokine receptors CCR5 and CXCR4 through activation of formyl peptide receptor-like 1. *Blood* 97:2941-2947.

Li, Q., Duan, L., Estes, J. D., Ma, Z. M., Rourke, T., Wang, Y., Reilly, C., Carlis, J., Miller, C. J. and Haase, A. T. (2005). Peak SIV replication in resting memory CD4$^+$ T cells depletes gut lamina propria CD4$^+$ T cells. *Nature* 434:1148-1152.

Liu, R., Paxton, W. A., Choe, S., Ceradini, D., Martin, S. R., Horuk, R., MacDonald, M. E., Stuhlmann, H., Koup, R. A. and Landau, N. R. (1996). Homozygous defect in HIV-1 coreceptor accounts for resistance of some multiply-exposed individuals to HIV-1 infection. *Cell* 86:367-377.

Liu, C. C., Perussia, B. and Young, J. D. (2000a). The emerging role of IL-15 in NK-cell development. *Immunol. Today* 21:113-116.

Liu, Q. H., Williams, D. A., McManus, C., Baribaud, F., Doms, R. W., Schols, D., De Clercq, E., Kotlikoff, M. I., Collman, R. G. and Freedman, B. D. (2000b). HIV-1 gp120 and chemokines activate ion channels in primary macrophages through CCR5 and CXCR4 stimulation. *Proc. Natl. Acad. Sci. U. S. A.* 97:4832-4837.

Liu, N. Q., Lossinsky, A. S., Popik, W., Li, X., Gujuluva, C., Kriederman, B., Roberts, J., Pushkarsky, T., Bukrinsky, M., Witte, M., Weinand, M. and Fiala, M. (2002). Human immunodeficiency virus type 1 enters brain microvascular endothelia by macropinocytosis dependent on lipid rafts and the mitogen-activated protein kinase signaling pathway. *J. Virol.* 76:6689-6700.

Lopalco, L., Barassi, C., Pastori, C., Longhi, R., Burastero, S. E., Tambussi, G., Mazzotta, F., Lazzarin, A., Clerici, M. and Siccardi, A. G. (2000). CCR5-reactive antibodies in seronegative partners of HIV-seropositive individuals down-modulate surface CCR5 *in vivo* and neutralize the infectivity of R5 strains of HIV-1 *In vitro. J. Immunol.* 164:3426-3433.

Lores, P., Boucher, V., Mackay, C., Pla, M., Von Boehmer, H., Jami, J., Barre-Sinoussi, F. and Weill, J. C. (1992). Expression of human CD4 in transgenic mice does not confer sensitivity to human immunodeficiency virus infection. *AIDS Res. Hum. Retroviruses* 8:2063-2071.

Luban, J., Bossolt, K. L., Franke, E. K., Kalpana, G. V. and Goff, S. P. (1993). Human immunodeficiency virus type 1 Gag protein binds to cyclophilins A and B. *Cell* 73:1067-1078.

Malim, M. H., Hauber, J., Le, S. Y., Maizel, J. V. and Cullen, B. R. (1989). The HIV-1 rev trans-activator acts through a structured target sequence to activate nuclear export of unspliced viral mRNA. *Nature* 338:254-257.

Malim, M. H., Tiley, L. S., McCarn, D. F., Rusche, J. R., Hauber, J. and Cullen, B. R. (1990). HIV-1 structural gene expression requires binding of the Rev trans-activator to its RNA target sequence. *Cell* 60:675-683.

Mangeat, B., Turelli, P., Caron, G., Friedli, M., Perrin, L. and Trono, D. (2003). Broad antiretroviral defence by human APOBEC3G through lethal editing of nascent reverse transcripts. *Nature* 424:99-103.

Marchetti, G., Meroni, L., Molteni, C., Bandera, A., Franzetti, F., Galli, M., Moroni, M., Clerici, M. and Gori, A. (2004). Interleukin-2 immunotherapy exerts a differential effect on CD4 and CD8 T cell dynamics. *AIDS* 18:211-216.

Marshall, J. D., Chehimi, J., Gri, G., Kostman, J. R., Montaner, L. J. and Trinchieri, G. (1999). The interleukin-12-mediated pathway of immune events is dysfunctional in human immunodeficiency virus-infected individuals. *Blood* 94:1003-1011.

Mattapallil, J. J., Douek, D. C., Hill, B., Nishimura, Y., Martin, M. and Roederer, M. (2005). Massive infection and loss of memory CD4+ T cells in multiple tissues during acute SIV infection. *Nature* 434:1093-1097.

McArthur, J. C., Haughey, N., Gartner, S., Conant, K., Pardo, C., Nath, A. and Sacktor, N. (2003). Human immunodeficiency virus-associated dementia: an evolving disease. *J. Neurovirol.* 9:205-221.

McDougal, J. S., Maddon, P. J., Dalgleish, A. G., Clapham, P. R., Littman, D. R., Godfrey, M., Maddon, D. E., Chess, L., Weiss, R. A. and Axel, R. (1986).

The T4 glycoprotein is a cell-surface receptor for the AIDS virus. *Cold Spring Harb. Symp. Quant. Biol.* 51:703-711.

Meissner, E. G., Duus, K. M., Loomis, R., D'Agostin, R. and Su, L. (2003). HIV-1 replication and pathogenesis in the human thymus. *Curr. HIV Res.* 1:275-285.

Mengozzi, M., De Filippi, C., Transidico, P., Biswas, P., Cota, M., Ghezzi, S., Vicenzi, E., Mantovani, A., Sozzani, S. and Poli, G. (1999). Human immunodeficiency virus replication induces monocyte chemotactic protein-1 in human macrophages and U937 promonocytic cells. *Blood* 93:1851-1857.

Menten, P., Struyf, S., Schutyser, E., Wuyts, A., De Clercq, E., Schols, D., Proost, P. and Van Damme, J. (1999). The LD78beta isoform of MIP-1alpha is the most potent CCR5 agonist and HIV-1-inhibiting chemokine. *J. Clin. Invest.* 104:R1-5.

Michael, N. L., Louie, L. G., Rohrbaugh, A. L., Schultz, K. A., Dayhoff, D. E., Wang, C. E. and Sheppard, H. W. (1997). The role of CCR5 and CCR2 polymorphisms in HIV-1 transmission and disease progression [see comments]. *Nat. Med.* 3:1160-1162.

Moll, H. (2003). Dendritic cells and host resistance to infection. *Cell. Microbiol.* 5:493-500.

Montaner, L. J., Doyle, A. G., Collin, M., Georges, H., James, W., Minty, A., Caput, D., Ferrara, P. and Gordon, S. (1993). Interleukin 13 inhibits human immunodeficiency virus type 1 production in primary blood-derived human macrophages *in vitro. J. Exp. Med.* 178:743-747.

Montaner, L. J., Bailer, R. T. and Gordon, S. (1997). IL-13 acts on macrophages to block the completion of reverse transcription, inhibit virus production, and reduce virus infectivity. *J. Leukoc. Biol.* 62:126-132.

Montes de Oca, R., Lee, K. K. and Wilson, K. L. (2005). Binding of barrier-to-autointegration factor (BAF) to histone H3 and selected linker histones including H1.1. *J. Biol. Chem.* 280:42252-42262.

Mosier, D. E., Gulizia, R. J., Baird, S. M., Wilson, D. B., Spector, D. H. and Spector, S. A. (1991). Human immunodeficiency virus infection of human-PBL-SCID mice. *Science* 251:791-794.

Nabel, G. and Baltimore, D. (1987). An inducible transcription factor activates expression of human immunodeficiency virus in T cells. *Nature* 326:711-713.

Naif, H., Ho-Shon, M., Chang, J. and Cunningham, A. L. (1994). Molecular mechanisms of IL-4 effect on HIV expression in promonocytic cell lines and primary human monocytes. *J. Leukoc. Biol.* 56:335-339.

Namikawa, R., Kaneshima, H., Lieberman, M., Weissman, I. L. and McCune, J. M. (1988). Infection of the SCID-hu mouse by HIV-1. *Science* 242:1684-1686.

Navarro, F. and Landau, N. R. (2004). Recent insights into HIV-1 Vif. *Curr. Opin. Immunol.* 16:477-482.

Neville, M., Stutz, F., Lee, L., Davis, L. I. and Rosbash, M. (1997). The importin-beta family member Crm1p bridges the interaction between Rev and the nuclear pore complex during nuclear export. *Curr. Biol.* 7:767-775.

Newman, E. N., Holmes, R. K., Craig, H. M., Klein, K. C., Lingappa, J. R., Malim, M. H. and Sheehy, A. M. (2005). Antiviral function of APOBEC3G can be dissociated from cytidine deaminase activity. *Curr. Biol.* 15:166-170.

Nibbs, R. J., Yang, J., Landau, N. R., Mao, J. H. and Graham, G. J. (1999). LD78beta, a non-allelic variant of human MIP-1alpha (LD78alpha), has enhanced receptor interactions and potent HIV suppressive activity. *J. Biol. Chem.* 274:17478-17483.

O'Garra, A. and Murphy, K. (1996). Role of cytokines in development of Th1 and Th2 cells. *Chem. Immunol.* 63:1-13.

Oberlin, E., Amara, A., Bachelerie, F., Bessia, C., Virelizier, J. L., Arenzana-Seisdedos, F., Schwartz, O., Heard, J. M., Clark-Lewis, I., Legler, D. F., Loetscher, M., Baggiolini, M. and Moser, B. (1996). The CXC chemokine SDF-1 is the ligand for LESTR/fusin and prevents infection by T-cell-line-adapted HIV-1 [Erratum: *Nature* 1996;384(6606):288]. *Nature* 382:833-835.

Obregon, E., Punzon, C., Fernandez-Cruz, E., Fresno, M. and Munoz-Fernandez, M. A. (1999). HIV-1 infection induces differentiation of immature neural cells through autocrine tumor necrosis factor and nitric oxide production. *Virology* 261:193-204.

Orenstein, J. M. and Jannotta, F. (1988). Human immunodeficiency virus and papovavirus infections in acquired immunodeficiency syndrome: an ultrastructural study of three cases. *Hum. Pathol.* 19:350-361.

Pal, R., Garzino-Demo, A., Markham, P. D., Burns, J., Brown, M., Gallo, R. C. and DeVico, A. L. (1997). Inhibition of HIV-1 infection by the beta-chemokine MDC [see comments]. *Science* 278:695-698.

Pantaleo, G., Butini, L., Graziosi, C., Poli, G., Schnittman, S. M., Greenhouse, J. J., Gallin, J. I. and Fauci, A. S. (1991a). Human immunodeficiency virus (HIV) infection in CD4$^+$ T lymphocytes genetically deficient in LFA-1: LFA-1 is required for HIV-mediated cell fusion but not for viral transmission. *J. Exp. Med.* 173:511-514.

Pantaleo, G., Poli, G., Butini, L., Fox, C., Dayton, A. I. and Fauci, A. S. (1991b). Dissociation between syncytia formation and HIV spreading. Suppression of syncytia formation does not necessarily reflect inhibition of HIV infection. *Eur. J. Immunol.* 21:1771-1774.

Pantaleo, G., Cohen, O. J., Schacker, T., Vaccarezza, M., Graziosi, C., Rizzardi, G. P., Kahn, J., Fox, C. H., Schnittman, S. M., Schwartz, D. H., Corey, L. and Fauci, A. S. (1998). Evolutionary pattern of human immunodeficiency virus (HIV) replication and distribution in lymph nodes following primary infection: implications for antiviral therapy. *Nat. Med.* 4:341-345.

Park, I. W., Wang, J. F. and Groopman, J. E. (2001). HIV-1 Tat promotes monocyte chemoattractant protein-1 secretion followed by transmigration of monocytes. *Blood* 97:352-358.

Patella, V., Florio, G., Petraroli, A. and Marone, G. (2000). HIV-1 gp120 induces IL-4 and IL-13 release from human Fc epsilon RI$^+$ cells through interaction with the VH3 region of IgE. *J. Immunol.* 164:589-595.

Patterson, B. K., Behbahani, H., Kabat, W. J., Sullivan, Y., O'Gorman, M. R., Landay, A., Flener, Z., Khan, N., Yogev, R. and Andersson, J. (2001).

Leukemia inhibitory factor inhibits HIV-1 replication and is upregulated in placentae from nontransmitting women. *J. Clin. Invest.* 107:287-294.

Paul, W. E. and Seder, R. A. (1994). Lymphocyte responses and cytokines. *Cell* 76:241-251.

Paxton, W. A., Martin, S. R., Tse, D., O'Brien, T. R., Skurnick, J., VanDevanter, N. L., Padian, N., Braun, J. F., Kotler, D. P., Wolinsky, S. M. and Koup, R. A. (1996). Relative resistance to HIV-1 infection of CD4 lymphocytes from persons who remain uninfected despite multiple high-risk sexual exposure. *Nat. Med.* 2:412-417.

Pelchen-Matthews, A., Kramer, B. and Marsh, M. (2003). Infectious HIV-1 assembles in late endosomes in primary macrophages. *J. Cell. Biol.* 162: 443-455.

Perrella, O., Carreiri, P. B., Perrella, A., Sbreglia, C., Gorga, F., Guarnaccia, D. and Tarantino, G. (2001). Transforming growth factor beta-1 and interferon-alpha in the AIDS dementia complex (ADC): possible relationship with cerebral viral load? *Eur. Cytokine Netw.* 12:51-55.

Perrin, L., Yerly, S., Marchal, F., Schockmel, G. A., Hirschel, B., Fox, C. H. and Pantaleo, G. (1998). Virus burden in lymph nodes and blood of subjects with primary human immunodeficiency virus type 1 infection on bitherapy. *J. Infect. Dis.* 177:1497-1501.

Peterson, P. K., Gekker, G., Chao, C. C., Schut, R., Molitor, T. W. and Balfour, H. H., Jr. (1991). Cocaine potentiates HIV-1 replication in human peripheral blood mononuclear cell cocultures. Involvement of transforming growth factor- beta. *J. Immunol.* 146:81-84.

Picchio, G. R., Gulizia, R. J. and Mosier, D. E. (1997). Chemokine receptor CCR5 genotype influences the kinetics of human immunodeficiency virus type 1 infection in human PBL-SCID mice. *J. Virol.* 71:7124-7127.

Pido-Lopez, J., Burton, C., Hardy, G., Pires, A., Sullivan, A., Gazzard, B., Aspinall, R., Gotch, F. and Imami, N. (2003). Thymic output during initial highly active antiretroviral therapy (HAART) and during HAART supplementation with interleukin 2 and/or with HIV type 1 immunogen (Remune). *AIDS Res. Hum. Retroviruses* 19:103-109.

Pitha, P. M. (1994). Multiple effects of interferon on the replication of human immunodeficiency virus type 1. *Antiviral. Res.* 24:205-219.

Poli, G., Orenstein, J. M., Kinter, A., Folks, T. M. and Fauci, A. S. (1989). Interferon-alpha but not AZT suppresses HIV expression in chronically infected cell lines. *Science* 244:575-577.

Poli, G., Bressler, P., Kinter, A., Duh, E., Timmer, W. C., Rabson, A., Justement, J. S., Stanley, S. and Fauci, A. S. (1990). Interleukin 6 induces human immunodeficiency virus expression in infected monocytic cells alone and in synergy with tumor necrosis factor alpha by transcriptional and post-transcriptional mechanisms. *J. Exp. Med.* 172:151-158.

Poli, G., Kinter, A. L., Justement, J. S., Bressler, P., Kehrl, J. H. and Fauci, A. S. (1992). Retinoic acid mimics transforming growth factor beta in the regulation of human immunodeficiency virus expression in monocytic cells. *Proc. Natl. Acad. Sci. U.S.A.* 89:2689-2693.

Poli, G., Kinter, A. L. and Fauci, A. S. (1994). Interleukin 1 induces expression of the human immunodeficiency virus alone and in synergy with interleukin 6 in chronically infected U1 cells: inhibition of inductive effects by the interleukin 1 receptor antagonist. *Proc. Natl. Acad. Sci. U. S. A.* 91:108-112.

Quaranta, M. G., Tritarelli, E., Giordani, L. and Viora, M. (2002). HIV-1 Nef induces dendritic cell differentiation: a possible mechanism of uninfected CD4($+$) T cell activation. *Exp. Cell. Res.* 275:243-254.

Quaranta, M. G., Mattioli, B., Spadaro, F., Straface, E., Giordani, L., Ramoni, C., Malorni, W. and Viora, M. (2003). HIV-1 Nef triggers Vav-mediated signaling pathway leading to functional and morphological differentiation of dendritic cells. *FASEB J.* 17:2025-2036.

Refaeli, Y., Van Parijs, L., London, C. A., Tschopp, J. and Abbas, A. K. (1998). Biochemical mechanisms of IL-2-regulated Fas-mediated T cell apoptosis. *Immunity* 8:615-623.

Rieckmann, P., Poli, G., Fox, C. H., Kehrl, J. H. and Fauci, A. S. (1991). Recombinant gp120 specifically enhances tumor necrosis factor-alpha production and Ig secretion in B lymphocytes from HIV-infected individuals but not from seronegative donors. *J. Immunol.* 147:2922-2927.

Rizzardi, G. P., Morawetz, R. A., Vicenzi, E., Ghezzi, S., Poli, G., Lazzarin, A. and Pantaleo, G. (1998). CCR2 polymorphism and HIV disease. Swiss HIV Cohort [letter]. *Nat. Med.* 4:252-253.

Rizzardi, G. P., Harari, A., Capiluppi, B., Tambussi, G., Ellefsen, K., Ciuffreda, D., Champagne, P., Bart, P. A., Chave, J. P., Lazzarin, A. and Pantaleo, G. (2002). Treatment of primary HIV-1 infection with cyclosporin A coupled with highly active antiretroviral therapy. *J. Clin. Invest.* 109:681-688.

Saksela, K., Muchmore, E., Girard, M., Fultz, P. and Baltimore, D. (1993). High viral load in lymph nodes and latent human immunodeficiency virus (HIV) in peripheral blood cells of HIV-1 infected chimpanzees. *J. Virol.* 67:7423-7427.

Sawaya, B. E., Thatikunta, P., Denisova, L., Brady, J., Khalili, K. and Amini, S. (1998). Regulation of TNFalpha and TGFbeta-1 gene transcription by HIV-1 Tat in CNS cells. *J. Neuroimmunol.* 87:33-42.

Sayah, D. M., Sokolskaja, E., Berthoux, L. and Luban, J. (2004). Cyclophilin A retrotransposition into TRIM5 explains owl monkey resistance to HIV-1. *Nature* 430:569-573.

Scarlatti, G., Tresoldi, E., Bjorndal, A., Fredriksson, R., Colognesi, C., Deng, H. K., Malnati, M. S., Plebani, A., Siccardi, A. G., Littman, D. R., Fenyo, E. M. and Lusso, P. (1997). *In vivo* evolution of HIV-1 co-receptor usage and sensitivity to chemokine-mediated suppression. *Nat. Med.* 3:1259-1265.

Schmidtmayerova, H., Sherry, B. and Bukrinsky, M. (1996). Chemokines and HIV replication [letter]. *Nature* 382:767.

Schroder, A. R., Shinn, P., Chen, H., Berry, C., Ecker, J. R. and Bushman, F. (2002). HIV-1 integration in the human genome favors active genes and local hotspots. *Cell* 110:521-529.

Schuitemaker, H., Kootstra, N. A., Koppelman, M. H., Bruisten, S. M., Huisman, H. G., Tersmette, M. and Miedema, F. (1992). Proliferation-dependent HIV-1 infection of monocytes occurs during differentiation into macrophages. *J. Clin. Invest.* 89:1154-1160.

Scoggins, R. M., Taylor, J. R., Jr., Patrie, J., van't Wout, A. B., Schuitemaker, H. and Camerini, D. (2000). Pathogenesis of primary R5 human immuno-deficiency virus type 1 clones in SCID-hu mice. *J. Virol.* 74:3205-3216.

Scripture-Adams, D. D., Brooks, D. G., Korin, Y. D. and Zack, J. A. (2002). Interleukin-7 induces expression of latent human immunodeficiency virus type 1 with minimal effects on T-cell phenotype. *J. Virol.* 76:13077-13082.

Sheehy, A. M., Gaddis, N. C., Choi, J. D. and Malim, M. H. (2002). Isolation of a human gene that inhibits HIV-1 infection and is suppressed by the viral Vif protein. *Nature* 418:646-650.

Sheehy, A. M., Gaddis, N. C. and Malim, M. H. (2003). The antiretroviral enzyme APOBEC3G is degraded by the proteasome in response to HIV-1 Vif. *Nat. Med.* 9:1404-1407.

Sherman, M. P. and Greene, W. C. (2002). Slipping through the door: HIV entry into the nucleus. *Microbes Infect.* 4:67-73.

Silvestri, G., Sodora, D. L., Koup, R. A., Paiardini, M., O'Neil, S. P., McClure, H. M., Staprans, S. I. and Feinberg, M. B. (2003). Nonpathogenic SIV infection of sooty mangabeys is characterized by limited bystander immunopathology despite chronic high-level viremia. *Immunity* 18:441-452.

Slifka, M. K. and Whitton, J. L. (2000). Clinical implications of dysregulated cytokine production. *J. Mol. Med.* 78:74-80.

Smithgall, M. D., Wong, J. G., Critchett, K. E. and Haffar, O. K. (1996). IL-7 up-regulates HIV-1 replication in naturally infected peripheral blood mononuclear cells. *J. Immunol.* 156:2324-2330.

Sova, P. and Volsky, D. J. (1993). Efficiency of viral DNA synthesis during infection of permissive and nonpermissive cells with vif-negative human immunodeficiency virus type 1. *J. Virol.* 67:6322-6326.

Sozzani, S., Ghezzi, S., Iannolo, G., Luini, W., Borsatti, A., Polentarutti, N., Sica, A., Locati, M., Mackay, C., Wells, T. N., Biswas, P., Vicenzi, E., Poli, G. and Mantovani, A. (1998). Interleukin 10 increases CCR5 expression and HIV infection in human monocytes. *J. Exp. Med.* 187:439-444.

Strebel, K. (2003). Virus-host interactions: role of HIV proteins Vif, Tat, and Rev. *AIDS* 17 Suppl 4:S25-34.

Strebel, K., Daugherty, D., Clouse, K., Cohen, D., Folks, T. and Martin, M. A. (1987). The HIV 'A' (sor) gene product is essential for virus infectivity. *Nature* 328:728-730.

Stremlau, M., Owens, C. M., Perron, M. J., Kiessling, M., Autissier, P. and Sodroski, J. (2004). The cytoplasmic body component TRIM5alpha restricts HIV-1 infection in Old World monkeys. *Nature* 427:848-853.

Stremlau, M., Perron, M., Welikala, S. and Sodroski, J. (2005). Species-specific variation in the B30.2(SPRY) domain of TRIM5alpha determines the potency of human immunodeficiency virus restriction. *J. Virol.* 79:3139-3145.

Su, L., Kaneshima, H., Bonyhadi, M., Salimi, S., Kraft, D., Rabin, L. and McCune, J. M. (1995). HIV-1-induced thymocyte depletion is associated with indirect cytopathogenicity and infection of progenitor cells *in vivo*. *Immunity* 2:25-36.

Sunila, I., Vaccarezza, M., Pantaleo, G., Fauci, A. S. and Orenstein, J. M. (1997). gp120 is present on the plasma membrane of apoptotic CD4 cells prepared from lymph nodes of HIV-1-infected individuals: an immunoelectron microscopic study. *AIDS* 11:27-32.

Swingler, S., Mann, A., Jacque, J., Brichacek, B., Sasseville, V. G., Williams, K., Lackner, A. A., Janoff, E. N., Wang, R., Fisher, D. and Stevenson, M. (1999). HIV-1 Nef mediates lymphocyte chemotaxis and activation by infected macrophages [see comments]. *Nat. Med.* 5:997-103.

Takahashi, K., Wesselingh, S. L., Griffin, D. E., McArthur, J. C., Johnson, R. T. and Glass, J. D. (1996). Localization of HIV-1 in human brain using polymcrase chain reaction/in situ hybridization and immunocytochemistry. *Ann. Neurol.* 39:705-711.

Trkola, A., Ketas, T. J., Nagashima, K. A., Zhao, L., Cilliers, T., Morris, L., Moore, J. P., Maddon, P. J. and Olson, W. C. (2001). Potent, broad-spectrum inhibition of human immunodeficiency virus type 1 by the CCR5 monoclonal antibody PRO 140. *J. Virol.* 75:579-588.

Truant, R. and Cullen, B. R. (1999). The arginine-rich domains present in human immunodeficiency virus type 1 Tat and Rev function as direct importin beta-dependent nuclear localization signals. *Mol. Cell. Biol.* 19:1210-1217.

Valentin, A., Lu, W., Rosati, M., Schneider, R., Albert, J., Karlsson, A. and Pavlakis, G. N. (1998). Dual effect of interleukin 4 on HIV-1 expression: implications for viral phenotypic switch and disease progression. *Proc. Natl. Acad. Sci. U. S. A.* 95:8886-8891.

Van Lint, C., Ghysdael, J., Paras, P., Jr., Burny, A., and Verdin, E. (1994). A transcriptional regulatory element is associated with a nuclease-hypersensitive site in the pol gene of human immunodeficiency virus type 1. *J. Virol.* 68:2632-2648.

Vartanian, J. P., Sommer, P. and Wain-Hobson, S. (2003). Death and the retrovirus. *Trends Mol. Med.* 9:409-413.

Varthakavi, V., Smith, R. M., Bour, S. P., Strebel, K. and Spearman, P. (2003). Viral protein U counteracts a human host cell restriction that inhibits HIV-1 particle production. *Proc. Natl. Acad. Sci. U. S. A.* 100:15154-15159.

Veazey, R. S., DeMaria, M., Chalifoux, L. V., Shvetz, D. E., Pauley, D. R., Knight, H. L., Rosenzweig, M., Johnson, R. P., Desrosiers, R. C. and Lackner, A. A. (1998). Gastrointestinal tract as a major site of CD4$^+$ T cell depletion and viral replication in SIV infection. *Science* 280:427-731.

Verani, A., Pesenti, E., Polo, S., Tresoldi, E., Scarlatti, G., Lusso, P., Siccardi, A. G. and Vercelli, D. (1998). CXCR4 is a functional coreceptor for infection of human macrophages by CXCR4-dependent primary HIV-1 isolates. *J. Immunol.* 161:2084-2088.

Vicenzi, E. and Poli, G. (2005). Infection of CD4$^+$ primary T cells and cell lines, generation of chronically infected cell lines, and induction of HIV expression. In *Current Protocols in Immunology.* W. Strober, ed. Hoboken, NJ: John WIley & Sons; 12.3.1-12.3.17.

Vicenzi, E., Bordignon, P. P., Biswas, P., Brambilla, A., Bovolenta, C., Cota, M., Sinigaglia, F. and Poli, G. (1999). Envelope-dependent restriction of human immunodeficiency virus type 1 spreading in CD4($^+$) T lymphocytes: R5 but not X4 viruses replicate in the absence of T-cell receptor restimulation [In Process Citation]. *J. Virol.* 73:7515-7523.

Vicenzi, E., Alfano, M., Ghezzi, S., Gatti, A., Veglia, F., Lazzarin, A., Sozzani, S., Mantovani, A. and Poli, G. (2000). Divergent regulation of HIV-1 replication in PBMC of infected individuals by CC chemokines: suppression by RANTES, MIP-1alpha, and MCP-3, and enhancement by MCP-1. *J. Leukoc. Biol.* 68:405-412.

Vicenzi, E., Panina-Bodignon, P., Vallanti, G., Di Lucia, P. and Poli, G. (2002). Restricted replication of primary HIV-1 isolates using both CCR5 and CXCR4 in Th2 but not in Th1 CD4($^+$) T cells. *J. Leukoc. Biol.* 72:913-920.

Villinger, F., Brar, S. S., Brice, G. T., Chikkala, N. F., Novembre, F. J., Mayne, A. E., Bucur, S., Hillyer, C. D. and Ansari, A. A. (1997). Immune and hematopoietic parameters in HIV-1-infected chimpanzees during clinical progression toward AIDS. *J. Med. Primatol.* 26:11-18.

Vitale, M., Caruso, A., De Francesco, M. A., Rodella, L., Bozzo, L., Garrafa, E., Grassi, M., Gobbi, G., Cacchioli, A. and Fiorentini, S. (2003). HIV-1 matrix protein p17 enhances the proliferative activity of natural killer cells and increases their ability to secrete proinflammatory cytokines. *Br. J. Haematol.* 120:337-343.

Wang, J., Harada, A., Matsushita, S., Matsumi, S., Zhang, Y., Shioda, T., Nagai, Y. and Matsushima, K. (1998). IL-4 and a glucocorticoid up-regulate CXCR4 expression on human CD4$^+$ T lymphocytes and enhance HIV-1 replication. *J. Leukoc. Biol.* 64:642-649.

Wang, J., Guan, E., Roderiquez, G. and Norcross, M. A. (1999). Inhibition of CCR5 expression by IL-12 through induction of beta- chemokines in human T lymphocytes. *J. Immunol.* 163:5763-5769.

Wang, J., Guan, E., Roderiquez, G. and Norcross, M. A. (2001). Synergistic induction of apoptosis in primary CD4($^+$) T cells by macrophage-tropic HIV-1 and TGF-beta1. *J. Immunol.* 167:3360-3366.

Wang, F. X., Xu, Y., Sullivan, J., Souder, E., Argyris, E. G., Acheampong, E. A., Fisher, J., Sierra, M., Thomson, M. M., Najera, R., Frank, I., Kulkosky, J., Pomerantz, R. J. and Nunnari, G. (2005). IL-7 is a potent and proviral strain-specific inducer of latent HIV-1 cellular reservoirs of infected individuals on virally suppressive HAART. *J. Clin. Invest.* 115:128-137.

Wei, X., Ghosh, S. K., Taylor, M. E., Johnson, V. A., Emini, E. A., Deutsch, P., Lifson, J. D., Bonhoeffer, S., Nowak, M. A., Hahn, B. H., Saag, M. S. and Shaw, G. M. (1995). Viral dynamics in human immunodeficiency virus type 1 infection. *Nature (London)* 373:117-122.

Wei, P., Garber, M. E., Fang, S. M., Fischer, W. H. and Jones, K. A. (1998). A novel CDK9-associated C-type cyclin interacts directly with HIV-1 Tat and mediates its high-affinity, loop-specific binding to TAR RNA. *Cell* 92:451-462.

Weissman, D., Poli, G. and Fauci, A. S. (1994). Interleukin 10 blocks HIV replication in macrophages by inhibiting the autocrine loop of tumor necrosis factor alpha and interleukin 6 induction of virus. *AIDS Res. Hum. Retroviruses* 10:1199-1206.

Weissman, D., Poli, G. and Fauci, A. S. (1995). IL-10 synergizes with multiple cytokines in enhancing HIV production in cells of monocytic lineage. *J. Acquir. Immune Defic. Syndr. Hum. Retrovirol.* 9:442-449.

Weissman, D., Rabin, R. L., Arthos, J., Rubbert, A., Dybul, M., Swofford, R., Venkatesan, S., Farber, J. M. and Fauci, A. S. (1997). Macrophage-tropic HIV and SIV envelope proteins induce a signal through the CCR5 chemokine receptor. *Nature* 389:981-985.

Wiley, C. A. and Nelson, J. A. (1988). Role of human immunodeficiency virus and cytomegalovirus in AIDS encephalitis. *Am. J. Pathol.* 133:73-81.

Xiao, H., Neuveut, C., Tiffany, H. L., Benkirane, M., Rich, E. A., Murphy, P. M. and Jeang, K. T. (2000). Selective CXCR4 antagonism by tat: implications for *in vivo* expansion of coreceptor use by HIV-1 [In Process Citation]. *Proc. Natl. Acad. Sci. U. S. A.* 97:11466-11471.

Yi, Y., Rana, S., Turner, J. D., Gaddis, N. and Collman, R. G. (1998). CXCR-4 is expressed by primary macrophages and supports CCR5- independent infection by dual-tropic but not T-tropic isolates of human immunodeficiency virus type 1. *J. Virol.* 72:772-777.

Yu, Q., Chen, D., Konig, R., Mariani, R., Unutmaz, D. and Landau, N. R. (2004a). APOBEC3B and APOBEC3C are potent inhibitors of simian immunodeficiency virus replication. *J. Biol. Chem.* 279:53379-53386.

Yu, Q., Konig, R., Pillai, S., Chiles, K., Kearney, M., Palmer, S., Richman, D., Coffin, J. M. and Landau, N. R. (2004b). Single-strand specificity of APOBEC3G accounts for minus-strand deamination of the HIV genome. *Nat. Struct. Mol. Biol.* 11:435-442.

Zagury, D., Lachgar, A., Chams, V., Fall, L. S., Bernard, J., Zagury, J. F., Bizzini, B., Gringeri, A., Santagostino, E., Rappaport, J., Feldman, M., Burny, A. and Gallo, R. C. (1998). Interferon alpha and Tat involvement in the immunosuppression of uninfected T cells and C-C chemokine decline in AIDS. *Proc. Natl. Acad. Sci. U. S. A.* 95:3851-3856.

Zhang, Z. Q., Wietgrefe, S. W., Li, Q., Shore, M. D., Duan, L., Reilly, C., Lifson, J. D. and Haase, A. T. (2004). Roles of substrate availability and infection of resting and activated CD4$^+$ T cells in transmission and acute simian immunodeficiency virus infection. *Proc. Natl. Acad. Sci. U. S. A.* 101: 5640-5645.

Zheng, Y. H., Irwin, D., Kurosu, T., Tokunaga, K., Sata, T. and Peterlin, B. M. (2004). Human APOBEC3F is another host factor that blocks human immunodeficiency virus type 1 replication. *J. Virol.* 78:6073-6076.

Zhou, Q. and Sharp, P. A. (1995). Novel mechanism and factor for regulation by HIV-1 Tat. *EMBO J.* 14:321-328.

Zhu, T., Muthui, D., Holte, S., Nickle, D., Feng, F., Brodie, S., Hwangbo, Y., Mullins, J. I. and Corey, L. (2002). Evidence for human immunodeficiency virus type 1 replication *in vivo* in CD14($^+$) monocytes and its potential role as a source of virus in patients on highly active antiretroviral therapy. *J. Virol.* 76:707-716.

Zinkernagel, R. M. and Hengartner, H. (1994). T-cell-mediated immunopathology versus direct cytolysis by virus: implications for HIV and AIDS. *Immunol. Today* 15:262-268.

Zocchi, M. R., Contini, P., Alfano, M. and Poggi, A. (2005). Pertussis toxin (PTX) B subunit and the nontoxic PTX mutant PT9K/129G inhibit Tat-induced TGF-{beta} production by NK cells and TGF-{beta}-mediated NK cell apoptosis. *J. Immunol.* 174:6054-6061.

Chapter 8

Innate Cellular Immune Responses in HIV Infection

Barbara Schmidt, Nicolai A. Kittan, Sabrina Haupt, and Jay A. Levy

8.1 Characteristics of Innate Immunity

Immune responses against pathogens are mediated by innate and adaptive immunity (Table 8.1). The innate immune system is evolutionary conserved over many species and was first described in *Drosophila*, whereas the adaptive immune system can be detected in vertebrates only. Major time differences are observed between stimulation and the type of immunologic response: the innate system usually reacts within minutes to days, whereas the adaptive system shows a more delayed response. The recognition of pathogens is also different between the two systems. Whereas innate immunity responds to pathogen-associated molecular patterns (PAMPs) such as unmethylated CpG DNA or single- and double-stranded RNA [reviewed in Beutler (2004)], adaptive immunity recognizes pathogens as distinct peptides presented in the context of MHC class I and class II molecules by professional antigen-presenting cells. Memory and booster

TABLE 8.1. Characteristics of innate and adaptive immune responses.

Innate immune response	Adaptive immune response
Evolutionary conserved	Only detected in vertebrates
Quick response within minutes to days	Delayed response within days to weeks
Recognition of molecular patterns	Recognition of specific antigens in the context of MHC class I and II presentation
No memory, no booster response	Memory and booster response

responses are well-known for adaptive but not innate immunity. These descriptions are not meant to be exclusive, as overlap features have been recognized (e.g., memory) (Kurtz, 2005), and many interactions exist between these two major branches of the immune system (Hoebe *et al.*, 2004).

Although the adaptive immune system has been studied in great detail during the past decades, the innate immunity has only recently received attention. This appreciation of innate immune responses came with the discovery of specific cellular receptors that respond to various foreign substances presented to the immune system. In this regard, the recognition of pathogens by the innate immune system is usually mediated via PAMP-recognition receptors (PRRs) the most notable of which are called Toll-like receptors (TLRs) [reviewed in Beutler (2004)]. These molecules were first identified as an embryonic dorsoventral regulatory gene cassette and later as a potent antifungal response in *Drosophila* (Lemaitre *et al.*, 1996). Human homologues of this type I transmembrane protein were shortly identified thereafter. They are characterized by an extracellular leucin-rich repeat region and a cytoplasmic domain homologous to the IL-1 receptor (Medzhitov *et al.*, 1997). The TLRs can be found on macrophages, dendritic cells, neutrophils, B cells, epithelial cells, and endothelial cells (Takeda and Akira, 2005). The cellular proteins produced in response to TLR signaling can elicit migration of new immune cells to an area of infection or inflammation and help in the maturation of dendritic cells to present antigens to the adaptive immune system. Although the recognition of self versus non-self has been based on the classic MHC concept of adaptive immunity, some immunologic phenomena cannot be explained by these mechanisms. In this regard, Matzinger has put forward a model for the recognition of danger and damage instead of non-self (Matzinger, 2002). This approach fits well into the recognition of PAMPs by TLRs.

Another innate immune system recognition process includes nucleotide-binding oligomerization domain (NOD) proteins and receptors that involve phagocytosis. Examples are the mannose receptors that interact with mannose-containing organisms (e.g., HIV or SIV) and encourage engulfment by macrophages. Such a ligand could be a mannose-binding lectin (Ezekowitz *et al.*, 1989).

The TLR signaling process allows intracellular events that permit the liberation of NF-κB and the transcription of cytokines [reviewed in Akira and Takeda (2004)]. All TLRs except TLR-3 seem to have a common signaling pathway that uses the adaptor protein MyD88 (myeloid differentiation factor 88). In some cases there is direct signaling through the Toll IL-1 receptor domain (TIR); in others (e.g., TLR-2, TLR-4), intracellular adaptor molecules such as TIRAP (TIR domain-containing adaptor protein) can be involved [reviewed in O'Neill *et al.* (2003); Pashine *et al.* (2005)]. The activation of the innate immune system, particularly at areas of infection or inflammation, can prepare the immune system for the induction of appropriate adaptive immune responses (Hoebe *et al.*, 2004; Iwasaki and Medzhitov, 2004).

8.2 Cells and Soluble Factors Involved in Innate Immunity Against HIV

Both cells (Table 8.2) and soluble cellular factors are involved in innate and adaptive immune responses against HIV. Dendritic cells can play a role in both immune systems (Medzhitov and Janeway, 2000). The plasmacytoid dendritic cells (PDCs) are a major component in the innate immune defense against HIV, whereas myeloid dendritic cells are more involved in antiviral adaptive immune responses (see Section 8.3). In terms of immune cells, CD8$^+$ cells can also be divided into two different functioning cells: cytotoxic T cells (CTLs) and non-cytotoxic T lymphocytes, which are part of the adaptive and innate immune system, respectively (Levy, 2001) (see Section 8.4). Another innate cell, the natural killer (NK) cell, can have an anti-HIV function by recognizing the downregulated expression of MHC class I molecules on HIV-infected cells (see Section 8.5). Natural killer T (NK-T) cell counts also decrease in HIV infection concomitant with the CD4$^+$ cell decline, whereas subsets of $\gamma\delta$ T cells increase in HIV-infected individuals (see Section 8.5). These cellular changes most likely reflect innate immune responses against HIV. The role of macrophages in HIV infection will not be discussed in this chapter, because they are often considered more active in adaptive immunity.

Soluble factors released by all these cells can have important effects on the immune responses in HIV infection. Most cytokines are involved in the innate immune defense [reviewed in Levy (2003)] and include the interferons, interleukins, and the chemokines such as RANTES, MIP-1α, and MIP-1β, which bind to the CCR5 coreceptor and can block HIV infection. The β-chemokines are produced by CD4$^+$ and CD8$^+$ T cells (Cocchi *et al.*, 1995; Greco *et al.*, 1998). The CXC-chemokine stromal-cell derived factor-1 (SDF-1) binds to the CXCR4 coreceptor and can inhibit the infection by T-tropic or X4 HIV-1 strains (Bleul *et al.*, 1996; Oberlin *et al.*, 1996). The β-chemokine macrophage-derived chemokine (MDC), which is expressed by activated PBMCs in some studies, was reported to suppress HIV infection of PBMCs (Pal *et al.*, 1997). Moreover, a potent HIV suppression *in vitro* was noted with IL-16 that is produced by T cells (Baier *et al.*, 1995). Leukemia inhibitory factor (LIF) was reported to be a potent endogenous HIV-1 suppressive factor produced locally in the

TABLE 8.2. Cells involved in the innate immune defense.

Cell type	Frequency in PBMC (%)	Pattern recognized	Pattern presented by	Characteristic surface receptors	Cytokine secretion upon activation	Immune polarization
PDC	0.2–0.5	Single-stranded RNA, small antiviral compounds via TLR7 Unmethylated CpG motifs via TLR9	Virus or virus-infected cells	BDCA2, BDCA4	IFN-α, IFN-β, TNF-α, IL-12	Th1
Non-cytotoxic CD8⁺ cells	20–40	Unclear	Virus or virus-infected cells	CD8⁺ CD28⁺ HLA-DR⁺	CAF	Th1
NK	Up to 15	Downregulated expression of MHC-I molecules on HIV-infected cells	Virus-infected cells	NKp46, NKp44, NKp30	IFN-γ, TNF-α, MIP-1α (CCL3), MIP-1β (CCL4), RANTES (CCL5)	Th1
iNK-T*	0.01-1	Glycolipids (α-GalCer)	CD1d	Vα24, Vβ11	IFN-γ, TNF-α, IL-4, IL-13	Th1, Th2
γδ T cells	3-6	Nonpeptidic phosphoantigens (alkylamines, amino-bisphonates)	Cell:cell contact required	Vγ2, Vδ2	IFN-γ, TNF-α	Th1

Th, T helper cell; IFN, interferon; TNF, tumor necrosis factor; MIP, macrophage inflammatory protein; RANTES, regulated upon activation normal T-cell expressed and secreted; IL, interleukin; CAF, CD8⁺ cell antiviral factor.
* The invariant NK-T cell.

placenta (Patterson *et al.*, 2001b). These findings require further evaluation for their clinical relevance (Mackewicz *et al.*, 1997; Levy, 2003).

Other molecules primarily localized within the secretory, azurophil, and cytosolic compartment of leukocytes can also exert anti-HIV activity [reviewed in Ganz and Lehrer (1997)], in particular cathelicidins (Lehrer and Ganz, 2002) and defensins (Lehrer and Ganz, 2002). In this chapter, we will focus on dendritic cells, particularly PDCs and their production of IFN-α as well as the CD8⁺ cell noncytotoxic response (CNAR), and its associated CD8$^+$ cell antiviral factor (CAF). Studies of other cells of the innate system and their role in HIV infection are also briefly reviewed.

8.3 Plasmacytoid and Myeloid Dendritic Cells and Their Association with HIV Infection

8.3.1 Overview

Nearly one-half century ago, Isaacs and Lindenmann (1957) discovered a soluble factor produced by influenza virus–infected chicken chorioallantois membrane cell cultures that prevented the infection of uninfected cells. They termed it "interferon" due to its activity as "virus interference." Although many cells can produce interferon, a single cell population was postulated to be a major producer of IFN in the body. These natural interferon-producing cells (NIPCs) [reviewed in Fitzgerald-Bocarsly (1993); Fitzgerald-Bocarsly (2002)] remained elusive until recently when PDCs were identified as cells that produce 200 to 1000 times more IFN than other cells in the body (Cella et al., 1999; Siegal et al., 1999). PDCs exhibit a plasma cell-like phenotype shortly after isolation but mature into dendritic cells after exposure to IL-3 and CD40L (Grouard et al., 1997). They migrate to secondary lymphatic tissue (Cyster, 1999), and reciprocally interact with T helper cells (Rissoan et al., 1999). Virus-stimulated PDCs drive primarily a potent Th1 polarization in the presence of CD40 ligand (Cella et al., 2000) and produce chemokines that induce migration of T and NK cells (Megjugorac et al., 2004). Under some conditions, they can elicit a Th2 type response (Ito et al., 2004). Thus, PDCs are an important link between innate and adaptive immunity (Kadowaki et al., 2000). PDCs can be generated in vivo using Flt3 ligand (Maraskovsky et al., 2000) or in vitro from human hematopoietic progenitors using a combination of Flt3 ligand and thrombopoietin (Chen et al., 2004).

PDCs make up one of the major two subpopulations of dendritic cells in the blood and are derived from lymphoid precursors. The myeloid dendritic cells (MDCs) develop from common myeloid precursors and represent a different lineage of dendritic cells (Liu, 2001). They express CD11c and Toll-like receptors (TLRs) 1, 2, 4, 5, and 8 (Kadowaki et al., 2001a) reviewed in (McKenna et al., 2005). PDCs are involved in innate and adaptive immune responses (Kadowaki et al., 2000; Kadowaki et al., 2001a; Kadowaki et al., 2001b). MDCs are major antigen presenting cells (APCs) in the adaptive immune system. MDCs are activated and induced to maturation as APCs by IFN-α (Kadowaki et al., 2000) and IFN-γ production (e.g., via NK cells)

(Sallusto *et al.*, 2004). Most recently, it has been shown that PDCs as well as MDCs interact with NK cells, which links both DC types to innate and adaptive immune responses in (viral) infections (Gerosa *et al.*, 2005).

8.3.2 PDC Characteristics

PDC are characterized as negative for lineage markers for T cells (CD3), B cells (CD19, CD20), monocytes (CD14), and NK cells (CD16, CD56), but positive for CD4, CXCR4, CCR5, and HLA-DR [reviewed in Fitzgerald-Bocarsly (2002)]. They do not seem to express DC-SIGN (Soumelis *et al.*, 2001; Schmidt *et al.*, 2004), although one group reported that subsets of PDCs express this surface molecule (Soilleux *et al.*, 2002). That finding needs to be further evaluated. Specific surface molecules, BDCA 2 and 4, have been identified on these dendritic cells by the use of monoclonal antibodies (Dzionek *et al.*, 2000). Blood dendritic cell antigen (BDCA) 2, also designated as dendritic cell lectin (DLEC), is a novel type II calcium-dependent lectin (Arce *et al.*, 2001; Dzionek *et al.*, 2001). It could play a role in ligand internalization and presentation; a fluorescently labeled antibody against BDCA2 is rapidly internalized and efficiently presented to T cells (Dzionek *et al.*, 2000). BDCA4 is identical to neuropilin-1 (NP-1), a neuronal receptor for the axon guidance factors and a receptor on endothelial and tumor cells for vascular endothelial growth factor (VEGF-A) (Dzionek *et al.*, 2002).

PDCs express TLR7 and 9 (Kadowaki *et al.*, 2001) and perhaps TLR10 (Hornung *et al.*, 2002). The natural ligand of TLR7 has been identified as single-stranded RNA (Diebold *et al.*, 2004; Heil *et al.*, 2004), which can be mimicked by small antiviral compounds (e.g., imiquimod) resulting in cytokine production and maturation of PDCs (Gibson *et al.*, 2002; Hemmi *et al.*, 2002). In contrast, TLR9 recognizes unmethylated CpG motifs of bacterial DNA (Hemmi *et al.*, 2000) and is a potent signal for growth, activation, and maturation of human dendritic cells (Hartmann *et al.*, 1999). TLR9 also synergizes with CD40L to induce high amounts of IL-12 (Krug *et al.*, 2001). It is not localized at the cell surface, but intracellularly (Ahmad-Nejad *et al.*, 2002) and signals after translocating from the endoplasmic reticulum to CpG DNA in the endo-/lysosome (Latz *et al.*, 2004).

As noted above, PDCs are the major producers of type 1 IFN. Type 1 IFN can play an important role in HIV infection by directly inhibiting virus replication (Yamamoto *et al.*, 1986; Michaelis and Levy, 1989). They can also induce strong innate immunity via NK cells (Biron *et al.*, 1999) or adaptive immune response by upregulating MHC expression (Biron, 1998; Bogdan, 2000; Hiroishi *et al.*, 2000; Kadowaki *et al.*, 2000; Kuchtey *et al.*, 2005) (Fig. 8.1).

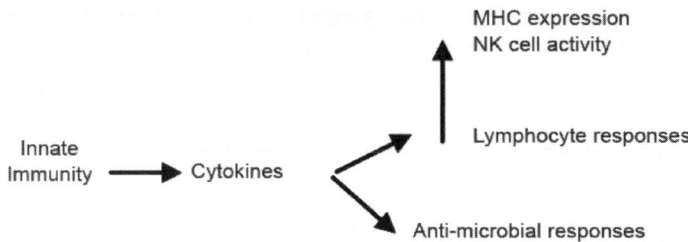

FIGURE 8.1. Importance of innate immunity. Activation of innate immunity leads to the production of cytokines that can have a direct antimicrobial effect (e.g., interferon) as well as enhance the innate (NK cell) and adaptive (T lymphocyte) immune responses (see text).

8.3.3 Relationship of Dendritic Cells to the HIV Clinical State

Early studies reported a decreased frequency of NIPCs in HIV infection (Howell *et al.*, 1994). Both PDC and MDC numbers are reduced in primary HIV-1 infection and these cell numbers can increase in parallel to the CD4$^+$ cells when viral replication is suppressed by antiretroviral therapy (Pacanowski *et al.*, 2001). Early PDC changes in response to antiretroviral therapy (ART) appear to be predictive of the viral rebound after interruption of HAART; a high number of PDCs and low viral load suggest that PDCs play an important role in the immunologic control of HIV replication (Siegal *et al.*, 2001; Soumelis *et al.*, 2001; Pacanowski *et al.*, 2004; Schmidt *et al.*, 2006; Killian *et al.*, 2006) (Table 8.3).

In acute or primary HIV infection, PDC numbers are reduced particularly in those with high viral loads. In time, further declines in PDC occur along with reduction in CD4$^+$ T cell levels (Killian, 2006). Subjects on therapy show a significant increase (p < 0.01) in PDC number along with CD4$^+$ T cell counts indicating concomitant changes in these two cell types during primary infection and after treatment (Killian, 2006).

In chronic HIV infection, reduced PDC numbers have been found in individuals with progressive disease (Table 8.3). This decrease is directly correlated with CD4$^+$ cell numbers and inversely correlated with HIV viral load (Donaghy *et al.*, 2001; Feldman *et al.*, 2001; Soumelis *et al.*, 2001; Chehimi *et al.*, 2002; Barron *et al.*, 2003; Finke *et al.*, 2004; Almeida *et al.*, 2005; Schmidt *et al.*, 2006). Several studies have also shown a concomitant decrease in MDC counts in HIV-infected patients with active viral replication (Grassi *et al.*, 1999; Donaghy *et al.*, 2001; Feldman *et al.*, 2001; Chehimi *et al.*, 2002; Barron *et al.*, 2003; Finke *et al.*, 2004; Almeida *et al.*, 2005). Both PDC and MDC numbers recover at least partially under ART (Barron *et al.*, 2003; Finke *et al.*, 2004; Schmidt *et al.*,

2006), although one study reported persistent decreases in PDC numbers and function despite HAART (Chehimi *et al.*, 2002).

TABLE 8.3. Numerical and functional deficiencies of cells involved in the innate immune defense in HIV–1 infected patients.

Cell type	Numerical deficiencies	References	Functional deficiencies	References
PDC	Reduced counts in HIV primary infection	(Pacanowski *et al.*, 2001 and 2004)	Lower frequency of IFN-α producing cells	(Feldman *et al.*, 2001; Siegal *et al.*, 2001; Chehimi *et al.*, 2002; Finke *et al.*, 2004)
			Enhanced expression of costimulatory molecules	(Barron *et al.*, 2003)
MDC	Reduced counts in HIV chronic infection	(Donaghy *et al.*, 2001; Feldman *et al.*, 2001; Soumelis *et al.*, 2001; Chehimi *et al.*, 2002; Barron *et al.*, 2003; Finke *et al.*, 2004; Almeida *et al.*, 2005; Schmidt *et al.*, 2006)	Downregulation of CXCR4/CCR5	(Almeida *et al.*, 2005)
			Impaired stimulation of allogeneic T-cell proliferation	(Donaghy *et al.*, 2003)
	Reduced counts in HIV chronic infection	(Donaghy *et al.*, 2001; Chehimi *et al.*, 2002; Barron *et al.*, 2003; Finke *et al.*, 2004; Almeida *et al.*, 2005)	Impaired stimulation of allogeneic T-cell proliferation	(Donaghy *et al.*, 2003)
			Downregulation of CXCR2/CXCR4	(Almeida *et al.*, 2005)
Non-cytotoxic CD8[+] cells	Increase in LTNP, decrease in progressive disease	(Walker and Levy, 1989; Walker *et al.*, 1991a; Chen *et al.*, 1993; Gomez *et al.*, 1994; Blackbourn *et al.*, 1996; Castelli *et al.*, 2002)	CAF activity only present in asymptomatic individuals and LTNP	(Walker and Levy, 1989; Brinchmann *et al.*, 1990; Walker *et al.*, 1991a; Landay *et al.*, 1993)

			Decreased expression of NCR	(De Maria et al., 2003; Mavilio et al., 2003)
NK	Reduction of CD16+ CD56+ cells, increase in CD16+ CD56- cells	(Hu et al., 1995; Ahmad et al., 2001; Tarazona et al., 2002; Mavilio et al., 2003; Alter et al., 2005; Azzoni et al., 2005)	Reduced lytic activity	(Ullum et al., 1995; Weber et al., 2000; Tasca et al., 2003)
			Low expression of perforin and granzyme A, cytokines (IFN-γ) and CC-chemokines	(Azzoni et al., 2002; Kottilil et al., 2003; Portales et al., 2003)
			NK-T1→NK-T2 shift	(Chan et al., 2003)
NK-T	NK-T cells either undetectable or severely reduced in cell numbers	(Motsinger et al., 2002; Sandberg et al., 2002; van der Vliet et al., 2002)	Downregulation of CD1d by nef, reducing the activation of NK-T cells	(Cho et al., 2005)
			Reduced production of IFN-γ	(Martini et al., 2000 and 2002)
γδ T cells	Increase of Vδ1 cells with concomitant decrease of Vδ2 cells	(Autran et al., 1989; Martini et al., 2000; Poles et al., 2003)	Reduction of proliferative responses	(Poccia et al., 1996; Wallace et al., 1997; Martini et al., 2000 and 2002)

NCR, natural cytotoxicity receptors; LTNP long-term nonprogressors.

Functional PDC deficits in HIV infection were observed by lower frequency of IFN-α responding cells in HIV-infected individuals (Feldman et al., 2001; Siegal et al., 2001; Chehimi et al., 2002; Finke et al., 2004) (Table 8.3). Moreover, the expression of costimulatory molecules CD80 and CD86 on PDCs was shown to be enhanced in HIV infection (Barron et al., 2003), whereas the expression of CXCR4/CCR5 and of CXCR2/CXCR4 were downregulated in PDCs and MDCs, respectively (Almeida et al., 2005). Moreover, PDCs and MDCs from HIV-infected individuals have shown a severely impaired stimulation of allogeneic T-cell proliferation in a mixed lymphocyte reaction (Donaghy et al., 2003).

8.3.4 HIV Interaction with Dendritic Cells and IFN Production

Both MDCs and PDCs are targets for HIV infection (Table 8.4). The first
evidence came from studies in which CD11c-negative peripheral blood
dendritic cells were shown to be susceptible to infection with an X4-tropic

TABLE 8.4. Effect of HIV infection on cells involved in the innate immune defense.

Cell type	HIV infection	References	Consequences of HIV infection	References
PDC	PDC can be infected by X4- and R5-tropic viruses	(Patterson *et al.*, 1999; Patterson *et al.*, 2001; Fong *et al.*, 2002; Donaghy *et al.*, 2003; Schmidt *et al.*, 2004)	Enhanced virus replication, cytopathic effects and cells death upon exposure to CD40L	(Fong *et al.*, 2002; Schmidt *et al.*, 2004)
			Maturation of PDC	(Yonezawa *et al.*, 2003; Fonteneau *et al.*, 2004)
			Upregulation of CCR7	(Fonteneau *et al.*, 2004) (Schmidt *et al.*)
MDC	MDC can be infected by HIV	(Donaghy *et al.*, 2003)		
Non-cytotoxic CD8+ cells	Not observed unless CD4 receptor is simultaneously expressed	NA	NA	NA
NK	CD4$^+$ CXCR4$^+$ and CCR5$^+$ NK cells can be infected by X4- and R5-tropic virus strains	(Gurney *et al.*, 2002; Valentin *et al.*, 2002)	Persistently-infected NK cells supporting virus propagation	(Gurney *et al.*, 2002)
			Depletion of HIV-infected NK cells	(Valentin *et al.*, 2002)
iNK-T*	CD4$^-$ CXCR4$^+$ and CCR5$^+$ NK-T cell lines can be infected by X4- and R5-tropic strains	(Motsinger *et al.*, 2002; Sandberg *et al.*, 2002; van der Vliet *et al.*, 2002)	Subsequent cell death	(Motsinger *et al.*, 2002; Sandberg *et al.*, 2002; van der Vliet *et al.*, 2002)
γδ T cells	CD4+ CXCR4+ and CCR5+ γδ T cells can be targeted by HIV-1	(Gurney *et al.*, 2002; Imlach *et al.*, 2003)	Productive infection with subsequent cell death	(Gurney *et al.*, 2002; Imlach *et al.*, 2003)

NA, not applicable.
* The invariant NK-T cell.

strain of HIV-1 (Patterson *et al.*, 1999; Patterson *et al.*, 2001a). These findings were confirmed with highly purified populations of PDCs and MDCs showing the presence of provirus in the majority of these cells isolated from untreated HIV-infected individuals (Donaghy *et al.*, 2003). However, PDCs were not reported to be a major reservoir for HIV-1 in infected individuals on virally suppressive HAART (Otero *et al.*, 2003).

HIV-1–infected PDCs only show a limited degree of virus replication; productive infection can be triggered by CD40 ligation that reduces the capacity of PDCs for IFN-α production (Fong *et al.*, 2002; Schmidt *et al.*, 2004). Under these conditions, cytopathic effects and cell death were observed *in vitro* and *in vivo* (Fong *et al.*, 2002; Schmidt *et al.*, 2004). PDCs respond to HIV-1 infection with IFN-α production and maturation (Yonezawa *et al.*, 2003), which concomitantly induces bystander maturation of MDCs (Fonteneau *et al.*, 2004). This sensitivity to HIV has been shown to be primarily in response to virus-infected cells (Schmidt *et al.*, 2005). The interaction of HIV and HIV-infected cells with PDCs resulting in the release of IFN-α appears to be mediated by high-affinity interaction of CD4 and gp120 (Beignon *et al.*, 2005; Schmidt *et al.*, 2005). It is blocked by chloroquine, which prevents endosomal acidification (Schmidt *et al.*, 2005). A recent report suggested that TLR7 was involved in this induction of IFN by HIV (Beignon *et al.*, 2005); results in our laboratory suggest a role of TLR9 (Schmidt *et al.*, 2005). For both TLRs, endocytosis is required. Exposure of PDCs to HIV-1 also induces upregulation of CCR7 (CD197), which mediates migration of cells to secondary lymphatic tissue and contributes to the loss of circulating PDCs (Fonteneau *et al.*, 2004; Schmidt *et al.*, 2005).

8.4 The Role of Noncytotoxic CD8$^+$ T Cells in Anti-HIV Responses

8.4.1 Overview

Two activities of CD8$^+$ T cells are involved in the immune defense against HIV: the cytotoxic response (CTLs) that destroys HIV-infected cells after recognition of the antigen in the MHC (an adaptive immune activity; McMichael and Rowland-Jones, 2001), and the noncytotoxic response, which suppresses HIV replication without killing the infected cells. The latter is characteristic of an innate immune response (Levy, 2003) (see below).

The CNAR was first detected when the isolation and replication of HIV in CD4$^+$ cells obtained from asymptomatic HIV-infected subjects was found to be suppressed by their CD8$^+$ cells in a dose-dependent manner

(Walker *et al.*, 1986). CNAR varied among different HIV-infected individuals depending on their clinical state and was active against all HIV
strains tested (Walker *et al.*, 1989; Walker *et al.*, 1991a). The activity was
also observed when CD8[+] cells from HIV-infected individuals were
cocultured with HIV-acutely infected CD4[+] cells from HIV-negative
donors (Walker *et al.*, 1991a). This finding indicated that the antiviral
response was not MHC restricted and thus differed from classic CTLs.
Nevertheless, optimal control is observed when CD4[+] and CD8[+] cells are
derived from the same donor (Mackewicz *et al.*, 1998). CNAR does not
affect the expression of various activation markers on CD4[+] cells or their
proliferation (Levy *et al.*, 1996).

This CD8[+] cell non-cytotoxic anti-HIV activity was confirmed by others
using T lymphocytes from asymptomatic HIV-1–infected individuals
(Chen *et al.*, 1993; Gomez *et al.*, 1994). CNAR correlates with high CD4[+]
cell counts and an asymptomatic clinical state in HIV infection (Chen
et al., 1993; Landay *et al.*, 1993; Gomez *et al.*, 1994; Castelli *et al.*, 2002).

CNAR appears to be part of the innate immune system (Table 8.5)
(Levy, 2003). It acts against all HIV-1, HIV-2, and the simian immunodeficiency virus (SIV). Its activity is not restricted by MHC recognition
and is an early response to HIV infection occurring before seroconversion
(Mackewicz *et al.*, 1994b). CNAR-like activity has been observed as well
with human T cell leukemia virus (HTLV)-infected cultures (Moriuchi and
Moriuchi, 2000) but not with infection of cells with any of the non-
retroviruses (Levy, 1996).

TABLE 8.5. CD8[+] cell non-cytotoxic anti-HIV response is part of the innate
immune system.

- Active against other retroviruses besides HIV
- Not restricted by class I or class II molecules
- Rapid, early response to HIV infection
- Found in exposed uninfected individuals and decreases after time of exposure
- No memory
- Mediated by a secreted cytokine

A subset of CD8[+] CD28[+] HLA-DR[+] cells has been reported to be
responsible for this antiviral effect (Landay *et al.*, 1993). Similarly, HIV
replication was suppressed optimally when CD8[+] cells were treated with
anti-CD3 and anti-CD28 antibodies during stimulation (Barker *et al.*,
1997). The effect most likely reflected the secretion of IL-2, because CD8[+]
cells recover anti-HIV activity in the presence of IL-2 (Barker *et al.*,
1999). A similar recovery was observed after cocultivation of CD8[+] cells
with dendritic cells; this effect was primarily mediated by the production
of IL-15 (Castelli *et al.*, 2004).

Subtractive hybridization approaches conducted using CD8$^+$ cells from HIV-discordant twins (Diaz *et al.*, 2003) have helped to identify another CNAR cell marker. The one infected twin suppressed HIV replication very well (CNAR$^+$); the other uninfected twin did not (CNAR-). In this comparative screen of mRNA expression in CD8$^+$ cells with low and high suppressing activity, upregulation of VCAM-1, a vascular endothelial molecule, correlated with the presence of antiviral activity (Diaz *et al.*, 2005). The observation uncovered a marker on CD8$^+$ cells, besides CD28, that is associated with strong anti-HIV activity.

CNAR has also been observed by several groups in animal models. As noted with HIV, SIV was more readily isolated from lymphocyte populations when CD8$^+$ cells were depleted. Moreover, SIV replication was blocked *in vitro* by CD8$^+$ T lymphocytes from infected, but not uninfected rhesus macaques (Kannagi *et al.*, 1988). The HIV-1 recovery in infected chimpanzees and the virus load was dramatically enhanced after enrichment of CD4$^+$ cells or depletion of CD8$^+$ cells from PBMCs (Castro *et al.*, 1991). Moreover, HIV was recovered from infected chimpanzees after they received anti-CD8 antibodies (Castro *et al.*, 1992). Further evidence came from HIV-2–infected baboons; CD8$^+$ cells isolated from these animals inhibited the HIV-1 replication in acutely infected autologous CD4$^+$ cells substantially more than CD8$^+$ cells from uninfected baboons (Blackbourn *et al.*, 1997). Finally, this type of CD8$^+$ cell antiviral activity was found as well with feline immunodeficiency (FIV)-infected cats (Jeng *et al.*, 1996).

In evaluating the CD8$^+$ cell inhibitory effect in acute or primary HIV infection, the activity was shown to precede neutralizing antibodies by several days (Mackewicz *et al.*, 1994b). CNAR was also found in exposed-uninfected individuals where CD8$^+$ cells from these individuals placed in cell culture prevented HIV infection of PBMCs (Stranford *et al.*, 1999; Furci *et al.*, 2002). This anti-HIV activity was also evident with CD8$^+$ T lymphocytes obtained from secondary lymphoid tissue of HIV-infected patients. Its strength correlated directly with less histopathology and low viral loads in the tissues and blood (Blackbourn *et al.*, 1996).

8.4.2 The CD8$^+$ T-Cell Antiviral Factor (CAF)

Early reports indicated that the suppression of virus replication by CD8$^+$ lymphocytes was mediated at least in part by a CD8$^+$ cell secreted antiviral factor (CAF) (Walker and Levy, 1989). The antiviral activity was observed when transwell inserts were used to prevent direct contact between CD4$^+$ and CD8$^+$ cells (Walker and Levy, 1989; Walker *et al.*, 1991b). These findings were confirmed by other groups reporting that HIV replication was inhibited by cell-free supernatants from autologous and allogeneic CD8[+] cells, which were obtained from asymptomatic HIV-infected as well as some HIV-uninfected individuals (Brinchmann *et al.*, 1990). Addition of fluids from CD8[+] cells reduced the number of cells expressing HIV antigens and RNA by more than 80% (Mackewicz *et al.*, 1995). Further studies revealed that CAF and CNAR can be blocked by up to 70–95% using protease inhibitors, in particular leupeptin (Mackewicz *et al.*, 2003). This finding suggested that cleavage of a precursor protein may be necessary to elicit CAF function (Levy, 2003).

8.4.3 Mechanism of Action

In early studies of CNAR, cell killing was shown not to be involved (Walker *et al.*, 1986; Walker *et al.*, 1991b; Wiviott *et al.*, 1990). Some studies reported that Tat-mediated transcription of the HIV-1 long terminal repeats (LTRs) was suppressed in CD4[+] T lymphocytes by CNAR (Chen *et al.*, 1993). Moreover, fluids obtained from CD8[+] T lymphocyte cultures directly inhibited the LTR-driven transcription in HIV-infected 1G5 cells carrying an LTR-luciferase construct (Mackewicz *et al.*, 1995). It suppressed the virus after integration (Mackewicz *et al.*, 1995; Butera *et al.*, 2001). CD8[+] cells and their fluids did not affect reverse transcription and integration or other early steps in the HIV replication cycle; however, the expression of viral RNA from the integrated provirus was specifically interrupted (Mackewicz *et al.*, 2000b). The suppression was not lentivirus-specific; it has been observed as well with LTR constructs of HTLV-1 and murine and avian retroviruses (Copeland *et al.*, 1995).

8.4.4 Characteristics of CAF

The exact nature of CAF remains undefined. Its activity is stable at high temperatures and low pH as well as after ether extraction and lyophilization (Levy *et al.*, 1996). Size chromatography experiments have revealed an activity between 10 and 30 kDa (Levy, 2003). CAF appears to be a unique molecule as CAF lacks identity with other cytokines and its activity cannot be blocked by antibodies against other soluble factors such as IFN-α and -β, TNF-α, TGF-β, IL-4, IL-6, and IL-8 (Mackewicz *et al.*, 1994a; Levy *et al.*, 1996; Levy, 2003). CAF was also shown not to be related to the chemokines RANTES, MIP-1α, and MIP-1β (Mackewicz *et al.*, 1997), which have been identified as anti-HIV factors (Cocchi *et al.*, 1995). Soluble factors present in CD8[+] cell granules were similarly excluded, for example, human granzyme A and B (Mackewicz *et al.*, 2000a), perforin (Mackewicz *et al.*, 2000b), and granulolysin (Mackewicz *et al.*, 2000c). Likewise, this activity was not linked to α-defensins, which are not produced by CD8$^+$ cells (Mackewicz *et al.*, 2003).

In the search for CAF, other natural anti-HIV factors have been found including the chemokines (Cocchi *et al.*, 1995) and α-defensins (Zhang *et al.*, 2002). CD8$^+$ cell lines were reported to secrete proteins at sizes between 3 and 10 kDa, which efficiently inhibited replication of M-tropic and T-tropic HIV-1 isolates through a non-cytolytic mechanism (Mosoian *et al.*, 2000). Another group described a heat-labile CD8$^+$ cell suppressive activity of >50 kDa that could bind to heparin; it was different from chemokines, cytokines, and the interleukins (Geiben-Lynn *et al.*, 2001). Subsequently, the same group reported that serum bovine antithrombin III was modified into an HIV-inhibiting factor by activated CD8$^+$ cells from HIV-infected individuals (Geiben-Lynn *et al.*, 2002). In other studies, the peroxiredoxins NKEF-A and NKEF-B were shown to be upregulated in activated CD8$^+$ cells; and these proteins in high concentrations inhibited HIV-1 replication (Geiben-Lynn *et al.*, 2003). A proteolytically cleaved form of the hemofiltrate CC chemokine (HCC)-1 was also reported to block entry of HIV-1 strains using CCR5 as coreceptor (Detheux *et al.*, 2000). However, these other proteins are not present in sufficient amounts in CAF-containing fluids to be responsible for the clinically relevant anti-HIV activity observed (Levy, 2003).

8.4.5 Relationship of CNAR to Clinical State

Non-cytotoxic anti-HIV activity of CD8$^+$ cells has been reported to play an important role in the early control of perinatal HIV infection (Pollack *et al.*, 1997). As noted above, a potent suppressive effect of CD8$^+$ cells has been observed in asymptomatic HIV-infected individuals with high CD4$^+$ cell counts (Chen *et al.*, 1993; Landay *et al.*, 1993; Gomez *et al.*, 1994; Barker *et al.*, 1998). Similarly, CNAR measured in blood and lymphoid tissue has correlated with lower viral loads (Blackbourn *et al.*, 1996). In infected subjects, CNAR was found decreased with antiretroviral therapy (Stranford *et al.*, 2001), but restored by administration of IL-2 (Martinez-Marino *et al.*, 2004). Recently, rapid screening assays for the detection of CNAR were developed that will allow a broader application in clinical studies by simplifying the experimental procedures (Castillo *et al.*, 2000; Killian *et al.*, 2005).

8.5 Other Innate Immune Cells

8.5.1 Natural Killer Cells

NK cells are an important component of the innate host antimicrobial defense system through their production of immunoregulatory cytokines and cytotoxic capacity (Trinchieri, 1989). Comprising about 15% of all circulating lymphocytes, two subsets of NK cells can be distinguished according to their expression of cell surface markers: (1) $CD56^{bright}$ $CD16^-$ subset, which accounts for 10% of NK cells. This subset is a major producer of cytokines upon stimulation but exhibits low natural cytotoxicity. (2) The $CD56^{dim}$ $CD16^+$ subset, which represents the majority of NK cells and is cytotoxic [reviewed in Cooper *et al.*, (2001)]. Upon activation, NK cells produce IFN-γ, TNF-α, MIP-1α (CCL3), MIP-1β (CCL4), and RANTES (CCL5) (Fehniger *et al.*, 1998; Oliva *et al.*, 1998). These cytokines can promote the activation of other immune effector cells (Sallusto *et al.*, 2004). NK cells preferentially lyse virus-infected cells or tumor cells whose MHC molecules are downregulated. In this regard, their function is blocked when HLA-A, -B, and -C are recognized by the killer cell Ig-like receptors (KIRs), HLA-E by the C-type lectin superfamily including CD94 and NKG2, and unknown ligands by natural cytotoxicity receptors (Lanier, 2005).

HIV interacts with NK cells in different ways [reviewed in Jacobs *et al.*, (2005)]. HIV proteins reduce the expression of MHC molecules from the cell surface [reviewed in Kamp *et al.*, (2000)], contributing to the viral escape from the host adaptive immune surveillance. As examples, Tat represses the transcription of MHC-I molecules, Vpu retains nascent MHC-1 chains in the endoplasmic reticulum, and Nef causes internalization of MHC molecules (Kamp *et al.*, 2000). Nevertheless, despite HLA-A and HLA-B downregulation, HLA-C or HLA-E are still expressed and help HIV avoid NK cell recognition and lysis (Cohen *et al.*, 1999; Bonaparte and Barker, 2004; Ward *et al.*, 2004). Recently, HLA-G expression induced on cells by HIV infection has been shown to inhibit NK killing of infected $CD4^+$ cells (Bonaparte and Barker, 2004) (Barker, E, pers. commun.).

The $CD4^+$ subset of NK cells which also expresses CCR5 or CXCR4 has been shown to be sensitive to HIV infection, with persistently infected cells supporting virus propagation (Valentin *et al.*, 2002) (Table 8.4). The SCID-hu mouse model has similarly shown that $CD4^+$ (NK) thymocytes expressing CCR5 and CXCR4 can be productively infected by HIV-1, which leads to depletion of these cells in this model (Gurney *et al.*, 2002).

In chronically HIV-infected individuals, CD16$^+$ CD56$^+$ cell numbers are generally reduced (Ahmad *et al.*, 2001; Tarazona *et al.*, 2002; Alter *et al.*, 2004; Azzoni *et al.*, 2005), whereas CD16$^+$ CD56$^-$ cell numbers are increased (Mavilio *et al.*, 2003) (Table 8.3); the latter cell type reflects a reduced expression of CD56 on NK cells (Hu *et al.*, 1995). In acute HIV-1 infection, an early loss of immune regulatory cytokine-secreting CD56^{bright} cells has been reported (Alter *et al.*, 2005). This loss is associated with a reduction in the cytolytic CD56^{dim} cells, paralleled by an accumulation of functionally anergic CD56^{neg} cells. A sequential loss of NK cell subsets results. In contrast, NK cell numbers and cytotoxicity were found preserved in a group of long-term nonprogressors with low CD4$^+$ cell counts (Ironson *et al.*, 2001).

In addition to numerical deficits, functional impairments of NK cell activities in HIV-infected subjects have been described (Table 8.3). The decreased surface expression of activating natural cytotoxicity receptors NKp30, NKp44, and NKp46 is associated with a reduction in NK-cell mediated killing of tumor cells *in vitro* (De Maria *et al.*, 2003; Mavilio *et al.*, 2003). Furthermore, the capacity of NK cells to lyse immature dendritic cells has been reported to be impaired (Tasca *et al.*, 2003). The expressions of perforin and granzyme A were low in HIV viremic individuals and restored by treatment with IFN-α (Portales *et al.*, 2003). Reduced lytic activity of NK cells (Weber *et al.*, 2000) and lymphokine-activated killer cells (Ullum *et al.*, 1995) was paralleled by a sustained impairment of IFN-γ secretion (Azzoni *et al.*, 2002). HIV replication mediated by NK cell–secreted CC-chemokines was also considerably reduced in patients with high viremia (Kottilil *et al.*, 2003).

8.5.2 NK-T Cells

Natural killer T (NK-T) cells, accounting for 0.01% to 1% of PBMCs, are a subset of T lymphocytes sharing NK and T cell markers [reviewed in Kim *et al.* (2002); Godfrey and Kronenberg (2004)]. Up to four sub-populations of cells can be identified by differential expression of CD4, CD8, or the absence of both markers (Wilson and Delovitch, 2003). In humans, the major NK-T cell subset studied expresses a restricted T-cell receptor repertoire of Vα24 preferentially paired with Vβ11 and is called the invariant NK-T cell (iNK-T) (Wilson and Delovitch, 2003). In contrast with conventional lymphocytes, NK-T cells do not recognize peptides presented in the context of MHC class I or class II, but recognize glycolipids (e.g., α-galactosylceramide) presented by the MHC class I-like molecule CD1d that is expressed on MDCs and other cells. About 50% of NK-T cells are positive for the CD4 receptor; some subsets express the

HIV coreceptors CXCR4 and CCR5 (Kim *et al.*, 2002). Among other chemokine and adhesion molecules, CCR7 is found on about 20% of NK-T cells that traffic to secondary lymphatic tissues (Kim *et al.*, 2002).

Upon activation, NK-T cells release IFN-γ, TNF-α, IL-4, and IL-13 within hours. NK-T cells can, therefore, polarize the immune response into a Th1 and Th2 direction. Recent observations suggest the CD4$^+$ NK-T cells release IL-4 and IL-13 favoring a Th2 response whereas the CD4$^-$ NK-T cells produce IFN-γ and TNF-α (Crough *et al.*, 2004). In addition, NK-T cells display cytolytic activities via perforin and granzyme B (Smyth *et al.*, 2000) as well as Fas targets via Fas ligand expression (Arase *et al.*, 1994).

NK-T cells have been implicated in the regulation of immune responses in autoimmune diseases, cancer, allograft transplant rejection, and infectious diseases. In HIV infection, NK-T cells with high expression levels of CCR5 and CD4 can be targeted by HIV (Table 8.4). NK-T cells have been found more susceptible to infection with R5-tropic than X4-tropic viruses; HIV replication in these cells is cytolytic (Motsinger *et al.*, 2002; Sandberg *et al.*, 2002). In this regard, when human NK-T cell clones were infected with different viral strains, susceptibility to R5 and X4-tropic isolates was consistent with the expression of the respective coreceptors, although the production of progeny virus was delayed in comparison with virus replication in PHA-stimulated PBMCs (Fleuridor *et al.*, 2003). These data suggest that CD4$^+$ NK-T cells could represent a gateway for establishment of HIV during or after initial transmission (Unutmaz, 2003).

Depletion of CD4$^+$ NK-T cells in HIV-infected individuals has been reported (Table 8.3). Reduced percentages of these cells have shown an inverse correlation with viral load in HIV-infected adults (Motsinger *et al.*, 2002) and a direct correlation with the CD4$^+$ T-cell depletion in HIV-infected pediatric subjects (Sandberg *et al.*, 2002). In contrast, some investigators have found a decrease in NK-T cells independent of the CD4$^+$ T lymphocyte count, CD4$^+$:CD8$^+$ cell ratio, and viral load (van der Vliet *et al.*, 2002). Because the depletion was largely observed during the first year after seroconversion, the authors attribute the depletion primarily to Fas-mediated apoptosis rather than HIV infection of these cells.

In HIV infection, functional deficiencies for NK-T cells have also been reported (Table 8.3). In this regard, when NK cells were characterized as Th1 and Th2 cells by the expression of IL-18R and ST2L, respectively, a clear NK Th-1:NK-Th-2 shift of these cells was observed in HIV-infected individuals (Chan *et al.*, 2001). A similar shift in Th response has been observed for NK-T cells (Chan *et al.*, 2003). Recent publications describe the downregulation of CD1d expressed on the cell surface of HIV-infected cells by the physical interaction of Nef with the cytoplasmic domain of CD1d. This process concomitantly reduces the activation of NK-T cells (Chan *et al.*, 2003; Cho *et al.*, 2005) and presumably their function.

8.5.3 γδ T Cells

The majority of circulating human γδ cells are Vγ2Vδ2 cells and mostly reside in epithelial mucosae [reviewed in Kabelitz and Wesch (2003)]. These cells recognize non-peptidic microbial phosphoantigens in a TCR-dependent, but MHC-independent manner [reviewed in Chen and Letvin (2003)]. Major expansions of these cells are observed in the blood of patients with bacterial infections where they primarily contribute to the antimicrobial host defense. Upon activation, they produce large amounts of IFN-γ and TNF-α and exert cytotoxicity via the perforin/granzyme pathway or induction of apoptosis via the Fas/FasL pathway (Boullier et al., 1997).

Several groups have reported a considerable increase in the proportion of Vδ1 cells in HIV-infected individuals (De Paoli et al., 1991; De Maria et al., 1992; Sindhu et al., 2003; Poggi et al., 2004), accompanied by a decrease of Vδ2 cells (Autran et al., 1989; Boullier et al., 1995; Martini et al., 2000; Dobmeyer et al., 2002; Poles et al., 2003) (Table 8.3). Conflicting data have been reported whether the ratio of Vδ1 to Vδ2 cells is restored during HAART: some groups have not seen this effect (Poles et al., 2003), whereas others observed at least a partial recovery with reinitiation of HAART after treatment interruption (Martini et al., 2002). In addition, the TCR repertoire on these cells was partially restored with a longer duration of HAART. The largest effect was observed in patients with viral loads below the detection level (Bordon et al., 2004).

Certain γδ cells expressing CD4, CCR5, and CXCR4 have been reported to be targeted by HIV-1, resulting in productive infection of these cells (Imlach et al., 2003) (Table 8.4). Similar data were reported using the SCID-hu mouse model: TCR$^+$ γδ thymocytes were susceptible to infection with X4- and R5-tropic viruses, and productive infection led to the depletion of these cells (Gurney et al., 2002). The viral Tat protein has been reported to interfere with the migration and transcytosis of CXCR3 and CXCR4 expressing Vδ2 and Vδ1 cells in response to IP-10 (CXCL10) and SDF-1 (CXCL12) (Poggi et al., 2004).

Functional abnormalities have also been reported in γδ cells of HIV-infected individuals (Table 8.3). Proliferative responses of these cells to Daudi cells and mycobacterial antigen were substantially reduced compared with HIV-uninfected controls (Poccia et al., 1996; Wallace et al., 1997). A substantial unresponsiveness was observed with respect to production of IFN-γ by Vδ2 cells of HIV-infected patients with opportunistic infections, which was restored by HAART (Martini et al., 2000; Martini et al., 2002). The expansion of cytotoxic Vδ1 cells, which directly correlated with viral load, could contribute to the HIV-immunopathogenesis via depletion of bystander CD4$^+$ T cells (Sindhu et al., 2003). Moreover, γδ cells have been found to produce an anti-HIV factor(s) that suppresses virus replication in CD4$^+$ cells (Poccia *et al.*, 2002). The nature of this factor has not yet been reported.

8.6 Concluding Remarks

The innate immune system as the first line of defense against pathogens has an important role in the defense against HIV infection and disease. With the high production of interferons and other cytokines and their function as APCs, the dendritic cells contribute to both innate and adaptive immune responses (Sections 8.3 and 8.4). Other cellular activities, particularly the CD8$^+$ cell noncytotoxic antiviral response and the potential attack on virus-infected cells by NK, NK-T, and γδ T cells, need to be better appreciated. They represent the earliest type of antimicrobial activity that could prevent HIV transmission and certainly limit its pathogenic course. At the same time, there are multiple effects of HIV on cells involved in this innate immune defense. HIV infection of these cells that express the CD4 receptor and coreceptor(s) can lead to cell death, as described for PDC, NK, and NK-T cells. If these cells do not support virus replication as has been described for immature PDCs and MDCs, they can be viral reservoirs and carry the virus to lymph nodes and other tissues. A reduction in cell number also results from the upregulation of cell surface markers such as CCR7, contributing to homing of these cells to secondary lymphatic tissue, as has been shown for PDCs. Numerical deficiencies can also result from the downregulation of cell surface markers used to identify certain cells, for example, CD56 on NK cells.

Reduced expression of surface markers also contributes to functional deficiencies as has been reported for the downregulation of natural cyto-toxicity receptors on NK cells. Equally important are other functional defects that not only affect the innate, but the adaptive immune response as well. Thus, impaired stimulation of allogeneic T-cell proliferation has been observed in PDCs and MDCs obtained from HIV-infected individuals. Reduced lytic activity as well as low expression of IFN-γ in NK cells impairs the interaction between the innate and adaptive immune system.

HIV has developed effective mechanisms to escape, survive in, and even profit from cells involved in the innate immune defense. It is very likely that increasing subsets of innate cells will be identified that are affected in their number and function in HIV-infected individuals. In this regard, MDCs are now separated into MDC1 and MDC2 cells according to their capability of differentiating into Langerhans cells (Ito *et al.*, 1999;

Dzionek *et al.*, 2000). Very low numbers of circulating MDC2 cells are difficult for measuring numbers and function. Similarly, increasing receptor molecules will be identified on innate and adaptive effector cells. These developments will contribute to a more detailed understanding of HIV immunopathogenicity and may result in new therapeutic approaches. It is particularly noteworthy that the first-line innate immune response is usually non-cytolytic (except for NK cells) and thus by principle should not cause any harm, whereas the second-line defense is cytotoxic and induces damage to infected as well as bystander cells. In this respect, agonists for Toll-like receptors 7 and 9 are already in use for enhancing innate immune responses in vaccination, tumors, and infectious diseases (Vollmer *et al.*, 2004; Craft *et al.*, 2005; Vollmer, 2005; Wille-Reece *et al.*, 2005). Some of these compounds as well as others could add a new face to anti-HIV therapy.

Acknowledgments: The authors would like to acknowledge the support of the German Research Foundation (grant SCHM 1702/1-1) (B.S.) as well as a grant from the Universitywide AIDS Research Program of the State of California (CC02-SF-002) (J.A.L.). Nicolai Kittan was supported by the graduate college GRK1071 ("Viruses of the Immune System"). We thank Kaylynn Peter for assistance in the preparation of the manuscript.

References

Ahmad, R., Sindhu, S. T., Toma, E., Morisset, R., Vincelette, J., Menezes, J. and Ahmad, A. (2001). Evidence for a correlation between antibody-dependent cellular cytotoxicity-mediating anti-HIV-1 antibodies and prognostic predictors of HIV infection. *J. Clin. Immunol.* 21:227-233.

Ahmad-Nejad, P., Hacker, H., Rutz, M., Bauer, S., Vabulas, R. M. and Wagner, H. (2002). Bacterial CpG-DNA and lipopolysaccharides activate Toll-like receptors at distinct cellular compartments. *Eur. J. Immunol.* 32:1958-1968.

Akira, S. and Takeda, K. (2004). Toll-like receptor signalling. *Nat. Rev. Immunol.* 4:499-511.

Almeida, M., Cordero, M., Almeida, J. and Orfao, A. (2005). Different subsets of peripheral blood dendritic cells show distinct phenotypic and functional abnormalities in HIV-1 infection. *AIDS* 19:261-271.

Alter, G., Malenfant, J. M., Delabre, R. M., Burgett, N. D., Yu, X. G., Lichterfeld, M., Zaunders, J. and Altfeld, M. (2004). Increased natural killer cell activity in viremic HIV-1 infection. *J. Immunol.* 173:5305-5311.

Alter, G., Teigen, N., Davis, B. T., Addo, M. M., Suscovich, T. J., Waring, M. T., Streeck, H., Johnston, M. N., Staller, K. D., Zaman, M. T., Yu, X. G., Lichterfeld, M., Basgoz, N., Rosenberg, E. S. and Altfeld, M. (2005).

Sequential deregulation of NK cell subset distribution and function starting in acute HIV-1 infection. *Blood* 106:3366-3369.

Arase, H., Arase, N., Kobayashi, Y., Nishimura, Y., Yonehara, S. and Onoe, K. (1994). Cytotoxicity of fresh NK1.1[+] T cell receptor alpha/beta[+] thymocytes against a CD4[+]8[+] thymocyte population associated with intact Fas antigen expression on the target. *J. Exp. Med.* 180:423-432.

Arce, I., Roda-Navarro, P., Montoya, M. C., Hernanz-Falcon, P., Puig-Kroger, A. and Fernandez-Ruiz, E. (2001). Molecular and genomic characterization of human DLEC, a novel member of the C-type lectin receptor gene family preferentially expressed on monocyte-derived dendritic cells. *Eur. J. Immunol.* 31:2733-2740.

Autran, B., Triebel, F., Katlama, C., Rozenbaum, W., Hercend, T. and Debre, P. (1989). T cell receptor g/d[+] lymphocyte subsets during HIV infection. *Clin. Exp. Immunol.* 75:206-210.

Azzoni, L., Papasavvas, E., Chehimi, J., Kostman, J. R., Mouner, K., Ondercin, J., Perussia, B. and Montaner, L. J. (2002). Sustained impairment of IFN-g secretion in suppressed HIV-infected patients despite mature NK cell recovery: evidence for a defective reconsititution of innate immunity. *J. Immunol.* 168:5764-5770.

Azzoni, L., Rutstein, R. M., Chehimi, J., Farabaugh, M. A. and Montaner, L. J. (2005). Dendritic and natural killer cell subsets associated with stable or declining CD4[+] cell counts in treated HIV-1-infected children. *J. Infect. Dis.* 191:1451-1459.

Baier, M., Werner, A., Bannert, N., Metzner, K. and Kurth, R. (1995). HIV suppression by interleukin-16. *Nature* 378:563.

Barker, E., Bossart, K. N., Fujimura, S. H. and Levy, J. A. (1997). CD28-costimulation increases CD8[+] cell suppression of HIV replication. *J. Immunol.* 159:5123-5131.

Barker, E., Bossart, K. N. and Levy, J. A. (1998). Primary CD8[+] cells from HIV-infected individuals can suppress productive infection of macrophages independent of b-chemokines. *Proc. Natl. Acad. Sci. U. S. A.* 95:1725-1729.

Barker, E., Bossart, K. N., Fujimura, S. H. and Levy, J. A. (1999). The Role of CD80 and CD86 in enhancing CD8([+]) cell suppression of HIV replication. *Cell. Immunol.* 196:95-103.

Barron, M. A., Blyveis, N., Palmer, B. E., MaWhinney, S. and Wilson, C. C. (2003). Influence of plasma viremia on defects in number and immunophenotype of blood dendritic cell subsets in human immunodeficiency virus 1-infected individuals. *J. Infect. Dis.* 187:26-37.

Beignon, A. S., McKenna, K., Skoberne, M., Manches, O., Dasilva, I., Kavanagh, D. G., Larsson, M., Gorelick, R. J., Lifson, J. D. and Bhardwaj, N. (2005). Endocytosis of HIV-1 activates plasmacytoid dendritic cells via Toll-like receptor-viral RNA interactions. *J. Clin. Invest.* 115:3265-3275.

Beutler, B. (2004). Innate immunity: an overview. *Mol. Immunol.* 40:845-59.

Biron, C. A. (1998). Role of early cytokines, including alpha and beta interferons (IFN- alpha/beta), in innate and adaptive immune responses to viral infections. *Semin. Immunol.* 10:383-90.

Biron, C. A., Nguyen, K. B., Pien, G. C., Cousens, L. P. and Salazar-Mather, T. P. (1999). Natural killer cells in antiviral defense: function and regulation by innate cytokines. *Annu. Rev. Immunol.* 17:189-220.

Blackbourn, D. J., Mackewicz, C. E., Barker, E., Hunt, T. K., Herndier, B., Haase, A. T. and Levy, J. A. (1996). Suppression of HIV replication by lymphoid tissue CD8[+] cells correlates with the clinical state of HIV-infected individuals. *Proc. Natl. Acad. Sci. U. S. A.* 93:13125-13130.

Blackbourn, D. J., Locher, C. P., Ramachandran, B., Barnett, S. W., Murthy, K. K., Carey, K. D., Brasky, K. M. and Levy, J. A. (1997). CD8[+] cells from HIV-2-infected baboons control HIV replication. *AIDS* 11:737-746.

Bleul, C. C., Farzan, M., Choe, H., Parolin, C., Clark-Lewis, I., Sodroski, J. and Springer, T. A. (1996). The lymphocyte chemoattractant SDF-1 is a ligand for LESTR/fusin and blocks HIV-1 entry. *Nature* 382:829-832.

Bogdan, C. (2000). The function of type I interferons in antimicrobial immunity. *Curr. Opin. Immunol.* 12:419-424.

Bonaparte, M. I. and Barker, E. (2004). Killing of human immunodeficiency virus-infected primary T-cells by autologous natural killer cells is dependent on the ability of the virus to alter the expression of major histocompatibility complex class I molecules. *Blood* 104:2087-2094.

Bordon, J., Evans, P. S., Propp, N., Davis, C. E., Jr., Redfield, R. R. and Pauza, C. D. (2004). Association between longer duration of HIV-suppressive therapy and partial recovery of the V gamma 2 T cell receptor repertoire. *J. Infect. Dis.* 189:1482-6.

Boullier, S., Cochet, M., Poccia, F. and Gougeon, M. L. (1995). CDR3-independent gamma delta V delta 1[+] T cell expansion in the peripheral blood of HIV-infected persons. *J. Immunol.* 154:1418-31.

Boullier, S., Dadaglio, G., Lafeuillade, A., Debord, T. and Gougeon, M. L. (1997). V delta 1 T cells expanded in the blood throughout HIV infection display a cytotoxic activity and are primed for TNF-alpha and IFN-gamma production but are not selected in lymph nodes. *J. Immunol.* 159:3629-37.

Brinchmann, J. E., Gaudernack, G. and Vartdal, F. (1990). CD8[+] T cells inhibit HIV replication in naturally infected CD4[+] T cells: evidence for a soluble inhibitor. *J. Immunol.* 144:2961-2966.

Butera, S. T., Pisell, T. L., Limpakarnjanarat, K., Young, N. L., Hodge, T. W., Mastro, T. D. and Folks, T. M. (2001). Production of a novel viral suppressive activity associated with resistance to infection amoung female sex workers exposed to HIV type 1. *AIDS Res. Hum. Retroviruses* 17:735-744.

Castelli, J., Thomas, E., Liu, Y.-J. and Levy, J. A. (2004). Mature dendritic cells can enhance CD8[+] cell noncytotoxic anti-HIV responses: The role of IL-15. *Blood* 103:2699-2704.

Castelli, J. C., Deeks, S. G., Shiboski, S. and Levy, J. A. (2002). Relationship of CD8[+] T cell non-cytotoxic anti-HIV response to CD4[+] T cell number in untreated asymptomatic HIV-infected individuals. *Blood* 99:4225-4227.

Castillo, R. C., Arango-Jaramillo, S., John, R., Weinhold, K., Kanki, P., Carruth, L. and Schwartz, D. H. (2000). Resistance to human immunodeficiency virus type 1 *in vitro* as a surrogate of vaccine-induced protective immunity. *J. Infect. Dis.* 181:897-903.

Castro, B. A., Walker, C. M., Eichberg, J. W. and Levy, J. A. (1991). Suppression of human immunodeficiency virus replication by CD8⁺ cells from infected and uninfected chimpanzees. *Cell. Immunol.* 132:246-255.

Castro, B. A., Homsy, J., Lennette, E., Murthy, K. K., Eichberg, J. W. and Levy, J. A. (1992). HIV-1 expression in chimpanzees can be activated by CD8⁺ cell depletion or CMV infection. *Clin. Immunol. Immunopathol.* 65:227-233.

Cella, M., Facchetti, F., Lanzavecchia, A. and Colonna, M. (2000). Plasmacytoid dendritic cells activated by influenza virus and CD40L drive a potent TH1 polarization. *Nat. Immunol.* 1:305-310.

Cella, M., Jarrossay, D., Facchetti, F., Alebardi, O., Nakajima, H., Lanzavecchia, A. and Colonna, M. (1999). Plasmacytoid monocytes migrate to inflamed lymph nodes and produce large amounts of type I interferon. *Nat. Med.* 5: 919-23.

Chan, W. L., Pejnovic, N., Lee, C. A. and Al-Ali, N. A. (2001). Human IL-18 receptor and ST2L are stable and selective markers for the respective type 1 and type 2 circulating lymphocytes. *J. Immunol.* 167:1238-44.

Chan, W. L., Pejnovic, N., Liew, T. V., Lee, C. A., Groves, R. and Hamilton, H. (2003). NKT cell subsets in infection and inflammation. *Immunol. Lett.* 85:159-63.

Chehimi, J., Campbell, D. E., Azzoni, L., Bacheller, D., Papasavvas, E., Jerandi, G., Mounzer, K., Kostman, J., Trinchieri, G. and Montaner, L. J. (2002). Persistent decreases in blood plasmacytoid dendritic cell number and function despite effective highly active antiretroviral therapy and increased blood myeloid dendritic cells in HIV-infected individuals. *J. Immunol.* 168: 4796-4801.

Chen, Z. W. and Letvin, N. L. (2003). Vgamma2Vdelta2⁺ T cells and anti-microbial immune responses. *Microbes Infect.* 5:491-498.

Chen, C. H., Weinhold, K. J., Bartlett, J. A., Bolognesi, D. P. and Greenberg, M. L. (1993). CD8⁺ T lymphocyte-mediated inhibition of HIV-1 long terminal repeat transcription: a novel antiviral mechanism. *AIDS Res. Hum. Retroviruses* 9:1079-1086.

Chen, W., Antonenko, S., Sederstrom, J. M., Liang, X., Chan, A. S., Kanzler, H., Blom, B., Blazar, B. R. and Liu, Y.-J. (2004). Thrombopoietin cooperates with FLT3-ligand in the generation of plasmacytoid dendritic cell precursors from human hematopoietic progenitors. *Blood* 103:2547-2553.

Cho, S., Knox, K. S., Kohli, L. M., He, J. J., Exley, M. A., Wilson, S. B. and Brutkiewicz, R. R. (2005). Impaired cell surface expression of human CD1d by the formation of an HIV-1 Nef/CD1d complex. *Virology* 337:242-52.

Cocchi, F., DeVico, A. L., Garzino-Demo, A., Arya, S. K., Gallo, R. C. and Lusso, P. (1995). Identification of RANTES, MIP-1alpha, and MIP-1beta as the major HIV-suppressive factors produced by CD8⁺ T cells. *Science* 270:1811-1815.

Cohen, G. B., Gandhi, R. T., Davis, D. M., Mandelboim, O., Chen, B. K., Strominger, J. L. and Baltimore, D. (1999). The selective downregulation of class 1 major histocompatibility complex proteins by HIV-1 protects HIV-infected cells from NK cells. *Immunity* 10.

Cooper, M. A., Fehniger, T. A. and Caligiuri, M. A. (2001). The biology of human natural killer-cell subsets. *Trends Immunol.* 22:633-40.

Copeland, K. F. T., McKay, P. J. and Rosenthal, K. L. (1995). Suppression of activation of the human immunodeficiency virus long terminal repeat by CD8+ T cells is not lentivirus specific. *AIDS Res. Hum. Retroviruses* 11: 1321-1326.

Craft, N., Bruhn, K. W., Nguyen, B. D., Prins, R., Lin, J. W., Liau, L. M. and Miller, J. F. (2005). The TLR7 agonist imiquimod enhances the anti-melanoma effects of a recombinant Listeria monocytogenes vaccine. *J. Immunol.* 175:1983-90.

Crough, T., Nieda, M. and Nicol, A. J. (2004). Granulocyte colony-stimulating factor modulates a-galactosylceramide-responsive human Va24+Vb11+ NKT cells. *J. Immunol.* 173:4960-4966.

Cyster, J. G. (1999). Chemokines and cell migration in secondary lymphoid organs. *Science* 286:2098-2102.

De Maria, A., Ferrazin, A., Ferrini, S., Ciccone, E., Terragna, A. and Moretta, L. (1992). Selective increase of a subset of T cell receptor gd T lymphocytes in the peripheral blood of patients with human immunodeficiency virus type 1 infection. *J. Infect. Dis.* 165:917-919.

De Maria, A., Fogli, M., Costa, P., Murdaca, G., Puppo, F., Mavilio, D., Moretta, A. and Moretta, L. (2003). The impaired NK cell cytolytic function in viremic HIV-1 infection is associated with a reduced surface expression of natural cytotoxicity receptors (NKp46, NKp30 and NKp44). *Eur. J. Immunol.* 33:2410-8.

De Paoli, P., Gennari, D., Martelli, P., Basaglia, G., Crovatto, M., Battistin, S. and Santini, G. (1991). A subset of gamma delta lymphocytes is increased during HIV-1 infection. *Clin. Exp. Immunol.* 83:187-91.

Detheux, M., Standker, L., Vakili, J., Munch, J., Forssmann, U., Adermann, K., Pohlmann, S., Vassart, G., Kirchhoff, F., Parmentier, M. and Forssmann, W. G. (2000). Natural proteolytic processing of hemofiltrate CC chemokine 1 generates a potent CC chemokine receptor (CCR)1 and CCR5 agonist with anti-HIV properties. *J. Exp. Med.* 192:1501-8.

Diaz, L. S., Stone, M., Mackewicz, C. E. and Levy, J. A. (2003). Differential gene expression in CD8+ cells exhibiting non-cytotoxic anti-HIV activity. *Virology* 311:400-409.

Diaz, L. S., Foster, H., Stone, M. R., Fujimura, S., Relman, D. A. and Levy, J. A. (2005). VCAM-1 expression on CD8+ cells correlates with enhanced anti-HIV suppressing activity. *J. Immunol.* 174:1574-9.

Diebold, S. S., Kaisho, T., Hemmi, H., Akira, S. and Reis e Sousa, C. (2004). Innate antiviral responses by means of TLR7-mediated recognition of single-stranded RNA. *Science* 303:1529-31.

Dobmeyer, T. S., Dobmeyer, R., Wesch, D., Helm, E. B., Hoelzer, D. and Kabelitz, D. (2002). Reciprocal alterations of Th1/Th2 function in gammadelta T-cell subsets of human immunodeficiency virus-1-infected patients. *Br. J. Haematol.* 118:282-8.

Donaghy, H., Gazzard, B., Gotch, F. and Patterson, S. (2003). Dysfunction and infection of freshly isolated blood myeloid and plasmacytoid dendritic cells in patients infected with HIV-1. *Blood* 101:4505-4511.

Donaghy, H., Pozniak, A., Gazzard, B., Qazi, N., Gilmour, J., Gotch, F. and Patterson, S. (2001). Loss of blood CD11c($^+$) myeloid and CD11c(-) plasmacytoid dendritic cells in patients with HIV-1 infection correlates with HIV-1 RNA virus load. *Blood* 98:2574-2576.

Dzionek, A., Fuchs, A., Schmidt, P., Cremer, S., Zysk, M., Miltenyi, S., Buck, D. W. and Schmitz, J. (2000). BDCA-2, BDCA-3, and BDCA-4: Three markers for distinct subsets of dendritic cells in human peripheral blood. *J. Immunol.* 165:6037-6046.

Dzionek, A., Inagaki, Y., Okawa, K., Nagafune, J., Rock, J., Sohma, Y., Winkels, G., Zysk, M., Yamaguchi, Y. and Schmitz, J. (2002). Plasmacytoid dendritic cells: from specific surface markers to specific cellular functions. *Hum. Immunol.* 63:1133-1148.

Dzionek, A., Sohma, Y., Nagafune, J., Cella, M., Colonna, M., Facchetti, F., Günther, G., Johnston, I., Lanzavecchia, A., Nagasaka, T., Okada, T., Vermi, W., Winkels, G., Yamamoto, T., Zysk, M., Yamaguchi, Y. and Schmitz, J. (2001). BDCA-2, a novel plasmacytoid dendritic cell-specific type II C-lectin mediates antigen capture and is a potent inhibitor on interferon a/b induction. *J. Exp. Med.* 194:1823-1934.

Ezekowitz, A. B., Kuhlman, M., Groopman, J. E. and Byrn, R. A. (1989). A human serum mannose-binding protein inhibits *in vitro* infection by the human immunodeficiency virus. *J. Exp. Med.* 169:185-196.

Fehniger, T. A., Herbein, G., Yu, H. X., Para, M. I., Bernstein, Z. P., Obrien, W. A. and Caligiuri, M. A. (1998). Natural killer cells from HIV-1($^+$) patients produce C-C chemokines and inhibit HIV-1 infection. *J. Immunol.* 161: 6433-6438.

Feldman, S., Stein, D., Amrute, S., Denny, T., Garcia, Z., Kloser, P., Sun, Y., Megjugorac, N. and Fitzgerald-Bocarsly, P. (2001). Decreased interferon-alpha production in HIV-infected patients correlates with numerical and functional deficiencies in circulating type 2 dendritic cell precursors. *Clin. Immunol.* 101:201-210.

Finke, J. S., Shodell, M., Shah, K., Siegal, F. P. and Steinman, R. M. (2004). Dendritic cell numbers in the blood of HIV-1 infected patients before and after changes in antiretroviral therapy. *J. Clin. Immunol.* 24:647-52.

Fitzgerald-Bocarsly, P. (1993). Human natural interferon-alpha producing cells. *Pharmacol. Ther.* 60:39-62.

Fitzgerald-Bocarsly, P. (2002). Natural interferon-a producing cells: the plasmacytoid dendritic cells. *Biotechniques* 33:16-29.

Fleuridor, R., Wilson, B., Hou, R., Landay, A., Kessler, H. and Al-Harthi, L. (2003). CD1d-restricted natural killer T cells are potent targets for human immunodeficiency virus infection. *Immunology* 108:3-9.

Fong, L., Mengozzi, M., Abbey, N. W., Herdier, B. G. and Engleman, E. G. (2002). Productive infection of plasmacytoid dendritic cells with human immunodeficiency virus 1 is triggered by CD40 ligation. *J. Virol.* 76: 11033-11041.

Fonteneau, J. F., Larsson, M., Beignon, A. S., McKenna, K., Dasilva, I., Amara, A., Liu, Y. J., Lifson, J. D., Littman, D. R. and Bhardwaj, N. (2004). Human Immunodeficiency Virus type 1 activates plasmacytoid dendritic cells and concomitantly induces the bystander maturation of myeloid dendritic cells. *J. Virol.* 78:5223-5232.

Furci, L., Lopalco, L., Loverro, P., Sinnone, M., Tambussi, G., Lazzarin, A. and Lusso, P. (2002). Non-cytotoxic inhibition of HIV-1 infection by unstimulated CD8$^+$ T lymphocytes from HIV-exposed-uninfected individuals. *AIDS* 16:1003-1008.

Ganz, T. and Lehrer, R. I. (1997). Antimicrobial peptides of leukocytes. *Curr. Opin. Hematol.* 4:53-8.

Geiben-Lynn, R., Brown, N., Walker, B. D. and Luster, A. D. (2002). Purification of a modified form of bovine antithrombin III as an HIV-1 CD8$^+$ T-cell antiviral factor. *J. Biol. Chem.* 277:42352-42357.

Geiben-Lynn, R., Kursar, M., Brown, N. V., Kerr, E. L., Luster, A. D. and Walker, B. D. (2001). Noncytolytic inhibition of X4 virus by bulk CD8$^+$ cells from human immunodefiency virus type 1 (HIV-1)-infected persons and HIV-1 specific cytotoxic T lymphocytes is not mediated by b-chemokines. *J. Virol.* 75:8306-8316.

Geiben-Lynn, R., Kursar, M., Brown, N. W., Addo, M. M., Shau, H., Lieberman, J., Luster, A. D. and Walker, B. D. (2003). HIV-1 antiviral activity of recombinant natural killer cell enhancing factors NKEF-A and NKEF-B, Members of the peroxiredoxin family. *The J. Biol. Chem.* 278:1569-1574.

Gerosa, F., Gobbi, A., Zorzi, P., Burg, S., Briere, F., Carra, G. and Trinchieri, G. (2005). The reciprocal interaction of NK cells with plasmacytoid or myeloid dendritic cells profoundly affects innate resistance functions. *J. Immunol.* 174:727-734.

Gibson, S. J., Lindh, J. M., Riter, T. R., Gleason, R. M., Rogers, L. M., Fuller, A. E., Oesterich, J. L., Gorden, K. B., Qiu, X., McKane, S. W., Noelle, R. J., Miller, R. L., Kedl, R. M., Fitzgerald-Bocarsly, P., Tomai, M. A. and Vasilakos, J. P. (2002). Plasmacytoid dendritic cells produce cytokines and mature in response to the TLR7 agonists, imiquimod and resiquimod. *Cell. Immunol.* 218:74-86.

Godfrey, D. I. and Kronenberg, M. (2004). Going both ways: immune regulation via CD1d-dependent NKT cells. *J. Clin. Invest.* 114:1379-1388.

Gomez, A. M., Smaill, F. M. and Rosenthal, K. L. (1994). Inhibition of HIV replication by CD8$^+$ T cells correlates with CD4 counts and clinical stage of disease. *Clin. Exp. Immunol.* 97:68-75.

Grassi, F., Hosmalin, A., McIlroy, D., Calvez, V., Debre, P. and Autran, B. (1999). Depletion in blood CD11c-positive dendritic cells from HIV-infected patients. *AIDS* 13:759-766.

Greco, G., Barker, E. and Levy, J. A. (1998). Differences in HIV replication in CD4$^+$ lymphocytes are not related to b-chemokine production. *AIDS Res. Hum. Retroviruses* 14:1407-1411.

Grouard, G., Rissoan, M. C., Filgueira, L., Durand, I., Banchereau, J. and Liu, Y. J. (1997). The enigmatic plasmacytoid T cells develop into dendritic cells with interleukin (IL)-3 and CD40-ligand. *J. Exp. Med.* 185:1101-1111.

Gurney, K. B., Yang, O. O., Wilson, S. B. and Uittenbogaart, C. H. (2002). TCR gamma delta[+] and CD161[+] thymocytes express HIV-1 in the SCID-hu mouse, potentially contributing to immune dysfunction in HIV infection. *J. Immunol.* 169:5338-5346.

Hartmann, G., Weiner, G. J. and Krieg, A. M. (1999). CpG DNA: a potent signal for growth, activation, and maturation of human dendritic cells. *Proc. Natl. Acad. Sci. U. S. A.* 96:9305-9310.

Heil, F., Hemmi, H., Hochrein, H., Ampenberger, F., Kirschning, C., Akira, S., Lipford, G., Wagner, H. and Bauer, S. (2004). Species-specific recognition of single-stranded RNA via toll-like receptor 7 and 8. *Science* 303:1526-1529.

Hemmi, H., Takeuchi, O., Kawai, T., Kaisho, T., Sato, S., Sanjo, H., Matsumoto, M., Hoshino, K., Wagner, H., Takeda, K. and Akira, S. (2000). A Toll-like receptor recognizes bacterial DNA. *Nature* 408:740.

Hemmi, H., Kaisho, T., Takeuchi, O., Sato, S., Sanjo, H., Hoshino, K., Horiuchi, T., Tomizawa, H., Takeda, K. and Akira, S. (2002). Small anti-viral compounds activate immune cells via the TLR7 MyD88- dependent signaling pathway. *Nat. Immunol.* 22:22.

Hiroishi, K., Tuting, T. and Lotze, M. T. (2000). IFN-alpha-expressing tumor cells enhance generation and promote survival of tumor-specific CTLs. *J. Immunol.* 164:567-572.

Hoebe, K., Janssen, E. and Beutler, B. (2004). The interface between innate and adaptive immunity. *Nat Immunol* 5:971-974.

Hornung, V., Rothenfusser, S., Britsch, S., Krug, A., Jahrsdorfer, B., Giese, T., Endres, S. and Hartmann, G. (2002). Quantitative expression of toll-like receptor 1-10 mRNA in cellular subsets of human peripheral blood mononuclear cells and sensitivity to CpG oligodeoxynucleotides. *J. Immunol.* 168:4531-4537.

Howell, D. M., Feldman, S. B., Kloser, P. and Fitzgerald-Bocarsly, P. (1994). Decreased frequency of functional natural interferon-producing cells in peripheral blood of patients with the acquired immune deficiency syndrome. *Clin. Immunol. Immunopathol.* 71:223-30.

Hu, P. F., Hultin, L. E., Hultin, P., Hausner, M. A., Jirji, K., Jewett, A., Bonavida, B., Detels, R. and Giorgi, J. V. (1995). Natural killer cell immunodeficiency in HIV disease is manifest by profoundly decreased numbers of CD16[+]CD56[+] cells and expansion of a population of CD16dimCD56- cells with low lytic activity. *J. Acquir. Immune Defic. Syndr. Hum. Retrovirol.* 10:331-340.

Imlach, S., Leen, C. and Simmonds, P. (2003). Phenotypic analysis of peripheral blood ga T lymphocytes and their targeting by human immunodeficiency virus type 1 *in vivo*. *Virology* 305:415-427.

Ironson, G., Balbin, E., Solomon, G., Fahey, J., Klimas, N., Schneiderman, N. and Fletcher, M. A. (2001). Relative preservation of natural killer cell cytotoxicity and number in healthy AIDS patients with low CD4 cell counts. *AIDS* 15:2065-2073.

Isaacs, A. and Lindenmann, J. (1957). Virus interference. I. The interferon. *Proc. R. Soc. Lond. B. Biol. Sci.* 147:258-267.

Ito, T., Amakawa, R., Inaba, M., Hori, T., Ota, M., Nakamura, K., Takebayashi, M., Miyaji, M., Yoshimura, T., Inaba, K. and Fukuhara, S. (2004).

Plasmacytoid dendritic cells regulate Th cell responses through OX40 ligand and type I IFNs. *J. Immunol.* 172:4253-4259.

Ito, T., Inaba, M., Inaba, K., Toki, J., Sogo, S., Iguchi, T., Adachi, Y., Yamaguchi, K., Amakawa, R., Valladeau, J., Saeland, S., Fukuhara, S. and Ikehara, S. (1999). A CD1a$^+$/CD11c$^+$ subset of human blood dendritic cells is a direct precursor of Langerhans cells. *J. Immunol.* 163:1409-1419.

Iwasaki, A. and Medzhitov, R. (2004). Toll-like receptor control of the adaptive immune responses. *Nat. Immunol.* 5:987-995.

Jacobs, R., Heiken, H. and Schmidt, R. E. (2005). Mutual interference of HIV and natural killer cell-mediated immune response. *Mol. Immunol.* 42:239-49.

Jeng, C. R., English, R. V., Childers, T., Tompkins, M. B. and Tompkins, W. A. F. (1996). Evidence for CD8$^+$ antiviral activity in cats infected with feline immunodeficiency virus. *J. Virol.* 70:2474-2480.

Kabelitz, D. and Wesch, D. (2003). Features and functions of gamma delta T lymphocytes: focus on chemokines and their receptors. *Crit. Rev. Immunol.* 23:339-370.

Kadowaki, N., Antonenko, S., Lau, J. Y. and Liu, Y. J. (2000). Natural interferon alpha/beta-producing cells link innate and adaptive immunity. *J. Exp. Med.* 192:219-226.

Kadowaki, N., Antonenko, S. and Liu, Y. J. (2001a). Distinct CpG DNA and polyinosinic-polycytidylic acid double-stranded RNA, respectively, stimulate CD11c- type 2 dendritic cell precursors and CD11c$^+$ dendritic cells to produce type I IFN. *J. Immunol.* 166:2291-2295.

Kadowaki, N., Ho, S., Antonenko, S., Malefyt, R. W., Kastelein, R. A., Bazan, F. and Liu, Y. J. (2001b). Subsets of human dendritic cell precursors express different toll-like receptors and respond to different microbial antigens. *J. Exp. Med.* 194:863-869.

Kamp, W., Berk, M. B., Visser, C. J. and Nottet, H. S. (2000). Mechanisms of HIV-1 to escape from the host immune surveillance. *Eur. J. Clin. Invest.* 30:740-746.

Kannagi, M., Chalifoux, L. V., Lord, C. I. and Letvin, N. L. (1988). Suppression of simian immunodeficiency virus replication *in vitro* by CD8$^+$ lymphocytes. *J. Immunol.* 140:2237-2242.

Killian, M. S., Ng, S., Mackewicz, C. E. and Levy, J. A. (2005). A screening assay for detecting CD8($^+$) cell non-cytotoxic anti-HIV responses. *J. Immunol. Methods* 304:137-150.

Killian, M. S., Fujimura, S. H., Hecht, F. M., and Levy, J. A. (2006). Similar changes in plasmacytoid dendritic cell and CD4 T-cell counts during primary HIV-1 infection and treatment. AIDS 20:1247-1252.

Kim, C. H., Butcher, E. C. and Johnston, B. (2002). Distinct subsets of human Valpha24-invariant NKT cells: cytokine responses and chemokine receptor expression. *Trends Immunol.* 23:516-519.

Kim, C. H., Johnston, B. and Butcher, E. C. (2002). Trafficking machinery of NKT cells: shared and differential chemokine receptor expression among V alpha 24($^+$)V beta 11($^+$) NKT cell subsets with distinct cytokine-producing capacity. *Blood* 100:11-16.

Kottilil, S., Chun, T. S., Moir, S., Liu, S., McLaughlin, M., Hallahan, C. W., Maldarelli, F., Corey, L. and Fauci, A. S. (2003). Innate immunity in HIV infection: effect of HIV viremia on natural killer cell function. *J. Infect. Dis.* 187:1038-1045.

Krug, A., Towarowski, A., Britsch, S., Rothenfusser, S., Hornung, V., Bals, R., Giese, T., Engelmann, H., Endres, S., Krieg, A. M. and Hartmann, G. (2001). Toll-like receptor expression reveals CpG DNA as a unique microbial stimulus for plasmacytoid dendritic cells which synergizes with CD40 ligand to induce high amounts of IL-12. *Eur. J. Immunol.* 31:3026-3037.

Kuchtey, J., Chefalo, P. J., Gray, R. C., Ramachandra, L. and Harding, C. V. (2005). Enhancement of dendritic cell antigen cross-presentation by CpG DNA involves type I IFN and stabilization of class I MHC mRNA. *J. Immunol.* 175:2244-2251.

Kurtz, J. (2005). Specific memory within innate immune systems. *Trends Immunol* 26:186-192.

Landay, A. L., Mackewicz, C. and Levy, J. A. (1993). An activated CD8$^+$ T cell phenotype correlates with anti-HIV activity and asymptomatic clinical status. *Clin. Immunol. Immunopathol.* 69:106-116.

Lanier, L. L. (2005). NK cell recognition. *Annu. Rev. Immunol.* 23:225-274.

Latz, E., Schoenemeyer, A., Visintin, A., Fitzgerald, K. A., Monks, B. G., Knetter, C. F., Lien, E., Nilsen, N. J., Espevik, T. and Golenbock, D. T. (2004). TLR9 signals after translocating from the ER to CpG DNA in the lysosome. *Nat. Immunol.* 5:190-198.

Lehrer, R. I. and Ganz, T. (2002). Defensins of vertebrate animals. *Curr. Opin. Immunol.* 14:96-102.

Lemaitre, B., Nicolas, E., Michaut, L., Reichhart, J. M. and Hoffmann, J. A. (1996). The dorsoventral regulatory gene cassette spatzle/Toll/Cactus controls the potent antifungal response in Drosophila adults. *Cell* 1996:973-983.

Levy, J. A. (1996). Infection by human immunodeficiency virus - CD4 is not enough. *N. Engl. J. Med.* 335:1528-1530.

Levy, J. A. (2001). The importance of the innate immune system in controlling HIV infection and disease. *Trends Immunol.* 22:312-316.

Levy, J. A. (2003). The search for the CD8$^+$ cell anti-HIV factor (CAF). *Trends Immunol.* 24:628-632.

Levy, J. A., Mackewicz, C. E. and Barker, E. (1996). Controlling HIV pathogenesis: the role of noncytotoxic anti-HIV activity of CD8$^+$ cells. *Immunol. Today* 17:217-224.

Liu, Y. J. (2001). Dendritic cell subsets and lineages, and their functions in innate and adaptive immunity. *Cell* 106:259-262.

Mackewicz, C. E., Ortega, H. and Levy, J. A. (1994a). Effect of cytokines on HIV replication in CD4$^+$ lymphocytes: lack of identity with the CD8$^+$ cell antiviral factor. *Cell. Immunol.* 153:329-343.

Mackewicz, C. E., Yang, L. C., Lifson, J. D. and Levy, J. A. (1994b). Non-cytolytic CD8 T-cell anti-HIV responses in primary infection. *Lancet* 344:1671-1673.

Mackewicz, C. E., Blackbourn, D. J. and Levy, J. A. (1995). CD8$^+$ cells suppress human immunodeficiency virus replication by inhibiting viral transcription. *Proc. Natl. Acad. Sci. U. S. A.* 92:2308-2312.

Mackewicz, C. E., Barker, E., Greco, G., Reyes-Teran, G. and Levy, J. A. (1997). Do b-chemokines have clinical relevance in HIV infection? *J. Clin. Invest.* 100:921-930.

Mackewicz, C. E., Garovoy, M. R. and Levy, J. A. (1998). HLA compatibility requirements for CD8$^+$ T-cell-mediated suppression of human immunodeficiency virus replication. *J. Virol.* 72:10165-10170.

Mackewicz, C. E., Lieberman, J., Froelich, C. and Levy, J. A. (2000a). HIV virions and HIV infection *in vitro* are unaffected by human granzymes A and B. *AIDS Res. Hum. Retroviruses* 16:367-372.

Mackewicz, C. E., Patterson, B. K., Lee, S. A. and Levy, J. A. (2000b). CD8($^+$) cell noncytotoxic anti-human immunodeficiency virus response inhibits expression of viral RNA but not reverse transcription or provirus integration. *J. Gen. Virol.* 81:1261-1264.

Mackewicz, C. E., Ridha, S. and Levy, J. A. (2000c). HIV virions and HIV replication are unaffected by granulysin. *AIDS* 14:328-330.

Mackewicz, C. E., Craik, C. S. and Levy, J. A. (2003). The CD8$^+$ cell noncytotoxic anti-HIV response can be blocked by protease inhibitors. *Proc. Natl. Acad. Sci. U. S. A.* 100:3433-3438.

Mackewicz, C. E., Yuan, J., Tran, P., Diaz, L., Mack, E., Selsted, M. E. and Levy, J. A. (2003). a-defensins can have anti-HIV activity but are not CD8$^+$ cell anti-HIV factors. *AIDS* 17:F23-F32.

Maraskovsky, E., Daro, E., Roux, E., Teepe, M., Maliszewski, C. R., Hoek, J., Caron, D., Lebsack, M. E. and McKenna, H. J. (2000). *In vivo* generation of human dendritic cell subsets by flt3 ligand. *Blood* 96:878-884.

Martinez-Marino, B., Shiboski, S., Hecht, F. M., Kahn, J. O. and Levy, J. A. (2004). Interleukin-2 therapy restores CD8 cell non-cyotoxic anti-HIV responses in primary infection subjects receiving HAART. *AIDS* 18: 1991-1999.

Martini, F., Urso, R., Gioia, C., De Felici, A., Narciso, P., Amendola, A., Paglia, M. G., Colizzi, V. and Poccia, F. (2000). gammadelta T-cell anergy in human immunodeficiency virus-infected persons with opportunistic infections and recovery after highly active antiretroviral therapy. *Immunology* 100:481-486.

Martini, F., Poccia, F., Goletti, D., Carrara, S., Vincenti, D., D'Offizi, G., Agrati, C., Ippolito, G., Colizzi, V., Pucillo, L. and Montesano, C. (2002). Acute human immunodeficiency virus replication causes a rapid and persistent impairment of Vg9Vd2 T cells in chronically infected patients undergoing structured treatment interruption. *J. Infect. Dis.* 186:847-850.

Matzinger, P. (2002). The danger model: a renewed sense of self. *Science* 296:301-305.

Mavilio, D., Benjamin, J., Daucher, M., Lombardo, G., Kottilil, S., Planta, M. A., Marcenaro, E., Bottino, C., Moretta, L., Moretta, A. and Fauci, A. S. (2003). Natural killer cells in HIV-1 infection: dichotomous effects of viremia on inhibitory and activating receptors and their functional correlates. *Proc. Natl. Acad. Sci. U. S. A.* 100:15011-15016.

McKenna, K., Beignon, A. S. and Bhardwaj, N. (2005). Plasmacytoid dendritic cells: linking innate and adaptive immunity. *J. Virol.* 79:17-27.

McMichael, A. J. and Rowland-Jones, S. L. (2001). Cellular immune responses to HIV. *Nature* 410:980-7.

Medzhitov, R. and Janeway, C., Jr. (2000). Innate immunity. *N. Engl. J. Med.* 343:338-44.

Medzhitov, R., Preston-Hurlburt, P. and Janeway Jr., C. A. (1997). A human homologue of the Drosophila Toll protein signals activation of adaptive immunity. *Nature* 388:394-397.

Megjugorac, N. J., Young, H. A., Amrute, S. B., Olshalsky, S. L. and Fitzgerald-Bocarsly, P. (2004). Virally stimulated plasmacytoid dendritic cells produce chemokines and induce migration of T and NK cells. *J. Leukoc. Biol.* 75: 504-514.

Michaelis, B. and Levy, J. A. (1989). HIV replication can be blocked by recombinant human interferon beta. *AIDS* 3:27-31.

Moriuchi, M. and Moriuchi, H. (2000). Class I-unrestricted noncytotoxic anti-HTLV-I activity of CD8($^+$) T cells. *Blood* 96:1994-1995.

Mosoian, A., Teixeira, A., Caron, E., Piqoz, J. and Klotman, M. E. (2000). CD8$^+$ cell lines isolated from HIV-1-infected children have potent soluble HIV-1 inhibitory activity that differs from b-chemokines. *Viral Immunol.* 13: 481-495.

Motsinger, A., Haas, D. W., Stanic, A. K., Van Kaer, L., Joyce, S. and Unutmaz, D. (2002). CD1d-restricted human natural killer T cells are highly susceptible to human immunodeficiency virus 1 infection. *J. Exp. Med.* 195:869.

O'Neill, L. A., Fitzgerald, K. A. and Bowie, A. G. (2003). The Toll-IL-1 receptor adaptor family grows to five members. *Trends Immunol.* 24:286-290.

Oberlin, E., Amara, A., Bachelerie, F., Bessia, C., Virelizier, J. L., Arenzana-Seisdedos, F., Schwartz, O., Heard, J. M., Clark-Lewis, I., Loetscher, M., Baggiolini, M. and Moser, B. (1996). The CXC chemokine SDF-1 is the ligand for LESTR/fusin and prevents infection by T-cell-line adapted HIV-1. *Nature* 382:833-835.

Oliva, A., Kinter, A. L., Vaccarezza, M., Rubbert, A., Catanzaro, A., Moir, S., Monaco, J., Ehler, L., Mizell, S., Jackson, R., Li, Y., Romano, J. W. and Fauci, A. S. (1998). Natural killer cells from human immunodeficiency virus (HIV)-infected individuals are an important source of CC-chemokines and suppress HIV-1 entry and replication *in vitro*. *J. Clin. Invest.* 102:223-231.

Otero, M., Nunnari, G., Leto, D., Sullivan, J., Wang, F.-X., Frank, I., Xu, Y., Patel, C., Dornadula, G., Kulkosky, J. and Pomerantz, R. J. (2003). Peripheral blood dendritic cells are not a major reservoir for HIV type 1 in infected individuals on virally suppressive HAART. *AIDS Res. Hum. Retroviruses* 19:1097-1103.

Pacanowski, J., Develioglu, L., Kamga, I., Sinet, M., Desvarieux, M., Girard, P. M. and Hosmalin, A. (2004). Early plasmacytoid dendritic cell changes predict plasma HIV load rebound during primary infection. *J. Infect. Dis.* 190:1889-1892.

Pacanowski, J., Kahi, S., Baillet, M., Lebon, P., Deveau, C., Goujard, C., Meyer, L., Oksenhendler, E., Sinet, M. and Hosmalin, A. (2001). Reduced blood CD123$^+$ (lymphoid) and CD11c$^+$ (myeloid) dendritic cell numbers in primary HIV-1 infection. *Blood* 98:3016-3021.

Pal, R., Garzino-Demo, A., Markham, P. D., Burns, J., Brown, M., Gallo, R. C. and DeVico, A. L. (1997). Inhibition of HIV-1 infection by the beta-chemokine MDC. *Science* 278:695-698.

Pashine, A., Valiante, N. M. and Ulmer, J. B. (2005). Targeting the innate immune response with improved vaccine adjuvants. *Nat. Med.* 11:s63-68.

Patterson, S., Robinson, S. P., English, N. R. and Knight, S. C. (1999). Subpopulations of peripheral blood dendritic cells show differential susceptibility to infection with a lymphotropic strain of HIV-1. *Immunol. Lett.* 66:111-116.

Patterson, S., Rae, A., Hockey, N., Gilmour, J. and Gotch, F. (2001a). Plasmacytoid dendritic cells are highly susceptible to human immunodeficiency virus type 1 infection and release infectious virus. *J. Virol.* 75:6710-6713.

Patterson, B. K., Behbahani, H., Kabat, W. J., Sullivan, Y., O'Gorman, M. R., Landay, A., Flener, Z., Khan, N., Yogev, R. and Andersson, J. (2001b). Leukemia inhibitory factor inhibits HIV-1 replication and is upregulated in placentae from nontransmitting women. *J. Clin. Invest.* 107:287-294.

Poccia, F., Boullier, S., Lecoeur, H., Cochet, M., Poquet, Y., Colizzi, V., Fournie, J. J. and Gougeon, M. L. (1996). Peripheral V gamma 9/V delta 2 T cell deletion and anergy to nonpeptidic mycobacterial antigens in asymptomatic HIV-1-infected persons. *J. Immunol.* 157:449-461.

Poccia, F., Gougeon, M. L., Agrati, C., Montesano, C., Martini, F., Pauza, C. D., Fisch, P., Wallace, M. and Malkovsky, M. (2002). Innate T-cell immunity in HIV infection: the role of Vgamma9Vdelta2 T lymphocytes. *Curr. Mol. Med.* 2:769-781.

Poggi, A., Carosio, R., Fenoglio, D., Brenci, S., Murdaca, G., Setti, M., Indiveri, F., Scabini, S., Ferrero, E. and Zocchi, M. R. (2004). Migration of V delta 1 and V delta 2 T cells in response to CXCR3 and CXCR4 ligands in healthy donors and HIV-1-infected patients: competition by HIV-1 Tat. *Blood* 103:2205-2213.

Poles, M. A., Barsoum, S., Yu, W., Yu, J., Sun, P., Daly, J., He, T., Mehandru, S., Talal, A., Markowitz, M., Hurley, A., Ho, D. and Zhang, L. (2003). Human immunodeficiency virus type 1 induces persistent changes in mucosal and blood gd T cells desptive suppressive therapy. *J. Virol.* 77:10456-10467.

Pollack, H., Zhan, M. X., Safrit, J. T., Chen, S. H., Rochford, G., Tao, P. Z., Koup, R., Krasinski, K. and Borkowsky, W. (1997). CD8[+] T-cell-mediated suppression of HIV replication in the first year of life: association with lower viral load and favorable early survival. *AIDS* 11:F9-F13.

Portales, P., Reynes, J., Pinet, V., Rouzier-Panis, R., Baillat, V., Clot, J. and Corbeau, P. (2003). Interferon-a restores HIV-induced alteration of natural killer cell perforin expression *in vivo*. *AIDS* 17:495-504.

Rissoan, M. C., Soumelis, V., Kadowaki, N., Grouard, G., Briere, F., de Waal Malefyt, R. and Liu, Y. J. (1999). Reciprocal control of T helper cell and dendritic cell differentiation. *Science* 283:1183-1186.

Sallusto, F., Geginat, J. and Lanzavecchia, A. (2004). Central memory and effector memory T cell subsets: function, generation, and maintenance. *Annu. Rev. Immunol.* 22:745-763.

Sandberg, J. K., Fast, N. M., Palacios, E. H., Fennelly, G., Dobroszycki, J., Palumbo, P., Wiznia, A., Grant, R. M., Bhardwaj, N., Rosenberg, M. G. and Nixon, D. F. (2002). Selective loss of innate CD4($^+$) V alpha 24 natural killer T cells in human immunodeficiency virus infection. *J. Virol.* 76: 7528-7534.

Schmidt, B., Scott, I., Whitmore, R. G., Foster, H., Fujimura, S., Schmitz, J. and Levy, J. A. (2004). Low-level HIV infection of plasmacytoid dendritic cells: onset of cytopathic effects and cell death after PDC maturation. *Virology* 329:280-288.

Schmidt, B., Ashlock, B. M., Foster, H., Fujimura, S. and Levy, J. A. (2005). HIV-infected cells are major inducers of plasmacytoid dendritic cells (PDC) interferon production, maturation and migration. *Virology* 343:256-266.

Schmidt, B., Fujimura, S. H., Martin, J. M. and Levy, J. A. (2006). Variations in plasmacytoid dendritic cell (PDC) and myeloid dendritic cell (MDC) levels in HIV-infected subjects on and off antiretroviral therapy. *J. Clin. Immunol.* 26:55-64.

Siegal, F. P., Kadowaki, N., Shodell, M., Fitzgerald-Bocarsly, P. A., Shah, K., Ho, S., Antonenko, S. and Liu, Y. J. (1999). The nature of the principal type 1 interferon-producing cells in human blood. *Science* 284:1835-1837.

Siegal, F. P., Fitzgerald-Bocarsly, P., Holland, B. K. and Shodell, M. (2001). Interferon-a generation and immune reconstitution during antiretroviral therapy for human immunodeficiency virus infection. *AIDS* 15:1603-1612.

Sindhu, S., Ahmad, R., Morisset, R., Ahmad, A. and Menezes, J. (2003). Peripheral blood cytotoxic gamma delta T lymphocytes from patients with human immunodeficiency virus type 1 infection and AIDS lyse uninfected CD4$^+$ T cells, and their cytocidal potential correlates with viral load. *J. Virol.* 77:1848-1855.

Smyth, M. J., Thia, K. Y., Street, S. E., Cretney, E., Trapani, J. A., Taniguchi, M., Kawano, T., Pelikan, S. B., Crowe, N. Y. and Godfrey, D. I. (2000). Differential tumor surveillance by natural killer (NK) and NKT cells. *J. Exp. Med.* 191:661-668.

Soumelis, V., Scott, I., Gheyas, F., Bouhour, D., Cozon, G., Cotte, L., Huang, L., Levy, J. and Liu, Y. J. (2001). Depletion of circulating natural type 1 interferon-producing cells in HIV-infected AIDS patients. *Blood* 98:906-912.

Soilleux, E. J., Morris, L. S., Leslie, G., Chehimi, J., Luo, Q., Levroney, E., Trowsdale, J., Montaner, L. J., Doms, R. W., Weissman, D., Coleman, N. and Lee, B. (2002). Constitutive and induced expression of DC-SIGN on dendritic cell and macrophage subpopulations in situ and *in vitro*. *J. Leukoc. Biol.* 71:445-457.

Stranford, S., Skurnick, J., Louria, D., Osmond, D., Chang, S., Sninsky, J., Ferrari, G., Weinhold, K., Lindquist, C. and Levy, J. (1999). Lack of infection in HIV-exposed individuals is associated with a strong CD8$^+$ cell noncytotoxic anti-HIV response. *Proc. Natl. Acad. Sci. U. S. A.* 96:1030-1035.

Stranford, S. A., Ong, J. C., Martinez-Marino, B., Busch, M., Hecht, F. M., Kahn, J. and Levy, J. A. (2001). Reduction in CD8$^+$ cell noncytotoxic anti-HIV activity in individuals receiving highly active antiretroviral therapy during primary infection. *Proc. Natl. Acad. Sci. U. S. A.* 98:597-602.

Takeda, K. and Akira, S. (2005). Toll-like receptors in innate immunity. *Int. Immunol.* 17:1-14.

Tarazona, R., Casado, J. G., Delarosa, O., Torre-Cisneros, J., Villanueva, J. L., Sanchez, B., Galiani, M. D., Gonzalez, R., Solana, R. and Pena, J. (2002). Selective depletion of CD56(dim) NK cell subsets and maintenance of CD56(bright) NK cells in treatment-naive HIV-1-seropositive individuals. *J. Clin. Immunol.* 22:176-183.

Tasca, S., Tambussi, G., Nozza, S., Capiluppi, B., Zocchi, M. R., Soldini, L., Veglia, F., Poli, G., Lazzarin, A. and Fortis, C. (2003). Escape of monocyte-derived dendritic cells of HIV-1 infected individuals from natural killer cell-mediated lysis. *AIDS* 17:2291-2298.

Trinchieri, G. (1989). Biology of natural killer cells. *Adv. Immunol.* 47:187-376.

Ullum, H., Gotzsche, P. C., Victor, J., Dickmeiss, E., Skinhoj, P. and Pedersen, B. K. (1995). Defective natural immunity: an early manifestation of human immunodeficiency virus infection. *J. Exp. Med.* 182:789-799.

Unutmaz, D. (2003). NKT cells and HIV infection. *Microbes Infect.* 5:1041-1047.

Valentin, A., Rosati, M., Patenaude, D. J., Hatzakis, A., Kostrikis, L. G., Lazanas, M., Wyvill, K. M., Yarchoan, R. and Pavlakis, G. N. (2002). Persistent HIV-1 infection of natural killer cells in patients receiving highly active antiretroviral therapy. *Proc. Natl. Acad. Sci. U. S. A.* 99:7015-7020.

van der Vliet, H. J., von Blomberg, B. M., Hazenberg, M. D., Nishi, N., Otto, S. A., van Benthem, B. H., Prins, M., Claessen, F. A., van den Eertwegh, A. J., Giaccone, G., Miedema, F., Scheper, R. J. and Pinedo, H. M. (2002). Selective decrease in circulating V alpha 24[+]V beta 11[+] NKT cells during HIV type 1 infection. *J. Immunol.* 168:1490-1495.

Vollmer, J. (2005). Progress in drug development of immunostimulatory CpG oligodeoxynucleotide ligands for TLR9. *Expert. Opin. Biol. Ther.* 5:673-682.

Vollmer, J., Weeratna, R., Payette, P., Jurk, M., Schetter, C., Laucht, M., Wader, T., Tluk, S., Liu, M., Davis, H. L. and Krieg, A. M. (2004). Characterization of three CpG oligodeoxynucleotide classes with distinct immunostimulatory activities. *Eur. J. Immunol.* 34:251-262.

Walker, C. M. and Levy, J. A. (1989). A diffusible lymphokine produced by CD8[+] T lymphocytes suppresses HIV replication. *Immunology* 66:628-630.

Walker, C. M., Moody, D. J., Stites, D. P. and Levy, J. A. (1986). CD8[+] lymphocytes can control HIV infection *in vitro* by suppressing virus replication. *Science* 234:1563-1566.

Walker, C. M., Moody, D. J., Stites, D. P. and Levy, J. A. (1989). CD8[+] T lymphocyte control of HIV replication in cultured CD4[+] cells varies among infected individuals. *Cell. Immunol.* 119:470-475.

Walker, C. M., Thomson-Honnebier, G. A., Hsueh, F. C., Erickson, A. L., Pan, L.-Z. and Levy, J. A. (1991a). CD8[+] T cells from HIV-1-infected individuals inhibit acute infection by human and primate immunodeficiency viruses. *Cell. Immunol.* 137:420-428.

Walker, C. M., Erikson, A. L., Hsueh, F. C. and Levy, J. A. (1991b). Inhibition of human immunodeficiency virus replication in acutely infected CD4[+] cells by CD8[+] cells involves a noncytotoxic mechanism. *J. Virol.* 65:5921-5927.

Wallace, M., Scharko, A. M., Pauza, C. D., Fisch, P., Imaoka, K., Kawabata, S., Fujihashi, K., Kiyono, H., Tanaka, Y., Bloom, B. R. and Malkovsky, M. (1997). Functional gamma delta T-lymphocyte defect associated with human immunodeficiency virus infections. *Mol. Med.* 3:60-71.

Ward, J. P., Bonaparte, M. I. and Barker, E. (2004). HLA-C and HLA-E reduce antibody-dependent natural killer cell-mediated cytotoxicity of HIV-infected primary T cell blasts. *AIDS* 18:1769-1779.

Weber, K., Meyer, D., Grosse, V., Stoll, M., Schmidt, R. E. and Heiken, H. (2000). Reconstitution of NK cell activity in HIV-1 infected individuals receiving antiretroviral therapy. *Immunobiology* 202:172-178.

Wille-Reece, U., Flynn, B. J., Lore, K., Koup, R. A., Kedl, R. M., Mattapallil, J. J., Weiss, W. R., Roederer, M. and Seder, R. A. (2005). HIV Gag protein conjugated to a Toll-like receptor 7/8 agonist improves the magnitude and quality of Th1 and CD8$^+$ T cell responses in nonhuman primates. *Proc. Natl. Acad. Sci. U. S. A.* 102:15190-15194.

Wilson, S. B. and Delovitch, T. L. (2003). Janus-like role of regulatory iNKT cells in autoimmune disease and tumour immunity. *Nat. Rev. Immunol.* 3:211-222.

Wiviott, L. D., Walker, C. M. and Levy, J. A. (1990). CD8$^+$ lymphocytes suppress HIV production by autologous CD4$^+$ cells without eliminating the infected cells from culture. *Cell. Immunol.* 128:628-634.

Yamamoto, J. K., Barre-Sinoussi, F., Bolton, V., Pedersen, N. C. and Gardner, M. B. (1986). Human alpha- and beta-interferon but not gamma- suppress the *in vitro* replication of LAV, HTLV-III, and ARV-2. *J. Interferon Res.* 6:143-152.

Yonezawa, A., Morita, R., Takaori-Kondo, A., Kadowaki, N., Kitawaki, T., Hori, T. and Uchiyama, T. (2003). Natural alpha interferon-producing cells respond to human immunodeficiency virus type 1 with alpha interferon production and maturation into dendritic cells. *J. Virol.* 77:3777-3784.

Zhang, L., Yu, W., He, T., Yu, J., Caffrey, R. E., Dalmasso, E. A., Fu, S., Pham, T., Mei, J., Ho, J. J., Zhang, W., Lopez, P. and Ho, D. D. (2002). Contribution of human alpha defensin 1, 2 and 3 to the anti-HIV-1 activity of CD8 antiviral factor. *Science* 298:995-1000.

Chapter 9

Adaptative Immune Responses in HIV-1 Infection

Mara Biasin and Mario Clerici

9.1 Introduction to the Characteristics of Specific Immunity

Adaptative immune responses to infectious agents have traditionally been classified into humoral and cellular immune responses. Humoral immunity, mediated by antibodies produced by B cells, is mostly finalized at the clearance of extracellular pathogens and toxins released by them, whereas cellular immunity, driven by T cells, is mainly directed at intracellular pathogens, such as viruses, that are inaccessible to circulating antibodies. There is nevertheless a considerable degree of overlap as these two types of immune response are not independent of each other. T cells produce cytokines that activate B cells playing a role in humoral immunity. B cells, in turn, influence cellular immunity through antibody-dependent cytotoxicity (ADCC) and function as highly efficient antigen-presenting cells (APCs), especially for proteins that bind to their surface immunoglobulins (Abdullah et al., 1999; Patil and Gupte, 1995).

Cellular immunity is largely a function of T lymphocytes and involves multiple mechanisms and cell types whose ultimate goal is to remove the invading pathogen. They recognize only peptide antigens presented within a binary complex with host proteins that are encoded by genes in the major histocompatibility complex (MHC) and are expressed on the surface of APCs. Lymphocytes consist of functionally distinct populations, the best defined of which are $CD4^+$ T helper cells and $CD8^+$ cytolytic, or cytotoxic, T lymphocytes (CTLs). In response to antigenic stimulation, T helper cells secrete cytokines such as interferons (IFN) and interleukines (IL), whose function is to stimulate the proliferation and differentiation of the T cells as well as other cells, including B cells, macrophages, and other leukocytes (Shearer and Clerici, 1996). CTLs and natural killer (NK) cells are responsible for the granule-exocytosis pathway killing that involves the release of several lysosomal enzymes and granules such as perforin, granzymes, and proteoglycans (Catalfamo and Henkart, 2003). The role of CTLs in human immunodeficency virus infection (HIV) is well-defined (Rowland-Jones *et al.*, 1998). Many data suggest that CD8$^+$ T cells play a fundamental role in controlling HIV replication, especially early in the disease (Autran *et al.*, 1996; Gruters *et al.*, 2002; Chouquet *et al.*, 2002). However, other studies have been unable to show a clear association between viral load and breath or magnitude of CTL responses in humans (Addo *et al.*, 2003; Betts *et al.*, 2001; Ogg *et al.*, 1998). This has raised questions about the clinical relevance of CTLs in HIV progression (Walker and Plata, 1990).

Humoral immunity is based on the effector function exerted by antibodies, and its role in HIV infection is still debated. Although various evidence sustains that humoral responses do not contribute to the clearing of the infection (Parren *et al.*, 1999; Poignard *et al.*, 1999), recent findings have highlighted how the presence of antibodies, belonging to the IgA subtype, represents a key factor in resistance to HIV-1 infection (Clerici *et al.*, 2002; Devito *et al.*, 2000; Lopalco *et al.*, 2005).

In this chapter, we will discuss our current knowledge on the effector functions used by cellular and humoral immunity in response to HIV-1 infection as well as the strategies employed by the virus to subvert the immune system. In the last section of the chapter, we will also consider the immune responses developed by HIV-exposed uninfected individuals (ESNs) and long-term nonprogressors (LTNPs) in response to HIV-1 so as to characterize and analyze the immunologic factors involved in the protection from HIV infection and its progression to acquired imunodeficiency sindrome (AIDS), which could be exploited in the setting-up of vaccine vaccine or therapeutic strategies.

9.2 Functional Immune Disregulation in the Different Phases of HIV-1 Infection

HIV-1 infection is associated with a complex pattern of alterations that profoundly affect the immune response (Cohen and Walker, 2001; Shearer, 1998). HIV-induced immune dysregulation impairs both the quantitative and qualitative homeostasis of the immune system, and this impairment is manifested as an array of multiple, characteristic defects that can be summarized as follows:

– A progressive decline in CD4$^+$ T lymphocytes numbers and a profound impairment of the functionality of these cells.

– A decrease in IL-2, IL-12, IL-15, and IFN-γ (type 1 cytokines that mainly stimulate cell-mediated immunity and immune defenses against intracellular pathogens) and an increase in IL-4, IL-5, IL-6, and IL-10 production (type-2 cytokines that mainly stimulate antibody production and humoral immunity and are finalized at the activation of immune defences against intracellular pathogens).

Changes in the immunophenotype of HIV-1 infected cells:

- CD4$^+$ T cells: increase in the subpopulations expressing CD45RO, Fas (CD95), CD25, CD38, HLA-DR, CD57, and CD69 and decrease in the subpopulations expressing CD45RA, CD7, and signaling lymphocytic activation molecule (SLAM).
- CD8$^+$ T cells: increase in the subpopulations expressing CD38, CD45RO, CD57, and CD69 and decrease in the subpopulations expressing CD45RA and CD28.
- NK cells: increase in the subpopulations expressing CD56 and decrease in the subpopulations expressing CD16/CD56.
- B lymphocytes: abnormal activation of humoral immunity with hypergammaglobulinemia, circulating immune complexes, hyper-production of IgE, and hypereosinophilia.

Global research efforts have led to detailed characterization not only of the effects of the virus on the host but also to an understanding of the ultimate failure of the immune system to contain the infection. The complex interactions occurring between virus and immune system have been deeply analyzed and characterized starting from the early phase of HIV-infection.

Primary HIV infection can be asymptomatic or associated with a flu-like syndrome characterized by fever, malaise, and weakness (Kahn and Walker, 1998). The early phase of HIV infection is also correlated with the presence of extremely elevated titers of HIV in the plasma (Ho *et al.*, 1994). A potent immune response is rapidly induced within a few days after primary infection. In this phase, both humoral immunity and cell-mediated immunity are promptly stimulated. It was nevertheless convincingly demonstrated that a significant reduction in HIV plasma viremia depends on the generation of a HIV-specific cell-mediated immune response (Borrow *et al.*, 1994, Poignard *et al.*, 1999). This concept is mostly based on the observation that the detection of HIV-specific CTLs precedes reduction in viral load, whereas the generation of neutralizing antibodies is delayed and is observed after the changes in HIV viremia have taken place (Borrow *et al.*, 1994). The idea that the modulation of HIV replication is associated with an intact and powerful HIV-specific cellular immunity (Clerici and Shearer, 1993, 1994) at least in part derives from the observation that HIV replication is relatively controlled in the initial phase of the infection, when the immune system is strong and still relatively undamaged.

A long period of clinical asymptomaticity follows primary HIV infection. This phase is initially characterized by a 100- to 1000-fold fall in viral load, reduced HIV replication in peripheral blood mononuclear cells (PBMCs), low HIV isolability from PBMCs, and a partial rise in peripheral blood CD4$^+$ T cell count (Levy, 1993). However, HIV replicates vigorously within the lymphatic tissues (Pantaleo et al., 1993a) and CD4 T-helper cell functions are significantly disrupted even in the earlier phases of asymptomatic HIV-1 infection (Clerici et al., 1989) as described in the following paragraphs.

From a virological point of view, the remarkably fast rate of HIV replication, and the enormous ability of the virus to modify in response to exogenous (pharmacologic) and endogenous (immune system) selective pressure, is underlined by the observation that a wild-type, drug-susceptible HIV strain can be totally replaced by a mutant, drug-resistant strain within 2–4 weeks after initiation of anti-retroviral therapy (Ho et al., 1995).

The end of the period of clinical latency coincides with the decline of CD4$^+$ T cells to less than 200 cells /μl. When the total number of CD4$^+$ T cells in the body has been reduced by at least half (Haase, 1999), the opportunistic tumors and infections that characterize AIDS beset the patient. This often occurs concomitantly with a precipitous rise in viremia and a crash in CD4$^+$ T cell count (Pantaleo et al., 1993b). Although these phases are very different in terms of viral and T-cell dynamics, they are obviously interdependent in many of their pathogenic mechanisms, and one cannot be considered in the absence of the other.

9.3 Cellular Immune Response to HIV-1

Both humoral and cell-mediated immune responses specific for HIV gene products occur in HIV-1–infected patients. The early response to HIV-1 infection is, in fact, similar in many ways to the immune response to other viruses and serves effectively to clear most of the virus present in the blood and in circulating T cells. Nonetheless, it is evident that these immune responses fail to eradicate the virus, and the infection eventually overwhelms the immune system in most individuals. Despite the poor effectiveness of immune responses to the virus, it is fundamental to characterize such responses, as the design of effective vaccines for immunization against HIV requires knowledge of the viral epitopes that are most likely to stimulate protective immunity. In this concern, cellular immune response to HIV infection, largely sustained by virus-specific CTLs, is probably the most powerful weapon at disposal of the immune system. Thus, the analyses of the mechanism of action of CTLs in HIV-1 infection as well as viral escape from cellular immunity have a basic value in the setting-up of efficient therapeutic/vaccine strategies.

9.3.1 Differentiation and Functions of CTLs

T-cell differentiation and selection in the thymus involves several stages culminating in the emergence of immature, differentiated, naïve $CD3^+CD8^+$ cells bearing the $\alpha\beta$ T-cell receptors. These naïve cells are released into circulation and typically move between blood and the secondary lymphoid organs, particularly the lymph nodes. Encounters with MHC class I–expressing cells, followed by costimulation signals such as those mediated by the CD80/86 pathway, leads to the activation and development of a primary response as well as further differentiation of these cells into effectors (CTLs) and memory cells (Mehta-Damani *et al.*, 1994). The most effective APCs include dendritic cells (DCs), which are capable of picking up pathogens (or proteins) and processing its proteins to produce MCH-restricted specific antigenic determinants (epitopes) that are recognized by T cells (Inaba and Inaba, 2005).

Memory and naïve T cells can be differentiated based on the expression of a number of surface markers, including the costimulatory molecules CD27 and CD28, the RO and RA isoforms of CD45, and CCR7, a homing receptor allowing cells to migrate to secondary lymphoid organs versus nonlymphoid tissues (Fig. 9.1). The expression pattern of CD27, CD45, and CCR7 reflects stages of gradual, linear differentiation from a typical central memory T cell (CM), with high proliferative capacity and low cytolytic activity, through an intermediate stage, including effector memory (EM) T cells, displaying immediate effector function, to a highly or terminally differentiated (TD) stage, represented by effector cells highly enriched in perforin that are specialized in the lysis of viral-infected targets (Geginat *et al.*, 2001; Lanzavecchia *et al.*, 2002; Sallusto *et al.*, 1999). Expansion potential decreases from CM to EM T cells and is very low in TD T cells, whereas effector function and cytolytic activity progressively increase during the differentiation process (Geginat *et al.*, 2001; Lanzavecchia *et al.*, 2002; Sallusto *et al.*, 1999). Of note, the relative proportion of these lymphocyte subsets was convincingly shown to be driven by antigenic exposure in the presence of cytokines such as IL-7 and IL-15 (Badovinac *et al.*, 2003; Lanzavecchia *et al.*, 2002; Manjunath *et al.*, 2001; Wherry *et al.*, 2003).

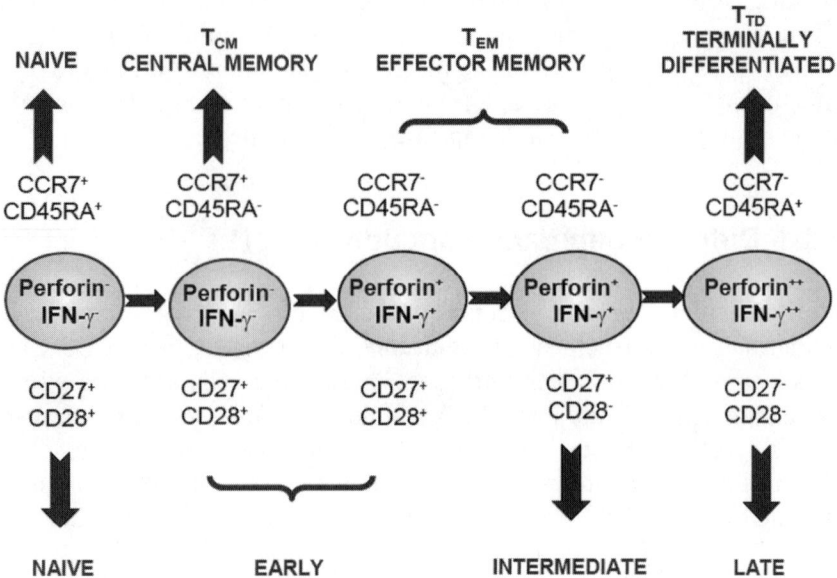

FIGURE 9.1. Memory and naïve T cells. Memory and naïve T cells can be differentiated based on the expression of a number of surface markers, including the costimulatory molecules CD27 and CD28, the RO and RA isoforms of CD45, and CCR7. Their different expression relates to stages of gradual, linear differentiation from a typical central memory T cell (CM), with low cytolytic activity, through an intermediate stage, including effector memory (EM) T cells, displaying immediate effector function, to a terminally differentiated (TD) stage, represented by effector cells highly enriched in perforin that are specialized in the lysis of viral infected targets.

CD8 T lymphocytes can exert their cytotoxic function, destroying the target cell, through two main mechanisms (Fig. 9.2). The first and well-known mechanism is based on the action of lytic enzymes (perforins) that, in the presence of high calcium concentration, polymerize on the cellular surface of the target cell favoring the formation of pores, because of which the osmotic gradient is lost (osmotic lysis) (Catalfamo and Henkart, 2003). A further mechanism of lysis involves the apoptosis of the target cell (fragmentation of genome into repetitive 200-bp fragments). Apoptosis can be secondary to the interaction of surface receptors expressed on the target cell with molecules released by CTLs (FasL/Fas) or can be due to the production of particular molecules, such as granzymes, "injected" by CTLs into the cytoplasm of the target cell (Kajino et al., 1998). Both mechanisms are induced by the release of cytokines (IL-2, IL-12, IFN-γ) produced by T helper cells, which "arm" cytotoxic lymphocytes allowing them to accomplish the above described function.

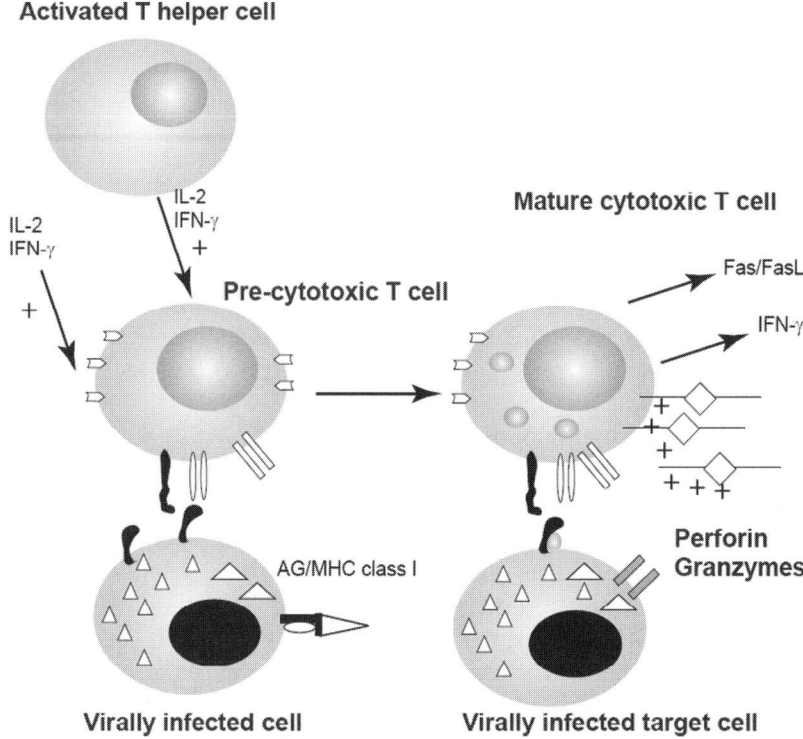

FIGURE 9.2. Activation of CD8[+] T lymphocytes. CD8[+] T lymphocytes activation is dependent on the presentation of processed antigenic peptides in association with MHC class I molecules expressed on the surface of a target cell and on the coordinated action of cytokines released by activated T helper cells (IL-2, IL-12, IFN-γ). CD8[+] T lymphocytes can exert their cytotoxic function through granule-dependent mechanisms (perforin, granzyme) or granule-independent mechanisms (TNF-α).

9.3.2 CTLs in HIV-1 Infection

Cellular immune response to HIV-infection is largely sustained by virus-specific CTLs. There is early circumstantial evidence in humans showing a temporal correlation between the initial drop in viremia and the appearance of HIV-1–specific CD8[+] T cells but not neutralizing antibodies. That is why the emergence of vigorous, HIV-specific CTLs in most individuals has been implicated as a reason for the declining of viremia (Borrow *et al.*, 1994, 1997; Koup *et al.*, 1994; Price *et al.*, 1997). Consistent with this observation, it has been verified that oligoclonal population of T lymphocytes expand markedly in the peripheral blood of infected individuals at the time of virus containment in the early weeks after HIV-1 infection (Pantaleo *et al.*, 1994). These populations of T cells are likely to represent clonally restricted CTLs.

Studies have also been done using assays to evaluate populations of CD8$^+$ T cells for killing, clonality, and tetramer binding in SIV-infected models. These studies have shown a clear temporal association between the expansion of CTLs and virus clearance (Chen *et al.*, 1995; Kuroda *et al.*, 1999; Yasutomi *et al.*, 1993). In further studies, however, vigorous CTL immune responses have been reported in HIV-1–infected people with global impairment of immune functions, whereas the absence of strong HIV-specific CTLs in individuals with fairly good immune response function has also been reported (Bernard *et al.*, 1998). This has raised questions as to the clinical relevance of CTLs in HIV progression.

Probably the most convincing evidence for the role played by CD8$^+$ CTLs in controlling HIV-1 replication derives from experimental studies in SIV-infected rhesus monkeys. *In vivo* depletion of CD8$^+$ cells in monkeys, achieved by infusion of monoclonal antibodies to CD8, deeply affected SIV replication, while prior to the depletion of CD8$^+$ T cells SIV replication was well-controlled in all of the monkeys (Jin *et al.*, 1999; Schmitz *et al.*, 1999). When the depletion was greater than 28 days, primary viremia was never cleared after infection, and the monkeys died with a rapidly progressive AIDS-like syndrome. Further, the transient depletion of CD8$^+$ T lymphocytes in SIV-infected rhesus monkeys coincided with an evident rise in viral load that returned to basal levels with the reappearance of CD8$^+$ cell population. A number of other experimental studies sustain the fundamental role played by CTL activation in the modulation of HIV-1 infection: (1) progression to AIDS is faster in patients who do not develop gag-specific CTL (Riviere *et al.*, 1995); (2) a powerful CTL function characterizes ESNs and LTNPs (Clerici and Shearer, 1994; Propato *et al.*, 2001); (3) the presence of strong HIV-specific CTLs characterizes children who did not progress to AIDS within the first year of birth, whereas those who progressed to AIDS had a lower CTL response (Buseyne *et al.*, 2002).

Consistent with the importance of CTLs in controlling HIV and SIV replication, MHC class I haplotypes of infected individuals have a significant predictive value for the rate of clinical disease progression. The rationale for these considerations is that human leukocyte antigen (HLA) genes are extremely polymorphic and that the host's HLA genotype is an important factor in determining the severity of other infectious diseases. The molecular basis for these differences is that each allelic variant of a class I or class II HLA molecule presents a restrained and specific repertoire of peptides to CD8$^+$ or CD4$^+$ T lymphocytes, which in turn mount a specific response against the pathogen. In other words, APCs are more or less capable of presenting a specific antigenic peptide to T lymphocytes, depending on whether the HLA variants that efficiently present peptides critical for the survival of the virus also mount a more efficient immune response against the virus. Because of HLA extreme polymorphism, several groups have correlated disease-free survival to the presence or absence of a specific HLA background. In this respect, patients fully heterozygous at class I alleles progress less rapidly to AIDS, probably because they are able to present in optimal fashion a wider repertoire of HIV-derived antigenic peptides (Carrington *et al.*, 1999). Similarly, the expression of the MHC I molecules HLA-B27 and HLA-B57 in HIV-1–infected individuals has been associated with better clinical outcomes after HIV-1 infection (Magierowska *et al.*, 1999), whereas the expression of a particular haplotype of HLA-A1, B8, B35, and Cw4 has been associated with worse outcome and rapid disease progression (Kaslow *et al.*, 1996). Specific HLA alleles have also been associated with vaccine responsiveness in HIV vaccine trials (Kaslow *et al.*, 2001). These observations underscore the importance of CTLs in containing HIV-1 replication and highlight the genetic constraints of immune control, the mechanism of which remains poorly understood.

9.3.3 Viral Escape from CTLs

The frequency of HIV-specific CTLs in HIV-infected patients is significantly elevated, representing 1% of total CD8 T cells circulating in peripheral blood. Furthermore, the quantity of HIV-specific CTLs is 10- to 50-fold higher compared with the HIV-infected target cells that are able to sustain HIV-1 replication. Despite these relevant findings and correlations associating CTLs with the control of viral replication, nearly all individuals ultimately develop AIDS within a decade if untreated, thus a central unanswered question is why replication of HIV-1, despite the induction of these powerful immune weapons, is not contained and leads to the progressive and ultimately profound immune suppression (Table 9.1).

TABLE 9.1. Possible causes responsible for functional defects of cytotoxic $CD8^-$ T lymphocytes (CTLs) observed in HIV-infected patients.

Reduced production of CTL-stimulating cytokines (IL-2, IL-12, IFN-γ)
Increased production of type 2 cytokines (IL-4, IL-10)
Reduced expression of MHC I molecules induced by viral antigens (nef)
Alteration in the signal transduction pathway
High mutation rate in HIV genome
Alteration in the pattern of accessory molecule expression
Reduction in the intracellular concentration of lytic enzymes (perforin, granzymes)
Alteration in the maturation pathway of $CD8^-$ T lymphocytes
Alteration in the production of propoptotic molecules

Viral escape from CTL responses has been documented during both acute (Allen *et al.*, 2000; Borrow *et al.*, 1997; Price *et al.*, 1997) and chronic infection (Goulder *et al.*, 1997; Phillips *et al.*, 1991), and a number of possible explanations to this phenomenon have been suggested and can be summarized as follows: (1) cytolytic capacity by CTLs is reduced as a consequence of functional alteration of $CD4^+$ T-helper lymphocytes (McNeil *et al.*, 2001; Pitcher *et al.*, 1999; Sieg *et al.*, 2002); (2) HIV-1 could escape CTL-mediated response thanks to a high mutation rate (Klenerman *et al.*, 2002); (3) the expression of costimulatory molecules on cell surface can be altered (Dudhane *et al.*, 1996; Esser *et al.*, 2001; Trabattoni *et al.*, 2002); (4) the concentration of perforin, CD95, and other lytic molecules could be reduced in HIV-1–infected patients (Amendola *et al.*, 2000; Trabattoni *et al.*, 2004a); (5) the process leading to CTLs maturation could be altered, preventing the assumption of their fully cytolytic potential (Champagne *et al.*, 2001; Schenal *et al.*, 2005). Among the many reasons proposed to justify the lack of immune control, more insight has been given to immune escape through the generation of mutations in targeted epitopes of the virus. When effective selection pressure is applied, the error-prone reverse transcriptase and high replication rate of HIV-1 allow for rapid replacement of circulating virus by those carrying resistant mutations, as was first observed with administration of potent anti-retroviral therapy. Escape occurs even through single amino acid mutations in an epitope, at site essential for MHC binding or T cell receptor recognition, but may be also affected by mutations in flanking regions that affect antigen processing. CTL escape has been well documented in a number of studies. A few years ago, Goulder *et al.* (2001) showed that in transmission studies an escape mutation within an HLA-B27 restricted epitope in the mother, associated with long-term nonprogressing HIV-infection, can be transmitted to the $B27^+$ child who then cannot make an immune response toward this epitope. Conversely, children who inherited HLA-B27 from their fathers and HIV from their mothers received a virus that has not been under B27-restricted selection pressure and was able to mount vigorous CTL responses and control viral replication. Studies conducted in SIV-infected monkeys developing an initial strong CTL response against an epitope in Tat have also demonstrated that the establishment of chronic uncontrolled infection is characterized by the emergence of new viruses with mutations in Tat (Allen *et al.*, 2000). Evidence supporting the influence of CTL selection pressure on this virus also comes from population studies examining association between HLA alleles and specific mutations (Moore *et al.*, 2002; Yusim *et al.*, 2002). Rare HLA alleles represent an advantage in controlling HIV infection as those individuals expressing rare alleles would be less likely to encounter viruses that had already developed fixed mutation in the dominant epitopes presented by that allele (Trachtenberg *et al.*, 2003).

Immune dysfunction may be another cause for viral escape, and it is likely that lack of HIV-specific CD4$^+$ T helper cell proliferation and expansion is a crucial feature of this impairment (McNeil *et al.*, 2001; Pitcher *et al.*, 1999). In macaque models, there is a clear loss of the capacity to express cytokines, beginning as early as the time of peak viremia in acute infection (McKay *et al.*, 2003), a feature that has been associated with the progression of HIV-infection in humans as illustrated in the next paragraph. The selective infection of HIV-1–specific CD4$^+$ T cells in infected individuals provides a mechanistic explanation for loss of these cells early in infection (Douek *et al.*, 2002) and explains why these responses are restored with early treatment of acute infection (Malhotra *et al.*, 2000; Oxenius *et al.*, 2000; Rosenberg *et al.*, 1997).

Several recent publications suggest how CTL inability to control HIV-1 replication could be secondary to their reduced perforin and granzyme intracellular concentration (Lieberman *et al.*, 2001; Yang *et al.*, 2005). The potential importance of these findings is reinforced by a study showing that HIV-specific CTLs lyse virus-infected targets via the granule-dependent pathway and extend previous reports indicating that reduced perforin expression is detected in CD8$^+$ T cells that infiltrate tissue in HIV-infected individuals (Andersson *et al.*, 1999). These findings could be related to the maturation process leading to the generation of EM and TD lymphocytes. As a matter of fact, it has been proven that HIV-1 inhibits certain steps of T-cell differentiation as a strategy to subvert immune response (Champagne *et al.*, 2001). In HIV-1–infected patients, the majority of HIV-specific T cells, in fact, appear to belong to the EM subset, which contains less perforin and has poor *ex vivo* killing compared with CMV-specific T cells, isolated from the same patients, enriched in T lymphocytes with TD phenotype. Thus, CTL functional defects observed in HIV-1–infected subjects could be secondary to an impaired maturation process leading to the accumulation of immature lymphocytes (CD45RA$^-$ CCR7$^-$CD8$^+$ and CD27$^+$CD28$^-$CD8$^+$) to the prejudice of functionally active CD27$^-$ CD28$^-$ CD8$^+$ T cells. Of note, as anti-retroviral therapy (ART) would be unable to restore functional impairment in CTLs of HIV-1–infected patients, it is possible to speculate on a direct anti-retroviral drug interference with the synthesis of perforin and granzymes. Preliminary results, obtained by culturing *in vitro* PBMCs of healthy individuals with anti-retroviral drugs, suggest this mechanism to be probably involved in CTL impairment. Additional data obtained in HIV-exposed health care workers undergoing ART-based prophylaxis show that perforin and granzyme are reduced in CD8$^+$ T lymphocytes after 1 month of therapy (Biasin *et al.*, unpublished data). These findings could explain why, after therapy interruption, a strong and uncontrolled rebound in HIV replication is observed even in those ART-treated patients in whom viral load has been undetectable for years.

9.3.4 Antiviral Soluble Factors

The role played by $CD8^+$ T lymphocytes in HIV infection, as well as in other viral infections, is far more complex. As a matter of fact, in addition to the previously described cytolytic activity, $CD8^+$ T lymphocytes can control and modulate HIV replication and infectivity through non-cytolytic mechanisms involving the production and release of antiviral soluble factors. The first of these factors, cell antiviral factor (CAF), was described in 1986 by Levy and his group. CAF suppresses virus replication at the level of the long terminal repeat–driven gene expression and the effect does not imply the lysis of the target cell (Levy et al., 1996). Interestingly, a higher production of CAF has been observed by $CD8^+$ T lymphocytes from LTNPs and ESNs (Bartnof, 1995; Stranford et al., 1999) and its expression seems to be positively modulated by type 1 cytokines (IL-2, IL-12, IFN-γ) (Chang et al., 2002). On the whole, these findings suggest a key role played by CAF in the control of HIV replication and progression to AIDS.

Other soluble factors exerting antiviral function include β-chemokines. Chemokines are chemoattractant substances secreted at sites of infection or injury (Luster, 1998). It has been known since the mid-1980s that presence of CD4 on a cell surface was necessary but not sufficient for entry of HIV into the cell. In addition, it was known that $CD8^+$ T cells secrete substances that interfere with the ability of HIV to infect cells (Clerici et al., 1992a). In 1995, Cocchi and colleagues identified these substances as RANTES, macrophage inflammatory protein-1α (MIP-1α), and MIP-1β. It was hypothesized that these substances bind to a receptor that the macrophage-tropic (M-tropic) strains of the virus requires for cell entry. In 1996, Feng and associates isolated CXCR4 (originally referred to as "fusin"), a chemokine receptor located on T cells that T-cell–tropic (T-tropic) HIV uses as a coreceptor along with CD4. Nowadays, it has been demonstrated that the interaction between the virus envelope protein gp120 and CD4 induces a conformational change that allows interaction between the virus and the chemokine receptor and ultimate fusion of the virus and host cell membrane (Kwong et al., 1998; Rizzuto et al., 1998; Wu et al., 1996; Wyatt et al., 1998). Thus, the currently held model is that M-tropic HIV strains (termed R5 viruses) infect macrophages, monocytes, and T cells by using the host's expression of CD4 and CCR5 as coreceptors. T-tropic HIV strains (termed X4 viruses) infect T cells by using CD4 and CXCR4 as coreceptors (Berger *et al.*, 1998). The host's natural ligands for these coreceptors are relevant because they may interfere with HIV entry into target cells blocking viral entry by interfering with viral binding to the receptor or by downregulating the receptor (Amara *et al.*, 1997). However, some *in vitro* studies have even suggested that RANTES, MIP-1α, and MIP-1β may upregulate replication of HIV in macrophages and monocytes by recruiting activated target cells (Canque *et al.*, 1996; Schmidtmayerova *et al.*, 1996a, 1996b). Furthermore, *in vivo* levels of RANTES in HIV-infected people have been shown not to be correlated with HIV-1 viral load (Li *et al.*, 2000). Further studies are needed to verify the role played by these soluble factors in HIV infection (Pal *et al.*, 1997).

More recently, the effects exerted by α-defensin 1, 2, and 3 in controlling HIV-replication and AIDS progression have been taken into consideration. α-Defensins are cationic, arginin-rich peptides secreted by a number of cells and have a broad antimicrobial spectrum (Fellerman and Stage, 2001). Cells of LTNP HIV-infected individuals were recently shown to produce relatively high quantities of α-defensins 1, 2, and 3 (Zhang *et al.*, 2002). This finding has further been extended to lymphocytes and cervical biopsies isolated from ESNs (Trabattoni *et al.*, 2004b) raising a global interest toward their potential employment as anti-retroviral compounds (Fig. 9.3).

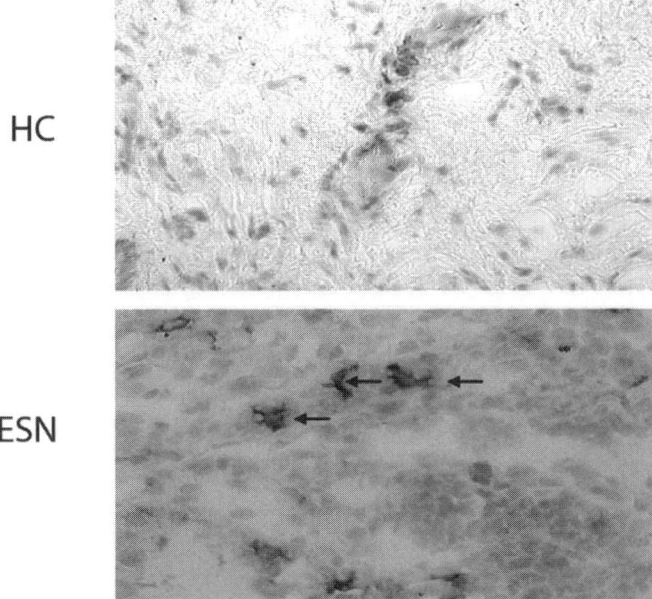

FIGURE 9.3. α-Defensin expression in HIV-infected individuals. α-Defensin detection by immunohistochemistry staining in cervical biopsies of a representative ESN and a representative healthy control (HC). The arrows indicate positive cells. (See Plate 2).

9.3.5 Alteration of T-Helper Functions in HIV-1 Infection

Virus-specific $CD4^+$ T lymphocytes also have an important role in controlling HIV replication mainly because of their role in facilitating CTLs and antibody responses, a helper function mediated by the release of proteins (cytokines) after antigen recognition. Interestingly, $CD4^+$ T helper functions are often disrupted even in the earlier phase of asymptomatic HIV infection (Clerici *et al.*, 1989) (Table 9.2). The T helper defects observed in this phase of the disease are complex, sequential, and independent of CD4 counts (Clerici *et al.*, 1989). Thus, when PBMCs of asymptomatic HIV-seropositive individuals are stimulated *in vitro* with recall antigens, HLA alloantigens, and mitogens, and IL-2 production is measured, different degrees of impairment in T-helper function are observed. To summarize: (1) IL-2 can be produced in response to all antigens; (2) recall-antigen stimulated IL-2 can be selectively defective; (3) only phytohemagglutinin (PHA)-stimulated IL-2 production can be observed; (4) the ability to produce IL-2 can be completely lost (Clerici *et al.*, 1993 and 1994). Approximately two-thirds of HIV-seropositive asymptomatic individuals show one or more of these T helper functional defects without exhibiting a critical reduction in the number of $CD4^+$ T lymphocytes. Even more important, in 1995 Dolan and colleagues demonstrated that these defects are predictive for the subsequent rate of decline in the number of $CD4^+$ T lymphocytes, time to diagnosis of AIDS, and time to death (Dolan *et al.*, 1995). In fact, at a parity of $CD4^+$ T cells and clinic stadiation, patients capable of producing IL-2 in response to soluble antigens, alloantigens, and mitogens showed a delayed progression to AIDS in comparison with patients with different anergy levels to the same stimuli. Therefore, defects in IL-2 production are predictive progression to AIDS (Dolan *et al.*, 1995).

TABLE 9.2. Possible causes responsible for functional defects of $CD4^-$ T helper lymphocytes observed in HIV-infected patients.

Direct cytophatic effect exerted by HIV

Reduced production of cytokines acting with an autocrine mechanism (IL-2)

Increased production of immunosuppressive soluble factors

Alteration in the signal transduction pathway

Increased production of type 2 cytokines (IL-4, IL-10)

Increased production of anti-CD4 self-antibodies

Alteration in the pattern of accessory molecule expression

Reduced expression of MHC and/or TCR molecules

Alteration in the antigen-processing pathway

Alteration in the formation of the MHC/Ag binary complex

Downregulation of the expression of the α-chain of the T-cell receptor

Besides the defects in IL-2 production, abnormalities in T helper integrity include a series of other alterations, among which reduced IL-2 receptor expression (Prince *et al.*, 1984), preferential loss of the naïve T-cell subset (Roeder *et al.*, 1995), and downregulation of the expression of the αchain of the T-cell receptor (Trimble and Liberman, 1998) may have an important role in disease progression. Furthermore, as T helper cells activation requires antigen recognition on the surface of an APC, defects in this cell population (MHC molecule expression, APC ability to process the antigens, biosynthesis of antigen/MHC complex, or reduction in costimulatory molecules expression) could be responsible for T helper functional impairment. However, experiments based on the use of cellular population (T helper and APC) derived from monozigous twins, discordant for HIV-1 serostatus, have rejected this hypothesis, suggesting how the different degrees of T helper functional impairment are secondary to the different T helper–APC interactions required for the diverse antigens (Clerici *et al.*, 1992).

Another key factor associating the progression of HIV-1 infection with T helper function is cytokine production and the T helper-1/T helper-2 (Th1/Th2) balance. The Th1/Th2 balance hypothesis emerged in the late 1980s, stemming from observations in mice of two subtypes of T helper cells differing in cytokine secretion patterns and other functions (Mosmann and Coffman, 1987, 1989). Because cytokines are produced by cell types other than T lymphocytes, a functional definition was introduced. Thus, type 1 cytokines, such as IL-2, IL-12, and IFN-γ, mainly stimulate cellular immunity and are more prone to fight viruses and other intracellular pathogens, eliminate cancerous cells, and stimulate delayed-type hypersensitivity (DTH) skin reaction, whereas type 2 cytokines, including IL-4, IL-5, IL-6, IL-10, and IL-13, mainly stimulate humoral immunity and upregulate antibody production to fight extracellular organisms (Clerici *et al.*, 1993, 1994) (Fig. 9.4). Currently, much of the literature elevates the type1/type2 balance concept to the level of paradigm according to which overactivation of either pattern can cause disease, and either pathway can downregulate the other (Kidd, 2003).

One of the most notable immunologic defects in HIV-1–infected individuals is poor HIV-1 T helper cell activity at all stages of the disease. Beginning with the period of clinical asymptomaticity, the ability of T lymphocytes to produce cytokines other than IL-2 upon antigenic stimulation is impaired as well, and this impairment seems to be pivotal in the progression of the infection. To summarize, *in vitro* stimulation of blood leukocytes from HIV-infected patients results in decreased production of IL-2, IFN-γ, and IL-12 and in increased production of IL-4, IL-5, IL-6, and IL-10 (Clerici and Shearer, 1993, 1994). These data led to the hypothesis that a type 1 to type 2 cytokine shift is present in HIV-1

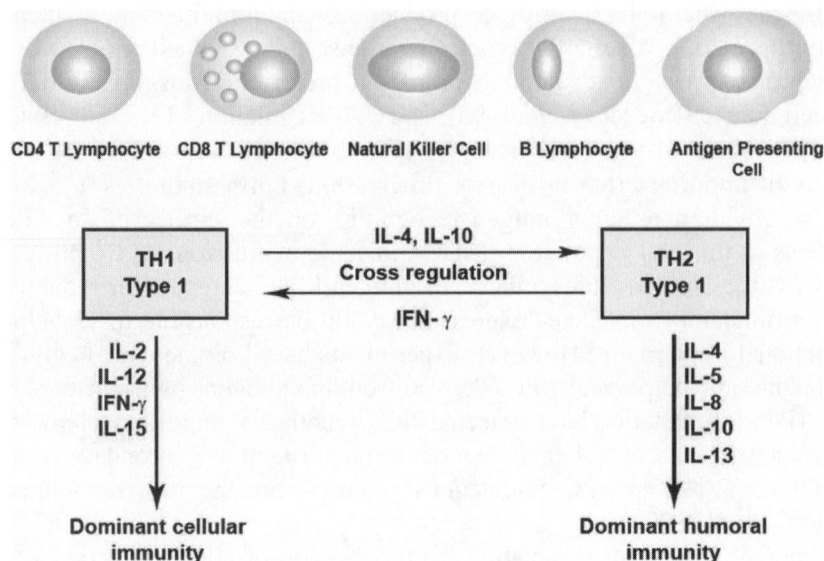

CD4 T Lymphocyte   CD8 T Lymphocyte   Natural Killer Cell   B Lymphocyte   Antigen Presenting
                                                                                      Cell

FIGURE 9.4. Production of cytokines by different cell types of the immune system. Type 1 cytokines (IL-2, IL-12, IFN-γ, IL-15) mainly stimulate cellular immunity and are more prone to fight viruses and other intracellular pathogens. Type 2 cytokines (IL-4, IL-5, IL-6, IL-10, IL-13) mainly stimulate humoral immunity and upregulate antibody production to fight extracellular organisms.

infection. Of note, this alteration in cytokine production plays a central role in the progression of HIV infection favoring the development and isolation of syncytium-inducing (X4) strains of HIV, whereas the opposite scenario is correlated with a condition of long-term nonprogression (Galli et al., 2001). Thus, even in the period in which clinical latency is observed, HIV-infection is immunologically and virologically extremely active and destructive. Accordingly to this scenario, Autran and colleagues (2005) recently demonstrated that HIV-1–specific CD4[+] Th1 responses combined with IgG2 antibodies and IFN-γ–producing CD4[+] Th1 cells are better predictors of long-term nonprogression than are virus parameters, host genes, or HIV-1–specific CD4[+] Th1 or CD8[+] T-cell proliferation.

   Given the role played by cytokines in the polarization of the immune response, polymorphism in their codifying sequences, capable of altering their rate of expression, could be related to resistance/progression of HIV-1 infection. Consistent with this hypothesis, cytokine gene polymorphism and particularly single nucleotide polymorphisms in the promoters of IL-4 and IL-10 have been shown to influence HIV-1 disease progression in seroconverter cohorts. The molecular basis of this difference is thought to be the modulation of mRNA production of IL-4 and IL-10, two Th2-type cytokines with pleiotropic effects on the immune system. However, both studies are single cohort studies and confirmation in other cohorts is necessary to really appreciate the impact of these findings (Nakayama *et al.*, 2000, 2002; Shin *et al.*, 2000).

The activation of CD4$^+$ T helper lymphocytes is dependent on the presentation of processed antigenic peptides in association with MCH class II molecules to lymphocytes that express a T-cell receptor specific for that binary complex. However, optimal lymphocyte activation requires a second signal that is delivered by the interaction between costimulatory, accessory molecules (Fig. 9.5). CD28 for example binds to a family of ligands on the surface of non–T cells collectively known as B7 molecules (Freeman *et al.*, 1989; Lencschow *et al.*, 1996), leading to the release of the cell growth factor IL-2. Besides B7.1 and B7.2, a number of other B7-like ligands, including B7-H1, have been recently described (Dong *et al.*, 1999; Chapoval *et al.*, 2001; Wang *et al.*, 2000). Interestingly, ligation of B7-H1 to T cells results in the preferential production of IL-10 (Dong *et al.*, 1999) and in increased T helper–dependent synthesis of trinitrophenyl (TNP)-specific immunoglobulin G2a (IgG2a) in mice (Tamura *et al.*, 2000). These results suggest that ligation of B7-H1 may be responsible for promoting type 2 cytokine–biased responses. Of note, in a

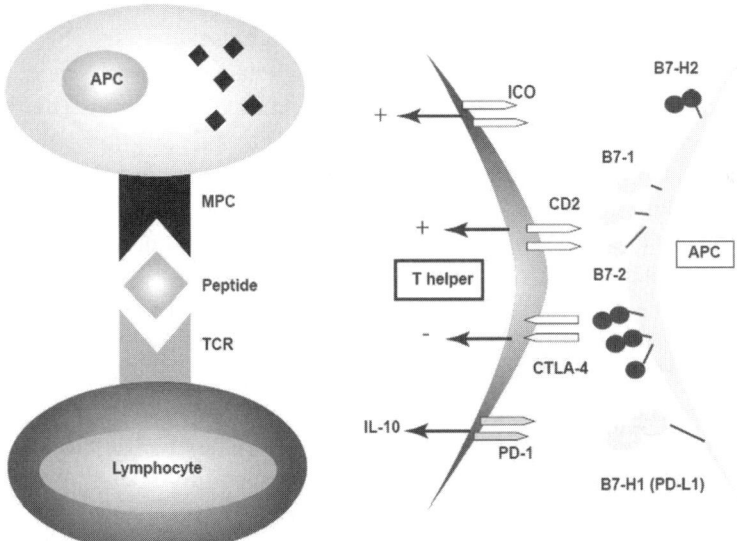

FIGURE 9.5. T-lymphocyte activation. Activation of T lymphocytes is dependent on the interaction between the T-cell receptor (TCR) with the antigen/MHC binary complex expressed on the surface of an antigen-presenting cell (APC). APCs also give costimulatory signals (interaction between membrane receptors) that are indispensable for the correct T-lymphocyte activation.

recent publication it has been demonstrated that B7-H1 synthesis and expression are upregulated in HIV-infected subjects, and the degree of dysregulation correlates with the severity of disease (Trabattoni *et al.*, 2003). Aberrant antigen presentation by APCs that exhibit increased B7-H1 expression and IL-10 production in HIV-1 infection could therefore be responsible for T-lymphocyte unresponsiveness and loss of protective immunity, and B7-H1 can be considered as a surrogate marker potentially involved in AIDS disease progression.

9.3.6 Immunologic Profile of People Living in Different Areas of the World

Immunologic results collected over time in the Gulu district of northern Uganda, where prevalence of HIV infection ranges between 14% and 25% (Fabiani *et al.*, 1998), show that lymphocytes from African HIV-infected individuals are abnormally activated (Bentwich *et al.*, 1995; Clerici *et al.*, 2000, 2001; Rizzardini *et al.*, 1996, 1998). Thus, production of IFN-γ and IL-10 by HIV-antigen-stimulted PBMCs of African HIV-infected individuals is higher than in European patients (Clerici *et al.*, 2001, 2000; Rizzardini *et al.*, 1996, 1998). Furthermore, immune activation in the African setting is not limited to HIV-infected individuals as IFN-γ and IL-10 is greatly augmented in HIV-uninfected individuals as well. This abnormal activation is associated with environmental conditions (hygienic conditions, parasitic and nonparasitic infections) and dietary limitations, as the immune response resumes a Th1-dominant pattern in Africans who move to Western countries (Clerici *et al.*, 2000).

Susceptibility to HIV infection is thought to be higher, and disease progression faster, in African individuals than in European individuals, and this conditions seems to be at least in part dependent on the immunologic profile of these subjects. As previously described *in vitro*, IFN-γ and IL-10 increase expression of CCR5, one of the main coreceptors for HIV on the surface of immune cells (Sozzani *et al.*, 1998). Data show that CCR5 is indeed upregulated *in vivo* in African individuals (Clerici *et al.*, 2000). The potential importance of this observation is underlined by results indicating that HIV infection in Africa is mostly supported by R5 viruses, the viral strain that uses CCR5 as its coreceptor (Abebe *et al.*, 1999). Thus, immune activation would provoke upregulation of CCR5 on target cells, and this would result in an evolutionary pressure on the viral quasispecies, leading to the preferential selection of R5 HIV strains (Clerici *et al.*, 2000, 2002a). The net result is prevalence of R5 viruses within a biological scenario (upregulation of CCR5) that facilitates infection of target cells.

Because this activation is associated with environmental factors, it has been proposed that other countries, undergoing a similar environmental burden, could share a comparable immune activation and alterations of the virus-host interaction. In particular, the analyses of functional and phenotypic parameters in HIV-infected and HIV-uninfected individuals from the Maharashtra-Mumbai region of India, where recent projection indicates an alarming rate of increase of HIV infection, have brought to light an interesting and unexpected immunologic profile. Thus, immune activation in Indian individuals is associated with increase in IFN-γ and IL-4 specific mRNA in both HIV-infected and -uninfected individuals (Biasin et al., 2003). This cytokine profile is different compared with that seen in Ugandan individuals, in whom IL-4 was not augmented. Even more important, IL-4 expression increase was associated with (1) a higher percentage of both CD4$^+$ and CD14$^+$ CXCR4-expressing cells and increased levels of CXCR4 mRNA in Indian compared with Italian subjects; (2) a higher rate of the T allele (twofold increase) in the -589 IL-4 polymorphic site, associated with increased expression of IL-4 (Nakayama et al., 2002). Of note, as HIV infection in India is mainly sustained by HIV-1 clade C, a strain of HIV that prevalently uses CCR5, irrespective of HIV disease status, the discrepancy between the predominant HIV strain (R5) and viral coreceptor (CXCR4) in India could partly justify the slower progression of HIV infection observed in Maharashtra-Mumbai patients.

9.4. Humoral Immune Responses in HIV-1 Infection

Infection with HIV-1 results in a wide range of clinical manifestations. HIV-1 preferentially infects CD4$^+$ cells including monocytes, T cells, and cells from the central nervous system, but peripheral B cells of HIV-infected patients, though not expressing CD4 molecule on their cell surface, show an abnormal phenotype (Amadori and Chieco-Bianchi, 1990; Lane et al., 1983; Martinez-Maza et al., 1987) and spontaneously produce high levels of antibodies (Amadori et al., 1989). Soon after HIV-1 infection, strong follicular hyperplasia is detectable within the lymphoid organs of HIV-1–infected patients and persists throughout the asymptomatic phase, whereas it rapidly decreases during normal immune responses (Cohen et al., 1995). In addition to an increase in size and disorganization of the follicular dendritic cells (FDCs) network, HIV-1 infection is accompanied by a thin and frequently disrupted mantle zone, a loss of germinal center (GC) polarization, and a decrease in CD80 expression on

GC B lymphocytes of hyperplastic HIV-1 seropositive lymph nodes (Legendre *et al.*, 1998). Thus, persistent HIV-1 infection is accompanied by progressive and irreversible destruction of lymphoid organs, a process involving not only CD4$^+$ T cells but also B cells and FDCs (Heat *et al.*, 1995; Spiegel *et al.*, 1992).

These data show that HIV-1 induces strong B-cell activation and a profound histologic alteration of lymphoid tissues but, as we will discuss in the following paragraphs, in spite of this strong and sustained activation of humoral immunity, it fails in containing and controlling HIV-1 infection and replication.

9.4.1 Differentiation and Functions of B Cells

B and T lymphocytes arise from a common bone marrow–derived precursor that becomes committed to the lymphocyte lineage. In the absence of antigen stimulation, mature B and T cells continuously recirculate throughout blood and lymphoid organs. Contact with the antigen induces migration of immature DCs from epithelia skin or mucosa into the draining lymphoid organ. The binding and uptake of the antigen results in the maturation of DCs and their relocation to the junction between the extrafollicular zone of the lymphoid tissue, rich in T lymphocytes, and the external area of the follicles, rich in B lymphocytes. At this site, antigen processed and presented by DCs independently activates B and T lymphocytes with the same antigenic specificity preventing their further recirculation (Liu, 1997; Melchers *et al.*, 1999). An efficient T-dependent humoral response is achieved only if these antigen-activated B and T lymphocytes rapidly establish direct cell-cell interactions, allowing B cells to survive, to differentiate into plasmocytes producing IgM with low affinity for the antigen, and to generate GC founder cell (Liu, 1997; Melchers *et al.*, 1999). Within GCs, B cells undergo a complex sequence of events: a clonal expansion phase, during which variable diversity joining (VDJ) hypermutation generates antibody variants, followed by a second phase involving the arrest of cell proliferation, isotype switching, and the selection of B lymphocytes expressing B-cell antigen receptor (BCR) with high affinity for the antigen (Kelsoe, 1996; Liu *et al.*, 1992). This process, known as affinity maturation, results in a progressive increase in antibody affinity selecting variants that bind antigen strongly (French *et al.*, 1989; Griffiths *et al.*, 1984). Selected centrocytes than differentiated into antigen-specific memory B cells and long-lived plasmoblast producing IgM, G, A, and E with a high affinity for the antigen. Plasmoblast and memory B cells then leave the follicle, migrating to specific areas in which they complete their differentiation (Ollila and Vihinen, 2005).

Antibodies specifically bind antigens in both the recognition phase and effector phase of humoral immunity. In the effector phase, antibodies secreted by B cells bind to antigens and trigger several effector mechanisms whose ultimate goal is to eliminate the antigen. Antibody-mediated effector functions include neutralization of microbes or toxic microbia products, activation of the complement system, opsonization of antigens for enhanced phagocytosis, ADCC, by which antibodies target microbes for lysis by cells of the innate immune system, and immediate hypersensitivity, by which antibodies trigger mast cells activation. According to its structural nature, each Ig isotype is specialized in performing one or more of these functions.

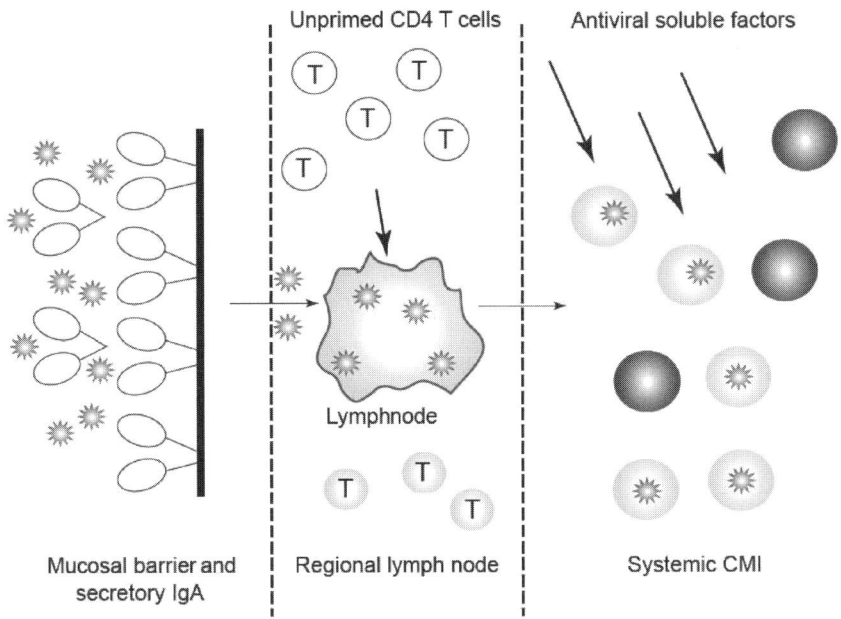

FIGURE 9.6. Diagram of the three-barrier model. First barrier: Mucosal HIV-specific Ig that mechanically prevents binding and penetration of HIV into the cellular targets. Second barrier: HIV is trapped in the lymph nodes where the IL-2–secreting T helper lymphocytes observed in the periphery are primed. Third barrier: HIV-specific CTLs and antiviral soluble factors.

As mucosa of female and male genital tracts are the portals of entry for sexually transmitted diseases of bacterial, fungal, parasitic, and viral origin, including HIV-1, and humoral defenses of mucosal surfaces is provided predominately by antibodies of the IgA isotype (McKanzie et al.,

2004; Woof and Kerr, 2004), the global research interest has been largely focused on this immunoglobulin class. IgA are synthesized locally by plasma cells, dimerized intracellularly together with a cysteine-rich polypeptide, called J chain, and then are selectively transported by a receptor-mediated mechanism into external secretions (Cyster, 2004; Fagarasan and Honjo, 2004). Mucosal IgA have been proposed to inhibit adherence and absorption of antigens and pathogens in the intestinal lumen, to neutralize intracellular organisms within epithelial cells, and to bind and excrete these foreign agents in the lamina propria to limit systemic exposure and spread. As illustrated in the next paragraphs, these effector mechanisms play a key role in the containment of HIV-1 infection and represent the first line of defense used by the immune system to prevent HIV-1 entry into the organism (Fig. 9.6).

9.4.2 Antibody Responses in HIV-1 Infection

Although antibody responses play a key role in clearing many viral infections, accumulating data suggest that this may not be true for HIV-1 infection. Peripheral B cells spontaneously produce large amounts of commutated Ig of various isotypes. Antibody responses to a variety of HIV-1 antigens are detectable within six to nine weeks after infection, and their appearance marks the seroconversion of the HIV-1–infected patient (D'Souza and Mathieson, 1996), but this HIV-specific part of the humoral response then progressively decreases in favor of the production of HIV-unrelated antibodies and autoantibodies. This evidence suggests that the process of affinity maturation or the survival of the specific memory pool is impaired during disease progression. Even more important, it has been convincingly demonstrated that antibodies isolated from sera of HIV-infected individuals have only a weak neutralizing activity for primary HIV-1 isolates (Montefiori et al., 1996; Moog et al., 1997; Moore et al., 1995), with most of the antibodies being non-neutralizing and directed at virion debris (Poignard et al., 1999). In addition, the burst of HIV-1 replication that occurs in the first days after initial infection is contained in the infected individuals well before the development of antibodies that are able to neutralize the virus (Pilgrim et al., 1997). While HIV-specific CD8[+] T lymphocytes can be isolated from HIV-infected patients soon after the infection, HIV-specific antibodies are usually evident a few months postinfection. Accordingly, in rhesus monkeys, despite a significant delay in the emergence of a virus-neutralizing antibody response, the kinetics of early viral clearance was not modified by the depletion of B lymphocytes after SIV infection (Schmits et al., 2003). These findings suggest that neutralizing antibodies may not be important in the early control of HIV-1 replication.

The most immunogenic HIV-1 molecules that elicit neutralizing antibody responses appear to be three distinct neutralizing domains within the HIV-1 envelope: the third hypervariable (V3) loop of the envelope glicoproteins, the CD4 binding site of the envelope, and the trans-membrane gp41protein. Given the role played by V3 loop in the interaction between HIV-1 envelope with chemokine receptors, it is not surprising that antibodies binding to this domain can successfully inhibit viral infection of cells (Choe *et al.*, 1996). Analogously, anti-CCR5 antibodies that cause antigen down-modulation and a CCR5-minus phenotype have been associated with resistance to HIV-1 infection (Lopalco *et al.*, 2000). Furthermore, antibodies directed to the V3 loop are the first neutralizing antibodies that arise in HIV-1–infected individuals (Javaherian *et al.*, 1989). In spite of this promising observation, this domain is problematic as a target for vaccine-elicited, broadly neutralizing antibodies, for at least two main reasons: (1) usually, V3 loop-specific antibodies are isolate-specific in their neutralizing ability; (2) the extensive glycosylation of the envelope characterizing primary HIV-1 isolates interferes with antibody access to that site (Reitter *et al.*, 1998). CD4 binding domains of HIV-1 envelope are highly conserved among viral isolates, and antibodies that bind to these domains are therefore reactive with a variety of viruses. Unfortunately, antibodies specific for CD4 binding site are only weakly neutralizing. The sequence of the trans-membrane gp41 protein is highly conserved among HIV-1 isolates, and monoclonal antibodies to gp41 that neutralize a variety of HIV-1 isolates have been described suggesting this region to be a good target for vaccine-induced antibody (Trkola *et al.*, 1995).

Several studies suggest that the contribution of neutralizing antibodies to the control of HIV-1 replication in subjects with established infection is marginal at best, despite data proving that they exert immune selection pressure (Richman *et al.*, 2003; Wei *et al.*, 2003). Two different studies, in particular, have proved the inability of humoral immunity to contain virus replication. In the first one, immunodeficient mice reconstituted with human lymphoid tissue have been infected with HIV-1 and then evaluated after infusion with neutralizing HIV-1–specific monoclonal antibodies. Antibodies had little effect on viral replication in this model (Poignard *et al.*, 1999). Similarly, in HIV-1–infected individuals, intravenous infusion of hyperimmune globulin with high titers of HIV-1–specific antibodies had little effect on viral load or disease progression (Jacobson, 1998).

Even more important, Smith-Franklin and colleagues (2002) have recently demonstrated that antibodies could play a negative role in the protection to HIV infection. Because FDCs trap antigens (and virus) in the form of immune complexes and are rich in Fcγ receptors (CD16), it has been proved *in vitro* that antibody and CD16 contribute to FDC-mediated maintenance of HIV infectivity. Accordingly, blocking FDC-Fcγ receptors or killing the FDCs dramatically reduced their ability to preserve virus infectivity.

Conversely, data stemming from studies performed after treatment interruption in HIV-1–infected individuals in whom highly active anti-retroviral therapy (HAART) was initiated during early seroconversion and who remained on therapy for 1 to 3 years suggest antibodies to be protective and to control viral rebound. As a matter of fact, autologous neutralizing antibodies emerged rapidly and correlated with a spontaneous downregulation in rebound viremia after treatment interruption, thus suggesting that potent neutralizing-antibody responses to autologous virus are able to mature and that in some people these responses contribute to the control of plasma viremia after treatment cessation (Montefiori et al., 2001). Accordingly, studies performed on SIV and HIV/SIV hybrid (SHIV) models have demonstrated that preexisting circulating neutralizing antibodies can modify the clinical outcome of infection in macaques. Injection of serum IgG or combinations of monoclonal antibodies that neutralize these viruses attenuates the phatogenicity or even blocks the establishment of infection by these lentiviruses (Baba et al., 2000; Haigwood et al., 1996; Mascola et al., 2000). The fact that neutralizing antibodies seem to be able to entirely protect against initial infection suggests that such antibodies will be very important in the setting-up of any strategy to prevent HIV-1 infection.

9.4.3 Viral Escape from Antibodies

As previously reported, viral escape is mainly due to the generation of mutations in targeted epitopes of the virus. In this regard, selection pressure exerted by humoral immune responses to HIV-1 is well documented, even if its precise contribution to immune failure is still debated. Selection pressure by neutralizing antibodies can be observed in vitro (Reits et al., 1988) and is apparent in vivo in early infection, as shown by the emergence of virus that is able to evade early autologous neutralizing antibodies even though it remains sensitive to neutralization by control sera (Albert et al., 1990). These data argue that neutralizing antibody responses account for the extensive variation in the envelope gene that is observed in the early months after primary HIV-1 infection. Studies performed using recombinant virus assay have shown that the rate of neutralizing-antibodies escape exceeds the rapid rate of change observed with drug selection pressure and can account for the extensive variability in the envelope protein compared with other genes (Richman *et al.*, 2003; Wei *et al.*, 1998). The mechanism of escape may involve changes in envelope glycans that shield antibody binding sites by steric hindrance (Wei *et al.*, 1998). These studies clearly show that neutralizing antibodies exert considerable selection pressure and that fully functional envelope variants that escape immune detection continuously emerge and become the dominant circulating species. Despite the clear induction of antibody escape, a direct link between the degree of antibody escape and disease progression remains to be shown.

9.5 Immune Responses in HIV-Exposed Seronegative Individuals

Two phenomena have been extensively investigated in HIV-1 infection: the absence of clinical progression in some HIV-infected patients and the ability of some individuals to resist HIV infection despite multiple and repeated exposure to HIV. Resistance to HIV infection is a relatively frequent condition that is observed in several different categories of individuals at high risk of infection including prostitutes, sexually exposed gay men, neonates of HIV-seropositive mothers, accidentally exposed health care workers, and heterosexual partners of HIV-seopositive subjects (Rowland-Jones and McMichael, 1995; Shearer and Clerici, 1996). Absence of HIV infection after exposure to HIV is associated, in a minority of ESN individuals, with mutation in the CCR5 coreceptor (Cocchi *et al.*, 1995; Liu *et al.*, 1996). It has nevertheless become apparent that actual exposure to HIV in these ESN individuals is linked to immune mechanisms of defense largely dependent on the activation of HIV-specific systemic and mucosal T helper and cytotoxic T cells (Clerici *et al.*, 1992b, 1993, 1994; Ranki *et al.*, 1989). Thus, IL-2 secretion by PBMCs, stimulated with synthetic peptides of the envelope proteins of HIV (Env), has been demonstrated in all the categories of ESN subjects mentioned above demonstrating that these individuals have HIV-specific memory T lymphocytes in peripheral blood. Biologically more important are results showing that HIV-specific IFN-γ–producing, HLA class I–restricted CTLs, evaluated with an ELISPOT assay, are present in ESN individuals. In fact, because peptide presentation in association with HLA class I molecules is dependent upon intracellular replication of the antigen, the detection of CTLs in ESN subjects strongly suggests that a subclinical infection with HIV has occurred in these individuals and allows speculation on the ability of the immune system to contain and/or reject HIV infection. More recently, it has been shown that exogenous antigen can also be processed for presentation by MHC class I molecules via an alternate antigen/processing mechanism pathway known as cross-priming presentation (Ackerman and Cresswell, 2004; Albert *et al.*, 1998). Thus, the detection of HIV-specific CTLs in ESN individuals might not be a consequence of abortive infection but rather could result from alternate processing mechanisms by DCs. This explanation is apparently reinforced by the observation that HIV cDNA was not detected in blood, seminal fluid, or cervical biopsies from ESN individuals (Mazzoli *et al.*, 1997).

The above observations have convincingly shown that resistance to HIV infection is associated with the detection of HIV-specific cellular immunity. Nevertheless, it is known that the majority of worldwide HIV infections are sexually transmitted and in this context, the first site of exposure is the genital mucosa (Kresina and Mathieson, 1999; Overbaugh *et al.*, 1999). That is why cellular immunity and mucosal immunity at this entry site have been deeply analyzed in recent years. The rationale for these analyses stems from a series of data including the detection of urinary HIV-specific antibodies in HIV-seronegative individuals with possible exposure to HIV (Urnovitz *et al.*, 1993) and the observations that (1) mucosal SIV-specific IgA are present in SIV-vaccinated macaques resisting intrarectal challenge with live, infectious SIV (Lehner *et al.*, 1995); and (2) neutralization of HIV is achieved by secretory (fecal) IgA induced by oral immunization (Bukawa *et al.*, 1995). The results so far obtained have shown that mucosal HIV-specific IgA, but not HIV-specific IgG, is present in cervical biopsies of heterosexual Italian women, African and Thai commercial sex workers, and in the ejaculates of HIV-uninfected male partners of HIV-infected women (Beyrer *et al.*, 1999; Lo Caputo *et al.*, 2003; Mazzoli *et al.*, 1997). Conversely, both HIV-specific IgG and IgA are detected in HIV-seropositive individuals, thus suggesting that in the absence of IgG, mucosal IgA acts as a mucosal surrogate marker of exposure to HIV not resulting in HIV infection. Three more considerations have largely contributed to strengthen the role of mucosal HIV-specific IgA in resistance to HIV infection: (1) HIV-specific IgA can neutralize HIV-1 isolates in an assay using PBMCs as target cells (Devito *et al.*, 2000); (2) mucosal and plasma HIV-specific IgA isolated from ESN individuals inhibits HIV-1 transcytosis across human epithelial cells (Devito *et al.*, 2000); (3) IgA from ESN subjects recognizes conserved epitopes (QUARILAV) on HIV-1 gp41 that differ from epitopes recognized by HIV-1–positive individuals (Clerici *et al.*, 2002b). These findings would explain why ESN individuals could resists HIV infection but do not clarify the mechanisms responsible for the production of IgA capable of recognizing different epitopes in different subjects. The answer to this basic question should probably be looked for in the genetic background of the individual, and studies based on the analyses of the murine model of Friend leukemia virus (FV) have already provided important results in this direction (Hasenkrug and Chesebro, 1997; Ney and D'Andrea, 2000). In this concern, chromosome 22, the orthologue of murine chromosome 15, seems to play a fundamental role in the resistance to HIV infection, and genes (i.e., APOBEC3G and APOBEC3F) mapping in a specific segment of this chromosome (22q13.1) (Jarmuz *et al.*, 2002) are under evaluation to prove their involvement in the defense against HIV entrance and/or replication.

Another recently discovered resistance factor to HIV infection is represented by α-defensins, cationic peptides with a broad antimicrobial activity whose concentration has been found significantly higher in both LTNP and ESN individuals (Trabattoni *et al.*, 2004a; Zhang *et al.*, 2002). Interestingly, the number of α-defensin–producing cells was 10-fold higher in ESNs and HIV-infected subjects than in low-risk HIV-unexposed controls in cells that were not stimulated *in vitro*, probably as a reflection of the chronic immune stimulation detected in ESNs and in HIV-infected partners (Trabattoni *et al.*, 2004a). The maintenance of a high production of α-defensins independently of antigenic stimulation is characteristic of mechanisms belonging to natural immunity and is in contrast with a number of observations showing that such exposure is necessary to maintain HIV-specific T and B lymphocytes in ESNs.

9.6 Immune Responses in Long-Term Nonprogressors

The course of HIV infection varies widely among individuals. The median time from infection to development of AIDS is 8 to 10 years, but a small proportion of patients has been characterized as LTNPs. These subjects have been identified among HIV-infected people whose risk factors include sexual exposure, intravenous drug use, and transfusions (Cao *et al.*, 1995, Hayrles *et al.*, 1996; Sheppard *et al.*, 1993). Strictly defined, LTNPs are those patients who, in the absence of treatment, do not show signs of the disease and maintain stable CD4 cell counts and low or undetectable viral loads despite prolonged periods of infection with HIV. Identification of these patients has led to comparison of people with different rates of disease progression as it is likely to hold information of paramount importance on how to successfully treat HIV infection. The interaction of numerous viral and host factors, such as viral virulence, host genetics, host immune response, and cytokine milieu, is believed to influence the course of disease.

As previously described, cellular immune response is a key factor in HIV pathogenesis, and studies of LTNPs as well as ESNs have suggested that certain aspects of host immunity may protect against HIV infection or slow disease progression. Studies have revealed that many LTNPs maintain vigorous HIV-specific cytotoxic T-cell activity (Haynes et al., 1996; Klein et al., 1995; Miedema and Klein, 1996; Rinaldo et al., 1995) thanks to highly activated and broadly directed HIV specific CTLs in the setting of very low viral loads, absence of viral escape mutants, and stable CD4 cell counts. In this concern, studies performed in different laboratories proved that (1) T cells of LTNPs, whose viral burdens were orders of magnitude lower than those typically found in progressors, were as susceptible to HIV infection as those of progressors but they had quantitatively greater $CD8^+$ T cell–mediated HIV-suppressive activity compared with that generally seen in progressors (Cao et al., 1995); (2) untreated HIV-infected patients at varying stages of disease show an inverse correlation between HIV-specific CTLs and viral load (Ogg et al., 1998); (3) HIV-specific CD8 CTLs and $CD4^+$ T cells were present in the cervical secretions of 63% of HIV-positive women, and CTL activity was more often observed in patients with higher CD4 counts and slower progression to HIV (Musey et al., 1997). Thus, the combination of low viral load, persistent CTL activity, and lack of disease progression is consistent with the hypothesis that CTLs have a protective effect (Harrer et al., 1996).

Another aspect of HIV-specific cell-mediated immunity is the T helper cell response, which has been shown to be important in maintenance of $CD8^+$ T-cell responses in other persistent viral infections (Matloubian et al., 1994; Zajac et al., 1998). A recent study investigated T-helper responses and CTL activity in untreated HIV-infected people with a wide range of viral loads and disease progression. T helper proliferative responses were positively correlated with Gag-specific CTL activity and negatively correlated with viral load. In addition, levels of CTL precursors seemed to depend on the presence of T helper function (Kalams et al., 1999). Thus, both HIV-specific T helper cells and HIV-specific CTLs may be vital in controlling progression of disease, and persistence of functional CTLs may depend in part on preservation of the T helper response. Furthermore, studies have shown that some LTNPs with undetectable viral loads demonstrate vigorous $CD4^+$ T helper responses in addition to persistent CTL responses (Rosenberg et al., 1997; Valentine et al., 1998), indicating ongoing viral replication and antigenic stimulation. These findings are supported by the following observations: (1) PBMCs from LTNPs show a higher production of type 1 cytokines (IL-2 and IFN-γ), and a lower production of type 2 cytokines (IL-10); (2) immunophenotypical analyses demonstrated a higher percentage of $CD4^+CD7^-$ (IL-10–producing lymphocytes) in progressors compared with LTNPs as well as an increase in $CD4^+SLAM^+$ (type 1 cytokine–producing lymphocytes) in LTNPs; (3) in the chimpanzee/HIV animal model, characterized by seroconversion in the absence of disease progression, CD4$^+$ T-cell proliferation is not compromised, type 1 cytokine production is prevalent, and HIV-specific CTL activity is strong and sustained.

High levels of CCR5-using chemokines have also been associated with slower disease progression (Ullum *et al.*, 1998). CD4$^+$ T cells from ESNs have been shown to produce increased levels of RANTES, MIP-1α, and MIP-1β and to suppress replication of M-tropic strains of HIV-1 (Furci *et al.*, 1997; Paxton *et al.*, 1996). Further research is needed to clarify the true clinical relevance and regulatory roles of these and other chemokines in HIV infection (Lee *et al.*, 1999; Pal *et al.*, 1997).

As for humoral immunity, several investigators have reported increased frequency and breadth of neutralizing antibody titers to HIV in LTNPs, but the ability of these neutralizing antibodies to neutralize clinically relevant primary or autologous viral isolates varies (Cao *et al.*, 1995, 1995; Loomis-price *et al.*, 1998; Pilgrim *et al.*, 1997). Other studies have found no difference in neutralizing antibodies between progressors and LTNPs (Barker *et al.*, 1998), and some studies of LTNPs have found weak or undetectable neutralizing antibody responses (Harrer *et al.*, 1996).

9.7 Concluding Remarks

Research efforts during the past several years have provided insight into the complex host response to HIV exposure and infection. Specifically, immunologic and genetic studies of LTNPs and ESNs have helped to elucidate the mechanisms by which some people have slow rates of disease progression or are protected from HIV acquisition. Current evidence suggests an important role of cytotoxic T cells and T helper cells in controlling viremia, slowing disease progression, and perhaps in preventing establishment of infection. The role of humoral immunity and mucosal immunity in preventing transmission and slowing disease progression remains unclear, and further studies are needed to elucidate their role in preventing infection and to set up new therapeutic and vaccine strategies.

Currently, this knowledge is being used to identify the effective *in vivo* immune responses that control HIV replication in infected people and to mimic these responses in HIV-negative individuals in order to develop a truly effective HIV vaccine.

References

Abdullah, N., Greenman, J., Pimenidou, A., Topping, K.P., Monson, J.R. (1999). The role of monocytes and natural killer cells in mediating antibody-dependent lysis of colorectal tumour cells. Cancer Immunol. Immunother. 48:517-524.

Abebe, A., Demissie, D., Goudsmit, J., Brouwer, M., Kuiken, C.L., Pollakis, G., Schuitemaker, H., Fontanet, A.L., Rinke de Wit, T.F. (1999). HIV-1 subtype C syncytium- and non-syncytium-inducing phenotypes and coreceptor usage among Ethiopian patients with AIDS. AIDS. 13:1305-1311.

Ackerman, A.L., Cresswell, P. (2004). Cellular mechanisms governing cross-presentation of exogenous antigens. Nat. Immunol. 5:678-684.

Addo, M.M., Yu, X.G., Rathod, A., Cohen, D., Eldridge, R.L., Strick, D., Johnston, M.N., Corcoran, C., Wurcel, A.G., Fitzpatrick, C.A., Feeney, M.E., Rodriguez, W.R., Basgoz, N., Draenert, R., Stone, D.R., Brander, C., Goulder, P.J., Rosenberg, E.S., Altfeld, M., Walker, B.D. (2003). Comprehensive epitope analysis of human immunodeficiency virus type 1 (HIV-1)-specific T-cell responses directed against the entire expressed HIV-1 genome demonstrate broadly directed responses, but no correlation to viral load. J. Virol. 77:2081-2092.

Albert, J., Abrahamsson, B., Nagy, K., Aurelius, E., Gaines, H., Nystrom, G., Fenyo, E.M. (1990). Rapid development of isolate-specific neutralizing antibodies after primary HIV-1 infection and consequent emergence of virus variants which resist neutralization by autologous sera. AIDS. 4:107-112.

Albert, M.L., Sauter, B., Bhardwaj, N. (1998). Dendritic cells acquire antigen from apoptotic cells and induce class I-restricted CTLs. Nature. 392:86-89.

Allen, T.M., O'Connor, D.H., Jing, P., Dzuris, J.L., Mothe, B.R., Vogel, T.U., Dunphy, E., Liebl, M.E., Emerson, C., Wilson, N., Kunstman, K.J., Wang, X., Allison, D.B., Hughes, A.L., Desrosiers, R.C., Altman, J.D., Wolinsky, S.M., Sette, A., Watkins, D.I. (2000). Tat-specific cytotoxic T lymphocytes select for SIV escape variants during resolution of primary viraemia. Nature. 407:386-390.

Amadori, A., Chieco-Bianchi, L. (1990). B-cell activation and HIV-1 infection: deeds and misdeeds. Immunol. Today. 11:374-379.

Amadori, A., Zamarchi, R, Ciminale, V., Del Mistro, A., Siervo, S., Alberti, A., Colombatti, M., Chieco-Bianchi, L. (1989). HIV-1-specific B cell activation. A major constituent of spontaneous B cell activation during HIV-1 infection. J. Immunol. 143:2146-2152.

Amendola, A., Poccia, F., Martini, F., Gioia, C., Galati, V., Pierdominici, M., Marziali, M., Pandolfi, F., Colizzi, V., Piacentini, M., Girardi, E., D'offizi, G. (2000). Decreased CD95 expression on naive T cells from HIV-infected persons undergoing highly active anti-retroviral therapy (HAART) and the influence of IL-2 low dose administration. Irhan Study Group. Clin. Exp. Immunol. 120:324-332.

Andersson, J., Behbahani, H., Lieberman, J., Connick, E., Landay, A., Patterson, B., Sonnerborg, A., Lore, K., Uccini, S., Fehniger Fehniger, T.E. (1999).

Perforin is not co-expressed with granzyme A within cytotoxic granules in CD8 T lymphocytes present in lymphoid tissue during chronic HIV infection. AIDS. 13:1295-1303.

Autran, B., Hadida, F., Haas, G. (1996). Evolution and plasticity of CTL responses against HIV. Curr. Opin. Immunol. 8:546-553.

Baba, T.W., Liska, V., Hofmann-Lehmann, R., Vlasak, J., Xu, W., Ayehunie, S., Cavacini, L.A., Posner, M.R., Katinger, H., Stiegler, G., Bernacky, B.J., Rizvi, T.A., Schmidt, R., Hill, L.R., Keeling, M.E., Lu, Y., Wright, J.E., Chou, T.C., Ruprecht, R.M. (2000). Human neutralizing monoclonal antibodies of the IgG1 subtype protect against mucosal simian-human immunodeficiency virus infection. Nat. Med. 6:200-206.

Badovinac, V.P., Harty, J.T. (2003). Memory lanes. Nat. Immunol. 2003; 4:212–213.

Belec, L., Ghys, P.D., Hocini, H., Nkengasong, J.N., Tranchot-Diallo, J., Diallo, M.O., Ettiegne-Traore, V., Maurice, C., Becquart, P., Matta, M., Si-Mohamed, A., Chomont, N., Coulibaly, I.M., Wiktor, S.Z., Kazatchkine, M.D. (2001). Cervicovaginal secretory antibodies to human immunodeficiency virus type 1 (HIV-1) that block viral transcytosis through tight epithelial barriers in highly exposed HIV-1-seronegative African women. J. Infect. Dis. 184:1412-1422.

Bentwich, Z., Kalinkovich, A., Weisman, Z. (1995). Immune activation is a dominant factor in the pathogenesis of African AIDS. Immunol. Today. 16:187-191.

Berger, E.A., Doms, R.W., Fenyo", E.M., Korber, B.T., Littman, D.R., Moore, J.P. (1998). A new classification for HIV-1 [Letter]. Nature. 391:240.

Bernard, N.F., Pederson, K., Chung, F., Ouellet, L., Wainberg, M.A., Tsoukas, C.M. (1998). HIV-specific cytotoxic T-lymphocyte activity in immunologically normal HIV-infected persons. AIDS. 12:2125-2139.

Betts, M.R., Ambrozak, D.R., Douek, D.C., Bonhoeffer, S., Brenchley, J.M., Casazza, J.P., Koup, R.A., Picker, L.J. (2001). Analysis of total human immunodeficiency virus (HIV)-specific CD4($+$) and CD8($+$) T-cell responses: relationship to viral load in untreated HIV infection. J. Virol. 75:11983-11991.

Beyrer, C., Artenstein, A.W., Rugpao, S., Stephens, H., VanCott, T.C., Robb, M.L., Rinkaew, M., Birx, D.L., Khamboonruang, C., Zimmerman, P.A., Nelson, K.E., Natpratan, C. (1999). Epidemiologic and biologic characterization of a cohort of human immunodeficiency virus type 1 highly exposed, persistently seronegative female sex workers in northern Thailand. Chiang Mai HEPS Working Group. J. Infect. Dis. 179:59-67.

Biasin, M., Boasso, A., Piacentini, L., Trabattoni, D., Magri, G., Deshmuks, R., Deshpande, A., Clerici, M. (2003). IL-4 and CXCR4 upregulation in HIV-infected and uninfected individuals from Maharashtra-Mumbai. AIDS. 17:1563-1565.

Borrow, P., Lewicki, H., Hahn, B.H., Shaw, G.M., Oldstone, M.B. (1994). Virus-specific CD8$^+$ cytotoxic T-lymphocyte activity associated with control of viremia in primary human immunodeficiency virus type 1 infection. J. Virol. 68:6103-6110.

Borrow, P., Lewicki, H., Wei, X., Horwitz, M.S., Peffer, N., Meyers, H., Nelson, J.A., Gairin, J.E., Hahn, B.H., Oldstone, M.B., Shaw, G.M. (1997). Antiviral pressure exerted by HIV-1-specific cytotoxic T lymphocytes (CTLs) during primary infection demonstrated by rapid selection of CTL escape virus. Nat. Med. 3:205-211.

Bukawa, H., Sekigawa, K., Hamajima, K., Fukushima, J., Yamada, Y., Kiyono, H., Okuda, K. (1995). Neutralization of HIV-1 by secretory IgA induced by oral immunization with a new macromolecular multicomponent peptide vaccine candidate. Nat. Med. 1:681-685.

Buseyne Buseyne, F., Le Chenadec, J., Corre, B., Porrot, F., Burgard, M., Rouzioux, C., Blanche, S., Mayaux, M.J., Riviere, Y. (2002). Inverse correlation between memory Gag-specific cytotoxic T lymphocytes and viral replication in human immunodeficiency virus-infected children. J. Infect. Dis. 186:1589-1596.

Canque, B., Rosenzwajg, M., Gey, A., Tartour, E., Fridman, W.H., Gluckman, J.C. (1996). Macrophage inflammatory protein-1alpha is induced by human immunodeficiency virus infection of monocyte-derived macrophages. Blood. 87:2011-2019.

Cao, Y., Qin, L., Zhang, L., Safrit, J., Ho, D.D. (1995). Virologic and immunologic characterization of long-term survivors of human immuno-deficiency virus type 1 infection. N. Engl. J. Med. 332:201-208.

Carrington, M., Nelson, G.W., Martin, M.P., Kissner, T., Vlahov, D., Goedert, J.J., Kaslow, R., Buchbinder, S., Hoots, K., O'Brien, S.J. (1999). HLA and HIV-1: heterozygote advantage and B*35-Cw*04 disadvantage. Science. 283:1748-1752.

Catalfamo, M., Henkart, P.A. Perforin and the granule exocytosis cytotoxicity pathway. Curr. Opin. Immunol. 15:522-7.

Catalfamo, M., Henkart, P.A. (2003). Perforin and the granule exocytosis cytotoxicity pathway. Curr. Opin. Immunol. 15:522-527.

Champagne, P., Ogg, G.S., King, A.S., Knabenhans, C., Ellefsen, K., Nobile, M., Appay, V., Rizzardi, G.P., Fleury, S., Lipp, M., Forster, R., Rowland-Jones, S., Sekaly, R.P., McMichael McMichael, A.J., Pantaleo, G. (2001). Skewed maturation of memory HIV-specific CD8 T lymphocytes. Nature. 410:106-111.

Chang, T.L., Mosoian, A., Pine, R., Klotman, M.E., Moore, J.P. (2002). A soluble factor(s) secreted from CD8($+$) T lymphocytes inhibits human immuno-deficiency virus type 1 replication through STAT1 activation. J. Virol. 76:569-581.

Chapoval, A.I., Ni, J., Lau, J.S., Wilcox, R.A., Flies, D.B., Liu, D., Dong, H., Sica, G.L., Zhu, G., Tamada, K., Chen, L. (2001). B7-H3: a costimulatory molecule for T cell activation and IFN-gamma production. Nat. Immunol. 2:269-274.

Chen, Z.W., Kou, Z.C., Lekutis, C., Shen, L., Zhou, D., Halloran, M., Li, J., Sodroski, J., Lee-Parritz, D., Letvin, N.L. (1995). T cell receptor V beta repertoire in an acute infection of rhesus monkeys with simian immuno-deficiency viruses and a chimeric simian-human immunodeficiency virus. J. Exp. Med. 182:21-31.

Choe, H., Farzan, M., Sun, Y., Sullivan, N., Rollins, B., Ponath, P.D., Wu, L., Mackay, C.R., LaRosa, G., Newman, W., Gerard, N., Gerard, C., Sodroski, J. (1996). The beta-chemokine receptors CCR3 and CCR5 facilitate infection by primary HIV-1 isolates. Cell. 85:1135-1148.

Chouquet, C., Autran, B., Gomard, E., Bouley, J.M., Calvez, V., Katlama, C., Costagliola, D., Riviere, Y.; IMMUNOCO Study Group. (2002). Correlation between breadth of memory HIV-specific cytotoxic T cells, viral load and disease progression in HIV infection. AIDS. 16:2399-2407.

Clerici, M., Shearer, G.M. (1993). A TH1→TH2 switch is a critical step in the etiology of HIV infection. Immunol. Today. 14:107-111.

Clerici, M., Shearer, G.M. (1994). The Th1-Th2 hypothesis of HIV infection: new insights. Immunol. Today. 15:575-581.

Clerici, M., Shearer, G.M. (1996). Correlates of protection in HIV infection and the progression of HIV infection to AIDS. Immunol. Lett. 51:69-73.

Clerici, M., Stocks, N.I., Zajac, R.A., Boswell, R.N., Lucey, D.R., Via, C.S., Shearer, G.M. (1989). Detection of three distinct patterns of T helper cell dysfunction in asymptomatic, human immunodeficiency virus-seropositive patients. Independence of $CD4^+$ cell numbers and clinical staging. J. Clin. Invest. 84:1892-1899.

Clerici, M., Giorgi, J.V., Chou, C.C., Gudeman, V.K., Zack, J.A., Gupta, P., Ho, H.N., Nishanian, P.G., Berzofsky, J.A., Shearer, G.M. (1992b). Cell-mediated immune response to human immunodeficiency virus (HIV) type 1 in seronegative homosexual men with recent sexual exposure to HIV-1. J. Infect. Dis. 165:1012-1019.

Clerici, M., Roilides, E., Via, C.S., Pizzo, P.A., Shearer, G.M. (1992a). A factor from CD8 cells of human immunodeficiency virus-infected patients suppresses HLA self-restricted T helper cell responses. Proc. Natl. Acad. Sci. U. S. A. 89:8424-8428.

Clerici, M., Sison, A.V., Berzofsky, J.A., Rakusan, T.A., Brandt, C.D., Ellaurie, M., Villa, M., Colie, C., Venzon, D.J., Sever, J.L. (1993). Cellular immune factors associated with mother-to-infant transmission of HIV. AIDS. 7:1427-1433.

Clerici, M., Levin, J.M., Kessler, H.A., Harris, A., Berzofsky, J.A., Landay, A.L., Shearer, G.M. (1994). HIV-specific T-helper activity in seronegative health care workers exposed to contaminated blood. JAMA. 271:42-46.

Clerici, M., Butto, S., Lukwiya, M., Saresella, M., Declich, S., Trabattoni, D., Pastori, C., Piconi, S., Fracasso, C., Fabiani, M., Ferrante, P., Rizzardini, G., Lopalco, L. (2000). Immune activation in africa is environmentally-driven and is associated with upregulation of CCR5. Italian-Ugandan AIDS Project. AIDS. 14:2083-2092.

Clerici, M., Boasso, A., Rizzardini, G., Deshpande, A., Biasin, M. (2002a). AIDS in Africa. Lancet. 360:1424.

Clerici, M., Barassi, C., Devito, C., Pastori, C., Piconi, S., Trabattoni, D., Longhi, R., Hinkula, J., Broliden, K., Lopalco, L. (2002b). Serum IgA of HIV-exposed uninfected individuals inhibit HIV through recognition of a region within the alpha-helix of gp41. AIDS. 16:1731-1741.

Cocchi, F., DeVico, A.L., Garzino-Demo, A., Arya, S.K., Gallo, R.C., Lusso, P. (1995). Identification of RANTES, MIP-1 alpha, and MIP-1 beta as the major HIV-suppressive factors produced by CD8+ T cells. Science. 270:1811-1815.

Cohen, D.E., Walker, B.D. (2001). Human immunodeficiency virus pathogenesis and prospects for immune control in patients with established infection. Clin. Infect. Dis. 32:1756-1768.

Cohen, O.J., Pantaleo, G., Schwartzentruber, D.J., Graziosi, C., Vaccarezza, M., Fauci, A.S. (1995). Pathogenic insights from studies of lymphoid tissue from HIV-infected individuals. J. Acquir. Immune. Defic. Syndr. Hum. Retrovirol. 10 Suppl 1:S6-14.

Cyster, J.G. (2004) Chemokines and cell migration in secondary lymphoid organs. Science. 286:2098-2102.

Devito, C., Broliden, K., Kaul, R., Svensson, L., Johansen, K., Kiama, P., Kimani, J., Lopalco, L., Piconi, S., Bwayo, J.J., Plummer, F., Clerici, M., Hinkula, J. (2000). Mucosal and plasma IgA from HIV-1-exposed uninfected individuals inhibit HIV-1 transcytosis across human epithelial cells. J. Immunol. 165:5170-5176.

Devito, C., Hinkula, J., Kaul, R., Lopalco, L., Bwayo, J.J., Plummer, F., Clerici, M., Broliden, K. (2000). Mucosal and plasma IgA from HIV-exposed seronegative individuals neutralize a primary HIV-1 isolate. AIDS. 14:1917-1920.

Dolan, M.J., Clerici, M., Blatt, S.P., Hendrix, C.W., Melcher, G.P., Boswell, R.N., Freeman, T.M., Ward, W., Hensley, R., Shearer, G.M. (1995). *In vitro* T cell function, delayed-type hypersensitivity skin testing, and CD4+ T cell subset phenotyping independently predict survival time in patients infected with human immunodeficiency virus. J. Infect. Dis. 172:79-87.

Dong, H., Zhu, G., Tamada, K., Chen, L. (1999). B7-H1, a third member of the B7 family, co-stimulates T-cell proliferation and interleukin-10 secretion. Nat. Med. 5(12):1365-1369.

Douek, D.C., Brenchley, J.M., Betts, M.R., Ambrozak, D.R., Hill, B.J., Okamoto, Y., Casazza, J.P., Kuruppu, J., Kunstman, K., Wolinsky, S., Grossman, Z., Dybul, M., Oxenius, A., Price, D.A., Connors, M., Koup, R.A. (2002). HIV preferentially infects HIV-specific CD4+ T cells. Nature. 417:95-98.

D'Souza, M.P., Mathieson, B.J. (1996). Early phases of HIV type 1 infection. AIDS Res. Hum. Retroviruses. 12:1-9.

Dudhane, A., Conti, B., Orlikowsky, T., Wang, Z.Q., Mangla, N., Gupta, A., Wormser, G.P., Hoffmann, M.K. (1996). Monocytes in HIV type 1-infected individuals lose expression of costimulatory B7 molecules and acquire cytotoxic activity. AIDS. Res. Hum. Retroviruses. 12:885-892.

Esser, M.T., Graham, D.R., Coren, L.V., Trubey, C.M., Bess, J.W., Jr, Arthur, L.O., Ott, D.E., Lifson, J.D. (2001). Differential incorporation of CD45, CD80 (B7-1), CD86 (B7-2), and major histocompatibility complex class I and II molecules into human immunodeficiency virus type 1 virions and microvesicles: implications for viral pathogenesis and immune regulation. J. Virol. 75:6173-6182.

Fabiani, M., Ble, C., Grivel, P., Lukwiya, M., Declich, S. (1998). 1989-1996 HIV-1 prevalence trends among different risk groups in Gulu District, North Uganda. J. Acquir. Immune. Defic. Syndr. Hum. Retrovirol. 18:514.

Fagarasan, S., Honjo, T. (2004). Regulation of IgA synthesis at mucosal surfaces. Curr. Opin. Immunol. 16:277-283.

Fellermann, K., Stange, E.F. (2001). Defensins - innate immunity at the epithelial frontier. Eur. J. Gastroenterol. Hepatol. 13:771-776.

Feng, Y., Broder, C.C., Kennedy, P.E., Berger, E.A. (1996). HIV-1 entry cofactor: functional cDNA cloning of a seven-transmembrane, G protein-coupled receptor. Science. 272:872-7.

Freeman, G.J., Freedman, A.S., Segil, J.M., Lee, G., Whitman, J.F., Nadler, L.M. (1989). B7, a new member of the Ig superfamily with unique expression on activated and neoplastic B cells. J. Immunol. 143:2714-2722.

French French, D.L., Laskov, R., Scharff, M.D. (1989). The role of somatic hypermutation in the generation of antibody diversity. Science. 244:1152-1157.

Furci, L., Scarlatti, G., Burastero, S., Tambussi, G., Colognesi, C., Quillent, C., Longhi, R., Loverro, P., Borgonovo, B., Gaffi, D., Carrow, E., Malnati, M., Lusso, P., Siccardi, A.G., Lazzarin, A., Beretta, A. (1997). Antigen-driven C-C chemokine-mediated HIV-1 suppression by CD4 T cells from exposed uninfected individuals expressing the wild-type CCR-5 allele. J. Exp. Med. 186:455-460.

Galli, G., Annunziato, F., Cosmi, L., Manetti, R., Maggi, E., Romagnani, S. (2001). Th1 and th2 responses, HIV-1 coreceptors, and HIV-1 infection. J. Biol. Regul. Homeost. Agents. 15:308-313.

Geginat, J., Sallusto, F., Lanzavecchia, A. (2001). Cytokine-driven proliferation and differentiation of human naive, central memory, and effector memory CD4(R) T cells. J. Exp. Med . 194:1711-1719.

Goulder, P.J., Brander, C., Tang, Y., Tremblay, C., Colbert, R.A., Addo, M.M., Rosenberg, E.S., Nguyen, T., Allen, R., Trocha, A., Altfeld, M., He, S., Bunce, M., Funkhouser, R., Goulder, P.J., Phillips, R.E., Colbert, R.A., McAdam, S., Ogg, G., Nowak, M.A., Giangrande, P., Luzzi, G., Morgan, B., Edwards, A., McMichael, A.J., Rowland-Jones, S. (1997). Late escape from an immunodominant cytotoxic T-lymphocyte response associated with progression to AIDS. Nat. Med. 3:212-217.

Griffiths, G.M., Berek, C., Kaartinen, M., Milstein, C. (1984). Somatic mutation and the maturation of immune response to 2-phenyl oxazolone. Nature. 312(5991):271-275.

Gruters, R.A., Vvan Baalen, C.A., Osterhaus, A.D. (2002). The advantage of early recognition of HIV-infected cells by cytotoxic T-lymphocytes. Vaccine. 20:2011-2015.

Haase, A.T. (1999). Population biology of HIV-1 infection: viral and CD4$^+$ T cell demographics and dynamics in lymphatic tissues. Annu. Rev. Immunol. 17:625-656.

Haigwood, N.L., Watson, A., Sutton, W.F., McClure, J., Lewis, A., Ranchalis, J., Travis, B., Voss, G., Letvin, N.L., Hu, S.L., Hirsch, V.M., Johnson, P.R.

(1996). Passive immune globulin therapy in the SIV/macaque model: early intervention can alter disease profile. Immunol. Lett. 51:107-114.

Harrer, T., Harrer, E., Kalams, S.A., Barbosa, P., Trocha, A., Johnson, R.P., Elbeik, T., Feinberg, M.B., Buchbinder, S.P., Walker, B.D. (1996). Cytotoxic T lymphocytes in asymptomatic long-term nonprogressing HIV-1 infection. Breadth and specificity of the response and relation to *in vivo* viral quasispecies in a person with prolonged infection and low viral load. J. Immunol. 156:2616-2623.

Hasenkrug, K.J., Chesebro, B. (1997). Immunity to retroviral infection: the Friend virus model. Proc. Natl. Acad. Sci. U. S. A. 94:7811-7816.

Haynes, B.F., Pantaleo, G., Fauci, A.S. (1996). Toward an understanding of the correlates of protective immunity to HIV infection. Science. 271:324-328.

Heath, S.L., Tew, J.G., Tew, J.G., Szakal, A.K., Burton, G.F. (1995). Follicular dendritic cells and human immunodeficiency virus infectivity. Nature. 377:740-744.

Ho, D.D., Neumann, A.U., Perelson, A.S., Chen, W., Leonard, J.M., Markowitz, M. (1995). Rapid turnover of plasma virions and CD4 lymphocytes in HIV-1 infection. Nature. 373:123-126.

Inaba, K., Inaba, M. (2005). Antigen recognition and presentation by dendritic cells. Int. J. Hematol. 81:181-187.

Jacobson, J.M. (1998). Passive immunization for the treatment of HIV infection. Mt. Sinai J. Med. 65:22-26.

Jarmuz, A., Chester, A., Bayliss, J., Gisbourne, J., Dunham, I., Scott, J., Navaratnam, N. (2002). An anthropoid-specific locus of orphan C to U RNA-editing enzymes on chromosome 22. Genomics. 79:285-296.

Javaherian, K., Langlois, A.J., McDanal, C., Ross, K.L., Eckler, L.I., Jellis, C.L., Profy, A.T., Rusche, J.R., Bolognesi, D.P., Putney, S.D., *et al.* (1989). Principal neutralizing domain of the human immunodeficiency virus type 1 envelope protein. Proc. Natl. Acad. Sci. U. S. A. 86:6768-6772.

Jin, X., Bauer, D.E., Tuttleton, S.E., Lewin, S., Gettie, A., Blanchard, J., Irwin, C.E., Safrit, J.T., Mittler Mittle, J., Weinberger, L., Kostrikis, L.G., Zhang, L., Perelson, A.S., Ho, D.D. (1999). Dramatic rise in plasma viremia after CD8($^+$) T cell depletion in simian immunodeficiency virus-infected macaques. J. Exp. Med. 189:991-998.

Kahn, J.O., Walker, B.D. (1998). Acute human immunodeficiency virus type 1 infection. N. Engl. J. Med. 339:33-39.

Kajino, K., Kajino, Y., Greene, M.I. (1998). Fas- and perforin-independent mechanism of cytotoxic T lymphocyte. Immunol. Res. 17(1-2):89-93.

Kalams, S.A., Buchbinder, S.P., Rosenberg, E.S., Billingsley, J.M., Colbert, D.S., Jones, N.G., Shea, A.K., Trocha, A.K., Walker, B.D. (1999). Association between virus-specific cytotoxic T-lymphocyte and helper responses in human immunodeficiency virus type 1 infection. J. Virol. 73:6715-6720.

Kaslow, R.A., Carrington, M., Apple, R., Park, L., Munoz, A., Saah, A.J., Goedert, J.J., Winkler, C., O'Brien, S.J., Rinaldo, C., Detels, R., Blattner, W., Phair, J., Erlich, H., Mann, D.L. (1996). Influence of combinations of human major histocompatibility complex genes on the course of HIV-1 infection. Nat. Med. 2:405-411.

Kaslow, R.A., Rivers, C., Tang, J., Bender, T.J., Goepfert, P.A., El Habib, R., Weinhold, K., Mulligan, M.J.; NIAID AIDS vaccine evaluation group. (2001). Polymorphisms in HLA class I genes associated with both favorable prognosis of human immunodeficiency virus (HIV) type 1 infection and positive cytotoxic T-lymphocyte responses to ALVAC-HIV recombinant canarypox vaccines. J. Virol. 75:8681-8689.

Kelsoe, G. (1996). Life and death in germinal centers (redux). Immunity. 4(2):107-111.

Kidd, P. (2003). Th1/Th2 balance: the hypothesis, its limitations, and implications for health and disease. Altern. Med. Rev. 8:223-246.

Klein, M.R., Vvan Baalen, C.A., Holwerda, A.M., Kerkhof Garde, S.R., Bende, R.J., Keet, I.P., Eeftinck-Schattenkerk, J.K., Osterhaus, A.D., Schuitemaker, H., Miedema, F. (1995). Kinetics of Gag-specific cytotoxic T lymphocyte responses during the clinical course of HIV- 1 infection: a longitudinal analysis of rapid progressors and long-term asymptomatics. J. Exp. Med.

Klenerman, P., Wu, Y., Phillips Phillip, R. (2002). HIV: current opinion in escapology. Curr. Opin. Microbiol. 5:408-13.

Koup, R.A., Safrit, J.T., Cao, Y., Andrews, C.A., McLeod, G., Borkowsky, W., Farthing, C., Ho, D.D. (1994). Temporal association of cellular immune responses with the initial control of viremia in primary human immunodeficiency virus type 1 syndrome. J. Virol. 68:4650-5.

Kresina, T.F., Mathieson, B. (1999). Human immunodeficiency virus type 1 infection, mucosal immunity, and pathogenesis and extramural research programs at the National Institutes of Health. J. Infect. Dis. 179 Suppl 3:S392-396.

Kuroda, M.J., Schmitz, J.E., Charini, W.A., Nickerson, C.E., Lifton, M.A., Lord, C.I., Forman, M.A., Letvin, N.L. (1999). Emergence of CTL coincides with clearance of virus during primary simian immunodeficiency virus infection in rhesus monkeys. J. Immunol. 162:5127-5133.

Kwong, P.D., Wyatt, R., Robinson, J., Sweet, R.W,, Sodroski, J., Hendrickson, W.A. (1998). Structure of an HIV gp120 envelope glycoprotein in complex with the CD4 receptor and a neutralizing human antibody. Nature. 393:648-659.

Lane, H.C., Masur, H., Edgar, L.C., Whalen, G., Rook, A.H., Fauci, A.S. (1983). Abnormalities of B-cell activation and immunoregulation in patients with the acquired immunodeficiency syndrome. N. Engl. J. Med. 309:453-458.

Lanzavecchia, A., Sallusto, F. (2002). Progressive differentiation and selection of the fittest in the immune response. Nat. Rev. Immunol. 2:982-2987.

Legendre, C., Raphael, M., Gras, G., Lefevre, E.A., Feuillard, J., Dormont, D., Richard Richard, Y. (1998) CD80 expression is decreased in hyperplastic lymph nodes of HIV$^+$ patients. Int. Immunol. 10:1847-1851.

Lehner, T., Wang, Y., Cranage, M., Bergmeier, L.A., Mitchell, E., Tao, L., Hall, G., Dennis, M., Cook, N., Jones, I., Doyle, C. (1998). Protective mucosal immunity elicited by targeted lymph node immunization with a subunit SIV envelope and core vaccine in macaques. Dev. Biol. Stand. 92:225-235.

Lenschow, D.J., Walunas, T.L., Bluestone, J.A. (1996). CD28/B7 system of T cell costimulation. Annu. Rev. Immunol. 14:233-258.

Levy, J.A. (1993). HIV pathogenesis and long-term survival. AIDS. 7(11):1401-1410.

Levy, J.A., Mackewicz, C.E., Barker, E. (1996). Controlling HIV pathogenesis: the role of the noncytotoxic anti-HIV response of $CD8^+$ T cells. Immunol. Today. 17:217-24.

Lieberman, J., Shankar, P., Manjunath, N., Andersson, J. (2001). Dressed to kill? A review of why antiviral CD8 T lymphocytes fail to prevent progressive immunodeficiency in HIV-1 infection. Blood. 98:1667-1677.

Liu, Y.J. (1997). Sites of B lymphocyte selection, activation, and tolerance in spleen. J. Exp. Med. 186:625-629.

Liu, Y.J., Johnson, G.D., Gordon, J., MacLennan, I.C. (1992). Germinal centres in T-cell-dependent antibody responses. Immunol. Today. 13:17-21.

Liu, R., Paxton, W.A., Choe, S., Ceradini, D., Martin, S.R., Horuk, R., MacDonald, M.E., Stuhlmann, H., Koup, R.A., Landau, N.R. (1996). Homozygous defect in HIV-1 coreceptor accounts for resistance of some multiply-exposed individuals to HIV-1 infection. Cell. 86:367-377.

Lo Caputo, S., Trabattoni, D., Vichi, F., Piconi, S., Lopalco, L., Villa, M.L., Mazzotta, F., Clerici, M. (2003). Mucosal and systemic HIV-1-specific immunity in HIV-1-exposed but uninfected heterosexual men. AIDS. 17531-539.

Loomis-Price, L.D., Cox, J.H., Mascola, J.R., VanCott, T.C., Michael, N.L., Fouts, T.R., Redfield, R.R., Robb, M.L., Wahren, B., Sheppard, H.W., Birx, D.L. (1998). Correlation between humoral responses to human immuno-deficiency virus type 1 envelope and disease progression in early-stage infection. J. Infect. Dis. 178:1306-1316.

Lopalco, L., Barassi, C., Pastori, C., Longhi, R., Burastero, S.E., Tambussi, G., Mazzotta, F., Lazzarin, A., Clerici, M., Siccardi, A.G. (2000). CCR5-reactive antibodies in seronegative partners of HIV-seropositive individuals down-modulate surface CCR5 *in vivo* and neutralize the infectivity of R5 strains of HIV-1 *In vitro*. J. Immunol. 164:3426-3433.

Lopalco, L., Barassi, C., Paolucci, C., Breda, D., Brunelli, D., Nguyen, M., Nouhin, J., Luong, T.T., Truong, L.X., Clerici, M., Calori, G., Lazzarin, A., Pancino, G., Burastero, S.E. (2005). Predictive value of anti-cell and anti-human immunodeficiency virus (HIV) humoral responses in HIV-1-exposed seronegative cohorts of European and Asian origin. J. Gen. Virol. 86:339-348.

Luster, A.D. (1998). Chemokines—chemotactic cytokines that mediate inflame-mation. N. Engl. J. Med. 338:436-445.

Magierowska, M., Theodorou, I., Debre, P., Sanson, F., Autran, B., Riviere, Y., Charron, D., Costagliola, D. (1999). Combined genotypes of CCR5, CCR2, SDF1, and HLA genes can predict the long-term nonprogressor status in human immunodeficiency virus-1-infected individuals. Blood. 93:936-941.

Malhotra, U., Berrey, M.M., Huang, Y., Markee, J., Brown, D.J., Ap, S., Musey, L., Schacker, T., Corey, L., McElrath, M.J. (2000). Effect of combination antiretroviral therapy on T-cell immunity in acute human immunodeficiency virus type 1 infection. J. Infect. Dis. 181:121-131.

Manjunath, N., Shankar, P., Wan, J., Weninger, W., Crowley, M.A., Hieshima, K., (2001). Effector differentiation is not prerequisite for generation of memory cytotoxic T lymphocytes. J. Clin. Invest. 108:871-878.

Martinez, V., Costagliola, D., Bonduelle, O., N'go, N., Schnuriger, A., Theodorou, I., Clauvel, J.P., Sicard, D., Agut, H., Debre, P., Rouzioux, C., Autran, B.; Asymptomatiques a Long Terme Study Group. (2005). Combination of HIV-1-specific CD4 Th1 cell responses and IgG2 antibodies is the best predictor for persistence of long-term nonprogression. J. Infect. Dis. 191:2053-2063.

Martinez-Maza, O., Crabb, E., Mitsuyasu, R.T., Fahey, J.L., Giorgi, J.V. (1987). Infection with the human immunodeficiency virus (HIV) is associated with an *in vivo* increase in B lymphocyte activation and immaturity. J. Immunol. 138:3720-3724.

Mascola, J.R., Stiegler, G., VanCott, T.C., Katinger, H., Carpenter, C.B., Hanson, C.E., Beary, H., Hayes, D., Frankel, S.S., Birx, D.L., Lewis, M.G. (2000). Protection of macaques against vaginal transmission of a pathogenic HIV-1/SIV chimeric virus by passive infusion of neutralizing antibodies. Nat. Med. 6:207-210.

Matloubian, M., Concepcion, R.J., Ahmed, R. (1994). CD4 T cells are required to sustain CD8 cytotoxic T-cell responses during chronic viral infection. J. Virol. 68:8056-8063.

Mazzoli, S., Trabattoni, D., Lo Caputo, S., Piconi, S., Ble, C., Meacci, F., Ruzzante, S., Salvi, A., Semplici, F., Longhi, R., Fusi, M.L., Tofani, N., Biasin, M., Villa, M.L., Mazzotta, F., Clerici, M. (1997). HIV-specific mucosal and cellular immunity in HIV-seronegative partners of HIV seropositive individuals. Nat. Med. 3:1250-1257.

McKay, P.F., Barouch, D.H., Schmitz, J.E., Veazey, R.S., Gorgone, D.A., Lifton, M.A., Williams, K.C., Letvin, N.L. (2003). Global dysfunction of CD4 T-lymphocyte cytokine expression in simian-human immunodeficiency virus/SIV-infected monkeys is prevented by vaccination. J. Virol. 77:4695-4702.

McKenzie, S.W., Dallalio, G., North, M., Frame, P., Means, R.T., Jr. (1996). Serum chemokine levels in patients with non-progressing HIV infection. AIDS. 10:F29-33.

McKenzie, B.S., Brady, J.L., Lew, A.M. (2004). Mucosal immunity: overcoming the barrier for induction of proximal responses. Immunol. Res. 30:35-71.

McNeil, A.C., Shupert, W.L., Iyasere, C.A., Hallahan, C.W., Mican, J.A., Davey, R.T., Jr, Connors, M. (2001). High-level HIV-1 viremia suppresses viral antigen-specific CD4(+) T cell proliferation. Proc. Natl. Acad. Sci. U. S. A. 98:13878-13883.

Mehta-Damani, A., Markowicz, S., Engleman, E.G. (1994). Generation of antigen-specific CD8+ CTLs from naive precursors. J. Immunol. 1994; 153:996-1003.

Melchers, F., Rolink, A.G., Schaniel, C. (1999). The role of chemokines in regulating cell migration during humoral immune responses. Cell. 99:351-354.

Miedema, F., Klein, M.R. (1996). AIDS pathogenesis: a finite immune response to blame? Science. 272:505-506.

Montefiori, D.C., Pantaleo, G., Fink, L.M., Zhou, J.T., Zhou, J.Y., Bilska, M., Miralles, G.D., Fauci, A.S. (1996). Neutralizing and infection-enhancing antibody responses to human immunodeficiency virus type 1 in long-term nonprogressors. J. Infect. Dis. 173:60-67.

Montefiori, D.C., Hill, T.S., Vo, H.T., Walker, B.D., Rosenberg, E.S. (2001). Neutralizing antibodies associated with viremia control in a subset of individuals after treatment of acute human immunodeficiency virus type 1 infection. J. Virol. 75(21):10200-10207.

Moog, C., Fleury, H.J., Pellegrin, I., Kirn, A., Aubertin, A.M. (1997). Autologous and heterologous neutralizing antibody responses following initial seroconversion in human immunodeficiency virus type 1-infected individuals. J. Virol. 71:3734-3741.

Moore, J.P., Cao, Y., Qing, L., Sattentau, Q.J., Pyati, J., Koduri, R., Robinson, J., Barbas, C.F., 3rd, Burton, D.R., Ho, D.D. (1995). Primary isolates of human immunodeficiency virus type 1 are relatively resistant to neutralization by monoclonal antibodies to gp120, and their neutralization is not predicted by studies with monomeric gp120. J. Virol. 69:101-109.

Moore, C.B., John, M., James, I.R., Christiansen, F.T., Witt, C.S., Mallal, S.A. (2002). Evidence of HIV-1 adaptation to HLA-restricted immune responses at a population level. Science. 296:1439-1443.

Mosmann, T.R., Coffman, R.L. (1987). Two types of mouse T helper cell clone: implication for immune regulation. Immunol. Today. 8:223-226.

Mosmann, T.R., Coffman, R.L. (1989). TH1 and TH2 cells: different patterns of lymphokine secretion lead to different functional properties. Annu. Rev. Immunol. 7:145-173.

Musey, L., Hu, Y., Eckert, L., Christensen, M., Karchmer, T., McElrath, M.J. (1997). HIV-1 induces cytotoxic T lymphocytes in the cervix of infected women. J. Exp. Med. 185:293-303.

Nakayama, E.E., Hoshino, Y., Xin, X., Liu, H., Goto, M., Watanabe, N., Taguchi, H., Hitani, A., Kawana-Tachikawa, A., Fukushima, M., Yamada, K., Sugiura, W., Oka, S.I., Ajisawa, A., Sato, H., Takebe, Y., Nakamura, T., Nagai, Y., Iwamoto, A., Shioda, T. (2000). Polymorphism in the interleukin-4 promoter affects acquisition of human immunodeficiency virus type 1 syncytium-inducing phenotype. J. Virol. 74:5452-5459.

Nakayama, E.E., Meyer, L., Iwamoto, A., Persoz, A., Nagai, Y., Rouzioux, C., Delfraissy, J.F., Debre, P., McIlroy, D., Theodorou, I., Shioda, T.; SEROCO Study Group. (2002). Protective effect of interleukin-4 -589T polymorphism on human immunodeficiency virus type 1 disease progression: relationship with virus load. J. Infect. Dis. 185(8):1183-1186.

Ney, P.A., D'Andrea, A.D. (2000). Friend erythroleukemia revisited. Blood. 96:3675-3680.

Ogg, G.S., Jin, X., Bonhoeffer, S., Dunbar, P.R., Nowak, M.A., Monard, S., Segal, J.P., Cao, Y., Rowland-Jones, S.L., Cerundolo, V., Hurley, A., Markowitz, M., Ho, D.D., Nixon, D.F., McMichael, A.J. (1998). Quantitation of HIV-1-specific cytotoxic T lymphocytes and plasma load of viral RNA. Science. 279:2103-2106.

Ollila, J., Vihinen, M. (2005). B cells. Int. J. Biochem. Cell. Biol. 37:518-523.

Overbaugh, J., Kreiss, J., Poss, M., Lewis, P., Mostad, S., John, G., Nduati, R., Mbori-Ngacha, D., Martin, H., Jr., Richardson, B., Jackson, S., Neilson, J., Long, E.M., Panteleeff, D., Welch, M., Rakwar, J., Jackson, D., Chohan, B., Lavreys, L., Mandaliya, K., Ndinya-Achola, J., Bwayo, J. (1999). Studies of human immunodeficiency virus type 1 mucosal viral shedding and transmission in Kenya. J. Infect. Dis. 179 Suppl 3:S401-404.

Oxenius, A., Price, D.A., Easterbrook, P.J., O'Callaghan, C.A., Kelleher, A.D., Whelan, J.A., Sontag, G., Sewell, A.K., Phillips, R.E. (2000). Early highly active antiretroviral therapy for acute HIV-1 infection preserves immune function of CD8$^+$ and CD4$^+$ T lymphocytes. Proc. Natl. Acad. Sci. U. S. A. 97:3382-3387.

Pal, R., Garzino-Demo, A., Markham, P.D., Burns, J., Brown, M., Gallo, R.C., DeVico, A.L. (1997) Inhibition of HIV-1 infection by the beta-chemokine MDC. Science. 278:695-698.

Pal, R., Garzino-Demo, A., Markham, P.D., Burns, J., Brown, M., Gallo, R. (1997). Inhibition of HIV-1 infection by the beta-chemokine MDC. Science. 278:695-698.

Pantaleo, G., Graziosi, C., Demarest, J.F., Butini, L., Montroni, M., Fox, C.H., Orenstein, J.M., Kotler, D.P., Fauci, A.S. (1993). HIV infection is active and progressive in lymphoid tissue during the clinically latent stage of disease. Nature. 362:355-358.

Pantaleo, G., Graziosi, C., Fauci, A.S. (1993). New concepts in the immunopathogenesis of human immunodeficiency virus infection. N. Engl. J. Med. 328:327-335.

Pantaleo, G., Demarest, J.F., Soudeyns, H., Graziosi, C., Denis, F., Adelsberger, J.W., Borrow, P., Saag, M.S., Shaw, G.M., Sekaly, R.P., et al. (1994). Major expansion of CD8$^+$ T cells with a predominant V beta usage during the primary immune response to HIV. Nature. 370:463-467.

Parren, P.W., Moore, J.P., Burton, D.R., Sattentau, Q.J. (1999). The neutralizing antibody response to HIV-1: viral evasion and escape from humoral immunity. AIDS. 13 Suppl A:S137-162.

Patil, J.S., Gupte, S.C. (1995). Role of antibody dependent cell mediated cytotoxicity in ABO hemolytic disease of the newborn. Indian. J. Pediatr. 62:587-592.

Paxton, W.A., Martin, S.R., Tse, D., O'Brien, T.R., Skurnick, J., VanDevanter, N.L., Padian, N., Braun, J.F., Kotler, D.P., Wolinsky, S.M., Koup, R.A. (1996). Relative resistance to HIV-1 infection of CD4 lymphocytes from persons who remain uninfected despite multiple high-risk sexual exposure. Nat. Med. 2:412-417.

Pelton, S.I., Burchett, S.K., McIntosh, K., Korber, B.T., Walker, B.D. (2001). Evolution and transmission of stable CTL escape mutations in HIV infection. Nature. 412(6844):334-338.

Phillips, R.E., Rowland-Jones, S., Nixon, D.F., Gotch, F.M., Edwards, J.P., Ogunlesi, A.O., Elvin, J.G., Rothbard, J.A., Bangham, C.R., Rizza, C.R., et al. (1991). Human immunodeficiency virus genetic variation that can escape cytotoxic T cell recognition. Nature. 354:453-459.

Pilgrim, A.K., Pantaleo, G., Cohen, O.J., Fink, L.M., Zhou, J.Y., Zhou, J.T., Bolognesi, D.P., Fauci, A.S., Montefiori, D.C. (1997). Neutralizing antibody responses to human immunodeficiency virus type 1 in primary infection and long-term-nonprogressive infection. J. Infect. Dis. 176:924-932.

Pitcher, C.J., Quittner, C., Peterson, D.M., Connors, M., Koup, R.A., Maino, V.C., Picker, L.J. (1999). HIV-1-specific CD4[+] T cells are detectable in most individuals with active HIV-1 infection, but decline with prolonged viral suppression. Nat. Med. 5(5):518-525.

Poignard, P., Sabbe, R., Picchio, G.R., Wang, M., Gulizia, R.J., Katinger, H., Parren, P.W., Mosier, D.E., Burton, D.R. (1999). Neutralizing antibodies have limited effects on the control of established HIV-1 infection *in vivo*. Immunity. 10:431-438.

Price, D.A., Goulder, P.J., Klenerman, P., Sewell, A.K., Easterbrook, P.J., Troop, M., Bangham, C.R., Phillips, R.E. (1997). Positive selection of HIV-1 cytotoxic T lymphocyte escape variants during primary infection. Proc. Natl. Acad. Sci. U. S. A. 94:1890-1895.

Prince, H.E., Kermani-Arab, V., Fahey, J.L. (1984). Depressed interleukin 2 receptor expression in acquired immune deficiency and lymphadenopathy syndromes. J. Immunol. 133:1313-1317.

Propato, A., Schiaffella, E., Vicenzi, E., Francavilla, V., Baloni, L., Paroli, M., Finocchi, L., Tanigaki, N., Ghezzi, S., Ferrara, R., Chesnut, R., Livingston, B., Sette, A., Paganelli, R., Aiuti, F., Poli, G., Barnaba Barnaba,V. (2001). Spreading of HIV-specific CD8[+] T-cell repertoire in long-term nonprogressors and its role in the control of viral load and disease activity. Hum. Immunol. 62:561-576.

Ranki, A., Mattinen, S., Yarchoan, R., Broder, S., Ghrayeb, J., Lahdevirta, J., Krohn, K. (1989). T-cell response towards HIV in infected individuals with and without zidovudine therapy, and in HIV-exposed sexual partners. AIDS. 3:63-69.

Reitter, J.N., Means, R.E., Desrosiers, R.C. (1998). A role for carbohydrates in immune evasion in AIDS. Nat. Med. 4:679-684.

Reitz, M.S., Jr., Wilson, C., Naugle, C., Gallo, R.C., Robert-Guroff, M. (1988). Generation of a neutralization-resistant variant of HIV-1 is due to selection for a point mutation in the envelope gene. Cell. 54:57-63.

Richman, D.D., Wrin, T., Little, S.J., Petropoulos, C.J. (2003). Rapid evolution of the neutralizing antibody response to HIV type 1 infection. Proc. Natl. Acad. Sci. U. S. A. 100:4144-4149.

Rinaldo, C., Huang, X.L., Fan, Z.F., Ding, M., Beltz, L., Logar, A., Liebmann, J., Cottrill, M. (1995). High levels of anti-human immunodeficiency virus type 1 (HIV-1) memory cytotoxic T lymphocyte activity and low viral load are associated with lack of disease in HIV-1-infected long-term nonprogressors. J. Virol. 69:5838-5842.

Riviere, Y., McChesney, M.B., Porrot, F., Tanneau-Salvadori, F., Sansonetti, P., Lopez, O., Pialoux, G., Feuillie, V., Mollereau, M., Chamaret, S., *et al.* (1995). Gag-specific cytotoxic responses to HIV type 1 are associated with a decreased risk of progression to AIDS-related complex or AIDS. AIDS Res. Hum. Retroviruses. 11:903-907.

Rizzardini, G., Piconi, S., Ruzzante, S., Fusi, M.L., Lukwiya, M., Declich, S., Tamburini, M., Villa, M.L., Fabiani, M., Milazzo, F., Clerici, M. (1996). Immunological activation markers in the serum of African and European HIV-seropositive and seronegative individuals. AIDS. 10:1535-1542.

Rizzardini, G., Trabattoni, D., Saresella, M., Piconi, S., Lukwiya, M., Declich, S., Fabiani, M., Ferrante, P., Clerici, M. (1998). Immune activation in HIV-infected African individuals. Italian-Ugandan AIDS cooperation program. AIDS. 12:2387-2396.

Rizzuto, C.D., Wyatt, R., Herna´ndez-Ramos, N., Sun, Y., Kwong, P.D., Hendrickson, W.A., (1998). A conserved HIV gp120 glycoprotein structure involved in chemokine receptor binding. Science. 280:1949-1953.

Roederer, M., Dubs, J.G., Anderson, M.T., Raju, P.A., Herzenberg, L.A., Herzenberg, L.A. (1995). CD8 naive T cell counts decrease progressively in HIV-infected adults. J. Clin. Invest. 95:2061-2066.

Rosenberg, E.S., Billingsley, J.M., Caliendo, A.M., Boswell, S.L., Sax, P.E., Kalams, S.A., Walker, B.D. (1997). Vigorous HIV-1-specific CD4$^+$ T cell responses associated with control of viremia. Science. 278:1447-1450.

Rowland-Jones, S.L., McMichael, A. (1995). Immune responses in HIV-exposed seronegatives: have they repelled the virus? Curr. Opin. Immunol. 7:448-455.

Rowland-Jones, S., Dong, T., Krausa, P., Sutton, J., Newell, H., Ariyoshi, K., Gotch, F., Sabally, S., Corrah, T., Kimani, J., MacDonald, K., Plummer, F., Ndinya-Achola, J., Whittle, H., McMichael, A. (1998). The role of cytotoxic T-cells in HIV infection. Dev. Biol. Stand. 92:209-214.

Sallusto, F., Lenig, D., Forster, R., Lipp, M., Lanzavecchia, A. (2001). Two subsets of memory T lymphocytes with distinct homing potentials and effector functions. Nature. 401:708-712.

Schenal, M., Lo Caputo, S., Fasano, F., Vichi, F., Saresella, M., Pierotti, P., Villa, M.L., Mazzotta, F., Trabattoni, D., Clerici, M. (2005). Distinct patterns of HIV-specific memory T lymphocytes in HIV-exposed uninfected individuals and in HIV-infected patients. AIDS. 19:653-661.

Schmidtmayerova, H., Nottet, H.S., Nuovo, G., Raabe, T., Flanagan, C.R., Dubrovsky, L. (1996a). Human immunodeficiency virus type 1 infection alters chemokine beta peptide expression in human monocytes: implications for recruitment of leukocytes into brain and lymph nodes. Proc. Natl. Acad. Sci. U. S. A. 93:700-704.

Schmidtmayerova, H., Sherry, B., Bukrinsky, M. (1996b). Chemokines and HIV replication [Letter]. Nature. 382:767.

Schmitz, J.E., Kuroda, M.J., Santra, S., Sasseville, V.G., Simon, M.A., Lifton, M.A., Racz, P., Tenner-Racz, K., Dalesandro, M., Scallon, B.J., Ghrayeb, J., Forman, M.A., Montefiori, D.C., Rieber, E.P., Letvin, N.L., Reimann, K.A. (1999). Control of viremia in simian immunodeficiency virus infection by CD8$^+$ lymphocytes. Science. 283:857-860.

Schmitz, J.E., Kuroda, M.J., Santra, S., Simon, M.A, Lifton, M.A., Lin, W., Khunkhun, R., Piatak, M., Lifson, J.D., Grosschupff, G., Gelman, R.S., Racz, P., Tenner-Racz, K., Mansfield, K.A., Letvin, N.L., Montefiori, D.C., Reimann, K.A. (2003). Effect of humoral immune responses on controlling viremia during primary infection of rhesus monkeys with simian immunodeficiency virus. J. Virol. 77:2165-2173.

Shearer, G.M. (1998). HIV-induced immunopathogenesis. Immunity. 9(5):587-593.

Shearer, G.M., Clerici, M. (1994). *In vitro* analysis of cell-mediated immunity: clinical relevance. Clin. Chem. 40:2162-2165.

Shearer, G.M., Clerici, M. (1996). Protective immunity against HIV infection: has nature done the experiment for us? Immunol. Today. 17:21-24.

Sheppard, H.W., Lang, W., Ascher, M.S., Vittinghoff, E., Winkelstein, W. (1993). The characterization of non-progressors: long-term HIV-1 infection with stable CD4 T-cell levels. AIDS. 7:1159-66.

Shin, H.D., Winkler, C., Stephens, J.C., Bream, J., Young, H., Goedert, J.J., O'Brien, T.R., Vlahov, D., Buchbinder, S., Giorgi, J., Rinaldo, C., Donfield, S., Willoughby, A., O'Brien, S.J., Smith, M.W. (2000). Genetic restriction of HIV-1 pathogenesis to AIDS by promoter alleles of IL10. Proc. Natl. Acad. Sci. U. S. A. 97:14467-14472.

Sieg, S.F., Mitchem, J.B., Bazdar, D.A., Lederman, M.M. (2002). Close link between CD4$^+$ and CD8$^+$ T cell proliferation defects in patients with human immunodeficiency virus disease and relationship to extended periods of CD4$^+$ lymphopenia. J. Infect. Dis. 185:1401-1416.

Smith-Franklin, B.A., Keele, B.F., Tew, J.G., Gartner, S., Szakal, A.K., Estes, J.D., Thacker, T.C., Burton, G.F. (2002). Follicular dendritic cells and the persistence of HIV infectivity: the role of antibodies and Fcgamma receptors. J. Immunol. 168:2408-2414.

Sozzani, S., Ghezzi, S., Iannolo, G., Luini, W., Borsatti, A., Polentarutti, N., Sica, A., Locati, M., Mackay, C., Wells, T.N., Biswas, P., Vicenzi, E., Poli, G., Mantovani, A. (1998). Interleukin 10 increases CCR5 expression and HIV infection in human monocytes. J. Exp. Med. 187:439-444.

Spiegel, H., Herbst, H., Niedobitek, G., Foss, H.D., Stein, H. (1992). Follicular dendritic cells are a major reservoir for human immunodeficiency virus type 1 in lymphoid tissues facilitating infection of CD4$^+$ T-helper cells. Am. J. Pathol. 140:15-22.

Stranford, S.A., Skurnick, J., Louria, D., Osmond, D., Chang, S.Y., Sninsky, J., Ferrari, G., Weinhold, K., Lindquist, C., Levy, J.A. (1999). Lack of infection in HIV-exposed individuals is associated with a strong CD8($^+$) cell noncytotoxic anti-HIV response. Proc. Natl. Acad. Sci. U. S. A. 96:1030-1035.

Tamura, H., Dong, H., Zhu, G., Sica, G.L., Flies, D.B., Tamada, K., Chen, L. (2001). B7-H1 costimulation preferentially enhances CD28-independent T-helper cell function. Blood. 97:1809-1816.

Trabattoni, D., Saresella, M., Biasin, M., Boasso, A., Piacentini, L., Ferrante, P., Dong, H., Maserati, R., Shearer, G.M., Chen, L., Clerici, M. (2003). B7-H1 is up-regulated in HIV infection and is a novel surrogate marker of disease progression. Blood. 101:2514-2520.

Trabattoni, D., Caputo, S.L., Maffeis, G., Vichi, F., Biasin, M., Pierotti, P., Fasano, F., Saresella, M., Franchini, M., Ferrante, P., Mazzotta, F., Clerici, M.

(2004a). Human alpha Defensin in HIV-Exposed But Uninfected Individuals. J. Acquir. Immune. Defic. Syndr. 35:455-463.

Trabattoni, D., Piconi, S., Biasin, M, Rizzardini, G., Migliorino, M., Seminari, E., Boasso, A., Piacentini, L., Villa, M.L., Maserati, R., Clerici, M. (2004b). Granule-dependent mechanisms of lysis are defective in CD8 T cells of HIV-infected, antiretroviral therapy-treated individuals. AIDS. 18:859-869.

Trachtenberg, E., Korber, B., Sollars, C., Kepler, T.B., Hraber, P.T., Hayes, E., Funkhouser, R., Fugate, M., Theiler, J., Hsu, Y.S., Kunstman, K., Wu, S., Phair, J., Erlich, H., Wolinsky, S. (2003). Advantage of rare HLA supertype in HIV disease progression. Nat. Med. 9:928-935.

Trimble, L.A., Lieberman, J. (1998). Circulating CD8 T lymphocytes in human immunodeficiency virus-infected individuals have impaired function and downmodulate CD3 zeta, the signaling chain of the T-cell receptor complex. Blood. 91:585-594.

Trkola, A., Pomales, A.B., Yuan, H., Korber, B., Maddon, P.J., Allaway, G.P., Katinger, H., Barbas, C.F., 3rd, Burton, D.R., Ho, D.D. (1995). Cross-clade neutralization of primary isolates of human immunodeficiency virus type 1 by human monoclonal antibodies and tetrameric CD4-IgG. J. Virol. 69:6609-6617.

Ullum, H., Cozzi Lepri, A., Victor, J., Aladdin, H., Phillips, A.N., Gerstoft, J., Skinhoj, P., Pedersen, B.K. (1998). Production of beta-chemokines in human immunodeficiency virus (HIV) infection: evidence that high levels of macrophage inflammatory protein-1beta are associated with a decreased risk of HIV disease progression. J. Infect. Dis. 177:331-336.

Urnovitz, H.B., Clerici, M., Shearer, G.M., Gottfried, T.D., Robison, D.J., Lutwick, L.I., Montagnier, L., Landers, D.V. (1993). HIV-1 antibody serum negativity with urine positivity. Lancet. 342:1458-1459.

Valentine, F.T., Paolino, A., Saito, A., Holzman, R.S. (1998). Lymphocyte-proliferative responses to HIV antigens as a potential measure of immunological reconstitution in HIV disease. AIDS Res. Hum. Retroviruses. 14(Suppl 2):S161-166.

Walker, B.D., Plata, F. (1990). Cytotoxic T lymphocytes against HIV. AIDS. 4(3):177-184.

Wang, S., Zhu, G., Chapoval, A.I., Dong, H., Tamada, K., Ni, J., Chen, L. (2000). Costimulation of T cells by B7-H2, a B7-like molecule that binds ICOS. Blood. 96:2808-2813.

Wei, X., Decker, J.M., Wang, S., Hui, H., Kappes, J.C., Wu, X., Salazar-Gonzalez, J.F., Salazar, M.G., Kilby, J.M., Saag, M.S., Komarova, N.L., Nowak, M.A., Hahn, B.H., Kwong, P.D., Shaw, G.M. (2003). Antibody neutralization and escape by HIV-1. Nature. 422:307-312.

Wherry, E.J., Teichgraber, V., Becker, T.C., Masopust, D., Kaech, S.M., Antia, R. (2003). Lineage relationship and protective immunity of memory CD8 T cell subsets. Nat. Immunol. 4:225-234.

Woof, J.M., Kerr, M.A. (2004). IgA function—variations on a theme. Immunology. 113:175-177.

Wu, L., Gerard. N.P., Wyatt. R., Choe, H., Parolin, C., Ruffing, N. (1996). CD4-induced interaction of primary HIV-1 gp120 glycoproteins with the chemokine receptor CCR5. Nature. 384:179-183.

Wyatt, R., Sodroski. J. (1998). The HIV-1 envelope glycoproteins: fusogens, antigens, and immunogens. Science. 280:1884-1888.

Yang, O.O., Lin, H., Dagarag, M., Ng, H.L., (200). Effros RB, Uittenbogaart CH. Decreased perforin and granzyme B expression in senescent HIV-1-specific cytotoxic T lymphocytes. Virology. 332:16-19.

Yasutomi, Y., Reimann, K.A., Lord, C.I., Miller, M.D., Letvin, N.L. (1993). Simian immunodeficiency virus-specific CD8[+] lymphocyte response in acutely infected rhesus monkeys. J. Virol. 67:1707-1711.

Yusim, K., Kesmir, C., Gaschen, B., Addo, M.M., Altfeld, M., Brunak, S., Chigaev, A., Detours, V., Korber, B.T. (2002). Clustering patterns of cytotoxic T-lymphocyte epitopes in human immunodeficiency virus type 1 (HIV-1) proteins reveal imprints of immune evasion on HIV-1 global variation. J. Virol. 76:8757-8768.

Zajac, A.J., Blattman, J.N., Murali-Krishna, K., Sourdive, D.J., Suresh, M., Altman, J.D., Ahmed, R. (1998). Viral immune evasion due to persistence of activated T cells without effector function. J. Exp. Med. 188:2205-2213.

Zhang, L., Yu, W., He, T., Yu, J., Caffrey, R.E., Dalmasso, E.A., Fu, S., Pham, T., Mei, J., Ho, J.J., Zhang, W., Lopez, P., Ho, D.D. (2002). Contribution of human alpha-defensin 1, 2, and 3 to the anti-HIV-1 activity of CD8 antiviral factor. Science. 298:995-1000.

Part III

Dendritic Cells and HIV Interactions and Their Role in Pathogenesis and Immunity

Chapter 10

Binding and Uptake of HIV by Dendritic Cells and Transfer to T Lymphocytes: Implications for Pathogenesis

Anthony L. Cunningham, John Wilkinson, Stuart Turville, and Melissa Pope

10.1 Introduction

More than 40 million people are now estimated to be living with HIV worldwide, of whom 47% are women. Five million people were infected in 2004, including approximately 650,000 children. Approximately 3 million died of AIDS in the same year. Seventy-five percent of cases are still present in sub-Saharan Africa. HIV is spreading rapidly in Asia, especially in India and China and in part of the Western Pacific, such as Papua New Guinea, and also in Eastern Europe. In sub-Saharan Africa, 58% of HIV-infected people are women (Annan, 2002).

10.2 Transmission of HIV

It is estimated that more than 80% of cases of HIV infections occur via heterosexual transmission worldwide. In sub-Saharan Africa and some of the parts of the developing world, HIV is often spread to the monogamous female partner within a family setting. This and other aspects of gender inequality (especially in young women) lead to difficulties in negotiation of safe sex or use of condoms.

Coinfection with other sexually transmitted pathogens especially those causing genital ulcerating disease, such as herpes simplex virus type 2 (HSV-2), enhances the likelihood of sexual transmission. This is most important in Africa, India, and other parts of the developing world, as well as in male-to-male sex in developed countries. For example, a recent meta-analysis showed a two- to threefold enhancement in risk of HIV transmission in both men and women with prior HSV-2 infection (Wald, 2004). Recent HSV-2 infection is most important (Reynolds *et al.*, 2003).

10.3 HIV Infection of Female Genital Tract

The human female genital tract is mostly composed of stratified squamous epithelia in the vagina and a small proportion of cuboidal epithelium extending from the endocervix onto the central ectocervix. This extension fluctuates in size throughout the menstrual cycle. When intact, together with the innate defenses of vaginal mucus, this barrier usually prevents the transmission of HIV present in seminal fluid (Greenhead *et al.*, 2000). An exception to this was the demonstration of infection of four of eight women undergoing artificial insemination by donor where the donor was HIV seropositive with low blood CD4 counts (and presumably a high seminal fluid viral load). Review of the patients' records showed no macroscopic evidence of vaginal ulceration or trauma at the time of artificial insemination although some patients had erosions, extensions of the endocervical epithelium onto the ectocervix (Stewart *et al.*, 1985).

In the macaque model, when progesterone is used to thin the vaginal and ectocervix epithelium, there is an increase in susceptibility to infection when vaginally challenged with SIV (Marx *et al.*, 1996; Miller *et al.*, 2005). Conversely, estrogen treatment was reported to afford resistance to vaginal SIV infection (Smith *et al.*, 2000, 2004). These findings suggest that fluctuating estrogen and progesterone levels during the menstrual cycle affects both the susceptibility of the female genital tract to SIV infection and the level of cervical virus shedding (Benki *et al.*, 2004; Sodora *et al.*, 1998). As recently shown in the SIV-macaque model, only minute proportions of (artificially) high titer inocula penetrate the epithelium to form small foci of infected CD4 lymphocytes in the sub-mucosa and occasionally in draining lymph nodes. The mucus, genital fluids, and intact epithelium provide a substantial barrier to HIV penetration. Viral growth in the scattered subepithelial nests of infected (mainly resting) CD4 lymphocytes is the major source of wider dissemination to other lymphoid tissue (Miller *et al.*, 2005).

Although several mechanisms have been proposed to transfer virus across the mucous membranes, such as direct infection of epithelial cells, transcytosis and transmigration of infected macrophages and T cells from the lumen, there is evidence in macaque and mouse models that uptake of HIV/SIV by dendritic cells (DCs) may be important in both intact and breached mucosa (Hladik *et al.*, 1999; Miller *et al.*, 2005). Ulceration or abrasion resulting in breaches in epithelial integrity allow direct access of the virus to underlying DCs in the stratified squamous epithelium or in the lamina propria whereby they can transport virus to lymph nodes (Fig. 10.1). Such breaches in the mucosa are commonly caused by other genital tract pathogens, especially HSV 1 or 2 or physical trauma. Genital herpes inflammation also leads to infiltration of macrophages and activated CD4 lymphocytes, target cells for HIV (Cunningham *et al.*, 1985).

Recent studies in the SIV-macaque model using cell-free virus that suggest augmentation of the innate defenses of vaginal mucus fluids and mucosa could further reduce the likelihood of viral penetration, increase the viral threshold required for such penetration, and perhaps reduce the size and frequency of the small submucosal and lymph node populations of infected T cells below levels required for further dissemination (Miller *et al.*, 2005).

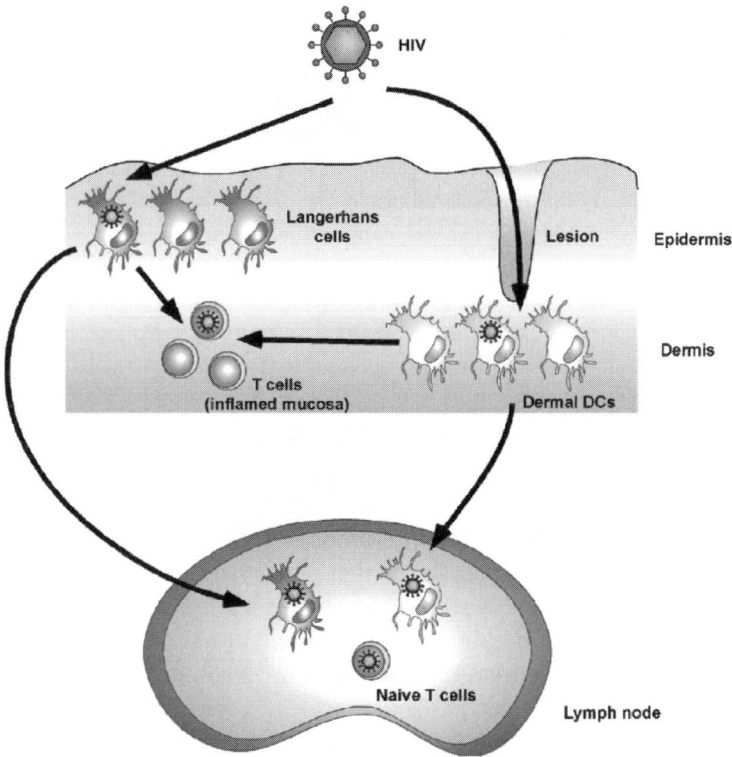

FIGURE 10.1. Role of DCs in HIV entry by sexual transmission.

10.4 The Role of Dendritic Cells in HIV Infection

DCs have a number of roles in HIV pathogenesis, probably including initial HIV uptake and infection in the genital and other epithelium (Fig. 10.1), transport to lymphoid tissue where they transfer HIV to and stimulate explosive HIV replication in T cells, and, conversely, priming of HIV-specific CD4 and CD8 cell-mediated immunity. Sessile immature DCs, such as Langerhans cells (LCs) in the epidermis and interstitial DCs in the dermis, lamina propria, or submucosa, are probably among the leukocytes infected by or capturing HIV (or SIV) after mucosal exposure (Hu *et al.*, 2000). DCs may also play a role in HIV infection and dissemination in other lymphoid tissue of thymus, gut, tonsil, spleen, and lymph nodes.

DCs are the major antigen-presenting cells within the body. They can be divided into several classes, the precursor DCs in blood, immature and mature plasmacytoid DCs in blood, and inflamed tissues and immature and mature myeloid DCs in blood and tissues. Immature myeloid DCs act as sentinels in various tissues to capture microbial antigens. In the stratified epithelium of the skin and anogenital mucosa, the major DCs are the epidermal LCs and the dermal DCs. In the rectal cuboidal epithelium, the DCs of the lamina propria are probably equivalent to dermal DCs. Access of viruses to LCs in stratified squamous epithelium of the skin is restricted by the overlying stratum corneum, but this is absent in the anal and vaginal epithelium and in the inner surface of the foreskin, which is rich in LCs. The latter may explain a recent report of marked reduction of HIV transmission by circumcision (incidence rate = 0.85 per 100 person-years) versus 49 (2.1 per 100 person-years) in the control group, corresponding with 60% protection (RR 0.40 95% CI: 0.24–0.68%; p <0.001) (Auvert *et al.*, 2005).

LCs are likely to be the first DCs to contact HIV. They are found in higher frequencies in the ectocervical epithelium than the vaginal epithelium with dendrites that reach toward the luminal surfaces. Whether these dendrites are exposed in the lumen of the vagina is debatable, but at least only minor tissue damage is required to expose these cells to HIV (Fig. 10.2). Support for dendrite exposure within the lumen comes from studies in the mouse intestine. DC processes can extend between epithelial cells of an intact barrier to reach the luminal surface (Rescigno *et al.*, 2001), allowing DCs the ability to directly entrap pathogens at a mucosal surface.

The most commonly used DC is derived from peripheral blood monocytes, monocyte-derived dendritic cells (MDDCs), by stimulation with IL-4 and GM-CSF (Romani *et al.*, 1994; Sallusto and Lanzavecchia, 1994). MDDCs resemble the immature CD14 dermal DC subset *in vivo*. However, they should really be considered a unique *in vitro* DC phenotype because they

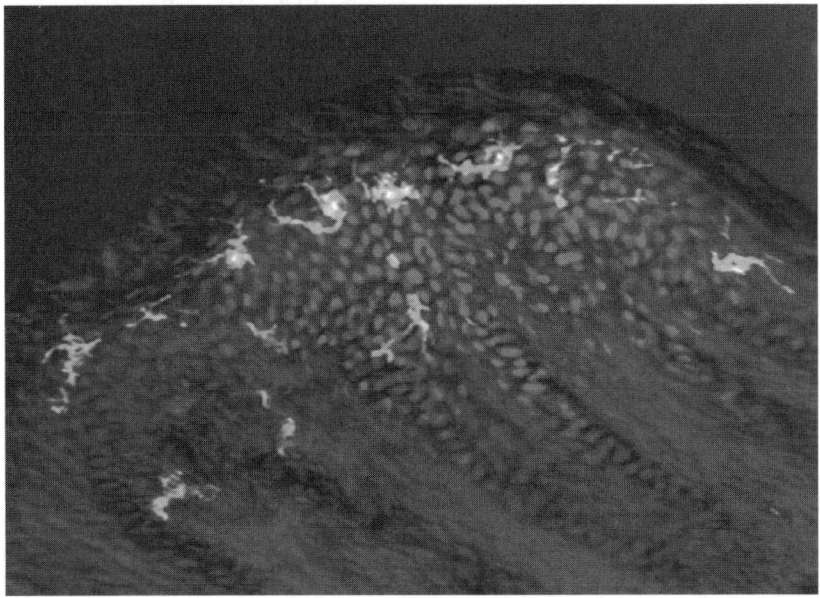

FIGURE 10.2. Langerhan's cells in human foreskin with processes extending to the tissue surface (image kindly provided by Scott McCoombe). (See Plate 3)

are derived from monocytes by treatment with IL-4 and GM-CSF. However a DC with such a phenotype and derived from monocytes has been proposed *in vivo* under conditions of inflammation (MacDonald *et al.*, 2002; Osugi *et al.*, 2002; Randolph *et al.*, 1999).

10.4.1 Immature versus Mature DCs

The maturation status of the DCs alters susceptibility to infection with different strains of HIV because the surface receptor expression changes. Immature DCs are specialized for antigen uptake and are capable of high levels of microbial (antigen) binding, endocytosis, macropinocytosis, and phagocytosis and therefore act as sentinels in the periphery. Exposure of DCs to "danger signals" in inflamed areas results in maturation and migration of DCs from the epithelium (after antigen uptake). These signals may be microbial products such as LPS or combinations of proinflammatory stimuli like TNF-α, IL-6, IL-1β, and PGE2 (Mehlhop *et al.*, 2002) or monocyte conditioned media. Maturation results in reduced endocytic activity, export of MHC II from endosomal compartments to the surface, and enhanced antigen presentation to T cells in local lymph nodes. The upregulation of adhesion molecules facilitates DC–T cell interactions. The C-type lectin receptors (CLRs) are downregulated on the cell surface. In addition, the CCR5 surface expression decreases while CXCR4 increases, thus altering susceptibility from R5 to X4 strains of HIV. Although most high CXCR4 expressing DCs will already be in the process of migration, microbicides targeting DCs may need to account for this diverse receptor expression and be able to prevent entry of both X4 and R5 strains of HIV (Peretti *et al.*, 2005; Wilkinson *et al.*, 2005).

10.4.2 HIV Receptors on DCs

The major surface receptors for microbial antigens on DCs are the Toll-like receptors (TLRs), mainly involved in signaling rather than viral entry, and the CLRs. More than 15 CLRs have now been described. The HIV binding receptors on DCs are CD4, the chemokine coreceptors (CCR5 or CXCR4), and a variety of CLRs. We have recently shown that different subsets of tissue DCs express a diversity of CLRs (Turville *et al.*, 2002, 2003), with the virus able to use specific CLRs on each subset thus enabling capture, infection, and/or transfer to T cells (see below).

10.4.2.1 C-type Lectin Receptors

Mannose receptor (CD206) was the first CLR identified but dendritic cell ICAM-3 grabbing non-integrin (DC-SIGN; CD209), a CLR found on the model of MDDC, has been most intensively studied because it was shown to bind HIV via the highly glycosylated gp120.

CLRs recognize oligosaccharides (usually high mannose) on microbial glycoproteins in a calcium-dependent manner via highly conserved carbohydrate recognition domains (CRDs). Ligands for the CLRs are still poorly characterized. There are two types of CLRs characterized by the positioning of their amino (N) terminus. Type I CLRs such as mannose receptors (MR) CD206 have their N terminus distal to the membrane (facing externally) and contain multiple CRDs. Type II CLRs have the N terminus within the cytoplasm and a single CRD, examples being DC-SIGN (CD209) and langerin (CD207) (Figdor *et al.*, 2002). Studies of recombinant fragments of the external domains of DC-SIGN showed it to be a tetramer. The DC-SIGN and DC-SIGNR family also have a neck domain between the TM and CRD that mediates oligomerization. Recently, our studies of DC-SIGN on the surface of MDDCs using short-range cross-linkers and mass spectrometry (MS/MS) not only confirmed its tetrameric tertiary structure, but also that it does not constitutively associate with CD4. As a result, lateral transfer of gp120 from DC-SIGN to CD4 is not likely to occur. It was also shown that only the tetramer had bound the natural ligand mannan and (monomeric) gp120 (Bernhard *et al.*, 2004) (Table 10.1).

TABLE 10.1. Structure of C-type lectins in HIV binding to dendritic cells.

Family	Type of protein	Oligomerization status	Domain structure (N-C)	CRDs per monomer
Mannose receptor	Type I transmembrane	Monomer/ dimmer*	cys rich-fibronectin- CRDs-TM-cyto	9 (4 are CHO binding)
DC-SIGN	Type II	Tetramer	cyto-TM-neck-CRD	One
Langerin	Type II transmembrane	Oligomer*	cyto-TM-neck-CRD	One

* Unpublished. cyto = cytoplastic tail; TM = Transmembrance domain; CRD = Carbohydrate recognition.

MR expressed on macrophages, DCs subsets, and epithelial and endothelial cells is a multidomain membrane associated receptor that selectively binds terminal sugars such as mannose, *N*-acetyl-glucosamine, and fucose. The latter is uncommon in mammalian oligosaccharides but common on the surfaces of microorganisms. Interaction with these ligands, as well as terminally mannosylated sugars on gp120 (which accounts for approximately half of its glycosylation sites) occurs via eight CRDs (Geyer *et al.*, 1988). It has been reported to be monomeric (Taylor and Drickamer, 1993) although we have recently found evidence for the presence of both monomers and dimers on the surface of MDDCs (Lai *et al.*, unpublished data).

Langerin on LCs is the least understood CLR of the trio, particularly with regard to its functional role, which as yet is to be elucidated. It is an important constituent of Birbeck granules, the organelle characterizing LCs (Valladeau *et al.*, 2000). It binds HIV probably by its mannose-bearing ligands (Turville *et al.*, 2002) and probably also binds other microbial glycoproteins (Stambach and Taylor, 2003). Ligand specificity also differs among these C-type lectins. DC-SIGN and DC-SIGNR bind with high affinity to branched tri-mannose glycosylation structures present in N-linked high mannose oligosaccharides on mammalian and pathogen cell surface, in particular on T cells and HIV envelope proteins respectively (Feinberg *et al.*, 2001). MR binds to L-fucose and D-mannose, especially to the latter as α 1-3 or α 1-6 linked branched oligomers (Kery *et al.*, 1992). MR is postulated to be a clearance molecule for microbes, with subsequent antigen uptake and presentation by antigen-presenting cells, like DCs and macro-phages (Sallusto *et al.*, 1995; Stahl and Ezekowitz, 1998). DC-SIGN has also recently been suggested to have a similar role to MR in antigen acquisition and processing, especially in DCs and perhaps also in some tissue macro-phages (Engering *et al.*, 2002; Sollieux *et al.*, 2002; Turville *et al.*, 2002).

HIV envelope gp120 is heavily glycosylated with a mixture of complex, high mannose and intermediate oligosaccharides constituting half of the molecular weight of the molecule. Of the 24 N-linked glycosylation sites on gp120 of the IIIB strain expressed in mammalian systems, approximately half consists of complex or intermediate and half consist of high mannose oligosaccharides. From mass spectrometric analyses of glycosylation sites and structural modeling, the high mannose glycans appear to cluster together on the binding surface, whereas the complex glycans form another more peripheral cluster with little structural overlap (Zhu *et al.*, 2000). However, HIV produced by macrophages has been shown to express more *N*-acetyl-lactosamine repeats on complex glycan chains, thus resulting in a higher degree of heterogeneity in glycosylation compared with T cell–derived HIV.

Therefore, HIV gp120 binding to CLRs may depend on the host cell from which HIV is produced. Indeed, HIV produced from T lymphocytes, but not macrophages, has been shown to bind DC-SIGN and DC-SIGNR. The presence of acetyl-lactosamine repeats on macrophage-derived HIV seems to hinder effective HIV gp120 attachment to these CLR (Lin *et al.*, 2000).

10.4.2.2 Diversity of CLR Expression on DCs

DC-SIGN has been strongly promoted as the key CLR on DCs for many functions, including HIV binding [reviewed by van Kooyk and Geijtenbeek (2003); Su *et al.* (2003)] but in uninflamed tissues *in vivo* expression is limited to dermal and lamina propria DCs not LCs in superficial epithet-lium. DC-SIGN is also expressed on MDDCs but not on fresh myeloid or plasmacytoid DCs in blood or uninflamed lymph nodes. MR is expressed on dermal DCs and MDDCs but not LCs. Macrophages express MR and some subsets of tissue macrophages express DC-SIGN. Epidermal LCs express langerin (Turville *et al.*, 2003, 2004). During monocyte differen-tiation into MDDCs, GM-CSF and IL-4 upregulate surface expression of MR and DC-SIGN, respectively. This may also occur *in vivo* in inflamed organs.

Soluble (monomeric) recombinant gp120 and HIV itself (with authentic trimeric gp120) bind to immature MDDCs and epithelial DCs primarily by their dominantly expressed CLRs, with minor binding to CD4, but to blood and tonsil DCs via CD4 alone (Turville *et al.*, 2002). Migrating LCs and dermal DCs bind gp120 via CD4 to a greater extent than CLRs, reflecting the change in expression of CLRs during maturation and migration (Turville *et al.*, 2001, 2002). Binding to the CLRs may result either in endocytosis or, to a lesser degree, subsequent interaction with CD4/CCR5, fusion and viral entry and *de novo* infection of the MDDCs.

CLRs probably concentrate HIV virions on the DC surface thus facilitating uptake by the cells. Transfer of virus to CD4 and a chemokine coreceptor exposed on the surface of the same cell results in membrane fusion and infection of the DC either on the cell surface or perhaps a minority after endocytosis. It is speculated that low viral titers can be concentrated in this way, thus resulting in endocytosis and transfer of more virus to T cells or in facilitation of infection of the DCs (see below).

On LCs, gp120 may also bind to at least one additional CLR, which may be more markedly downregulated by migration than langerin, correlating with the observation that gp120 binding is reduced on migrating LCs (Turville *et al.*, 2001 and 2002). LCs can be isolated from skin by collagenase digestion and magnetic bead isolation or flow sorting, but this is labor intensive and yields are low. Alternatively, LCs can be isolated by the easier "crawl out" method from epidermal explants but these LCs are partially mature and express lower levels of the CLRs that bind gp120 (Tchou *et al.*, 2003). Convenient accurate *in vitro* LC models are required that can be used to study the quaternary structure of langerin on intact cells and its interactions with HIV.

Assuming DCs are conclusively proved to play a key role in sexual transmission of HIV, blockade of infection will require to inhibit not just DC-SIGN but several other CLRs on genital tract DCs including at least langerin on LCs as well as MR and DC-SIGN on dermal or lamina propria DCs. Recent strategies for microbicide development have not been designed to cope with such diversity. A dual approach to target both the incoming virus and the resident DC population(s) could be used with soluble receptors or their ligands. The reasons for targeting the CLRs on DCs rather than CD4 and CCR5 on macrophages and T cells are threefold: (i) HIV binds CLRs with a higher affinity than CD4 and CCR5, (ii) DCs capture and transfer virus, and (iii) DCs stimulate high replication in CD4$^+$ T-cell populations.

Although CLRs on DC subsets may be considered primary targets, it is worth repeating that MR is also present on macrophages and provides an entry receptor for HIV on these cells that can facilitate transfer to T cells. (Ezekowitz *et al.*, 1989; Nguyen and Hildreth, 2003). Macrophages present in the genital subepithelial stromal tissues and at a higher frequency in inflamed or breached mucosa provide a reservoir of HIV that is difficult to access by current anti-retroviral therapies. Strategies that also reduce the size of this reservoir have obvious advantages in the delayed clinical development of the disease.

10.5 Transmission of HIV to T Cells by DCs

Recent studies of genital tract SIV infection in the macaque model suggest there are two targets for HIV transferred across epithelium either for underlying foci of lamina propria lymphatic tissue or draining lymph nodes. In both sites, HIV can replicate in CD4 lymphocytes to high levels and cause rapid and severe CD4 lymphocyte depletion. As mentioned above, in SIV-macaque models, transfer of virus from DCs to CD4 lymphocytes *in vitro* has been shown to be biphasic, with both immature and mature DCs able to transfer virus across a "viral" synapse that is formed when DCs and T cells meet (Turville *et al.*, 2004). (This is not identical in antigen expression to the "immunologic" synapse between DCs transferring antigen to T cells.) DCs can transfer virus after binding to CLRs on DCs without becoming infected, a process known as trans enhancement (Geijtenbeek *et al.*, 2000). It was initially suggested that HIV was retained on the surface of DC-SIGN transfected THP1 cells prior to transfer to T cells in a process entirely independent of DC infection (Geijtenbeek *et al.*, 2000). However, these THP1 cells have subsequently been shown to be B lymphocytes, and subsequent work by several groups including ours showed that DC-SIGN binding (Turville *et al.*, 2002) results in HIV internalization in MDDCs. Macrophages also transfer HIV to T cells and the initial capture and internalization is mediated by MR (Nguyen and Hildreth, 2003).

With immature MDDCs, HIV bound to DC-SIGN and MR is internalized through endocytosis into endosomal compartments within the cell. The HIV envelope protein gp120 trafficked into endosomes and was completely degraded in lysosomes but whole HIV/SIV, although degraded, is initially localized in smaller superficial endosomes that are negative for the early endosomal marker EEA-1 beyond 1 h (Table 10.2). In comparison with CD4 lymphocytes and macrophages, infection of permissive DC subsets is of low productivity. However, even DC infection levels of less than 1% are more than sufficient to provide an explosive viral infection in CD4 lymphocytes (Kawamura *et al.*, 2000).

TABLE 10.2. Characterization of HIV contained in DC compartments.

Marker	Immature DC	Mature DC	Viral synapse
Immunoflorescence			
EEA1	–		NA
LAMP1	–	–	NA
Tetraspannins			
CD9		++	
CD61		++	+
CD83		+	+
Electron microscopy			
Clathrin-coated vesicles		+	

Furthermore, as shown by Lee and colleagues (Lee *et al.*, 2001) and lately ourselves (Turville *et al.*, 2002), capture of HIV by the CLRs, DC-SIGN and MR facilitates entry of HIV into the cytoplasm by the conventional CD4/CCR5-mediated neutral membrane fusion. This process, described above, is known as "cis" infection. Therefore, CLRs facilitate entry into two possible pathways, the endocytic vesicular or neutral fusion/infection into the cytoplasm (Fig. 10.3). Indeed, after HIV binds to its main target cells—CD4 lymphocytes, macrophages, and DCs—there is a gradation in the importance of neutral fusion versus endocytosis mode of entry. Neutral fusion of HIV (at the cell membrane) predominates in CD4 lymphocytes, and endocytic uptake predominates in DCs. The latter is clearly facilitated by the initial contact of HIV with CLRs rather than CD4. Some infection in DCs could occur at the early stages within endosomes, especially as CD4, CCR5, and CXCR4 recycle in this compartment. Frederickson and colleagues (Frederickson *et al.*, 2002) have demonstrated that neutralization of endosomal acidification with proton pump inhibitors like chloroquine and bafilomycin leads to an increase in infection. We have indeed observed a positive effect on DC infection after chloroquine and bafilomycin exposure (unpublished data).

Contact of the DC with a CD4$^+$ T cell causes the internalized vesicles containing the virus to migrate to the cell-cell junction, and virions can be transferred across this synapse (McDonald *et al.*, 2003). This appears to

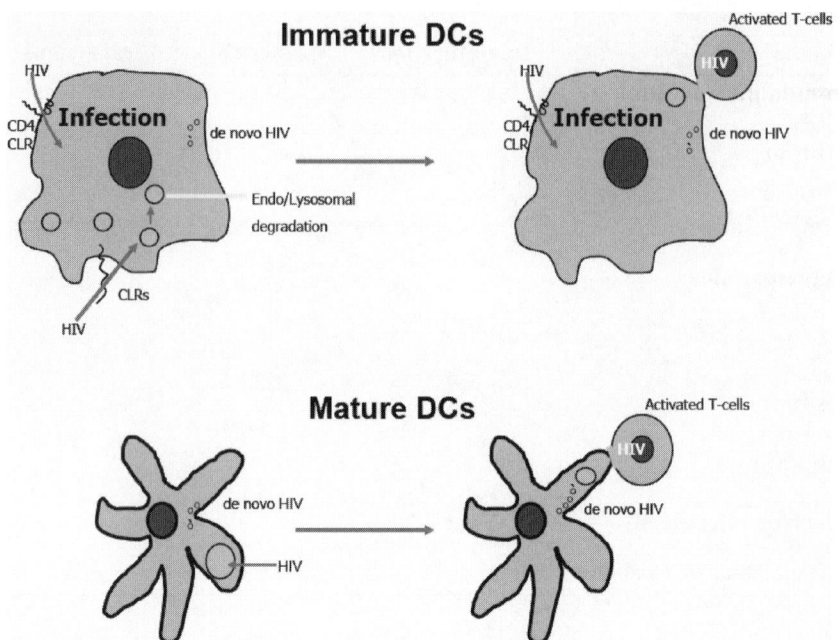

FIGURE 10.3. Uptake of HIV by monocyte-derived dendridic cells and transfer to T cells.

be more efficient with mature than immature MDDCs (see below) (Turville *et al.*, 2004). This trans enhancement, which persists for approximately 6 to 12 h with a tail to 24 h after HIV binding, is likely to be an important factor in the local transfer of virus to intraepithelial lymphocytes and possibly at lower levels to lamina propria T lymphocytes. However, because DC migration to lymph nodes requires approximately 12 to 24 h, trans enhancement may not result in an effective "Trojan horse" (Lin *et al.*, 2000; Sewell and Price, 2001) as the majority of virus is likely to be degraded during this time via the endosomal acidic proteolytic environment in the DC. However, cis infection or neutral fusion results in slowly developing *de novo* virus production more than 24 h after HIV binding. This is the period when DCs would certainly have reached underlying lamina propria T lymphocytes and also the lymph nodes. Therefore, transfer of *de novo* infecting virus is more likely to be important for viral dissemination than trans infection. Furthermore, *in vitro* cis transfer is relatively more efficient than trans transfer and occurs at low viral input multiplicity (Turville *et al.*, 2005). The reasons for this may be a higher peaking and longer duration of viral transfer (over 72 to 96 h) (Wilkinson *et al.*, unpublished), viral transfer capacity up to 40 days (Popov *et al.*, 2005), and the need for only a very small infected DC population (<0.1%). Thus, the prevention of viral attachment and inhibition of viral entry is of paramount importance in preventing both trans and cis infection. However, when sexually transmitted copathogens are present in the vaginal mucosa, such as HSV, and induce inflammation, numerous activated CD4$^+$ T lymphocytes or macrophages infiltrating the epithelium and lamina propria might be infected within the first few hours by the trans mechanism.

10.5.1 Effect of Maturation

Maturation of epithelial DCs is often associated with their emigration to lymphatic tissues and lymph nodes. Maturation stimuli may be endogenous such as antigen uptake or viral replication in DCs or exogenous "danger signals" including lipopolysaccharide or cytokines such as TNF-α and prostaglandins (E2). These stimuli induce downregulation of the CLRs, DC-SIGN, MR, and to a lesser degree langerin concordantly with upregulation of CD83, the costimulatory molecules CD80 and 86, and CCR7, which binds to the chemokines resulting in migration. HIV binds and enters a different compartment in mature to immature DCs (Frank *et al.*, 2002). This mature DC compartment is single, large, coated with clathrin, and recently shown to be labeled for CD9, CD63, and CD81 (Garcia *et al.*, 2005). Virus can be transferred to CD4 cells from this compartment in trans. Mature DCs are less endocytic and phagocytic than iDCs although they retain intact infectious particles within the endosome for longer periods and are more efficient at antigenic stimulation of T cells and HIV transfer to these cells. Indeed, maturation enhances the formation of both immunologic and viral synapses with T cells. Of note, mature DCs stimulate bold CD4 and CD8 HIV/SIV specific lymphocytes whereas immature DCs trigger only CD4 lymphocyte responses (Frank *et al.*, 2003). LFA-ICAM interactions and the tetraspannins are involved in this synapse formation but the CLRs appear not to be concentrated at the synapse (Garcia *et al.*, 2005). Obviously, CLRs are essential for uptake of HIV into the endocytic compartment in iDCs. Maturation downregulates CCR5 and upregulates CXCR4. Therefore, replication of R5 strains of HIV is inhibited in mature DCs, and indeed replication is greatly reduced owing to other stages of restriction. However, after maturation and/or migration endocytosed virus is transferred with greater efficiency to T cells than by immature DCs.

10.5.2 HIV, CLRs, and Nonepithelial DCs

Myeloid or plasmacytoid DCs in noninflamed tonsil, and probably lymph nodes, express low levels of CLRs (DC-SIGN and MR). However, in inflammatory conditions, cytokine production may result in CLR upregulation, that is, GM-CSF for MR and IL-4 for DC-SIGN (Soilleux *et al.*, 2002). A potential role of CLRs and DCs in mucosal lymphatic tissue, lymph nodes, thymus, and placenta in enhancing HIV replication in such inflammatory conditions needs to be examined. DC-SIGN and possibly other CLRs are expressed on DCs in normal uninflamed placenta, thymus, and lung (Soilleux *et al.*, 2002).

10.5.3 Control of HIV Spread and Vaginal Microbicides

The traditional tools of education, abstinence, and use of safe sex techniques including condoms have worked well in selected countries, including the United States, Europe, Thailand, and Uganda, but in parts of sub-Saharan Africa and India, condom use is extremely low as evident by the continued growth of new adult cases in these regions. Currently, the 3×5 initiative to enable the use of anti-retrovirals by 3 million people in sub-Saharan Africa and other parts the developing world by 2005 is faltering. However, the recent Group of Eight announcement on targeting HIV in Africa may bolster the use of anti-retrovirals in assisting the control of transmission. Progress in vaccines is slow with many recent setbacks. Therefore, an interim mechanism of control of HIV transmission, particularly heterosexual transmission in the developing world, and which empowers women is urgently required. Microbicides might also be used to simultaneously target other genital pathogens such as HSV, which increase susceptibility to HIV (Cunningham and Dwyer, 2004).

There are two basic approaches for microbicide development. The anti-HIV agent must either target the virus while it is in the lumen of the vagina or else target the epithelial cells, by preventing virus attachment and fusion or spread from the host cell [reviewed by Shattock and Moore (2003); Moore (2005)]. A combination of both approaches may lead to the greatest success, as complete inhibition of viral entry is so difficult.

10.5.4 Targeting the Virus

Successful microbicide candidates must specifically inhibit binding, entry, or replication and spread of a wide variety of HIV-1 and HIV-2 strains, especially clinical isolates and probably both R5 and X4 strains in different cell types. Conventional anti-retrovirals such as nucleoside reverse transcriptase inhibitors with poor oral bioavailability (e.g., MIV-150 and UC781) may still be useful as topical inhibitors (Fletcher *et al.*, 2005). Advances in the understanding of the process of viral attachment to target cells, leading to envelope-mediated membrane fusion and infection, have provided greater opportunity for the development of drugs that interfere with each of these stages. Many of the new drugs targeting the attachment/ fusion stage of HIV replication, such as T-20, are worth testing as topical microbicides. The two main targets on the surface of HIV are the viral membrane and the envelope glycoproteins. However, nonspecific reagents that disrupt the virus membrane will also have an effect on the host cells resulting in the ulceration and increased infection observed with nonoxynol-9. Similar approaches with other detergents and drugs that deplete the cholesterol content of the virus and cellular membranes must be approached with caution. A more specific approach is the development of drugs that target the envelope glycoproteins. For examples, cyanovirin and the neutralizing antibody b12 (which both recognize high mannose oligosaccharides on gp120) have been shown to be effective *in vivo* but at quite high concentrations (Tsai *et al.*, 2003, 2004; Veazey *et al.*, 2003). Plant lectins are also potent inhibitors of HIV infection of DCs and subsequent transfer to T cells (Turville *et al.*, 2005) (They are entering *in vivo* microbicide studies as a part of the European collaborative initiative for micobicide development.) Such inhibitors which bind to gp120 or gp41 and impair the subsequent binding of virus to target cell must be highly potent (Poignard *et al.*, 2001), as DCs taking up virus migrate away from high local concentrations of microbicide and if residual, HIV will be disseminated to T cells in a more permissive setting with little or no microbicide present.

10.5.5 Targeting the Cell

The alternative to targeting the virus is to prevent HIV binding to its target cell(s). HIV is capable of binding to a wide variety of cells in the body including CD4 and CD8 lymphocytes, macrophages, astrocytes, bone marrow precursors, and DCs. In the context of the genital mucosa, DCs, CD4 lymphocytes, and macrophages are the main targets for HIV infection. The obvious way of preventing infection of the target cell is to block the cell-surface receptors for HIV binding and entry, that is, CD4, chemokine coreceptors CCR5 (Lederman *et al.*, 2004) and CXCR4 and the CLRs present on DCs. Much effort has been directed at the development of CD4 and coreceptor antagonists CCR5, many of which are in early stages of clinical trials, but little attention has been paid to the CLRs, apart from DC-SIGN.

10.6 Models for Examining HIV Infection and Testing of Vaginal Microbicides in the Genital Mucosa

10.6.1 Monocyte-derived Dendritic Cells and Monocyte-derived Langerhans Cells

Immature MDDCs serve as a convenient *in vitro* model for some of the epithelial DCs encountered by HIV during sexual transmission. They express the CLRs MR and DC-SIGN on their surface and display most similarity to dermal or lamina propria DCs but not langerin-expressing DCs. They can be cultured in greater numbers than isolated myeloid DCs, which are present in blood at very low frequencies. MDDCs can also be generated *in vitro* as both immature and mature cells allowing both phenotypes to be studied.

Epidermal and dermal DCs derived freshly from human epidermal explants are difficult to isolate in high numbers (Turville *et al.*, 2002). The isolation of immature epithelial DCs using collagenase digestion is laborious and yields relatively few cells, with a 1 cm^2 piece of skin only yielding approximately 10,000 DCs (and yields vary). "Crawl-out" methods of isolation where emigrant DCs collected from explants *in vitro* have a partially mature phenotype including downregulation of their CLRs and reflect cells that have been activated by tissue trauma or another pathogen. These *ex vivo* partially mature DCs provide a more representative model of *in vitro* LCs than the traditional *in vitro* derived MDDCs.

In vivo immature LCs do not express MR or DC-SIGN (Turville *et al.*, 2002). Studies of immature epithelial LCs *in situ* would be greatly facilitated by the development of an accurate *in vitro* model of this cell. Many groups have defined an LC model from CD34$^+$ stem cells or CD14$^+$ monocytes with a variety of cytokines (Gatti *et al.*, 2000; Geissmann *et al.*, 1998; Jaksits *et al.*, 1999; Mohamadzadeh *et al.*, 2001; Strunk *et al.*, 1996) yet to date there is no model defined by the phenotype as langerin$^+$ DC-SIGN$^-$MR$^-$E-cadherin$^+$CCR6$^+$Birbeck Granule$^+$CD83$^-$. Such a model would be extremely useful for studying HIV binding and entry allowing direct comparisons with MDDCs for expression of other unknown CLRs on the surface likely to bind HIV/gp120.

10.6.2 *Ex Vivo* Cervical Explants

The *ex vivo* cervical explant model has the advantage that it is the most accurate *in vitro* model of infection of uninflamed genital tract tissues with resting cells of the immune system being present. However, data from these organ cultures of the cervix must be considered with caution as there are likely to be lower levels of DCs in the mucosa than that found *in vivo*, probably because of emigration after trauma (Fletcher *et al.*, 2005). This model has highlighted the importance of CD4 and the HIV coreceptors CCR5 and CXCR4 in localized HIV infection within the cervix and MR expression on emigrant DCs. While MR was found to have minimal impact on localized infection within cervical tissue, MR-dependent uptake and virus dissemination by migratory DCs was found to be more efficient. Again these studies found that while DC-SIGN was responsible for a considerable fraction of HIV uptake, other CLRs were found to be equally important as the simultaneous blocking of both CD4 and DC-SIGN did not completely abolish HIV transmission (Hu *et al.*, 2004).

Although DCs are likely to be responsible for the majority of HIV dissemination across the mucosa, *ex vivo* studies using human cervical tissue indicate that HIV may also be disseminated to draining lymph nodes by migrating CD4$^+$ T cells (Shattock and Moore, 2003) possibly from submucosal sites (Miller *et al.*, 2005). LCs and T cells are often observed in direct contact in the human cervical and vaginal epithelium (Parr and Parr, 1991), which could facilitate trans infection of the T cells with HIV. Therefore, inhibition of HIV uptake by microbicides may have to inhibit the entire receptor repertoire of all HIV target cells present within the mucosa. If such inhibitors cannot prevent transport of residual virus across the mucosa, then microbicides that cross the mucosal barrier and inhibit initial spread in CD4 lymphocytes might (Fletcher *et al.*, 2005).

10.6.3 Macaque Models

The macaque model of vaginal transmission using SIV is the main pre-clinical model for microbicide testing (Moore, 2005). Progesterone is used to thin the stratified squamous vaginal epithelium to uniformly enhance infection with doses of virus 100 times that found in human semen (see above). Many of the SHIV strains, chimeric simian/HIV viruses, have a preference for the CXCR4 coreceptor, while the HIV strains involved in the sexual transmission of HIV are predominately CCR5 utilizing. As a result, SHIV SF162 is now widely used in the macaque vaginal challenge model as it utilizes the CCR5 coreceptor (Lederman *et al.*, 2004). It is important to consider that the concentration of inhibitors of HIV binding and entry required to block vaginal transmission *in vivo* was 1,000-fold higher than that required to prevent virus infection *in vitro* (Veazey *et al.*, 2003). Therefore, there may be differences in successful inhibitors *in vitro* compared with those that are successful *in vivo*. Testing in macaque models is essential before selecting candidates for human trials. In the future, the effect of microbicides on HIV transmission in the setting of coinfection with other sexually transmitted pathogens (especially those causing genital ulceration) must also be tested using these models.

10.7 Concluding Remarks

In this chapter, we have demonstrated that HIV utilizes key biological features of DCs in order to optimize access to and replication within its principle targets, the CD4 lymphocytes in the mucosa or lymph nodes. The virus is able to use the heterogeneity of its heavily glycosylated envelope molecule gp120 to bind to the most important CLRs on the surface of epithelial DCs, langerin, DC-SIGN, and MR, through different structural motifs in its high mannose sugars. This capacity to bind to all of these three receptors would allow it to bind to any of the epithelial DCs whether in minimal abrasions or deeper lacerations of stratified squamous epithelium or cuboidal or columnar epithelium of the anorectal and genital tract. After concentration by these molecules on the surface of DCs, the virus is able to enter and utilize the endosomal system of the DC for transfer to T cells slightly altering and subverting pathways of antigen presentation. Furthermore, these pathways allow HIV to infect activated or inactivated CD4 lymphocytes simultaneously presenting antigen so there may be specific infection and eventual knockout of CD4 lymphocytes that respond to HIV antigen (Lore *et al.*, 2005). Apart from the importance of this DC uptake pathway in epithelial cells as a potential mechanism for HIV transport to resting CD4 lymphocytes in focal rests in the lamina propria or in lymph nodes, such mechanisms may operate elsewhere in the body, particularly where inflammation may upregulate CLRs on the surface of DCs. CLRs of epithelial DCs should be considered as targets for vaginal (and perhaps anorectal) microbicides.

References

Annan, K.A. (2002). In Africa, AIDS has a women's face. New York Times 29 December: 9.

Auvert, B., Taljaard, D., Lagarde, E., Sobngwi-Tambekou, J., Sitta, R., and Puren, A. (2005). Randomized, controlled intervention trial of male circumcision for reduction of HIV infection risk: The ANRS 1265 Trial. PLoS Med 2:e298

Benki, S., Mostad, S.B., Richardson, B.A., Mandaliya, K., Kreiss, J.K., and Overbaugh, J. (2004). Cyclic shedding of HIV-1 RNA in cervical secretions during the menstrual cycle. J. Infect. Dis. 189:2192-2201.

Bernhard, O.K., Lai, J., Wilkinson, J., Sheil, M.M., and Cunningham, A.L. (2004). Proteomic analysis of DC-SIGN on dendritic cells detects tetramers required for ligand binding but no association with CD4. J. Biol. Chem. 279:51828-51835.

Cunningham, A.L., Turner, R.R., Miller, A.C., Para, M.F., and Merigan, T.C. (1985). Evolution of recurrent herpes simplex lesions. An immunohistologic study. J. Clin. Invest. 75:226-233.

Cunningham, A.L., and Dwyer, D.E. (2004). The pathogenesis underlying the interaction of HIV and herpes simplex virus after co-infection. J. HIV Ther. 9:9-13.

Engering, A., Geijtenbeek, T.B., van Vliet, S.J., Wijers, M., van Liempt, E., Demaurex, N., Lanzavecchia, A., Fransen, J., Figdor, C.G., Piguet, V., and van Kooyk, Y. (2002). The dendritic cell-specific adhesion receptor DC-SIGN internalizes antigen for presentation to T cells. J. Immunol. 168:2118-2126.

Ezekowitz, R.A., Kuhlman, M., Groopman, J.E., and Byrn, R.A. (1989). A human serum mannose-binding protein inhibits *in vitro* infection by the human immunodeficiency virus. J. Exp. Med. 169:185-196.

Feinberg, H., Mitchell, D.A., Drickamer, K., and Weis, W.I. (2001). Structural basis for selective recognition of oligosaccharides by DC-SIGN and DC-SIGNR. Science 294:2163-2166.

Figdor, C.G., van Kooyk, Y., and Adema, G.J. (2002). C-type lectin receptors on dendritic cells and Langerhans cells. Nat. Rev. Immunol. 2:77-84.

Fletcher, P., Kiselyeva, Y., Wallace, G., Romano, J., Griffin, G., Margolis, L., and Shattock, R. (2005). The nonnucleoside reverse transcriptase inhibitor UC-781 inhibits human immunodeficiency virus type 1 infection of human cervical tissue and dissemination by migratory cells. J. Virol. 79:11179-11186.

Frank, I., Piatak, M. Jr., Stoessel, H., Romani, N., Bonnyay, D., Lifson, J.D., and Pope, M. (2002). Infectious and whole inactivated simian immunodeficiency viruses interact similarly with primate dendritic cells (DCs): differential intracellular fate of virions in mature and immature DCs. J. Virol. 76: 2936-2951.

Frank, I., Santos, J.J., Mehlhop, E., Villamide-Herrera, L., Santisteban, C., Gettie, A., Ignatius, R., Lifson, J.D., and Pope, M. (2003). Presentation of exogenous whole inactivated simian immunodeficiency virus by mature dendritic cells induces CD4$^+$ and CD8$^+$ T-cell responses. J. Acquir. Immune. Defic. Syndr. 34:7-19.

Fredericksen, B.L., Wei, B.L., Yao, J., Luo, T., and Garcia, J.V. (2002). Inhibition of endosomal/lysosomal degradation increases the infectivity of human immunodeficiency virus. J. Virol. 76:11440-11446.

Garcia, E., Pion, M., Pelchen-Matthews, A., Collinson, L., Arrighi, J.F., Blot, G., Leuba, F., Escola, J.M., Demaurex, N., Marsh, M., and Piguet, V. (2005). HIV-1 trafficking to the dendritic cell-T-cell infectious synapse uses a pathway of tetraspanin sorting to the immunological synapse. Traffic 6: 488-501.

Gatti, E., Velleca, M.A., Biedermann, B.C., Ma, W., Unternaehrer, J., Ebersold, M.W., Medzhitov, R., Pober, J.S., and Mellman, I. (2000). Large-scale culture and selective maturation of human Langerhans cells from granulocyte colony-stimulating factor-mobilized CD34$^+$ progenitors. J. Immunol. 164: 3600-3607.

Geijtenbeek, T.B., Kwon, D.S., Torensma, R., van Vliet, S.J., van Duijnhoven, G.C., Middel, J., Cornelissen, I.L., Nottet, H.S., KewalRamani, V.N., Littman, D.R., Figdor, C.G., and van Kooyk, Y. (2000). DC-SIGN, a dendritic cell-specific HIV-1-binding protein that enhances trans-infection of T cells. Cell 100:587-597.

Geissmann, F., Prost, C., Monnet, J.P., Dy, M., Brousse, N., and Hermine, O. (1998). Transforming growth factor beta1, in the presence of granulocyte/ macrophage colony-stimulating factor and interleukin 4, induces differen-tiation of human peripheral blood monocytes into dendritic Langerhans cells. J. Exp. Med. 187:961-966.

Geyer, H., Holschbach, C., Hunsmann, G., and Schneider, J. (1988). Carbo-hydrates of human immunodeficiency virus. Structures of oligosaccharides linked to the envelope glycoprotein 120. J. Biol. Chem. 263:11760-11767.

Greenhead, P., Hayes, P., Watts, P.S., Laing, K.G., Griffin, G.E., and Shattock, R.J. (2000). Parameters of human immunodeficiency virus infection of human cervical tissue and inhibition by vaginal virucides. J. Virol. 74:5577-5586.

Hladik, F., Lentz, G., Akridge, R.E., Peterson, G., Kelley, H., McElroy, A., and McElrath, M.J. (1999). Dendritic cell-T-cell interactions support coreceptor-independent human immunodeficiency virus type 1 transmission in the human genital tract. J. Virol. 73:5833-5842.

Hu, J., Gardner, M.B., and Miller, C.J. (2000). Simian immunodeficiency virus rapidly penetrates the cervicovaginal mucosa after intravaginal inoculation and infects intraepithelial dendritic cells. J. Virol. 74:6087-6095.

Hu, Q., Frank, I., Williams, V., Santos, J.J., Watts, P., Griffin, G.E., Moore, J.P., Pope, M., and Shattock, R.J. (2004). Blockade of attachment and fusion receptors inhibits HIV-1 infection of human cervical tissue. J. Exp. Med. 199:1065-1075.

Jaksits, S., Kriehuber, E., Charbonnier, A.S., Rappersberger, K., Stingl, G., and Maurer, D. (1999). CD34$^+$ cell-derived CD14$^+$ precursor cells develop into Langerhans cells in a TGF-beta 1-dependent manner. J. Immunol. 163: 4869-4877.

Kawamura, T., Cohen, S.S., Borris, D.L., Aquilino, E.A., Glushakova, S., Margolis, L.B., Orenstein, J.M., Offord, R.E., Neurath, A.R., and Blauvelt, A.

(2000). Candidate microbicides block HIV-1 infection of human immature Langerhans cells within epithelial tissue explants. J. Exp. Med. 192: 1491-1500.

Kery, V., Krepinsky, J.J., Warren, C.D., Capek, P., and Stahl, P.D. (1992). Ligand recognition by purified human mannose receptor. Arch. Biochem. Biophys. 298:49-55.

Kwon, D.S., Gregorio, G., Bitton, N., Hendrickson, W.A., and Littman, D.R. (2002). DC-SIGN-mediacted internalization of HIV is required for trans-enhancement of T-cell infection. Immunity 16:135-144.

Lederman, M.M., Veazey, R.S., Offord, R., Mosier, D.E., Dufour, J., Mefford, M., Piatak, M. Jr., Lifson, J.D., Salkowitz, J.R., Rodriguez, B., Blauvelt, A., Hartley, O. (2004). Prevention of vaginal SHIV transmission in rhesus macaques through inhibition of CCR5. Science 306:485-487.

Lee, B., Leslie, G., Soilleux, E., O'Doherty, U., Baik, S., Levroney, E., Flummerfelt, K., Swiggard, W., Coleman, N., Malim, M., Doms, R.W. (2001). Cis expression of DC-SIGN allows for more efficient entry of human and simian immunodeficiency viruses via CD4 and a coreceptor. J. Virol. 75:12028-12038.

Lin, C.L., Sewell, A.K., Gao, G.F., Whelan, K.T., Phillips, R.E., Austyn, J.M. (2000). Macrophage-tropic HIV induces and exploits dendritic cell chemotaxis. J. Exp. Med. 192:587-594.

Lore, K., Smed-Sorensen, A., Vasudevan, J., Mascola, J.R., and Koup, R.A. (2005). Myeloid and plasmacytoid dendritic cells transfer HIV-1 preferentially to antigen-specific CD4$^+$ T cells. J. Exp. Med. 201:2023-2033.

MacDonald, K.P., Munster, D.J., Clark, G.J., Dzionek, A., Schmitz, J., and Hart, D.N. (2002). Characterization of human blood dendritic cell subsets. Blood 100:4512-4520.

Marx, P.A., Spira, A.I., Gettie, A., Dailey, P.J., Veazey, R.S., Lackner, A.A., Mahoney, C.J., Miller, C.J., Claypool, L.E., Ho, D.D., and Alexander, N.J. (1996). Progesterone implants enhance SIV vaginal transmission and early virus load. Nat. Med. 2:1084-1089.

McDonald, D., Wu, L., Bohks, S.M., KewalRamani, V.N., Unutmaz, D., and Hope, T.J. (2003). Recruitment of HIV and its receptors to dendritic cell-T cell junctions. Science 300:1295-1297.

Mehlhop, E., Villamide, L.A., Frank, I., Gettie, A., Santisteban, C., Messmer, D., Ignatius, R., Lifson, J.D., and Pope, M. (2002). Enhanced *in vitro* stimulation of rhesus macaque dendritic cells for activation of SIV-specific T cell responses. J. Immunol. Methods 260:219-234.

Miller, C.J., Li, Q., Abel, K., Kim, E.Y., Ma, Z.M., Wietgrefe, S., La Franco-Scheuch, L., Compton, L., Duan, L., Shore, M.D., Zupancic, M., Busch, M., Carlis, J., Wolinsky, S., and Haase, A.T. (2005). Propagation and dissemination of infection after vaginal transmission of simian immuno-deficiency virus. J. Virol. 79:9217-9227.

Mohamadzadeh, M., Berard, F., Essert, G., Chalouni, C., Pulendran, B., Davoust, J., Bridges, G., Palucka, A.K., and Banchereau, J. (2001). Interleukin 15 skews monocyte differentiation into dendritic cells with features of Langerhans cells. J. Exp. Med. 194:1013-1020.

Moore, J.P. (2005). Topical microbicides become topical. N. Engl. J. Med. 352:298-300.

Nguyen, D.G., and Hildreth, J.E. (2003). Involvement of macrophage mannose receptor in the binding and transmission of HIV by macrophages. Eur. J. Immunol. 33:483-493.

Nobile, C., Moris, A., Porrot, F., Sol-Foulon, N., and Schwartz, O. (2003). Inhibition of human immunodeficiency virus type 1 Env-mediated fusion by DC-SIGN. J. Virol. 77:5313-5323.

Osugi, Y., Vuckovic, S., and Hart, D.N. (2002). Myeloid blood CD11c(+) dendritic cells and monocyte-derived dendritic cells differ in their ability to stimulate T lymphocytes. Blood 100:2858-2866.

Parr, M.B., and Parr, E.L. (1991). Langerhans cells and T lymphocyte subsets in the murine vagina and cervix. Biol. Reprod. 44:491-498.

Peretti, S., Schiavoni, I., Pugliese, K., and Federico, M. (2006). Selective elimination of HIV-1-infected cells by Env-directed, HIV-1-based virus-like particles. Virology 345:115-126.

Poignard, P., Saphire, E.O., Parren, P.W., and Burton, D.R. (2001). gp120: Biologic aspects of structural features. Ann. Rev. Immunol. 19:253-274.

Popov, S., Chenine, A.L., Gruber, A., Li, P.L., and Ruprecht, R.M. (2005). Long-term productive human immunodeficiency virus infection of CD1a-sorted myeloid dendritic cells. J. Virol. 79:602-608.

Randolph, G.J., Inaba, K., Robbiani, D.F., Steinman, R.M., and Muller, W.A. (1999). Differentiation of phagocytic monocytes into lymph node dendritic cells *in vivo*. Immunity 11:753-761.

Rescigno, M., Urbano, M., Valzasina, B., Francolini, M., Rotta, G., Bonario, R., Granucci, F., Kraehenbuhl, J.P., Ricciardi-Castagnoli, P. (2001). Dendritic cells express tight junction proteins and penetrate gut epithelial monolayers to sample bacteria. Nat. Immunol. 2:361-367.

Reynolds, S.J., Risbud, A.R., Shepherd, M.E., Zenilman, J.M., Brookmeyer, R.S., Paranjape, R.S., Divekar, A.D., Gangakhedkar, R.R., Ghate, M.V., Bollinger, R.C., Mehendale, S.M. (2003). Recent herpes simplex virus type 2 infection and the risk of human immunodeficiency virus type 1 acquisition in India. J. Infect. Dis. 87:1513-1521.

Romani, N., Gruner, S., Brang, D., Kampgen, E., Lenz, A., Trockenbacher, B., Konwalinka, G., Fritsch, P.O., Steinman, R.M., and Schuler, G. (1994). Proliferating dendritic cell progenitors in human blood. J. Exp. Med. 180: 83-93.

Sallusto, F., and Lanzavecchia, A. (1994). Efficient presentation of soluble antigen by cultured human dendritic cells is maintained by granulocyte/macrophage colony-stimulating factor plus interleukin 4 and downregulated by tumor necrosis factor alpha. J. Exp. Med. 179:1109-1118.

Sallusto, F., Cella, M., Danieli, C., and Lanzavecchia, A. (1995). Dendritic cells use macropinocytosis and the mannose receptor to concentrate macromolecules in the major histocompatibility complex class II compartment: downregulation by cytokines and bacterial products. J. Exp. Med. 182:389-400.

Sewell, A.K., and Price, D.A. (2001). Dendritic cells and transmission of HIV-1. Trends Immunol. 22:173-175.

Shattock, R.J., and Moore, J.P. (2003). Inhibiting sexual transmission of HIV-1 infection. Nat. Rev. Microbiol. 1:25-34.

Smith, S.M., Baskin, G.B., and Marx, P.A. (2000). Estrogen protects against vaginal transmission of simian immunodeficiency virus. J. Infect. Dis. 182:708-715.

Smith, S.M., Mefford, M., Sodora, D., Klase, Z., Singh, M., Alexander, N., Hess, D., and Marx, P.A. (2004). Topical estrogen protects against SIV vaginal transmission without evidence of systemic effect. AIDS 18:1637-1643.

Sodora, D.L., Gettie, A., Miller, C.J., and Marx, P.A. (1998). Vaginal transmission of SIV: assessing infectivity and hormonal influences in macaques inoculated with cell-free and cell-associated viral stocks. AIDS Res. Hum. Retroviruses 14 Suppl 1:S119-23.

Soilleux, E.J., Morris, L.S., Leslie, G., Chehimi, J., Luo, Q., Levroney, E., Trowsdale, J., Montaner, L.J., Doms, R.W., Weissman, D., Coleman, N., and Lee, B. (2002). Constitutive and induced expression of DC-SIGN on dendritic cell and macrophage subpopulations in situ and *in vitro*. J. Leukoc. Biol. 71:445-457.

Stahl, P.D., and Ezekowitz, R.A. (1998). The mannose receptor is a pattern recognition receptor involved in host defense. Curr. Opin. Immunol. 10:50-55.

Stambach, N.S., and Taylor, M.E. (2003). Characterization of carbohydrate recognition by langerin, a C-type lectin of Langerhans cells. Glycobiology 13:401-410.

Stewart, G.J., Tyler, J.P., Cunningham, A.L., Barr, J.A., Driscoll, G.L., Gold, J., and Lamont, B.J. (1985). Transmission of human T-cell lymphotropic virus type III (HTLV-III) by artificial insemination by donor. Lancet 2:581-585.

Strunk, D., Rappersberger, K., Egger, C., Strobl, H., Kromer, E., Elbe, A., Maurer, D., and Stingl, G. (1996). Generation of human dendritic cells/Langerhans cells from circulating CD34$^+$ hematopoietic progenitor cells. Blood 87:1292-1302.

Su, S.V., Gurney, K.B., and Lee, B. (2003). Sugar and spice: viral envelope-DC-SIGN interactions in HIV pathogenesis. Curr. HIV Res. 1:87-99.

Taylor, M.E., Drickamer, K. (1993). Structural requirements for high affinity binding of complex ligands by the macrophage mannose receptor. J. Biol. Chem. 268:399-404.

Tchou, I., Sabido, O., Lambert, C., Misery, L., Garraud, O., and Genin, C. (2003). Technique for obtaining highly enriched, quiescent immature Langerhans cells suitable for *ex vivo* assays. Immunol. Lett. 86:7-14.

Tsai, C.C., Emau, P., Jiang, Y., Tian, B., Morton, W.R., Gustafson, K.R., and Boyd, M.R. (2003). Cyanovirin-N gel as a topical microbicide prevents rectal transmission of SHIV89.6P in macaques. AIDS Res. Hum. Retroviruses 19:535-541.

Tsai, C.C., Emau, P., Jiang, Y., Agy, M.B., Shattock, R.J., Schmidt, A., Morton, W.R., Gustafson, K.R., and Boyd, M.R. (2004). Cyanovirin-N inhibits AIDS virus infections in vaginal transmission models. AIDS Res. Hum. Retroviruses 20:11-18.

Turville, S.G., Arthos, J., Donald, K.M., Lynch, G., Naif, H., Clark, G., Hart, D., Cunningham, A.L. (2001). HIV gp120 receptors on human dendritic cells. Blood 98:2482-2488.

Turville, S.G., Cameron P.U., Handley, A., Lin, G., Pöhlmann, S., Doms, R.W., and Cunningham, A.L. (2002). Diversity of receptor binding HIV on dendritic cell subsets. Nat. Immunol. 3:975-983.

Turville, S.G., Wilkinson, J., Cameron, P.U., Dable, J., and Cunningham, A.L. (2003). The role of dendritic cell C-type lectin receptors in HIV pathogenesis. J. Leukoc. Biol. 74:710-718.

Turville, S.G., Santos, J.J., Frank, I., Cameron, P.U., Wilkinson, J., Miranda-Saksena, M., Dable, J., Stossel, H., Romani, N., Piatak, M. Jr., Lifson, J.D., Pope, M., Cunningham, A.L. (2004). Dendritic cells internalize and transfer immunodeficiency viruses to CD4$^+$ lymphocytes in two distinct phases. Blood 103: 2170-2179.

Turville, S.G., Vermeire, K., Balzarini, J., and Schols, D. (2005). Sugar-binding proteins potently inhibit dendritic cell human immunodeficiency virus type 1 (HIV-1) infection and dendritic-cell-directed HIV-1 transfer. J. Virol. 79:13519-13527.

Valladeau, J., Ravel, O., Dezutter-Dambuyant, C., Moore, K., Kleijmeer, M., Liu,Y., Duvert-Frances, V., Vincent, C., Schmitt, D., Davoust, J., Caux, C., Lebecque, S., and Saeland, S. (2000). Langerin, a novel C-type lectin specific to Langerhans cells, is an endocytic receptor that induces the formation of Birbeck granules. Immunity 12:71-81.

van Kooyk, Y., and Geijtenbeek, T.B. (2003). DC-SIGN: escape mechanism for pathogens. Nat. Rev. Immunol. 3:697-709.

Veazey, R.S., Shattock, R.J., Pope, M., Kirijan, J.C., Jones, J., Hu, Q., Ketas, T., Marx, P.A., Klasse, P.J., Burton, D.R., and Moore, J.P. (2003). Prevention of virus transmission to macaque monkeys by a vaginally applied monoclonal antibody to HIV-1 gp120. Nat. Med. 9:343-346.

Wald, A. (2004). Synergistic interactions between herpes simplex virus type-2 and human immunodeficiency virus epidemics. Herpes 11:70-76.

Wilkinson, J. and Cunningham, A.L. (2006). Mucosal transmission of HIV-1: first stop dendritic cells. Curr. Drug. Targets (still in press).

Zhu, X., Borchers, C., Bienstock, R.J., and Tomer, K.B. (2000). Mass spectro-metric characterization of the glycosylation pattern of HIV-gp120 expressed in CHO cells. Biochemistry 39:11194-11204.

Chapter 11

Loss, Infection, and Dysfunction of Dendritic Cells in HIV Infection

Steven Patterson, Heather Donaghy, and Peter Kelleher

11.1 Overview

Presentation of antigen by dendritic cells (DCs) to T cells is essential for the initiation of primary adaptive immune responses. DCs also play a key role in innate immunity through secretion of cytokines that activate NK cells and production of interferons with antiviral activity. After their discovery 30 years ago by Steinman and colleagues (Steinman *et al.*, 1975), DCs were shown to be highly efficient at stimulating CD4 T-cell proliferation in response to cognate peptides presented on MHC class II molecules. Subsequently, they were found to stimulate CD8 T cells recognizing MHC class I peptide complexes while more recently they were shown to cluster B cells and play a role in immunoglobulin class switching (Banchereau and Steinman, 1998; Banchereau *et al.*, 2000). Given the numerous roles that have now been ascribed to DCs, it is clear that loss or impaired functioning of these cells could be highly detrimental to the host immune system.

The complexity of the DC system has been revealed during the past 10 years with the discovery of a number of phenotypically and functionally different DC populations in both human and mouse (Shortman and Liu, 2002). The property that unifies all DC populations is their ability to acquire, process, and present antigen initiating the proliferation of T cells recognizing cognate peptide complexed with MHC class I and II molecules. Other leukocytes, including B lymphocytes and macrophages, can also stimulate T cells but are less efficient than DCs. DCs are unique in being the only kind of antigen-presenting cell that can efficiently stimulate the proliferation of naïve T cells. They are derived from CD34[+] stem cells that originate in the bone marrow and have been identified in virtually every body tissue, with the possible exception of the brain. In humans, the precursors of tissue DC constitute approximately 1% of blood mono-nuclear cells and can be separated into two phenotypically and functionally distinct populations, myeloid DCs (mDCs), which are the precursors of the classical peripheral tissue DCs, and the more recently identified plasmacytoid DCs (pDCs), shown to be identical to the "natural interferon producing cell" (Rissoan *et al.*, 1999; Robinson *et al.*, 1999). They are routinely identified by the expression of MHC class II HLA-DR and the lack of labeling with a cocktail of lineage-specific antibodies that identifies other mononuclear cell populations and includes anti-CD3, -CD14, -CD16, -CD19, and -CD56. The β-integrin, CD11c and the receptor for IL-3, CD123, are differentially expressed by the mDCs and pDCs, respectively, and are used to differentiate the two populations. This chapter will focus on human mDCs and pDCs and their role in HIV infection. To help appreciate the consequences of infection, loss, and dysfunction of DCs mediated by HIV infection, we give a brief overview of DC biology.

11.2 Introduction to Myeloid and Plasmacytoid DCs

11.2.1 Myeloid DCs

11.2.1.1 Phenotype and Location

Cells identifiable as myeloid DCs are present in blood, peripheral tissues, and secondary lymphoid tissues at different stages of development. CD11c[+] DCs present in blood are thought to represent the precursors of immature myeloid tissue DCs and are seeded to peripheral tissues where their main function is to detect, internalize, process, and present antigen from invading pathogens. They are thus found in larger numbers in tissues where pathogens gain access to the body such as the skin, mucosal tissue in the airways, gastrointestinal and genital tracts. Expression of the CXCR1, CCR1, CCR2, and CCR5 on the blood precursors facilitates recruitment to inflamed tissues through recognition of inflammatory chemokines (Sallusto *et al.*, 1998; Sozzani *et al.*, 1998). At least two main subpopulations of myeloid DCs have been identified in peripheral tissues: Langerhans cells and dermal/interstitial type DCs. Langerhans cells are the more superficially located DCs and are found in the epidermal layers of the skin and adjacent to the epithelial cells in the genital tract. Dermal/interstitial DCs are found in relatively large numbers in the dermis, in the less superficial layers of mucosal tissue, and in other peripheral tissues. Interestingly, although Langerhans cells have been described in rectal tissue (Hussain and Lehner, 1995), there are no reports of these cells in other mucosal compartments in the gut. Langerhans uniquely express Birbeck granules and can be further identified by membrane labeling for langerin, however, they lack expression of CD11b and DC-SIGN. Dermal/interstitial DCs, on the other hand, do not have Birbeck granules but are positive for DC-SIGN and CD11b. Both these myeloid DCs express CD1a at the immature tissue stage of differentiation. DCs can be generated from monocytes by culturing for 5–7 days in GM-CSF and IL-4 and phenotypically resemble immature dermal DCs (Romani *et al.*, 1994; Sallusto and Lanzavecchia, 1994). Currently, it is not clear to what extent monocyte-derived DCs (mdDCs) contribute to the pool of immature tissue DCs *in vivo*. In addition to mature DCs, immature DCs are also found in the secondary lymphoid tissue and may arise from a slow turnover of tissue DCs.

11.2.1.2 Maturation

Myeloid dendritic cells in peripheral tissue are immature in that they express lower levels of MHC and the costimulatory molecules, CD80 and CD86, required for the activation of T cells through interaction with CD28 and are generally negative for the maturation marker CD83. Immature DCs show high levels of endocytosis, which enables them to constantly sample the environment to detect invading pathogens. Maturation of myeloid DCs is triggered by pathogen-derived molecules such as bacterial lipopoly-saccharide, flagellin, or double-stranded RNA (Pasare and Medzhitov, 2004). These molecules are recognized by evolutionary conserved pattern recognition receptors, particularly Toll-like receptors (TLRs), expressed by mDCs. Maturation is also induced by molecules associated with inflammation and necrosis including TNF-α and heat shock protein (Srivastava, 2002). In addition to upregulation of MHC and costimulatory molecules, maturation is accompanied by upregulation of CCR7, which mediates homing to lymphoid tissue in response to CCL19 (MIP-3β) and CCL21 (6Ckine) (Sallusto *et al.*, 1999) and a downregulation of endocytic capacity (Sallusto *et al.*, 1995). In the T-cell areas of the lymph node, mature DCs presenting peptides complexed with MHC class II and class I molecules cluster, activate, and stimulate proliferation of CD4 and CD8 T cells. DCs then undergo further activation in lymphoid tissue through signaling by ligands of the TNF superfamily such as CD40L and TRANCE expressed on antigen-specific T cells. CD4[+] T-cell signaling induces secretion of high levels of IL-12p70 (Cella *et al.*, 1996; Snijders *et al.*, 1998) and promote DC survival (Wong *et al.*, 1997; Josien *et al.*, 2000), which is probably necessary for sustained CD4 T-cell proliferation and the development of Th1-mediated immunity.

11.2.1.3 Development

Myeloid DCs are derived from CD34[+] bone marrow stem cells under the influence of GM-CSF and TNF-α (Caux *et al.*, 1992; Reid *et al.*, 1992). *In vitro* studies have shown that in the presence of these cytokines, stem cells differentiate into two phenotypically distinct populations that are either CD1a[+] CD14[-] or CD1a[-] CD14[+]. The former differentiate into Langerhans cells and the latter into dermal DCs. Precursor cells destined to become Langerhans cells have been shown to express CLA, a molecule that facilitates homing to the skin (Strunk *et al.*, 1997). Skin Langerhans cells have a much longer life in tissue than other DC types and under noninflammatory conditions are replaced by precursors in the skin that have yet to be fully characterized (Merad *et al.*, 2002). There is an absolute requirement for TGF-β in the differentiation of Langerhans cells because these cells are absent in TGF-β knockout mice but other types of DCs are present (Borkowski *et al.*, 1996). Sufficient levels of TGF-β are probably present in FCS (Foetal calf serum) to support differentiation of Langerhans cells *in vitro*. As discussed above, dermal-type DCs can be derived from monocytes and, interestingly, addition of TGF-β to these cultures leads to the generation of cells with Birbeck granules, a marker uniquely associated with Langerhans cells (Geissmann *et al.*, 1998). Dendritic cells in blood are sometimes termed "pre-DCs" as they can be identified as belonging to the DC family based on the absence of lineage markers of other cell types and expression of MHC class II but do not express CD1a, Langerin, or DC-SIGN associated with immature tissue DCs. Recent studies of blood mDC differentiation are discussed in Section 11.3.

11.2.1.4 Function

Myeloid DCs initiate adaptive immunity by capturing pathogens and presenting their antigens to both CD4 and CD8 T cell (Banchereau *et al.*, 1998). However, only dermal-type DCs are able to secrete IL-10, cluster B cells, and mediate immunoglobulin class switching (Caux *et al.*, 1997; Dubois *et al.*, 1997; Fayette *et al.*, 1997; Saint-Vis *et al.*, 1998). Langerhans cells, on the other hand, are more potent stimulators of CD8 cytotoxic T cells (CTLs) than dermal or mdDCs and whereas on stimulation with CD40 ligand (CD40L) mdDCs produce large amounts of IL-12, a cytokine that promotes Th1 responses, Langerhans cells produce little or no Il-12 (Cella *et al.*, 1996; Ratzinger *et al.*, 2004). It is increasingly apparent that DCs display marked flexibiliy in their response depending on the nature, duration, timing, and concentration of stimulation (Langenkamp *et al.*, 2000), a feature that enables the DC to orchestrate the type of immune response that is generated.

The function of DCs depends on their maturational state with immature and mature DCs inducing tolerance and immunity, respectively (Dhodapkar *et al.*, 2001). Under noninflammatory conditions, the steady migration of immature tissue DCs to lymphoid tissue is considered to maintain peripheral tolerance (Hawiger *et al.*, 2001). Myeloid DCs also play a role in innate immunity through the production of IL-12, which induces interferon-γ production by NK cells (Yu *et al.*, 2001).

11.2.2 Plasmacytoid DCs

11.2.2.1 Location and Development

Plasmacytoid DCs are phenotypically and functionally distinct from mDCs. They were first observed by histologists in the T-cell areas of lymphoid tissue and termed either plasmacytoid T cells or plasmacytoid monocytes because of their well-developed endoplasmic reticulum reminiscent of plasma cells (Lennert and Remmele, 1958; Facchetti *et al.*, 1988). During the past 10 years, these cells have been shown to be derived from lineage-negative CD11c- blood cells that express CD4, MHC class II, and CD123, the receptor for IL-3, and to belong to the DC family. Generally, these cells do take up residence in peripheral tissues but migrate directly from the blood to secondary lymphoid tissue. There is an absence of pDCs from normal noninflamed tissues but they have been detected in the nasal mucosal tissue in experimentally induced human allergic rhinitis (Jahnsen *et al.*, 2000) as well as in the skin during certain inflammatory diseases (Wollenberg *et al.*, 2002). However, CD123 positive cells were not detected in DCs isolated from normal or inflamed intestinal tissue in Crohn disease patients (Bell *et al.*, 2001). Thus it is unclear to what extent pDCs migrate to inflamed tissue and is an area that warrants further investigation. Less is known about the development pathway of pDCs, but Flt-3 ligand has been shown to increase pDC numbers in human and is thought to play a key role on generation of pDCs (Blom *et al.*, 2000; Chen *et al.*, 2004). There is debate whether all DCs are derived from a common precursor or whether pDCs differentiate along a separate lymphoid pathway and is clearly an area requiring further studies (del Hoyo *et al.*, 2002; Corcoran *et al.*, 2003).

11.2.2.2 Maturation and Function

pDCs express the chemokine receptor CXCR3 that facilitates migration to inflamed lymph nodes in response to CXCL9 and CXCL10. Activation in the blood by microbial ligands induces the maturation of pDCs with the upregulation of CCR7 that mediates homing to secondary lymphoid tissue under the influence of CCL19 and CCL21. CD62L (L-selectin), which binds peripheral lymph node addressin on high endothelial venules, is constitutively expressed by pDCs and facilitates steady migration to lymph nodes under noninflammatory conditions (Nakano *et al.*, 2001).

A remarkable property of these cells is their ability to produce large amounts of interferon-α (IFN-α), 10- to 100-fold more than any other type of leukocyte, on stimulation with viruses or appropriate TLR ligands such as unmethylated CpG DNA through TLR9 (Liu, 2005). In addition to the direct role played by IFN-α in innate immunity to virus infections, this cytokine also influences adaptive immunity by promoting Th1 immunity (Parronchi *et al.*, 1996; Rogge *et al.*, 1997). Freshly isolated blood pDCs are not strong stimulators of allogenic T cells, but after culture with IL-3 and CD40L they become as potent stimulators of naïve allogeneic T cells as mdDCs (Grouard *et al.*, 1997). As immature pDCs are not located at sites where pathogens commonly enter the body, it is not clear whether they play a significant direct role in antigen presentation. That they could potentially contribute to adaptive immunity is supported by studies demonstrating that pDCs pulsed with either live or heat-inactivated (56°C) influenza virus can expand antigen-specific CD4 and CD8 responses (Fonteneau *et al.*, 2003). The efficiency of T-cell expansion mediated by pDCs is similar to that induced by CD11c[+] myeloid blood DCs. However, mDCs but not pDCs are able to present boiled virus, which has no hemagglutinin-mediated fusion capacity, to CD4 T cells and presumably reflects differences in endocytic properties of the two DC types. Interestingly, pDCs have recently been shown to have a role in the generation of HSV-specific CTLs (Yoneyama *et al.*, 2005). For more detailed descriptions of pDC biology, several recent reviews are available (Colonna *et al.*, 2004; Liu, 2005; McKenna *et al.*, 2005).

11.3 Practical Issues in DC Research

The low numbers of blood DCs and lack of specific markers has hampered human DC research for many years. In the HIV field there are conflicting reports, particularly the question whether these cells are targets for HIV infection, which are probably partly due to the difficulty of isolating relatively pure populations of cells. The pDC population has only recently been identified, and some of the early studies were undoubtedly performed with mixed populations of mDCs and pDCs (Patterson *et al.*, 1991). Different isolation procedures for blood DCs probably resulted in different cell populations and findings. A number of developments have helped resolve problems of conflicting data; the identification of the two blood DC populations is probably the most significant (Rissoan *et al.*, 1999; Robinson *et al.*, 1999). Early studies used morphologic criteria alone to identify and enumerate blood DCs (Macatonia *et al.*, 1990). Based on morphology, it can be difficult to differentiate between monocytes and DCs and this method fails to take into account possible changes in morphology due to infection. The commonly used current approach of identifying DCs, based on expression of MHC class II and absence of lineage markers and further differentiation into pDC and mDC populations using antibodies to CD123 and CD11c, has greatly improved reliability. However, there is scope for further improvements in methodology. Although most DC enumeration studies have estimated DC number after separation of mononuclear cells by centrifugation over ficoll, a whole blood assay should lead to greater consistency. Because mDCs and pDCs together only constitute about 1% of the total mononuclear population, consensus is needed regarding the number of cells that need to be acquired to gain an accurate estimate.

Two approaches have been employed to acquire purified human DC populations for *in vitro* studies of HIV-DC interactions, either direct isolation from blood or tissues or generation *in vitro* from precursors. Myeloid and plasmacytoid DCs can be isolated from blood by flow cytometry cell sorting using the same cocktail of antibodies used for identification. To obtain sufficient cells, it is necessary start with buffy coats derived from up to 500 ml of blood. For reasons of cost and speed, most protocols involve preliminary depletion of non-DC populations by density gradient separation and or immunomagnetic beads (Patterson *et al.*, 2001a). This approach has not been used to isolate DCs from HIV patients and probably reflects safety concerns and the low numbers of cells that can be isolated from the small volume of blood normally obtained from patients. These problems have been largely alleviated by the discovery of specific or semispecific BDCA (Blood dendritic cell antigen) markers for blood pDCs and mDCs (Dzionek *et al.*, 2000). Employing BDCA antibodies conjugated to magnetic beads, it is usually possible to isolate between 50×10^3 and 100×10^3 each of mDCs and pDCs from 50 ml of HIV patient blood and more than 10^6 from a buffy coat.

The study of human DCs was revolutionised by the discovery that large numbers of DCs could be generated from monocytes by culturing in GM-CSF and IL-4 for 5–7 days (Romani et al., 1994; Sallusto et al., 1994). This procedure gives rise to an immature population of DCs expressing CD11b, CD11c, MHC class II, CD1a, DC-SIGN, and low levels of the costimulatory molecules CD80 and CD86. Although mdDCs have greatly facilitated HIV studies, it is not clear how closely representative they are of tissue DCs. They most closely resemble dermal type DCs, however, studies comparing in vitro–generated and DCs isolated from the dermal tissues have found differences in the levels of expression of CD14, CD1a, and DC-SIGN (Turville et al., 2002). mdDCs also differ from DCs isolated from peripheral blood in terms of migratory capacity, ability to make cytokines (IL-12), and induction of effector T-cell subsets (Luft et al., 2002). DCs can also be generated from $CD34^+$ stem cells isolated from umbilical cord blood. Culture of these cells for 5 days in GM-CSF, TNF-α, SCF, and Flt-3 ligand generates two cell populations, a $CD1a^+ CD14^-$ and a $CD1a^- CD14^+$ population, which can be separated and further differentiated into Langerhans and dermal-type DCs, respectively (Caux et al., 1996; Rozis et al., 2005). Langerhans can also be isolated directly from skin biopsies after separating the epidermal and dermal by protease digestion (Richters et al., 2001). Langerhans cells migrate out of the epidermal layer during overnight culture, however, the main drawback of this procedure is that the isolated cells invariably mature during isolation.

11.4 Differentiation of Blood Myeloid DCs

The consensus view is that $CD11c^+$ mDCs are the precursors of tissue DCs, principally Langerhans cells and dermal/interstitial type DCs (Ito et al., 1999). To further investigate the differentiation of blood DCs, we have used BDCA-1 immunomagnetic beads to isolate mDCs from the blood of normal healthy controls and studied their differentiation and function during in vitro culture (Patterson et al., 2005). Freshly isolated BDCA-1 positive DCs are either negative for or express low to moderate levels of CD11b.

On culture with of GM-CSF and IL-4, they differentiate into major $CD11b^+$ and minor $CD11b^-$ populations (Fig. 11.1). After 24 h and subsequent time points, the $CD11b^-$ fraction express higher levels of MHC

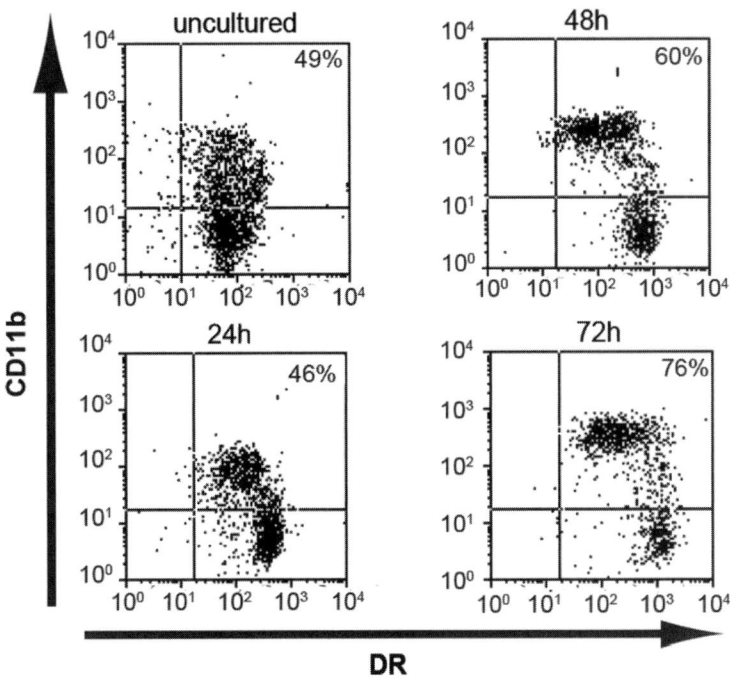

FIGURE 11.1. CD11c$^+$ blood mDCs differentiate into CD11b$^+$ and CD11b$^-$ populations on culture with GM-CSF and IL-4.

class II DR. Freshly isolated CD11b$^+$ and CD11b$^-$ DCs have a similar phenotype showing low levels of the costimulatory molecule CD86 and no or little expression of other maturation-associated markers including CD80, CD83, and CD40. In addition, they do not express DC-SIGN or CD1a, which are found on immature tissue DCs. After 48 h in culture, the major population of CD11b$^+$ cells express CD1a and DC-SIGN but express low levels or are negative for costimulatory molecules. This phenotype is similar to that of immature mdDCs generated *in vitro*. By contrast, the CD11b$^-$ DCs show a highly mature phenotype with high expression of MHC class II, CD80, CD83, and CD86 (Fig. 11.2). Cultured CD11b$^-$ DCs also express the chemokine lymph node homing receptor CCR7, consistent with their more mature phenotype, whereas this receptor is absent on the CD11b$^+$ DCs.

Several lines of evidence suggest that the two DC populations represent separate differentiation pathways. First the CD11b$^-$ cells, unlike the CD11b$^+$ population, never express CD1a or DC-SIGN. Second, if the two populations are separated by cell sorting and then cultured, the CD11b phenotype is maintained. Third, on stimulation with various maturation stimuli including LPS, poly I:C, and CD40L, the CD11b DCs mature and

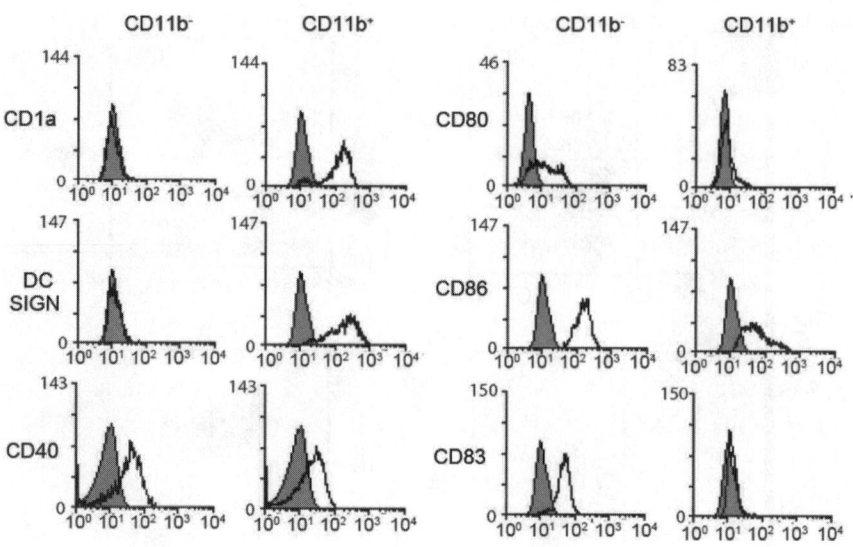

FIGURE 11.2. Phenotype of CD11b$^+$ and CD11b$^-$ blood mDCs after 48 h culture. The filled histograms represent labeling with an isotype control.

upregulate expression of MHC class II and costimulatory molecules but continue to express CD11b, CD1a, and DC-SIGN. That the two populations are *bona fide* DCs is confirmed by the fact that both efficiently stimulate T-cell proliferation. Stimulation of the CD11b$^-$ DCs induced no or minimal further phenotypic maturation suggesting that they are programmed to mature constitutively. We hypothesize that the CD11b$^+$ DCs are precursors of immature dermal/interstitial DCs based on the expression of DC-SIGN and CD1a, but the CD11b$^-$ cells cannot be readily placed into any known existing subpopulation of DCs in peripheral tissue. They do not develop into Langerhans cells even in the presence of TGF-β, a cytokine that is essential for Langerhans cell differentiation (Borkowski *et al.*, 1996; Caux *et al.*, 1999).

Intriguingly, two distinct CD11c positive DC populations have been described in human thymus, a minor CD11b$^+$ and a major CD11b$^-$ population (Vandenabeele *et al.*, 2001). The former is negative for CD83 while the latter expresses high levels of CD80, CD83, and CD86 and thus has a similar phenotype to the CD11b$^-$ DCs. This CD11b$^-$ DC population is found adjacent to Hassall's corpuscles in the thymic medulla. Hassall's corpuscles secrete thymic stromal lymphopoietin (TSLP), which activates thymic DCs to express high levels of CD80 and CD86. The TSLP-conditioned DCs promote the differentiation of CD4$^+$CD25$^+$ FoxP3$^+$

T-regulatory cells (Watanabe, 2005). Further studies are needed to determine if this DC subset expresses CD11b. It has been proposed that thymic DCs differentiate *in situ* rather than being seeded from the blood (Vandenabeele *et al.*, 2001), however, in view of the similarity between the blood and thymic CD11b⁻ populations, it is conceivable that thymic DCs may be derived from mDCs in the blood. Given the role of mature thymic DCs in the development of $CD4^+CD25^+FoxP3^+$ regulatory T cells, the myeloid CD11b⁻ DC subset may also stimulate the proliferation of this T-cell subset in peripheral blood. It seems unlikely that thymic CD11b⁻ DCs would depend on inflammatory or pathogenic signals to mature but may be programmed to develop directly into mature cells as appears to be the case with the CD11b⁻ blood DCs. Later in this chapter, the different-tiation of blood mDCs in HIV infection will be discussed and possible consequences explored.

11.5 DCs and Transmission of HIV

Heterosexual transmission accounts for the majority of HIV transmission events. The consensus view is that DCs play an important role in the early stages of infection but there is a lack of agreement on the precise details. Studies in macaques exposed to SIV intravaginally detected virus in DCs in the vaginal mucosa adjacent to the epithelial layer and in lymph nodes at early times after infection. (Spira *et al.*, 1996; Hu *et al.*, 2000). Based on these observations, it was proposed that Langerhans cells, the most superficially located DCs, become infected early after transmission and then transmit the virus to T cells after migrating to the lymph node. This hypothesis would explain the preferential transmission of R5 viruses as tissue Langerhans cells are reported to express CCR5 but not CXCR4 (Zaitseva *et al.*, 1997). The discovery of the C-type lectin DC-SIGN and its ability to mediate binding of virus to dermal DCs via the envelope glycoprotein generated a second hypothesis (Geijtenbeek *et al.*, 2000a; Geijtenbeek *et al.*, 2000b). Attachment of virus to DC-SIGN does not result in infection but to entry into a cytoplasmic compartment where it can remain in an infectious state for several days until migration to the lymph node where transmission to T cells occurs. There are two concerns with this hypothesis: the first is that DC-SIGN is not expressed by the superficially located Langerhans cells and the second is that DC-SIGN does not discriminate between binding of X4 and R5 viruses and thus cannot explain preferential transmission of the latter.

DCs undoubtedly facilitate transport of virus from the genital tract to the lymph node but may not be the only early targets in the genital tract. Analysis of cells obtained from the cervix with a cytobrush revealed that, unlike CD4 T cells in the blood, the majority of cervical CD4 T cells expressed activation markers including CD69, CD25, and MHC class II DR and thus are likely to be susceptible to productive virus infection (Stevenson *et al.*, 1990; Prakash *et al.*, 2001). This view is supported by studies in macaques infected intravaginally with SIV that found CD4 T cells in the intraepithelial compartment or lamina propria were major targets of infection at early time points (Zhang *et al.*, 1999). Thus early in infection there is local virus replication in the cervical mucosa, and DCs will be exposed either to transmitted or newly produced virus. Langerhans or dermal DCs could become productively infected and/or bind virus through lectin receptors and on migration to the lymph node transmit infection to T cells by an infectious or noninfectious route.

11.6 Loss of DCs in HIV Infection

11.6.1 Tissue Langerhans Cells

The earliest report of alterations in DCs in HIV infection appeared in 1984 when the loss of skin Langerhans cell in AIDS patients was described by Belsito and colleagues (Belsito *et al.*, 1984). However, these findings were not supported by a number of subsequent investigations that found no evidence for loss of skin Langerhans cells at any stage of disease (Kalter *et al.*, 1991; Compton *et al.*, 1996; Nandwani *et al.*, 1996). By contrast with skin Langerhans cells, in women with HIV there is a loss of Langerhans cells in cervical epithelium, which correlates with stage of disease development but is independent of infection with human papilloma viruses (Spinillo *et al.*, 1993; Rosini *et al.*, 1996; Levi *et al.*, 2005) Interestingly, a study of 29 AIDS patients also found a reduction in the number of Langerhans cells in the esophageal mucosa (Charton-Bain *et al.*, 1999). Loss of Langerhans cells at these sites may result in impaired responses to mucosal infections.

Differences in population maintenance pathways may explain why there is loss of mucosal but not skin Langerhans cells in HIV. Under noninflammatory conditions, there is a slow turnover of Langerhans cells in the skin with cells migrating to the lymph node being replaced by cells generated locally rather than by recruitment from the blood (Kamath *et al.*, 2002; Merad *et al.*, 2002). This contrasts with other tissues including intestinal mucosa and the dermis where DCs are replaced every few days by cells from the blood (Pugh *et al.*, 1983; Kamath *et al.*, 2002). In this context of Langerhans cell generation in the skin, it is interesting that *in vitro* CD34 stem cells can differentiate normally into Langerhans cells the presence of HIV (Canque *et al.*, 1996).

11.6.2 Blood DCs

The first evidence that DC numbers are reduced in HIV infection relied on enrichment by density gradient centrifugation followed by identification based on morphologic appearance in the light microscope (Macatonia *et al.*, 1990). Despite this rather unsophisticated approach and the fact that DC subpopulations were not differentiated, a depletion of DCs, particularly in AIDS patients, was described. Later work identified DCs by the absence of labeling for CD3, CD14, CD19, and CD56 and positive staining for MHC class II DR and confirmed the original findings by flow cytometry (Patterson *et al.*, 1996; Gompels *et al.*, 1998; Patterson *et al.*, 1998). Subsequently, Grassi *et al.* (1999) demonstrated that loss of DCs correlated with increasing virus load. It was shown by ELISPOT assays for IFN-α that there was a reduction in the number of IFN-α–producing cells in HIV infection (Howell *et al.*, 1994), however, it was not until several years later that these cells were shown to be DCs (Siegal *et al.*, 1999).

Further delineation of the extent of DC loss during HIV infection was possible after the identification and characterization of two distinct blood DC populations, and several groups have now reported alterations in these cells during HIV infection. Studies in chronically infected patients showed a severe depletion of both mDCs and pDCs in patients with AIDS (Donaghy *et al.*, 2001; Pacanowski *et al.*, 2001; Feldman *et al.*, 2001; Soumelis *et al.*, 2001; Chehimi *et al.*, 2002; Barron *et al.*, 2003). Most of these investigations were performed in Europe and North America where HIV clade B predominates but similar findings were made in a Ugandan cohort of patients where clades A and D are the major subtypes (Jones *et al.*, 2001). There are conflicting findings on the correlation between CD4 number and virus load and loss of DC subpopulations. Our studies on anti-retroviral drug naïve patients found a correlation between virus load and numbers of mDCs and pDCs (Fig. 11.3) (Donaghy *et al.*, 2001). A study that divided patients into two groups, more and less than 200 CD4 T cells per μl, found reduced numbers of pDCs and mDCs in both groups with the most severe reduction in patients with less than 200 CD4 T cells. However, this study did not find a close correlation between the number of either DC type and virus load or CD4 number (Feldman *et al.*, 2001). The

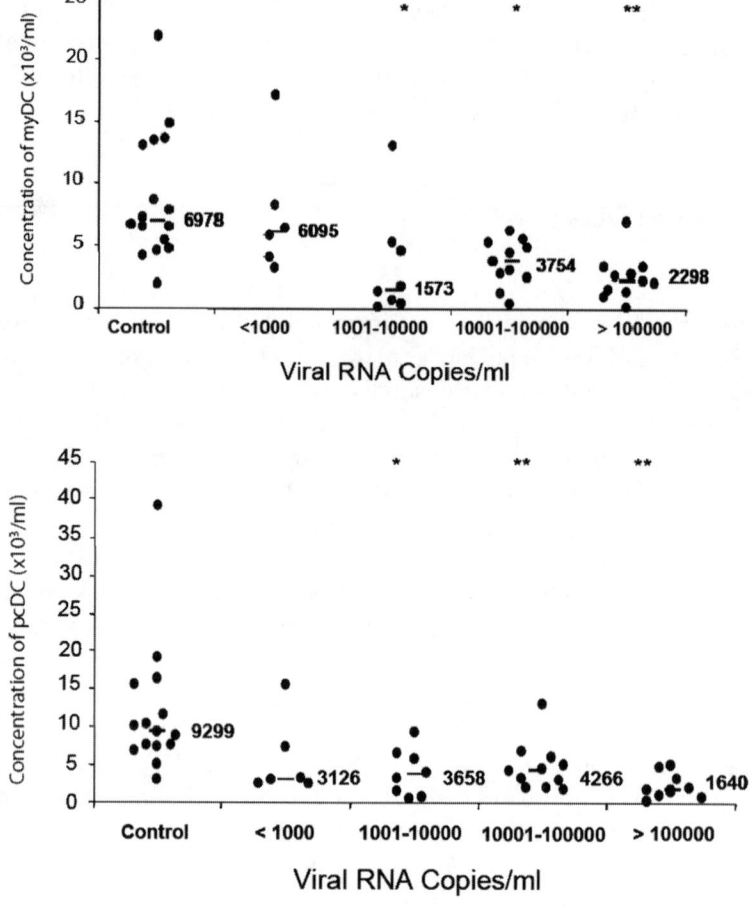

FIGURE 11.3. Loss of mDCs and pDCs in HIV-infected patients with increasing virus load.

percentage of pDCs able to produce IFN-α on stimulation with herpes simplex 1 is reduced in HIV infection suggesting functional impairment in addition to depletion (Feldman *et al.*, 2001). The investigation by Chehimi *et al.* (2002) described a correlation between virus load and depletion of mDCs whereas pDC loss did not correlate with virus load or CD4 number. A study by Soumelis *et al.* (2001) on pDCs observed reduced numbers in patients with AIDS. This latter study found that in long-term survivors, defined as individuals infected for more than 10 years with no clinical sign of disease, there were higher levels of pDCs than in healthy controls. This suggests that increased pDC numbers may help protect against disease development. Conversely, in patients with severely depleted pDCs, defined as below 2 per microliter of blood, the incidence of opportunistic infections and Kaposi sarcoma (KS) was increased. Interestingly, some patients with less than 2 pDCs per microliter and KS had relatively high CD4 counts whereas two patients with CD4 counts of less than 100 but with pDCs of more than 2 per microliter did not have AIDS-defining illnesses.

It has long been known that there is a defect in IFN-α production by PBMCs in HIV infection, both in symptomatic and asymptomatic patients (Lopez *et al.*, 1983; Ferbas *et al.*, 1995). It now seems clear that this may be explained by loss and possibly dysfunction of blood pDCs. In individuals acutely infected by HIV, a sharp reduction in mDC and pDC numbers was observed (Pacanowski *et al.*, 2001). Because all these patients were immediately treated with HAART, it is not known whether there is some recovery as patients progress to the chronic stage of disease in which there is at least partial control of virus. The underlying mechanisms of depletion may be different in acute infection as studies in SIV macaques have shown increase numbers of lymph node DCs in acute infection and a reduction in animals with AIDS (Zimmer *et al.*, 2002). The mechanism of blood DC loss is unknown but seems unlikely to be due to direct infection in blood because viral genome can be detected in very few blood DCs (see Section 11.7.3.1). Neither does it seem to be due to increased recruitment to the tissues or lymph nodes, as in patients who have had their virus load reduced to undetectable levels by HAART there is in most cases no return to normal DC numbers. Impaired generation of DCs from bone marrow precursors, increased rates of DC turnover, and activation-induced death are among the possible reasons for the loss but currently data to support these hypotheses are not yet available.

11.6.3 Anti-retroviral Drugs and DC Numbers

To date, there have been only a very limited number of studies on the effect of anti-retroviral therapy on blood DC number. However, it is clear that there are marked differences in response between individuals with only some patients showing improved DC numbers and then rarely returning to the levels seen in uninfected controls. In a study of 13 acutely infected individuals receiving therapy 26 to 57 days after infection, there was no increase in the mean number of either mDCs or pDCs after 6 to 12 months treatment (Pacanowski *et al.*, 2001). However, when individual patients were analyzed, two patients showed recovery of pDCs to normal levels while numbers decreased further in eight patients. Five patients showed an increase in mDCs, and in eight patients numbers were decreased. Interestingly, pDC but not mDC recovery correlated with virus load and increase in CD4 numbers. In a cross-sectional study of chronically infected patients, there was an increase in the mean number of mDCs and pDCs in patients responding well to HAART but levels were still significantly less than in controls (Barron *et al.*, 2003). HAART failures, on the other hand, showed little improvement in either cell population. In this study, five chronically infected patients were analyzed before and after 6 months into therapy. Four showed increased numbers of pDCs, but mDCs only improved in two. A second study of 26 chronically ill individuals in whom HAART reduced virus loads to less than 50 copies per milliliter found recovery of the mDC population (Chehimi *et al.*, 2002). By contrast, normal pDC numbers were only seen in six patients and despite recovery of the pDCs, IFN-α production by PBMCs remained low in these patients. It is difficult to reconcile these findings with a study of 294 HAART patients showing recovery of IFN-α production to levels sufficient to confer protection against opportunistic infections (Siegal *et al.*, 1999). Resolving these conflicting findings will require further cross-sectional and longitudinal studies to correlate DC numbers, function, and clinical outcome.

11.7 Infection of DCs by HIV

11.7.1 *In Vitro* Studies

The question of the susceptibility of DCs to infection by HIV and the role this might play in the pathogenesis of the disease has been controversial with the publication of conflicting data. The reasons for this are no doubt due, at least in part, to the difficulty in isolating pure populations of cells and the unwitting use of preparations comprising heterogeneous populations of DCs. Different isolation procedures employed in different laboratories may have preferentially selected different populations or induced variable levels of differentiation and maturation. The current consensus is that DCs can be infected by HIV but there remains lack of agreement on the role this plays in virus transmission and the pathogenesis of disease development.

11.7.1.1 HIV Receptor Expression on DCs

Early studies on blood DCs, which probably comprised mixed populations of mDCs and pDCs, demonstrated expression of the primary HIV receptor, CD4 (O'Doherty *et al.*, 1993; Ferbas *et al.*, 1994; Patterson *et al.*, 1995), and in subsequent studies mRNA for the chemokine coreceptors CCR5 and CXCR4 (Ayehunie *et al.*, 1997). Uncultured pDCs and mDCs both express CD4 but expression levels are slightly higher in pDCs (Fig. 11.4). By flow

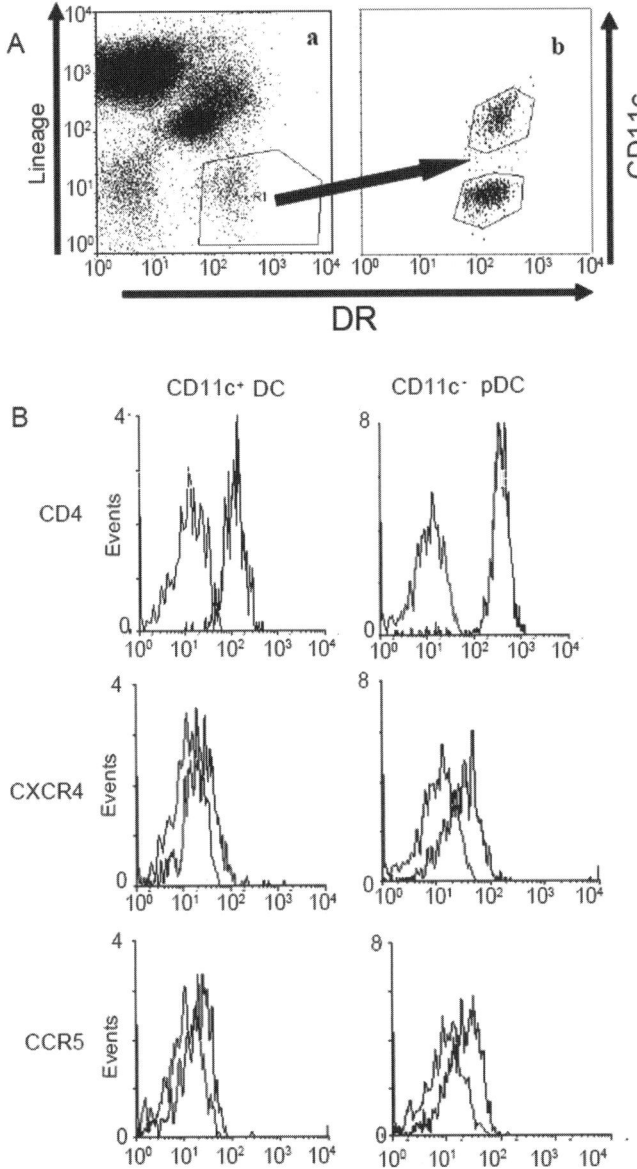

FIGURE 11.4. (A) Blood DCs can be differentiated by the absence of labeling with a cocktail of lineage markers and expression of HLA DR and further divided into CD11c⁺ myeloid and CD11⁻ plasmacytoid populations. (B) Expression of CD4, CXCR4, and CCR5 on CD11c⁺ myeloid and CD11c⁻ plasmacytoid DCs. The first histogram in each plot represents labeling with an isotype control antibody.

cytometry, low levels of CCR5 and CXR4, confirmed by RT-PCR, were demonstrated on mDCs whereas higher levels of these receptors are present on pDCs (Patterson *et al.*, 2001b; Fong *et al.*, 2002; Yonezawa *et al.*, 2003). mddDCs express CD4, CCR5, and CXCR4 (Granelli Piperno *et al.*, 1996; Bakri *et al.*, 2001) and the C-type lectin receptor, DC-SIGN, which has been shown to bind the HIV envelope glycoprotein (Geijtenbeek *et al.*, 2000a; Geijtenbeek *et al.*, 2000b). DC-SIGN messenger RNA can be detected in freshly isolated blood myeloid DCs but the receptor is not expressed on the cell surface until they are cultured (Patterson *et al.*, 2001b; Patterson *et al.*, 2005). Several independent reports have found no evidence for expression of DC SIGN on pDCs (Patterson *et al.*, 2001b; Baribaud *et al.*, 2002; Fong *et al.*, 2002; Yonezawa *et al.*, 2003).

Langerhans cells express CD4 and are the most superficially located population of DCs in the genital tract lying adjacent to epithelial cells. Thus they have been proposed to be among the first cells to be encountered and infected by the virus (Braathen, 1987; Niedecken *et al.*, 1987; Miller *et al.*, 1994). Studies on freshly isolated skin Langerhans cells have shown that they express CCR5 on the cell membrane whereas CXCR4 is present on the membrane of intracellular vacuoles and only appears on the cell surface after culture (Zaitseva *et al.*, 1997). These findings may explain the preferential transmission of CCR5-utilizing macrophage tropic (R5) rather than lymphotropic virus (X4) that requires CXCR4 for cellular entry. However, samples obtained with a cervical brush detected both CCR5 and CXCR4 on the surface of CD1a positive cells suggesting that Langerhans cells in the genital tract may not precisely mirror those in the skin (Prakash *et al.*, 2004). Langerhans cells do not express DC-SIGN thus questioning the role of this molecule in early transmission events (Geijtenbeek *et al.*,2000b). C-type lectins such as mannose receptor (CD206) and Langerin (CD207) are expressed by Langerhans cells and may be involved in initial attachment HIV-1 gp120 to this cell population (Turville *et al.*, 2003).

11.7.2 *In Vitro* Infection Studies

11.7.2.1 Blood Plasmacytoid DCs

Electron microscopy (EM) studies demonstrated virus budding from blood DCs infected *in vitro* with an X4 strain of HIV (Patterson and Knight, 1987). Further EM studies on the enriched populations of DCs used for these experiments identified two morphologically distinct populations of DCs that were termed type 1 and type 2 DCs (Patterson *et al.*, 1991). With the discovery and characterization of blood mDCs and pDCs, it became clear that the type 1 and type 2 DCs corresponded with mDCs and pDCs (Patterson *et al.*, 1999; Robinson *et al.*, 1999). Productive infection by X4 virus was only observed in DCs corresponding with pDCs. In subsequent studies, highly purified pDCs were isolated by cell sorting and exposed to R5 and X4 strains of virus (Patterson *et al.*, 2001b). Analysis of infected cells by limiting dilution nested PCR showed replication of both R5 and X4 strains of virus although higher levels of provirus are seen in cells infected with R5 virus (Fig. 11.5). Later studies confirmed productive infection of pDCs by R5 and X4 viruses (Fong *et al.*, 2002; Yonezawa *et al.*, 2003; Schmidt *et al.*, 2004) and also demonstrated integration of the proviral DNA into the pDC genome. Productive infection of pDCs in one study was found to be dependent on maturation by CD40L (Fong *et al.*, 2002). It is not clear whether this reflects differences in the degree of maturation induced during isolation or whether other factors account for the discrepancies. Infectious and inactivated HIV induce pDC maturation and secretion of IFN-α (Ferbas *et al.*, 1994). It is thus perhaps surprising that the virus is able to replicate in cells that produce copious amounts of IFN-α. The presence of this cytokine does mediate suppressive effects as levels of virus are increased when pDCs are infected in the presence of IFN-α neutralizing antibodies (Fong *et al.*, 2002; Yonezawa *et al.*, 2003). Infected pDCs also secrete RANTES/CCL5 and MIP-1α/CCL3, which may reduce transmission of virus from infected pDCs.

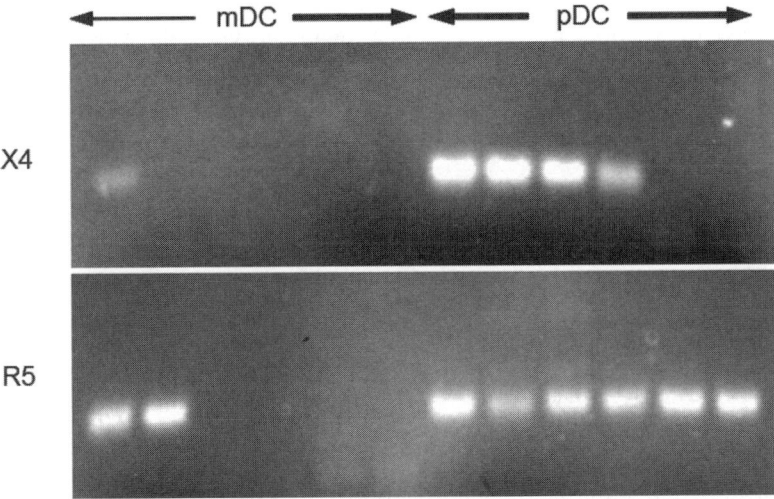

FIGURE 11.5. Blood mDCs and pDCs infected for 9 days with X4 and R5 virus. Provirus DNA estimated by nested PCR on fivefold dilutions of DNA; the highest concentration of DNA for each assay is equivalent to 5000 cells.

11.7.2.2 Blood Myeloid DCs

Although low levels of CXCR4 and CCR5 are expressed on mDCs, most studies have observed preferential replication of R5 and no or low-level infection by X4 viruses (Patterson *et al.*, 2001b; Yonezawa *et al.*, 2003; Ganesh *et al.*, 2004). Despite low levels of infection, mDCs very efficiently transmit virus to T cells in cocultures (Cameron *et al.*, 1992). In one series of experiments, conjugates of skin-derived DCs and T cells in the absence of added exogenous T-cell stimuli supported high-level productive virus replication whereas none was detected in DCs or T cells alone. However, in further experiments it was shown that virus was not transmitted by AZT-treated DCs indicating that at least some level of replication in DCs is required (Pope *et al.*, 1994; Pope *et al.*, 1995). On maturation, blood mDCs become more resistant to R5 virus infection but are enhanced in their ability to transmit virus to CD4 T cells, presumably reflecting stronger stimulation of T cells through higher level expression of MHC, adhesion and costimulatory molecules on the DCs. Transmission to T cells is not blocked by virus neutralizing antibody suggesting that virus transfer occurs through close contacts that develop between DC and T cell that exclude antibody (Ganesh *et al.*, 2004). The T-cell membrane in these contact zones contains higher levels of CD4 and chemokine receptors, which probably increase the efficiency of transmission and has been termed an infectious synapase (McDonald *et al.*, 2003). These latter studies employing mdDCs showed high concentrations of virus-containing vacuoles in close proximity to the synapse. The CD11b negative mDCs described in Section 11.4 have a mature phenotype and thus would be predicted to efficiently transmit virus to T cells although may be less likely to become infected. Further elucidation of the susceptibility of these cells to virus infection and their role in dissemination of HIV is required.

11.7.2.3 Monocyte-derived DCs

These cells are considered models for immature dermal-type DCs and are phenotypically similar to the CD11b$^+$ cells that differentiate from blood myeloid DCs. Some investigators propose they are resistant to HIV infection but bind and transmit virus to T cells whereas others report that they can be productively infected by HIV. The true picture probably includes elements of both hypotheses. A number of studies have demonstrated that mdDCs are susceptible to infection but produce low levels of virus that in some cases is only detected in coculture with T cells (Tsunetsuguyokota *et al.*, 1995; Granelli-Piperno *et al.*, 1998; Popov *et al.*, 2005). As with blood mDCs, there is preferential replication of R5 virus, however, immature mdDCs express CXCR4, albeit at a lower level than CCR5, and permit entry of X4 virus as indicated by the presence of early reverse-transcribed viral products (Bakri *et al.*, 2001). When the ratio of early to late (after the second template switch) reverse transcript products were compared in X4 and R5 infected immature mdDCs, 8-50 fold more reactions proceeded to completion in R5 infected cells. These findings correlate with differences in the levels of the two viruses released into the culture supernatants. Mature mdDCs show reduced ability to produce infectious virus. Analysis of infected mature mdDCs indicated that infection up to the stage of DNA integration is similar in mature and immature DCs but there is a marked reduction in transcription from the viral LTR (Long Terminal Repeat) in mature DCs (Bakri *et al.*, 2001). Interestingly, ultrastructural studies of infected mature mdDCs have shown that virus particles accumulate in deep perinuclear vacuoles and may represent a reservoir of virus that can be transmitted to T cells. By contrast, fewer intact particles were seen in immature mdDCs and were present in more peripherally located vesicles (Frank *et al.*, 2002).

One mechanism of virus transfer to T cells is through binding to a DC-specific C type lectin, DC-SIGN, that recognizes ICAM-2 and ICAM-3. Binding of virus to DC-SIGN occurs through the highly glycoslysated envelope protein gp120 but does not mediate infection (Geijtenbeek *et al.*, 2000a). Virus bound to DC-SIGN is internalized and can persist for several days without being degraded before initiating T-cell infection. Langerhans cells are negative for DC-SIGN but express another lectin receptor, langerin, which may play a similar role in virus transmission. *In vivo* dermal DCs were found to express DC-SIGN at lower levels than that found on DCs generated *in vitro* and thus the efficiency of virus transmission by this pathway may be lower than at first thought (Turville *et al.*, 2002). Depending on the nature of the pathogen encountered, DCs can promote Th1 or Th2 type responses (Kapsenberg, 2003). DCs primed to stimulate Th1 immunity are more efficient in transmitting HIV to T cells. This correlates with upregulation of ICAM-1, which would increase cell-cell contact (Sanders *et al.*, 2002). HIV-coded proteins may also increase the efficiency of transmission from DC to T cells. Nef augments DC–T cell clustering by upregulation of DC-SIGN (Sol-Foulon *et al.*, 2002). Tat can improve the efficiency of virus transmission to T cells and monocyte/macrophages by DCs through upregulating interferon-inducible protein 10 (IP10) and human monokine induced by interferon γ (HuMIG), two molecules that attract T cells via binding to CXCR3 and monocyte/macrophages by secretion of monocyte chemotactic proteins 2 and 3 (MCP-2, MCP-3) (Izmailova *et al.*, 2003).

11.7.2.4 Langerhans Cells

Because of their superficial location at sites where HIV enters the body, there is considerable interest in HIV infection of Langerhans cells. *In vitro* studies have either generated Langerhans cells from CD34 precursors or isolated them directly from skin. CD34 stem cells cultured with GM-CSF and TNF-α give rise to both dermal type DCs and Langerhans cells (Caux *et al.*, 1996), and as most studies have not separated these two populations, it is not always clear whether published findings are strictly restricted to Langerhans cells. Langerhans cells can be isolated from skin by separating the epidermal layer from the dermis by protease treatment and then collecting the cells as they migrate out of the epidermal layer during overnight culture. Although preparations free of dermal DCs can be acquired, the cells leaving the tissue are invariably mature and show upregulation of CXCR4. Thus findings may not necessarily fully mirror *in vivo* events. Bearing this in mind, several studies have shown productive infection of Langerhans cells with R5 and X4 viruses (Canque *et al.*, 1996; Blauvelt *et al.*, 1997; Sivard *et al.*, 2004). Infection of Langerhans cells by R5 and X4 virus is inhibited by the CCR5 and CXCR4 ligands, RANTES and SDF1, respectively. As with mDCs, infected Langerhans cells can cells can transmit virus to T cells (Ludewig *et al.*, 1996).

11.7.3 DC Infection *In Vivo*

11.7.3.1 Blood DCs

In situ hybridization studies detected HIV DNA in a population of lineage-negative PBMCs isolated from infected patients (Macatonia *et al.*, 1990). Viral DNA was also detected in DCs purified from the spleen of infected patients but at a low frequency, 1 in 3000 to 18,000 DCs (McIlroy *et al.*, 1995). With the development of antibody reagents specific for blood mDCs and pDCs, it became possible to purify sufficient DCs from 50 ml of blood and look for evidence of infection. By limiting dilution nested PCR using pol primers, infected mDCs from 13 of 14 HAART naïve patients were detected at frequencies ranging from 1 in 40 to 1 in 5000, and integrated provirus was detected in three samples (Donaghy *et al.*, 2003). The same study found HIV DNA in pDCs from 12 patients at frequencies of 1 in 200 to 1 in 5000 but no integrated virus was detected. However, when pDCs were cultured with allogeneic T cells, viral p24 was detected in 2 of 32 cocultures. Given that both DC populations express CD4, CCR5, and CXCR4 and that 10 of the patients had virus loads greater than 50,000 copies/ml, higher levels of infection and of integrated virus might have been expected. The explanation may be that whereas T cells recirculate, blood DCs are in transit from the blood to the tissues or lymphoid organs thus there would not be progressive accumulation of infected DCs as probably occurs with CD4 T cells. In addition, binding of virus to pDC induces IFN-α secretion and may also stimulate migration to the lymph node (Ferbas *et al.*, 1994). Future studies need to determine whether the virus induces upregulation of CCR7. That pDCs are targets *in vivo* was confirmed by the finding of p24 positive CD123 positive cells in tonsils from HIV-infected individual (Fong *et al.*, 2002).

11.7.3.2 Langerhans Cells

Some studies report that infection of skin Langerhans cells by HIV is rare (Kalter *et al.*, 1991) but the majority find that these cells are infected although there is marked variation in the level of infection reported in different investigations (Tschachler *et al.*, 1987; Rappersberger *et al.*, 1988; Zambruno *et al.*, 1991; Giannetti *et al.*, 1993; Cimarelli *et al.*, 1994; Henry *et al.*, 1994; Compton *et al.*, 1996). An immunohistochemical study of two patients detected p24 gag in 30–50% of skin Langerhans cells (Compton *et al.*, 1996) whereas other studies found 0.1–3.5% to be infected. In three independent studies, the number of patients with infected Langerhans cells was 7 of 40, 7 of 9, and 6 of 9 (Tschachler *et al.*, 1987; Zambruno *et al.*, 1991; Giannetti *et al.*, 1993). The finding of mRNA for regulatory and structural genes in highly purified skin Langerhans cells from infected individuals (Henry *et al.*, 1994; Compton *et al.*, 1996) together with EM studies showing virus particle associated with these cells (Tschachler *et al.*, 1987; Rappersberger *et al.*, 1988) is strong evidence that they are susceptible to productive infection *in vivo*.

11.8 Dysfunction of DCs in HIV Infection

DCs mediate both adaptive and innate immune responses and thus reduction in their numbers would impact on both arms of the immune response. Here we examine data indicating that the situation may be further exacerbated by impaired or altered function by the DCs that survive.

11.8.1 Blood Myeloid DCs

There have been relatively few studies analyzing the function of blood DCs in HIV, probably reflecting the difficulty in obtaining sufficient numbers of pure cells. Cell preparations from HIV-infected patients enriched for blood DCs, which would have contained both mDC and pDC populations, were found in early studies to be impaired in their ability to stimulate allogeneic and antigen-specific (influenza and tetanus toxoid) T-cell proliferation in asymptomatic and AIDS patients (Macatonia *et al.*, 1990; Macatonia *et al.*, 1992). More recently, highly purified preparations of blood mDCs isolated from HAART naïve patients were also found to be severely impaired in their ability to stimulate allogeneic T-cell proliferation (Donaghy *et al.*, 2003). The observed defect was not due to secondary infection of responding CD4 T cells as viral p24 was not detected in the majority of MLR (Mixed Leucocyte Reaction) supernatants. There are a number of possible explanations for the limited capacity of DCs to stimulate allogeneic T-cell responses including failure of DC differentiation (see below), enhanced or accelerated death of DCs in allogeneic T-cell cultures, and stimulation of T-cell subsets such as $CD4^+CD25^+$ regulatory T cells that may inhibit allogeneic T-cell responses.

As a possible explanation for impaired T stimulation capacity, blood mDCs were analyzed for their ability to differentiate in 48-h cultures into phenotypically distinct $CD11b^+$ and $CD11b^-$ populations as described in Section 11.4. For the majority of patients (14 of 19), there was a failure to develop into major $CD11b^+$, $CD1a^+$ and minor $CD11b^-$, $CD83^+$ populations (Fig. 11.6) and in addition in most of the cultures there was poor DC survival. These observations explain, at least in part, the impaired ability of these DCs to stimulate T-cell proliferation. However, even in three patients showing normal differentiation and reasonable viability after 48 h (54–58%) there was still substantial impairment of T-cell stimulation. IL-12 enhances T-cell proliferation (Perussia *et al.*, 1992) and IL-12 production is reduced in *Staphyloccus aureus* or CD40L stimulated PBMCs from HIV-infected patients (Chehimi *et al.*, 1994; Chougnet *et al.*, 1998). Thus reduced production of IL-12 may partly explain the impaired T-cell stimulating capacity of blood mDCs. However, as monocytes also secrete IL-12 (Dandrea *et al.*, 1992), studies on purified blood DCs are required to test this hypothesis.

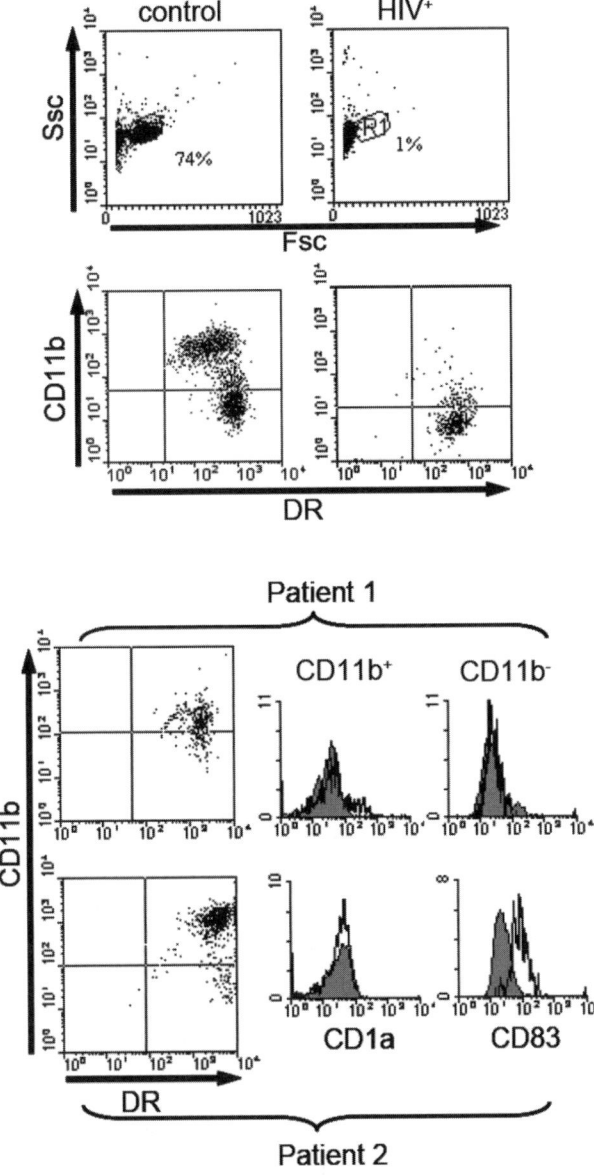

FIGURE 11.6. (A) Control and HIV blood mDCs cultured for 48 h. mDCs from HIV-infected individual show poor survival and impaired ability to differentiate into CD11b⁺ and CD11b⁻ populations. (B) mDCs from two HIV-infected patients showing limited differentiation into CD11b⁺ and CD11b⁻ populations. Cells from patient 1 fail to express CD1a or CD83, whereas CD11⁻ DCs in patient 2 do express CD83 but the CD11b⁺ DCs fail to express CD1a.

11.8.2 Monocyte-derived DCs

DCs can be derived from monocytes of HIV-infected patients and by contrast with freshly isolated blood mDCs show normal phenotype and capacity to stimulate T-cell proliferation (Fan *et al.*, 1997; Sapp *et al.*, 1999). These cells are not infected with HIV and can be generated from patients at all stages of disease, although there is some reduction in yield that correlates with CD4 T-cell numbers and virus load. Interestingly, DCs derived from monocytes at early or late stages of disease are able to stimulate virus-specific CD8 responses when transfected with vaccinia vectors coding for HIV proteins or after pulsing with HIV peptides. These observations have been translated into "proof of principle" experiments to test the efficacy of mdDCs as a therapeutic vaccine (Lu *et al.*, 2003). SIV-infected macaques were vaccinated with mdDCs that were matured after pulsing with chemically inactivated SIV. After three vaccinations, SIV-specific IFN-γ and CTL cell responses were augmented, CD4 T-cell counts were increased, and peripheral blood virus load was reduced by 1000-fold and maintained at this level for 34 weeks. This strategy was then tested in 18 HIV-infected patients using inactivated autologous HIV-1 (Lu *et al.*, 2004). Suppression of virus load by 90% for more than 12 months after vaccination was seen in eight patients and correlated with increased gag-specific effector CD8 cells and CD4 Th1 cells.

The mdDC vaccine trials are highly encouraging but raise a number of questions. The first is to what extent does differentiation of monocytes into DCs *in vitro* reflect *in vivo* events? If the *in vitro* model parallels events *in vivo,* is there a failure in the monocyte-DC differentiation pathway in HIV-infected patients? As functional DC can be generated from monocytes of infected individuals, it is possible that they become infected during differentiation, alternatively the cytokine environment in infected individuals may not be conducive to DC differentiation. HIV-specific CD8 cytotoxic T cells control but do not eradicate virus during the chronic stages of the disease (Rowland-Jones *et al.*, 1997). The continued replication of the virus, albeit partially controlled, may reflect impaired development of CD8 cells to a terminally differentiated phenotype with maximal cytolytic activity (Champagne *et al.*, 2001). It would be interesting to know whether the success of the DC vaccination studies is due to the generation of terminally differentiated CD8 cells. In HIV infection there is preferential infection of virus-specific CD4 T cells (Douek *et al.*, 2002), which probably reflects the close interactions that occur between infected DCs and antigen-specific CD4 T cells during antigen presentation. How then do HIV-specific CD4 T cells escape infection as they are activated by the injected DCs? Naïve CD4 T cells require contact with antigen on a DC for 6 h to stimulate cell division (Lezzi *et al.*, 1998) whereas naïve CD8 cells require a minimum of 2 h contact (van Stipdonk *et al.*, 2001). This period may be too short for autologous virus to infect the DC and disrupt stimulation of effector T cells. Patients and macaques were injected with mature mdDCs, which are less susceptible to productive HIV infection and may partly explain the success of this vaccination strategy (Bakri *et al.*, 2001; Granelli-Piperno *et al.*, 2004). The success of targeting DCs for therapeutic vaccination suggests that conventional prophylactic vaccines may be improved by strategies to target a vaccine to DCs *in vivo*. DC presentation may be more effective with viral peptides/ proteins non-replication-competent virus than replication-competent virus as the latter may directly kill HIV-1–specific CD4 T cells leading to loss of CD4 T-cell responses and impairment and loss of memory CD8 anti-HIV immune responses.

Functional defects are observed in mdDCs infected with HIV *in vitro*. Their ability to stimulate allogeneic T-cell proliferation is reduced but they have normal levels of MHC class II and costimulatory molecule expression and increase production of IL-12 (Kawamura *et al.*, 2003). Binding of HIV gp120 to CD4 on T cells inhibits signaling and proliferation (Diamond *et al.*, 1988; Cefai *et al.*, 1990; Oyaizu *et al.*, 1990) and could account for impaired stimulation by infected DCs as soluble CD4 alleviated suppression. Although normal levels of IL-12 production are seen in HIV-infected DC cultures stimulated with CD40L, intracellular labeling showed that only p24 gag negative but not p24 positive DCs produced IL-12 (Smed-Sorensen *et al.*, 2004).

11.8.3 Langerhans Cells

In mice infected with retrovirus Rauscher leukemia virus, there is impaired migration of skin Langerhans cells to the draining lymph node after skin painting with a contact sensitizer and is considered to be a contributory factor to the immunosuppressive properties of this virus (Gabrilovich *et al.*, 1994). Similarly, in late-stage infection of macaques with SIV there is impaired migration of Langerhans cells to lymph nodes (Zimmer *et al.*, 2002). In an elegant series of experiments by Blauvelt and colleagues on twins that were discordant for HIV infection, the functional properties of Langerhans cells were studied (Blauvelt *et al.*, 1995; Blauvelt *et al.*, 1996). At different stages of disease, Langerhans cells from infected twins pulsed with recall antigens were able to stimulate T cells from the uninfected twin but were unable to stimulate proliferation of allogeneic T cells. These findings suggest normal stimulation of memory responses but defective stimulation of naïve T cells.

11.8.4 Plasmacytoid DCs

As with blood mDCs there are relatively few studies on the function of pDCs in HIV infection. It has long been known that there is a defect in IFN-α production by PBMCs in HIV infection, both in symptomatic and asymptomatic patients (Lopez *et al.*, 1983; Ferbas *et al.*, 1995), which can now be explained by loss and possibly dysfunction of blood pDCs. The percentage of pDCs able to produce IFN-α on stimulation with virus was reduced in HIV infection suggesting functional impairment in addition to depletion (Feldman *et al.*, 2001). The ability of pDCs from HIV-infected individuals to stimulate allogeneic T-cell proliferation is suppressed (Donaghy *et al.*, 2003). The mechanism of suppression is unclear but, as with mDCs, is unlikely to be a direct effect of infection as less than 0.5% of cells are infected. Neither is it due to secondary infection of responding lymphocytes as p24 gag antigen is rarely detected in the supernatants of MLR cultures.

11.9 Concluding Remarks

DCs play a role at all stages of HIV pathogenesis from transmission of the virus through to the development of AIDS. In the genital tract, DCs become infected and bind the virus via lectin receptors and then transport the virus to the draining lymph node where efficient transmission to T cells occurs. DC–T cell clusters support high-level virus replication and the subsequent loss of pDCs leads to a reduction in innate immunity. The loss and dysfunction of mDCs would contribute to reduced IL-12 responses, affect NK function, and cause polarization toward Th2 responses. Alterations in the mDC population may also reduce the capacity to generate CTLs and may have some bearing on the impaired development of terminally differentiated CD8 cells. Assessment of DC numbers and function could facilitate better monitoring of the immune status of patients. This may be particularly valuable in HAART patients undergoing structured treatment interruption where there is recent evidence of a correlation between pDC numbers after treatment and control of virus after treatment cessation (Pacanowski *et al.*, 2004). Elucidating the relative contributions of DC-mediated heterosexual transmission of HIV by infectious and noninfectious routes will help in the rationale design of virucides. Our continued efforts to understand the role of DCs in HIV pathogenesis will hopefully lead to better patient care and may indicate new strategies for disease prevention.

References

Ayehunie, S., GarciaZepeda, E.A., Hoxie, J.A., Horuk, R., Kupper, T.S., Luster, A.D., and Ruprecht, R.M. (1997). Human immunodeficiency virus-1 entry into purified blood dendritic cells through CC and CXC chemokine coreceptors. Blood 90:1379-1386.

Bakri, Y., Schiffer, C., Zennou, V., Charneau, P., Kahn, E., Benjouad, A., Gluckman, J.C., and Canque, B. (2001). The maturation of dendritic cells results in postintegration inhibition of HIV-1 replication. J. Immunol. 166:3780-3788.

Banchereau, J., Briere, F., Caux, C., Davoust, J., Lebecque, S., Liu, Y.T., Pulendran, B., and Palucka, K. (2000). Immunobiology of dendritic cells. Annu. Rev. Immunol. 18:767-811.

Banchereau, J. and Steinman, R.M. (1998). Dendritic cells and the control of immunity. Nature 392:245-252.

Baribaud, F., Pohlmann, S., Leslie, G., Mortari, F., and Doms, R.W. (2002). Quantitative expression and virus transmission analysis of DC-SIGN on monocyte-derived dendritic cells. J. Virol. 76:9135-9142.

Barron, M.A., Blyveis, N., Palmer, B.E., MaWhinney, S., and Wilson, C.C. (2003). Influence of plasma viremia on defects in number and immunophenotype of blood dendritic cell subsets in human immunodeficiency virus 1-infected individuals. J. Infect. Dis. 187:26-37.

Bell, S.J., Rigby, R., English, N., Mann, S.D., Knight, S.C., Kamm, M.A., and Stagg, A.J. (2001). Migration and maturation of human colonic dendritic cells. J. Immunol. 166:4958-4967.

Belsito, D.V., Sanchez, M.R., Baer, R.L., Valentine, F., and Thorbecke, G.J. (1984). Reduced Langerhans cell Ia-antigen and ATPase activity in patients with the acquired immunodeficiency syndrome. N. Engl. J. Med. 310: 1279-1282.

Blauvelt, A., Asada, H., Saville, M.W., KlausKovtun, V., Altman, D.J., Yarchoan, R., and Katz, S.I. (1997). Productive infection of dendritic cells by HIV-1 and their ability to capture virus are mediated through separate pathways. J. Clin. Invest. 100:2043-2053.

Blauvelt, A., Clerici, M., Lucey, D.R., Steinberg, S.M., Yarchoan, R., Walker, R., Shearer, G.M., and Katz, S.I. (1995). Functional-studies of epidermal langerhans cells and blood monocytes in HIV-infected persons. J. Immunol. 154:3506-3515.

Blauvelt, A., Chougnet, C., Shearer, G.M., and Katz, S.I. (1996). Modulation of T cell responses to recall antigens presented by langerhans cells in HIV-discordant identical twins by anti- interleukin (IL)-10 antibodies and IL-12. J. Clin. Invest. 97:1550-1555.

Blom, B., Ho, S., Antonenko, S., and Liu, Y.J. (2000). Generation of interferon alpha-producing predendritic cell (pre-DC)2 from human CD34($^+$) hematopoietic stem cells. J. Exp. Med. 192:1785-1795.

Borkowski, T.A., Letterio, J.J., Farr, A.G., and Udey, M.C. (1996). A role for endogenous transforming growth factor beta 1 in Langerhans cell biology:

The skin of transforming growth factor beta 1 null mice is devoid of epidermal Langerhans cells. J. Exp. Med. 184:2417-2422.

Braathen, L.R. (1987). Langerhans cells as primary target cells for HIV infection. Lancet 2:1094.

Cameron, P.U., Freudenthal, P.S., Barker, J.M., Gezelter, S., Inaba, K., and Steinman, R.M. (1992). Dendritic cells exposed to human-immunodeficiency-virus type-1 transmit a vigorous cytopathic infection to CD4+ T-cells. Science 257:383-386.

Canque, B., Rosenzwajg, M., Camus, S., Yagello, M., Bonnet, M.L., Guigon, M., and Gluckman, J.C. (1996). The effect of *in vitro* human immunodeficiency virus infection on dendritic-cell differentiation and function. Blood 88: 4215-4228.

Caux, C., Dezutter Dambuyant, C., Schmitt, D., and Banchereau, J. (1992). GM-CSF and TNF-Alpha cooperate in the generation of dendritic Langerhans cells. Nature 360:258-261.

Caux, C., Vanbervliet, B., Massacrier, C., Dezutter Dambuyant, C., deSaintVis, B., Jacquet, C., Yoneda, K., Imamura, S., Schmitt, D., and Banchereau, J. (1996). CD34(+) hematopoietic progenitors from human cord blood differentiate along two independent dendritic cell pathways in response to GM-CSF+TNF alpha. J. Exp Med. 184:695-706.

Caux, C., Massacrier, C., Vanbervliet, B., Dubois, B., Durand, I., Cella, M., Lanzavecchia, A., and Banchereau, J. (1997). CD34+ hematopoietic progenitors from human cord blood differentiate along two independent dendritic cell pathways in response to granulocyte-macrophage colony-stimulating factor plus tumor necrosis factor a. 2. Functional analysis. Blood 90:1458-1470.

Caux, C., Massacrier, C., Dubois, B., Valladeau, J., Dezutter-Dambuyant, C., Durand, I., Schmitt, D., and Saeland, S. (1999). Respective involvement of TGF-beta and IL-4 in the development of Langerhans cells and non-Langerhans dendritic cells from CD34(+) progenitors. J. Leuk. Biol. 66: 781-791.

Cefai, D., Debre, P., Kaczorek, M., Idziorek, T., Autran, B., and Bismuth, G. (1990). Human Immunodeficiency virus-1 glycoproteins-Gp120 and glycoproteins-Gp160 specifically inhibit the Cd3/T-cell-antigen receptor phosphoinositide transduction pathway. J. Clin. Invest. 86:2117-2124.

Cella, M., Scheidegger, D., PalmerLehmann, K., Lane, P., Lanzavecchia, A., and Alber, G. (1996). Ligation of CD40 on dendritic cells triggers production of high levels of IL-12 and enhances T cell stimulatory capacity: T-T help via APC activation. J. Exp. Med. 184:747-752.

Champagne, P., Ogg, G.S., King, A.S., Knabenhans, C., Ellefsen, K., Nobile, M., Appay, V., Rizzardi, G.P., Fleury, S., Lipp, M., Forster, R., Rowland-Jones, S., Sekaly, R.P., McMichael, A.J., and Pantaleo, G. (2001). Skewed maturation of memory HIV-specific CD8 T lymphocytes. Nature 410: 106-111.

Charton-Bain, M.C., Terris, B., Dauge, M.C.H., Marche, C., Walker, F., Bouchaud, O., Xerri, L., and Potet, F. (1999). Reduced number of Langerhans cells in oesophageal mucosa from AIDS patients. Histopathology 34:399-404.

Chehimi, J., Starr, S.E., Frank, I., Dandrea, A., Ma, X.J., Macgregor, R.R., Sennelier, J., and Trinchieri, G. (1994). Impaired interleukin-12 production in human immunodeficiency virus-infected patients. J. Exp Med. 179:1361-1366.

Chehimi, J., Campbell, D.E., Azzoni, L., Bacheller, D., Papasavvas, E., Jerandi, G., Mounzer, K., Kostman, J., Trinchieri, G., and Montaner, L.J. (2002). Persistent decreases in blood plasmacytoid dendritic cell number and function despite effective highly active antiretroviral therapy and increased blood myeloid dendritic cells in HIV-infected individuals. J. Immunol. 168: 4796-4801.

Chen, W., Antonenko, S., Sederstrom, J.M., Liang, X.Q., Chan, A.S.H., Kanzler, H., Blom, B., Blazar, B.R., and Liu, Y.J. (2004). Thrombopoietin cooperates with FLT3-hgand in the generation of plasmacytoid dendritic cell precursors from human hematopoietic progenitors. Blood 103:2547-2553.

Chougnet, C., Thomas, E., Landay, A.L., Kessler, H.A., Buchbinder, S., Scheer, S., and Shearer, G.M. (1998). CD40 ligand and IFN-gamma synergistically restore IL-12 production in HIV-infected patients. Eur. J. Immunol. 28: 646-656.

Cimarelli, A., Zambruno, G., Marconi, A., Girolomoni, G., Bertazzoni, U., and Giannetti, A. (1994). Quantitation by competitive PCR of HIV-1 Proviral DNA in epidermal Langerhans cells of HIV-infected patients. J. Acquir. Immune Defic. Syndr. Hum. Retrovirol. 7:230-235.

Colonna, M., Trinchieri, G., and Liu, Y.J. (2004). Plasmacytoid dendritic cells in immunity. Nat. Immunol. 5:1219-1226.

Compton, C.C., Kupper, T.S., and Nadire, K.B. (1996). HIV-infected Langerhans cells constitute a significant proportion of the epidermal Langerhans cell population throughout the course of HIV disease. J. Invest. Dermatol. 107:822-826.

Corcoran, L., Ferrero, I., Vremec, D., Lucas, K., Waithman, J., O'Keeffe, M., Wu, L., Wilson, A., and Shortman, K. (2003). The lymphoid past of mouse plasmacytoid cells and thymic dendritic cells. J. Immunol. 170:4926-4932.

Dandrea, A., Rengaraju, M., Valiante, N.M., Chehimi, J., Kubin, M., Aste, M., Chan, S.H., Kobayashi, M., Young, D., Nickbarg, E., Chizzonite, R., Wolf, S.F., and Trinchieri, G. (1992). Production of natural-killer-cell stimulatory factor (interleukin-12) by peripheral-blood mononuclear-cells. J. Exp. Med. 176:1387-1398.

del Hoyo, G.M., Martin, P., Vargas, H.H., Ruiz, S., Arias, C.F., and Ardavin, C. (2002). Characterization of a common precursor population for dendritic cells. Nature 415:1043-1047.

Dhodapkar, M.V., Steinman, R.M., Krasovsky, J., Munz, C., and Bhardwaj, N. (2001). Antigen-specific inhibition of effector T cell function in humans after injection of immature dendritic cells. J. Exp. Med. 193:233-238.

Diamond, D.C., Sleckman, B.P., Gregory, T., Lasky, L.A., Greenstein, J.L., and Burakoff, S.J. (1988). Inhibition of Cd4[+] T-cell function by the HIV envelope protein, Gp120. J. Immunol. 141:3715-3717.

Donaghy, H., Gazzard, B., Gotch, F., and Patterson, S. (2003). Dysfunction and infection of freshly isolated blood myeloid and plasmacytoid dendritic cells in patients infected with HIV-1. Blood 101:4505-4511.

Donaghy, H., Pozniak, A., Gazzard, B., Qazi, N., Gilmour, J., Gotch, F., and Patterson, S. (2001). Loss of blood CD11c($^+$) myeloid and CD11c(-) plasmacytoid dendritic cells in patients with HIV-1 infection correlates with HIV-1 RNA virus load. Blood 98:2574-2576.

Douek, D.C., Brenchley, J.M., Betts, M.R., Ambrozak, D.R., Hill, B.J., Okamoto, Y., Casazza, J.P., Kuruppu, J., Kuntsman, K., Wolinsky, S., Grossman, Z., Dybul, M., Oxenius, A., Price, D.A., Connors, M., and Koup, R.A. (2002). HIV preferentially infects HIV-specific CD4($^+$) T cells. Nature 417:95-98.

Dubois, B., Vanbervliet, B., Fayette, J., Massacrier, C., VanKooten, C., Briere, F., Banchereau, J., and Caux, C. (1997). Dendritic cells enhance growth and differentiation of CD40- activated B lymphocytes. J. Exp. Med. 185:941-951.

Dzionek, A., Fuchs, A., Schmidt, P., Cremer, S., Zysk, M., Miltenyi, S., Buck, D.W., and Schmitz, J. (2000). BDCA-2, BDCA-3 and BDCA-4: Three markers for distinct subsets of dendritic cells in human peripheral blood. J. Immunol. 165:6037-6046.

Facchetti, F., Dewolfpeeters, C., Mason, D.Y., Pulford, K., Vandenoord, J.J., and Desmet, V.J. (1988). Plasmacytoid T-cells - Immunohistochemical evidence for their monocyte macrophage origin. Am. J. Pathol. 133:15-21.

Fan, Z., Huang, X.L., Zheng, L., Wilson, C., Borowski, L., Liebmann, J., Gupta, P., Margolick, J., and Rinaldo, C. (1997). Cultured blood dendritic cells retain HIV-1 antigen-presenting capacity for memory CTL during progressive HIV-1 infection. J. Immunol. 159:4973-4982.

Fayette, J., Dubois, B., Vandenabeele, S., Bridon, J.M., Vanbervliet, B., Durand, I., Banchereau, J., Caux, C., and Briere, F. (1997). Human dendritic cells skew isotype switching of CD40-activated naive B cells towards IgA(1) and IgA(2). J. Exp. Med. 185:1909-1918.

Feldman, S., Stein, D., Amrute, S., Denny, T., Garcia, Z., Kloser, P., Sun, Y., Megjugorac, N., and Fitzgerald-Bocarsly, P.(2001). Decreased interferon-alpha production in HIV-infected patients correlates with numerical and functional deficiencies in circulating type 2 dendritic cell precursors. Clin. Immunol. 101:201-210.

Ferbas, J., Navratil, J., Logar, A., and Rinaldo, C. (1995). Selective decrease in human-immunodeficiency-virus type-1 (HIV-1)-induced alpha-interferon production by peripheral-blood mononuclear-cells during HIV-1 infection. Clin. Diagn. Lab. Immunol. 2:138-142.

Ferbas, J.J., Toso, J.F., Logar, A.J., Navratil, J.S., and Rinaldo, C.R. (1994). CD4($^+$) blood dendritic cells are potent producers of IFN-alpha in response to in-vitro HIV-1 infection. J. Immunol. 152:4649-4662.

Fong, L., Mengozzi, M., Abbey, N.W., Herndier, B.G., and Engleman, E.G. (2002). Productive infection of plasmacytoid dendritic cells with human immunodeficiency virus type 1 is triggered by CD40 ligation. J. Virol. 76:11033-11041.

Fonteneau, J.F., Gilliet, M., Larsson, M., Dasilva, I., Munz, C., Liu, Y.J., and Bhardwaj, N. (2003). Activation of influenza virus-specific CD4($^+$) and

CD8($^+$) T cells: A new role for plasmacytoid denchitic cells in adaptive immunity. Blood 101:3520-3526.

Frank, I., Piatak, M., Stoessel, H., Romani, N., Bonnyay, D., Lifson, J.D., and Pope, M. (2002). Infectious and whole inactivated simian immunodeficiency viruses interact similarly with primate dendritic cells (DCs): Differential intracellular fate of virions in mature and immature DCs. J. Virol. 76: 2936-2951.

Gabrilovich, D.I., Woods, G.M., Patterson, S., Harvey, J.J., and Knight, S.C. (1994). Retrovirus-induced immunosuppression via blocking of dendritic cell-migration and down-regulation of adhesion molecules. Immunology 82:82-87.

Ganesh, L., Leung, K., Lore, K., Levin, R., Panet, A., Schwartz, O., Koup, R.A., and Nabel, G.J. (2004). Infection of specific dendritic cells by CCR5-tropic human immunodeficiency virus type 1 promotes cell-mediated transmission of virus resistant to broadly neutralizing antibodies. J. Virol. 78:11980-11987.

Geijtenbeek, T.B.H., Kwon, D.S., Torensma, R., van Vliet, S.J., van Duijnhoven, G.C.F., Middel, J., Cornelissen, I.L.M.H., Nottet, H.S.L.M., KewalRamani, V.N., Littman, D.R., Figdor, C.G., and van Kooyk, Y. (2000a). DC-SIGN, a dendritic cell-specific HIV-1-binding protein that enhances trans-infection of T cells. Cell 100:587-597.

Geijtenbeek, T.B.H., Torensma, R., van Vliet, S.J., van Duijnhoven, G.C.F., Adema, G.J., van Kooyk, Y., and Figdor, C.G. (2000b). Identification of DC-SIGN, a novel dendritic cell-specific ICAM-3 receptor that supports primary immune responses. Cell 100:575-585.

Geissmann, F., Prost, C., Monnet, J.P., Dy, M., Brousse, N., and Hermine, O. (1998). Transforming growth factor beta 1 in the presence of granulocyte/ macrophage colony-stimulating factor and interleukin 4, induces differentiation of human peripheral blood monocytes into dendritic Langerhans cells. J. Exp. Med. 187:961-966.

Giannetti, A., Zambruno, G., Cimarelli, A., Marconi, A., Negroni, M., Girolomoni, G., and Bertazzoni, U. (1993). Direct detection of HIV-1 RNA in epidermal Langerhans cells of HIV-infected patients. J. Acquir. Immune Defic. Syndr. Hum. Retrovirol. 6:329-333.

Gompels, M., Patterson, S., Roberts, M.S., Macatonia, S.E., Pinching, A.J., and Knight, S.C. (1998). Increase in dendritic cell numbers, their function and the proportion uninfected during AZT therapy. Clin. Exp. Immunol. 112:347-353.

Granelli Piperno, A., Moser, B., Pope, M., Chen, D.L., Wei, Y., Isdell, F., ODoherty, U., Paxton, W., Koup, R., Mojsov, S., Bhardwaj, N., Clark Lewis, I., Baggiolini, M., and Steinman, R.M. (1996). Efficient interaction of HIV-1 with purified dendritic cells via multiple chemokine coreceptors. J. Exp. Med. 184:2433-2438.

Granelli-Piperno, A., Delgado, E., Finkel, V., Paxton, W., and Steinman, R.M. (1998). Immature dendritic cells selectively replicate macrophagetropic (M-tropic) human immunodeficiency virus type 1, while mature cells efficiently transmit both M- and T-Tropic virus to T cells. J. Virol. 72:2733-2737.

Granelli-Piperno, A., Golebiowska, A., Trumpfheller, C., Siegal, F.P., and Steinman, R.M. (2004). HIV-1-infected monocyte-derived dendritic cells do not undergo maturation but can elicit IL-10 production and T cell regulation. Proc. Natl. Acad. Sci. U. S. A. 101:7669-7674.

Grassi, F., Hosmalin, A., McIlroy, D., Calvez, V., Debre, P., and Autran, B. (1999). Depletion in blood CD11c-positive dendritic cells from HIV- infected patients. AIDS 13:759-766.

Grouard, G., Rissoan, M.C., Filgueira, L., Durand, I., Banchereau, J., and Liu, Y.J. (1997). The enigmatic plasmacytoid T cells develop into dendritic cells with interleukin (IL)-3 and CD40-ligand. J. Exp. Med. 185:1101-1111.

Hawiger, D., Inaba, K., Dorsett, Y., Guo, M., Mahnke, K., Rivera, M., Ravetch, J.V., Steinman, R.M., and Nussenzweig, M.C. (2001). Dendritic cells induce peripheral T cell unresponsiveness under steady state conditions *in vivo*. J. Exp. Med. 194:769-779.

Henry, M., Uthman, A., Ballaun, C., Stingl, G., and Tschachler, E. (1994). Epidermal Langerhans cells of AIDS patients express HIV-1 regulatory and structural genes. J. Invest. Dermatol 103:593-596.

Howell, D.M., Feldman, S.B., Kloser, P., and Fitzgeraldbocarsly, P. (1994). Decreased frequency of functional natural interferon-producing cells in peripheral-blood of patients with the acquired-immune-deficiency-syndrome. Clin. Immunol. Immunopathol. 71:223-230.

Hu, J.J., Gardner, M.B., and Miller, C.J. (2000). Simian immunodeficiency virus rapidly penetrates the cervicovaginal mucosa after intravaginal inoculation and infects intraepithelial dendritic cells. J. Virol. 74:6087-6095.

Hussain, L.A. and Lehner, T. (1995). Comparative investigation of Langerhans cells and potential receptors for HIV in oral, genitourinary and rectal epithelia. Immunology 85:475-484.

Ito, T., Inaba, M., Inaba, K., Toki, J., Sogo, S., Iguchi, T., Adachi, Y., Yamaguchi, K., Amakawa, R., Valladeau, J., Saeland, S., Fukuhara, S., and Ikehara, S. (1999). A CD1a$(^+)$/CD11c$(^+)$ subset of human blood dendritic cells is a direct precursor of Langerhans cells. J. Immunol. 163:1409-1419.

Izmailova, E., Bertley, F.M.N., Huang, Q., Makori, N., Miller, C.J., Young, R.A., and Aldovini, A. (2003). HIV-1 Tat reprograms immature dendritic cells to express chemoattractants for activated T cells and macrophages. Nat. Med. 9:191-197.

Jahnsen, F.L., Lund-Johansen, F., Dunne, J.F., Farkas, L., Haye, R., and Brandtzaeg, P. (2000). Experimentally induced recruitment of plasmacytoid (CD123(high)) dendritic cells in human nasal allergy. J. Immunol. 165: 4062-4068.

Jones, G.J., Watera, C., Patterson, S., Rutebemberwa, A., Kaleebu, P., Whitworth, J.A., Gotch, F.M., and Gilmour, J.W. (2001). Comparative loss and maturation of peripheral blood dendritic cell subpopulations in African and non-African HIV-1-infected patients. AIDS 15:1657-1663.

Josien, R., Li, H.L., Ingulli, E., Sarma, S., Wong, B.R., Vologodskaia, M., Steinman, R.M., and Choi, Y. (2000). TRANCE, a tumor necrosis factor family member, enhances the longevity and adjuvant properties of dendritic cells *in vivo*. J. Exp. Med. 191:495-501.

Kalter, D.C., Greenhouse, J.J., Orenstein, J.M., Schnittman, S.M., Gendelman, H.E., and Meltzer, M.S. (1991). Epidermal Langerhans cells are not principal reservoirs of virus in HIV disease. J. Immunol. 146:3396-3404.

Kamath, A.T., Henri, S., Battye, F., Tough, D.F., and Shortman, K. (2002). Developmental kinetics and lifespan of dendritic cells in mouse lymphoid organs. Blood 100:1734-1741.

Kapsenberg, M.L. (2003). Dendritic-cell control of pathogen-driven T-cell polarization. Nat. Rev. Immunol. 3:984-993.

Kawamura, T., Gatanaga, H., Borris, D.L., Connors, M., Mitsuya, H., and Blauvelt, A. (2003). Decreased stimulation of CD4($^+$) T cell proliferation and IL-2 production by highly enriched populations of HIV-infected dendritic cells. J. Immunol. 170:4260-4266.

Langenkamp, A., Messi, M., Lanzavecchia, A., and Sallusto, F. (2000). Kinetics of dendritic cell activation: impact on priming of T(H)1, T(H)2 and nonpolarized T cells. Nat. Immunol. 1:311-316.

Lennert, K. and Remmele, W. (1958). Karyometric research on lymph node cells in man. I. Germinoblasts, lymphoblasts and lymphocytes. Acta Haematol. 19:99-113.

Levi, G., Feldman, J., Holman, S., Salarieh, A., Strickler, H.D., Alter, S., and Minkoff, H. (2005). Relationship between HIV viral load and Langerhans cells of the cervical epithelium. J. Obstet. Gynaecol. Res. 31:178-184.

Lezzi, G., Karjalainen, K., and Lanzavecchia, A. (1998). The duration of antigenic stimulation determines the fate of naive and effector T cells. Immunity 8: 89-95.

Liu, Y.J. (2005). IPC: Professional type 1 interferon-producing cells and plasmacytoid dendritic cell precursors. Annu. Rev. Immunol. 23:275-306.

Lopez, C., Fitzgerald, P.A., and Siegal, F.P. (1983). Severe acquired immune-deficiency syndrome in male-homosexuals - Diminished capacity to make interferon-alpha in vitro associated with severe opportunistic infections. J. Infect. Dis. 148:962-966.

Lu, W., Wu, X.X., Lu, Y.Z., Guo, W.Z., and Andrieu, J.M. (2003). Therapeutic dendritic-cell vaccine for simian AIDS. Nat. Med. 9:27-32.

Lu, W., Arraes, L.C., Ferreira, W.T., and Andrieu, J.M. (2004). Therapeutic dendritic-cell vaccine for chronic HIV-1 infection. Nat. Med. 10:1359-1365.

Ludewig, B., Gelderblom, H.R., Becker, Y., Schafer, A., and Pauli,G. (1996). Transmission of HIV-1 from productively infected mature Langerhans cells to primary CD4$^+$ T lymphocytes results in altered T cell responses with enhanced production of IFN-gamma and IL-10. Virology 215:51-60.

Luft, T., Jefford, M., Luetjens, P., Toy, T., Hochrein, H., Masterman, K.A., Maliszewski, C., Shortman, K., Cebon, J., and Maraskovsky, E. (2002). Functionally distinct dendritic cell (DC) populations induced by physiologic stimuli: prostaglandin E-2 regulates the migratory capacity of specific DC subsets. Blood 100:1362-1372.

Macatonia, S.E., Gompels, M., Pinching, A.J., Patterson, S., and Knight, S.C. (1992). Antigen presentation by macrophages but not by dendritic cells in human-immunodeficiency-virus (HIV) infection. Immunology 75:576-581.

Macatonia, S.E., Lau, R., Patterson, S., Pinching, A.J., and Knight, S.C. (1990). Dendritic cell infection, depletion and dysfunction in HIV-infected individuals. Immunology 71:38-45.

McDonald, D., Wu, L., Bohks, S.M., Kewal Ramani, V.N., Unutmaz, D., and Hope, T.J. (2003). Recruitment of HIV and its receptors to dendritic cell-T cell junctions. Science 300:1295-1297.

McIlroy, D., Autran, B., Cheynier, R., Wainhobson, S., Clauvel, J.P., Oksenhendler, E., Debre, P., and Hosmalin, A. (1995). Infection frequency of dendritic cells and Cd4($^+$) T-lymphocytes in spleens of human immuno-deficiency virus-positive patients. J. Virol. 69:4737-4745.

McKenna, K., Beignon, A.S., and Bhardwaj, N. (2005). Plasmacytoid dendritic cells: Linking innate and adaptive immunity. J. Virol. 79:17-27.

Merad, M., Manz, M.G., Karsunky, H., Wagers, A., Peters, W., Charo, I., Weissman, I.L., Cyster, J.G., and Engleman, E.G. (2002). Langerhans cells renew in the skin throughout life under steady-state conditions. Nat. Immunol. 3:1135-1141.

Miller, C.J., Vogel, P., Alexander, N.J., Dandekar, S., Hendrickx, A.G., and Marx, P.A. (1994). Pathology and localization of simian immunodeficiency virus in the reproductive-tract of chronically infected male rhesus macaques. Lab. Invest. 70:255-262.

Nakano, H., Yanagita, M., and Gunn, M.D. (2001). CD11c($^+$)B220($^+$)Gr-1($^+$) cells in mouse lymph nodes and spleen display characteristics of plasmacytoid dendritic cells. J. Exp. Med. 194:1171-1178.

Nandwani, R., Gazzard, B.G., Barton, S.E., Hawkins, D.A., Zemelman, V., and Staughton, R.C.D. (1996). Does HIV disease progression influence epidermal Langerhans cell density? Br. J. Dermatol. 134:1087-1092.

Niedecken, H., Lutz, G., Bauer, R., and Kreysel, H. W. (1987). Langerhans cell as primary target and vehicle for transmission of HIV. Lancet 2:519-520.

ODoherty, U., Steinman, R.M., Peng, M., Cameron, P.U., Gezelter, S., Kopeloff, I., Swiggard, W.J., Pope, M., and Bhardwaj, N. (1993). Dendritic cells freshly isolated from human blood express CD4 and mature into typical immunostimulatory dendritic cells after culture in monocyte-conditioned medium. J. Exp. Med. 178:1067-1078.

Oyaizu, N., Chirmule, N., Kalyanaraman, V.S., Hall, W.W., Good, R.A., and Pahwa, S. (1990). Human-immunodeficiency-virus type-1 envelope glycol-protein Gp120 produces immune defects in CD4$^+$ lymphocytes-T by inhibiting interleukin-2 messenger-RNA. Proc. Natl. Acad. Sci. U. S. A. 87:2379-2383.

Pacanowski, J., Kahi, S., Baillet, M., Lebon, P., Deveau, C., Goujard, C., Meyer, L., Oksenhendler, E., Sinet, M., and Hosmalin, A. (2001). Reduced blood CD123($^+$) (lymphoid) and CD11c($^+$) (myeloid) dendritic cell numbers in primary HIV-1 infection. Blood 98:3016-3021.

Pacanowski, M., Develioglu, L., Kamga, I., Sinet, M., Desvarieux, M.S., Girard, P.M., and Hosmalin, A. (2004). Early plasmacytoid dendritic cell changes predict plasma HIV load rebound during primary infection. J. Infect. Dis. 190:1889-1892.

Parronchi, P., Mohapatra, S., Sampognaro, S., Giannarini, L., Wahn, U., Chong, P.L., Mohapatra, S., Maggi, E., Ranz, H., and Romagnani, S. (1996). Effects of interferon-alpha on cytokine profile, T cell receptor repertoire and peptide reactivity of human allergen-specific T cells. Eur. J. Immunol. 26:697-703.

Pasare, C. and Medzhitov, R. (2004). Toll-like receptors: Linking innate and adaptive immunity. Microbes Infect. 6:1382-1387.

Patterson, S. and Knight, S.C. (1987). Susceptibility of human peripheral-blood dendritic cells to infection by human-immunodeficiency-virus. J. Gen. Virol. 68:1177-1181.

Patterson, S., Gross, J., Bedford, P., and Knight, S.C. (1991). Morphology and phenotype of dendritic cells from peripheral-blood and their productive and nonproductive infection with human-immunodeficiency-virus type-1. Immunology 72:361-367.

Patterson, S., Gross, J., English, N., Stackpoole, A., Bedford, P., and Knight, S.C. (1995). CD4 expression on dendritic cells and their infection by human-immunodeficiency-virus. J. Gen. Virol. 76:1155-1163.

Patterson, S., Helbert, M., English, N.R., Pinching, A.J., and Knight, S.C. (1996). The effect of AZT on dendritic cell number and provirus load in the peripheral blood of AIDS patients: A preliminary study. Res. Virol. 147:109-114.

Patterson, S., English, N.R., Longhurst, H., Balfe, P., Helbert, M., Pinching, A.J., and Knight, S.C. (1998). Analysis of human immunodeficiency virus type 1 (HIV-1) variants and levels of infection in dendritic and T cells from symptomatic HIV-1-infected patients. J. Gen. Virol. 79:247-257.

Patterson, S., Robinson, S.P., English, N.R., and Knight, S.C. (1999). Subpopulations of peripheral blood dendritic cells show differential susceptibility to infection with a lymphotropic strain of HIV-1. Immunol. Lett. 66:111-116.

Patterson, S., Rae, A., and Donaghy, H. (2001a). Purification of dendritic cells from peripheral blood. Dendritic Cell Protocols 64, 111-120. New Jersey, Humana Press Inc. Methods in Molecular Medicine. Robinson, S. P. and Stagg, A. J.

Patterson, S., Rae, A., Hockey, N., Gilmour, J., and Gotch, F. (2001b). Plasmacytoid dendritic cells are highly susceptible to human immuno-deficiency virus type 1 infection and release infectious virus. J. Virol. 75:6710-6713.

Patterson, S., Donaghy, H., Amjadi, P., Gazzard, B., Gotch, F., and Kelleher, P. (2005). Human BDCA-1-positive blood dendritic cells differentiate into phenotypically distinct immature and mature populations in the absence of exogenous maturational stimuli: Differentiation failure in HIV infection. J. Immunol. 174:8200-8209.

Perussia, B., Chan, S.H., Dandrea, A., Tsuji, K., Santoli, D., Pospisil, M., Young, D., Wolf, S.F., and Trinchieri, G. (1992). Natural-killer (NK) cell stimulatory factor or Il-12 has differential-effects on the proliferation of TCR-alpha-beta[+], TCR-gamma-delta[+] lymphocytes-T, and NK cells. J. Immunol. 149:3495-3502.

Pope, M., Betjes, M.G.H., Romani, N., Hirmand, H., Cameron, P.U., Hoffman, L., Gezelter, S., Schuler, G., and Steinman, R.M. (1994). Conjugates of dendritic cells and memory T-lymphocytes from skin facilitate productive infection with HIV-1. Cell 78:389-398.

Pope, M., Gezelter, S., Gallo, N., Hoffman, L., and Steinman, R.M. (1995). Low-level of HIV-1 infection in cutaneous dendritic cells promote extensive viral replication upon binding to memory CD4($^+$)T cells. J. Exp. Med. 182:2045-2056.

Popov, S., Chenine, A.L., Gruber, A., Li, P.L., and Ruprecht, R.M. (2005). Long-term productive human immunodeficiency virus infection of CD1a-sorted myeloid dendritic cells. J. Virol. 79:602-608.

Prakash, M., Kapembwa, M.S., Gotch, F., and Patterson, S. (2001). Higher levels of activation markers and chemokine receptors on T lymphocytes in the cervix than peripheral blood of normal healthy women. J. Reprod. Immunol. 52:101-111.

Prakash, M., Kapembwa, M.S., Gotch, F., and Patterson, S. (2004). Chemokine receptor expression on mucosal dendritic cells from the endocervix of healthy women. J. Infect. Dis. 190:246-250.

Pugh, C.W., Macpherson, G.G., and Steer, H.W. (1983). Characterization of non-lymphoid cells derived from rat peripheral lymph. J. Exp. Med. 157:1758-1779.

Rappersberger, K., Gartner, S., Schenk, P., Stingl, G., Groh, V., Tschachler, E., Mann, D.L., Wolff, K., Konrad, K., and Popovic, M. (1988). Langerhans cells are an actual site of HIV-1 replication. Intervirology 29:185-194.

Ratzinger, G., Baggers, J., de Cos, M.A., Yuan, J.D., Dao, T., Reagan, J.L., Munz, C., Heller, G., and Young, J.W. (2004). Mature human Langerhans cells derived from CD3($^+$) hematopoietic progenitors stimulate greater cytolytic T lymphocyte activity in the absence of bioactive IL-12p70, by either single peptide presentation or cross-priming, than do dermal-interstitial or monocyte-derived dendritic cells. J. Immunol. 173:2780-2791.

Reid, C.D.L., Stackpoole, A., Meager, A., and Tikerpae, J. (1992). Interactions of tumor-necrosis-factor with granulocyte- macrophage colony-stimulating factor and other cytokines in the regulation of dendritic cell-growth invitro from early bipotent CD34$^+$ progenitors in human bone-marrow. J. Immunol. 149:2681-2688.

Richters, C.D., Hoekstra, M.J., Du Pont, J.S., Kreis, R.W., and Kamperdijk, E.W.A. Isolation of human skin dendritic cells by *in vitro* migration. Dendritic Cell Protocols, 145-154. (2001). Methods in Molecular Medicine. Robinson, S. P. and Stagg, A. J., Humana Press, Totowa, New Jersey.

Rissoan, M.C., Soumelis, V., Kadowaki, N., Grouard, G., Briere, F., Malefyt, R.D., and Liu, Y.J. (1999). Reciprocal control of T helper cell and dendritic cell differentiation. Science 283:1183-1186.

Robinson, S.P., Patterson, S., English, N., Davies, D., Knight, S.C., and Reid, C.D.L. (1999). Human peripheral blood contains two distinct lineages of dendritic cells. Eur. J. Immunol. 29:2769-2778.

Rogge, L., BarberisMaino, L., Biffi, M., Passini, N., Presky, D.H., Gubler, U., and Sinigaglia, F. (1997). Selective expression of an interleukin-12 receptor component by human T helper 1 cells. J. Exp. Med. 185:825-831.

Romani, N., Gruner, S., Brang, D., Kampgen, E., Lenz, A., Trockenbacher, B., Konwalinka, G., Fritsch, P.O., Steinman, R.M., and Schuler, G. (1994). Proliferating dendritic cell progenitors in human blood. J. Exp. Med. 180: 83-93.

Rosini, S., Caltagirone, S., Tallini, G., Lattanzio, G., Demopoulos, R., Piantelli, M., and Musiani, P. (1996). Depletion of stromal and intraepithelial antigen-presenting cells in cervical neoplasia in human immunodeficiency virus infection. Hum. Pathol. 27:834-838.

Rowland-Jones, S., Tan, R.S., and McMichael, A. (1997). Role of cellular immunity in protection against HIV infection. Adv. Immunol. 65:277-346.

Rozis, G., De Silva, S., Benlahrech, A., Papagatsias, P., Harris, J., Gotch, F, Dickson, G., and Patterson, S. (2005). Langerhans cells are more efficiently transduced than dermal dendritic cells by adenovirus vectors expressing either group C or group B fibre protein: Implications for mucosal vaccines. Eur. J. Immunol. 35:2617-2626.

Saint-Vis, B., Fugier-Vivier, I., Massacrier, C., Gaillard, C., Vanbervliet, B., Ait-Yahia, S., Banchereau, J., Liu, Y.J., Lebecque, S., and Caux, C. (1998). The cytokine profile expressed by human dendritic cells is dependent on cell subtype and mode of activation. J. Immunol. 160:1666-1676.

Sallusto, F., Cella, M., Danieli, C., and Lanzavecchia, A. (1995). Dendritic cells use macropinocytosis and the mannose receptor to concentrate macro-molecules in the major histocompatibility complex class-ii compartment - down-regulation by cytokines and bacterial products. J. Exp. Med. 182:389-400.

Sallusto, F. and Lanzavecchia, A. (1994). Efficient presentation of soluble-antigen by cultured human dendritic cells is maintained by granulocyte-macrophage colony-stimulating factor plus interleukin-4 and down-regulated by tumor-necrosis-factor-alpha. J. Exp. Med. 179:1109-1118.

Sallusto, F., Palermo, B., Lenig, D., Miettinen, M., Matikainen, S., Julkunen, I., Forster, R., Burgstahler, R., Lipp, M., and Lanzavecchia, A. (1999). Distinct patterns and kinetics of chemokine production regulate dendritic cell function. Eur. J. Immunol. 29:1617-1625.

Sallusto, F., Schaerli, P., Loetscher, P., Schaniel, C., Lenig, D., Mackay, C.R., Qin, S.X., and Lanzavecchia, A. (1998). Rapid and coordinated switch in chemokine receptor expression during dendritic cell maturation. Eur. J. Immunol. 28:2760-2769.

Sanders, R.W., de Jong, E.C., Baldwin, C.E., Schuitemaker, J.H.N., Kapsenberg, M.L., and Berkhout, B. (2002). Differential transmission of human immunodeficiency virus type I by distinct subsets of effector dendritic cells. J. Virol. 76:7812-7821.

Sapp, M., Engelmayer, J., Larsson, M., Granelli-Piperno, A., Steinman, R., and Bhardwaj, N. (1999). Dendritic cells generated from blood monocytes of HIV-1 patients are not infected and act as competent antigen presenting cells eliciting potent T-cell responses. Immunol. Lett. 66:121-128.

Schmidt, B., Scott, I., Whitmore, R.G., Foster, H., Fujimura, S., Schmitz, J., and Levy, J.A. (2004). Low-level HIV infection of plasmacytoid dendritic cells:

onset of cytopathic effects and cell death after PDC maturation. Virology 329:280-288.

Shortman, K. and Liu, Y.J. (2002). Mouse and human dendritic cell subtypes. Nat. Rev. Immunol. 2:151-161.

Siegal, F.P., Kadowaki, N., Shodell, M., Fitzgerald-Bocarsly, P.A., Shah, K., Ho, S., Antonenko, S., and Liu, Y.J. (1999). The nature of the principal type 1 interferon-producing cells in human blood. Science 284:1835-1837.

Sivard, P., Berlier, W., Picard, B., Sabido, O., Genin, C., and Misery, L. (2004). HIV-1 infection of Langerhans cells in a reconstructed vaginal mucosa. J. Infect. Dis. 190:227-235.

Smed-Sorensen, A., Lore, K., Walther-Jallow, L., Andersson, J., and Spetz, A.L. (2004). HIV-1-infected dendritic cells up-regulate cell surface markers but fail to produce IL-12 p70 in response to CD40 ligand stimulation. Blood 104:2810-2817.

Snijders, A., Kalinski, P., Hilkens, C.M.U., and Kapsenberg, M.L. (1998). High-level IL-12 production by human dendritic cells requires two signals. Int. Immunol. 10:1593-1598.

Sol-Foulon, N., Moris, A., Nobile, C., Boccaccio, C., Engering, A., Abastado, J.P., Heard, J.M., van Kooyk, Y., and Schwartz, O. (2002). HIV-1 nef-induced upregulation of DC-SIGN in dendritic cells promotes lymphocyte clustering and viral spread. Immunity 16:145-155.

Soumelis, V., Scott, I., Gheyas, F., Bouhour, D., Cozon, G., Cotte, L., Huang, L., Levy, J.A., and Liu, Y.J. (2001). Depletion of circulating natural type I interferon-producing cells in HIV-infected AIDS patients. Blood 98:906-912.

Sozzani, S., Allavena, P., D'Amico, G., Luini, W., Bianchi, G., Kataura, M., Imai, T., Yoshie, O., Bonecchi, R., and Mantovani, A. (1998). Cutting edge: Differential regulation of chemokine receptors during dendritic cell maturation: A model for their trafficking properties. J. Immunol. 161: 1083-1086.

Spinillo, A., Tenti, P., Zappatore, R., Deseta, F., Silini, E., and Guaschino, S. (1993). Langerhans cell counts and cervical intraepithelial neoplasia in women with human-immunodeficiency-virus infection. Gynecol. Oncol. 48:210-213.

Spira, A.I., Marx, P.A., Patterson, B.K., Mahoney, J., Koup, R.A., Wolinsky, S.M., and Ho, D.D. (1996). Cellular targets of infection and route of viral dissemination after an intravaginal inoculation of simian immunodeficiency virus into rhesus macaques. J. Exp. Med. 183:215-225.

Srivastava, P. (2002.) Interaction of heat shock proteins with peptides and antigen presenting cells: Chaperoning of the innate and adaptive immune responses. Annu. Rev. Immunol. 20:395-425.

Steinman, R.M., Adams, J.C., and Cohn, Z.A. (1975). Identification of a novel cell type in peripheral lymphoid organs of mice .4. Identification and distribution in mouse spleen. J. Exp. Med. 141:804-820.

Stevenson, M., Stanwick, T.L., Dempsey, M.P., and Lamonica, C.A. (1990). HIV-1 replication is controlled at the level of T-cell activation and proviral integration. EMBO J. 9:1551-1560.

Strunk, D., Egger, C., Leitner, G., Hanau, D., and Stingl, G. (1997). A skin homing molecule defines the Langerhans cell progenitor in human peripheral blood. J. Exp. Med. 185:1131-1136.

Tschachler, E., Groh, V., Rappersberger, K., Gardner, S., Popovic, M., Schenk, P., Konrad, K., Mann, D.L., Wolff, K., and Stingl, G. (1987). Epidermal Langerhans cells (LC) - A virus reservoir in HIV-infection. J. Invest. Dermatol. 88:522.

Tsunetsuguyokota, Y., Akagawa, K., Kimoto, H., Suzuki, K., Iwasaki, M., Yasuda, S., Hausser, G., Hultgren, C., Meyerhans, A., and Takemori, T. (1995). Monocyte-derived cultured dendritic cells are susceptible to human-immunodeficiency-virus infection and transmit virus to resting T-cells in the process of nominal antigen presentation. J. Virol. 69:4544-4547.

Turville, S., Wilkinson, J., Cameron, P., Dable, J., and Cunningham, A.L. (2003). The role of dendritic cell C-type lectin receptors in HIV pathogenesis. J. Leuk. Biol. 74:710-718.

Turville, S.G., Cameron, P.U., Handley, A., Lin, G., Pohlmann, S., Doms, R.W., and Cunningham, A.L. (2002). Diversity of receptors binding HIV on dendritic cell subsets. Nat. Immunol. 3:975-983.

van Stipdonk, M.J.B., Lemmens, E.E., and Schoenberger, S.P. (2001). Naive CTLs require a single brief period of antigenic stimulation for clonal expansion and differentiation. Nat. Immunol. 2:423-429.

Vandenabeele, S., Hochrein, H., Mavaddat, N., Winkel, K., and Shortman,K. (2001). Human thymus contains 2 distinct dendritic cell populations. Blood 97:1733-1741.

Watanabe, N., Wang, Y.H., Lee, H.K., Ito, T., Wang, Y.H., Cao, W., and Liu, Y.J. (2005). Hassall's corpuscles instruct dendritic cells to induce CD4($^+$)CD25($^+$) regulatory T cells in human thymus. Nature 436:1181-1185.

Wollenberg, A., Wagner, M., Gunther, S., Towarowski, A., Tuma, E., Moderer, M., Rothenfusser, S., Wetzel, S., Endres, S., and Hartmann, G. (2002). Plasmacytoid dendritic cells: A new cutaneous dendritic cell subset with distinct role in inflammatory skin diseases. J. Invest. Dermatol. 119:1096-1102.

Wong, B.R., Josien, R., Lee, S.Y., Sauter, B., Li, H.L., Steinman, R.M., and Choi, Y.W. (1997). TRANCE (tumor necrosis factor [TNF]-related activation-induced cytokine), a new TNF family member predominantly expressed in T cells, is a dendritic cell-specific survival factor. J. Exp. Med. 186:2075-2080.

Yoneyama, H., Matsuno, K., Toda, E., Nishiwaki, T., Matsuo, N., Nakano, A., Narumi, S., Lu, B., Gerard, C., Ishikawa, S., and Matsushima, K. (2005). Plasmacytoid DCs help lymph node DCs to induce anti-HSV CTLs. J. Exp. Med. 202:425-435.

Yonezawa, A., Morita, R., Takaori-Kondo, A., Kadowaki, N., Kitawaki, T., Hori, T., and Uchiyama, T. (2003). Natural alpha interferon-producing cells respond to human immunodeficiency virus type 1 with alpha interferon production and maturation into dendritic cells. J. Virol. 77:3777-3784.

Yu, Y., Hagihara, M., Ando, K., Gansuvd, B., Matsuzawa, H., Tsuchiya, T., Ueda, Y., Inoue, H., Hotta, T., and Kato, S. (2001). Enhancement of human cord blood CD34($^+$) cell-derived NK cell cytotoxicity by dendritic cells. J. Immunol. 166:1590-1600.

Zaitseva, M., Blauvelt, A., Lee, S., Lapham, C.K., Klaus-Kovtun, V., Mostowski, H., Manischewitz, J., and Golding, H. (1997). Expression and function of CCR5 and CXCR4 on human Langerhans cells and macrophages: Implications for HIV primary infection. Nat. Med. 3:1369-1375.

Zambruno, G., Mori, L., Marconi, A., Mongiardo, N., Derienzo, B., Bertazzoni, U., and Giannetti, A. (1991). Detection of HIV-1 in epidermal Langerhans cells of HIV- infected patients using the polymerase chain-reaction. J. Invest. Dermatol. 96:979-982.

Zhang, Z.Q., Schuler, T., Zupancic, M., Wietgrefe, S., Staskus, K.A., Reimann, K.A., Reinhart, T.A., Rogan, M., Cavert, W., Miller, C.J., Veazey, R.S., Notermans, D., Little, S., Danner, S.A., Richman, D.D., Havlir, D., Wong, J., Jordan, H.L., Schacker, T.W., Racz, P., Tenner-Racz, K., Letvin, N.L., Wolinsky, S., and Haase, A.T. (1999). Sexual transmission and propagation of SIV and HIV in resting and activated CD4($^+$) T cells. Science 286:1353-1357.

Zimmer, M.I., Larregina, A.T., Castillo, C.M., Capuano, S., Falo, L.D., Murphey-Corb, M., Reinhart, T.A., and Barratt-Boyes, S.M. (2002). Disrupted homeostasis of Langerhans cells and interdigitating dendritic cells in monkeys with AIDS. Blood 99:2859-2868.

Chapter 12

HIV Exploitation of DC Biology to Subvert the Host Immune Response

Manuela Del Cornò, Lucia Conti, Maria Cristina Gauzzi, Laura Fantuzzi, and Sandra Gessani

12.1 Introduction

The potency of dendritic cells (DCs) as antigen-presenting cells (APCs) makes them pivotal in the initiation of host response to control and/or eliminate viral infections. It is now becoming clear that DC function is not fixed but is adaptable in response to signals from the environment, including cytokines, the nature of the pathogen and its ability or not to trigger pattern recognition receptors on DCs, as well as linked to the DC subset and the proximity of T cells. Moreover, increasing clinical evidence in humans indicates that natural infection with most viruses is deleterious to the host. This is because of the efficacy of the immune response that may be associated with substantial tissue destruction, but also because some viruses escape the immune response with various strategies or induce immunosuppression. Several studies in this respect have now revealed that although DCs play a pivotal role in inducing protective immunity against viral infections, viruses could use these cells as a tool for subverting the host immune response, thus inducing immune suppression. Furthermore, DCs can serve as latent viral reservoirs and prevent the complete elimination of the virus, therefore leading to immunopathology.

We have summarized the latest developments on the role of DCs in the immunopathogenesis of viral infections, focusing on human immunodeficiency virus type 1 (HIV-1) productive infection, as well as on bystander effects of its soluble products. The different mechanisms exploited by HIV-1 to subvert the host immune system to its own advantage, thus favoring its persistence and spread through the body, are extensively analyzed. We will begin the discussion with a brief overview on the general concepts of the antiviral immune response.

12.2 The Host Response to Viral Infection

A fundamental aspect for a successful virus infection is represented by the fact that the pathogen must have enough time in the host cell to replicate its genome, package it, and produce a new infectious progeny. The host-virus relationship is a dynamic process in which the virus attempts to minimize its visibility while the host attempts to prevent and eradicate infection with minimal collateral damage to itself. Initially, the virus must recognize, bind, and enter its target cell and migrate to the appropriate cellular compartment. Here, its genome is transcribed, translated, and replicated to permit the assembly and export of new virions so that the infection can spread to additional susceptible target cells and hosts. Conversely, the host must be able to recognize the presence of the virus and eliminate it as quickly and efficiently as possible. This usually occurs as a stepwise series of events in which the infected cell and the immune system each play a critical role.

12.2.1 Innate Immune Response

The first barrier that the virus has to overcome in order to successfully replicate and spread is the rapid host innate response (Fig. 12.1). Crossing the epithelial barrier by a pathogen, including viruses, stimulates an inflammatory response characterized by the rapid production by epithelial cells of inflammatory cytokines (IL-1β and TNF-α), chemokines (CXCL8 and CCL3), and vasoactive amines. These act in concert to stimulate vascular permeability for effector blood proteins (plasmin, thrombin, complement, and antibodies), to increase blood flow and recruit inflammatory cells to the site of infection, thus initiating inflammation and antiviral immune response. These soluble mediators also activate macrophages, DCs, and natural killer (NK) cells. The type of tissue infected and the nature of the cytokines and chemokines produced profoundly influences the type of cells recruited during this inflammatory phase and this in turn influences the type of adaptive immune response that follows. Among the cytokines produced, during the innate response, type I IFN are of crucial importance. Although initially identified as antiviral compounds, their potent immuno-modulatory activity on different immune cell populations, including DCs,

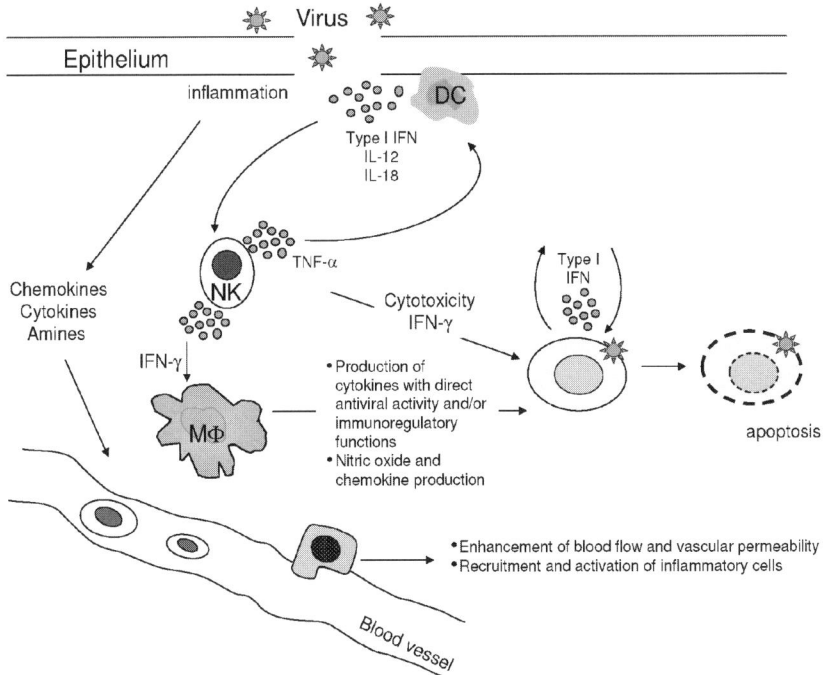

FIGURE 12.1. Innate immune response to viral infection. The inflammatory response that follows after crossing the epithelial barrier by a virus triggers the production of a variety of soluble factors coordinately acting to stimulate vascular permeability for effector blood proteins, increased blood flow, and recruitment/ activation of innate immunity effector cells, including macrophages (MΦ), DCs, and NK cells to the infection site. DCs and NK cells have the ability to reciprocally activate one another. This cross-talk involves both cell-to-cell contact and soluble mediators produced by the two cell types. Virus infection also activates type I and type II IFN production that act on the surrounding uninfected cells by restricting viral spread, regulating the generation/activation of other immune cells, or both. Furthermore, virus-infected cells can undergo apoptosis as a mechanism to prevent viral replication.

NK, CD4$^+$ and CD8$^+$ T cells, has been subsequently recognized (Belardelli, 1995; Santini *et al.*, 2000). The induction of type I IFN by virus-infected cells occurs very rapidly, within a few hours postinfection, and represents the earliest antiviral response in the host (Haller *et al.*, 2006). More importantly, the early virus-induced cytokine response may not be limited to type I IFN, as certain viruses, including HIV and lymphocytic choriomeningitis virus, have the potential to infect and rapidly activate macrophages to produce additional cytokines, such as TNF-α (Guidotti and Chisari, 1999), which can trigger independent antiviral activities. Nonetheless, the induction of type I IFN is widely accepted as the most immediate and important antiviral host response. Although every cell or tissue of the body can produce and secrete IFN upon stimulation, a specialized cell type, the blood plasmacytoid DC (PDC), has been shown to be the greatest IFN producer on viral challenge Siegal et al., 1999). The amount of IFN secreted by PDCs is 100 to 1000 times more than that produced by any other blood cell type (Liu, 2005; Siegal et al., 1999). IFN production is transcriptionally activated upon recognition of a virus, or its molecular patterns, by the Toll-like receptors (TLRs) of DCs (Reis e Sousa, 2004). In turn, the secreted IFN stimulates DC maturation and subsequent expression of proinflammatory cytokines and costimulatory molecules, leading to the transition to an adaptive antiviral immunity (Liu, 2005). The importance of type I IFN as antiviral effector molecule is further underlined by the multiple mechanisms that viruses have evolved in interfering with its biological activity. Several comprehensive reviews have been written on these subjects in recent years (Haller et al., 2006; Katze et al., 2002).

Another important cell component in the innate immune response to viruses is represented by NK cells (Biron et al., 1999; Hamerman et al., 2005; Lodoen and Lanier, 2005). NK cells can be activated by the direct engagement of activating receptors on their cell surface by ligands expressed on the surface of virus-infected cells. Furthermore, NK cells can be activated by exposure to cytokines, mainly type I IFN, produced by other immune cells in response to a viral infection. In this regard, it has been suggested that a range of cytokines produced by DCs (i.e., IL-12, IL-18, and type I IFN) are required for the induction of the various NK cell effector functions (Hamerman et al., 2005). On the other hand, NK cells release soluble factors (TNF-α, IFN-γ) that can promote DC activation. The end result of this reciprocal interaction is NK cell activation, pro-liferation, cytotoxicity, IFN-γ production, and DC maturation, leading to effector cytokine production and T cell activation (Munz et al., 2005). Activated NK cells, as well as macrophages, can directly kill virus-infected cells and also produce inflammatory cytokines at the sites of infection, which recruit or activate other immune cells. In this respect, the IFN-γ secreted by NK cells after activation is of crucial importance for the activation of macrophages and T cells, both of which are important effectors in the response to infection.

In the struggle between virus and host, control over the cell death machinery is of critical importance for survival. Viruses are obligatory intracellular parasites and, as such, must modulate apoptotic pathways to control the life span of their host in order to complete their replicative cycle. The innate response mounted by the immune system incorporates activation of the apoptic pathway, particularly by members of the TNF cytokine family, as a mechanism to restrict viral replication (Blaho, 2004;

Strater and Moller, 2004). However, for the host, success is harsh and potentially costly, as apoptosis often contributes to pathogenesis. Thus, apoptosis serves as a powerful selective pressure for the virus to evade, and a variety of strategies have been exploited by viruses to manipulate this aspect of the host innate response (Benedict et al., 2002). In fact, viruses encode several specific proteins to modulate apoptosis to their own profit. Prior to virus replication, these proteins serve to inhibit premature apoptosis of the virus-infected cells. However, after completion of the viral life cycle, viral-induced apoptosis has been suggested to represent a mechanism to disseminate the viral progeny without causing inflammatory responses (Everett and McFadden, 1999).

12.2.2 Adaptive Immune Response

DCs are highly specialized in capturing and presenting antigens to T cells and stimulating the differentiation and proliferation of B cells (Banchereau and Steinman, 1998; Steinman et al., 1999). Because of these functions, DCs represent key modulators linking the innate and adaptive arms of the immune response and allow the transition to the adaptive immune response in viral infections (Fig. 12.2). Although DCs have not yet been shown to exert any direct antiviral activity themselves, they secrete a large variety of antiviral and immunoregulatory cytokines (type I IFN, TNF-α, IL-1, IL-6, IL-12, and IL-18) after activation (Bender et al., 1998; Stockwin et al., 2000). It is clear that soluble factors produced by virus-infected cells, phagocytes, and APCs, or by other cells of the innate immunity system can play a direct role in controlling the replication of many viruses during the initial phase of the infection. However, due to their immunomodulatory activity, these factors can also contribute to the initiation of the adaptive immune response. The contribution of the latter to viral clearance relies on the antiviral functions of cytotoxic T cells (CTLs), T helper (Th) cells, and B cells. It is generally assumed that their relative importance relies on the particular type of viral infection. For instance, $CD8^+$ CTLs but not $CD4^+$ T cells or IFN-γ are thought to be important for clearing acute infections, particularly of non-cytophatic viruses (Tishon et al., 1995; Zinkernagel et al., 1996). Conversely, $CD4^+$ T cells, IFN-γ and neutralizing antibodies but not CTLs are thought to be crucial for clearing cytophatic viruses (Zinkernagel et al., 1996). Chronic, persistent virus infections are instead cleared by $CD4^+$ T cells and IFN-γ but not adequately by CTLs (Tishon et al., 1995).

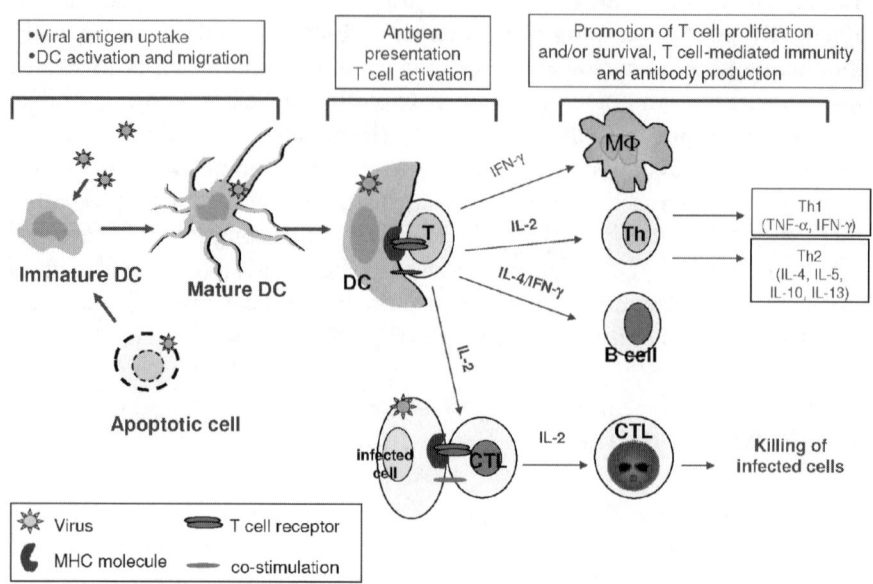

FIGURE 12.2. Adaptive immune response to viral infection. Activation of the innate immune system plays an instructive role for the induction of antigen-specific immune responses, which are initiated when DCs within or adjacent to infected epithelium come into contact with the infecting virus, process it into antigenic peptides, migrate to the lymph nodes, and present them to the T cells. The virus-specific CD4[+] T cells provide essential help for the induction of antiviral CTLs, antibodies, and memory T cells.

Antigen-specific Th cell, B cell, and antibody immune responses are initiated when DCs within or adjacent to epithelium come into contact with infecting viruses. DCs take up virus particles and process them into antigenic peptides that are bound to MHC class II molecules. Alternatively, DCs can take up virus antigens processed by other phagocytic/ APCs, such as macrophages. DCs then migrate to local lymph nodes and present the MHC-peptide complex to T and B cells with antigen receptors that can bind to the complex (see Chapter 13). While Th cells can be primed by APCs that have internalized viral antigens released by other cells, priming of CTLs usually requires the processing of viral proteins that are either endogenously produced within or phagocytosed by professional APCs (Steinman et al., 1999) in the context of MHC class I peptide-antigen complexes. For viruses that do not infect professional APCs, tissue-derived DCs that have processed apoptotic virus-infected cells and debris are likely to migrate to the regional lymph nodes to allow CTL priming to occur (Sallusto and Lanzavecchia, 1999). On clonal expansion of primed CTLs, these cells exit the lymph nodes and home-in to infected tissues, where they recognize viral antigens and perform their effector function by killing virus-infected cells. It is noteworthy that CTLs, as well as NK cells, use apoptosis as the principal killing mechanism. They do so by releasing certain cytotoxic molecules, such as TNF-α, perforin and granzyme. Hence, viral anti-apoptotic strategies also help the virus to evade CTL and NK cell-mediated killing of infected cells (Iannello et al., 2006). CTL responses can also be initiated by DCs and helped by Th cells, a major source of IL-2. The accepted dogma is that complete clearance of intracellular viruses by the immune response depends on the destruction of infected cell by the effectors of the innate and adaptive immunity system, namely CTLs and NK cells. This concept, however, has been recently challenged by experimental evidence showing that much of the antiviral potential of these cells reflects their ability to produce antiviral cytokines, such as IFN-γ and TNF-α at the site of infection (Guidotti et al., 1996). Indeed, these cytokines can purge viruses from infected cells noncytophatically as long as the cell is able to activate antiviral mechanisms and the virus is sensitive to them (Guidotti and Chisari, 1999). Importantly, the same cytokines also control virus infections indirectly, by modulating the induction, amplification, recruitment, and effector functions of the immune response and by upregulating antigen processing and display of viral epitopes at the surface of infected cell. In keeping with these concepts, it is not surprising that a number of viruses encode proteins that have the potential to inhibit the antiviral activity of cytokines (Katze et al., 2002).

Th cells play a key role in antiviral immunity (Mosmann and Coffman, 1989). They participate to the antiviral response both directly (by producing antiviral cytokines) and indirectly (by providing help for B cells and CTLs). Th cells can be divided into two main subsets, Th1 and Th2, based on their different cytokine profiles. Th1 cells produce IL-2, TNF-α and IFN-γ, which possess antiviral activities and regulate the cellular immune response, whereas Th2 cells produce IL-4, IL-5, IL-10, and IL-13, which are known to stimulate antibody production (de Jong et al., 2005; Kidd, 2003; Mosmann and Coffman, 1989). Th cells also promote B-cell differentiation in antibody-producing plasma cells. Antiviral antibodies contribute to viral clearance mainly by blocking virus entry into susceptible cells and by removing infectious virions from the circulation, thereby preventing extracellular viral spread. This is a complex process that involves not only antiviral antibodies but also the complement system and phagocytic cells (Bachmann and Zinkernagel, 1997; Burton et al., 2000). In addition, antibodies perform other antiviral functions besides the neutralization of extracellular virions. For instance, the deposition of antibodies on the surface of infected cells may prevent the release of virions from the infected cells. Finally, the virus-specific CD4[+] Th cells provide essential help not only for the induction of CTLs and antibodies but also for the generation of memory

T cells, critical for an efficient virus-specific anamnestic response (Welsh *et al.*, 2004).

The host invariably responds to infecting viruses by activating its innate immune system and mounting virus-specific humoral and cellular immune responses. These responses are aimed at controlling viral replication and eliminating the infecting virus from the host. However, viruses have evolved numerous and sophisticated strategies to counteract and evade host antiviral responses (Iannello *et al.*, 2006). Unraveling these viral strategies is necessary to gain insights into host-pathogen interactions and their co-evolution.

12.3 Virus-induced Phenotypic and Functional Alterations of Human DCs

Most viral infections are initiated by virus penetration through respiratory, gut, or genital mucosal surfaces that act as a barrier between the body and the external environment, although viruses also may directly enter the bloodstream via arthropod vectors or animal bites. Entry through the blood is likely to shunt the virus directly into the spleen, where T and B cells and APCs are readily available to initiate an immune response. In the more common peripheral routes of viral entry, the virus will likely encounter DCs, which develop from bone marrow progenitors, enter the circulation, and home-in to peripheral tissues (see Chapters 1 and 2), where they await activation by "danger signals" (Matzinger, 2002; Reis e Sousa *et al.*, 1999). These cells reside in pluristratified epithelia of the oral mucosae, vagina, cervix, and the rectum and are referred to as "immature DCs" (de Fraissinette *et al.*, 1989). Although there is as yet limited evidence that DCs represent the initial virus target during natural infection, several observations support the hypothesis that mucosal DCs become infected at the early stage of a viral infection. In this respect, studies in macaques exposed to the simian immunodeficiency virus (SIV) revealed that SIV enters the mucosae within 1 h of vaginal exposure, primarily infects intraepithelial DCs, and can be found in association with DCs in draining lymph nodes within 18 h (Hu *et al.*, 2000). In addition, epidermal and dermal interstitial immature DCs can be infected with dengue virus (DV), an arthropod-borne flavivirus, which is introduced into human skin by mosquito bites (Wu *et al.*, 2000). In keeping with *in vivo* studies, a tremendous amount of literature has documented the susceptibility of DCs, either derived *in vitro* from monocytes or CD34[+] progenitors as well as blood DCs and epidermal Langerhans cells, to *in vitro* infection by several viruses. These viruses include HIV (Rinaldo and Piazza, 2004), measles (MV) (Grosjean *et al.*, 1997), influenza (Bhardwaj *et al.*, 1994), DV (Ho *et al.*, 2001), Lassa and Ebola (Baize *et al.*, 2004; Mahanty *et al.*, 2003), and herpesviruses (Abendroth *et al.*, 2001; Asada *et al.*, 1999; Kruse *et al.*, 2000; Morrow *et al.*, 2003; Riegler *et al.*, 2000). However, the productivity of infection is generally low in DCs but it is highly enhanced by contact with T cells, as demonstrated for HIV (Cameron *et al.*, 1992) and MV (Fugier-Vivier *et al.*, 1997).

The interactions between viruses and DCs have recently gained considerable interest in the context of their possible importance in the pathogenesis of viral infection (Pollara *et al.*, 2005). Because of their central role in the induction of immune responses, modulation of DC maturation and functional activity represents a strategic mechanism for a pathogen to evade immune surveillance. Growing evidence is accumulating on the capacity of viruses to differently affect DC biology, thus interfering at several steps of DC-induced immune response, including DC generation and survival, DC morphology, DC maturation, antigen presentation/processing, migration, and T cell activation (Kaiserlian and Dubois, 2001; Palucka and Banchereau, 2002; Pollara *et al.*, 2005) (Table 12.1).

TABLE 12.1. Virus-induced phenotypic and functional alterations of human DCs.

Effect	Virus	Mechanism(s)
Interference with DC generation	HTLV-1, EBV	Monocyte infection
Interference with DC survival	HSV-1, VV, MV, HCV, influenza virus	Modulation of apoptosis, aberrant response to cytokines
Loss of DC morphology	HSV-1 and 2, MV	Disruption of Rho GTPase signaling pathway
Induction of aberrant maturation pathway	MV, HSV-1, DV	Uncoupling phenotypic from functional DC activation
Inhibition of DC maturation	HSV-1, HHV-6, VZV, EBV, CMV, VV, Ebola and Lassa viruses	Inhibition of maturation stimuli activity, secretion of regulatory cytokines, inhibition of costimulatory molecule and MHC-peptide complex expression
Modulation of DC migratory capacity	CMV, poxvirus, HSV, MV	Impaired response to chemokines, secretion of soluble chemokine receptor homologues

This effect is usually achieved by the expression of viral genes and virus replication in DCs or their precursors (Fugier-Vivier et al., 1997; Ho et al., 2001; Kakimoto et al., 2002; Li et al., 2002; Makino et al., 2000; Morrow et al., 2003; Riegler et al., 2000; Salio et al., 1999; Schnorr et al., 1997). However, modulation of DC functions in the absence of productive infection or upon exposure to viral soluble factors has also been reported for hepatitis C virus (HCV) (Dolganiuc et al., 2003), vaccinia virus (VV) (Engelmayer et al., 1999), and human herpes virus (HHV-6) (Smith et al., 2005), as well as HIV-1, whose effects are extensively discussed in Section 12.4 and therefore not included in Table 12.1.

12.3.1 Interference with DC Generation and Survival

Viruses can interfere with DC generation by infecting monocytes, the blood precursors of DCs, and impairing their differentiation into DCs, as described for human T cell leukemia virus (HTLV)-1 (Makino et al., 2000). Likewise, Epstein-Barr virus (EBV) infection of monocytes inhibits their differentiation into DCs, leading to an abnormal cellular response to GM-CSF and IL-4 and to apoptotic cell death (Li et al., 2002). In addition, infection of DCs with VV, MV, influenza virus, and herpes simplex virus (HSV)-1 (Cella et al., 1999; Engelmayer et al., 1999; Fugier-Vivier et al., 1997; Muller et al., 2004) or expression of HCV core proteins (Siavoshian et al., 2005) may affect their survival by activating proapoptotic (Siavoshian et al., 2005) or inhibiting anti-apoptotic pathways (Muller et al., 2004). This induced cell death is likely to represent a very efficient way to target the death of DCs at its most critical moment, during their interaction with T cells, as reported for MV (Servet-Delprat et al., 2000a).

12.3.2 Loss of DC Morphology

Cytokeleton organization and turnover play an important role in DC endocytosis and pinocytosis (Garrett et al., 2000; West et al., 2000), as well as in the formation of membrane ruffles and lamellipodia necessary to carry out these functions (Swetman et al., 2002). In addition, cytoskeleton rearrangement is important for chemokine-mediated migration and for antigen processing and presentation (Dong et al., 2003). Loss of DC morphology as a consequence of viral infection has been observed with several viruses, including HSV-1 and -2 and MV (Fugier-Vivier et al., 1997; Murata et al., 2000; Pollara et al., 2003). This effect may result from the disruption of the small Rho GTPase signaling pathway, key regulators of the actin cytoskeleton turnover (Murata *et al.*, 2000; Pollara *et al.*, 2003).

12.3.3 Interference with DC Maturation

The process of DC maturation represents a strategic target for viruses to escape the host immune response for different reasons. For example, viruses can directly induce an aberrant pattern of DC maturation, promoting migration and transmission of infection to T cells in lymphoid organs or, through their antigen-presenting capacity, the DCs could induce B-cell production of virus neutralizing antibodies, thereby facilitating Fc receptor–mediated infection by virus bound to immune complexes (Homsy *et al.*, 1989). For instance, MV infection of DCs leads to upregulation of activation markers (HLA-DR, CD86, and CD83), enhances IL-12 production (Servet-Delprat *et al.*, 2000b), and stimulates DC migration in response to CCL19 (Dubois *et al.*, 2001), indicative of their activation. DC maturation appears largely dependent on IFN-α which is induced during MV infection (Dubois *et al.*, 2001). Interestingly, not only MV-infected DCs but also bystander uninfected DCs undergo a maturation process that results from contact, engulfment of MV-induced apoptotic DCs by uninfected DCs, or both (Servet-Delprat *et al.*, 2000b), as well as from the IFN-α acting on surrounding uninfected DCs (Dubois *et al.*, 2001). Despite the activated phenotype, MV-infected DCs suppress mitogen-dependent proliferation of uninfected peripheral blood lymphocytes (PBLs), indicating that they may potentially contribute to immune suppression in acute measles (Fugier-Vivier *et al.*, 1997; Grosjean *et al.*, 1997; Schnorr *et al.*, 1997). Likewise, HSV-1–infected DCs showed a more mature phenotype than uninfected DCs, characterized by elevated MHC class II and CD86 expression. This upregulation was, however, unable to render the DCs potent T cell stimulators, due to the ability of the virus to disrupt the DC maturation process (Pollara *et al.*, 2003; Salio *et al.*, 1999). In addition, it has been reported that infection of immature DCs with DV causes DC expression of maturation markers, as well as of TNF-α and IFN-α, which could explain delayed DC apoptosis in the late phase of infection (Ho *et al.*, 2001). DV-infected DCs induce the interacting T cells to proliferate, to produce IL-2 as well as to express activation markers, and a cytokine pattern (IL-4, IL-10, IFN-γ) indicative of a Th0 phenotype (Ho *et al.*, 2001). Overall, these results indicate that at least some viruses can turn on an aberrant maturation program of DCs by uncoupling phenotypic from functional activation. This permits migration to lymph nodes of DCs that are strongly impaired in their capacity to activate T cell proliferation, thus resulting in immune suppression.

Alternatively, certain viruses including HSV-1, HHV-6, human cyto-megalovirus (HCMV), EBV, varicella zoster virus (VZV), VV, Ebola, and Lassa viruses may escape immune recognition by directly blocking DC maturation and migration, thus preventing both viral spreading and initiation of the immune response (Engelmayer et al., 1999; Jenne et al., 2000; Moutaftsi et al., 2002; Pollara et al., 2003; Salio et al., 1999; Smith et al., 2003). Inhibition can occur by various means including (i) secretion of soluble receptor homologues for TNF-α, a maturation-inducing stimulus, as well as homologues for factors inhibiting the activity of type I IFN and GM-CSF, as described for VV (Tortorella et al., 2000); (ii) secretion of regulatory cytokines such as the viral IL-10 from EBV or poxviruses (Spriggs, 1999); (iii) inactivation of intracellular pathways such as those targeted by CMV to prevent surface expression of costimulatory molecules and MHC-peptide complexes (Tortorella et al., 2000). Infection of already mature DCs by VZV has also been reported to downregulate maturation-associated surface molecules and to inhibit DC allostimulatory capacity, although the underlying mechanisms have not yet been defined (Morrow et al., 2003).

Independent of the mechanism used, the net result of interfering with DC maturation is an altered capacity of DCs to activate T cells. This defect could be due to inhibition of factors contributing to T cell proliferation and differentiation, as with IL-12. Inhibition of cytokine secretion may result from a complete block of secretion or alternatively from a skewing of the pattern of secreted cytokines. Furthermore, inhibition of T cell activation may also be partly due to the fact that virus-infected DCs become killer cells, as described for CMV and MV, that render DC cytotoxic through the upregulation of both FasL/CD95L and TRAIL (Raftery et al., 2001; Vidalain et al., 2000).

12.3.4 Modulation of DC Migration

In addition to the virus-induced impairment of DC functional activation, characterized by a reduced capacity to stimulate T cell proliferation and to produce IL-12, some viruses also exploit the migratory capacity of DCs as a mechanism to paralyze the early immune response of the host. In this respect, it has been shown that CMV impairs immature DC migration in response to the inflammatory chemokines CCL3 and CCL5 (Varani et al., 2005), as well as mature DC chemotactic response to CCL19 and CCL21 (Moutaftsi et al., 2004). Conversely, as described in Section 12.3.3, the MV-induced chemotactic response of mature DCs to CCL19, allowing migration to draining lymph nodes, could account for the success of the potent T cell immune suppression during acute measles (Dubois *et al.*, 2001), as the viral glycoproteins expressed on the surface of MV-infected DCs potently inhibit T cell proliferation. In addition, some viruses, including poxviruses and HSV, encode secreted homologues of chemokine receptors that act as chemokine antagonists to prevent the attraction of additional DCs at infection sites (McFadden and Murphy, 2000; Tortorella *et al.*, 2000).

12.4 DCs in HIV Pathogenesis: Protective or Defective?

HIV-1 infection is associated with a progressive and gradual loss of immune competence, ultimately leading to an increased susceptibility to neoplastic and infectious diseases. Although HIV infection correlates with abnormalities in most compartments of the immune system, defects in cell-mediated immunity appear to be of great clinical relevance (Douek, 2003; Letvin and Walker, 2003; Stevenson, 2003). Such defects include aberrant or absent $CD4^+$ T cell responses, inefficient $CD8^+$ T cell activity, and dysregulation of APC function. A general state of immune activation is associated with all stages of the disease. This is probably associated with the inability of the immune system to eliminate the virus after primary infection, so that the continuous viral replication during the entire course of the disease acts as a chronic stimulus for both sensitized and nonsensitized cells, thus providing a permanent source of virus-infected cells, free infectious virus, and soluble viral antigens.

APCs, including monocytes/macrophages and DCs, represent the principal cell targets for HIV infection and play important and well recognized roles in multiple aspects of AIDS pathogenesis (Collman *et al.*, 2003; Lore and Larsson, 2003). The study of APCs in HIV pathogenesis has been ongoing for almost two decades. The initial studies recognized the important role of monocytes/macrophages as targets for HIV infection and their ability to serve as reservoirs of virus, particularly in tissues (Crowe *et al.*, 2003). In addition, the impact of HIV infection on another class of APCs, the DCs, has been more recently characterized (Lore and Larsson, 2003). These cells have been identified to play a crucial role in the pathogenesis of HIV infection not only for their susceptibility to the virus but also because their exploitation as a viral target cell can interfere with the development of the host antiviral defense (Knight, 2001). Furthermore, DCs are actively involved in HIV spreading through the body by virtue of their expression of dendritic cell-specific ICAM-3-grabbing nonintegrin 1 (DC-SIGN), a DC-specific surface receptor capable of interacting with the viral gp120, thus allowing the transport of infectious virus from the periphery to secondary lymphoid organs (Geijtenbeek *et al.*, 2000). The observation that, in spite of the success of anti-retroviral therapy in the prolonged suppression of viral load in AIDS patients, infectious virus persists in virus reservoirs, including macrophages and DCs, makes their essential contribution to the pathogenesis of HIV infection even more evident (Aquaro *et al.*, 2002; Letvin and Walker, 2003).

An important aspect of the immunopathogenesis of AIDS is the observation that HIV target cells can also be exposed to different viral gene products, expressed at the surface of infected cells, secreted or released in the microenvironment as a consequence of the death of infected cells. Soluble HIV proteins have been shown to exert bystander effects on neighboring immune cell populations in the absence of productive infection and to interfere with the induction of protective immune response by altering the production of soluble immune mediators as well as the cellular differentiation/activation state (Stevenson, 2003).

The issue of whether HIV affects the functional competence of APCs is still controversial and not yet completely clarified. Although infection, depletion, and dysfunction of DCs may contribute to the immunopathogenesis associated with HIV disease (see Chapter 11), specific cellular and molecular mechanisms by which HIV alters DC functions have not yet been fully understood. Exposure of DCs to HIV-1 may initiate an aberrant pathway of DC maturation, which is associated with a profound impairment of DC functions critically important for the generation of a protective immune response. To address these issues, several studies have been carried out in the past few years by using either infectious or inactivated HIV (Table 12.2) or by using individual viral gene products either endogenously expressed or exogenously administered in soluble form (Table 12.3). Although these studies were successful in associating some of the initially described HIV-1–mediated effects on DC phenotypic and functional properties with the activity of at least some specific HIV-1 gene products (Nef, Tat, Vpr, and gp120), they also yielded a few contrasting results. The discrepancies observed are possibly due to the different populations of monocytes that were used to generate DCs, as well as to the different experimental conditions employed to differentiate and stimulate these cells. In some cases, the different and even opposite results obtained are also probably related to the different delivery/expression systems used to express specific HIV genes into DCs. An overview of the major effects observed on monocyte-derived DC (MDDC) exposure to HIV-1 soluble products is shown in Figure 12.3.

TABLE 12.2. Modulation of DC function by infectious and noninfectious virions.

Virion status and strain		DC response	DC subset	Reference
Inactivated virion	AT2 inactivated (X4 and R5)	- Induction of IFN-α	PDC	Herbeuveal, 2005
	AT2 inactivated (X4 and R5)	- Upregulation of CD80, CD86, CD40, CD83, and MHC class II	MDDC	Fantuzzi, 2004
	AT2 inactivated (X4 and R5)	- Induction of type I IFN - Upregulation of CCL2, CCL3, CCL4	PDC	Del Cornò, 2005
	AT2 inactivated (X4 and R5)	- Upregulation of CD80, CD86, CD83, CCR7 - Induction of IFN-α and TNF-α	PDC	Fonteneau, 2004
	Heat inactivated HIV	- Induction of IFN-α - Upregulation of CD80, CD86	PDC	Yonezawa, 2003
Viral infection	X4 and R5	- Partial upregulation of CD83, CD86, and CCR7 - Downregulation of DC-SIGN - ERK1/2 activation - p38 MAPK activation	iMDDC mMDDC	Wilflingseder, 2004
	X4 and R5	- Upregulation of CD80, CD86, CD83, CCR7 - Induction of IFN-α and TNF-α	PDC	Fonteneau, 2004
	X4 and R5	- Induction of IFN-α - Upregulation of CD80, CD86	PDC	Yonezawa, 2003
	R5 and X4	- Induction of IFN-α, CCL3, and CCL5	PDC	Fong, 2002
	R5	- Upregulation of costimulatory molecules - Impaired response to maturation stimuli	MDDC	Smed-Sorensen, 2004
	R5	- Impaired response to maturation stimuli	MDDC	Granelli-Piperno, 2004
	R5	- Impaired T cell stimulatory capacity and IL-12 secretion	MDDC	Kawamura, 2003
	R5	- Induction of DC cytotoxic activity	MDDC	Lichtner, 2004

TABLE 12.3. Modulation of DC function by HIV-1 soluble proteins.

Protein	Conditions	DC response	DC subset	Reference
gp120	Recombinant protein (R5 and X4)	- Induction of type I IFN - Upregulation of CCL2, CCL3, CCL4	PDC	Del Cornò, 2005
	Recombinant protein (R5 and X4)	- Upregulation of CD80, CD86, CD40, CD83, and MHC class II, downregulation of MHC class I - No induction of IL-12, TNF-α, CCL3, CCL4, CCL5 - No stimulation of T cell response - Impaired maturation of DC promoted by classical stimuli	iMDDC	Fantuzzi, 2004
	Recombinant protein (R5)	- Induction of migration	iMDDC	Lin, 2000
	Recombinant protein (X4)	- Upregulation of CD80, CD86, CD40, MHC class II, CD54, CXCR4, CCR7 - Downregulation of MR, CCR6 - Induction of IL-10, IL-12, IL-18, TNF-α	iMDDC iMDDC mMDDC	Williams, 2002
	Coated beads	- Binding to DC-SIGN induces recruitment of LFA-1	iMDDC	van Gisbergen, 2005
Nef	Adenovirus-Nef	- Induction of IL-6, IL-12, TNF-α, CCL3, CCL4, CXCL8 - No phenotypic changes - Increased T lymphocyte stimulatory capacity - Activation of STAT3 and NFkB	iMDDC	Messmer, 2002a and 2002b
	VSV/HIV pseudotypes	- Downregulation of MHC class I - Upregulation of DC-SIGN	iMDDC	Sol-Foulon, 2002
	VSV/HIV pseudotypes	- Downregulation of MHC class I and CD1a - No effect on DC activation markers	iMDDC	Shinya, 2004
	Adenovirus-Nef	- No change in membrane levels of MHC class I or CD4	iMDDC	Cramer, 2001
	Recombinant Protein	- Induction of IL-1β, IL-8, IL-12, IL-15, TNF-α, CCL3, CCL4 - Upregulation of CD1a, CD40, CXCR4, MHC class II, CD80, CD86 - Downregulation of MHC class I and MR	iMDDC	Quaranta, 2002 and 2003

		- Increased T lymphocyte stimulatory capacity - Decreased endocytic activity - Activation of NFkB and Vav - Cytoskeleton rearrangement - Inhibition of CD8$^+$ CTL cytotoxic activity and induction of CD8$^+$ T cell apoptosis	iMDDC	Quaranta, 2002 and 2003
	Vaccinia virus-Nef	- Downregulation of MHC class I - Impaired CD8$^+$ T cell–specific response	iMDDC	Andrieu, 2001
Tat	Adenovirus-Tat	- Expression of interferon-inducible genes. - Upregulation of CXCL9, CXCL10, CCL7, CCL8 - No change in membrane levels of CD1a, CD1c, CD40, CD62L, CD83, CD86, MHC class II	iMDDC	Izmailova, 2003
	Recombinant Protein	- Upregulation of CD40, CD80, CD83, CD86, MHC class I and II - Induction of IL-12, TNF-α, CCL3, CCL4, CCL5 - Increased T lymphocyte stimulatory capacity	iMDDC	Fanales-Belasio, 2002
	Recombinant Protein	- Induction of migration	iMDDC	Benelli, 1998
Vpr	VSV/HIV pseudotypes	- Downregulation of CD80, CD86, CD83 - Induction of IL-10, inhibition of IL-12 production - Impaired maturation of DC promoted by classical stimuli	iMDDC; mMDDC	Majumder, 2005
	Recombinant Protein	- Downregulation of CD80, CD86, CD83 - Impaired maturation of DC promoted by classical stimuli - Inhibition of IL-12 and TNF-α production	iMDDC; mMDDC	Muthumani, 2005

iMDDC, immature MDDC; mMDDC, mature MDDC.

The following sections will examine the mechanisms underlying the capacity of HIV-1 to dysregulate DC development and functions. Aspects will be discussed concerning both the effects related to the expression of specific viral components in the course of HIV productive infection and bystander effects exerted by soluble viral factors released at the sites of active viral replication on these cells.

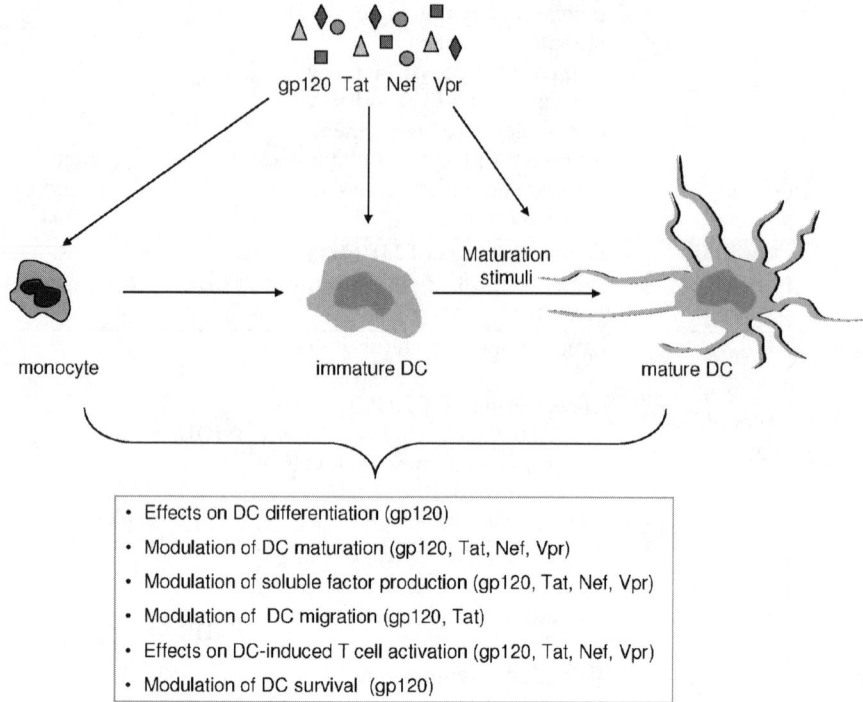

FIGURE 12.3. Bystander effects of HIV-1 soluble products on MDDCs. HIV gene products can be released in the serum and tissues as soluble factors or be exposed at the surface of infected cells (gp120) and interfere with DC differentiation and maturation, acting directly on monocytes and immature DCs or interfering with their response to maturation stimuli. Depending on the HIV product to which DCs are exposed, the net result of this interaction can be the modulation of soluble factor production, of T cell activation, as well as of DC survival.

12.4.1 HIV-1 Effects on DC Differentiation/Maturation

With respect to the effects of HIV infection of DCs generated from mono-cyte precursors (MDDCs), both *in vitro* and *in vivo* studies have revealed a profound alteration of their properties. Although MDDCs infected *in vitro* with R5 HIV-1 upregulate the expression of costimulatory molecules, their production of effector cytokines in response to CD40L is skewed as compared with control uninfected cells (Smed-Sorensen *et al.*, 2004).

In keeping with the hypothesis that HIV infection disables DC function, Granelli-Piperno and colleagues subsequently demonstrated that, although HIV infection itself does not induce DC phenotypic and functional maturation, infected DCs specifically fail to mature in response to different stimuli (Granelli-Piperno *et al.*, 2004). In addition, when DCs from HIV-infected patients were infected *in vitro* and cultured with autologous T cells, IL-10 was produced in 6 out of the 10 patients and, furthermore, the DC–T cell cocultures could suppress a further immune response, the allogeneic T cell proliferation, by secreting soluble factors, including IL-10. Finally, Kawamura and co-workers reported that although enriched populations of *in vitro* HIV-infected DCs secreted increased levels of IL-12 whereas IL-10 production was decreased, these cells were poor stimulators of allogeneic CD4$^+$ T cell proliferation and IL-2 production. Interestingly, HIV-infected DCs secreted gp120 that impairs normal CD4$^+$ T cell functions *in vitro* (Kawamura *et al.*, 2003).

Until a few years ago, little was known on the effect of HIV early interactions with DCs on the phenotypic and functional properties of these cells. The use of viruses inactivated with AT-2, a procedure that blocks infectivity without altering the conformation of the envelope proteins (Rossio *et al.*, 1998), as well as of the recombinant envelope glycoproteins has made it possible to demonstrate that most of the effects observed in HIV-exposed DCs are independent of virus replication and directly related to gp120 recognition. This further highlights the importance of bystander effects exerted by HIV soluble products on surrounding uninfected cells. In particular, it was shown that exposure of MDDCs to AT-2–inactivated virus or gp120 results in the acquisition of a mature phenotype, characterized by a significant upregulation of costimulatory molecules, MHC class II antigens, as well as the appearance of the CD83 maturation marker. However, despite the acquisition of an apparently activated phenotype, gp120-exposed DCs are functionally impaired in terms of effector cytokines (IL-12 and TNF-α) and chemokines (CCL2, CCL3, CCL4) production and capacity to induce T cell proliferative responses. gp120-exposed DCs also retained a significantly high capacity to up-take antigens (Fantuzzi *et al.*, 2004).

The finding that HIV exposure of DCs uncouples phenotypic from functional activation is consistent with the above described results achieved by Kawamura and colleagues (Kawamura *et al.*, 2003). Functional but not phenotypic impairment of DCs is also observed when immature DCs are generated from monocytes in the presence of gp120 and then stimulated to mature by LPS or CD40L (Fantuzzi *et al.*, 2004). However, it should be noted that, in contrast with the above-mentioned studies documenting DC impairment, Williams and colleagues reported that X4 gp120 exposure of DCs results in the acquisition of an activated phenotype correlating with increased IL-12 production and allostimulatory capacity (Williams *et al.*, 2002). These discrepancies are likely explained by the fact that different populations of monocytes were used to generate

DCs, as well as by the different experimental conditions used to differentiate and stimulate these cells.

HIV-1 gp120 has been shown to activate signal transduction pathways through both CD4 and chemokine receptors in several cell types, including DCs. Recent studies revealed that at least some of these signaling events are of crucial importance for the HIV-mediated effects on DC biology. In keeping with previous results obtained in T lymphocytes and macrophages, showing that HIV triggers a MAPK-mediated signal transduction cascade, acting as a central pathway in the signaling network of the host cell (Del Corno *et al.*, 2001; Popik and Pitha, 2000), the MAPK family members ERK 1/2 and p38 have been shown to be activated soon after the infection of DCs with HIV-1 (Wilflingseder *et al.*, 2004). In particular, it has been shown that short- and long-term exposure of immature MDDCs to different R5 and X4 HIV-1 strains results in alterations of DC phenotype associated with differential phosphorylation of the ERK1/2 and p38 MAPK signaling pathways. The phosphorylation of the ERK1/2 was most prominent soon after the addition of the virus, and short-term HIV-1 exposure also enhanced p38 MAPK phosphorylation in both immature and mature DCs. Activation of the MAPK pathway could be dependent on the interaction of the viral envelope protein(s) with cellular receptors, as the enhanced HIV-1–induced p38 MAPK phosphorylation was not affected by the reverse transcriptase inhibitor AZT (Wilflingseder *et al.*, 2004). Interestingly, this signal transduction pathway has been reported to play a role in DC maturation induced by a variety of agents (LPS, TNF-α, contact sensitizers, and ribotoxic stress) (Arrighi *et al.*, 2001; Puig-Kroger *et al.*, 2001; Rescigno *et al.*, 1998; Sato *et al.*, 1999).

In addition to the effect of early interactions of HIV with DCs, the effect of other viral gene products on DC biology has been also investigated. Concerning the role of Tat in the modulation of DC properties and functions, different and even opposite results have been reported, probably due to the different delivery/expression systems used. In particular, it has been shown that recombinant native Tat is efficiently taken up by MDDCs and that its uptake results in a significant enhancement of the expression of MHC class I and II, costimulatory molecules, and CD83 (Fanales-Belasio *et al.*, 2002). Furthermore, in contrast with what was observed for gp120, this mature phenotype was accompanied by full functional activation of DCs as assessed by a marked and dose-dependent induction of effector cytokines (IL-12 and TNF-α) and β-chemokines (CCL3, CCL4, and CCL5) and a significant enhancement of the DC capacity to stimulate the proliferation of lymphocytes in response to allogeneic and recall antigens. These results not only suggest a direct role for soluble Tat in inducing the maturation of DCs and the generation of T cell responses but also argue in favor of an adjuvant property of this protein in the induction of the immune response toward other antigens. In contrast with the results achieved by using recombinant native Tat, the endogenous expression of Tat through adenoviral vectors in immature DCs completely failed to induce their maturation in terms of acquisition of maturation-specific markers and production of effector cytokines. However, endogenous Tat expression in DCs led to marked changes in the transcriptional profile of these cells, including expression of transcription factors (STAT1, IRF-7) and chemokines (IP-10, MIG, CCL7, and CCL8) (Izmailova et al., 2003).

Similar to the above-described effects of Tat, a number of different Nef-specific effects have been observed in DCs, depending on whether recombinant or endogenously expressed Nef were used (Messmer et al., 2002a; Messmer et al., 2002b; Quaranta et al., 2002; Quaranta et al., 2003; Sol-Foulon et al., 2002). In particular, recombinant Nef exposure of immature MDDCs was shown to result in the upregulation of CD1a, costimulatory and MHC class II molecules, concomitantly with downmodulation of MHC class I and mannose receptor (Quaranta et al., 2002). The functional consequence of this phenotypic maturation was a decrease in endocytic and phagocytic activities of DCs, upregulation of the production of inflammatory cytokines and chemokines, and increase in the $CD4^+$ T cell stimulatory capacity. In keeping with these results, a specific downregulation of MHC class I was also observed upon DC transduction with Nef-expressing vaccinia virus vector (Andrieu et al., 2001) or infection with nef^+ VSVG-HIV-1 (Shinya et al., 2004), while a decrease in CD1a expression was conversely seen under these latter conditions. As mentioned above, in contrast with the results achieved by Quaranta and co-workers, the phenotypic properties of DCs were not found to be affected upon adenoviral vector- or VSVG-HIV-mediated delivery of Nef (Messmer et al., 2002b; Sol-Foulon et al., 2002). Furthermore, it was recently reported that recombinant Nef exposure of immature MDDCs results in a Vav-mediated rearrangement of actin microfilaments leading to the formation of uropods and ruffles (Quaranta et al., 2003). The imme-diate consequence of these morphologic changes is an increased capacity of Nef exposed DCs to form clusters with allogeneic $CD4^+$ T cells, in keeping with the hypothesis that the main goal of Nef-induced mani-pulation of DCs is to foster virus dissemination. Consistent with this hypothesis, a marked Nef-induced upregulation of DC-SIGN, whose expression is responsible for the DC-mediated transfer of infectious virions to T lymphocytes, has been described (Sol-Foulon et al., 2002). Infecting immature MDDCs with wild-type HIV-1 resulted in a marked accumulation of DC-SIGN at the surface of infected DCs. This effect was independent of cell maturation and directly associated with the expression of Nef, as no effect was observed on infection of MDDCs with Δnef-HIV-1. The functional consequence of DC-SIGN surface accumulation was again a dramatic increase in DC–T cell cluster formation and HIV transfer. One of the major issues of HIV pathogenesis is virus escape from CTL response. In this regard, Nef has been described to contribute to the impairment of CD8$^+$ T-cell functions by down-regulating the expression of MHC class I in HIV-infected cells (Schwartz *et al.*, 1996). Recently, a Nef-mediated decrease in MHC class I surface expression has also been shown in MDDCs, independently of the delivery system used (Andrieu *et al.*, 2001; Quaranta *et al.*, 2002; Shinya *et al.*, 2004). Moreover, this downmodulation has been correlated with the impairment of Nef-expressing immature DCs in their presentation to Nef-specific CD8$^+$ T cell clones (Andrieu *et al.*, 2001). In addition to this classical mechanism of CD8$^+$ T cell response impairment, a novel mechanism has also been proposed by which Nef induces anergy and apoptosis in CD8$^+$ T cells by manipulating DC functions (Quaranta *et al.*, 2004). In particular, Nef was shown not only to inhibit CD8$^+$ T lymphocyte cytotoxic activity and IFN-γ production via downregulation of MHC class I surface expression, but also to induce apoptosis of CD8$^+$ T cells by upregulating the expression of both TNF-α and FasL in DCs. According to these findings, Nef has acquired the capacity to subvert DC biology in order to affect both functional competence and survival of CD8$^+$ T cells, by modulating the expression and production of key signaling receptors and molecules.

The effects of Vpr on the biology and functions of DCs have been only recently investigated (Majumder *et al.*, 2005; Muthumani *et al.*, 2005). The use of two different experimental approaches, Vpr-expressing virus and recombinant Vpr, has allowed the study of both direct effects, possibly occurring in HIV-infected DCs, and bystander effects due to exposure of uninfected cells to soluble Vpr. In particular, it has been shown that exposure of immature MDDCs to recombinant Vpr results in down-modulation of the expression of CD83 and costimulatory molecules. Moreover, both in the context of virus replication and exogenous addition, this protein was shown to inhibit the upregulation of the same molecules on DC maturation induction by TNF-α, LPS, or CD40L. Concerning the effects of Vpr exposure on the functional properties of DCs, both cytokine production and antigen presentation to T cells were reported to be affected by the protein. In fact, recombinant Vpr-exposed DCs produced lower amounts of IL-12 and TNF-α compared with control or Gag-exposed DCs and exhibited a reduced capacity to stimulate the proliferation of allogeneic T lymphocytes. Similarly, DCs infected with Vpr-expressing HIV were reported to retain the antigen uptake capacity typical of immature cells, to produce lower amounts of IL-12, and to be less efficient in stimulating antigen-specific CD8$^+$ T cell clones as compared with DCs infected with the Vpr-minus virus. As already reported for other Vpr effects on different cell populations, the downmodulation of costimulatory molecules on DCs was shown to be blocked by the glucocorticoid receptor (GR) antagonist RU486, thus suggesting the involvement of GR-Vpr interaction in this phenomenon (Muthumani *et al.*, 2005). Interestingly, despite the well-known capacity of Vpr to induce apoptosis in different cell types, the survival of DCs does not seem to be influenced by this protein (Majumder *et al.*, 2005).

Initial studies carried out with DC subsets isolated from HIV-infected patients revealed that *ex vivo* PDCs and myeloid DCs (MDCs) exhibit a higher expression of the maturation markers CD86 and CD40 (Barron *et al.*, 2003; Grassi *et al.*, 1999). Likewise, *in vitro* studies showed that PDCs undergo maturation after direct exposure to HIV-1, as assessed by an increased expression of costimulatory molecules, MHC class II, CD83, and CCR7 (Fonteneau *et al.*, 2004; Yonezawa *et al.*, 2003). Conversely, although the phenotypic and functional properties of MDCs were not directly affected by the virus, the cytokines secreted by HIV-activated PDCs were able to induce their maturation (Fonteneau *et al.*, 2004). In contrast with the HIV-mediated activation of blood DC subsets, the DCs found in lymphoid tissues at late stages of HIV-1 infection failed to undergo full maturation and showed reduced expression of CD80 and CD86 (Hsieh *et al.*, 2003; Lore *et al.*, 2002). Because functional maturation of DCs is essential for activation and expansion of effector T cells, the decreased expression of costimulatory molecules is expected to have negative effects on antigen presentation and activation of T cell responses.

12.4.2 Modulation of Cytokine/Chemokine Secretion

Several studies investigated the functional consequences of HIV exposure of different DC subsets on their capacity to release cytokines and chemokines, immunomediators that play a fundamental role in controlling the homeostasis of the immune system. As a key cell type in innate antiviral immunity, PDCs have been studied in more detail in the context of HIV-1 infection and characterized for their capacity to directly recognize and respond to HIV-1 infection by producing large quantities of IFN-α (Beignon *et al.*, 2005; Fong *et al.*, 2002; Fonteneau *et al.*, 2004; Herbeuval *et al.*, 2005; Schmidt *et al.*, 2005; Yonezawa *et al.*, 2003). Virus-induced IFN-α contributes, at least in part, to the restriction of viral replication in PDCs themselves (Fong *et al.*, 2002; Yonezawa *et al.*, 2003) and CD4$^+$ T cells (Schmidt *et al.*, 2005) as well as to the PDC-induced MDC maturation (Fonteneau *et al.*, 2004) (Fig. 12.4).

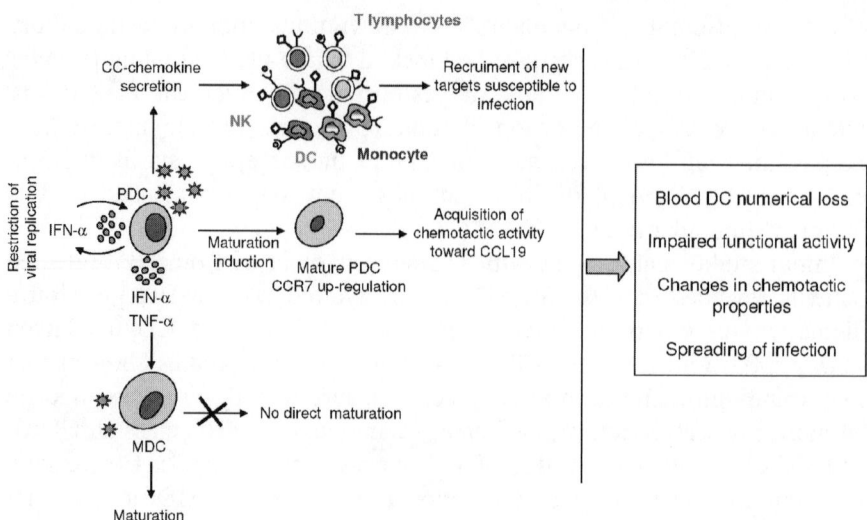

FIGURE 12.4. Immunoregulatory effects of HIV-1 exposure of blood DC subsets. HIV-exposed PDCs undergo a maturation process and produce IFN-α, which in turn acts as a restriction factor for HIV replication in these cells. PDC maturation is associated with CCR7 upmodulation and increased chemotactic response to CCL19. In contrast, MDCs do not directly respond to HIV, although they can be productively infected, but soluble factors released by infected PDCs may promote their maturation. HIV-exposed PDCs also release β-chemokines thus favoring the recruitment of other immune cells susceptible to HIV infection. HIV-1 encounter of PDCs and MDCs ultimately leads to their depletion, although the underlying mechanisms are not yet well defined.

There is general consensus that gp120 is required for IFN-α induction by PDCs and that this effect is mediated through its interaction with CD4 (Fonteneau et al., 2004; Beignon et al., 2005; Herbeuval et al., 2005; Schmidt et al., 2005). However, whether gp120 is sufficient to induce IFN-α production remains controversial. We have recently studied the relative importance of HIV-1 surface components in triggering IFN production in DCs. To address this issue, we comparatively studied IFN-α production by donor-matched blood DC subsets and in vitro generated MDDCs, exposed to either recombinant gp120 or gp41 (Del Cornò et al., 2005). Both X4 and R5 gp120 were sufficient to induce IFN-α secretion by PDCs, although to a lesser extent than the whole inactivated virus. Conversely, negligible IFN-α secretion was observed upon gp41 treatment (Del Cornò et al., 2005). In contrast, neither gp120 nor gp41 induced IFN-α production by MDCs or MDDCs. However, in other studies, gp120 and gp160 pre-parations from different viral strains failed to induce any detectable IFN production by PDCs (Beignon et al., 2005; Herbeuval et al., 2005;

Schmidt *et al.*, 2005). Interestingly, it was recently reported that IFN-α production by PDCs in response to HIV-1 requires at least two interactions between the virus and the cell. Initially, envelope-CD4 interactions mediate endocytosis of HIV. Subsequently, endosomally delivered viral nucleic acids stimulate PDCs through Toll-like receptor (TLR)7 (Beignon *et al.*, 2005). Additionally, a role for TLR9 signaling in this process was suggested by the inhibition of HIV-induced IFN production by low concentrations of chloroquine (Schmidt *et al.*, 2005), which are known to selectively block TLR9 (Lee *et al.*, 2003). These discrepant results between studies may indicate the existence of distinct mechanisms operating at different levels, that is, one at the level of the cellular membrane upon early DC-HIV-1 interaction followed by another event occurring within the PDC after viral internalization. It should be noted that different experimental conditions were used in the above-mentioned studies (notably, differences in the number of cells, in the gp120/gp160 preparations, or in the time of treatment), which may have favored one mechanism over another, providing thus an explanation for such discrepant results.

PDC interaction with HIV has also been described to result in the production of a panel of chemokines. In particular, comparative analysis of PDCs, MDCs, and MDDCs derived from the same donor demonstrated that gp120 upregulates the production of some inflammatory chemokines (CCL2, CCL3, and CCL4) only in PDCs, whereas MDCs seem to be completely refractory (Del Corno *et al.*, 2005). These results are in agreement with the data showing the induction, in PDCs, of the proinflammatory cytokine TNF-α by AT-2–inactivated HIV (Fonteneau *et al.*, 2004) as well as of CCL3 and CCL5 by infectious virus (Fong *et al.*, 2002).

In contrast with what was observed for gp120, Tat exposure of DCs results in a marked and dose-dependent induction of inflammatory cytokines (IL-12 and TNF-α) and β-chemokines (CCL3, CCL4, and CCL5) in MDDCs (Fanales-Belasio *et al.*, 2002). By using the same cellular model but a different experimental approach (endogenous expression of Tat through adenoviral vector), it was reported that Tat expression in immature MDDCs upmodulates the expression of a panel of mRNA coding for chemokines (Izmailova *et al.*, 2003). This Tat induced transcription correlates with secretion of CXCL10, CXCL9, CCL7 and CCL8 in the culture medium of Tat-expressing DCs and with the capacity of conditioned medium to promote the chemotaxis of target monocytes and activated T lymphocytes. The same chemokines were also found in the culture medium of DCs productively infected with different R5 strains of HIV-1 at early stages of infection, and their secretion was shown to be responsible for the chemotactic activity of supernatants from infected cultures. The induction of chemotactic factors acting either on monocytes or on activated T lymphocytes, both representing the main HIV target cell populations *in vitro*, may ultimately lead to an increased virus propagation through a DC-mediated recruitment of new target cells.

Similar to Tat, both endogenously produced and exogenously added Nef were found to upregulate the secretion of a panel of inflammatory cytokines (IL-1β, IL-6, IL-8, IL-12, IL-15, TNF-α) and monocyte/lymphocyte-recruiting β-chemokines (CCL3, CCL4, CCL5, CXCL8) in immature MDDCs (Messmer *et al.*, 2002b; Quaranta *et al.*, 2002; Teleshova *et al.*, 2003).

In contrast with the above-reported effects of other HIV-1 soluble products, Vpr has not yet been described to modulate chemokine secretion in DCs. However, cytokine profiling of MDDCs either infected with HIV-1 vpr[+] virus (Majumder *et al.*, 2005) or exposed to extracellular delivered Vpr (Muthumani *et al.*, 2005) indicated that this protein inhibits the production of IL-12 and TNF-α and upregulates IL-10, whereas IL-6 and IL-1β production is unaltered (Majumder *et al.*, 2005; Muthumani *et al.*, 2005).

12.4.3 Regulation of DC Chemotactic Functions

Leukocyte infiltration at the sites of infection or inflammation is a key event in host defense. The recruitment of leukocytes to an inflammatory site is a controlled event dependent on the binding of a chemotactic ligand to the cells, polarization of the cells in the direction of the gradient, and movement toward the signal source (Ridley *et al.*, 2003). Both native and recombinant gp120 are chemotactic for monocytes (Cruikshank *et al.*, 1990), suggesting that gp120 may recruit and activate monocytes during the course of infection. Parallel studies carried out in MDDCs have shown that immature DCs migrate toward R5 but not X4 HIV-1 strains. Furthermore, pre-exposure to R5 HIV-1 or its recombinant gp120 protein prevents migration toward CCR5 ligands (Lin *et al.*, 2000). This activation of APCs by gp120 can result in the internalization of a number of chemoattractant receptors, including HIV-1 fusion coreceptors, contributing to the reported viral interference after initial infection as well as to the suppression of APC-dependent inflammatory reactions.

As previously seen for gp120, in addition to its capacity to influence the endogenous production of chemokines by DCs, Tat was also reported to act as a direct chemoattractant for these cells (Benelli *et al.*, 1998). DC migration, induced in a dose-dependent manner by recombinant Tat, was shown to be mediated by both the basic and the RGD domains of the protein and to involve the expression of $\alpha_v\beta_3$ integrin. The capacity of Tat to directly recruit DCs at the active HIV replication sites would be of further advantage for the virus and might contribute to the overall increase in the number of potential cells becoming infected.

The capacity of HIV to affect DC migration is not only related to the ability of some viral genes to directly trigger DC chemotaxis (gp120 and Tat) but also to the fact that, at least in some DC subsets, HIV exposure results in major changes in the expression of chemokine receptors, thus profoundly modifying the response of DCs to specific chemoattractants. In this regard, it has been reported that HIV-1 markedly enhances CCR7 expression in PDCs (Fonteneau et al., 2004; Schmidt et al., 2005). It has been suggested that this HIV-mediated effect may account for the decrease of these cells in blood and for the increased number of DCs seen in lymph nodes in vivo during primary HIV-1 infection (Lore et al., 2002).

12.4.4 Effect on DC Survival

Several reports showed that significant changes in the number of DC subsets can occur during the course of HIV infection, as extensively described in Chapter 11. However, little is still known on the underling mechanisms responsible for this phenomenon and it is still debated whether DC numerical changes are the result of cell loss or redistribution of specific subsets from the blood to other areas of the body. In this regard, it has been reported that exposure of PDCs to AT-2–inactivated X4 and R5 HIV-1 affects the viability of these cells rendering them able to survive for a few days in the absence of IL-3 (Fonteneau et al., 2004). The increased survival of HIV-exposed PDCs as well as their enhanced expression of CCR7 enabling them to migrate in response to the nodal chemokine CCL19 provide some evidence in favor of the hypothesis that at least in part PDCs may relocate in different areas in HIV infection.

12.4.5 Effect on DC Cytotoxic Activity

MDDCs exposed to HIV-1 can become cytotoxic and kill interacting CD4[+] and CD8[+] T cells, instead of activating them, via induction of apoptosis in a FasL-, TRAIL-, or TNF-α–dependent manner (Geleziunas et al., 2001; Lichtner et al., 2004; Quaranta et al., 2004). The apoptosis of CD8[+] T cells occurs via activation of the death receptor pathway and is due to Nef-induced upregulation of TNF-α and expression of FasL by the DCs, thus making the DC cytotoxic (Quaranta et al., 2004). The characterization of this effect revealed an additional bystander effect by which noninfected T cells are killed during HIV-1 infection.

12.5 Concluding Remarks

The cellular immune response developed against HIV is complex. DCs are crucial for host immunity and, at the same time, HIV-1 interacts with DCs to facilitate pathogenesis. This puts the immune system in a dilemma: the cell required for priming of HIV-specific T cell, namely the DC, is also responsible for the transmission of the virus to the newly activated T cell.

Despite the extensive research on DC biology over the past decade, we are still at the early stages of understanding all aspects of HIV interaction with DCs. Recent findings show that HIV-1 exerts immunomodulatory effects on certain DC subsets, including impairment of maturation, altered cytokine profiles, defective stimulatory function, altered migration and acquisition of cytotoxic activity. It is now becoming evident that at least in part these effects do not require direct infection of DCs but can result from bystander effects exerted by HIV soluble products locally released at the sites of active viral replication. Because the different viral proteins have been shown to exert both similar and opposite effects on immune response through their action on DCs, it is not yet clear how these effects could work together *in vivo* and contribute to the establishment of infection and eventually to disease progression. It is expected that a delicate balance between the effects of single proteins, together with host-specific factors, would contribute to dictate the outcome of disease. Delineating the specific functions and pathways affected and their roles in immune escape will improve our knowledge of AIDS pathogenesis and enable further improvements in the design of therapeutic and/or prophylactic strategies for combating HIV infection.

References

Abendroth, A., Morrow, G., Cunningham, A. L. and Slobedman, B. (2001). Varicella-zoster virus infection of human dendritic cells and transmission to T cells: implications for virus dissemination in the host. J Virol. 75: 6183-92.

Andrieu, M., Chassin, D., Desoutter, J. F., Bouchaert, I., Baillet, M., Hanau, D., Guillet, J. G. and Hosmalin, A. (2001). Downregulation of major histo-compatibility class I on human dendritic cells by HIV Nef impairs antigen presentation to HIV-specific CD8[+] T lymphocytes. AIDS Res Hum Retroviruses. 17: 1365-70.

Aquaro, S., Calio, R., Balzarini, J., Bellocchi, M. C., Garaci, E. and Perno, C. F. (2002). Macrophages and HIV infection: therapeutical approaches toward this strategic virus reservoir. Antiviral Res. 55: 209-25.

Arrighi, J. F., Rebsamen, M., Rousset, F., Kindler, V. and Hauser, C. (2001). A critical role for p38 mitogen-activated protein kinase in the maturation of human blood-derived dendritic cells induced by lipopolysaccharide, TNF-alpha, and contact sensitizers. J Immunol. 166: 3837-45.

Asada, H., Klaus-Kovtun, V., Golding, H., Katz, S. I. and Blauvelt, A. (1999). Human herpesvirus 6 infects dendritic cells and suppresses human immunodeficiency virus type 1 replication in coinfected cultures. J Virol. 73: 4019-28.

Bachmann, M. F. and Zinkernagel, R. M. (1997). Neutralizing antiviral B cell responses. Annu Rev Immunol. 15: 235-70.

Baize, S., Kaplon, J., Faure, C., Pannetier, D., Georges-Courbot, M. C. and Deubel, V. (2004). Lassa virus infection of human dendritic cells and macrophages is productive but fails to activate cells. J Immunol. 172: 2861-9.

Banchereau, J. and Steinman, R. M. (1998). Dendritic cells and the control of immunity. Nature. 392: 245-52.

Barron, M. A., Blyveis, N., Palmer, B. E., MaWhinney, S. and Wilson, C. C. (2003). Influence of plasma viremia on defects in number and immuno-phenotype of blood dendritic cell subsets in human immunodeficiency virus 1-infected individuals. J Infect Dis. 187: 26-37.

Beignon, A. S., McKenna, K., Skoberne, M., Manches, O., DaSilva, I., Kavanagh, D. G., Larsson, M., Gorelick, R. J., Lifson, J. D. and Bhardwaj, N. (2005). Endocytosis of HIV-1 activates plasmacytoid dendritic cells via Toll-like receptor-viral RNA interactions. J Clin Invest. 115: 3265-75.

Belardelli, F. (1995). Role of interferons and other cytokines in the regulation of the immune response. APMIS. 103: 161-79.

Bender, A., Albert, M., Reddy, A., Feldman, M., Sauter, B., Kaplan, G., Hellman, W. and Bhardwaj, N. (1998). The distinctive features of influenza virus infection of dendritic cells. Immunobiology. 198: 552-67.

Benedict, C. A., Norris, P. S. and Ware, C. F. (2002). To kill or be killed: viral evasion of apoptosis. Nat Immunol. 3: 1013-8.

Benelli, R., Mortarini, R., Anichini, A., Giunciuglio, D., Noonan, D. M., Montalti, S., Tacchetti, C. and Albini, A. (1998). Monocyte-derived dendritic cells and monocytes migrate to HIV-Tat RGD and basic peptides. AIDS. 12: 261-8.

Bhardwaj, N., Bender, A., Gonzalez, N., Bui, L. K., Garrett, M. C. and Steinman, R. M. (1994). Influenza virus-infected dendritic cells stimulate strong proliferative and cytolytic responses from human CD8[+] T cells. J Clin Invest. 94: 797-807.

Biron, C. A., Nguyen, K. B., Pien, G. C., Cousens, L. P. and Salazar-Mather, T. P. (1999). Natural killer cells in antiviral defense: function and regulation by innate cytokines. Annu Rev Immunol. 17: 189-220.

Blaho, J. A. (2004). Virus infection and apoptosis (issue II) an introduction: cheating death or death as a fact of life? Int Rev Immunol. 23: 1-6.

Burton, D. R., Williamson, R. A. and Parren, P. W. (2000). Antibody and virus: binding and neutralization. Virology. 270: 1-3.

Cameron, P. U., Freudenthal, P. S., Barker, J. M., Gezelter, S., Inaba, K. and Steinman, R. M. (1992). Dendritic cells exposed to human immunodeficiency virus type-1 transmit a vigorous cytopathic infection to CD4$^+$ T cells. Science. 257: 383-7.

Cella, M., Salio, M., Sakakibara, Y., Langen, H., Julkunen, I. and Lanzavecchia, A. (1999). Maturation, activation, and protection of dendritic cells induced by double-stranded RNA. J Exp Med. 189: 821-9.

Collman, R. G., Perno, C. F., Crowe, S. M., Stevenson, M. and Montaner, L. J. (2003). HIV and cells of macrophage/dendritic lineage and other non-T cell reservoirs: new answers yield new questions. J Leukoc Biol. 74: 631-4.

Crowe, S., Zhu, T. and Muller, W. A. (2003). The contribution of monocyte infection and trafficking to viral persistence, and maintenance of the viral reservoir in HIV infection. J Leukoc Biol. 74: 635-41.

Cruikshank, W. W., Center, D. M., Pyle, S. W. and Kornfeld, H. (1990). Biologic activities of HIV-1 envelope glycoprotein: the effects of crosslinking. Biomed Pharmacother. 44: 5-11.

de Fraissinette, A., Schmitt, D. and Thivolet, J. (1989). Langerhans cells of human mucosa. J Dermatol. 16: 255-62.

de Jong, E. C., Smits, H. H. and Kapsenberg, M. L. (2005). Dendritic cell-mediated T cell polarization. Springer Semin Immunopathol. 26: 289-307.

Del Corno, M., Liu, Q. H., Schols, D., de Clercq, E., Gessani, S., Freedman, B. D. and Collman, R. G. (2001). HIV-1 gp120 and chemokine activation of Pyk2 and mitogen-activated protein kinases in primary macrophages mediated by calcium-dependent, pertussis toxin-insensitive chemokine receptor signaling. Blood. 98: 2909-16.

Del Corno, M., Gauzzi, M. C., Penna, G., Belardelli, F., Adorini, L. and Gessani, S. (2005). Human immunodeficiency virus type 1 gp120 and other activation stimuli are highly effective in triggering alpha interferon and CC chemokine production in circulating plasmacytoid but not myeloid dendritic cells. J Virol. 79: 12597-601.

Dolganiuc, A., Kodys, K., Kopasz, A., Marshall, C., Do, T., Romics, L., Jr., Mandrekar, P., Zapp, M. and Szabo, G. (2003). Hepatitis C virus core and nonstructural protein 3 proteins induce pro- and anti-inflammatory cytokines and inhibit dendritic cell differentiation. J Immunol. 170: 5615-24.

Dong, R., Cwynarski, K., Entwistle, A., Marelli-Berg, F., Dazzi, F., Simpson, E., Goldman, J. M., Melo, J. V., Lechler, R. I., Bellantuono, I., Ridley, A. and Lombardi, G. (2003). Dendritic cells from CML patients have altered actin organization, reduced antigen processing, and impaired migration. Blood. 101: 3560-7.

Douek, D. C. (2003). Disrupting T-cell homeostasis: how HIV-1 infection causes disease. AIDS Rev. 5: 172-7.

Dubois, B., Lamy, P. J., Chemin, K., Lachaux, A. and Kaiserlian, D. (2001). Measles virus exploits dendritic cells to suppress CD4$^+$ T-cell proliferation via expression of surface viral glycoproteins independently of T-cell trans-infection. Cell Immunol. 214: 173-83.

Engelmayer, J., Larsson, M., Subklewe, M., Chahroudi, A., Cox, W. I., Steinman, R. M. and Bhardwaj, N. (1999). Vaccinia virus inhibits the maturation of human dendritic cells: a novel mechanism of immune evasion. J Immunol. 163: 6762-8.

Everett, H. and McFadden, G. (1999). Apoptosis: an innate immune response to virus infection. Trends Microbiol. 7: 160-5.

Fanales-Belasio, E., Moretti, S., Nappi, F., Barillari, G., Micheletti, F., Cafaro, A. and Ensoli, B. (2002). Native HIV-1 Tat protein targets monocyte-derived dendritic cells and enhances their maturation, function, and antigen-specific T cell responses. J Immunol. 168: 197-206.

Fantuzzi, L., Purificato, C., Donato, K., Belardelli, F. and Gessani, S. (2004). Human immunodeficiency virus type 1 gp120 induces abnormal maturation and functional alterations of dendritic cells: a novel mechanism for AIDS pathogenesis. J Virol. 78: 9763-72.

Fong, L., Mengozzi, M., Abbey, N. W., Herndier, B. G. and Engleman, E. G. (2002). Productive infection of plasmacytoid dendritic cells with human immunodeficiency virus type 1 is triggered by CD40 ligation. J Virol. 76: 11033-41.

Fonteneau, J. F., Larsson, M., Beignon, A. S., McKenna, K., Dasilva, I., Amara, A., Liu, Y. J., Lifson, J. D., Littman, D. R. and Bhardwaj, N. (2004). Human immunodeficiency virus type 1 activates plasmacytoid dendritic cells and concomitantly induces the bystander maturation of myeloid dendritic cells. J Virol. 78: 5223-32.

Fugier-Vivier, I., Servet-Delprat, C., Rivailler, P., Rissoan, M. C., Liu, Y. J. and Rabourdin-Combe, C. (1997). Measles virus suppresses cell-mediated immunity by interfering with the survival and functions of dendritic and T cells. J Exp Med. 186: 813-23.

Garrett, W. S., Chen, L. M., Kroschewski, R., Ebersold, M., Turley, S., Trombetta, S., Galan, J. E. and Mellman, I. (2000). Developmental control of endocytosis in dendritic cells by Cdc42. Cell. 102: 325-34.

Geijtenbeek, T. B., Kwon, D. S., Torensma, R., van Vliet, S. J., van Duijnhoven, G. C., Middel, J., Cornelissen, I. L., Nottet, H. S., KewalRamani, V. N., Littman, D. R., Figdor, C. G. and van Kooyk, Y. (2000). DC-SIGN, a dendritic cell-specific HIV-1-binding protein that enhances trans-infection of T cells. Cell. 100: 587-97.

Geleziunas, R., Xu, W., Takeda, K., Ichijo, H. and Greene, W. C. (2001). HIV-1 Nef inhibits ASK1-dependent death signalling providing a potential mechanism for protecting the infected host cell. Nature. 410: 834-8.

Granelli-Piperno, A., Golebiowska, A., Trumpfheller, C., Siegal, F. P. and Steinman, R. M. (2004). HIV-1-infected monocyte-derived dendritic cells do not undergo maturation but can elicit IL-10 production and T cell regulation. Proc Natl Acad Sci U S A. 101: 7669-74.

Grassi, F., Hosmalin, A., McIlroy, D., Calvez, V., Debre, P. and Autran, B. (1999). Depletion in blood CD11c-positive dendritic cells from HIV-infected patients. AIDS. 13: 759-66.

Grosjean, I., Caux, C., Bella, C., Berger, I., Wild, F., Banchereau, J. and Kaiserlian, D. (1997). Measles virus infects human dendritic cells and blocks their allostimulatory properties for CD4$^+$ T cells. J Exp Med. 186: 801-12.

Guidotti, L. G. and Chisari, F. V. (1999). Cytokine-induced viral purging—role in viral pathogenesis. Curr Opin Microbiol. 2: 388-91.

Guidotti, L. G., Borrow, P., Hobbs, M. V., Matzke, B., Gresser, I., Oldstone, M. B. and Chisari, F. V. (1996). Viral cross talk: intracellular inactivation of the hepatitis B virus during an unrelated viral infection of the liver. Proc Natl Acad Sci U S A. 93: 4589-94.

Haller, O., Kochs, G. and Weber, F. (2006). The interferon response circuit: induction and suppression by pathogenic viruses. Virology. 344: 119-30.

Hamerman, J. A., Ogasawara, K. and Lanier, L. L. (2005). NK cells in innate immunity. Curr Opin Immunol. 17: 29-35.

Herbeuval, J. P., Hardy, A. W., Boasso, A., Anderson, S. A., Dolan, M. J., Dy, M. and Shearer, G. M. (2005). Regulation of TNF-related apoptosis-inducing ligand on primary CD4$^+$ T cells by HIV-1: role of type I IFN-producing plasmacytoid dendritic cells. Proc Natl Acad Sci U S A. 102: 13974-9.

Ho, L. J., Wang, J. J., Shaio, M. F., Kao, C. L., Chang, D. M., Han, S. W. and Lai, J. H. (2001). Infection of human dendritic cells by dengue virus causes cell maturation and cytokine production. J Immunol. 166: 1499-506.

Homsy, J., Meyer, M., Tateno, M., Clarkson, S. and Levy, J. A. (1989). The Fc and not CD4 receptor mediates antibody enhancement of HIV infection in human cells. Science. 244: 1357-60.

Hsieh, S. M., Pan, S. C., Hung, C. C., Chen, M. Y. and Chang, S. C. (2003). Differential impact of late-stage HIV-1 infection on *in vitro* and *in vivo* maturation of myeloid dendritic cells. J Acquir Immune Defic Syndr. 33: 413-9.

Hu, J., Gardner, M. B. and Miller, C. J. (2000). Simian immunodeficiency virus rapidly penetrates the cervicovaginal mucosa after intravaginal inoculation and infects intraepithelial dendritic cells. J Virol. 74: 6087-95.

Iannello, A., Debbeche, O., Martin, E., Attalah, L. H., Samarani, S. and Ahmad, A. (2006). Viral strategies for evading antiviral cellular immune responses of the host. J Leukoc Biol. 79: 16-35.

Izmailova, E., Bertley, F. M., Huang, Q., Makori, N., Miller, C. J., Young, R. A. and Aldovini, A. (2003). HIV-1 Tat reprograms immature dendritic cells to express chemoattractants for activated T cells and macrophages. Nat Med. 9: 191-7.

Jenne, L., Hauser, C., Arrighi, J. F., Saurat, J. H. and Hugin, A. W. (2000). Poxvirus as a vector to transduce human dendritic cells for immunotherapy: abortive infection but reduced APC function. Gene Ther. 7: 1575-83.

Kaiserlian, D. and Dubois, B. (2001). Dendritic cells and viral immunity: friends or foes? Semin Immunol. 13: 303-10.

Kakimoto, M., Hasegawa, A., Fujita, S. and Yasukawa, M. (2002). Phenotypic and functional alterations of dendritic cells induced by human herpesvirus 6 infection. J Virol. 76: 10338-45.

Katze, M. G., He, Y. and Gale, M., Jr. (2002). Viruses and interferon: a fight for supremacy. Nat Rev Immunol. 2: 675-87.

Kawamura, T., Gatanaga, H., Borris, D. L., Connors, M., Mitsuya, H. and Blauvelt, A. (2003). Decreased stimulation of CD4$^+$ T cell proliferation and IL-2 production by highly enriched populations of HIV-infected dendritic cells. J Immunol. 170: 4260-6.

Kidd, P. (2003). Th1/Th2 balance: the hypothesis, its limitations, and implications for health and disease. Altern Med Rev. 8: 223-46.

Knight, S. C. (2001). Dendritic cells and HIV infection; immunity with viral transmission versus compromised cellular immunity? Immunobiology. 204: 614-21.

Kruse, M., Rosorius, O., Kratzer, F., Stelz, G., Kuhnt, C., Schuler, G., Hauber, J. and Steinkasserer, A. (2000). Mature dendritic cells infected with herpes simplex virus type 1 exhibit inhibited T-cell stimulatory capacity. J Virol. 74: 7127-36.

Lee, J., Chuang, T.H., Redecke, V., She, L., Pitha, P.M., Carson, D.A., Raz, E., Cottam, H.B. (2003). Molecular basis for the immunostimulatory acrivity of guanine nucleoside analogs: activation of Toll-like receptor 7. Proc Nat Acad Sci U S A. 100: 6646-6651.

Letvin, N. L. and Walker, B. D. (2003). Immunopathogenesis and immunotherapy in AIDS virus infections. Nat Med. 9: 861-6.

Li, L., Liu, D., Hutt-Fletcher, L., Morgan, A., Masucci, M. G. and Levitsky, V. (2002). Epstein-Barr virus inhibits the development of dendritic cells by promoting apoptosis of their monocyte precursors in the presence of granulocyte macrophage-colony-stimulating factor and interleukin-4. Blood. 99: 3725-34.

Lichtner, M., Maranon, C., Vidalain, P. O., Azocar, O., Hanau, D., Lebon, P., Burgard, M., Rouzioux, C., Vullo, V., Yagita, H., Rabourdin-Combe, C., Servet, C. and Hosmalin, A. (2004). HIV type 1-infected dendritic cells induce apoptotic death in infected and uninfected primary CD4 T lymphocytes. AIDS Res Hum Retroviruses. 20: 175-82.

Lin, C. L., Sewell, A. K., Gao, G. F., Whelan, K. T., Phillips, R. E. and Austyn, J. M. (2000). Macrophage-tropic HIV induces and exploits dendritic cell chemotaxis. J Exp Med. 192: 587-94.

Liu, Y. J. (2005). IPC: professional type 1 interferon-producing cells and plasmacytoid dendritic cell precursors. Annu Rev Immunol. 23: 275-306.

Lodoen, M. B. and Lanier, L. L. (2005). Viral modulation of NK cell immunity. Nat Rev Microbiol. 3: 59-69.

Lore, K. and Larsson, M. (2003). The role of dendritic cells in the pathogenesis of HIV-1 infection. APMIS. 111: 776-88.

Lore, K., Sonnerborg, A., Brostrom, C., Goh, L. E., Perrin, L., McDade, H., Stellbrink, H. J., Gazzard, B., Weber, R., Napolitano, L. A., van Kooyk, Y. and Andersson, J. (2002). Accumulation of DC-SIGN$^+$CD40$^+$ dendritic cells with reduced CD80 and CD86 expression in lymphoid tissue during acute HIV-1 infection. AIDS. 16: 683-92.

Mahanty, S., Hutchinson, K., Agarwal, S., McRae, M., Rollin, P. E. and Pulendran, B. (2003). Cutting edge: impairment of dendritic cells and adaptive immunity by Ebola and Lassa viruses. J Immunol. 170: 2797-801.

Majumder, B., Janket, M. L., Schafer, E. A., Schaubert, K., Huang, X. L., Kan-Mitchell, J., Rinaldo, C. R., Jr. and Ayyavoo, V. (2005). Human immunodeficiency virus type 1 Vpr impairs dendritic cell maturation and T-cell activation: implications for viral immune escape. J Virol. 79: 7990-8003.

Makino, M., Wakamatsu, S., Shimokubo, S., Arima, N. and Baba, M. (2000). Production of functionally deficient dendritic cells from HTLV-I-infected monocytes: implications for the dendritic cell defect in adult T cell leukemia. Virology. 274: 140-8.

Matzinger, P. (2002). The danger model: a renewed sense of self. Science. 296: 301-5.

McFadden, G. and Murphy, P. M. (2000). Host-related immunomodulators encoded by poxviruses and herpesviruses. Curr Opin Microbiol. 3: 371-8.

Messmer, D., Bromberg, J., Devgan, G., Jacque, J. M., Granelli-Piperno, A. and Pope, M. (2002a). Human immunodeficiency virus type 1 Nef mediates activation of STAT3 in immature dendritic cells. AIDS Res Hum Retroviruses. 18: 1043-50.

Messmer, D., Jacque, J. M., Santisteban, C., Bristow, C., Han, S. Y., Villamide-Herrera, L., Mehlhop, E., Marx, P. A., Steinman, R. M., Gettie, A. and Pope, M. (2002b). Endogenously expressed nef uncouples cytokine and chemokine production from membrane phenotypic maturation in dendritic cells. J Immunol. 169: 4172-82.

Morrow, G., Slobedman, B., Cunningham, A. L. and Abendroth, A. (2003). Varicella-zoster virus productively infects mature dendritic cells and alters their immune function. J Virol. 77: 4950-9.

Mosmann, T. R. and Coffman, R. L. (1989). TH1 and TH2 cells: different patterns of lymphokine secretion lead to different functional properties. Annu Rev Immunol. 7: 145-73.

Moutaftsi, M., Mehl, A. M., Borysiewicz, L. K. and Tabi, Z. (2002). Human cytomegalovirus inhibits maturation and impairs function of monocyte-derived dendritic cells. Blood. 99: 2913-21.

Moutaftsi, M., Brennan, P., Spector, S. A. and Tabi, Z. (2004). Impaired lymphoid chemokine-mediated migration due to a block on the chemokine receptor switch in human cytomegalovirus-infected dendritic cells. J Virol. 78: 3046-54.

Muller, D. B., Raftery, M. J., Kather, A., Giese, T. and Schonrich, G. (2004). Frontline: Induction of apoptosis and modulation of c-FLIPL and p53 in immature dendritic cells infected with herpes simplex virus. Eur J Immunol. 34: 941-51.

Munz, C., Steinman, R. M. and Fujii, S. (2005). Dendritic cell maturation by innate lymphocytes: coordinated stimulation of innate and adaptive immunity. J Exp Med. 202: 203-7.

Murata, T., Goshima, F., Daikoku, T., Takakuwa, H. and Nishiyama, Y. (2000). Expression of herpes simplex virus type 2 US3 affects the Cdc42/Rac pathway and attenuates c-Jun N-terminal kinase activation. Genes Cells. 5: 1017-27.

Muthumani, K., Hwang, D. S., Choo, A. Y., Mayilvahanan, S., Dayes, N. S., Thieu, K. P. and Weiner, D. B. (2005). HIV-1 Vpr inhibits the maturation and activation of macrophages and dendritic cells *in vitro*. Int Immunol. 17: 103-16.

Palucka, K. and Banchereau, J. (2002). How dendritic cells and microbes interact to elicit or subvert protective immune responses. Curr Opin Immunol. 14: 420-31.

Pollara, G., Speidel, K., Samady, L., Rajpopat, M., McGrath, Y., Ledermann, J., Coffin, R. S., Katz, D. R. and Chain, B. (2003). Herpes simplex virus infection of dendritic cells: balance among activation, inhibition, and immunity. J Infect Dis. 187: 165-78.

Pollara, G., Kwan, A., Newton, P. J., Handley, M. E., Chain, B. M. and Katz, D. R. (2005). Dendritic cells in viral pathogenesis: protective or defective? Int J Exp Pathol. 86: 187-204.

Popik, W. and Pitha, P. M. (2000). Exploitation of cellular signaling by HIV-1: unwelcome guests with master keys that signal their entry. Virology. 276: 1-6.

Puig-Kroger, A., Relloso, M., Fernandez-Capetillo, O., Zubiaga, A., Silva, A., Bernabeu, C. and Corbi, A. L. (2001). Extracellular signal-regulated protein kinase signaling pathway negatively regulates the phenotypic and functional maturation of monocyte-derived human dendritic cells. Blood. 98: 2175-82.

Quaranta, M. G., Tritarelli, E., Giordani, L. and Viora, M. (2002). HIV-1 Nef induces dendritic cell differentiation: a possible mechanism of uninfected CD4($^+$) T cell activation. Exp Cell Res. 275: 243-54.

Quaranta, M. G., Mattioli, B., Spadaro, F., Straface, E., Giordani, L., Ramoni, C., Malorni, W. and Viora, M. (2003). HIV-1 Nef triggers Vav-mediated signaling pathway leading to functional and morphological differentiation of dendritic cells. FASEB J. 17: 2025-36.

Quaranta, M. G., Mattioli, B., Giordani, L. and Viora, M. (2004). HIV-1 Nef equips dendritic cells to reduce survival and function of CD8$^+$ T cells: a mechanism of immune evasion. FASEB J. 18: 1459-61.

Raftery, M. J., Schwab, M., Eibert, S. M., Samstag, Y., Walczak, H. and Schonrich, G. (2001). Targeting the function of mature dendritic cells by human cytomegalovirus: a multilayered viral defense strategy. Immunity. 15: 997-1009.

Reis e Sousa, C. (2004). Toll-like receptors and dendritic cells: for whom the bug tolls. Semin Immunol. 16: 27-34.

Reis e Sousa, C., Sher, A. and Kaye, P. (1999). The role of dendritic cells in the induction and regulation of immunity to microbial infection. Curr Opin Immunol. 11: 392-9.

Rescigno, M., Martino, M., Sutherland, C. L., Gold, M. R. and Ricciardi-Castagnoli, P. (1998). Dendritic cell survival and maturation are regulated by different signaling pathways. J Exp Med. 188: 2175-80.

Ridley, A. J., Schwartz, M. A., Burridge, K., Firtel, R. A., Ginsberg, M. H., Borisy, G., Parsons, J. T. and Horwitz, A. R. (2003). Cell migration: integrating signals from front to back. Science. 302: 1704-9.

Riegler, S., Hebart, H., Einsele, H., Brossart, P., Jahn, G. and Sinzger, C. (2000). Monocyte-derived dendritic cells are permissive to the complete replicative cycle of human cytomegalovirus. J Gen Virol. 81: 393-9.

Rinaldo, C. R., Jr. and Piazza, P. (2004). Virus infection of dendritic cells: portal for host invasion and host defense. Trends Microbiol. 12: 337-45.

Rossio, J. L., Esser, M. T., Suryanarayana, K., Schneider, D. K., Bess, J. W., Jr., Vasquez, G. M., Wiltrout, T. A., Chertova, E., Grimes, M. K., Sattentau, Q., Arthur, L. O., Henderson, L. E. and Lifson, J. D. (1998). Inactivation of human immunodeficiency virus type 1 infectivity with preservation of conformational and functional integrity of virion surface proteins. J Virol. 72: 7992-8001.

Salio, M., Cella, M., Suter, M. and Lanzavecchia, A. (1999). Inhibition of dendritic cell maturation by herpes simplex virus. Eur J Immunol. 29: 3245-53.

Sallusto, F. and Lanzavecchia, A. (1999). Mobilizing dendritic cells for tolerance, priming, and chronic inflammation. J Exp Med. 189: 611-4.

Santini, S. M., Lapenta, C., Logozzi, M., Parlato, S., Spada, M., Di Pucchio, T. and Belardelli, F. (2000). Type I interferon as a powerful adjuvant for monocyte-derived dendritic cell development and activity *in vitro* and in Hu-PBL-SCID mice. J Exp Med. 191: 1777-88.

Sato, K., Nagayama, H., Tadokoro, K., Juji, T. and Takahashi, T. A. (1999). Extracellular signal-regulated kinase, stress-activated protein kinase/c-Jun N-terminal kinase, and p38mapk are involved in IL-10-mediated selective repression of TNF-alpha-induced activation and maturation of human peripheral blood monocyte-derived dendritic cells. J Immunol. 162: 3865-72.

Schmidt, B., Ashlock, B. M., Foster, H., Fujimura, S. H. and Levy, J. A. (2005). HIV-infected cells are major inducers of plasmacytoid dendritic cell interferon production, maturation, and migration. Virology. 343: 256-66.

Schnorr, J. J., Xanthakos, S., Keikavoussi, P., Kampgen, E., ter Meulen, V. and Schneider-Schaulies, S. (1997). Induction of maturation of human blood dendritic cell precursors by measles virus is associated with immunosuppression. Proc Natl Acad Sci U S A. 94: 5326-31.

Schwartz, O., Marechal, V., Le Gall, S., Lemonnier, F. and Heard, J. M. (1996). Endocytosis of major histocompatibility complex class I molecules is induced by the HIV-1 Nef protein. Nat Med. 2: 338-42.

Servet-Delprat, C., Vidalain, P. O., Azocar, O., Le Deist, F., Fischer, A. and Rabourdin-Combe, C. (2000a). Consequences of Fas-mediated human dendritic cell apoptosis induced by measles virus. J Virol. 74: 4387-93.

Servet-Delprat, C., Vidalain, P. O., Bausinger, H., Manie, S., Le Deist, F., Azocar, O., Hanau, D., Fischer, A. and Rabourdin-Combe, C. (2000b). Measles virus induces abnormal differentiation of CD40 ligand-activated human dendritic cells. J Immunol. 164: 1753-60.

Shinya, E., Owaki, A., Shimizu, M., Takeuchi, J., Kawashima, T., Hidaka, C., Satomi, M., Watari, E., Sugita, M. and Takahashi, H. (2004). Endogenously expressed HIV-1 nef down-regulates antigen-presenting molecules, not only class I MHC but also CD1a, in immature dendritic cells. Virology. 326: 79-89.

Siavoshian, S., Abraham, J. D., Thumann, C., Kieny, M. P. and Schuster, C. (2005). Hepatitis C virus core, NS3, NS5A, NS5B proteins induce apoptosis in mature dendritic cells. J Med Virol. 75: 402-11.

Siegal, F. P., Kadowaki, N., Shodell, M., Fitzgerald-Bocarsly, P. A., Shah, K., Ho, S., Antonenko, S. and Liu, Y. J. (1999). The nature of the principal type 1 interferon-producing cells in human blood. Science. 284: 1835-7.

Smed-Sorensen, A., Lore, K., Walther-Jallow, L., Andersson, J. and Spetz, A. L. (2004). HIV-1-infected dendritic cells up-regulate cell surface markers but fail to produce IL-12 p70 in response to CD40 ligand stimulation. Blood. 104: 2810-7.

Smith, A., Santoro, F., Di Lullo, G., Dagna, L., Verani, A. and Lusso, P. (2003). Selective suppression of IL-12 production by human herpesvirus 6. Blood. 102: 2877-84.

Smith, A. P., Paolucci, C., Di Lullo, G., Burastero, S. E., Santoro, F. and Lusso, P. (2005). Viral replication-independent blockade of dendritic cell maturation and interleukin-12 production by human herpesvirus 6. J Virol. 79: 2807-13.

Sol-Foulon, N., Moris, A., Nobile, C., Boccaccio, C., Engering, A., Abastado, J. P., Heard, J. M., van Kooyk, Y. and Schwartz, O. (2002). HIV-1 Nef-induced upregulation of DC-SIGN in dendritic cells promotes lymphocyte clustering and viral spread. Immunity. 16: 145-55.

Spriggs, M. K. (1999). Virus-encoded modulators of cytokines and growth factors. Cytokine Growth Factor Rev. 10: 1-4.

Steinman, R. M., Inaba, K., Turley, S., Pierre, P. and Mellman, I. (1999). Antigen capture, processing, and presentation by dendritic cells: recent cell biological studies. Hum Immunol. 60: 562-7.

Stevenson, M. (2003). HIV-1 pathogenesis. Nat Med. 9: 853-60.

Stockwin, L. H., McGonagle, D., Martin, I. G. and Blair, G. E. (2000). Dendritic cells: immunological sentinels with a central role in health and disease. Immunol Cell Biol. 78: 91-102.

Strater, J. and Moller, P. (2004). TRAIL and viral infection. Vitam Horm. 67: 257-74.

Swetman, C. A., Leverrier, Y., Garg, R., Gan, C. H., Ridley, A. J., Katz, D. R. and Chain, B. M. (2002). Extension, retraction and contraction in the formation of a dendritic cell dendrite: distinct roles for Rho GTPases. Eur J Immunol. 32: 2074-83.

Teleshova, N., Frank, I. and Pope, M. (2003). Immunodeficiency virus exploitation of dendritic cells in the early steps of infection. J Leukoc Biol. 74: 683-90.

Tishon, A., Lewicki, H., Rall, G., Von Herrath, M. and Oldstone, M. B. (1995). An essential role for type 1 interferon-gamma in terminating persistent viral infection. Virology. 212: 244-50.

Tortorella, D., Gewurz, B. E., Furman, M. H., Schust, D. J. and Ploegh, H. L. (2000). Viral subversion of the immune system. Annu Rev Immunol. 18: 861-926.

Varani, S., Frascaroli, G., Homman-Loudiyi, M., Feld, S., Landini, M. P. and Soderberg-Naucler, C. (2005). Human cytomegalovirus inhibits the migration of immature dendritic cells by down-regulating cell-surface CCR1 and CCR5. J Leukoc Biol. 77: 219-28.

Vidalain, P. O., Azocar, O., Lamouille, B., Astier, A., Rabourdin-Combe, C. and Servet-Delprat, C. (2000). Measles virus induces functional TRAIL production by human dendritic cells. J Virol. 74: 556-9.

Welsh, R. M., Selin, L. K. and Szomolanyi-Tsuda, E. (2004). Immunological memory to viral infections. Annu Rev Immunol. 22: 711-43.

West, M. A., Prescott, A. R., Eskelinen, E. L., Ridley, A. J. and Watts, C. (2000). Rac is required for constitutive macropinocytosis by dendritic cells but does not control its downregulation. Curr Biol. 10: 839-48.

Wilflingseder, D., Mullauer, B., Schramek, H., Banki, Z., Pruenster, M., Dierich, M. P. and Stoiber, H. (2004). HIV-1-induced migration of monocyte-derived dendritic cells is associated with differential activation of MAPK pathways. J Immunol. 173: 7497-505.

Williams, M. A., Trout, R. and Spector, S. A. (2002). HIV-1 gp120 modulates the immunological function and expression of accessory and co-stimulatory molecules of monocyte-derived dendritic cells. J Hematother Stem Cell Res. 11: 829-47.

Wu, S. J., Grouard-Vogel, G., Sun, W., Mascola, J. R., Brachtel, E., Putvatana, R., Louder, M. K., Filgueira, L., Marovich, M. A., Wong, H. K., Blauvelt, A., Murphy, G. S., Robb, M. L., Innes, B. L., Birx, D. L., Hayes, C. G. and Frankel, S. S. (2000). Human skin Langerhans cells are targets of dengue virus infection. Nat Med. 6: 816-20.

Yonezawa, A., Morita, R., Takaori-Kondo, A., Kadowaki, N., Kitawaki, T., Hori, T. and Uchiyama, T. (2003). Natural alpha interferon-producing cells respond to human immunodeficiency virus type 1 with alpha interferon production and maturation into dendritic cells. J Virol. 77: 3777-84.

Zinkernagel, R. M., Bachmann, M. F., Kundig, T. M., Oehen, S., Pirchet, H. and Hengartner, H. (1996). On immunological memory. Annu Rev Immunol. 14: 333-67.

Chapter 13

Cross-Presentation by Dendritic Cells: Role in HIV Immunity and Pathogenesis

Concepción Marañón, Guillaume Hoeffel, Anne-Claire Ripoche, and Anne Hosmalin

13.1 Introduction

Dendritic cells (DCs) have a critical role in HIV infection as they are the only antigen-presenting cells that can activate naïve T cells and, therefore, elicit the strong specific T-cell responses that are crucial in controlling HIV replication. The antigen presentation pathways used by DCs for eliciting these responses are described in this chapter. DCs can present HIV directly after being infected themselves. They have also developed specific antigen cross-presentation pathways that allow them to present exogenous antigens in association with major histocompatibility complex (MHC) class I molecules to $CD8^+$ T lymphocytes. The different cross-presentation pathways used by DCs to present HIV antigens are described here. Cross-presentation from live infected $CD4^+$ T lymphocytes is compared with that from dead infected cells. The role of these cross-presentation mechanisms in HIV immunity and pathogenesis is discussed with the aim of understanding how to modulate the specific T-cell responses to control HIV replication and disease.

13.2 Antigen Presentation Pathways

T-lymphocyte receptors recognize complexes formed by MHC molecules and epitopes. Epitopes result from the enzymatic digestion of antigens. This allows for the restriction of the immune response to antigens presented by specialized antigen-presenting cells (Zinkernagel and Doherty,

1974). Antigen presentation includes antigen uptake, degradation, and loading onto MHC class I and II molecules, which are then expressed at the cell surface. DCs are basically the only presenting cells that can present antigens to naïve T lymphocytes. They are thus required to induce memory antigen-specific immunity, which is a requisite for the achievement of any effective vaccination. DCs are also required for the induction of tolerance, and therefore to avoid autoimmunity and graft rejection. The DC-stimulated lymphocytes expand and differentiate into memory lympho-cytes, with effector or regulator functions that are required in antitumor and antiviral defenses or in tolerance (Banchereau and Steinman, 1998; Banchereau *et al.*, 2000; Guermonprez *et al.*, 2002; Heath *et al.*, 2004). Antigen presentation is tightly regulated by DC maturation.

13.2.1 MHC Class II–restricted Antigen Presentation

CD4$^+$ T lymphocytes recognize epitopes of various lengths associated with an MHC class II molecule (Babbitt *et al.*, 1985). Digestion leading to these epitopes takes place in late endosomes and is regulated by DC maturation. Immature DCs express few MHC class II molecules at their surface. They accumulate internalized antigens in endocytic vesicles where MHC II molecules are bound to the invariant chain (Ii). During DC maturation, MHC class II molecule synthesis is upregulated, and protease activity of the cathepsins involved in antigen and in Ii degradation increases together with the acidification of the endocytic compartment. The Ii is cleaved and the chaperone CLIP peptide is released from the MHC class II molecule cleft, where newly processed epitopes bind (Germain *et al.*, 1996). MHC molecule-epitope complexes can then be exported to the surface (Boes *et al.*, 2004; Chow *et al.*, 2002). In mature DCs, endocytosis is reduced as well as the transport of internalized MHC class II molecules to lysosomes for degradation, while MHC class II molecule–epitope complexes are mainly expressed and stable at the cell surface, together with costimulatory and adhesion molecules. Therefore, during DC maturation, all concurs to accumulate antigen and MHC class II molecules in the same compart-ments, then to load these MHC molecules with epitopes, which are then presented optimally to CD4$^+$ T lymphocytes.

13.2.2 MHC Class I–restricted Antigen Presentation

CD8$^+$ T lymphocytes recognize 9- to 10-amino-acid epitopes that associate with an MHC class I molecule through anchor residues (Rammensee *et al.*,

1995; Townsend *et al.*, 1985). Classically, and in all cell types, digestion of antigens synthesized in the cytosol, and marked for rapid degradation by ubiquitin, is started by the proteasome (Kloetzel and Ossendorp, 2004; Niedermann *et al.*, 1999; York *et al.*, 1999). For one epitope in the HIV-1 Nef protein, the proteasome was shown to be replaced by another housekeeping cytosolic enzyme, the tripeptidyl-peptidase II (TPP-II) (Seifert *et al.*, 2003). Most epitopes may derive from defective ribosomal products (DRiPs) (Yewdell *et al.*, 2003). Some epitopes may derive from post-translational protein splicing or may be encoded by alternative reading frames (Cardinaud *et al.*, 2004; Hanada *et al.*, 2004). Epitope precursors are then further trimmed, especially at their N-termini, by aminopeptidases, before or after transport by the transporter associated with the antigen-processing (TAP) molecules into the endoplasmic reticulum (ER), where they bind to nascent MHC class I molecules and are exported to the cell surface (Saveanu *et al.*, 2002).

Most cells display class I–restricted epitopes originating from endogenous proteins encoded by the genes they express. This allows CD8$^+$ T lymphocytes to identify cells that express viral proteins or tumor cells and to eliminate them. However, DCs have developed specific antigen cross-presentation pathways that allow them to present exogenous antigens in association with MHC class I molecules (Rock *et al.*, 2005b). Cross-priming from antigen-bearing cells or soluble proteins *in vivo* was first evidenced by Bevan (Bevan, 1976; Bevan, 1987). Cross-presentation is required for the presentation of microorganisms that do not infect DCs or for tolerization toward self molecules specific for other cell types. Cross-presentation pathways can be exploited in strategies for the development of new-generation vaccines. They are also fundamental in graft rejection (Heath and Carbone, 2001; Kurts *et al.*, 1996). They may represent major presentation pathways compared with direct presentation pathways occurring during infection (Chen *et al.*, 2004; Lizee *et al.*, 2003; Sigal *et al.*, 1999). This may depend on whether DCs are mostly infected themselves or ingest microbial nucleic acids (Freigang *et al.*, 2003; Spetz *et al.*, 1999), proteins, or cell fragments (Rock and Shen, 2005b). During infection, the fact that cross-presentation pathways are only found in DCs (or activated macrophages) allows the of adaptive immunity or tolerance without the destruction of uninfected, non–antigen-presenting cells by cytotoxic T lymphocytes.

Cross-presentation requires 1000- to 10,000-fold lower concentrations of antigen when a soluble protein is associated with particles, cells, or bacteria, compared with soluble protein alone (Kovacsovics-Bankowski *et al.*, 1993; Nair *et al.*, 1992). After endocytosis, cross-presentation pathways mostly involve an endosome-to-cytosol pathway in DCs and macrophages (Rock and Shen, 2005b). Then, antigens join the proteasome- and the TAP-dependent pathway. When these cells have phagocytosed latex beads loaded with soluble antigen, they perform cross-presentation using a fusion compartment between phagosomes and the ER (Ackerman et al., 2003; Guermonprez et al., 2003; Houde et al., 2003). The conditions of formation of this compartment still need to be defined precisely (Touret et al., 2005). The transport mechanism to cytosolic degrading enzymes is not yet elucidated. It depends at least partly on the sec61 translocon, which exports misfolded proteins from the ER into the cytosol for degradation (Ackerman and Cresswell, 2004; Rock, 2003). In vitro data correlate with in vivo priming data, where cross-priming is made by hematopoietic cells and is mostly dependent on TAP and on the presence of a sequence addressing MHC class I molecules to endo-lysosomal compartments (Huang et al., 1994; Kurts et al., 1999; Lizee et al., 2003; Sigal et al., 1999). Cross-presentation can also involve a vacuolar pathway, independent of proteasome and TAP, but dependent on lysosomal acidification and on cathepsin S (Pfeifer et al., 1993; Shen et al., 2004b; Wick and Pfeifer, 1996). Epitopes then probably bind to MHC class I molecules in recycling endosomes or in the phagosome-ER compartment. In addition, in contrast with latex beads, soluble proteins escaping proteolysis gain access from macropinosomes to the perinuclear ER through retrograde mechanisms that still need to be characterized (Ackerman et al., 2005; Day et al., 1997; Mallard et al., 1998).

Formal evidence of the role of DCs in cross-presentation was brought by inducible depletion in mice, which could abrogate cross-priming of CD8[+] T cells (Jung et al., 2002). In mice, CD8[+] DCs were described as the subpopulation mostly responsible for cross-presentation (Belz et al., 2005; den Haan et al., 2000), but activated CD8[-] DCs as well as activated macrophages can also be involved (Bellone et al., 1997; den Haan and Bevan, 2002; Kovacsovics-Bankowski et al., 1993; Norbury et al., 2002; Pozzi et al., 2005; Scheinecker et al., 2002; Yrlid and Wick, 2002).

During DC maturation, MHC class I–restricted antigen presentation is regulated: in addition to transient antigen uptake enhancement, immunoproteasome subunit and TAP expression are upregulated (Delamarre et al., 2003; Gil-Torregrosa et al., 2004; Macagno et al., 1999). Moreover, after inflammatory stimulation of DCs, newly synthesized ubiquitinated proteins are stored within DC aggresome-like induced structures (DALIS), perhaps to optimize exogenous (infectious) antigen processing and loading onto

MHC I molecules (Lelouard *et al.*, 2002). Of note, IFN-α (from virally stimulated pDCs) can mediate cross-presentation *in vivo* during the course of viral infection (Le Bon *et al.*, 2003) and IFN-α (from activated T cells or NK cells) can enhance cross-presentation from Langerhans cells (LCs) *in vitro* (Matsuo *et al.*, 2004). Therefore, presentation of exogenous antigens by DCs to CD8$^+$ T cells, and induction of the effector mechanisms of CD8$^+$ T cells, which may disrupt homeostasis, can only occur when internalized antigens are accompanied by maturation signals linked to pathogens or to inflammatory signals.

MHC class I–restricted antigen presentation is thought to be of special importance for controlling the infection by the human immunodeficiency virus (HIV), because such infection is generally associated with CD8$^+$ T-cell responses. These responses are crucial in controlling HIV replication, as evidenced by converging data in patients and by CD8$^+$ T-cell depletion experiments in the simian immunodeficiency virus infection model, which is very close to HIV infection (Heeney *et al.*, 1999; Letvin and Walker, 2003; Pantaleo and Koup, 2004).

13.3 Direct Presentation of HIV from Infected DCs

The different pathways of HIV antigen presentation to CD8$^+$ T lymphocytes by DCs are illustrated in Figure 13.1. The most obvious pathway for viral antigen presentation is the direct, classical antigen presentation pathway from antigens synthesized in the cytosol. Specific killing by cytotoxic CD8$^+$ T lymphocytes of HIV infected CD4$^+$ T lymphocytes was evidenced by flow cytometry (Collins *et al.*, 1998). Specific killing of HIV-infected DCs can also be shown (Buseyne *et al.*, 2001). Given the low infection frequency of these cells *in vivo* (McIlroy *et al.*, 1995), this may be an unfrequent event. However, it may be extremely efficient in lymphoid organs, including the intestinal lymphoid tissue, where most of the virus is produced (Embretson *et al.*, 1993; Pantaleo *et al.*, 1991; Veazey and Lackner, 2004).

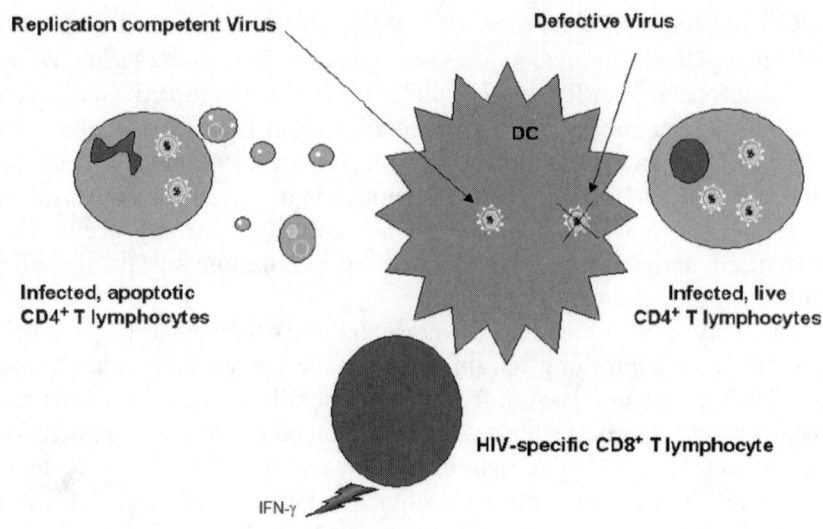

FIGURE 13.1. How do DCs present HIV to T lymphocytes? HIV antigens can be presented by DCs to CD8$^+$ T lymphocytes through different pathways. The most classical is direct presentation of replication-competent virus from infected DCs: HIV directs synthesis of its proteins in the cytosol of the DC and thus can be processed by cytosolic enzymes. Alternatively, HIV can be presented by several cross-presentation mechanisms. Defective virus that cannot replicate nor direct synthesis of its proteins can be internalized into DC through different mechanisms, then cross-presented. Virus from infected, apoptotic CD4$^+$ T lymphocytes can be cross-presented by DC. In addition, CD4$^+$ T lymphocytes do not need to be apoptotic or dead to give HIV antigens to DC for cross-presentation: DC can cross-present HIV antigens also from live, infected CD4$^+$ T lymphocytes.

13.4 Cross-Presentation of HIV

13.4.1 Cross-Presentation of Defective HIV

In addition to presenting directly epitopes from live virus, DCs can present epitopes from inactivated HIV without any viral protein synthesis to specific CD8$^+$ T lymphocytes (Buseyne *et al.*, 2001) (Fig. 13.1). This may be a relatively frequent event *in vivo* because the reverse transcriptase of HIV is error-prone and generates an excess of replication-deficient viral particles compared with the number of infectious particles. HIV is internalized into cells of the DC lineage either through CD4 and the coreceptors CCR5 or CXCR4, but also through DC-SIGN (for interstitial DC and macrophages), langerin (for LC), other lectins, and other molecules, which mediate endosomal entry (Turville *et al.*, 2002). Binding to DC-SIGN for instance leads to cross-presentation of inactivated HIV (Moris *et al.*, 2004).

13.4.2 Cross-Presentation from Apoptotic, HIV-infected Cells

During cell turnover and embryogenesis, homeostatic replacement of the cells occurs through programmed cell death or apoptosis. Apoptotic cells are phagocytosed by resident macrophages and by immature DCs. Then, DCs cross-present the antigens that are present in the apoptotic cells. These include virus-infected cells (Albert *et al.*, 1998b), tumor cells (Berard *et al.*, 2000; Masse *et al.*, 2002; Schuler and Steinman, 1997), transplanted cells, and autologous cells containing self antigens (Heath and Carbone, 2001).

The DC population responsible for apoptotic cell cross-priming *in vivo* is believed to be the $CD8^+$ $CD11^+$ population in the mouse (Iyoda *et al.*, 2002), although in kinetic studies $CD8^-$ cells were shown to be able to perform this function (Morelli *et al.*, 2003; Scheinecker *et al.*, 2002; Moron *et al.*, 2002).

Entry of apoptotic bodies into DCs is mediated by numerous specific receptors (Skoberne *et al.*, 2005) including scavenger receptors such as CD36 (Albert *et al.*, 1998a) or SRA; integrins such as $\alpha v\beta 5$ (Albert *et al.*, 1998a); receptors recognizing phosphatidylserine, which is exposed at the surface of apoptotic cells; CD14; CD91 (or α_2-macroglobulin receptor) and Lox1, which binds heat shock proteins loaded with antigens (Binder and Srivastava, 2005); and the complement receptors CR3 and CR4.

Two teams seeking the best methods for therapeutic vaccination using DCs from HIV^+ patients showed that human DCs cross-present HIV antigens from autologous infected, apoptotic, or even necrotic $CD4^+$ T lymphocytes (Larsson *et al.*, 2002; Zhao *et al.*, 2002) (Fig. 13.1). The pathway remains to be determined, even though the pathway for influenza virus antigen cross-presentation by human DCs from dead recombinant vaccinia virus–infected monocytes was found to depend on cathepsin D proteolysis in an acidic compartment as well as on the proteasome and on TAP, indicating a requirement for access to the cytosol and then to the ER (Fonteneau *et al.*, 2003). Interestingly, DCs that phagocytose dead infected $CD4^+$ T lymphocytes also phagocytose proviral DNA that encodes for HIV proteins, giving rise to presentation (Spetz *et al.*, 1999). This was of course controlled by anti-retroviral treatment of the cultures in the studies cited above (Larsson *et al.*, 2002; Zhao *et al.*, 2002).

Among the different HIV antigen-presentation pathways that had been described in DCs, the cross-presentation pathway from infected, apoptotic CD4$^+$ T lymphocytes could be regarded as the most efficient, if one could extrapolate from other antigenic systems showing potentiation at small antigen dose compared with soluble protein (Kovacsovics-Bankowski *et al.*, 1993; Regnault *et al.*, 1999). In addition, HIV infection induces a high susceptibility to apoptosis in T cells, including infected CD4$^+$ T lymphocytes (Estaquier *et al.*, 1994). Thus, apoptotic infected CD4$^+$ T lymphocytes probably represent a huge source of antigen (Ho *et al.*, 1995). To assess quantitatively the relative efficiencies of this and the other HIV antigen-presentation pathways in DCs, we performed antigen-presentation experiments to HIV-specific CD8$^+$ T-cell lines from HIV-infected patients using DCs loaded with different amounts of virus: either free, infectious HIV-1-lav, or free HIV-1-lav and AZT to inhibit virus replication (i.e., inactivated virus), or HIV-1-lav infected, apoptotic CD4$^+$ T lymphocytes. The amount of p24 was measured by ELISA either in the virus preparations or in the infected CD4$^+$ T-lymphocyte cultures. Indeed, cross-presentation by DCs from HIV-infected CD4$^+$ T lymphocytes was quantitatively by far the most efficient antigen-presentation pathway, compared with direct presentation from live or inactivated virus (Marañón *et al.*, 2004).

13.4.3 Cross-Presentation from Live, HIV-infected Cells

13.4.3.1 Evidence of Cross-Presentation from Live, HIV-infected Cells

To assess apoptotic CD4$^+$ T-lymphocyte uptake by DCs, we dyed these lymphocytes using the lipophilic membrane dye PKH, then we added them to the DCs, and we measured by flow cytometry the proportion of DCs labeled by PKH after incubation. Surprisingly, in our controls, the CD4$^+$ T lymphocytes did not need to be apoptotic for the PKH to be associated with DCs. Even live CD4$^+$ T lymphocytes lent their dye to the DCs. Were these lymphocytes really alive or had they undergone apoptosis during the overnight culture? This was controlled using caspase inhibitors and kinetic measurement of T-cell apoptosis using annexin V staining during the time of coculture. While the level of PKH associated with DCs increased steadily, no apoptosis increase was detected in the T cells in the coculture, thus showing that PKH was really acquired from live cells. Indeed, as shown in Figure 13.2, DCs

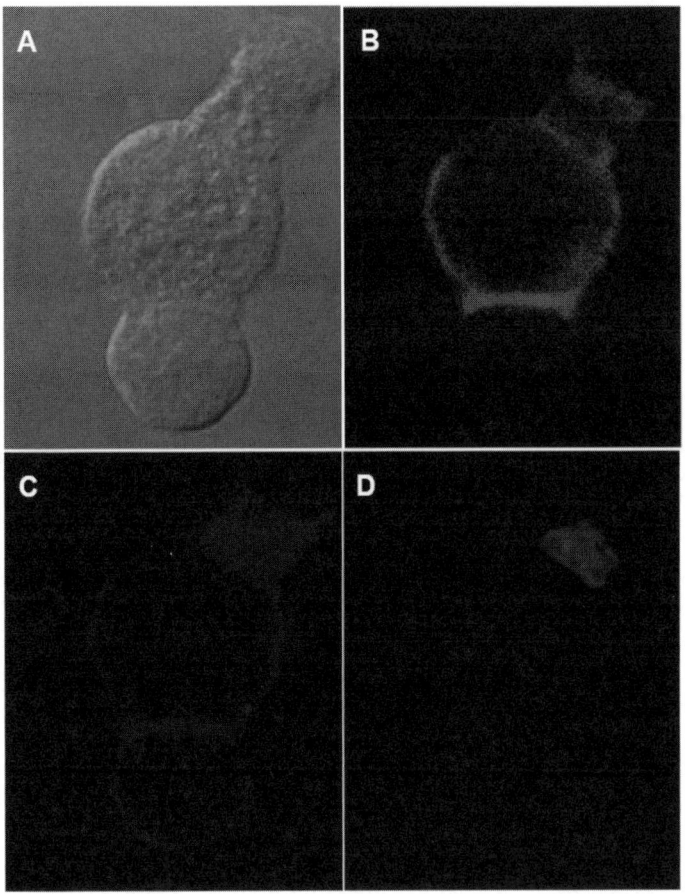

FIGURE 13.2. DCs acquire molecules from live as well as apoptotic infected CD4⁺ T lymphocytes. DCs and HIV-infected CD4⁺ T cells interact closely when cultured together, as shown by confocal transmission microscopy (A). DCs are labeled in magenta by immunofluorescence using anti-CD11c antibodies. (B) DCs acquire CD3 molecules (in blue; C) from the T-cell receptor complex. T cells (smaller, round, labeled in blue and not in magenta) have been irradiated. Some are apoptotic, as shown by TUNEL labeling in red (D), others are alive, but both are in close contact with DCs. After this contact and material transfer, DCs cross-present the HIV antigens from the infected CD4⁺ T lymphocytes to effector CD8⁺ T lymphocytes. (See Plate 4)

interacted closely with apoptotic as well as live CD4⁺ T lymphocytes. Was the PKH-labeled material derived from the T cells only associated to the DCs or was it internalized? Internalization into early and late endosomes was evidenced by confocal microscopy and colocalization with transferrin and acetyl-LDL, respectively. Did this internalization lead to HIV antigen cross-presentation? It led to HIV antigen presentation to CD8$^+$ T-cell lines with dose-response curves superimposable on those obtained after internalization of antigen from apoptotic, infected CD4$^+$ T lymphocytes. The use of AZT throughout these experiments precluded direct antigen presentation from live, infectious virus, which in any case proved to be very inefficient, even at very high concentrations, for recognition by our primary CD8$^+$ T-cell lines derived from HIV$^+$ patients. Presentation was restricted by MHC class I molecules from the DCs and not from the infected CD4$^+$ T lymphocytes (Marañón et al., 2004). Therefore, HIV antigen transfer and cross-presentation from live, infected CD4$^+$ T lymphocytes was evidenced, as in a tumor antigen model in macaques (Harshyne et al., 2001), and in a model using OVA-expressing recombinant vaccinia virus, where antigen production was restricted to the donor cell and antigen presentation to the recipient cell (Ramirez and Sigal, 2002).

13.4.3.2 Mechanisms of Antigen Internalization from Live Cells into DCs

After excluding direct virus infection, several mechanisms could account for antigen internalization from live cells into DCs. The first potential mechanism was the transfer of exosomes or of microparticles from the infected CD4$^+$ T lymphocytes. Exosomes are membrane vesicles contained within multivesicular bodies and secreted upon fusion of these compartments from the plasma membrane (Thery et al., 2002). They are generated by tumor cells and by various cells including T lymphocytes and DCs, and they are a source of tumor antigens for CD8$^+$ T-cell cross-priming (Wolfers et al., 2001). Transfer of fluorescein isothiocytanate ovalbumin loaded antigen between syngeneic DCs leading to cross-presentation of MHC class II molecules between allogeneic DCs was suggested to occur through exosomes (Andre et al., 2004; Bedford et al., 1999; Knight et al., 1998). To assess the exosome hypothesis, as shown in Figure 13.3, 450-nm-pore Transwells were used to separate infected CD4$^+$ T lymphocytes (in the top well) and DCs (in the bottom well). When labeled and infected CD4$^+$ T lymphocytes were cultured in the top wells, the HIV p24 protein was revealed in the bottom wells, and small particles with intermediate sizes between those of 100 and 430 nm beads were detected by flow cytometry. Despite these potential sources of HIV antigens in the bottom wells, DCs did not cross-present HIV antigens. A cell-to-cell contact in the same well was required for cross-presentation. If exosomes still mediated

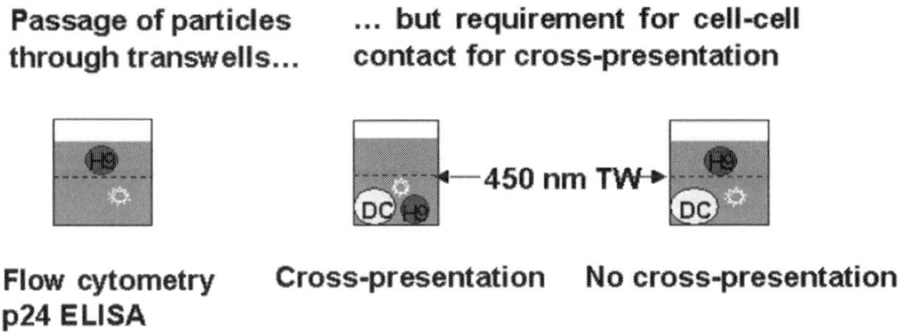

Passage of particles through transwells...   **... but requirement for cell-cell contact for cross-presentation**

Flow cytometry   Cross-presentation   No cross-presentation
p24 ELISA

FIGURE 13.3. HIV antigen cross-presentation by DC from live cells requires cell-cell contact. Dendritic cells were cultured in the bottom wells of 450 nm transwells while HIV-1 infected CD4$^+$ T cells (H9 cell line) were cultured either in the top wells or the bottom wells. Despite evidence of HIV p24 and microparticles through the transwell membrance, cross-presentation occureed only when DC and infected CD4$^+$ T cells were in the same well.

HIV antigen transfer from live cells, this would only be in the context of a close contact, like in an antigen-independent immunologic synapse between DC and T cell (Revy *et al.*, 2001).

The second potential mechanism was the transfer of CD4$^+$ T lymphocyte material by nibbling or trogocytosis. (Marañón *et al.*, 2004). Trogocytosis allows transfer of large molecules from live cell to live cell. This transfer occurs between different cell types, particularly cells of the immune system during the formation of antigen-dependent (Huang *et al.*, 1999; Hudrisier *et al.*, 2001; Stinchcombe *et al.*, 2001) or -independent (Tabiasco *et al.*, 2003) synapses. Its purpose may be to amplify lymphocyte activation, then to terminate immune responses by fratricide, or even to transfer cell metabolites (Joly and Hudrisier, 2003). In addition, the fact that T cells down-modulate peptide-MHC complexes by removing them from the surface of APCs may help to select activation of higher affinity T cells (Kedl *et al.*, 2002). Nibbling is a process previously shown in macaque DCs, which involves extraction of plasma membrane from DCs or other hematopoietic cells by DC endocytosis into vesicles. It gives rise to tumor antigen cross-presentation (Harshyne *et al.*, 2001). Real-time pictures obtained in our laboratory by Nomarsky microscopy and by fluorescence microscopy were indeed compatible with nibbling.

Additional mechanisms may include the transfer of small, preprocessed peptides through gap junctions. Gap junctions are assemblies of intercellular channels, each constituted by the junction of two hemichannels formed by the six connexin 43 molecules. Connexin 43 is expressed broadly, including on DCs and activated lymphocytes. Gap junctions can mediate cross-presentation by intercellular peptide transfer between adjacent cells (Neijssen *et al.*, 2005). Small molecules can also be transferred from live cell to live cell at a distance up to 100 μm through the formation of tunelling nanotubules, which connect DCs and allow calcium flux induction (Watkins and Salter, 2005); antigen transfer and cross-presentation were not demonstrated through this mechanism, but they are likely to occur. In these two mechanisms, the size of the molecules that can be transferred may be limited, therefore, preprocessing may be necessary in the donor cell.

13.5 Comparison of Cross-Presentation from Live or from Dead Cells

13.5.1 Uptake and Antigen-processing Pathways

The receptors involved in antigen transfer from live cells were compared with those involved from apoptotic cells. As in the previous nibbling data (Harshyne *et al.*, 2003), scavenger receptors, but not integrins or mannose receptors, seemed to be involved (Marañón *et al.*, 2004). The data indicated a redundancy of potential receptors as previously found for antigen transfer from apoptotic cells (Albert *et al.*, 1998a). Many different receptors and factors that mediate the uptake of apoptotic cells (reviewed in Skoberne *et al.* (2005)) need to be studied for the uptake from live cells. A different array of receptors is expected for live and apoptotic cells, with some more specific for apoptotic cells, like phosphatidyl serine receptors (although live cells can transiently expose phosphatidylserine at their surface during activation) (Henson *et al.*, 2001; Somersan and Bhardwaj, 2001), and others that would be common to both, including some scavenger receptors. Usage of different receptors leads to different processing and signaling pathways (Akakura *et al.*, 2004), thus to a different polarization of the T-cell responses elicited by the DCs.

Depending on experimental systems, the antigen transferred during cross-presentation may be in the form of full-length proteins (Norbury *et al.*, 2004; Shen and Rock, 2004a; Wolkers *et al.*, 2004), proteasomal products that are epitope precursors needing further degradation (Serna *et al.*, 2003), or mature epitopes bound or not to heat shock proteins (Binder and Srivastava, 2005). This depends on the availability of preprocessed epitope precursors in the source of antigen, as shown for apoptotic cells (Blachere *et al.*, 2005). Phagocytosis or macropinocytosis allow internalization into the endosomal pathway of whole protein as well as preprocessed antigen and passage through the still incompletely characterized "phagosome-to-cytosol" pathway. Conversely, the direct passage into the cytosol through gap junctions or tuneling nanotubules may only allow acquisition of preproceessed epitopes or epitope precursors, because of potential size constraints; these peptides ought then to be transported by TAP into the ER, with or without trimming by enzymes from the cytosol or the ER. These important differences are relevant to the mechanism of antigen acquisition from live as well as apoptotic cells.

13.5.2 Role of DC Maturation

When DCs are completely mature, but not "exhausted," and in the presence of sufficient amounts of antigen, they induce Th1-type and cytotoxic T-cell responses (Boonstra et al., 2003; Lanzavecchia and Sallusto, 2001; Sallusto et al., 2004). When they are semi-mature, they induce regulatory T-cell responses (Albert et al., 2001; Lutz and Schuler, 2002; O'Garra and Vieira, 2004; Steinman et al., 2003). Maturation is triggered by "danger" signals (Matzinger, 2002): contact with microbial pathogen-associated molecular patterns (PAMPs), signals from damaged peripheral tissues, inflammatory signals, and activated T-cell signals. PAMPs stimulate pattern-recognition receptors (PRRs), among which Toll-like receptors and nucleotide binding and oligomerization domains (NODs) (Philpott and Girardin, 2004) play an important role. Intracellular and plasma membrane TLR ligation synergize to induce DC maturation (Napolitani et al., 2005). Interestingly, inducible phagosome maturation (acidification, LAMP expression), which influences cross-presentation, is dependent on the presence of TLR ligands in their cargo (Blander and Medzhitov, 2004). Among activated T-cell signals, CD154 (CD40 ligand) signaling synergizes with PRR signaling to induce maximal stimulation of DCs, giving rise to IL-12 secretion, which is essential for Th1-type and cytotoxic responses (Reis e Sousa, 2004). Among inflamematory signals, IFN-α induces DC differentiation and maturation from monocytes, acts as a maturation stimulus, and is a powerful adjuvant for vaccination (Lapenta et al., 2003; Montoya et al., 2002).

It will be necessary to know (Fig. 13.4) whether live cells induce DC maturation by themselves or when they are infected by a pathogen or stressed, in the presence of exogenous or "endogenous adjuvants" (Gallucci et al., 1999). The degree of DC maturation should determine whether live cell cross-presentation gives rise to immunity or to tolerance.

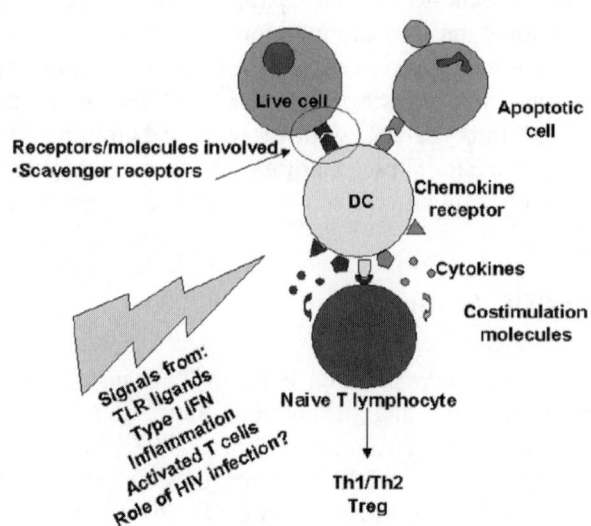

FIGURE 13.4. DC-pathogen interactions and cross-presentation of antigens associated with live or apoptotic cells: immunity or tolerance? After incubation with live compared to apoptotic cells, DC may be stimulated through a different array of receptors, leading to different expression of cytokines, chemokines and costimulation molecules, which in turn may polarize naïve T lymphocytes to express molecular profiles characteristic of either T helper 1, T helper 2 or regulatory T cells. Signals from pathogenic or endogenous adjuvant may influence DC maturation and T cell polarization. The role of HIV itself needs to be determined.

13.5.3 Is Death Necessary to Cross-Presentation?

In a philosophical perspective, the current concept of "death-defying immunity" (Albert, 2004) is fascinating and must somehow make sense. Not only do phagocytes clear efficiently the newly dead cells and recycle their elementary cell nutrients, but also they sample the molecules that were expressed in the cells immediately before death, thus transferring immunologic information to T lymphocytes, leading to immune tolerance or activation (Steinman *et al.*, 2000). The debate has been focused on the opposition between apoptosis and necrosis (Sauter *et al.*, 2000), then on the mechanism of apoptosis induction (McIlroy and Gregoire, 2003), which determines signal 3 that polarizes T-cell responses. Apoptotic cells per se most often do not induce DC maturation, and rather they have an inhibitory effect on DC activation (reviewed in Savill *et al.* (2003);

Skoberne *et al.*, (2005)). On the contrary, necrosis releases endogenous adjuvants like uric acid (Gallucci *et al.*, 1999; Rock *et al.*, 2005a; Sauter *et al.*, 2000). Sometimes, microbial signals have been pointed to as the inducers of immunogenicity that was formerly ascribed to death or to heat shock proteins (Bausinger *et al.*, 2002; Salio *et al.*, 2000).

The question is whether death is necessary to cross-presentation. *In vivo* data indicate that veiled DCs traffic in the thoracic duct from the gut, carrying apoptotic bodies from intestinal cells (Huang *et al.*, 2000). Cross-presentation of gastric epithelial cell–specific antigen by DCs from the draining lymph node was formally demonstrated in mice (Scheinecker *et al.*, 2002). But in that case, is apoptosis of gastric cells necessary to cross-presentation? The most convincing data on the necessity of apoptosis come from the elegant demonstration that apoptosis inhibition *in vivo* led to inhibition of cross-tolerization to islet antigens and to full development of diabetes (Hugues *et al.*, 2002; Turley *et al.*, 2003).

Conversely, the concept of antigen cross-presentation from live cells seems threatening for homeostasis. This may be the case if the immune response is polarized toward activation and effector mechanisms, but not if it is polarized toward tolerance. It may be difficult to evidence technically the cross-presentation from live cells *in vivo*, because live cells may be killed very fast after injection (Iyoda *et al.*, 2002). However, the major source of antigen is most often in live cells and might be a source for immune tolerance in the absence of full DC maturation (Rénia *et al.*, 2005).

Understanding the circumstances when cross-presentation from live versus apoptotic cells can happen will help to modulate better immune responses to different antigens, especially those from HIV.

13.6 Role of Cross-Presentation from Infected Cells in HIV Immunity and Pathogenesis

In vivo in HIV-infected patients, $CD8^+$ T-cell responses are not as strong as those found in other viral infections such as EBV or CMV infections, which are well controlled by the immune system (Dalod *et al.*, 1999; Letvin and Walker, 2003). The epitopic repertoire is limited, the proliferative and IL-2–producing capacities are impaired; cytotoxicity is reduced, as well as perforin secretion (Pantaleo and Koup, 2004). HIV-specific $CD4^+$ regulatory T cells participate in this $CD8^+$ T-cell response deficiency: indeed, *in vitro* depletion of these cells in cultures from infected patients restores higher HIV-specific $CD8^+$ T-cell responses

(Aandahl *et al.*, 2004; Kinter *et al.*, 2004; Weiss *et al.*, 2004). Similar to HIV-specific CD8$^+$ T-cell responses, which have a dual antiviral and immunopathogenic role (Hosmalin *et al.*, 2001), regulatory T cells probably have ambivalent functions: to inhibit the antiviral effects of CD8$^+$ T cells, but also to inhibit their pathogenic effects, especially the hyperactivation of CD4$^+$ T cells, which renders them permissive to HIV replication (Kinter *et al.*, 2004; Kornfeld *et al.*, 2005).

In vitro, HIV stimulates only incompletely the maturation of monocyte-derived DCs: costimulation molecules may be upregulated or not, depending on experimental conditions, but IL-12 p70 production is deficient, and regulatory T cells can be induced (Granelli-Piperno *et al.*, 2004; Smed-Sorensen *et al.*, 2004). HIV-1 gp120 induces partial maturation of DCs, with decreased IL-12 production and impaired allogeneic CD4$^+$ T-cell proliferation (Fantuzzi *et al.*, 2004; Kawamura *et al.*, 2003). Conversely to myeloid or monocyte-derived DCs, plasmacytoid DCs are directly activated by HIV-1, probably through TLR7 (Beignon *et al.*, 2005), and produce type I interferons (Schmidt *et al.*, 2004; Yonezawa *et al.*, 2003) that in turn induce the maturation of myeloid DCs (Fonteneau *et al.*, 2004).

The induction of CD154 (CD40 ligand) during T-cell activation is impaired in HIV$^+$ patients with advanced disease (Chougnet, 2003; Subauste *et al.*, 2004; Vanham *et al.*, 1999). *In vitro* data indicate that this is mediated at least in part by the interaction of gp120 with CD4 (Zhang *et al.*, 2004). IL-12 induction *in vitro* is impaired in the PBMCs from HIV$^+$ patients (Chougnet *et al.*, 2002; Ma and Montaner, 2000; Marshall *et al.*, 1999). HIV may impair DC maturation, IL-12 p70 secretion, and T-cell stimulatory function not only directly but also through impairment of CD40-CD154 interactions with CD4$^+$ T cells (Zhang *et al.*, 2006). These requirements would be interesting to test in the setting of cross-presentation from live cells.

In HIV patients, we were the first to find that the numbers of circulating myeloid DCs are lower than in healthy donors (Donaghy *et al.*, 2004; Servet *et al.*, 2002); this may also result in an impairment of HIV antigen cross-presentation. When DCs are infected themselves, microorganisms may evade direct presentation through escape mechanisms (Lilley and Ploegh, 2005; Yewdell and Hill, 2002). Cross-presentation from infected cells definitely allows to avoid escape mechanisms evolved in HIV such as class I and II MHC antigen down-modulation by Nef (Andrieu *et al.*, 2001; Schwartz *et al.*, 1996; Stumptner-Cuvelette *et al.*, 2001). These escape mechanisms affect only directly infected cells, not the DCs that cross-present HIV antigens. A good proportion of HIV-infected cells, especially CD4$^+$ T cells, undergoes apoptosis, but another pool returns to a resting state and builds the latent reservoir pool (Blankson *et al.*, 2002). Cross-presentation of HIV and autoantigens from apoptotic, infected CD4$^+$

T cells probably induces effector and regulatory T-cell responses, as indicated by data on vinculin, a molecule that is overexpressed on apoptotic cells: DCs can cross-prime vinculin-specific CTLs *in vitro*, if the apoptotic cells express CD154, otherwise additional T-cell help is required; in HIV-infected patients, the frequency of peripheral apoptotic T cells expressing CD154 was correlated with the frequency of vinculin-specific CTLs (Propato *et al.*, 2001). This does not exclude the possibility that DCs can also cross-prime HIV from live infected cells.

Cross-presentation from live infected cells may provide additional therapeutic opportunities during HIV infection. These cells represent an abundant source of antigen, and they comprise the latent reservoirs, which hamper viral eradication because they are not detectable by the immune system. Viral protein production might be induced at low levels in these reservoirs during HAART structured interruptions, by using different appropriate factors: thus, IL-2 would be useful for latently infected T cells (Chun *et al.*, 1999; Levy *et al.*, 2003), while colony-stimulating factors would be adequate for latently infected cells from the monocyte/macrophage lineage including cells from the microglia (Aladdin *et al.*, 2000; Armstrong and Kazanjian, 2001; Pires *et al.*, 2005). The use of additional T-cell helper signals like soluble CD154 or CD40 agonists, or IFN-α at an appropriate time (Le Bon *et al.*, 2003; Longman *et al.*, 2006), might help to increase cross-presentation to HIV-specific CD8$^+$ T cells. Conversely, if the hyperactivation of the immune system proves to be a major pathogenic factor, one might imagine in some cases to down-modulate CD40 expression to favor HIV-specific T-cell regulation (Martin *et al.*, 2003).

13.7 Concluding Remarks

In conclusion, cross-presentation from HIV-infected cells by DCs probably plays an important role to induce the effector or regulatory CD8$^+$ T-cell responses that are crucial in viral replication control. Cross-presentation of HIV antigens might be exploited for immune therapy, but the respective roles of apoptotic and live cells as HIV antigen sources remain to be established.

Acknowledgments: We thank all members of the Dendritic Cells Workgroup from the ANRS (AC19), as well as J.P. Abastado, M. Albert, S. Amigorena, C. Chougnet, R. Germain, G. Milon, L. Renia, R. Thomas, S. Gessani, A. Trautmann, L. Zitvogel, and others for helpful discussions.

References

Aandahl, E. M., Michaelsson, J., Moretto, W. J., Hecht, F. M., and Nixon, D. F. (2004). Human CD4$^+$ CD25$^+$ regulatory T cells control T-cell responses to human immunodeficiency virus and cytomegalovirus antigens. J. Virol. 78:2454-2459.

Ackerman, A. L., and Cresswell, P. (2004). Cellular mechanisms governing cross-presentation of exogenous antigens. Nat. Immunol. 5:678-684.

Ackerman, A. L., Kyritsis, C., Tampé, R., and Cresswell, P. (2003). Early phagosomes in dendritic cells form a cellular compartment sufficient for cross presentation of exogenous antigens. Proc. Natl. Acad. Sci. U. S. A. 100:12889-12894.

Ackerman, A. L., Kyritsis, C., Tampé, R., and Cresswell, P. (2005). Access of soluble antigens to the endoplasmic reticulum can explain cross-presentation by dendritic cells. Nat. Immunol. 6:107-113.

Akakura, S., Singh, S., Spataro, M., Akakura, R., Kim, J. I., Albert, M. L., and Birge, R. B. (2004). The opsonin MFG-E8 is a ligand for the alphavbeta5 integrin and triggers DOCK180-dependent Rac1 activation for the phago-cytosis of apoptotic cells. Exp. Cell. Res. 292:403-416.

Aladdin, H., Ullum, H., Dam Nielsen, S., Espersen, C., Mathiesen, L., Katzenstein, T. L., Gerstoft, J., Skinhoj, P., and Pedersen, B. K. (2000). Granulocyte colony-stimulating factor increases CD4$^+$ T cell counts of human immunodeficiency virus-infected patients receiving stable, highly active antiretroviral therapy: results from a randomized, placebo- controlled trial. J. Infect. Dis. 181:1148-1152.

Albert, M. L. (2004). Death-defying immunity: do apoptotic cells influence antigen processing and presentation? Nat. Rev. Immunol. 4:223-231.

Albert, M. L., Jegathesan, M., and Darnell, R. B. (2001). Dendritic cell maturation is required for the cross-tolerization of CD8$^+$ T cells. Nat. Immunol. 2:1010-1017.

Albert, M. L., Pearce, S. F., Francisco, L. M., Sauter, B., Roy, P., Silverstein, R. L., and Bhardwaj, N. (1998a). Immature dendritic cells phagocytose apoptotic cells via alphavbeta5 and CD36, and cross-present antigens to cytotoxic T lymphocytes. J. Exp. Med. 188:1359-1368.

Albert, M. L., Sauter, B., and Bhardwaj, N. (1998b). Dendritic cells acquire antigen from apoptotic cells and induce class I-restricted CTLs. Nature 392:86-89.

Andre, F., Chaput, N., Schartz, N. E., Flament, C., Aubert, N., Bernard, J., Lemonnier, F., Raposo, G., Escudier, B., Hsu, D. H., Tursz, T., Amigorena, S., Angevin, E., and Zitvogel, L. (2004). Exosomes as potent cell-free peptide-based vaccine. I. Dendritic cell-derived exosomes transfer functional MHC class I/peptide complexes to dendritic cells. J. Immunol. 172:2126-2136.

Andrieu, M., Chassin, D., Desoutter, J. F., Bouchaert, I., Hanau, D., Guillet, J. G., and Hosmalin, A. (2001). HIV-1 Nef protein induces a down-regulation of the dendritic cell surface expression of MHC class I and impairs antigen presentation to CTL. AIDS Res. Hum. Retroviruses 17:1365-1370.

Armstrong, W. S., and Kazanjian, P. (2001). Use of cytokines in human immunodeficiency virus-infected patients: colony-stimulating factors, erythropoietin, and interleukin-2. Clin. Infect. Dis. 32:766-773.

Babbitt, B. P., Allen, P. M., Matsueda, G., Haber, E., and Unanue, E. R. (1985). The binding of immunogenic peptides to Ia histocompatibility molecules. Nature 317:359-361.

Banchereau, J., and Steinman, R. M. (1998). Dendritic cells and the control of immunity. Nature 392:245-252.

Banchereau, J., Briere, F., Caux, C., Davoust, J., Lebecque, S., Liu, Y. J., Pulendran, B., and Palucka, K. (2000). Immunobiology of dendritic cells. Annu. Rev. Immunol. 18:767-811.

Bausinger, H., Lipsker, D., Ziylan, U., Manie, S., Briand, J. P., Cazenave, J. P., Muller, S., Haeuw, J. F., Ravanat, C., de la Salle, H., and Hanau, D. (2002). Endotoxin-free heat-shock protein 70 fails to induce APC activation. Eur. J. Immunol. 32:3708-3713.

Bedford, P., Garner, K., and Knight, S. C. (1999). MHC class II molecules transferred between allogeneic dendritic cells stimulate primary mixed leukocyte reactions. Int. Immunol. 11:1739-1744.

Beignon, A. S., McKenna, K., Skoberne, M., Manches, O., Dasilva, I., Kavanagh, D. G., Larsson, M., Gorelick, R. J., Lifson, J. D., and Bhardwaj, N. (2005). Endocytosis of HIV-1 activates plasmacytoid dendritic cells via Toll-like receptor-viral RNA interactions. J. Clin. Invest. 115:3265-3275.

Bellone, M., Iezzi, G., Rovere, P., Galati, G., Ronchetti, A., Protti, M. P., Davoust, J., Rugarli, C., and Manfredi, A. A. (1997). Processing of engulfed apoptotic bodies yields T cell epitopes. J. Immunol. 159:5391-5399.

Belz, G. T., Shortman, K., Bevan, M. J., and Heath, W. R. (2005). CD8{alpha}[+] dendritic cells selectively present MHC class I-restricted noncytolytic viral and intracellular bacterial antigens in vivo. J. Immunol. 175:196-200.

Berard, F., Blanco, P., Davoust, J., Neidhart-Berard, E. M., Nouri-Shirazi, M., Taquet, N., Rimoldi, D., Cerottini, J. C., Banchereau, J., and Palucka, A. K. (2000). Cross-priming of naive CD8 T cells against melanoma antigens using dendritic cells loaded with killed allogeneic melanoma cells. J. Exp. Med. 192:1535-1544.

Bevan, M. J. (1976). Cross-priming for a secondary cytotoxic response to minor H antigens with H-2 congenic cells which do not cross-react in the cytotoxic assay. J. Exp. Med. 143:1283-1288.

Bevan, M. J. (1987). Antigen recognition. Class discrimination in the world of immunology. Nature 325:192-194.

Binder, R. J., and Srivastava, P. K. (2005). Peptides chaperoned by heat-shock proteins are a necessary and sufficient source of antigen in the cross-priming of CD8[+] T cells. Nat. Immunol. 6:593-599.

Blachere, N. E., Darnell, R. B., and Albert, M. L. (2005). Apoptotic cells deliver processed antigen to dendritic cells for cross-presentation. PLoS Biol. 3:e185.

Blander, J. M., and Medzhitov, R. (2004). Regulation of phagosome maturation by signals from toll-like receptors. Science 304:1014-1018.

Blankson, J. N., Persaud, D., and Siliciano, R. F. (2002). The challenge of viral reservoirs in HIV-1 infection. Annu. Rev. Med. 53:557-593.

Boes, M., Cuvillier, A., and Ploegh, H. (2004). Membrane specializations and endosome maturation in dendritic cells and B cells. Trends Cell. Biol. 14:175-183.

Boonstra, A., Asselin-Paturel, C., Gilliet, M., Crain, C., Trinchieri, G., Liu, Y. J., and O'Garra, A. (2003). Flexibility of mouse classical and plasmacytoid-derived dendritic cells in directing T helper type 1 and 2 cell development: dependency on antigen dose and differential toll-like receptor ligation. J. Exp. Med. 197:101-109.

Buseyne, F., Le Gall, S., Boccaccio, C., Abastado, J. P., Lifson, J. D., Arthur, L. O., Riviere, Y., Heard, J. M., and Schwartz, O. (2001). MHC-I-restricted presentation of HIV-1 virion antigens without viral replication. Nat. Med. 7:344-349.

Cardinaud, S., Moris, A., Fevrier, M., Rohrlich, P. S., Weiss, L., Langlade-Demoyen, P., Lemonnier, F. A., Schwartz, O., and Habel, A. (2004). Identification of cryptic MHC I-restricted epitopes encoded by HIV-1 alternative reading frames. J. Exp. Med. 199:1053-1063.

Chen, W., Masterman, K. A., Basta, S., Mansour Haeryfar, S. M., Dimopoulos, N., Knowles, B., Bennink, J. R., and Yewdell, J. W. (2004). Cross-priming of CD8$^+$ T cells by viral and tumor antigens is a robust phenomenon. Eur. J. Immunol. 34:194-199.

Chougnet, C. (2003). Role of CD40 ligand dysregulation in HIV-associated dysfunction of antigen-presenting cells. J. Leukoc. Biol. 74:702-709.

Chougnet, C., Shearer, G. M., and Landay, A. L. (2002). The role of antigen-presenting cells in HIV pathogenesis. Curr. Infect. Dis. Rep. 4:266-271.

Chow, A., Toomre, D., Garrett, W., and Mellman, I. (2002). Dendritic cell maturation triggers retrograde MHC class II transport from lysosomes to the plasma membrane. Nature 418:988-994.

Chun, T. W., Engel, D., Mizell, S. B., Hallahan, C. W., Fischette, M., Park, S., Davey, R. T., Jr., Dybul, M., Kovacs, J. A., Metcalf, J. A., Mican, J. M., Berrey, M. M., Corey, L., Lane, H. C., and Fauci, A. S. (1999). Effect of interleukin-2 on the pool of latently infected, resting CD4$^+$ T cells in HIV-1-infected patients receiving highly active anti-retroviral therapy. Nat. Med. 5:651-655.

Collins, K. L., Chen, B. K., Kalams, S. A., Walker, B. D., and Baltimore, D. (1998). HIV-1 Nef protein protects infected primary cells against killing by cytotoxic T lymphocytes. Nature 391:397-401.

Dalod, M., Dupuis, M., Deschemin, J.-C., Goujard, C., Deveau, C., Meyer, L., Ngo, N., Rouzioux, C., Guillet, J.-G., Delfraissy, J.-F., Sinet, M., and Venet, A. (1999). Weak anti-HIV CD8$^+$ T-cell effector activity in HIV primary infection. J. Clin. Invest. 104:1431-1439.

Day, P. M., Yewdell, J. W., Porgador, A., Germain, R. N., and Bennink, J. R. (1997). Direct delivery of exogenous MHC class I molecule-binding oligopeptides to the endoplasmic reticulum of viable cells. Proc. Natl. Acad. Sci. U. S. A. 94:8064-8069.

Delamarre, L., Holcombe, H., and Mellman, I. (2003). Presentation of exogenous antigens on major histocompatibility complex (MHC) class I and MHC class II molecules is differentially regulated during dendritic cell maturation. J. Exp. Med. 198:111-122.

den Haan, J. M., and Bevan, M. J. (2002). Constitutive versus activation-dependent cross-presentation of immune complexes by CD8($^+$) and CD8(-) dendritic cells *in vivo*. J. Exp. Med. 196:817-827.

den Haan, J. M., Lehar, S. M., and Bevan, M. J. (2000). CD8($^+$) but not CD8(-) dendritic cells cross-prime cytotoxic T cells *in vivo*. J. Exp. Med. 192:1685-1696.

Donaghy, H., Stebbing, J., and Patterson, S. (2004). Antigen presentation and the role of dendritic cells in HIV. Curr. Opin. Infect. Dis. 17:1-6.

Embretson, J., Zupancic, M., Ribas, J. L., Burke, A., Racz, P., Tenner-Racz, K., and Haase, A. T. (1993). Massive covert infection of helper T lymphocytes and macrophages by HIV during the incubation period of AIDS. Nature 362:359-362.

Estaquier, J., Idziorek, T., De Bels, F., Barré-Sinoussi, F., Hurtrel, B., Aubertin, A.-M., Venet, A., Mehtali, M., Muchmore, E., Michel, P., Mouton, Y., Girard, M., and Ameisen, J.-C. (1994). Programmed cell death and AIDS: significance of T-cell apoptosis in pathogenic and nonpathogenic primate lentiviral infections. Proc. Natl. Acad. Sci. U. S. A. 91:9431-9435.

Fantuzzi, L., Purificato, C., Donato, K., Belardelli, F., and Gessani, S. (2004). Human immunodeficiency virus type 1 gp120 induces abnormal maturation and functional alterations of dendritic cells: a novel mechanism for AIDS pathogenesis. J. Virol. 78:9763-9772.

Fonteneau, J. F., Kavanagh, D. G., Lirvall, M., Sanders, C., Cover, T. L., Bhardwaj, N., and Larsson, M. (2003). Characterization of the MHC class I cross-presentation pathway for cell-associated antigens by human dendritic cells. Blood 102:4448-4455.

Fonteneau, J. F., Larsson, M., Beignon, A. S., McKenna, K., Dasilva, I., Amara, A., Liu, Y. J., Lifson, J. D., Littman, D. R., and Bhardwaj, N. (2004). Human immunodeficiency virus type 1 activates plasmacytoid dendritic cells and concomitantly induces the bystander maturation of myeloid dendritic cells. J. Virol. 78:5223-5232.

Freigang, S., Egger, D., Bienz, K., Hengartner, H., and Zinkernagel, R. M. (2003). Endogenous neosynthesis vs. cross-presentation of viral antigens for cytotoxic T cell priming. Proc. Natl. Acad. Sci. U. S. A. 100:13477-13482.

Gallucci, S., Lolkema, M., and Matzinger, P. (1999). Natural adjuvants: endogenous activators of dendritic cells. Nat. Med. 5:1249-1255.

Germain, R. N., Castellino, F., Han, R., Reis e Sousa, C., Romagnoli, P., Sadegh-Nasseri, S., and Zhong, G. M. (1996). Processing and presentation of endocytically acquired protein antigens by MHC class II and class I molecules. Immunol. Rev. 151:5-30.

Gil-Torregrosa, B. C., Lennon-Dumenil, A. M., Kessler, B., Guermonprez, P., Ploegh, H. L., Fruci, D., van Endert, P., and Amigorena, S. (2004). Control of cross-presentation during dendritic cell maturation. Eur. J. Immunol. 34:398-407.

Granelli-Piperno, A., Golebiowska, A., Trumpfheller, C., Siegal, F. P., and Steinman, R. M. (2004). HIV-1-infected monocyte-derived dendritic cells do not undergo maturation but can elicit IL-10 production and T cell regulation. Proc. Natl. Acad. Sci. U. S. A. 101:7669-7674.

Guermonprez, P., Saveanu, L., Kleijmeer, M., Davoust, J., Van Endert, P., and Amigorena, S. (2003). ER-phagosome fusion defines an MHC class I cross-presentation compartment in dendritic cells. Nature 425:397-402.

Guermonprez, P., Valladeau, J., Zitvogel, L., Thery, C., and Amigorena, S. (2002). Antigen presentation and T cell stimulation by dendritic cells. Annu. Rev. Immunol. 20:621-667.

Hanada, K., Yewdell, J. W., and Yang, J. C. (2004). Immune recognition of a human renal cancer antigen through post-translational protein splicing. Nature 427:252-256.

Harshyne, L. A., Watkins, S. C., Gambotto, A., and Barratt-Boyes, S. M. (2001). Dendritic cells acquire antigens from live cells for cross-presentation to CTL. J. Immunol. 166:3717-3723.

Harshyne, L. A., Zimmer, M. I., Watkins, S. C., and Barratt-Boyes, S. M. (2003). A role for class A scavenger receptor in dendritic cell nibbling from live cells. J. Immunol. 170:2302-2309.

Heath, W. R., and Carbone, F. R. (2001). Cross-presentation in viral immunity and self-tolerance. Nat. Rev. Immunol. 1:126-134.

Heath, W. R., Belz, G. T., Behrens, G. M., Smith, C. M., Forehan, S. P., Parish, I. A., Davey, G. M., Wilson, N. S., Carbone, F. R., and Villadangos, J. A. (2004). Cross-presentation, dendritic cell subsets, and the generation of immunity to cellular antigens. Immunol. Rev. 199:9-26.

Heeney, J. L., Beverley, P., McMichael, A., Shearer, G., Strominger, J., Wahren, B., Weber, J., and Gotch, F. (1999). Immune correlates of protection from HIV and AIDS - more answers but yet more questions. Immunol. Today 20:247-251.

Henson, P. M., Bratton, D. L., and Fadok, V. A. (2001). The phosphatidylserine receptor: a crucial molecular switch? Nat. Rev. Mol. Cell. Biol. 2:627-633.

Ho, D. D., Neumann, A. U., Perelson, A. S., Chen, W., Leonard, J. M., and Markowitz, M. (1995). Rapid turnover of plasma virions and CD4 lymphocytes in HIV-1 infection. Nature 373:123-126.

Hosmalin, A., Samri, A., Dumaurier, M. J., Dudoit, Y., Oksenhendler, E., Karmochkine, M., Autran, B., Wain-Hobson, S., and Cheynier, R. (2001). HIV-specific effector CTL and HIV-producing cells co-localize in white pulps and germinal centers from infected patients. Blood 97:2695-2701.

Houde, M., Bertholet, S., Gagnon, E., Brunet, S., Goyette, G., Laplante, A., Princiotta, M. F., Thibault, P., Sacks, D., and Desjardins, M. (2003). Phagosomes are competent organelles for antigen cross-presentation. Nature 425:402-406.

Huang, A. Y., Golumbek, P., Ahmadzadeh, M., Jaffee, E., Pardoll, D., and Levitsky, H. (1994). Role of bone marrow-derived cells in presenting MHC class I-restricted tumor antigens. Science 264:961-965.

Huang, J. F., Yang, Y., Sepulveda, H., Shi, W., Hwang, I., Peterson, P. A., Jackson, M. R., Sprent, J., and Cai, Z. (1999). TCR-mediated internalization of peptide-MHC complexes acquired by T cells. Science 286:952-954.

Huang, F. P., Platt, N., Wykes, M., Major, J. R., Powell, T. J., Jenkins, C. D., and MacPherson, G. G. (2000). A discrete subpopulation of dendritic cells transports apoptotic intestinal epithelial cells to T cell areas of mesenteric lymph nodes. J. Exp. Med. 191:435-444.

Hudrisier, D., Riond, J., Mazarguil, H., Gairin, J. E., and Joly, E. (2001). Cutting edge: CTLs rapidly capture membrane fragments from target cells in a TCR signaling-dependent manner. J. Immunol. 166:3645-3649.

Hugues, S., Mougneau, E., Ferlin, W., Jeste, D., Hofman, P., Homann, D., Beaudoin, L., Schrike, C., Von Herrath, M., Lehuen, A., and Glaichenhaus, N. (2002). Tolerance to islet antigens and prevention from diabetes induced by limited apoptosis of pancreatic beta cells. Immunity 16:169-181.

Iyoda, T., Shimoyama, S., Liu, K., Omatsu, Y., Akiyama, Y., Maeda, Y., Takahara, K., Steinman, R. M., and Inaba, K. (2002). The CD8$^+$ dendritic cell subset selectively endocytoses dying cells in culture and *in vivo*. J. Exp. Med. 195:1289-1302.

Joly, E., and Hudrisier, D. (2003). What is trogocytosis and what is its purpose? Nat. Immunol. 4:815.

Jung, S., Unutmaz, D., Wong, P., Sano, G., De los Santos, K., Sparwasser, T., Wu, S., Vuthoori, S., Ko, K., Zavala, F., Pamer, E. G., Littman, D. R., and Lang, R. A. (2002). *In vivo* depletion of CD11c($^+$) dendritic cells abrogates priming of CD8($^+$) T cells by exogenous cell-associated antigens. Immunity 17:211-220.

Kawamura, T., Gatanaga, H., Borris, D. L., Connors, M., Mitsuya, H., and Blauvelt, A. (2003). Decreased stimulation of CD4$^+$ T cell proliferation and IL-2 production by highly enriched populations of HIV-infected dendritic cells. J. Immunol. 170:4260-4266.

Kedl, R. M., Schaefer, B. C., Kappler, J. W., and Marrack, P. (2002). T cells down-modulate peptide-MHC complexes on APCs *in vivo*. Nat. Immunol. 3:27-32.

Kinter, A. L., Hennessey, M., Bell, A., Kern, S., Lin, Y., Daucher, M., Planta, M., McGlaughlin, M., Jackson, R., Ziegler, S. F., and Fauci, A. S. (2004). CD25($^+$)CD4($^+$) regulatory T cells from the peripheral blood of asymptomatic HIV-infected individuals regulate CD4($^+$) and CD8($^+$) HIV-specific T cell immune responses *in vitro* and are associated with favorable clinical markers of disease status. J. Exp. Med. 200:331-343.

Kloetzel, P. M., and Ossendorp, F. (2004). Proteasome and peptidase function in MHC-class-I-mediated antigen presentation. Curr. Opin. Immunol. 16:76-81.

Knight, S. C., Iqball, S., Roberts, M. S., Macatonia, S., and Bedford, P. A. (1998). Transfer of antigen between dendritic cells in the stimulation of primary T cell proliferation. Eur. J. Immunol. 28:1636-1644.

Kornfeld, C., Ploquin, M. J., Pandrea, I., Faye, A., Onanga, R., Apetrei, C., Poaty-Mavoungou, V., Rouquet, P., Estaquier, J., Mortara, L., Desoutter, J. F., Butor, C., Le Grand, R., Roques, P., Simon, F., Barre-Sinoussi, F., Diop, O. M., and Muller-Trutwin, M. C. (2005). Antiinflammatory profiles during primary SIV infection in African green monkeys are associated with protection against AIDS. J. Clin. Invest. 115:1082-1091.

Kovacsovics-Bankowski, M., Clark, K., Benacerraf, B., and Rock, K. L. (1993). Efficient major histocompatibility complex class I presentation of exogenous antigen upon phagocytosis by macrophages. Proc. Natl. Acad. Sci. U. S. A. 90:4942-4946.

Kurts, C., Heath, W. R., Carbone, F. R., Allison, J., Miller, J. F. A. P., and Kosaka, H. (1996). Constitutive class-I restricted exogenous presentation of self antigens *in vivo*. J. Exp. Med. 184:923-930.

Kurts, C., Heath, W. R., Carbone, F. R., Allison, J., and Miller, J. F. A. P. (1999). CD8 T cell ignorance of tolerance to islet Ags depends on Ag dose. Proc. Natl. Acad. Sci. U. S. A. 96:12703-12707.

Lanzavecchia, A., and Sallusto, F. (2001). The instructive role of dendritic cells on T cell responses: lineages, plasticity and kinetics. Curr. Opin. Immunol. 13:291-298.

Lapenta, C., Santini, S. M., Logozzi, M., Spada, M., Andreotti, M., Di Pucchio, T., Parlato, S., and Belardelli, F. (2003). Potent immune response against HIV-1 and protection from virus challenge in hu-PBL-SCID mice immunized with inactivated virus-pulsed dendritic cells generated in the presence of IFN-alpha. J. Exp. Med. 198:361-367.

Larsson, M., Fonteneau, J. F., Lirvall, M., Haslett, P., Lifson, J. D., and Bhardwaj, N. (2002). Activation of HIV-1 specific CD4 and CD8 T cells by human dendritic cells: roles for cross-presentation and non-infectious HIV-1 virus. AIDS 16:1319-1329.

Le Bon, A., Etchart, N., Rossmann, C., Ashton, M., Hou, S., Gewert, D., Borrow, P., and Tough, D. F. (2003). Cross-priming of CD8[+] T cells stimulated by virus-induced type I interferon. Nat. Immunol. 4:1009-1015.

Lelouard, H., Gatti, E., Cappello, F., Gresser, O., Camosseto, V., and Pierre, P. (2002). Transient aggregation of ubiquitinated proteins during dendritic cell maturation. Nature 417:177-182.

Letvin, N. L., and Walker, B. D. (2003). Immunopathogenesis and immuno-therapy in AIDS virus infections. Nat. Med. 9:861-866.

Levy, Y., Durier, C., Krzysiek, R., Rabian, C., Capitant, C., Lascaux, A. S., Michon, C., Oksenhendler, E., Weiss, L., Gastaut, J. A., Goujard, C., Rouzioux, C., Maral, J., Delfraissy, J. F., Emilie, D., and Aboulker, J. P. (2003). Effects of interleukin-2 therapy combined with highly active antiretroviral therapy on immune restoration in HIV-1 infection: a randomized controlled trial. AIDS 17:343-351.

Lilley, B. N., and Ploegh, H. L. (2005). Viral modulation of antigen presentation: manipulation of cellular targets in the ER and beyond. Immunol. Rev. 207:126-144.

Lizee, G., Basha, G., Tiong, J., Julien, J. P., Tian, M., Biron, K. E., and Jefferies, W. A. (2003). Control of dendritic cell cross-presentation by the major histocompatibility complex class I cytoplasmic domain. Nat. Immunol. 4:1065-1073.

Lutz, M. B., and Schuler, G. (2002). Immature, semi-mature and fully mature dendritic cells: which signals induce tolerance or immunity? Trends Immunol. 23:445-449.

Ma, X., and Montaner, L. J. (2000). Proinflammatory response and IL-12 expression in HIV-1 infection. J. Leukoc. Biol. 68:383-390.

Macagno, A., Gilliet, M., Sallusto, F., Lanzavecchia, A., Nestle, F. O., and Groettrup, M. (1999). Dendritic cells up-regulate immunoproteasomes and the proteasome regulator PA28 during maturation. Eur. J. Immunol. 29:4037-4042.

Mallard, F., Antony, C., Tenza, D., Salamero, J., Goud, B., and Johannes, L. (1998). Direct pathway from early/recycling endosomes to the Golgi apparatus revealed through the study of shiga toxin B-fragment transport. J. Cell. Biol. 143:973-990.

Marañón, C., Desoutter, J. F., Hoeffel, G., Cohen, W., Hanau, D., and Hosmalin, A. (2004). Dendritic cells cross-present HIV antigens from live as well as apoptotic infected CD4$^+$ T lymphocytes. Proc. Natl. Acad. Sci. U. S. A. 101:6092-6097.

Marshall, J., Chehimi, J., Gri, G., Kostman, J., Montaner, L., and Trinchieri, G. (1999). The IL-12-mediated pathway of immune events is dysfunctional in HIV-infected individuals. Blood 94:1003-1011.

Martin, E., O'Sullivan, B., Low, P., and Thomas, R. (2003). Antigen-specific suppression of a primed immune response by dendritic cells mediated by regulatory T cells secreting interleukin-10. Immunity 18:155-167.

Masse, D., Voisine, C., Henry, F., Cordel, S., Barbieux, I., Josien, R., Meflah, K., Gregoire, M., and Lieubeau, B. (2002). Increased vaccination efficiency with apoptotic cells by silica-induced, dendritic-like cells. Cancer Res. 62:1050-1056.

Matsuo, M., Nagata, Y., Sato, E., Atanackovic, D., Valmori, D., Chen, Y. T., Ritter, G., Mellman, I., Old, L. J., and Gnjatic, S. (2004). IFN-{gamma} enables cross-presentation of exogenous protein antigen in human Langerhans cells by potentiating maturation. Proc. Natl. Acad. Sci. U. S. A. 101:14467-14472.

Matzinger, P. (2002). The danger model: a renewed sense of self. Science 296:301-305.

McIlroy, D., and Gregoire, M. (2003). Optimizing dendritic cell-based anticancer immunotherapy: maturation state does have clinical impact. Cancer Immunol. Immunother. 52:583-591.

McIlroy, D., Autran, B., Cheynier, R., Wain-Hobson, S., Clauvel, J.-P., Oksenhendler, E., Debre, P., and Hosmalin, A. (1995). Dendritic cell and CD4$^+$ T lymphocyte infection frequency in spleens of HIV-positive patients. J. Virol. 69:4737-4745.

Montoya, M., Schiavoni, G., Mattei, F., Gresser, I., Belardelli, F., Borrow, P., and Tough, D. F. (2002). Type I interferons produced by dendritic cells promote their phenotypic and functional activation. Blood 99:3263-3271.

Morelli, A. E., Larregina, A. T., Shufesky, W. J., Zahorchak, A. F., Logar, A. J., Papworth, G. D., Wang, Z., Watkins, S. C., Falo, L. D., Jr., and Thomson, A. W. (2003). Internalization of circulating apoptotic cells by splenic marginal zone dendritic cells: dependence on complement receptors and effect on cytokine production. Blood 101:611-620.

Moris, A., Nobile, C., Buseyne, F., Porrot, F., Abastado, J. P., and Schwartz, O. (2004). DC-SIGN promotes exogenous MHC-I-restricted HIV-1 antigen presentation. Blood 103:2648-2654.

Nair, S., Zhou, F., Reddy, R., Huang, L., and Rouse, B. T. (1992). Soluble proteins delivered to dendritic cells via pH-sensitive liposomes induce primary cytotoxic T lymphocyte responses *in vitro*. J. Exp. Med. 175:609-612.

Napolitani, G., Rinaldi, A., Bertoni, F., Sallusto, F., and Lanzavecchia, A. (2005). Selected Toll-like receptor agonist combinations synergistically trigger a T helper type 1-polarizing program in dendritic cells. Nat. Immunol. 6:769-776.

Neijssen, J., Herberts, C., Drijfhout, J. W., Reits, E., Janssen, L., and Neefjes, J. (2005). Cross-presentation by intercellular peptide transfer through gap junctions. Nature 434:83-88.

Niedermann, G., Geier, E., Lucchiari-Hartz, M., Hitziger, N., Ramsperger, A., and Eichmann, K. (1999). The specificity of proteasomes: impact on MHC class I processing and presentation of antigens. Immunol. Rev. 172:29-48.

Norbury, C. C., Malide, D., Gibbs, J. S., Bennink, J. R., and Yewdell, J. W. (2002). Visualizing priming of virus-specific CD8$^+$ T cells by infected dendritic cells *in vivo*. Nat. Immunol. 3:265-271.

Norbury, C. C., Basta, S., Donohue, K. B., Tscharke, D. C., Princiotta, M. F., Berglund, P., Gibbs, J., Bennink, J. R., and Yewdell, J. W. (2004). CD8$^+$ T cell cross-priming via transfer of proteasome substrates. Science 304:1318-1321.

O'Garra, A., and Vieira, P. (2004). Regulatory T cells and mechanisms of immune system control. Nat. Med. 10:801-805.

Pantaleo, G., Graziosi, C., Butini, L., Pizzo, P. A., Schnittman, S. M., Kotler, D. P., and Fauci, A. S. (1991). Lymphoid organs function as major reservoirs for human immunodeficiency virus. Proc. Natl. Acad. Sci. U. S. A. 88:9838-9842.

Pantaleo, G., and Koup, R. A. (2004). Correlates of immune protection in HIV-1 infection: what we know, what we don't know, what we should know. Nat. Med. 10:806-810.

Pfeifer, J. D., Wick, M. J., Roberts, R. L., Findlay, K., Normark, S. J., and Harding, C. V. (1993). Phagocytic processing of bacterial antigens for class I MHC presentation to T cells. Nature 361:359-362.

Philpott, D. J., and Girardin, S. E. (2004). The role of Toll-like receptors and Nod proteins in bacterial infection. Mol. Immunol. 41:1099-1108.

Pires, A., Nelson, M., Pozniak, A. L., Fisher, M., Gazzard, B., Gotch, F., and Imami, N. (2005). Mycobacterial immune reconstitution inflammatory syndrome in HIV-1 infection after antiretroviral therapy is associated with deregulated specific T-cell responses: beneficial effect of IL-2 and GM-CSF immunotherapy. J. Immune Based Ther. Vaccines 3:7.

Pozzi, L. A., Maciaszek, J. W., and Rock, K. L. (2005). Both dendritic cells and macrophages can stimulate naive CD8 T cells *in vivo* to proliferate, develop effector function, and differentiate into memory cells. J. Immunol. 175:2071-2081.

Propato, A., Cutrona, G., Francavilla, V., Ulivi, M., Schiaffella, E., Landt, O., Dunbar, R., Cerundolo, V., Ferrarini, M., and Barnaba, V. (2001). Apoptotic cells overexpress vinculin and induce vinculin-specific cytotoxic T-cell cross-priming. Nat. Med. 7:807-813.

Ramirez, M. C., and Sigal, L. J. (2002). Macrophages and dendritic cells use the cytosolic pathway to rapidly cross-present antigen from live, vaccinia-infected cells. J. Immunol. 169:6733-6742.

Rammensee, H., Friede, T., and Stevanovic, S. (1995). MHC ligands and peptide motifs: first listing. Immunogenetics 41:178-228.

Regnault, A., Lankar, D., Lacabanne, V., Rodriguez, A., Thery, C., Rescigno, M., Saito, T., Verbeek, S., Bonnerot, C., Ricciardi-Castagnoli, P., and Amigorena, S. (1999). Fcgamma receptor-mediated induction of dendritic cell maturation and major histocompatibility complex class I-restricted antigen presentation after immune complex internalization. J. Exp. Med. 189:371-380.

Reis e Sousa, C. (2004). Activation of dendritic cells: translating innate into adaptive immunity. Curr. Opin. Immunol. 16:21-25.

Rénia, L., Marañón, C., Hosmalin, A., Grüner, A. C., Silvie, O., and Snounou, G. (2005). Do apoptotic malaria infected hepatocytes initiate protective immune responses? J. Infect. Dis. 193:163-164.

Revy, P., Sospedra, M., Barbour, B., and Trautmann, A. (2001). Functional antigen-independent synapses formed between T cells and dendritic cells. Nat. Immunol. 2:925-931.

Rock, K. L. (2003). The ins and outs of cross-presentation. Nat. Immunol. 4:941-943.

Rock, K. L., Hearn, A., Chen, C. J., and Shi, Y. (2005a). Natural endogenous adjuvants. Springer Semin. Immunopathol. 26:231-246.

Rock, K. L., and Shen, L. (2005b). Cross-presentation: underlying mechanisms and role in immune surveillance. Immunol. Rev. 207:166-183.

Salio, M., Cerundolo, V., and Lanzavecchia, A. (2000). Dendritic cell maturation is induced by mycoplasma infection but not by necrotic cells. Eur. J. Immunol. 30:705-708.

Sallusto, F., Geginat, J., and Lanzavecchia, A. (2004). Central memory and effector memory T cell subsets: function, generation, and maintenance. Annu. Rev. Immunol. 22:745-763.

Sauter, B., Albert, M., Francisco, L., Larsson, M., Somersan, S., and Bhardwaj, N. (2000). Consequences of cell death: exposure to necrotic tumor cells, but not primary tissue cells or apoptotic cells, induces the maturation of immunostimulatory dendritic cells. J. Exp. Med. 191:423-433.

Saveanu, L., Fruci, D., and van Endert, P. (2002). Beyond the proteasome: trimming, degradation and generation of MHC class I ligands by auxiliary proteases. Mol. Immunol. 39:203.

Savill, J., Gregory, C., and Haslett, C. (2003). Cell biology. Eat me or die. Science 302:1516-1517.

Scheinecker, C., McHugh, R., Shevach, E. M., and Germain, R. N. (2002). Constitutive presentation of a natural tissue autoantigen exclusively by dendritic cells in the draining lymph node. J. Exp. Med. 196:1079-1090.

Schmidt, B., Scott, I., Whitmore, R. G., Foster, H., Fujimura, S., Schmitz, J., and Levy, J. A. (2004). Low-level HIV infection of plasmacytoid dendritic cells: onset of cytopathic effects and cell death after PDC maturation. Virology 329:280-288.

Schuler, G., and Steinman, R. M. (1997). Dendritic cells as adjuvants for immune-mediated resistance to tumors. J. Exp. Med. 186:1183-1187.

Schwartz, O., Marechal, V., Le Gall, S., Lemonnier, F., and Heard, J. M. (1996). Endocytosis of major histocompatibility complex class I molecules is induced by the HIV-1 Nef protein. Nat. Med. 2:338-342.

Seifert, U., Maranon, C., Shmueli, A., Desoutter, J. F., Wesoloski, L., Janek, K., Henklein, P., Diescher, S., Andrieu, M., de la Salle, H., Weinschenk, T., Schild, H., Laderach, D., Galy, A., Haas, G., Kloetzel, P. M., Reiss, Y., and Hosmalin, A. (2003). An essential role for tripeptidyl peptidase in the generation of an MHC class I epitope. Nat. Immunol. 4:375-379.

Serna, A., Ramirez, M. C., Soukhanova, A., and Sigal, L. J. (2003). Cutting edge: efficient MHC class I cross-presentation during early vaccinia infection requires the transfer of proteasomal intermediates between antigen donor and presenting cells. J. Immunol. 171:5668-5672.

Servet, C., Zitvogel, L., and Hosmalin, A. (2002). Dendritic cells in innate immune responses against HIV. Curr. Mol. Med. 2:739-756.

Shen, L., and Rock, K. L. (2004a). Cellular protein is the source of cross-priming antigen *in vivo*. Proc. Natl. Acad. Sci. U. S. A. 101:3035-3040.

Shen, L., Sigal, L. J., Boes, M., and Rock, K. L. (2004b). Important role of cathepsin S in generating peptides for TAP-independent MHC class I crosspresentation *in vivo*. Immunity 21:155-165.

Sigal, L. J., Crotty, S., Andino, R., and Rock, K. L. (1999). Cytotoxic T-cell immunity to virus-infected non-haematopoietic cells requires presentation of exogenous antigen. Nature 398:77-80.

Skoberne, M., Beignon, A. S., Larsson, M., and Bhardwaj, N. (2005). Apoptotic cells at the crossroads of tolerance and immunity. Curr. Top. Microbiol. Immunol. 289:259-292.

Smed-Sorensen, A., Lore, K., Walther-Jallow, L., Andersson, J., and Spetz, A. L. (2004). HIV-1-infected dendritic cells up-regulate cell surface markers but fail to produce IL-12 p70 in response to CD40 ligand stimulation. Blood 104:2810-2817.

Somersan, S., and Bhardwaj, N. (2001). Tethering and tickling: a new role for the phosphatidylserine receptor. J. Cell. Biol. 155:501-504.

Spetz, A.-L., Patterson, B. K., Lore, K., Andersson, J., and Holmgren, L. (1999). Functional gene transfer of HIV DNA by an HIV receptor-independent mechanism. J. Immunol. 163:736-742.

Steinman, R., Turley, S., Melman, I., and Inaba, K. (2000). The induction of tolerance by dendritic cells that have captured apoptotic cells. J. Exp. Med. 191:411-416.

Steinman, R. M., Hawiger, D., and Nussenzweig, M. C. (2003). Tolerogenic dendritic cells. Annu. Rev. Immunol. 21:685-711.

Stinchcombe, J. C., Bossi, G., Booth, S., and Griffiths, G. M. (2001). The immunological synapse of CTL contains a secretory domain and membrane bridges. Immunity 15:751-761.

Stumptner-Cuvelette, P., Morchoisne, S., Dugast, M., Le Gall, S., Raposo, G., Schwartz, O., and Benaroch, P. (2001). HIV-1 Nef impairs MHC class II antigen presentation and surface expression. Proc. Natl. Acad. Sci. U. S. A. 98:12144-12149.

Subauste, C. S., Wessendarp, M., Portilllo, J. A., Andrade, R. M., Hinds, L. M., Gomez, F. J., Smulian, A. G., Grubbs, P. A., and Haglund, L. A. (2004). Pathogen-specific induction of CD154 is impaired in $CD4^+$ T cells from human immunodeficiency virus-infected patients. J. Infect. Dis. 189:61-70.

Tabiasco, J., Vercellone, A., Meggetto, F., Hudrisier, D., Brousset, P., and Fournie, J. J. (2003). Acquisition of viral receptor by NK cells through immunological synapse. J. Immunol. 170:5993-5998.

Thery, C., Zitvogel, L., and Amigorena, S. (2002). Exosomes: composition, biogenesis and function. Nat. Rev. Immunol. 2:569-579.

Touret, N., Paroutis, P., Terebiznik, M., Harrison, R. E., Trombetta, S., Pypaert, M., Chow, A., Jiang, A., Shaw, J., Yip, C., Moore, H. P., van der Wel, N., Houben, D., Peters, P. J., de Chastellier, C., Mellman, I., and Grinstein, S. (2005). Quantitative and dynamic assessment of the contribution of the ER to phagosome formation. Cell 123:157-170.

Townsend, A. R. M., Gotch, F. M., and Davey, J. (1985). Cytotoxic T cells recognize fragments of the influenza nucleoprotein. Cell 42:457-467.

Turley, S., Poirot, L., Hattori, M., Benoist, C., and Mathis, D. (2003). Physiological beta cell death triggers priming of self-reactive T cells by dendritic cells in a type-1 diabetes model. J. Exp. Med. 198:1527-1537.

Turville, S. G., Cameron, P. U., Handley, A., Lin, G., Pohlmann, S., Doms, R. W., and Cunningham, A. L. (2002). Diversity of receptors binding HIV on dendritic cell subsets. Nat. Immunol. 3:975-983.

Vanham, G., Penne, L., Devalck, J., Kestens, R., Colebunders, R., Bosmans, E., Thielemans, K., and Ceuppens, J. (1999). Decreased CD40 ligand induction in CD4 T cells and dysregulated IL-12 production during HIV infection. Clin. Exp. Immunol. 117:335-342.

Veazey, R. S., and Lackner, A. A. (2004). Getting to the guts of HIV pathogenesis. J. Exp. Med. 200:697-700.

Watkins, S. C., and Salter, R. D. (2005). Functional connectivity between immune cells mediated by tunneling nanotubules. Immunity 23:309-318.

Weiss, L., Donkova-Petrini, V., Caccavelli, L., Balbo, M., Carbonneil, C., and Levy, Y. (2004). Human immunodeficiency virus-driven expansion of $CD4^+CD25^+$ Regulatory T cells which suppress HIV-specific CD4 T-cell responses in HIV-infected patients. Blood 104:3249-3256.

Wick, M. J., and Pfeifer, J. D. (1996). Major histocompatibility complex class I presentation of ovalbumin peptide 257-264 from exogenous sources: protein context influences the degree of TAP-independent presentation. Eur. J. Immunol. 26:2790-2799.

Wolfers, J., Lozier, A., Raposo, G., Regnault, A., Thery, C., Masurier, C., Flament, C., Pouzieux, S., Faure, F., Tursz, T., Angevin, E., Amigorena, S., and Zitvogel, L. (2001). Tumor-derived exosomes are a source of shared tumor rejection antigens for CTL cross-priming. Nat. Med. 7:297-303.

Wolkers, M. C., Brouwenstijn, N., Bakker, A. H., Toebes, M., and Schumacher, T. N. (2004). Antigen bias in T cell cross-priming. Science 304:1314-1317.

Yewdell, J. W., and Hill, A. B. (2002). Viral interference with antigen presentation. Nat. Immunol. 3:1019-1025.

Yewdell, J. W., Reits, E., and Neefjes, J. (2003). Making sense of mass destruction: quantitating MHC class I antigen presentation. Nat. Rev. Immunol. 3:952-961.

Yonezawa, A., Morita, R., Takaori-Kondo, A., Kadowaki, N., Kitawaki, T., Hori, T., and Uchiyama, T. (2003). Natural alpha interferon-producing cells respond to human immunodeficiency virus type 1 with alpha interferon production and maturation into dendritic cells. J. Virol. 77:3777-3784.

York, I. A., Goldberg, A. L., Mo, X. Y., and Rock, K. L. (1999). Proteolysis and class I major histocompatibility complex antigen presentation. Immunol. Rev. 172:49-66.

Yrlid, U., and Wick, M. J. (2002). Antigen presentation capacity and cytokine production by murine splenic dendritic cell subsets upon Salmonella encounter. J. Immunol. 169:108-116.

Zhang, R., Fichtenbaum, C. J., Hildeman, D. A., Lifson, J. D., and Chougnet, C. (2004). CD40 ligand dysregulation in HIV infection: HIV glycoprotein 120 inhibits signaling cascades upstream of CD40 ligand transcription. J. Immunol. 172:2678-2686.

Zhang, R., Lifson, J. D., and Chougnet, C. (2006). Failure of HIV-exposed CD4[+] T cells to activate dendritic cells is reversed by restoration of CD40/CD154 interactions. Blood 107:1989-1995.

Zhao, X. Q., Huang, X. L., Gupta, P., Borowski, L., Fan, Z., Watkins, S. C., Thomas, E. K., and Rinaldo, C. R., Jr. (2002). Induction of anti-human immunodeficiency virus type 1 (HIV-1) CD8([+]) and CD4([+]) T-cell reactivity by dendritic cells loaded with HIV-1 X4- infected apoptotic cells. J. Virol. 76:3007-3014.

Zinkernagel, R. M., and Doherty, P. C. (1974). Restriction of *in vitro* T-cell mediated cytotoxicity in lymphocytic choriomeningitis within a syngeneic or semiallogenic system. Nature 248:701-702.

Chapter 14

Immunotherapy of HIV Infection: Dendritic Cells as Targets and Tools

Imerio Capone, Giuseppe Tambussi, Paola Rizza,
and Adriano Lazzarin

14.1 Introduction

Since the start of the HIV epidemic, more than 60 million people have been infected with the virus. Currently, 42 million individuals are living with AIDS worldwide, and more than 20 million people have died as a consequence of the disease. Highly active anti-retroviral therapy (HAART), defined as a combination therapy of drugs capable of inhibiting HIV replication and lowering plasma viral load to undetectable level, is currently used as the standard care for HIV-infected patients. Despite the recent success in the development of this drug therapy, several limitations, such as the failure in eradicating the infection, toxic side effects, the emergence of drug-resistant viruses, together with its high cost, represent the challenges for achieving a long-term control of the progression of the HIV infection in infected individuals. Although there is no doubt that anti-retroviral drugs remain the reference treatment for HIV-1 infection, it is now generally accepted that we must broaden our treatment strategies focusing on the improvement of safety and efficacy of therapies and the simplification of disease management. It has been demonstrated that short interruptions of drug treatment can act as a boost for HIV immunity, and this has raised the question whether immune-based therapies could be considered as useful and powerful tools for enhancing immune responses to HIV, thereby limiting the use of anti-retroviral drugs and improving the efficacy of the treatment in a long-term strategy? An increasing interest in immune-based therapies for HIV infections stems from emerging data suggesting that some immune correlates can play a key role in reducing virus replication. In particular, it has been shown in animal models as well as in humans that both T helper (Th) cells and cytotoxic T lymphocytes (CTLs) are important in containing HIV replication (Jin *et al.*, 1999; Rosenberg *et al.*, 1997; Schmitz *et al.*, 1999). Therefore, enhancing immune control of the virus in terms of the strength and extent of HIV-specific cellular responses might have a clinical benefit or enhance the efficacy when we use antiviral drugs.

Several strategies of immunotherapy have been assessed in HIV-infected patients. They include the use of cytokines, immune cells, drugs inhibiting T-cell activation, and vaccines. Therapeutic immunization may be intended as a complementary approach to anti-retroviral drug therapy to induce an effective control of HIV replication, thus preventing immunologic defects caused by virus spreading and, at the same time, permitting long periods of an absence of therapy. Currently, the development of HIV-1 vaccines is focusing on strategies capable of eliciting effective T-cell responses. An ensemble of studies has now revealed that dendritic cells (DCs) play a crucial role in linking innate and adaptive immunity against pathogens. Several factors are important in designing an effective therapeutic vaccine, but the capability of the antigen formulation targeting DCs in sufficient amounts to trigger an effective response represents one of the most significant prerequisites. The complex biology of DCs, the heterogenicity of these cells and of their functions in the regulation of the immune response, and the importance of recent progress of knowledge in the field of vaccine research have been extensively described in other chapters of this book. This chapter will focus on the state of the art of immunotherapy for the treatment of HIV-1 infection, with particular emphasis on DCs as targets and tools of HIV vaccines.

14.2 The Rationale for Combining HAART with Immunotherapy in the Treatment of HIV-1 Infection

14.2.1 Drug Therapy Limitations

The standard HAART combination of drugs includes a minimum of one protease inhibitor (PI) or non-nucleoside reverse-transcriptase inhibitor (NNRTI) associated with two nucleoside reverse-trancriptase inhibitors (NRTI). Since its introduction in 1996, HAART profoundly changed the course and management of HIV infection, reducing HIV-related morbidity and mortality and decreasing HIV transmission (Palella *et al.*, 1998). Despite this success, some intrinsic limitations have since emerged rendering HAART as a long-term toxicity risk together with its inefficacy. First, the drug therapy fails to eradicate the virus (Siliciano *et al.*, 2003), resulting, at best, in a long-term containment of viral load below the detectable level and a $CD4^+$ T-cell level that ensures a very low risk of progression to full-blown AIDS. However, a life-long treatment is required to prevent disease progression and this represents a major obstacle for a worldwide application of such treatments due to their high costs, which largely exceeds the pro-capita resources of the great majority of HIV-infected individuals, especially those in developing countries. Second, current drug therapies are associated with several toxic side effects, including lipid disorders and mitochondrial toxicity, thus often requiring treatment interruptions. These complications and the low patient compliance represent the main reasons for delaying the initiation of drug therapy, as recommended by the recent new therapeutic guidelines (Yeni *et al.*, 2004). Third, prolonged treatment with anti-retroviral drugs results in the selection of HIV-1 variants displaying a resistance to the drugs (Aleman *et al.*, 2002). New therapeutics, such as inhibitors of virus entry, are continuously under development, but there is little hope that there soon will be effective drugs available to overcome problems related to toxicity or resistance; therefore, there is no doubt that alternative or complementary strategies for HIV control are urgently needed.

14.2.2 Immune Control of HIV-1

The characterization of immune responses against HIV-1 has indicated that both Th cells and CTLs are involved in the control of viral replication in the early phases of acute infection (Rosenberg *et al.*, 1997; Schmitz *et al.*, 1999). In contrast, HIV-1 neutralizing antibodies appear much later, and it is likely that they do not contribute to the early control of HIV-1 infection (Parren *et al.*, 1999). Although unable to eradicate the virus, HIV-1 specific T-cell responses contain viral load until escape mechanisms prevail, thus subverting immune control and favoring the progression of infection and the gradual development of the disease. On this basis, enhancing the host immune responses against HIV could be a valid strategy for overcoming this lack of viral control, thus providing a rationale for immune-based therapy of HIV-1 infection. This rationale stems from first observations, that the treatment of HIV-1 infection with anti-retroviral drugs induced an increment of naïve T-cells as well as an improvement of their functional defects typically observed in advanced HIV-infected patients (Autran *et al.*, 1997). In fact, it has been shown that in these patients, HAART can restore the immunity to microbial antigens (Pontesilli *et al.*, 1999; Rinaldo *et al.*, 1999), indicating that HIV-induced immune suppression can be reversible. On the other hand, HIV-specific immunity is not fully restored in the course of HAART. In fact, although HIV-specific CD4$^+$ T cells are prevented from being destroyed when drugs are administered in the early phase of infection (Rosenberg *et al.*, 1997; Oxenius *et al.*, 2000), they can typically remain at low levels for years when drug treatment is initiated in the chronic phase (Lederman *et al.*, 1998; Li *et al.*, 1998). Other studies have shown that HIV-specific immunity is not enhanced but instead declines when viral replication is suppressed by drug administration (Ogg *et al.*, 1999; Pitcher *et al.*, 1999), suggesting that the antigenic stimulation is too weak for the induction as well as maintenance of the immune response. The concept that immunity to HIV is dependent on antigen exposure has been confirmed by several studies, which demonstrated that a transient restoration of HIV-specific CD4$^+$ and CD8$^+$ T-cell responses can occur when HIV replication recommences during HAART interruptions, thus re-exposing HIV-1 antigens to the immune system (autovaccination) (Ortiz *et al.*, 1999; Papasavvas *et al.*, 2000; Rosenberg *et al.*, 2000). Unfortunately, these responses fail to enhance viral control and to contain viral load (Carcelain *et al.*, 2001; Oxenius *et al.*, 2002), although a transient control was observed in some uncontrolled treatment interruption studies (Lisziewicz *et al.*, 1999; Rosenberg *et al.*, 2000). This could be due to the period of time elapsing between the antigenic stimulus and T-cell responses (Davenport *et al.*, 2004), which allows a rapid infection of newly generated HIV-specific CD4$^+$ T cells (Douek *et al.*, 2002) and a selection of mutant viruses escaping immune control (Altfeld *et al.*, 2002; Jost *et al.*, 2002). Furthermore, re-exposing HIV-infected subjects to the same virus antigens leads to the expansion of memory CTL clones, thus limiting the diversification of lymphocyte repertoire, consistently with the "antigenic-sin" theory (Autran and Carcelain, 2000). Hence, the major challenge of the research is to stimulate strong and durable T-cell responses capable of controlling HIV infection in individuals unable to develop such responses on their own, with the aim of achieving equilibrium between HIV replication and host immune responses, thus permitting long off-therapy periods. The strategy is defined as "therapeutic immunization" and consists of the vaccination of HIV-infected individuals when the virus is under the suppressive activity of HAART, so that an effective immunity can be stimulated, before rather than after the viral rebound occurs during the drug interruption, thus resulting in a better containment of virus replication. Figure 14.1 illustrates the advantage and limitations of the current HAART regimen and the rationale for it being combined with strategies of immunotherapy aimed at inducing a protective immune response. Several studies carried out in animal models support this concept, and some results from clinical trials suggest that this objective could be achieved also in humans.

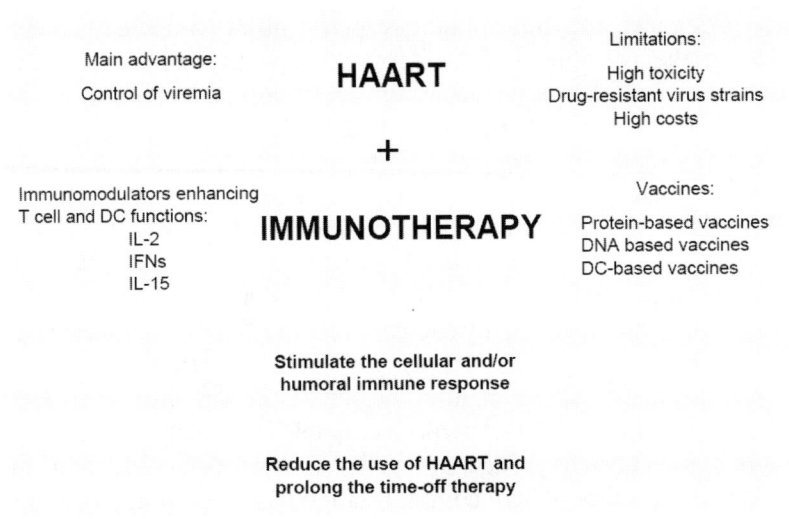

FIGURE 14.1. The principal challenge in combining HAART with immuno-modulators enhancing immune cell functions and/or vaccines for the management of HIV-1 infection. HAART can control viral replication but presents several limitations, such as high toxicity, costs, and emergence of drug-resistant virus strains. All this could be overcome by the combination of HAART with therapeutic vaccine immunomodulatory agents, with the aim of stimulating cellular and/or humoral immune responses capable of reducing the use of HAART and prolonging the time off-therapy.

14.3 Overview of the Clinical Trials of Immunotherapy for the Treatment of HIV-1 Infection

14.3.1 Introduction

The inadequate immune reconstitution after anti-retroviral therapy, the discordant immuno-virological response, and the aim to obtain a specific immunity against HIV leads to clinical immunotherapy applications. Several strategies of immunotherapy based on different types of rationale have been evaluated in clinical trials with HIV-infected patients. These strategies range from the use of some cytokines, such as IL-2, capable of promoting T-cell functions, to the use of substances suppressing T-cell activation. In addition to these strategies (all based on the use of factors acting on T-cell functions), several groups have tested protocols of active immunization, represented by different types of HIV vaccines, whose main goal is to prompt an antiviral immune response in HIV-infected patients. The issue of the possible usefulness of "therapeutic vaccines" (active immunization in patients with an established chronic infection) is still matter of debate. The results of pilot clinical studies in patients infected with HIV, HBV, and HPV support the interest for further clinical investigation on therapeutic vaccination strategies (Autran *et al.*, 2004). One major limitation in addressing the potential efficacy of therapeutic vaccination in HIV-infected individuals is represented by the lack of an effective prophylactic vaccine (Desrosiers, 2004). However, some data are available on the immunogenicity of certain types of HIV vaccines in infected patients (Gahery *et al.*, 2005; Gotch, 2005). In the following subsections, we review some main strategies of immunotherapy of HIV infection based on the use of factors/drugs capable of directly acting on the patient's immune cells (in particular T cells). In a subsequent section, we discuss some approaches of active immunization based on strategies aimed at the *in vivo* targeting of patients' DCs by using classical and novel categories of vaccine adjuvants. In the final sections of this chapter, we provide an overview of the results available to date, in both animal models and humans, supporting the concept that vaccines based on the use of autologous DCs pulsed with HIV antigens can signify a valuable approach for inducing a strong and potentially protective immune response. We conclude the chapter by emphasizing the importance of further investigations on the biology of DCs for the better evaluation of the role of these cells as targets and tools for HIV vaccines.

14.3.2 IL-2 Treatment in HIV-infected Patients

The failure of eradicating HIV infection by combining several antiretroviral agents has refocused most of the attention on the long-term management of HIV disease. In this context, the possibility of combining antiviral and immune-based approaches has a solid basis in the consolidated experience of IL-2 administration. Typically known as one cytokine of the adaptive immune response, IL-2 has been extensively evaluated in a large number of clinical trials designed to boost T and NK cell immunity in a therapeutic setting against cancer and chronic viral infections (Eklund and Kuzel, 2004; Anaya and Sias, 2005). Multiple trials on different cohorts of HIV-infected individuals have shown that intermittent administration of 3–12 MIU/d of IL-2 can restore normal levels of circulating $CD4^+$ T lymphocytes, a crucial parameter of the immune function.

The retrospective analysis of IL-2–treated HIV-infected individuals supports the hypothesis that the observed increases of $CD4^+$ T-cell counts are not a simple "cosmetic" change in the composition of PBMCs of these individuals. Multiple randomized trials have demonstrated that the administration of intermittent cycles of IL-2 to HIV-infected patients can lead to profound, sustained increases in CD4$^+$ cell number and percentage. After three to six 5-day cycles of IL-2 administered every 2 months, CD4$^+$ cell numbers may remain elevated for years without additional cycles. While this therapy causes marked changes in CD4$^+$ cell numbers, preferentially observed in cells with a naïve phenotype, only minimal changes are seen in CD8$^+$ or NK cell numbers.

The mechanisms leading to these increases have remained obscure. *Ex vivo* studies have documented increases in proliferation and death of CD8$^+$ as well as CD4$^+$ T cells during an IL-2 cycle, suggesting that, while increased proliferation may play a role, proliferation alone does not explain the preferential expansion of CD4$^+$ cells with respect to CD8$^+$ cells. It is hoped that ongoing long-term phase III trials with a clinical endpoint will also address this fundamental question.

An added value of IL-2 experimental therapy is represented by its broader implication: administration of a cytokine, a messenger protein that allows the intercellular communication among immune cells, as a drug given to immunodeficient individuals may lead to the reconstitution of their immune responses, thus increasing their capacity to fight opportunistic infections and tumors. Some concerns on the use of IL-2, however, have recently been raised as a consequence of the finding that, under some conditions, this cytokine can enhance the proliferation of CD25$^+$/CD4$^+$ regulatory T cells, which may increase in certain patients categories and can produce immune regulatory effects (Antony *et al.*, 2005). Other cytokines, such as IL-7 or IL-15, can be valuable candidates for new strategies of immunotherapy of HIV-infection (Ahmad *et al.*, 2005; Nunnari and Pomerantz, 2005; d'Ettorre *et al.*, 2006), but results from relevant clinical studies are still not available.

14.3.3 The Use of Immunomodulatory Drugs in Primary HIV Infection

Primary or acute HIV infection (PHI) refers to the specific period of HIV viral infection or viral inoculation, while early HIV infection refers to the period of time after recent infection. In the vast majority of adult cases (approximately 75%), early HIV infection is associated with an acute viral syndrome. The clinical signs of HIV infection include skin rash, lymphadenopathy, splenomegaly, arthritis, and, less frequently, aseptic meningitis. At present, it is widely accepted that primary HIV infection is possibly the most crucial period for HAART initiation. This is because a series of observations have revealed that the initial interaction between the virus and the host during primary infection determines the pattern and the rate of disease progression. Several studies have provided evidence that the initiation of HAART during primary infection is associated with effective suppression of virus replication and spread, a shorter symptomatic phase, restoration of $CD4^+$ T-cell counts, and the preservation of HIV-specific T-cell responses. Furthermore, recent data suggest that initiation of therapy during primary infection is associated with long-term control of virus replication after HAART discontinuation. However, it is still unclear whether early initiation of HAART is able to interfere with the formation of the pool of latently virus-infected $CD4^+$ T cells, which represents a major limitation for the eradication of HIV-1. The long-term, although partial, control of virus replication seems to require several cycles of a structured treatment interruption.

A principal feature of PHI is massive immune activation. Immune activation as a general mechanism of HIV-associated disease has been proposed in the past and has been substantiated by several observations (Lori and Lisziewicz, 2000). It has been assumed that the immune activation may be deleterious for several reasons. Although HIV-1 may replicate in both quiescent and activated $CD4^+$ T cells, proliferating and activated $CD4^+$ T cells support massive HIV replication and production. High levels of virus replication and massive stimulation of the immune system may lead to clonal exhaustion of HIV-specific $CD8^+$ T cells and rapid elimination of virus-specific $CD4^+$ T helper cells. Alternative immune-based interventions include anti-retroviral drugs in combination with compounds that are able to inhibit the function and reactivity of some cells of the immune system (Rizzardi *et al.*, 2002a; Fumero *et al.*, 2004). The goal of these strategies is to achieve the long-term control of virus replication with the use of minimal or possibly no antiretroviral drugs.

14.3.3.1 Use of Cyclosporin A Alongside HAART

In the past, some studies have tested the effects of therapeutic agents such as cyclosporin A (CsA) that selectively suppress T-cell activation in HIV-1 infection. CsA, a cyclic undecapeptide, is an immunosuppressive agent that has been hypothesized to suppress HIV replication indirectly, by limiting T-cell activation, and directly, by interference with HIV gag polyprotein processing, resulting in production of noninfectious particles (Thali *et al.*, 1994). CsA suppresses T-cell activation by directly blocking activation of the genes for IL-2, IL-4, and the IL-2 receptor in T cells, thus inhibiting IL-2–dependent T-cell proliferation and differentiation. The results from early clinical studies were quite disappointing, because there was no evidence of a beneficial effect. In this regard, it is important to note, however, that in these studies CsA was used as monotherapy in patients with chronic infection and at advanced stages of disease. In contrast, administration of CsA in monkeys acutely inoculated with SIV (simian immunodeficiency virus) showed a beneficial effect on the kinetics of CD4[+] T-cell depletion (Martin *et al.*, 1997).

In the past few years, the hypothesis has been that the rapid shutdown of immune activation associated with PHI may be beneficial for both immunologic and virologic measures in patients that have been tested. The effect of CsA administration in combination with HAART was investigated in a pilot clinical study (Rizzardi *et al.*, 2002b). Overall, these data suggest that decreasing T-cell activation in the very early phases of HIV disease has a beneficial impact on the long-term course of the infection, contributing to the establishment, following PHI, of a more favorable immunologic set-point that affects the ultimate pattern and rate of disease progression. It is also worth noting that this was the first study providing evidence on the benefits achieved with HAART during primary HIV-1 infection, via the use of an immune-modulating strategy interfering with early pathogenic events. These results provide the rationale for the evaluation of CsA and other therapeutic immune-modulating strategies in larger randomized clinical trials.

14.3.3.2 Mycophenolic Acid in Patients Undergoing Supervised Interruption of Therapy

Mycophenolate mofetil (MMF) is a well-known drug broadly used in renal transplantation for its ability to selectively inhibit lymphocyte division. MMF is the morpholinoethyl ester of the active compound mycophenolic acid (MPA). Interest in MPA as an immunosuppressive agent in renal transplantation has been renewed in clinical studies showing that the proliferation of T and B lymphocytes is selectively blocked by MPA, as their purine synthesis primarily relies on the *de novo* synthesis pathway, which is blocked by MPA.

The effect of MPA on cell activation and HIV-1 infection has been investigated in both *in vitro* and *in vivo* studies (Chapuis *et al.*, 2000). Overall, the available data indicate that: (a) MPA is a potent inhibitor of the proliferation of activated T cells, even in the presence of IL-2; (b) MPA suppresses cell proliferation by inducing apoptosis and cell death of a large proportion of activated T cells; (c) MPA is able to suppress HIV infection in CD4[+] T cells, because it is likely to cause the depletion of important substrates in order to develop an effective reverse transcription by impairing the *de novo* synthesis of guanosine nucleotides. Therefore,

MPA might inhibit HIV infection by a dual mechanism, including an antiviral mechanism exerted by depleting the substrate for the reverse transcriptase and an immunologic mechanism able to reduce the pool of activated $CD4^+$ T lymphocytes that may in turn support productive HIV-1 infection. The activity of MPA on T-cell proliferation and HIV-1 has been tested in pilot clinical studies. The results indicate that MMF is able to significantly reduce the *in vivo* size of the dividing $CD4^+$ T-cell pool (Chapuis *et al.*, 2000). It remains to be determined whether MMF exerts an effect on the pool of resting latently infected $CD4^+$ T cells. It is probable that MMF has no effect on a resting cell, thus not directly affecting the size of this pool of cells. However, once these cells are activated in the presence of MMF, MMF could in turn induce apoptosis and therefore cell death. These results, albeit preliminary, suggest that MMF may have an indirect impact on the pool of resting latently infected $CD4^+$ T cells, contributing to its depletion *in vivo*.

14.4 The Dendritic Cell as Target for the Development of HIV-1 Vaccines

The development of an effective anti-HIV vaccine requires not only the characterization of the relevant virus antigens potentially important in achieving immune protection but also the identification of potent adjuvants, which are necessary in inducing suitable levels of neutralizing antibodies as well as for ensuring the generation of vigorous antiviral T-cell responses. It is now commonly accepted that DCs are the target cells of any type of vaccine, and we have recently understood different strategies in enhancing vaccine efficacy by acting at various levels on these cells. The understanding of the signals necessary to convert DCs into immunostimulatory APCs capable of priming appropriate effector T cells has allowed the identification of key-molecules suitable for the development of novel and effective adjuvants, including immunopotentiators recognized as strong DC activators, as well as novel strategies of *in vivo* targeting of DCs. Figure 14.2 illustrates how DCs can be exploited as targets in immunotherapy strategies for the treatment of HIV infection. In this regard, there is an increasing interest in the development of new categories of vaccine adjuvants that specifically act on DCs. Depending on their mechanisms of action, two main categories of adjuvants can be distinguished, which act at different levels on DCs: (i) delivery systems, which ensure that the antigen properly targets the DCs; (ii) immunopotentiators, which directly exert an effect on DCs by interacting with specific receptors (O'Hagan and Valiante, 2003).

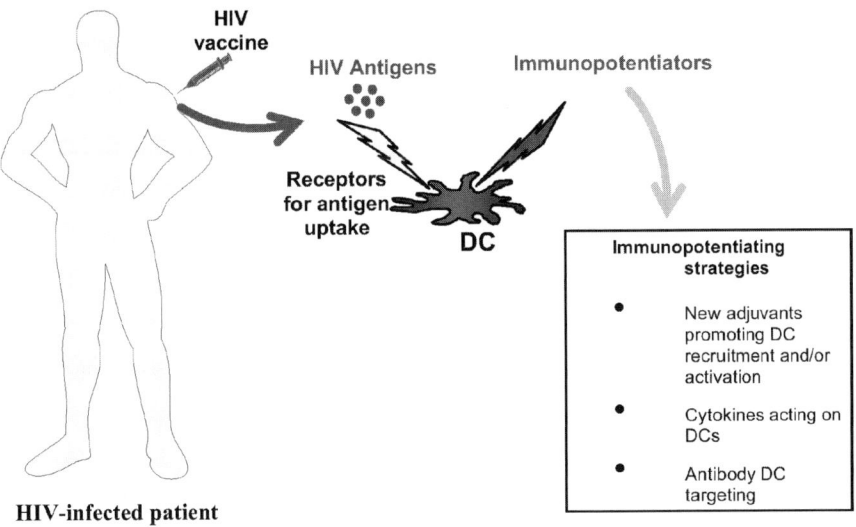

FIGURE 14.2. Dendritic cell targeting for the development of immunotherapeutic strategies against HIV, based on the recent knowledge of DC biology. DCs can be targeted *in vivo* by immunopotentiating agents (adjuvants, cytokines, antibodies, etc.), which may enhance their capability to take up and process antigens and/or to promote their activation and migration to lymphoid organs in order to efficiently prime immune responses.

These aspects have been extensively addressed in Chapter 5. Good examples of *in vivo* targeting of reference antigens to DCs, by the use of a specific antibody to the DEC-205 receptor, have been provided in mouse models by Bonifaz and colleagues (2002, 2004). These as well as other strategies of HIV targeting to DCs have been summarized in Chapter 6 of this book and are not further discussed here. Reference again is made of the well-known "theory of danger signals" proposed by Matzinger and co-workers (Gallucci *et al.*, 1999), because it is still generally important in the understanding how adjuvants should act on DCs for the development of effective therapeutic vaccines in patients with severe chronic diseases such as HIV infection and cancer. According to this theory, the immune system is structured to distinguish dangerous from harmless, rather than self from non-self. Figure 14.3 illustrates how immunopotentiators, including the natural adjuvants, can contribute to the activation of DCs and, therefore, to the generation of a potentially protective immune response. These immunopotentiators can include microbe-derived factors, such as CpG oligonucleotides, or virus-induced cytokines (IFN-α). These special type of adjuvants could be considered as valuable candidates for boosting antiviral innate immunity in HIV-infected patients, who generally show marked alterations in the phenotype and function of circulating DC population (Kamga, *et al.*, 2005; Almeida *et al.*, 2005).

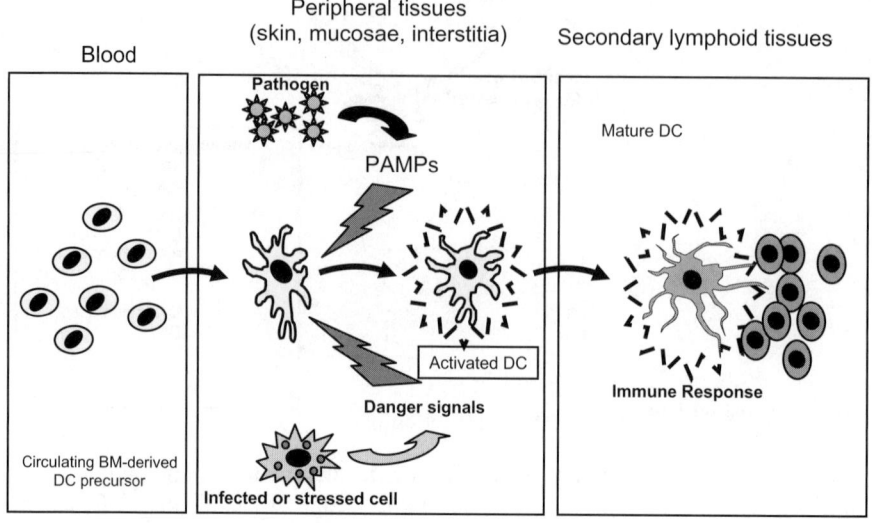

FIGURE 14.3. Role of immunopotentiators and natural adjuvants in the activation of DCs. Circulating DC precursors, in the presence of danger signals (inflammatory cytokines) derived from infected cells or molecular products derived from a family of evolutionary conserved structural elements shared by microbial pathogens (PAMPs), are activated and migrate to secondary lymphoid tissues, where they can prime immune responses. Cytokines (IFN-α), bacterial products such as CpG oligonucleotides, and synthetic adjuvants capable of inducing differentiation/activation of DCs can be valuable components in strategies of immunotherapy in HIV-infected patients, who generally show various types of DC dysfunctions (see Chapters 8 and 11).

14.5 Dendritic Cells as Tools for the Development of Therapeutic Vaccines Against HIV

14.5.1 Introduction

In recent years, special attention has been given to the use of DCs as potentially ideal cellular adjuvants for the development of therapeutic vaccines (Santini and Belardelli, 2003; Banchereau and Palucka, 2005). DCs are in fact the professional APCs capable of stimulating naïve T cells to initiate a primary immune response and of processing extracellular antigens for presentation by MHC class I molecules (Banchereau *et al.*, 2000). The use of DCs as cellular adjuvants for the preparation of therapeutic vaccines against some human malignancies has become a frequent experimental approach on the basis of promising results generated in animal tumor models. Table 14.1 summarizes some main clinical trials in which DC-based vaccines have been used. On the whole, the main lessons derived from the evaluation of the results of clinical trials performed so far with DC-based vaccines in cancer patients are the following. First, no major side effects have been reported, suggesting that a DC-based immunotherapy does not imply the toxicity often observed in patients treated with a high dose of IL-2 or IFN-α. Second, in cancer models the detected immune response appears to be superior to that achieved by the use of other types of vaccines, suggesting a potential advantage, in terms of efficacy, in using a DC-based vaccine (Banchereau and Palucka, 2005). The optimal type of DC for vaccine preparation, the number of cells to be injected and route of administration are still crucial issues to be defined. The principal methods for the preparation of DC-based vaccines for clinical trials have been extensively described in Chapter 6 of this book. With regard to patients with chronic infectious diseases, such as hepatitis B and C or HIV-1 infection, DC-based vaccination strategies are still at a very early stage of development. The following sections illustrate the main results regarding the use of DCs as tools for the development of therapeutic vaccines against HIV-1. Figure 14.4 summarizes the steps necessary to obtain *ex vivo* DCs suitable for therapeutic vaccination of HIV patients. We first review the results of some *in vivo* studies based on the use of human cells transplanted in immuno-deficient mice and injected with HIV. We then provide the reader with a brief overview of the results on the DC-based vaccines in monkeys. Finally, we report on the very few studies in the use of DC-based vaccines in HIV-infected patients.

TABLE 14.1. Main clinical studies with DCs in the most common tumors and AIDS.

Disease	Vaccine	No of patients	Immune response[*]	Clinical response[*]	Authors
Melanoma	iDC pulsed with tumor peptides or lysate	16	11/16	5/16	Nestle, 1998
Melanoma	mDC pulsed with MAGE-3 peptide	13	8/11	6/11	Thurner, 1999
Melanoma	mDC pulsed with tumor peptides	14	4/14	8/14	Mackensen, 2000
Melanoma	mDC pulsed with tumor peptides	18	16/18	10/17	Banchereau, 2001
Melanoma	mDC pulsed with tumor peptides	28	12/16	9/16	Schuler-Thurner, 2002
Melanoma	mDC pulsed with tumor cells	19	3/10	6/12	O'Rourke, 2003
Melanoma	iDC pulsed with MART peptide	18	18/18	3/18	Butterfield, 2003
Melanoma	iDC pulsed with tumor peptides or lysate	33	5/15	9/33	Hersey, 2004
Prostate cancer	mDC pulsed with PSM-P1/PSM-P2	33	—	9/25	Murphy, 1999
Prostate cancer	mDC pulsed with PAP	21	21/21	6/21	Fong, 2001a
Prostate cancer	mDC pulsed with tumor antigen RNA	13	13/13	7/13	Heiser, 2002
RCC	mDC pulsed with tumor lysate	35	5/6	10/27	Holtl, 2002
RCC	mDC pulsed with tumor RNA	15	6/7	—	Su, 2003
RCC	mDC fused with tumor cells	12	7/12	4/12	Marten, 2003
Colon cancer	mDC pulsed with CEA peptide	12	7/12	5/12	Fong, 2001b
Cervical cancer	mDC pulsed with HPV E7 protein	1	1	1	Santin, 2002
AIDS	mDC pulsed with inactivated HIV	18	18/18	18/18	Lu, 2004
AIDS	mDC pulsed with inactivated HIV	12	0/12	0/12	Garcia, 2005

[*]Immune response and clinical response are reported according to the number of patients considered fully evaluable.
iDC, immature dendritic cells: mDC, mature dendritic cells.

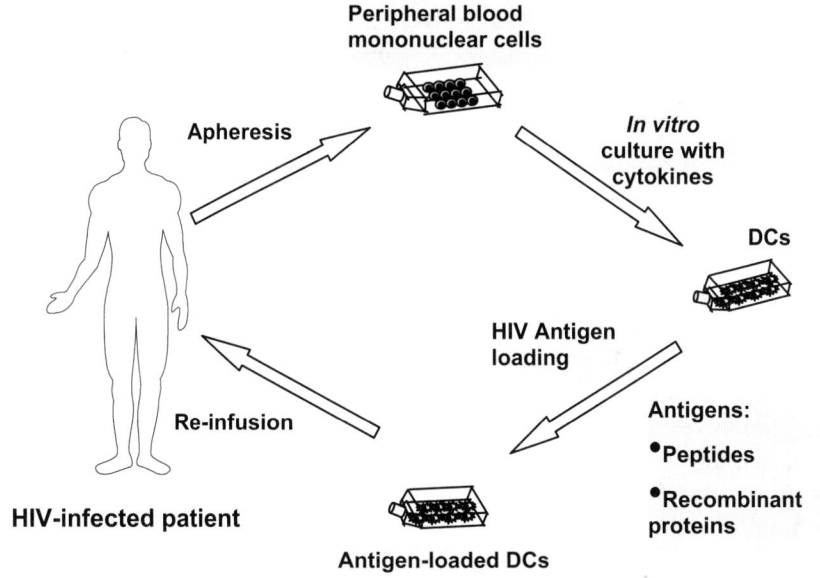

FIGURE 14.4. DCs as tools in the immunotherapy of HIV infections. DCs can be generated *ex vivo* by culturing leukapheresis products (PBMCs, CD14$^+$ monocytes, or CD34$^+$ cells) from HIV-infected patients in cytokine medium and, after antigen loading, are re-injected in patients for therapeutic purposes. Chapter 6 summarizes the main methods for the generation of DCs for their antigen loading and discusses some crucial issues for the development of DC-based vaccines. DC-based vaccines have largely been tested in clinical studies in cancer patients (Table 14.1), and very little information thus far is available for the potential efficacy of DC-based vaccines in HIV-infected patients (Lu et al., 2004; Garcia et al., 2005).

14.5.2 Studies in Hu-PBL-SCID Mouse Model

The results of a set of studies performed in severe combined immunodeficient (SCID) mice reconstituted with human peripheral blood cells (PBLs) have shown that DCs pulsed with AT-2–inactivated HIV are capable of inducing *in vivo* a human immune response against HIV. In particular, partially mature DCs generated within 3 days after stimulation of CD14$^+$ monocytes with type I IFN and GM-CSF (Santini *et al.*, 2000) have proven to be particularly effective in this *in vivo* model. These cells (IFN-DCs) were characterized by a marked upregulation of HLA-DR and costimulatory molecules CD80 and CD86, as well as the appearance of certain markers of activated DCs, such as CD83 and CD25. When injected intravenously into SCID mice, IFN-DCs exhibited an enhanced migratory behavior with respect to immature DCs generated in the presence of IL-4 (Parlato *et al.*, 2001). *In vitro* experiments have not only demonstrated that IFN-DCs are better stimulators of autologous PBLs than IL-4/GM-CSF DCs, but that they are also capable of inducing a stronger Th1-oriented response, as assessed by higher number of IFN-α–producing cells in primary cultures stimulated with IFN-DCs compared with cultures stimulated with IL-4/GM-CSF DCs (Santini *et al.*, 2000). The strong response elicited *in vitro* by IFN-DCs was confirmed in the hu-PBL-SCID mouse model. In fact, it was found that, when hu-PBL-SCID mice were immunized with donor's autologous DCs pulsed with AT-2–inactivated HIV SF162 strain, high levels of anti-HIV antibodies can be elicited (Santini *et al.*, 2000). Mice immunized with IFN-DCs displayed higher levels of anti-HIV antibodies than those exhibited by xenochimeras immunized with IL-4/GM-CSF DCs. Antibodies belonging to the IgG1 isotype were detected especially in mice immunized with IFN-DCs, thus suggesting a Th1-biased response. It was then evaluated whether the immunization of hu-PBL-SCID mice with AT-2/HIV-pulsed DCs could result in the generation of human anti-HIV CD8$^+$ T cells. Vaccination of hu-PBL-SCID mice with virus-pulsed DCs resulted in a clear-cut generation of HIV-1–specific CD8$^+$ T cells (mainly against gag and pol proteins). Furthermore, the extent of response of the animals immunized with IL-4-DCs was lower than that detected in mice immunized with IFN-DCs (Lapenta *et al.*, 2003). Of particular note, the characterization of CD8$^+$ T-cell response has revealed the presence of CTLs specific for the immunodominant gag-derived SL9 epitope, which is highly conserved even in the presence of strong selective pressure to viral escape (Brander *et al.*, 1998; Brander *et al.*, 1999). In a second set of experiments, hu-PBL-SCID mice were challenged with HIV-1 SF162 after immunization, and then the animals were assayed for the extent of viral infection/replication. The results have shown that mice immunized with virus-pulsed IFN-DCs were protected from the infection, in contrast with what has been observed in mice injected with unpulsed IFN-DCs (Lapenta *et al.*, 2003). Interestingly, only the hu-PBL-SCID mice vaccinated with AT-2/HIV-pulsed DCs showed the presence of an anti-HIV CD8$^+$ T-cell response. On the whole, these results may suggest that vaccines based on the use of patient's DCs can be effective in inducing a robust potentially protective humoral and cellular immune response against HIV.

14.5.3 Studies in Non-Human Primate Model

Encouraging results, supporting the findings obtained in the hu-PBL-SCID mouse model and the potential interest for a DC-based immune intervention in HIV-infected individuals, have been obtained in macaques (Bhardwaj and Walker, 2003). Using the simian immunodeficiency virus (SIV) as an AIDS animal model, Lu and co-workers (2003) have demonstrated that immunotherapy of HIV infection can be effective in controlling virus replication. Macaques previously infected with the pathogenic strain SIVmac251 were immunized with autologous mature DCs pulsed with AT-2–inactivated SIV or with unpulsed DCs as a control. A total of five injections every 2 weeks were performed. Within ten days after the first immunization, both blood SIV cellular DNA and plasma RNA levels significantly decreased. After three immunizations, 7 out of 10 animals vaccinated with SIV-pulsed DCs exhibited a 1000-fold decrease of SIV plasma RNA, while the other three animals showed a lower rate of decrease. No effect was observed in all four animals vaccinated with unpulsed DCs. Of note, low viremia was maintained for the whole duration of the study (up to 300 days postvaccination) in all seven responder animals. The authors have also monitored the immune responses induced by vaccination. In all the monkeys of the control group, no increase in $CD4^+$ T-cell count was observed. Whereas a significant raise of $CD4^+$ T cells was detected in all animals vaccinated with SIV-pulsed DCs, however no significant differences were observed in $CD4^+$ T-cell count evolution by comparing the seven monkeys with low viremia and the three animals with higher viremia. The $CD8^+$ T-cell response was also increased in the vaccinated monkeys; interestingly, the seven monkeys with lower viremia showed a significantly higher CTL activity. Of note, the same animals showed higher levels of neutralizing antibodies than those observed in the three vaccinated animals exhibiting higher viremia, while neutralizing antibodies titers of the four control monkeys remained low and unchanged.

14.5.4 Clinical Studies with DC-based Vaccines in HIV-infected Individuals

To date, only two phase I clinical studies have been carried out for evaluating the safety and efficacy of DC-based HIV-1 vaccines. Both studies have been designed for therapeutic treatment of patients with chronic HIV-1 infection.

The first study (Lu et al., 2004) included 18 untreated infected subjects who were vaccinated with autologous DCs. These DCs had been generated from monocytes according to the IL-4/GM-CSF standard method, matured in the presence of TNF-α, and pulsed with AT-2–inactivated HIV-1 derived from the same patient. Each subject received a total of three doses of the DC-based vaccine and was observed for 1 year thereafter without any antiviral therapy. The vaccine was generally well tolerated and no local or systemic side effect was developed by any of the patients, with the exception of an increase in the size of peripheral lymph nodes. The authors then monitored both clinical responses and the generation of HIV immunity. After immunization, the median plasma viral load decreased by 80%, eight individuals showed a decrease of more than 90% over the period of the study (1 year), while for the other 10 the reduction was weaker and transient. The median HIV-1 cellular DNA also decreased by 50%, remaining at this level for almost the whole period of the study. Blood $CD4^+$ T-cell count increased significantly for a short period of time (3 months), returning progressively to baseline, while no significant changes were observed in the $CD8^+$ T-cell count. The authors have also characterized HIV-1 immunity induced by vaccination in terms of humoral as well as cellular responses. The total antibodies to HIV-1 remained unchanged after the vaccination and the neutralizing antibodies were detected at low level (1/10 titers). All 18 vaccinated patients showed a threefold increase of IL-2 and a twofold increase of IFN-α secreting HIV-1–specific $CD4^+$ T cells, and this effect strongly correlated with the change in the plasma viral load observed after vaccination. Interestingly, HIV-1-gag–specific $CD8^+$ T cells also augmented by more than threefold and remained at this level until the end of the study. Of note, perforin-expressing $CD8^+$ T cells (effector cells) were detected in the $CD8^+$ cell population induced by vaccination, and a significant linear correlation was observed between the frequency of these cells and the plasma viral load reduction. Moreover, a strong correlation between the frequency of HIV-1–specific perforin-expressing $CD8^+$ T cells and the frequency of IL-2- or IFN-α–expressing $CD4^+$ T cells was observed.

The second study (Garcia *et al.*, 2005) was carried out on 18 patients with chronic HIV-1 infection undergoing HAART, who were randomized either to be vaccinated with autologous monocyte-derived DCs loaded with autologous heat-inactivated HIV-1 (12 subjects) or to represent a control group (six subjects). After treatment (five immunizations at 6-week intervals), HAART was interrupted and the patients were observed for at least 24 weeks to monitor safety and both the immune and clinical responses. Overall, the vaccine was well tolerated and no local or systemic reactions were observed, with the exception of two patients who experienced mild flu-like symptoms 24 hours after immunization. There was no significant change in plasma viral load between baseline and the levels measured at the end of the follow-up period, although a moderate decrease ($\geq$0.5 log10) was observed in 4 of the 12 vaccinated subjects. No significant increase in neutralizing antibody titers was detected. The authors also monitored HIV-1–specific $CD8^+$ and $CD4^+$ T-cell responses.

Surprisingly, CD8$^+$ T cells decreased during the treatment period, and after stopping HAART there was a progressive recovery, as assessed by the frequency of CD8$^+$ T cells specific for Gag-derived pools of overlapping peptides. This decline of CD8$^+$ T-cell response was similar for both, the patients showing a reduction in the plasma viral load and those not exhibiting any change of this parameter. A moderate increase of HIV-1–specific CD4$^+$ lymphoproliferative response to the p24 HIV antigen was observed after two vaccine administrations, although this was not significant.

14.6 Concluding Remarks

It has been argued that the next major advance in the management of HIV-infected patients will likely occur as a result of the development of immunotherapy protocols capable of inducing a long-term immune control of HIV replication. We have reviewed different types of immunotherapy strategies. In most cases, the general approach is based on attempts to stimulate nonspecifically (by the use of cytokines such as IL-12) or specifically (by vaccines) the immune response in patients. These types of immune intervention, as well as those based on the use of substances capable of inhibiting T-cell activation, can be successfully integrated with HAART in different phases of the natural history of the infection, provided that we are able to clearly identify the purpose, the optimal timing, and the treatment modalities.

The recent knowledge on the biology of DCs and the progress in biotechnology has opened up new perspectives in the immunotherapy of HIV infection. We can now design strategies for the *in vivo* targeting of the relevant HIV antigens to the DCs. Likewise, new natural and synthetic molecules capable of restoring and/or enhancing DC activities, often impaired in patients, have recently been identified and can be tested for their possible role in strategies of immunotherapy of HIV infection. In addition to this, a considerable interest has focused on the use of patients' DCs loaded with HIV antigens as a potentially more effective strategy of therapeutic vaccination in HIV-infected individuals. While the results of a recent pilot trial (Lu *et al.*, 2004) appear to be encouraging in pursuing this therapeutic vaccination strategy, the overall information available at the present time is poor and not conclusive. Of particular note, DCs are important targets of HIV infection, and attention should be paid to the choice of DCs used in clinical studies. Different types of DCs may exhibit not only a different potential in inducing antiviral immunity but also a different degree of susceptibility to HIV infection and capability to transfer the virus to the target cell. Thus, both preclinical and clinical studies are needed in order to evaluate the effectiveness of DC-based vaccines in the immunotherapy of HIV infection. We conclude this chapter by emphasizing that although the possible future validation of DC-based vaccines for the immunotherapy of HIV infection will certainly not solve the drastic needs of HIV-infected individuals in the developing countries, the progress of the research in this field will help us to identify novel and practical strategies for the *in vivo* targeting of the relevant HIV antigens to the right DCs. All this will lead to the definition of new cost-effective immunotherapy protocols to be combined with HAART in the management of HIV-infected patients.

Acknowledgments: Work in the authors' laboratories has been partially supported by grants from the Italian National Project on AIDS.

References

Ahmad, A., Ahmad, R., Iannello, A., Toma, E., Morisset, R., and Sindhu, S. T. (2005). IL-15 and HIV infection: lessons for immunotherapy and vaccination. Curr. HIV Res. 3:261-270.

Aleman, S., Soderbarg, K., Visco-Comandini, U., Sitbon, G., and Sonnerborg, A. (2002). Drug resistance at low viraemia in HIV-1-infected patients with antiretroviral combination therapy. AIDS 16:1039-1044.

Almeida, M., Corsero, M., Almeida, J., and Orfao, A. (2005). Different subsets of peripheral blood dendritic cells show distinct phenotypic and functional abnormalities in HIV-1 infection. AIDS 19:261-271.

Altfeld, M., Allen, T. M., Yu, X. G., Johnston, M. N., Agrawal, D., Korber, B. T., Montefiori, D. C., O'Connor, D. H., Davis, B. T., Lee, P. K., Maier, E. L., Harlow, J., Goulder, P. J., Brander, C., Rosenberg, E. S., and Walker, B. D. (2002). HIV-1 superinfection despite broad CD8[+] T-cell responses containing replication of the primary virus. Nature 420:434-439.

Anaya, J. P., and Sias, J. J. (2005). The use of interleukin-2 in human immunodeficiency virus infection. Pharmacotherapy 25:86-95.

Antony, P. A., and Restifo, N. P. (2005). CD4[+]CD25[+] T regulatory cells, immunotherapy of cancer, and interleukin-2. J. Immunother. 28:120-128.

Autran, B., Carcelain, G., Li, T. S., Blanc, C., Mathez, D., Tubiana, R., Katlama, C., Debre, P., and Leibowitch, J. (1997). Positive effects of combined antiretroviral therapy on CD4[+] T cell homeostasis and function in advanced HIV disease. Science 277:112-116.

Autran, B., and Carcelain, G. (2000). AIDS. Boosting immunity to HIV-can the virus help? Science 290:946-949.

Autran, B., Carcelain, G., Combadiere, B., and Debre, P. (2004). Therapeutic vaccines for chronic infections. Science 305:205-8.

Banchereau, J., Briere, F., Caux, C., Davoust, J., Lebecque, S., Liu, Y. J., Pulendran, B., and Palucka, K. (2000). Immunobiology of dendritic cells. Annu. Rev. Immunol. 18:767-811.

Banchereau, J., Palucka, A. K., Dhodapkar, M., Burkeholder, S., Taquet, N., Rolland, A., Taquet, S., Coquery, S., Wittkowski, K. M., Bhardwaj, N., Pineiro, L., Steinman, R., and Fay, J. (2001). Immune and clinical responses in patients with metastatic melanoma to CD34($+$) progenitor-derived dendritic cell vaccine. Cancer Res. 61:6451-6458.

Banchereau, J., and Palucka, A. K. (2005). Dendritic cells as therapeutic vaccines against cancer. Nat. Rev. Immunol. 5:296-306.

Bhardwaj, N., and Walzer, B. D. (2003). Immunotherapy for AIDS virus infections: cautious optimism for cell-based vaccine. Nat. Med. 9:13-14.

Bonifaz, L., Bonnyay, D., Mahnke, K., Rivera, M., Nussenzweig, M. C., and Steinman, R. M. (2002). Efficient targeting of protein antigen to the dendritic cell receptor DEC-205 in the steady state leads to antigen presentation on major histocompatibility complex class I products and peripheral CD8$^+$ T cell tolerance. J. Exp. Med. 196:1627-1638.

Bonifaz, L. C., Bonnyay, D. P., Charalambous, A., Darguste, D. I., Fujii, S., Soares, H., Brimnes, M. K., Moltedo, B., Moran, T. M., and Steinman, R. M. (2004). *In vivo* targeting of antigens to maturing dendritic cells via the DEC-205 receptor improves T cell vaccination. J. Exp. Med. 199:815-824.

Brander, C., Hartman, K. E., Trocha, A. K., Jones, N. G., Johnson, R. P., Korber, B., Wentworth, P., Buchbinder, S. P., Wolinsky, S., Walker, B. D., and Kalams, S. A. (1998). Lack of strong immune selection pressure by the immuno-dominant, HLA-A*0201-restricted cytotoxic T lymphocyte response in chronic human immunodeficiency virus-1 infection. J. Clin. Invest. 101:2559-2566.

Brander, C., Yang, O. O., Jones, N. G., Lee, Y., Goulder, P., Johnson, R. P., Trocha, A., Colbert, D., Hay, C., Buchbinder, S., Bergmann, C. C., Zweerink, H. J., Wolinsky, S., Blattner, W. A., Kalams, S. A., and Walker, B. D. (1999). Efficient processing of the immunodominant, HLA-A*0201-restricted human immunodeficiency virus type 1 cytotoxic T-lymphocyte epitope despite multiple variations in the epitope flanking sequences. J. Virol. 73:10191-10198.

Butterfield, L. H., Ribas, A., Dissette, V. B., Amarnani, S. N., Vu, H. T., Oseguera, D., Wang, H. J., Elashoff, R. M., McBride, W. H., Mukherji, B., Cochran, A. J., Glaspy, J. A., and Economou, J. S. (2003). Determinant spreading associated with clinical response in dendritic cell-based immu-notherapy for malignant melanoma. Clin. Cancer Res. 9:998-1008.

Carcelain, G., Tubiana, R., Samri, A., Calvez, V., Delaugerre, C., Agut, H., Katlama, C., and Autran, B. (2001). Transient mobilization of human immunodeficiency virus (HIV)-specific CD4 T-helper cells fails to control virus rebounds during intermittent antiretroviral therapy in chronic HIV type 1 infection. J. Virol. 75:234-241.

Chapius, A. G., Rizzardi, G. P., D'Agostino, C., Attinger, A., Knabenhans, C., Fleury, S., Acha-Orbea, H., and Pantaleo, G. (2000). Effects of mycophanolic acid on human immunodeficiency virus infection *in vitro* and *in vivo*. Nat. Med. 6:735-736.

D'Ettorre, G., Andreotti, M., Carnevalini, M., Andreoni, C., Zaffiri, L., Vullo, V., Vella, S., and Mastroianni, C. M. (2006). Interleukin-15 enhances the secretion of IFN-gamma and CC chemokines by natural killer cells from HIV viremic and aviremic patients. Immunol. Lett. 103:192-195.

Davenport, M. P., Ribeiro, R. M, and Perelson, A. S. (2004). Kinetics of virus-specific CD8$^+$ T cells and the control of human immunodeficiency virus infection. J. Virol. 78:10096-10103.

Desrosiers, R. C. (2004). Prospects for an AIDS vaccine. Nat. Med. 10:221-223.

Douek, D. C., Brenchley, J. M., Betts, M. R., Ambrozak, D. R., Hill, B. J., Okamoto, Y., Casazza, J. P., Kuruppu, J., Kunstman, K., Wolinsky, S., Grossman, Z., Dybul, M., Oxenius, A., Price, D. A., Connors, M., and Koup, R. A. (2002). HIV preferentially infects HIV-specific CD4$^+$ T cells. Nature 417:95-98.

Eklund, J. W., and Kuzel, T. M. (2004). A review of recent findings involving interleukin-2-based cancer therapy. Curr. Opin. Oncol. 16:542-546.

Fong, L., Brockstedt, D., Benike, C., Breen, J. K., Strang, G., Ruegg, C. L., and Engleman, E. G. (2001a). Dendritic cell-based xenoantigen vaccination for prostate cancer immunotherapy. J. Immunol. 167:7150-7156.

Fong, L., Hou, Y., Rivas, A., Benike, C., Yuen, A., Fisher, G. A., Davis, M. M., and Engleman, E. G. (2001b). Altered peptide ligand vaccination with Flt3 ligand expanded dendritic cells for tumor immunotherapy. Proc. Natl. Acad. Sci. U. S. A. 98:8809-8814.

Fumero, E., Garcia, F., and Gatell, J. M. (2004). Immunosuppressive drugs as an adjuvant to HIV treatment. J. Antimicrob. Chemother. 53:415-417.

Gahery, H., Choppin, J., Bourgault, I., Fischer, E., Maillere, B., and Guillet, J. G. (2005). HIV preventive vaccine research at the ANRS: the lipopeptide vaccine approach. Therapie 60:243-248.

Gallucci, S., Lolkema, M., and Matzinger, P. (1999). Natural adjuvants: endogenous activators of dendritic cells. Nat. Med. 5:1249-1255.

Garcia, F., Lejeune, M., Climent, N., Gil, C., Alcami, J., Morente, V., Alos, L., Ruiz, A., Setoain, J., Fumero, E., Castro, P., Lopez, A., Cruceta, A., Piera, C., Florence, E., Pereira, A., Libois, A., Gonzalez, N., Guila, M., Caballero, M., Lomena, F., Joseph, J., Miro, J. M., Pumarola, T., Plana, M., Gatell, J. M., and Gallart, T. (2005). Therapeutic immunization with dendritic cells loaded with heat-inactivated autologous HIV-1 in patients with chronic HIV-1 infection. J. Infect Dis. 191:1680-1685.

Gotch, F. (2005). Therapeutic vaccines and immunotherapy revisited. J. HIV Ther. 10:48-50.

Heiser, A., Coleman, D., Dannull, J., Yancey, D., Maurice, M. A., Lallas, C. D., Dahm, P., Niedzwiecki, D., Gilboa, E., and Vieweg, J. (2002). Autologous dendritic cells transfected with prostate-specific antigen RNA stimulate CTL responses against metastatic prostate tumors. J. Clin. Invest. 109:409-417.

Hersey, P., Menzies, S. W., Halliday, G. M., Nguyen, T., Farrelly, M. L., De Silva, C., and Lett, M. (2004). Phase I/II study of treatment with dendritic cell vaccines in patients with disseminated melanoma. Cancer Immunol. Immunother. 53:125-134.

Holtl, L., Zelle-Rieser, C., Gander, H., Papesh, C., Ramoner, R., Bartsch, G., Rogatsch, H., Barsoum, A. L., Coggin, J. H. Jr., and Thurnher, M. (2002). Immunotherapy of metastatic renal cell carcinoma with tumor lysate-pulsed autologous dendritic cells. Clin. Cancer Res. 8:3369-3376.

Jin, X., Bauer, D. E., Tuttleton, S. E., Lewin, S., Gettie, A., Blanchard, J., Irwin, C. E., Safrit, J. T., Mittler, J., Weinberger, L., Kostrikis, L. G., Zhang, L., Perelson, A. S., and Ho, D. D. (1999). Dramatic rise in plasma viremia after CD8($^+$) T cell depletion in simian immunodeficiency virus-infected macaques. J. Exp. Med. 189:991-998.

Jost, S., Bernard, M. C., Kaiser, L., Yerly, S., Hirschel, B., Samri, A., Autran, B., Goh, L. E., and Perrin, L. (2002). A patient with HIV-1 superinfection. N. Engl. J. Med. 347:731-736.

Kamga, I., Kahi, S., Develioglu, L., Lichtner, M., Maranon, C., Deveau, C., Meyer, L., Goujard, C., Lebon, P., Sinet, M., and Hosmalin, A. (2005). Type I interferon production is profoundly and transiently impaired in primary HIV-1 infection. J. Infect. Dis. 192:303-310.

Lapenta, C., Santini, S. M., Logozzi, M., Spada, M., Andreotti, M., Di Pucchio, T., Parlato, S., and Belardelli, F. (2003). Potent immune response against HIV-1 and protection from virus challenge in hu-PBL-SCID mice immunized with inactivated virus-pulsed dendritic cells generated in the presence of IFN-alpha. J. Exp. Med. 198:361-367.

Lederman, M. M., Connick, E., Landay, A., Kuritzkes, D. R., Spritzler, J., St. Clair, M., Kotzin, B. L., Fox, L., Chiozzi, M. H., Leonard, J. M., Rousseau, F., Wade, M., Roe, J. D., Martinez, A., and Kessler, H. (1998). Immunologic responses associated with 12 weeks of combination antiretroviral therapy consisting of zidovudine, lamivudine, and ritonavir: results of AIDS Clinical Trials Group Protocol 315. J. Infect. Dis. 178:70-79.

Li, T. S., Tubiana, R., Katlama, C., Calvez, V., Ait Mohand, H., and Autran, B. (1998). Long-lasting recovery in CD4 T-cell function and viral-load reduction after highly active antiretroviral therapy in advanced HIV-1 disease. Lancet 351:1682-1686.

Lisziewicz, J., Rosenberg, E., Lieberman, J., Jessen, H., Lopalco, L., Siliciano, R., Walker, B., and Lori, F. (1999). Control of HIV despite the discontinuation of antiretroviral therapy. N. Engl. J. Med. 340:1683-1684.

Lori, F., and Lisziewicz, J. (2000) Role of immune modulation in primary HIV infection. J. Biol. Regul. Homeost. Agents 14:45-48.

Lu, W., Wu, X., Lu, Y., Guo, W., and Andrieu, J.M. (2003). Therapeutic dendritic-cell vaccine for simian AIDS. Nat. Med. 9:27-32.

Lu, W., Arraes, L. C., Ferreira, W. T., and Andrieu, J. M. (2004). Therapeutic dendritic-cell vaccine for chronic HIV-1 infection. Nat. Med. 10:1359-1365.

Mackensen, A., Herbst, B., Chen, J. L., Kohler, G., Noppen, C., Herr, W., Spagnoli, G. C., Cerundolo, V., and Lindemann, A. (2000). Phase I study in melanoma patients of a vaccine with peptide-pulsed dendritic cells generated *in vitro* from CD34($^+$) hematopoietic progenitor cells. Int. J. Cancer 86:385-392.

Marten, A., Renoth, S., Heinicke, T., Albers, P., Pauli, A., Mey, U., Caspari, R., Flieger, D., Hanfland, P., Von Ruecker, A., Eis-Hubinger, A. M., Muller, S., Schwaner, I., Lohmann, U., Heylmann, G., Sauerbruch, T., and Schmidt-Wolf, I. G. (2003). Allogeneic dendritic cells fused with tumor cells: preclinical results and outcome of a clinical phase I/II trial in patients with metastatic renal cell carcinoma. Hum. Gene Ther. 14:483-494.

Martin, L. N., Murphey-Corb, M., Mack, P., Baskin, G. B., Pantaleo, G., Vaccarezza, M., Fox, C. H., and Fauci, A. S. (1997). Cyclosporin A modulation of early virologic and immunologic events during ptimary simian immunodeficiency virus infection in rhesus monkeys. J. Infect. Dis. 176:374-383.

Murphy, G. P., Tjoa, B. A., Simmons, S. J., Jarisch, J., Bowes, V. A., Ragde, H., Rogers, M., Elgamal, A., Kenny, G. M., Cobb, O. E., Ireton, R. C., Troychak, M. J., Salgaller, M. L., and Boynton, A. L. (1999). Infusion of dendritic cells pulsed with HLA-A2-specific prostate-specific membrane antigen peptides: a phase II prostate cancer vaccine trial involving patients with hormone-refractory metastatic disease. Prostate 38:73-78.

Nestle, F. O., Alijagic, S., Gilliet, M., Sun, Y., Grabbe, S., Dummer, R., Burg, G., and Schadendorf, D. (1998). Vaccination of melanoma patients with peptide-or tumor lysate-pulsed dendritic cells. Nat. Med. 4:328-332.

Nunnari, G., and Pomerantz, R. J. (2005). IL-7 as a potential therapy for HIV-1-infected individuals. Expert. Opin. Biol. Ther. 5:1421-1426.

Ogg, G. S., Jin, X., Bonhoeffer, S., Moss, P., Nowak, M. A., Monard, S., Segal, J. P., Cao, Y., Rowland-Jones, S. L., Hurley, A., Markowitz, M., Ho, D. D., McMichael, A. J., and Nixon, D. F. (1999). Decay kinetics of human immunodeficiency virus-specific effector cytotoxic T lymphocytes after combination antiretroviral therapy. J. Virol. 73:797-800.

O'Hagan, D. T., and Valiante, N. M. (2003). Recent advances in the discovery and delivery of vaccine adjuvants. Nat. Rev. Drug Discov. 2:727-735.

O'Rourke, M. G., Johnson, M., Lanagan, C., See, J., Yang, J., Bell, J. R., Slater, G. J., Kerr, B. M., Crowe, B., Purdie, D. M., Elliott, S. L., Ellem, K. A., and Schmidt, C. W. (2003). Durable complete clinical responses in a phase I/II trial using an autologous melanoma cell/dendritic cell vaccine. Cancer Immunol. Immunother. 52:387-395.

Ortiz, G. M., Nixon, D. F., Trkola, A., Binley, J., Jin, X., Bonhoeffer, S., Kuebler, P. J., Donahoe, S. M., Demoitie, M. A., Kakimoto, W. M, Ketas, T., Clas, B., Heymann, J. J., Zhang, L., Cao, Y., Hurley, A., Moore, J. P., Ho, D. D., and Markowitz, M. (1999). HIV-1-specific immune responses in subjects who temporarily contain virus replication after discontinuation of highly active antiretroviral therapy. J. Clin. Invest. 104:R13-18.

Oxenius, A., Price, D. A., Easterbrook, P. J., O'Callaghan, C. A., Kelleher, A. D., Whelan, J. A., Sontag, G., Sewell, A. K., and Phillips, R. E. (2000). Early highly active antiretroviral therapy for acute HIV-1 infection preserves immune function of CD8$^+$ and CD4$^+$ T lymphocytes. Proc. Natl. Acad. Sci. U. S. A. 97:3382-3387.

Oxenius, A., Price, D. A., Gunthard, H. F., Dawson, S. J., Fagard, C., Perrin, L., Fischer, M., Weber, R., Plana, M., Garcia, F., Hirschel, B., McLean, A., and Phillips, R. E. (2002). Stimulation of HIV-specific cellular immunity by structured treatment interruption fails to enhance viral control in chronic HIV infection. Proc. Natl. Acad. Sci. U. S. A. 99:13747-13752.

Palella, F. J., Jr., Delaney, K. M., Moorman, A. C., Loveless, M. O., Fuhrer, J., Satten, G. A., Aschman, D. J., and Holmberg, S. D. (1998). Declining morbidity and mortality among patients with advanced human immunodeficiency virus infection. HIV Outpatient Study Investigators. N. Engl. J. Med. 338: 853-860.

Papasavvas, E., Ortiz, G. M., Gross, R., Sun, J., Moore, E. C., Heymann, J. J., Moonis, M., Sandberg, J. K., Drohan, L. A., Gallagher, B., Shull, J., Nixon, D. F., Kostman, J. R., and Montaner, L. J. (2000). Enhancement of human immunodeficiency virus type 1-specific CD4 and CD8 T cell responses in chronically infected persons after temporary treatment interruption. J. Infect. Dis. 182:766-775.

Parlato, S., Santini, S. M., Lapenta, C., Di Pucchio, T., Logozzi, M., Spada, M., Giammarioli, A. M., Malori, W., Fais, S., and Belardelli, F. (2001). Expression of CCR-7, MIP-3beta, and Th-1 chemokines in type I IFN-induced monocyte-derived dendritic cells: importance for the rapid acquisition of potent migratory and functional activities. Blood 98:3022-3029.

Parren, P. W., Moore, J. P., Burton, D. R., and Sattentau, Q. J. (1999). The neutralizing antibody response to HIV-1: viral evasion and escape from humoral immunity. AIDS 13 Suppl A: S137-162.

Pitcher, C. J., Quittner, C., Peterson, D. M., Connors, M., Koup, R. A., Maino, V. C., and Picker, L. J. (1999). HIV-1-specific CD4[+] T cells are detectable in most individuals with active HIV-1 infection, but decline with prolonged viral suppression. Nat. Med. 5:518-525.

Pontesilli, O., Kerkhof-Garde, S., Notermans, D. W., Foudraine, N. A., Roos, M. T., Klein, M. R., Danner, S. A., Lange, J. M., and Miedema, F. (1999). Functional T cell reconstitution and human immunodeficiency virus-1-specific cell-mediated immunity during highly active antiretroviral therapy. J. Infect. Dis. 180:76-86.

Rinaldo, C. R., Jr., Liebmann, J. M., Huang, X. L., Fan, Z., Al-Shboul, Q., McMahon, D. K., Day, R. D., Riddler, S. A., and Mellors, J. W. (1999). Prolonged suppression of human immunodeficiency virus type 1 (HIV-1) viremia in persons with advanced disease results in enhancement of CD4 T cell reactivity to microbial antigens but not to HIV-1 antigens. J. Infect. Dis. 179:329-336.

Rizzardi, G. P., Lazzarin, A., and Pantaleo, G. (2002a). Potential role of immune modulation in the effective long-term control of HIV-1 infection. J. Biol. Regul. Homeost. Agents 16:83-90.

Rizzardi, G. P., Larari, A., Capiluppi, B., Tambussi, G., Ellefsen, K., Ciuffreda, D., Champagne, P., Bart, P. A., Chave, J. P., Lazzarin, A., and Pantaleo, G. (2002b). Treatment of primari HIV-1 infection with cyclosporin A couplet with highly active antiretroviral therapy. J. Clin. Invest. 109:681-688.

Rosenberg, E. S., Billingsley, J. M., Caliendo, A. M., Boswell, S. L., Sax, P. E., Kalams, S. A., and Walker, B. D. (1997). Vigorous HIV-1-specific CD4[+] T cell responses associated with control of viremia. Science 278:1447-1450.

Rosenberg, E. S., Altfeld, M., Poon, S. H., Phillips, M. N., Wilkes, B. M., Eldridge, R. L., Robbins, G. K., D'Aquila, R. T., Goulder, P. J., and Walker, B. D. (2000). Immune control of HIV-1 after early treatment of acute infection. Nature 407:523-526.

Santin, A. D., Bellone, S., Gokden, M., Cannon, M. J., and Parham, G. P. (2002). Vaccination with HPV-18 E7-pulsed dendritic cells in a patient with metastatic cervical cancer. N. Engl. J. Med. 346:1752-1753.

Santini, S. M., Lapenta, C., Logozzi, M., Parlato, S., Spada, M., Di Pucchio, T., and Belardelli, F. (2000). Type I interferon as a powerful adjuvant for monocyte-derived dendritic cell development and activity *in vitro* and in Hu-PBL-SCID mice. J. Exp. Med. 191:1777-1788.

Santini, S. M., and Belardelli, F. (2003). Advances in the use of dendritic cells and new adjuvants for the development of therapeutic vaccines. Stem Cells 21:495-505.

Schmitz, J. E., Kuroda, M. J., Santra, S., Sasseville, V. G., Simon, M. A., Lifton, M. A., Racz, P., Tenner-Racz, K., Dalesandro, M., Scallon, B. J., Ghrayeb, J., Forman, M. A., Montefiori, D. C., Rieber, E. P., Letvin, N. L., and Reimann, K. A. (1999). Control of viremia in simian immunodeficiency virus infection by CD8$^+$ lymphocytes. Science 283:857-860.

Schuler-Thurner, B., Schultz, E. S., Berger, T. G., Weinlich, G., Ebner, S., Woerl, P., Bender, A., Feuerstein, B., Fritsch, P. O., Romani, N., and Schuler, G. (2002) Rapid induction of tumor-specific type 1 T helper cells in metastatic melanoma patients by vaccination with mature, cryopreserved, peptide-loaded monocyte-derived dendritic cells. J. Exp. Med. 195:1279-1288.

Siliciano, J. D., Kajdas, J., Finzi, D., Quinn, T. C., Chadwick, K., Margolick, J. B., Kovacs, C., Gange, S. J., and Siliciano, R. F. (2003). Long-term follow-up studies confirm the stability of the latent reservoir for HIV-1 in resting CD4$^+$ T cells. Nat. Med. 9:727-728.

Su, Z., Dannull, J., Heiser, A., Yancey, D., Pruitt, S., Madden, J., Coleman, D., Niedzwiecki, D., Gilboa, E., and Vieweg, J. (2003). Immunological and clinical responses in metastatic renal cancer patients vaccinated with tumor RNA-transfected dendritic cells. Cancer Res. 63:2127-2133.

Thali, M., Bukovsky, A., Kondo, E., Rosenwirth, B., Walsh, C. T., Sodroski, J., and Gottlinger, H. G. (1994). Functional association of cyclophilin A with HIV-1 virions. Nature 372:319-320.

Thurner, B., Haendle, I., Roder, C., Dieckmann, D., Keikavoussi, P., Jonuleit, H., Bender, A., Maczek, C., Schreiner, D., von den Driesch, P., Brocker, E. B., Steinman, R. M., Enk, A., Kampgen, E., and Schuler, G. (1999). Vaccination with mage-3A1 peptide-pulsed mature, monocyte-derived dendritic cells expands specific cytotoxic T cells and induces regression of some metastases in advanced stage IV melanoma. J. Exp. Med. 190:1669-1678.

Yeni, P. G., Hammer, S. M., Hirsch, M. S., Saag, M. S., Schechter, M., Carpenter, C. C., Fischl, M. A., Gatell, J. M., Gazzard, B. G., Jacobsen, D. M., Katzenstein, D. A., Montaner, J. S., Richman, D. D., Schooley, R. T., Thompson, M. A., Vella, S., and Volberding, P. A. (2004). Treatment for adult HIV infection: 2004 recommendations of the International AIDS Society-USA Panel. JAMA 292:251-265.

Index

A

Acquired immunodeficiency syndrome (AIDS), 245–246, 252, 256–257, 262, 264, 267

Activation markers, 61, 95, 308, 416, 457

Adaptive immunity
DC-mediated Th1/Th2 polarization and molecular basis, 22–23
MDCs, 18–20
PDCs, 20–22
viral infection, 451–454

Adjuvants, 171–187, 189–195

Antigen(s)
presentation, 213, 221, 224, 226–229, 485–489, 492, 494
processing, 46, 56–58, 63, 99–105
targeting, 134, 227

Antiviral soluble factors, 300–301, 309–310, 344–345, 353

APOlipoprotein B mRNA-Editing enzyme Catalytic polypeptide-like editing complex (APOBEC), 245, 259–260, 273

AT-2 inactivated HIV-1, 465, 471, 473

B

B cell function, 69

Blood DCs. *See* DC subsets; MDCs; PDCs

C

C type lectin receptors and HIV binding and uptake, 386–390, 391, 393–394, 396–397

CCR7, 53–55

Chemokines, 9, 11, 14–17, 19, 344, 361

Common lymphoid progenitors (CLP), 4–5, 7

Common myeloid progenitors (CMP), 4–5, 7

Costimulation, 59–60

Cross-presentation, 210, 219, 224, 228, 485–501

Cytokines, 5, 6, 11, 14–23, 171, 174–177, 179–181, 183, 187–194

Cytotoxic T Lymphocytes (CTL), 261, 451–454

D

DC *See* Dendritic cells

DC subsets. *See also* MDCs, PDCs, Langerhans cells, MoDC, MDDC, follicular DC, 4–10, 49–52
antigen uptake and processing, 14–15
cross-talk and innate immunity, 16
cytokine and chemokine production, 15–16
invariant receptors, 11–14
trafficking pathways, 51–55

DC-SIGN, 12–14, 16

Dendritic cells (DC), 3–24, 85–87, 96–98, 107, 171, 175–180, 203–229, 381–398, 405–432, 485–501, 515–534
development, 4–7
differentiation, 46, 48–51, 55, 60–61, 69
functional plasticity, 3–4, 18–21, 49, 51–56
HIV-1 pathogenesis, 447, 459–473
maturation, 17, 130–131, 135–143, 146–148, 153–156

migration, 17
origin, 4 7, 46–51
tolerance, central and peripheral,
  23–25
viral infection, 448–456, 461
"Designer DC," 154–157

E
Envelope (Env), 245–246, 248–250,
  253, 258, 262, 265, 267–268,
  270–271. *See also* gp120

F
Fc receptors, 11, 13–14, 17
Follicular dendritic cells (FDC),
  265–266. *See also* DC subsets

G
Gamma/delta T cells, 299–300,
  305–306, 315–316
Genital mucosa, 395–398
Germinal center DCs (GCDC), 10
gp120, 460, 462, 464–466, 470–473.
  See also Envelope (Env)
Granulocyte-macrophage colony-
  stimulating factor (GM-CSF), 5,
  12, 17
Gut-Associated Lymphoid Tissue
  (GALT), 245–246, 264–265

H
Hematopoietic stem cells (HSC), 6
Highly active anti-retroviral therapy
  (HAART), 264–267, 269, 273
HIV-1
  Binding and uptake, 381–398
  Class I MHC (or Major
    histocompatibility complex),
    485–489, 494, 500
  DC cytotoxic activity, 461, 463,
    468, 473
  DC differentiation, 451, 456,
    464–469

DC maturation, 450, 455,
  457–458, 464–469
DC migration, 457–459, 464,
  472–473
DC secretion of
  cytokines/chemokines, 455,
  458, 469–472
DC subsets
  MDDC, 461–464, 466–468,
    470–473
  MDC, 469–471
  PDC, 450, 461–462, 469–471,
    473
DC survival, 455–456, 464–465,
  469, 473
See also Human Immunodeficiency
  Virus
HIV-exposed seronegative individuals
  (ESN), 345, 357–359
Human immunodeficiency virus
  (HIV), 203–206, 208–209,
  211–213, 216–217, 219,
  221–222, 245–273, 297–317,
  381–398, 405–432, 515–534
Human T lymphotropic virus
  (HTLV), 257, 261, 455–456

I
Immune evasion, 87–88, 105
Immunomonitoring, 204, 216–218
Immunopotentiator, 177, 179
Immune suppression, 59–60, 63, 67
Immunotherapy, 203, 206, 515–534
Indoleamine 2,3-dioxygenase (IDO),
  148–151
Innate immunity, 16, 52, 297–303
  viral infection, 448–451
Interferon, 299–301, 303, 316, 333,
  335, 338–339, 341, 344,
  347–348, 350–351, 357, 360,
  448–454, 469–472
Interleukin (IL), 4–6, 12–13, 15–24,
  48, 106, 137, 256, 299, 300,
  311, 333

L

Langerhans cells (LC), 4–5, 9, 12, 406–409, 412, 414–417, 422, 425–427, 431. *See also* DC subsets

Lipopolysaccharide (LPS), 5, 10, 13, 16, 18, 20, 22, 52, 93, .134, 173–174, 208, 393

Long terminal repeats (LTR), 247–249, 261, 271

Long-term non progressors (LTNP), 254, 257–258, 265, 334, 359–361

M

Macrophage colony-stimulating factor (M-CSF), 5

Maturation, 85–87, 89, 91–94, 98, 101–104, 106–110. *See also* DC maturation

Monocyte-derived DCs (MoDC or MDDC) *See also* dendritic cells (DC)

Mucosa-associated invariant T (MAIT) cells, 8

Mucosal DCs, 9

Myeloid dendritic cells (MDC), 5, 7–10, 12–25, 46–51, 299, 301–307, 406–415, 421–422, 424, 428–429
Th2 development, 19
See also DC subsets

N

Natural killer (NK) cells, 11, 15–16, 22, 312–313

nef, 460, 462–464, 467–468, 472

NK-T cells, 305, 313–314, 316

NOD2 (nucleotide oligomerization domain 2), 13

Non-cytotoxic CD8+ cells, 300–301, 304, 306–311, 316

P

Pathogen-associated molecular patterns (PAMP), 11

Pattern-recognition receptors (PRR), 11–13

Perforin, 334, 337–339, 341–343

Peripheral tolerance, 129–130, 134, 149, 151–153

Plasmacytoid dendritic cells (PDC), 5–10, 12–18, 20–25, 46–48, 299, 409–410, 411, 422–423, 432
IL-3–dependent pathway, 21–22
TLR-dependent pathway, 21
tolerogenic functions, 24–25
See also DC subsets

Pulmonary immunity, 68–69

R

Regulatory T cells, 133–134, 138, 142, 144, 148, 152–154

S

Secondary lymphoid organs (SLO), 10

Severe combined immunodeficiency (SCID), 258, 263, 267

Skin DCs, 9

SLO DCs, 10–11

Solid organ DCs, 9–10

Systemic lupus erythematosus (SLE), 21

T

T lymphocytes
CD4+ T lymphocytes, 335, 340, 346
CD8+ T lymphocytes, 339–340, 343–344, 354
Cytotoxic T lymphocytes (CTL), 333–335, 337–344, 346, 353, 357–358, 360–361

tat, 460, 463–464, 466–467, 471–473

Thymic DCs, 7

Thymic stromal lymphopoietin (TSLP), 10, 19–20, 23

Tolerogenic DCs, 96, 98, 109

Toll-like receptors (TLR), 5–6, 9,
    11–13, 15, 17, 20–21, 52, 58, 64,
    66, 298, 371
Transforming growth factor (TGF)-β,
    4–5, 18, 20, 23,
TRIpartite Motif protein (TRIM5), 260

Tumor necrosis factor (TNF)-α, 5,
    15–17, 19–22
V
Vitamin D₃, 139–140, 147
vpr, 460, 463–464, 468–469, 472

Printed in the United States of America